DATE DUE			
NOV 12 1984			

Developmental Biology of Freshwater Invertebrates

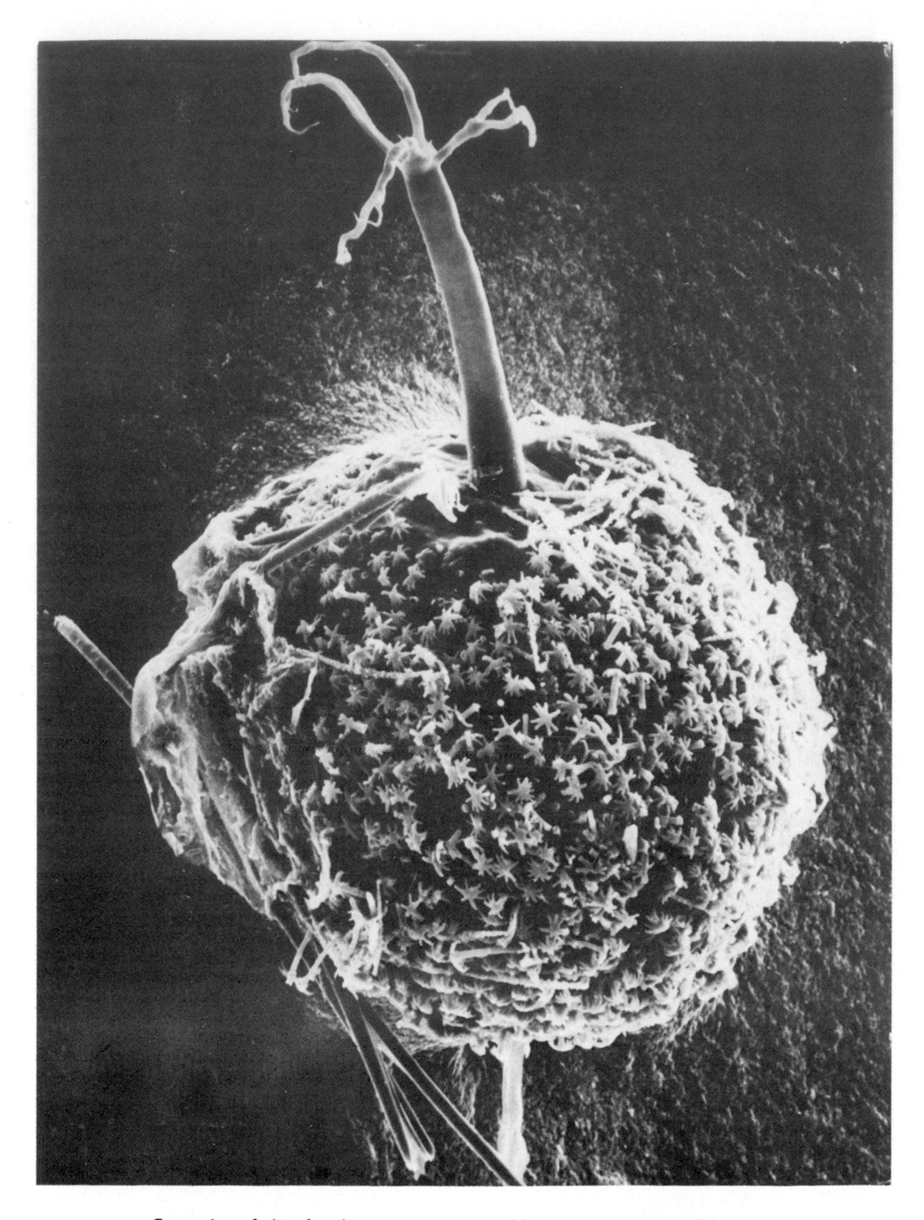

Gemule of the freshwater sponge, *Heteromeyenia tubisperma.*
From Harrison, F.W. (1981) Hydrobiologia *77*:257–259.

Developmental Biology of Freshwater Invertebrates

Editors

Frederick W. Harrison

Western Carolina University
Cullowhee, North Carolina

Ronald R. Cowden

Quillen-Dishner College of Medicine
East Tennessee State University
Johnson City, Tennessee

Alan R. Liss, Inc., New York

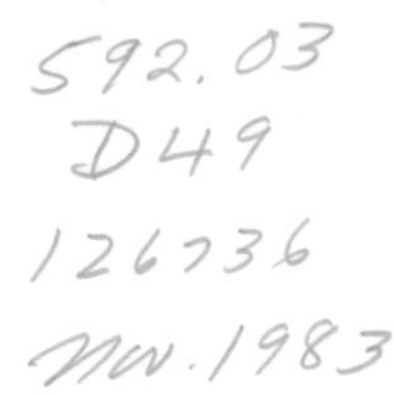

Address all Inquiries to the Publisher
Alan R. Liss, Inc., 150 Fifth Avenue, New York, NY 10011

Printed in the United States of America.

Library of Congress Cataloging in Publication Data
Main entry under title:

Developmental biology of freshwater invertebrates.
Includes bibliographies and index.
1. Freshwater invertebrates—Development. I. Harrison, Frederick Williams, 1938– . II. Cowden, Ronald R.
QL141.D48 1982 592'.03 82-14964
ISBN 0-8451-0222-2

Dedication

The Editors would like to dedicate this book to our wives who have tolerated a cumulative 43 years of systematic neglect because of their husbands' fascination with biology; endured the sudden descent for hours, days, or even weeks of "Visiting Firemen"—some enjoyable and some not so; survived sojourns of weeks at luxurious field research stations; expended vacations on meetings and congresses; made homes for us and our dependents in foreign lands on "sabbatical years;" and listened with patience and sympathy to our concerns.

To: Marion Godfrey Boyd Harrison
Beverly Marie Louise Sherwood Cowden

Contents

Contributors

Mario Benazzi [151]
Istituto di Zoologia e Anatomia Comparata, University of Pisa, Via Volta 4, 56100 Pisa, Italy

Burton J. Bogitsh [221]
Department of General Biology, Vanderbilt University, Nashville, Tennessee 37235

O. Stephen Carter [221]
Department of Physiology, Graduate School of Nursing, Vanderbilt University, Nashville, Tennessee 37235

C. C. Chinnappa [485]
Department of Biology, University of Calgary, Calgary, Alberta T2N 1N4, Canada

Ronald R. Cowden [xi, xiii, 213]
Department of Biophysics, Quillen-Dishner College of Medicine, East Tennessee State University, Johnson City, Tennessee 37601

Juan Fernández [317]
Departamento de Biología, Facultad de Ciencias Básicas y Farmacéuticas, Universidad de Chile, Casilla 653, Santiago, Chile

Vittorio Gremigni [151]
Istituto di Zoologia e Anatomia Comparata, University of Pisa, Via Volta 4, 56100 Pisa, Italy

Frederick W. Harrison [xi, xiii, 1]
Department of Biology, Western Carolina University, Cullowhee, North Carolina 28723

Georgia E. Lesh-Laurie [69]
College of Graduate Studies, Cleveland State University, Cleveland, Ohio 44115

Charles F. Lytle [129]
Zoology Department, North Carolina State University, Raleigh, North Carolina 27650

The bold face number in brackets following each contributor's name is the opening page number of that contributor's paper.

John B. Morrill [399]
Division of Natural Sciences, New College of the University of South Florida, Sarasota, Florida 33580

Hideo Mukai [535]
Department of Biology, Faculty of Education, Gunma University, Aramaki, Maebashi 371, Japan

Diane R. Nelson [363]
Department of Biological Sciences, East Tennessee State University, Johnson City, Tennessee 37614

Nancy Olea [317]
Departamento de Biología, Facultad de Ciencias Básicas y Farmacéuticas, Universidad de Chile, Casilla 653, Santiago, Chile

Einhard Schierenberg [249]
Department of Molecular Biology, Max-Planck-Institute for Experimental Medicine, D-3400 Göttingen, Federal Republic of Germany. Currently affiliated with Department of Molecular, Cellular and Developmental Biology, University of Colorado, Boulder, Colorado 80390

Takashi Shimizu [283]
Zoological Institute, Faculty of Science, Hokkaido University, Sapporo 060, Japan

Grace A. Wyngaard [485]
Department of Zoology, University of Maryland, College Park, Maryland 20742

Preface

Two years ago we were sitting on a hill in the North Carolina mountains and started to muse on the possibility of developing an invertebrate embryology course based at one of the inland freshwater laboratories. This, in turn, led to a tentative listing of forms that might be used, and speculation about who might be the available authority on each group. Since contemporary higher educational economics would prohibit a course of this sort based on the old Woods Hole marine invertebrate embryology course format, we decided that it would be useful to have the details of collection, maintenance, normal development, experimental management, and the main problems in developmental biology posed by each group or organism collected into a single volume. Telephone calls to colleagues elicited some volunteers and the identification of further potential authors. Although it was possible to cover a number of species or groups, qualified authors were not available for others. However, the decision was made that the book we had in mind need not be encyclopedic to be useful.

We hope that the reader will find, as we did, that a surprising number of groups are represented, and that the investigator interested in developmental biology need not feel constrained to set up an imperfect miniature ocean in order to work with interesting creatures and problems in developmental biology.

We would like to confess to one significant lapse from purity in terms of freshwater organisms. *Caenorhabditis elegans* is a soil nematode that may be cultured in liquid media. Dr. von Schierenberg protested as much, but we felt that the importance of this organism is so great that its omission would have been an even greater lapse.

We hope you will be as amazed and gratified as we were with the quality of the information published in this volume, and that this information will cause some of you to use the species covered in your research.

Frederick W. Harrison and Ronald R. Cowden
May 27, 1982

Introduction

The chapters that form this volume demonstrate that a substantial segment of the metazoan invertebrates evolved at least some representatives which are adapted to freshwater habitats. Close inspection reveals three major categories: (1) evolutionary primitive diploblastic or triploblastic animals (sponges, coelenterates, and flatworms); (2) triploblastic representatives (flatworms and nematodes) of the main protostomic branch leading to coelomate or pseudocoelomate body plans; (3) schizocoelous or hemocoelous coelomates (oligochaetes, leeches, pulmonate gastropods, and arthropods); or if none of these three, evolutionary side-branches that appear to be dead ends (bryozoans and tardigrades).

Although early development in most arthropods is very unlike that of the determinate meroblastic spirally cleaving lineage, within some of these groups there are forms that deviate substantially from the expected pattern. In cephalopod development a large yolky egg offers a striking example of departure from the idealized cleavage pattern characteristic of the group. At the other end of the spectrum, triclad flatworms form an exception in the sense that they do not seem to adhere to the rigid formula of spiral determinate cleavage. Polyclads, however, which represent the next recognized evolutionary progression in the turbellaria, do adhere to the pattern and form Müller's larva, a modified trochophore. The persistence of the spiral cleavage pattern and the similarity in organization of embryos up to the gastrula stage is remarkable, but there are exceptions and specializations along the way. Development in arthropods is an extreme example of specialization. It would be interesting to speculate on why representatives of this particular branch of the animal kingdom adapted themselves to freshwater habitats while the deuterostomic forms exhibiting regular or irregular indeterminate cleavage and archenteron formation did not. Most of the forms recognized in the deuterostome evolutionary pathway are reasonably advanced and specialized, but evolutionary age and phylogenetic status are relative concepts. It seems that there would have been geologic time enough and opportunity enough to have freshwater representatives of these groups. Though more detailed exploration of the fresh waters of the world might turn up some representatives of the deuterostome lineage, deuterostomes are not represented in the animals we know. It is at least curious that the main metazoans adapted to life in freshwater are either primitive forms, divergent offshoots, or representatives of a specific major evolutionary branch.

Unquestionably, the embryologist with access to the marine environment finds a richer lode of animal diversity than he could expect from freshwater forms. Even within the groups represented, the options are relatively restricted compared to the species adapted to the marine environment. There are notable exceptions. The parasitic forms or forms that may have a parasitic life history and the arthropods might be rated at equivalence, but even diversity within the freshwater leeches pales beside the abundance and diversity of marine polychaete species.

Though the experimental use to which one might put any of these organisms is conditioned by the special interests and skills of the investigator, each of these groups offers some special experimental options that might be examined with advantage. One of the problems one must face in the consideration of cellular differentiation of a vertebrate is the astounding number of specialized classes or types of cells, not to speak of the transitional variants that participate in the structure of each organ or tissue system. In contrast, the more restricted total number of cell types encountered in sponges, coelenterates, flatworms, and even nematodes very significantly simplifies this problem. However, with the exception of freshwater sponge cells which are of respectable size, the cells of most of these representatives are relatively small and far from ideal light microscopic optical objects. In spite of this, there is an abundance of experience that assures us that, though it is often difficult or challenging to work with these cells, it is far from impossible. Methods for maceration and dissociation have demonstrated that the individual cell types can be identified, and that even automated cytochemical techniques may be used to define the various subpopulations of cells. As a group they offer options for the investigation of developmental phenomena in less complex systems.

Freshwater sponges, hydra, and planarians have been used mainly in experiments designed to study regeneration and morphogenesis. The options are now multiplying. Dissociated sponge cells can be isolated on Percoll gradients, and virtually pure populations of archeocytes and other specialized cell types can be obtained, recombined, and considered in culture. Cultured archeocytes can form a functional, if somewhat imperfect, sponge. In coelenterates the development of nematocysts from interstitial cells represents an excellent example of progressive clonal cellular differentiation. With the development of specific antibodies to neuropeptides and neuropeptidelike substances, it has been possible to gain a better appreciation of functional diversity and organization in these organisms. In flatworms, or at least in planarians, it has become possible to dissociate cells from a normal worm and maintain these in active cell division for up to twelve days. This maintenance of cell division appears to be achieved by retarding differentiation. Regeneration as well as cell division can be affected by a variety of biogenic amines such as histamine, serotonin, and norepinephrine. These substances are associated with the nervous system and appear to have a major role in the establishment of polarity as well as the rate of regeneration. In freshwater sponges and hydra, it is possible to selectively deplete the organism of interstitial cells or archeocytes. Grafting experiments can be used with hydra to perturb normal organization.

Our knowledge of the cytogenetics of both sponges and coelenterates rests mainly on older publications that made use of sectioned material. Because the chromosomes are small and relatively numerous, cytogenetic study of these groups has been considered difficult to impossible. The methods available to contemporary cytogenetics should allow one to overcome these problems. There are reasons to believe that a surprising diversity of karyological conditions might be found within and among these populations, and that this information would affect our appreciation of chromosomal function in development. This has clearly been the case in free-living flatworms, which have more accessible chromosomes. The presence of diploid somatic cells, triploid male germ cells, and hexaploid female germ cells in a race of *Dugesia lugubris* has allowed the unequivocal demonstration that differentiated germ cells can dedifferentiate and participate in the regeneration of somatic structures.

Trematodes and other helminth parasites require intermediate hosts,which are, in the case of trematodes, mollusks. The primary larva has evolved into a mobile, protected bag of germinal cells that seeks and penetrates an appropriate host. The larva must defend and protect itself from the host's natural defense systems while becoming established in the host tissue. It then must amplify itself by growth and development of yet additional asexual intermediate reproductive units—i.e., the daughter sporocysts, which ultimately form the cercaria that penetrate the definitive mammalian hosts. The ability to perform the intermediate portion of this life-cycle step *in vitro* represents a major step toward exploration of the process of developmental amplification, growth, and differentiation in these intermediate stages. Since mollusks exhibit significant cellular responses to foreign tissues or macromolecules, the evolution of selective tolerance to the parasitic intermediates continues to be a question of major practical and theoretical importance.

Examining the bryozoans, one finds that they are neither overly simple nor as complex as coelomate organisms. They offer options for the examination of alternative paths of asexual development in a more complex system than that found in coelenterates. Bryozoans allow study of adaptation of a gemmulelike form of asexual reproduction, and afford opportunities for examination of the developmental genetics of colonial systems.

In light of advances in genetics and cell, molecular and developmental biology, the insight that the soil nematode, *Caenorhabditis elegans,* provided the ideal vehicle for combining all of these aspects in the consideration of development of a particular organism might well have been correct. It is small and easily cultivated, and displays a constant cell number achieved by spiral determinate cleavage and subsequent determinate development of those specific lineages. A sufficient amount of spadework had been done on its biology and genetics before it became *the* organism of major interest so that reasonably well-characterized mutants were available. There were other candidates for the ideal vehicle, but the existence of known mutants seemed to be the decisive factor. In a heroic effort, investigators have determined the fates and lineages of virtually every cell in the organism. The animal has even had the good grace to accumulate specific fluorescent substances in certain tissues of which genetic mutants have been obtained. It is probably no exaggeration to contend that more is known concerning the total biology of development in this organism than in any other. As noted, since this organism is clearly a triploblastic pseudocoelomate member of the protostomic branch of metazoan evolution, it occupies a transitional position between the more primitive and more complex forms in this lineage. All these considerations prompted us to include *Caenorhabditis,* a non-freshwater organism, in this book.

Three major groups of coelomate spiralians are represented in this volume: oligochaete annelids, leeches, and pulmonate snails. They differ considerably in the extent to which they have been studied and in the experimental uses to which they have been put. The leeches were "rediscovered" about a half decade ago after sporadic use in neurobiology and neuroendocrinology. The technology allowing pressure microinjection of either horseradish peroxidase or a fluorescent-labeled synthetic peptide into cells became possible. Microinjection allows direct visualization of cells belonging to those lineages. The relatively large, tough leech blastomeres are technically more easily injected than some smaller, more fragile blastomeres. Thus leeches have been extensively used in the investigation of cell lineage in a determinate system, particularly in studies of the nervous system. In addition, the Cold Spring Harbor group prepared monoclonal antibodies against various specific components in whole-leech CNS tissue, and was then able to demonstrate the specific locations of neurons that share particular peptides. Though the leech nervous system is complex when compared to that of a planarian or hydra, it is simple in comparison to a vertebrate nervous system. Just as *Caenorhab-*

ditis seems ideal for the study of all aspects of development at this time, the leech CNS seems to be of the proper order of complexity for state-of-the-art investigation of the nervous system.

Tubifex has been a "classical object" of morphological and experimental investigation of spiralian development, but its limited use in developmental biology is probably due to the fact that maintaining these animals in proper condition and obtaining embryos require some special facilities and careful attention to detail. Even under the best conditions, the number of eggs packaged in each cocoon is not sufficient to stimulate the projection of large-scale experimental biochemical investigations. This organism serves as an excellent example of spiralian development leading to direct development of the definitive animal. It has been useful in blastomere deletion experiments, centrifugation experiments, and for light-microscopic and ultrastructural cytological investigations of development. It could be profitably studied cytochemically, and it might also serve for comparative studies of cell lineage or of regulative capacities in later phases of development.

Development in snails of the genus *Lymnaea* has been more thoroughly studied than development of any coelomate animal presented in this volume. An immense body of detailed information about development in *Lymnaea* and related genera has been accumulated; classical morphology of development and gametogenesis; centrifugation experiments on mature and immature eggs and embryos at various stages of development; the composition and fate of surrounding fluids in egg capsules; the physical conditions of intrachorionic development; cell-surface morphology of cells and whole embryos at different stages; effects of blastomere deletions; and biochemical changes in development and during tissue histogenesis. These are just a few examples; there are many more. However, a striking feature of these embryos is that after reasonably rigid determinate development to the gastrula stage, there is a short period of time during which these embryos display a substantial capacity for regulation. Then once more the system abruptly becomes restricted. The full significance of this is not completely understood, and the problem should attract the attention of researchers.

The freshwater spiralians offer a large continuum of options, particularly when combined with the large assortment of marine organisms that exhibit similar patterns of early development. However, each presents options in detail that might be exploited by an insightful investigator. Cell lineage and the effects of blastomere deletion are known for the three groups presented and for *Caenorhabditis*. It is an advantage to have comparative information concerning virtually the total developmental history of a series of organisms that share a common pattern of early development.

Development of the cyclopid micro-arthropods and of the tardigrades, which have an ambiguous taxonomic position, was selected for presentation for quite different reasons. The tardigrades are microfauna that inhabit microaquatic environments. In spite of the fact that these animals are largely invisible to the unaided human eye, a surprising amount of published work is available concerning their embryology. The cells are of respectable size, the chromosomes are of a reasonable size and number, and some species have been studied cytogenetically. The material presented should mainly make the reader aware of this invisible biological world that is just starting to be appreciated since the advent of scanning electron microscopes. Of necessity, most approaches to this group have to be in the form of descriptive embryology, but the opportunities for experimentation will certainly improve as more is known about development in this group. In contrast to the other groups that have been considered, cleavage is regular and coeloms develop from structures described as archenterons. However, because of the tardigrades uncertain taxonomic status, the significance of this aspect of tardigrade development has not been firmly established.

The copepod genera *Cyclops, Mesocyclops,* and other cyclopid genera present another situation. About thirty years ago it was recognized that cleavage stages offer favorable material for the study of mitotic spindles and nuclear changes during the cell cycle. This was mainly the work of a single investigator, and, with the lapse of his interest in the system, the group became the focus of a cytogenetic effort that, until recently, was again virtually the work of a single scientist. Development in these animals shares with some other forms the characteristic of eliminating some chromatin from nongerminal (somatic) cells during the later cleavage divisions. Depending on the species in question, chromatin diminution may start as early as the third cleavage, but beyond this point of initial diminution there is some confusion about both the amount and kind of chromatin that is lost or retained. The developmental significance of chromatin diminution is unknown. Conflicts in the literature may arise from differences in the equipment and technical details of measurements, or from studies of different species or populations. Cytogenetics of the group has received a good deal of attention because the chromosomes are relatively large and few in number. The cytogenetics of this group has been reviewed in some detail. There has been some recent work on the cytochemistry of nuclei during *Cyclops* development that addresses both DNA content and qualitative changes in chromatin and other intranuclear structures. One would hope that interest in the study of development in this interesting group might be reawakened. However, serious study of these organisms had to wait until suitable laboratory culture methods became available. These culture methods, now available, were outgrowths of population ecology and genetic studies, and should open these organisms to broader experimental exploitation.

There are other groups or organisms that might have been included in this volume, but there are obvious space limitations and limitations of the availability of potential authors familiar with the groups that might be represented. More emphasis might have been placed on parasitic helminths: while they continue to receive the attention of parasitologists, aspects of their development that are of more basic biological interest might be considered. The account of mother-and daughter sporocyst development in schistosomes should be considered, as an example. There are freshwater prosobranchs and lamellibranchs, and it would be advantageous to have access to the details of development of some of these molluscan species, but appropriate authors were not identified.

Beyond those obvious forms that might have escaped this particular net, there is an ambiguous group of mainly marine organisms that inhabit estuaries but which have become adapted to variable concentrations of brackish water. In addition there are some terrestrial forms that experience some part of their life cycles in fresh water. However, the organisms covered in this volume seemed to present a reasonable sample, and the coverage is not, and was not intended to be, encyclopedic. The editors hope that the accounts presented will stimulate the imaginations of some graduate students in developmental biology and indicate new options to more experienced investigators.

Frederick W. Harrison and Ronald R. Cowden

Developmental Biology of Freshwater Invertebrates, pages 1–67

Developmental Biology of Freshwater Sponges

Frederick W. Harrison

ABSTRACT The freshwater sponges offer an extremely useful and challenging model of differentiation, morphogenesis, and cell-to-cell communication. This chapter includes an extensive treatment of basic sponge biology including discussions of structure, cell types, natural history of adult sponges, and general normal embryology. Collecting techniques, methods of maintaining sponges in the laboratory, and methods of obtaining and rearing embryos are discussed. All features of development, including current investigations, are presented.

The freshwater sponges are members of the great acalcareous class Demospongiae. While the vast majority of demosponges are marine, a number of sponges, derived from different ancestors, inhabit freshwater environments. Recent studies (Brien '67a, '70, '73a; Racek and Harrison, '75) have provided insights into the possible pathways of spongillid evolution during the past 100 million years. Gemmule-producing freshwater sponges existed in inland waters long before the isolation of nonspongillid Porifera of rather marine facies in the so-called "ancient freshwater lakes." Recent studies of a number of gemmuleless Porifera from a range of thalassoid environments (Racek and Harrison, unpublished data) show that they must have arisen not only from different ancestors but also in much more recent geological times. The polyphyletic origin of the freshwater sponges, assumed by Marshall (1883), commented on by Penney and Racek ('68), Racek ('69), and discussed by Brien ('66a, '66b, '69, '73b), has thus been well documented.

Freshwater sponges are all leuconoid, i.e., the flagellated cells, choanocytes, are restricted to choanocytic chambers. The sponge is separated from the external environment by a simple squamous epithelium, the pinacoderm. The exopinacoderm covers the free surface of the sponge (Fig. 1), while the basopinacoderm delimits the attachment surface. Internally, canals are lined by an endopinacoderm. Anatomically and physiologically, freshwater sponges are organized around an elaborate water-conducting system, the aquiferous

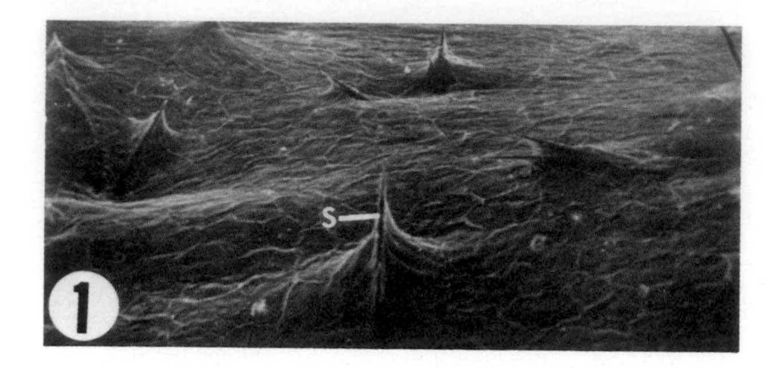

Fig. 1. The surface of the sponge, formed by the exopinacoderm, is supported in tent-like fashion by fasicles of spicules (s.). The borders of pinacocytes are visible. Scanning electron microscopy (SEM) × 270. (Scanning electron micrograph courtesy of Dr. Louis De Vos.)

system. Water enters the sponge by passing through openings, ostia, in the exopinacoderm. Some of these openings may lie within porocytes, cells that enclose a pore. Large subepithelial spaces may lie below the exopinacoderm. The remainder of the inhalent system is composed of incurrent canals that lead to choanocytic chambers. Water enters the choanocytic chambers by small apertures, prosopyles, that lie between choanocytes. The choanocytes form the somewhat spherical choanocytic chambers. Water leaves the choanocytic chambers by apertures called the apopyles (Fig. 2). Central cells may lie within the chamber or at the apopyle with pseudopodial extensions in contact with choanocytes, thus affecting the rate and direction of water current flow (Connes et al., '71a). Water passes from the choanocytic chambers into excurrent canals (Fig. 3). These lead into oscula, openings by which water leaves the sponge. These may be located on epidermal "chimneys." The entire region lying between the pinacoderm and the choanocytic chambers is the mesohyl. The mesohyl contains various ameboid cell types that move freely throughout this region of loosely textured connective tissue (De Vos, '79), which is rich in collagen and glycoproteins (refer to Garrone, '78, for an excellent review of sponge connective tissue). The freshwater sponge skeleton is composed of siliceous spicules, which may be found free within the mesoglea or may be formed into fasicles that support the exopinacoderm in a tent-like fashion.

FRESHWATER SPONGE CELL TYPES

There are approximately ten cell types in freshwater sponges. However, as Pavans de Ceccatty ('79) notes, their classification is difficult because of

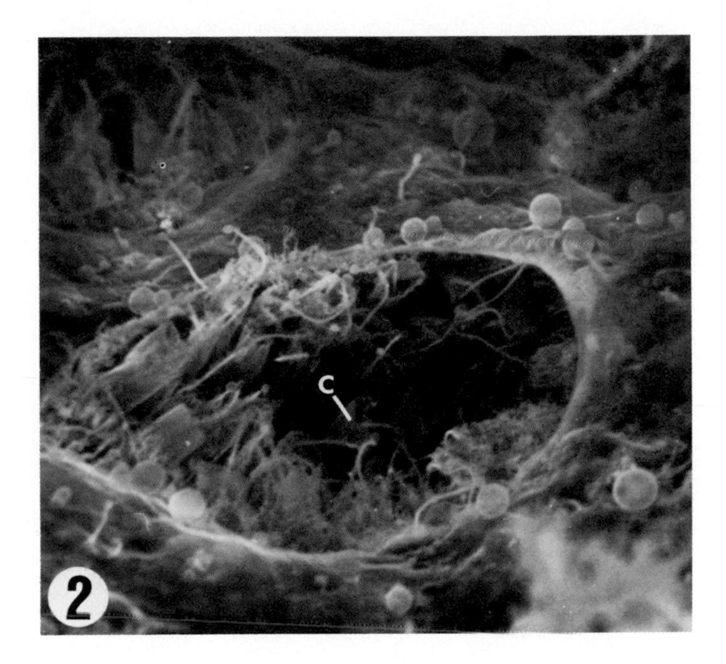

Fig. 2. Choanocyte collars and flagellae project from the apopyle of a choanocytic chamber. c, collar. SEM × 2,650. (Courtesy of Dr. Louis De Vos, De Vos, '79).

their mobility and because their position within the sponge may determine their function. It seems possible that several of the "cell types" recognized in freshwater sponges represent different phases in the modulation of a given kind of cell rather than a permanent state of differentiation. Comprehension of developmental processes in sponges depends upon a clear understanding of the structure and function of the various kinds of cells (for reviews see Lévi, '70; Brien, '73b; Bergquist, '78; Garrone, '78; Hartman et al., '80). The following is subdivided into descriptions of cells comprising an epithelium, i.e., pinacoderm and choanoderm, and the cell types of the mesohyl.

The two fundamental cell types of sponges are pinacocytes and choanocytes (Lévi, '70); their presence alone is sufficient to define the group. Brien ('67b), in fact, considered that there were two sponge cell lines, pinacocytes and mesohyl cells constituting the "ectomesenchyme," and choanocytes comprising the "choanoderm." The concept of sponges as a diploblastic group stems from the idea of two distinct cell lines. However, choanocytes are derived from archeocyte progenitors from the mesohyl and, in fact, can be transformed into endopinacocytes or germ cells (Diaz, '74; Diaz and

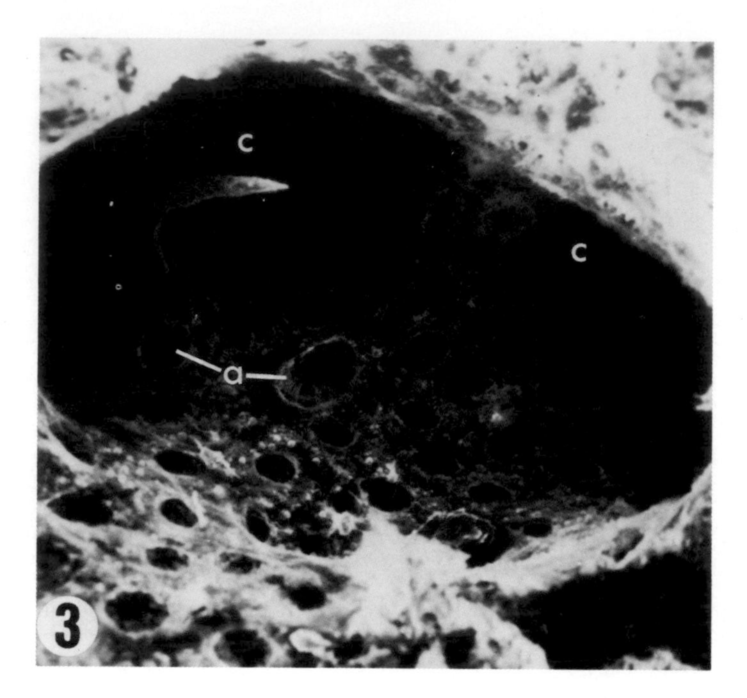

Fig. 3. Several exhalent canals (c) are evident in this micrograph. Numerous apopyles (a) arranged in regular fashion, empty into the canals. SEM × 500. (Courtesy of Dr. Louis De Vos.)

Connes, '80). For discussion of the diploblastic condition refer to Hyman ('40), Brien ('72), Tuzet ('73), and Rasmont ('79).

Pinacocytes

Pinacocytes are the predominant cell of the sponge surface (exopinacocytes), line the aquiferous canals (endopinacocytes), and form the basal attachment surface (basopinacocytes). The sponge pinacoderm differs from epithelia in higher animals in that it lacks a basement lamina (Lévi, '70) (although Garrone, '78, reported subepithelial laminae resembling typical basal laminae in some species) and in the absence of specialized cell junctions (Ledger, '75). Although localized increases in membrane density, which are sometimes associated with bundles of tonofilaments and which approach the structure of macula adherens, have been reported in *Ephydatia fluviatilis* (Feige, '69) and in *E. mulleri* (Pottu-Boumendil, '75), true desmosomes have never been described in sponges. The fact that basopinacocytes become ameboid during wound healing (Harrison, '72a) is attributable to the lack of tight junctions. Most cell junctions are simple, although complex interdigitations may exist.

Pinacocytes of freshwater sponges form a simple squamous epithelium in which nuclei lie in the same plane as the cytoplasm. This differs from the situation in most other sponges in which exopinacocytes form a "T" with the nucleus contained within a clublike extension protruding into the mesohyl (see Bagby, '70; Boury-Esnault, '73, for reviews).

Pinacocyte nuclei may possess nucleoli but this is variable. Basopinacocytes, particularly in growing sponges, actively synthesize protein during secretion of the spongin attachment to the substratum. During attachment, the rough endoplasmic reticulum is extended with expanded cisternae filled with a cloudy or granular material (Garrone, '78). At this time, basopinacocytes also incorporate ^{14}C-phenylalanine, a characteristic amino acid occurring in spongin (Efremova, '70). They also exhibit considerable red fluorescence following vital fluorochroming with acridine orange used at 10^{-4} M concentration (Harrison and Davis, in press). Following the period of synthesis, these cells become quiescent and exhibit a low level of secretory activity (Harrison, '72a). They continue to function in osmoregulation, containing numerous perinuclear contractile vacuoles (Harrison, '72a; Brauer, '75). Basopinacocytes at the sponge periphery actively phagocytize bacteria and demonstrate considerable levels of acid phosphatase activity in cytoplasmic inclusions (Harrison '72a) (Figs. 4,5).

Exopinacocytes contain small amounts of nonglycogenic carbohydrate, acid phosphatase, lipid, and scattered protein inclusions in the cytoplasm. Little RNA is present in these cells.

Porocytes

The surface epithelium of freshwater sponges is pierced by incurrent openings of two types. The least complex of these is a separation between exopinacocytes to form a circular or elliptical simple ostium (Wintermann-Kilian et al., '69; Harrison, '74a). Incurrent ostia are also found in porocytes. Porocytes are derived from exopinacocytes and resemble them histochemically and in general morphology (Brien, '43; Wintermann-Kilian et al., '69; Harrison, '74a; De Vos, '79). The cell types differ in that porocytes contain a central pore during the dilation phase (Fig. 6). Studies of living porocytes (Harrison, '72b) demonstrate varying degrees of pore dilation, which is accompanied by changes in the number of contractile vacuoles (Figs. 7–9).

Choanocytes

Freshwater-sponge choanocytes are found in eurypylous choanocytic chambers, i.e., chambers in which the large apopyle opens directly into an excurrent canal. Although choanocytes are characteristic of sponges, they

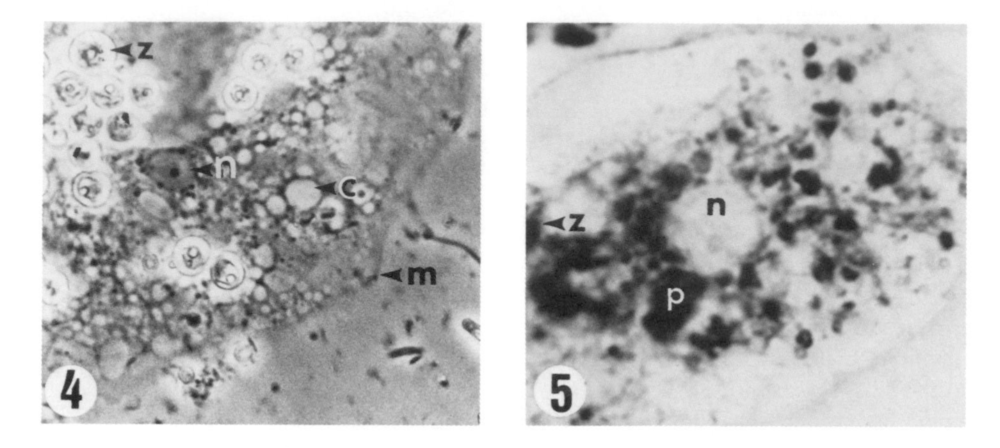

Fig. 4. Periphery of a sponge showing numerous contractile vacuoles. c, contractile vacuole; m, sponge margin; *n*, basopinacocyte nucleus with typically small nucleolus; z, intracellular zoochlorella. Phase contrast of the living sponge. × 600. (Harrison, '72a.)

Fig. 5. Acid phosphatase activity in a basopinacocyte. Enzyme activity (p) is seen largely in granules near the nucleus (n), but the reactive sites are present throughout the cytoplasm z, intracellular zoochlorella. Simultaneous coupling azo dye method. × 1,200. (Harrison, '72a.)

are also found in other phyla (Brien, '73b; Tuzet, '73). Choanocytes are polarized cells with the free (luminal) surface bearing a collar of microvilli that surrounds a single flagellum (Figs. 10,11). Recent studies by Brill ('73), De Vos ('77, '79), Garrone ('78), and Garrone et al. ('80) have amplified the earlier studies of Rasmont ('59) and Fjerdingstad ('61) defining choanocyte structure. The nucleus is central, lacks a nucleolus, and contains highly condensed chromatin, which in turn suggests low levels of transcriptional activity. A well-developed Golgi apparatus is located near the apical nuclear surface, and numerous vacuoles, many containing considerable acid phosphatase activity (Harrison, '74b), are present in the cytoplasm. The low levels of cytoplasmic RNA suggest reduced synthetic activity, which is reasonable in view of the nuclear morphology. Collar microvilli are interconnected by a meshwork of glycoprotein filaments (Brill, '73; Garrone, '78). Choanocytes do not rest upon a basement membrane and are attached to each other by interdigitations.

In addition to their role in creation of a water current, choanocytes also capture food. Studies by Kilian ('52), Schmidt ('70), and Weissenfels ('76) demonstrated that, following bacterial capture by the collar, most of the particles were transferred to mesohyl cells for digestion and final excretion

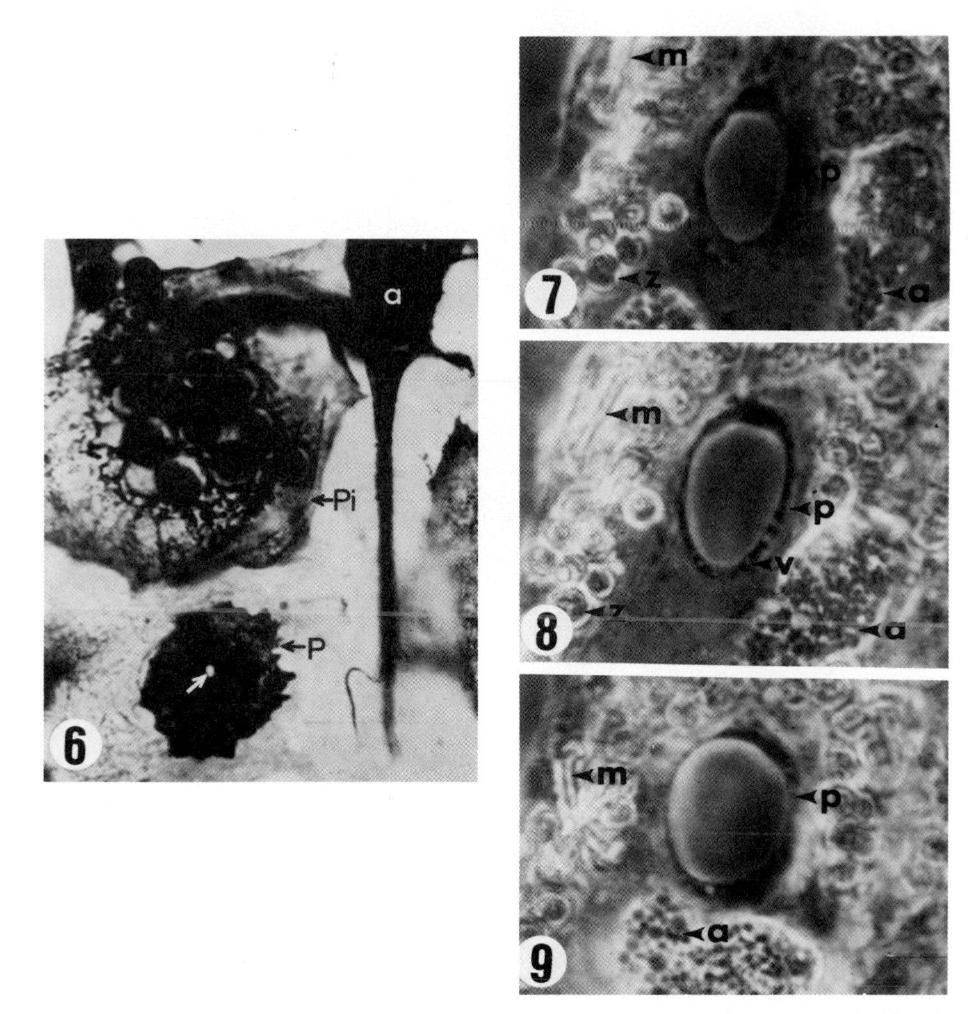

Fig. 6. The superior surface of the sponge *Corvomeyenia carolinensis* is characterized by exopinacocytes (Pi) and porocytes (p), which are separated from each other by a noncellular layer. The porocyte exhibits partial pore dilation (white arrow). An ameboid cell (a) migrates over the outer surface of the sponge. Protargol impregnation. × 750. (Harrison, '74b.)

Figs. 7–9. Living porocytes of *Corvomeyenia carolinensis*. A single porocyte is seen in the process of dilation. Porocyte cytoplasm is seen as a dark ring surrounding the elliptical central pore. Various ameboid cells (a) are seen in the series as they migrate near the porocyte. Cytoplasmic vacuolation (v) increases as the porocyte dilates. m, microsclere; p, porocyte; z, intracellular zoochlorella. Phase contrast of the living sponge. × 750. (Harrison, '72b.)

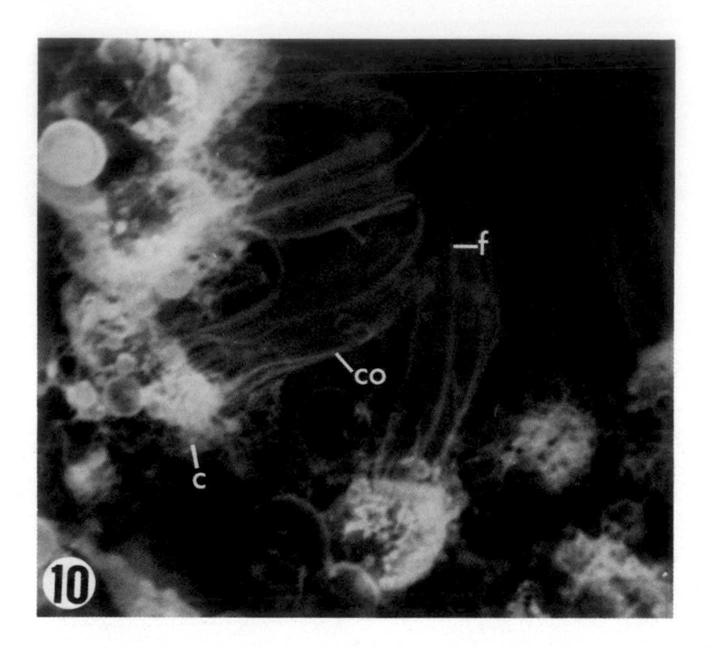

Fig. 10. Choanocytes (c) of *Ephydatia fluviatilis* project their collars (co) and flagellae (f) into the lumen of a choanocytic chamber. Observe that the choanocytes do not rest upon a basement membrane. SEM × 5,700. (Courtesy of Dr. Louis De Vos.)

of indigestible debris into the excurrent system. [For further information on freshwater sponge feeding, clearance rates, algal symbioses, etc., see Gilbert and Allen ('73a, '73b), Frost ('78, '80a, '80b), Frost and Williamson ('80), Willenz ('80), and Williamson ('79).] In some marine sponges, gametes are derived from choanocytes (for review see Simpson, '80). Leveaux ('42) reported that spermatocytes develop from choanocytes in *S. lacustris* and *E. fluviatilis*. Since freshwater sponge choanocyte nuclear differentiation closely approaches that of terminally specialized larval epithelial cells (Harrison and Cowden, '75c) and as Leveaux's observations were based upon static material, the question warrants further study.

Cells of the Mesohyl

Archeocytes. Archeocytes are totipotent ameboid macrophagelike cells that occur throughout the mesohyl (Fig. 57). They have a prominent nucleolus within an enlarged, vesicular nucleus. The cytoplasm, which typically contains considerable RNA, is often filled with inclusions, many of which are phagosomes exhibiting acid phosphatase activity. Archeocytes, with their

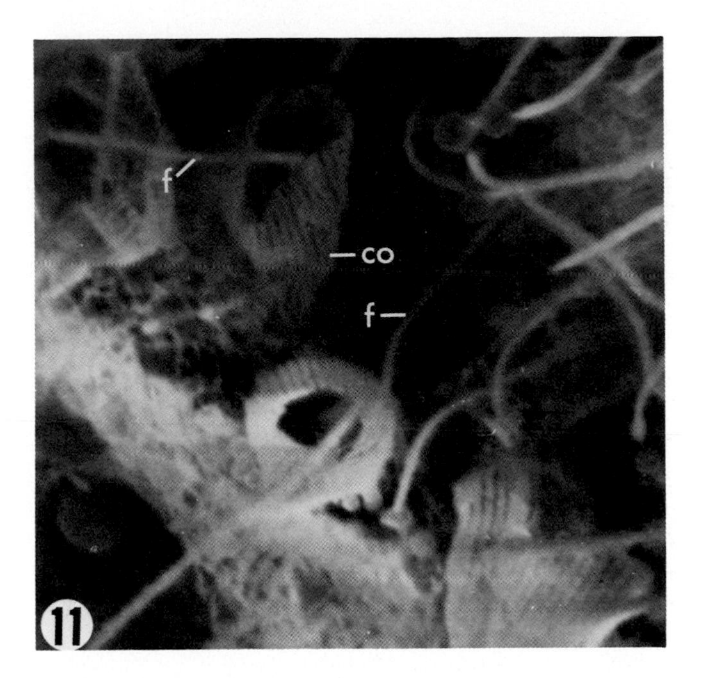

Fig. 11. Apopyle of a choanocytic chamber. The microvillar nature of the choanocyte's collar is evident. co, collar; f, flagellum. SEM × 10,000. (Courtesy of Dr. Louis De Vos; De Vos, '79.)

virtual blastomeric potential for differentiation, play a major role in sexual reproduction, since in all likelihood they develop into either ova or spermatocytes. They also persist as a significant cell population in larvae and undergo cytodifferentiation, either prior to or following larval attachment to the substratum (Brien and Meewis, '38; Harrison and Cowden, '75b, '75c). A number of asexual processes are also based upon the developmental potential of archeocytes. As they become thesocytes of gemmules, archeocytes form a true stem-cell population from which all the cells of the newly hatched sponge will be derived.

Collencytes, lophocytes, and spongocytes. Collencytes are somewhat nondescript cells, recognizable in the mesohyl only by their apparent lack of a nucleolus and by their stellate shape with branching pseudopods. Along with the more easily recognized lophocytes, collencytes actively synthesize and secrete collagen fibrils. Lophocytes (Figs. 12, 56) are polarized ameboid cells that trail a tuft of collagen fibrils. The nucleus contains a nucleolus. The rough endoplasmic reticulum is highly developed and the cytoplasm contains

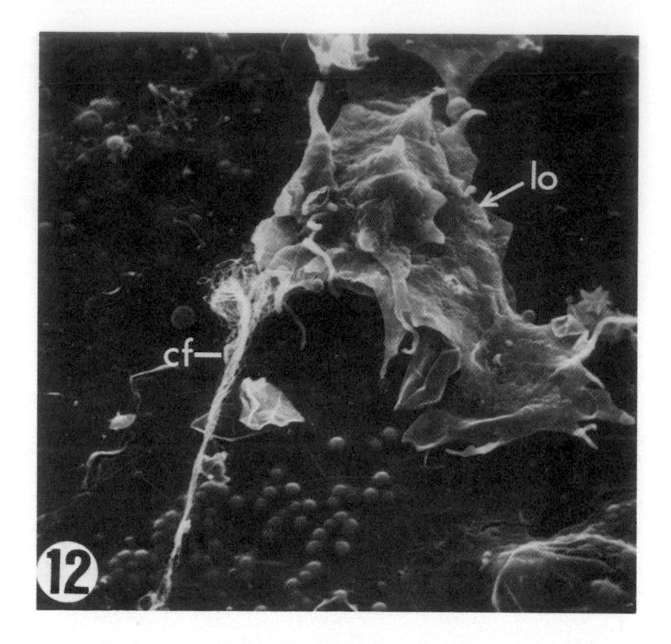

Fig. 12. A lophocyte (lo) migrating over basopinacocytes trails a tuft of collagen fibrils (cf). SEM × 3,400. (Courtesy of Dr. Louis De Vos.)

inclusions that may be homogenous or may contain filaments but are not phagosomes (Garrone, '78).

Spongocytes are archeocytelike cells that secrete spongin, a special form of collagen. They are recognizable by their position in the sponge and by their morphology, at least during the period of most active secretory activity. Garrone ('78) defines three forms of spongocytes: spongocytes secreting the perispicular spongin (which is only seen in freshwater sponges as a fine spicular sheath or as a tuft at junctions in fasicles of spicules), spongocytes secreting the gemmule coat, and basopinacocytes secreting the basal spongin. All contain a very highly developed rough endoplasmic reticulum but, at least in basopinacocytes, the period of synthetic activity, and thus the period during which the cells exhibit typical morphological characteristics, may be quite transitory. All aspects of collencytes, lophocytes, and spongocytes have been examined in detail by Garrone ('78).

Sclerocytes. Sclerocytes secrete spicules (Figs. 13,14) intracellularly. In early stages of spicule formation their nucleus contains a nucleolus and they contain considerable cytoplasmic RNA, which seems to be associated with the synthesis of an internal proteinaceous axial filament. This filament lies

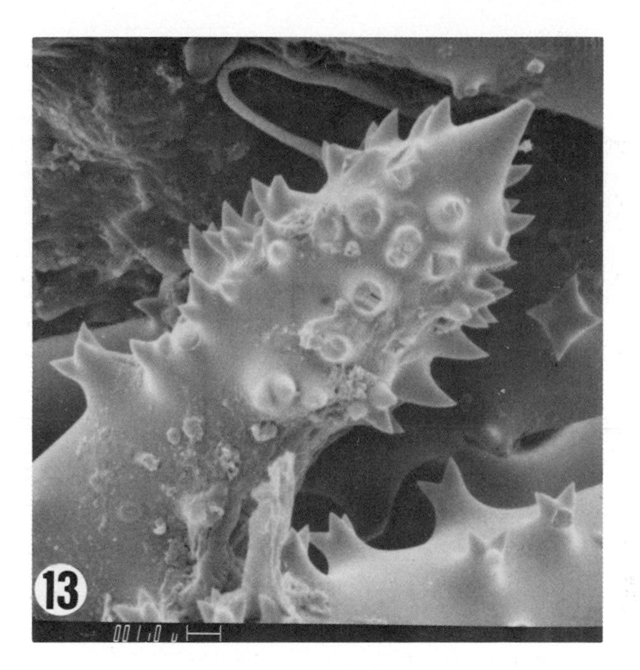

Fig. 13. Siliceous megascleres of *Trochospongilla pennsylvanica* bear multiple spines at the spicule tip. SEM. Scale bar represents 1 μm.

within a vacuole bounded by a membrane, the silicalemma, that functions in the transport and deposition of silica upon the axial filament (for further details of spicule formation see Simpson and Vaccaro, '74; Simpson et al., '79; Ledger and Jones, '77; Garrone, '69; Drum, '68; and for reviews refer to Brien, '73b; Jones, '79; Hartman, '81; Garrone et al., '81; Simpson, '81).

Cells containing inclusions. Besides sclerocytes, a number of freshwater sponge cells contain characteristic cytoplasmic inclusions. Trophocytes, or "nurse cells," are nutritive cells involved in oogenesis and gemmule development (Figs. 15–21). Cystencytes, seen particularly in *E. mülleri* (Tessenow, '69), contain a very large spherical inclusion while the "yolk cells" (Harrison, '74b) of *Corvomeyenia carolinensis* are filled with numerous inclusions of the same general size range. The inclusions of all these cells except thesocytes (binucleate archeocytes of dormant gemmules) appear to consist mainly of glycoproteins. Stylocytes of *C. carolinensis* contain an intranuclear protein crystal in addition to cytoplasmic glycoprotein inclusions (Harrison et al., '74). The function of most of these cells is not understood, although Bretting ('79) demonstrated that lectins were stored inside vesicles in the "spherulous cells" of several species of marine sponges.

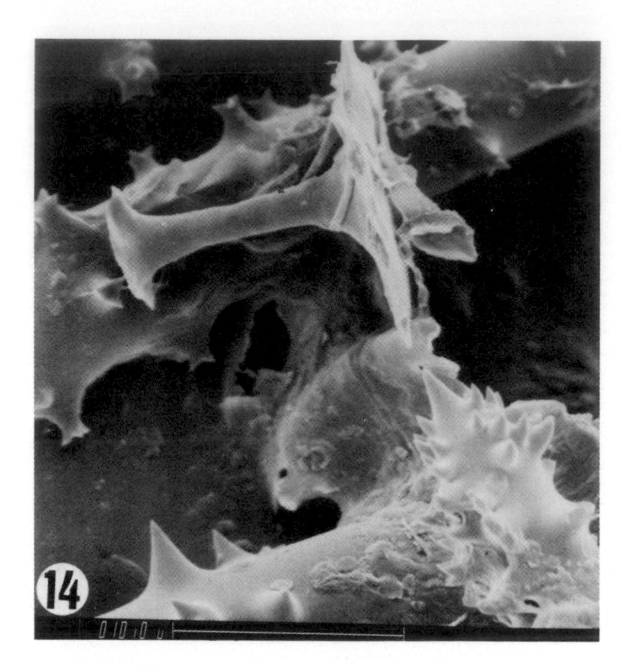

Fig. 14. A characteristic *T. pennsylvanica* gemmosclere, a minute birotulate with a slender shaft and terminal rotules of greatly differing diameters, lies among spined megoscleres. SEM. Scale bar represents 10 μm.

Fig. 15. Earliest recognizable oocyte. This phase of oocyte growth involves direct assimilation of reserve materials from the sponge mesohyl. Although sialic acid-enriched carbohydrates (s) are associated with the cell surface and with perinuclear patches (p), fluorescence does not exceed that of perispicular spongin. a, archeocyte; o, oocyte. × 1,500. Dansylhydrazine method for sialic acid in Figures 15–22, 24, and 25. (Harrison and Cowden, in press.)

Fig. 16. As development has proceeded, cells from the sponge mesohyl have migrated to the periphery of the oocyte (o). Inclusions rich in sialic acid in these cells suggest that they are trophocytes (t). Accumulations of sialic-acid-enriched carbohydrate are visible in the oocyte cytoplasm and at the cell periphery. n, oocyte nucleus × 1,500. (Harrison and Cowden, in press.)

Fig. 17. During the second phase of oocyte growth, the developing oocyte phagocytizes trophocytes and synthesizes vitelline platelets. The oocyte is surrounded by trophocytes (t). Note sialic acid at the oocyte periphery and in cytoplasmic inclusions (i). n, oocyte nucleus. × 1,500. (Harrison and Cowden, in press.)

Fig. 18. Phagocytized trophocytes (t) within the cytoplasm of the oocyte are recognized by their considerable cytoplasmic and nucleolar fluorescence following specific fluorochroming for sialic acid. Vitelline platelets (p) fluoresce in the oocyte cytoplasm and sialic acid (s) is present as a layer at the oocyte surface. The oocyte nucleolus is not in the plane of section although the nucleus (n) is seen as a dark region. × 1,500. (Harrison and Cowden, in press.)

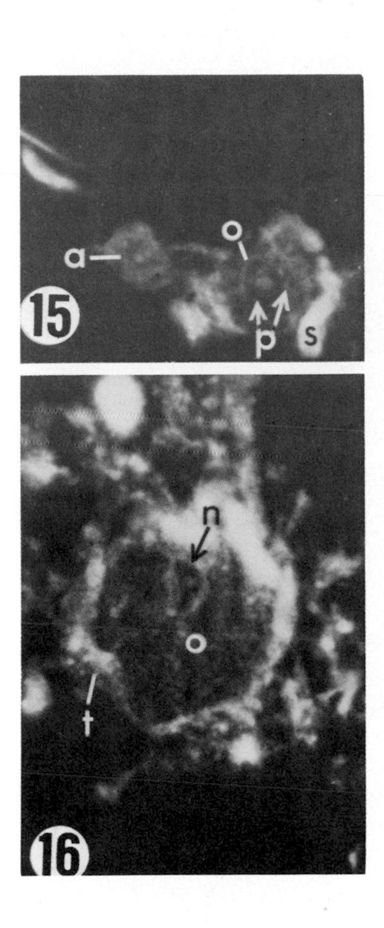
a
o
p
s
15
n
o
t
16

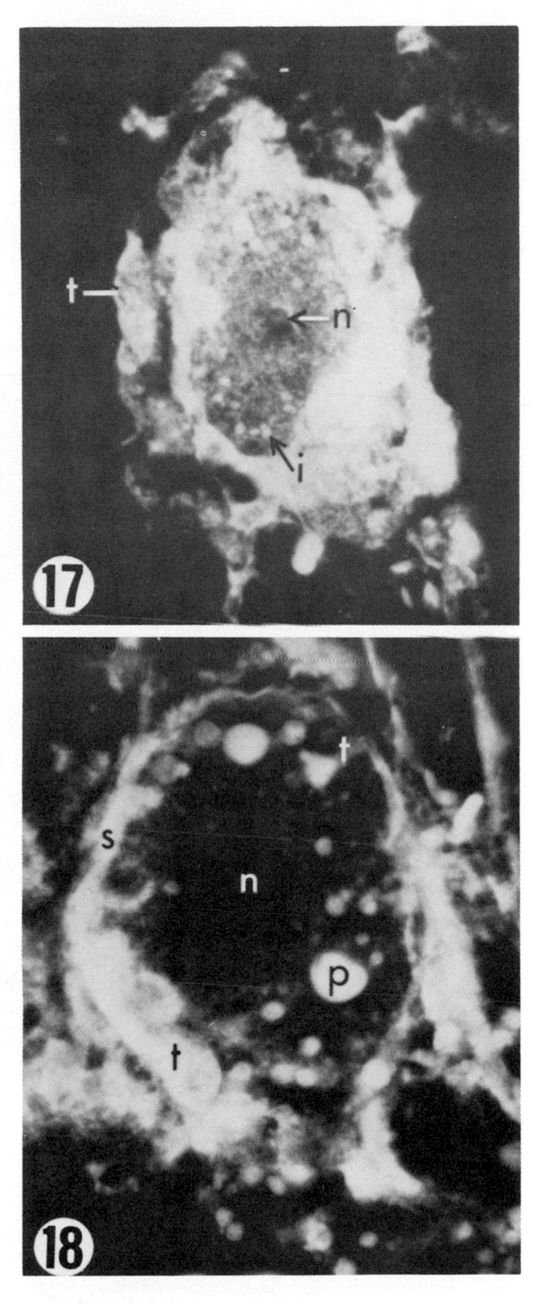
t
n
i
17
t
s
n
p
t
18

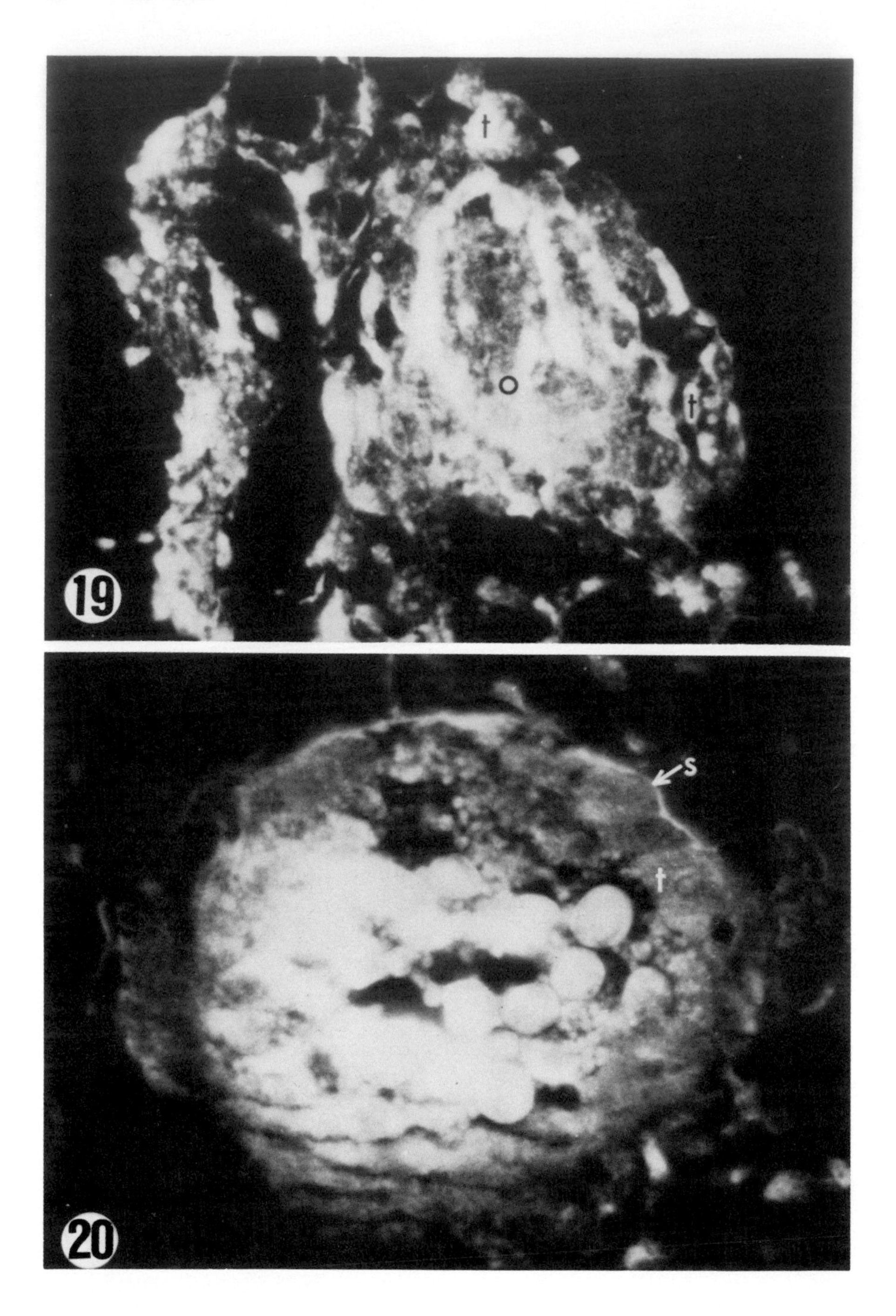
t
o
t
19
s
t
20

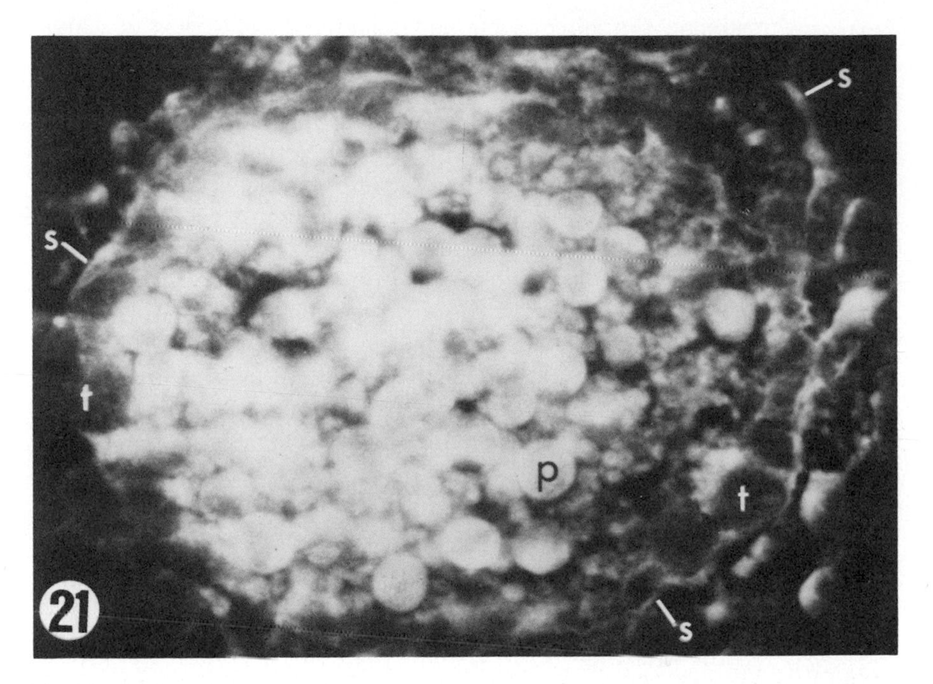

Fig. 21. Vitelline platelets (p) obscure most internal detail in this mature oocyte. Phagocytized trophocytes (t) exhibit polarized cytoplasmic fluorescence as they apparently continue to pass carbohydrate reserves to the oocyte. Sialic acid (s) is also present on the surface of the oocyte. × 1,250. (Harrison and Cowden, in press.)

NATURAL HISTORY OF ADULT FRESHWATER SPONGES

Systematics is central to understanding the natural history of freshwater sponges but, until recently, spongillid systematics was hopelessly confused. The revisionary work of Penney and Racek ('68) brought some degree of order into this area and demonstrated global evolutionary patterns within the gemmule-forming spongillids. Studies by Poirrier ('69, '74, '76) showed ecomorphic variation in freshwater sponges and established the interrelationship of environmental parameters and classification.

The natural history and ecology of freshwater sponges, particularly the North American species, have been reviewed by Harrison ('74a) in a species-

---

Figs. 19, 20. As the oocyte (o) grows it contains and is surrounded by trophocytes (t). Note the phagocytized trophocyte (t) in Figure 20. Sialic acid (s) is still present on the surface of the oocyte. × 1,250. (Harrison and Cowden, in press.)

by-species treatment. For an excellent and comprehensive review of demosponge ecology, refer to Sarà and Vacelet ('73). Considerable work still needs to be done to define the basic life events and natural history of many freshwater sponge species, particularly those occurring in the Southern Hemisphere and in subtropical and tropical zones of the Northern Hemisphere.

Freshwater sponges will attach and grow upon almost any substrate. There are definite species preferences for lentic or lotic environments. Although excessive turbulence with accompanying buffeting and scouring tends to eliminate sponges from habitats subject to severe flooding, sponges benefit from mild current flow that brings food and oxygen, removes waste, and washes off sediments. The effects of light intensity upon sponges vary, as species tolerances are not uniform. The effects of light intensity are always modified by water color and transparency. Water temperature, in addition to influencing life cycle events, may also influence morphological features and may actually be a limiting factor.

Sponges may be regularly collected from water of temperatures up to 37°C although the combination of higher water temperatures and declining water levels is frequently associated with sponge degeneration and/or gemmule formation (Poirrier, '69). While a good many freshwater sponge species survive only as gemmules during the winter, some may undergo limited regression (Van de Vyver and Willenz, '75) similar to that reported in some marine sponges (Simpson, '68). The distribution of freshwater sponge species can be correlated with the chemical properties of the habitat and the individual tolerances of the species to these factors (Harrison, '74a, '77, '79; Harrison and Harrison, '77, '79; Poirrier, '69, '76).

General Normal Embryology

Sexuality and sex determination. Freshwater sponges have evolved a number of sexual and asexual reproductive mechanisms. Within the Porifera a wide variety of sexual conditions exists (Sarà '74, for a review of sexuality in the phylum). Among the freshwater sponges, there have been few detailed studies of life cycles. However, these studies indicate a wide degree of variability in the form of sexuality and sex determination. Freshwater sponges may exhibit contemporaneous hermaphrodism (*Ephydatia fluviatilis,* Leveaux, '42; *Spongilla lacustris,* Leveaux, '42; *Eunapius fragilis,* Harrison, unpublished observations; *Trochospongilla pennsylvanica,* Simpson and Gilbert, '73); may be dioecious (*Ephydatia fluviatilis,* Van de Vyver and Willenz, '75; *Spongilla lacustris,* Gilbert, '74); or periodically dioecious (*Ochridaspongia rotunda,* Gilbert and Hadzisce, '77). There are also cases of sex reversal in which individual sponges undergo annual sex reversal (*S. lacustris,* Gilbert and Simpson, '76) or a relatively long-standing sex reversal of a strain of one species may occur. Van de Vyver and Willenz ('75) observed that the β strain of *E. fluviatilis,* after undergoing oogenesis and producing larvae in 1969, was exclusively male for the next 3 years.

There is need for further study to clarify the nature of sexuality and sex determination within the freshwater sponges. This will entail detailed microscopic examination of tissue collected from the same individual throughout the entire reproductive phase of the life cycle. Whenever possible, sponges used in such studies should be reared in the field from individual gemmules. In those species that undergo total seasonal degeneration, leaving only gemmules in the skeletal framework of the parent sponge, small pieces (1 mm^2) of tissue must be excised weekly for microscopic examination throughout the complete growth period. On the other hand, in species that undergo seasonal regression with the onset of winter or of the hot season but that survive in a disorganized condition, samples should be taken throughout the entire year.

Gametogenesis. The relatively few studies of life cycle events of spongillid freshwater sponges indicate that, at least in the northern temperate zone, embryogenesis occurs during the early summer months. Brien ('67c) described embryogenesis in specimens of the potamolepid African sponge *Potamolepis stendelli* collected in December and in March. Efremova and Efremova ('79) collected larvae in July and August from the endemic lubomirskiid sponge *Baikalospongia bacillifera.*

Gilbert and Hadzisce ('77) found reproductive activity occurring throughout the year in *Ochridospongia rotunda* a benthic species subjected to little variation in water temperature. In *Ephydatia fluviatilis,* which undergoes limited tissue regression during overwintering in colder areas of the northern temperate zone, Van de Vyver and Willenz ('75) observed oocytes as early as November and throughout the winter months. Vitellogenesis occurred in May and was closely correlated with the timing of spermatogenesis in the male β strain. Only those sponges that had overwintered entered into sexual reproduction. This report is in marked contrast to Gilbert's observations ('74) that oocyte production in *S. lacustris* requires no environmental stimulus specific to a particular time of year. Gilbert considers that oocyte formation is under endogenous control, most probably associated with gemmule hatching. In Gilbert's study, oocytes differentiated under conditions of increasing and decreasing water temperatures and day lengths, and were always formed in implanted gemmule-bearing sponge skeletal frameworks about 1 week after gemmule hatching. Earlier, Simpson and Gilbert ('73) found that in natural populations of *S. lacustris* and *Trochospongilla (Tubella) pennsylvanica* egg formation occurred over a 6–7 week period in May and June following hatching from gemmules during late April and early May. Spermatogenesis occurred over a shorter period during May–June, beginning 3 to 4 weeks after initiation of egg production, and was correlated with an increase in temperature to 20°C. Egg production began in *S. lacustris* when water temperatures were 5°–10°C, whereas water temperatures were between 15° and 20°C in late May when eggs were observed in *T. pennsylvanica.* Simpson's study ('68) of life cycle events in the marine poecilosclerid sponge *Microciona prolifera* indicates that gametogenesis in that species is closely related with the rise in water temperatures. Eggs appear when temperatures range between 10° and 18°C and spermatozoa appear when temperatures are between 16° and 20°C.

The nature of the control of life cycle events, including gametogenesis, is simply not known at this time. It seems possible that, at least in some species, phases of life cycle may be endogenously controlled or that there may be requirements associated with prior overwintering (Van de Vyer and Willenz, '75) gemmule hatching (Gilbert, '74), or a possible induction of spermatogenesis by eggs (Simpson, '68). Recent studies by Rasmont ('74) and Harrison et al. ('81) implicating cyclic nucleotides in sponge developmental processes suggest that the control of life-cycle involves mechanisms that are neither straightforward nor simple.

COLLECTING TECHNIQUES

The methods employed in collecting sponges will often be determined by the collector's proposed use of the material. To assure species identification one must collect gemmule-bearing specimens since most taxonomic schemes in use today employ gemmule and gemmosclere (gemmule spicule) morphology as diagnostic criteria. It is not necessary to collect entire specimens.

For developmental studies, it is often useful to collect gemmules rather than collecting the entire sponge. Following gemmule production and deterioration of the parent sponge, gemmules can be found as a basal mat or, in branching species, retained within the parent sponge skeletal framework. These mats or skeletal branches can be gathered and transported to the laboratory in habitat water in Zip-lock plastic freezer bags placed into an ice-filled cooler. Following transfer of the gemmules to capped vials, they should be stored in the dark at 4°C in Millipore-Q–filtered habitat water or in an appropriate artificial medium. It will be necessary to change the fluid a number of times until complete decomposition of the parent sponge and associated debris. Gemmules can be stored under these conditions for a number of months and can even be shipped air freight if packed with dry ice in an insulated container. Gemmules can be carefully removed from the mat or from the skeletal framework by dissection over ice using watchmakers' forceps. If individual gemmules are to be returned to storage it is imperative that they not be allowed to warm during the dissection as dormancy will be broken in many species.

There are few commercial sources of freshwater sponges. Carolina Biological Supply Company in Burlington, North Carolina, has advertised them in the past.

MAINTENANCE OF SPONGES IN THE LABORATORY

Methodology

Although adult sponges can be brought into the laboratory tied to microscope slides and reared in dishes containing filtered pond water (Penney, '32; Harrison, '74b), sponges used in laboratory studies are in most cases reared from gemmules (see above) that have been stored refrigerated in the dark at 0°–4°C. Sponges may be maintained in the laboratory in continuous flow (Imlay and Paige, '72; Poirrier et al., '81) or in static systems (Rasmont, '61; Mank and Kilian, '79; Rozenfeld and Curtis, '80). Because of the accumulation of bacteria and fungal contamination on the gemmule surface, it may be necessary to treat the gemmules with antiseptic agents prior to any experimental study. Rozenfeld and Curtis ('80) applied successive treatments, with agitation, of 2% hydrogen peroxide for 1 minute and 2% sodium hypochlorite for 2 minutes, followed by five washes in sterile mineral medium, and completely freed gemmules from bacterial and fungal contamination and grew sponges in sterile culture. Although hatching was retarded there was no decrease in the percentage of gemmules hatching or in rates of normal development.

Double sterilization of gemmules appears to be most critical when sterile cell culture is planned, particularly following hydroxyurea inhibition of choanocyte differentiation (Rozenfeld and Rasmont, '76). However, with usual culture procedures, cleansing with only hydrogen peroxide (Rasmont, '61, '75) is sufficient as this will cleanse gemmules of all contaminants except bacteria. During development into normal sponges, the bacterial count remains surprisingly low, i.e., 100–1,000 bacteria/ml in a 25-ml Petri dish containing 120 gemmules (Rasmont, '75; Rozenfeld and Curtis, '80). This is apparently because of the bacterial diet of sponges.

Sponges have been maintained in other laboratories without prior sterilization of gemmules, but in these cases the medium was changed each day (Mank and Kilian, '79) or a continuous flow system was used with addition of streptomycin, penicillin, and gentamycin at initial concentrations of 50 μM to control growth of filamentous bacteria (Poirrier et al., '81).

Sponges have been reared successfully in a variety of media. Ostrom and Simpson ('79) grew *S. lacustris* from gemmules in water that had been distilled and subsequently purified in a Millipore Milli-Q_2 system. Rasmont ('61) raised sponges, particularly *E. fluviatilis*, in "M" medium, a dilute ionic solution[1]. The optimum concentration and ionic balance of mineral media, often reflecting habitat conditions, may vary even for the same species. Stock solutions for mineral media should be heat sterilized, as heating the complete medium precipitates carbonates and silicate (Rozenfeld and Curtis, '80). Mank and Kilian ('79), growing gemmule-derived *S. lacustris* in either pond water or nonsterile Eagle's basal medium plus Earle's salts (BME), successfully reared sponges for over a year. Media were prepared (with and without silicates; 2.1 mg SiO_2/liter) by dissolving 0.1 gm BME and 0.022 gm $NaHCO_3$ in 1 liter of double-distilled water. They found it essential to change the medium daily as the sponges' growth rate was significantly reduced or the sponges died if left in the same medium for more than 2 days. They attributed this to increased bacterial and fungal contamination and to the increase in metabolic products. Poirrier et al. ('81), using a continuous-flow system, reared *E. fluviatilis* and *Spongilla alba* in habitat water filtered through Whatman #1 filters and filtered again through 0.45 μm-pore–diameter Millipore filters to remove microorganisms. Millipore-filtered habitat water was diluted with deionized water to a salinity of 1 ppt. Culture vessel water was renewed by continuous flow at a rate of approximately 100 ml/hour, exchanging the water in the vessel every hour. Water was constantly stirred by a magnetic stir-bar. An air flow of about 0.5 liters/minute was directed at the water surface in the vessel.

Although sponges hatched from gemmules will survive for some time while they use nutrient reserves stored in the cells, one must supply food for long-term culture of sponges. Rasmont ('61) fed sponge cultures suspensions of *Escherichia coli* bacteria that had been thoroughly washed with "M" medium, killed by tyndallization, and kept deep-frozen until use. Mank and Kilian ('79) fed sponges daily a variety of materials, i.e., commercial fish food (Liquifry No. 1, Bio Min 66, etc.), bacteria, and the yeast *Saccharomyces cerevisiae*. Liquifry was fed at a rate of 0.05 ml/liter of moderately filtered pond water. Bio Min was fed at 20 mg/liter. Liquifry produced flourishing growth, and sponges grew well with Bio Min. Sponges fed 15 mg of *S. cerevisiae* also did quite well. Similar growth was obtained when these nutrients or *E. coli* (0.5 μl suspended in 0.2 ml distilled water) were added to BME. Sponges held in a mixture of pond water and BME (3:1) and fed on Liquifry or *E. coli* also produced gemmules and lived for several months. Imlay and Paige ('72) fed sponges trout fry food (Glencoe starter granules) at 0.5 gm/day in a continuous flow multichamber system. Food was placed in the prechamber to avoid contamination of the sponges with suspended

[1] Ca^{++}, 2.00 mEq per liter; Mg^{++}, 1.00 mEq per liter; Na^{+}, 1.00 mEq per liter; K^{+}, 0.05 mEq2 per liter; Cl^{-}, 2.05 mEq per liter; SO_4^{--}, 1.00 mEq per liter; SiO_3^{--}, 0.50 mEq per liter; and HCO_3^{-}, 0.50 mEq per liter.

sediment. Accumulated food was removed every 7–10 days. Poirrier et al. ('81) fed sponges a pulse of live *E. coli*, introduced into a continuous flow of filtered habitat water every 12 hours. Sponge growth rate was a function of bacterial pulse concentration.

It is apparent that freshwater sponges can be maintained in the laboratory for long periods of time and will form gemmules and possibly larvae. I suggest a culture system organized around the following procedures:

1. Cleansing or, for cell culture, sterilization of gemmules following removal from the refrigerator (Rozenfeld and Curtis, '80);
2. Maintenance of gemmule-derived sponges in, preferably, a continuous flow system or, alternately, a static system in which the medium is changed daily;
3. Utilization of any appropriate medium (above) following experimental determination of the optimum medium for the particular species to be cultured; and
4. Feeding with Liquifry (Mank and Kilian, '79) or with bacteria (Rasmont, '61; Poirrier et al., '81).

Use of a magnetic stirrer and stir-bar to generate current and use of directed air flow (Poirrier et al., '81) appear to be quite beneficial. Developing sponges can be grown in the culture on Petri dishes (Rasmont's group uses clusters of gemmules, 120 total gemmules/25-ml Petri dish); between two glass slides (Ankel and Eigenbrodt, '50) in "sandwich culture," which is particularly suitable for time-lapse cinematography; on microscope slides or cover slips for vital staining or phase-contrast studies (Harrison, '72b); on Parafilm which facilitates sectioning; or on Epon or other plastics.

Obtaining and Rearing Embryos

Freshwater sponge larvae are retained in the parent sponge until they exhibit many characteristics of adult sponges. In spite of this long period of incubation, there have been only isolated studies in which larvae were collected and manipulated experimentally. In these (Van de Vyver, '70; Van de Vyver and Willenz, '75), sponges hatched from gemmules were reared on glass plates for 2 weeks in the laboratory and were subsequently transferred to the field, i.e., a pond outflow, for further growth. Plates bearing the sponges were brought into the laboratory during the period of larval release, allowing examination (a hundred free swimming larvae in 1 ml of "M" medium).

FEATURES OF DEVELOPMENT

Sexual Reproduction

Origin of germ cells. Germ cells in sponges develop from either ameboid archeocytes of the mesohyl or from choanocytes, depending upon the species (for reviews see Tuzet, '64; Brien, '73b; Fell, '74a; Diaz, '79). Freshwater sponge oocytes apparently develop from mesohyl archeocytes (Leveaux, '42), although in other groups of sponges oocytes may arise through transformation of choanocytes (Duboscq and Tuzet, '44; Tuzet, '47; Sarà, '55; Tuzet and Pavans de Ceccatty, '58; Diaz et al., '73). Spermatocytes were reported (Leveaux, '42) to develop from choanocytes in *S. lacustris* and *E. fluviatilis*. There are numerous reports of similar origins of spermatocytes in other sponges (Dendy, '14; Gatenby, '20, '27; Tuzet and Pavans de Ceccatty, '58; Tuzet et al., '70; Diaz and Connes, '80). However, there have been as many reports that mesohyl archeocytes give rise to spermatocytes (Okada, '28; Tuzet, '30; Fincher, '40; Lévi, '56). In related studies, Harrison and Cowden ('75b, '76b) demonstrated cytophotometrically that deoxyribonucleoprotein organization in the condensed nuclei of choanocytes and terminally differentiated larval epitheliocytes of *Ephydatia fluviatilis* and *Eunapius fragilis* was quite similar. The similarity of high extinction values suggested a correspondingly low level of transcriptional activity and raised the question of the likelihood of a choanocytic origin of germ cells in these sponges. There have been no experimental studies that conclusively demonstrate the origin of germ cells and, since all extant interpretations are based upon static observations of mostly fixed material, a conclusive statement cannot be made at present.

Oogenesis. Early stages of oocyte growth involve assimilation of reserve materials, probably by pinocytosis, directly from the mesohyl of the sponge. The second phase of oogenesis begins as trophocytes containing reserve material in discrete cytoplasmic inclusions migrate to the surface of the oocyte. The oocyte engulfs trophocyte nurse cells that have migrated through the surrounding sheath of follicle cells and synthesizes vitelline platelets. Following this apparent phagocytosis, or possibly partial surrounding of trophocytes by the oocyte, the plasma membranes and intercelluar structures of these cells appear to remain intact and relatively undegraded throughout the period of oogenesis (Figs. 15–21). Fell ('69) also reported the presence and persistence of trophocytes in several species of haliclonid marine sponges. As Fell ('74a) noted, the physiological relationship of trophocytes to oocyte growth and development in sponges has not been satisfactorily defined, although a number of investigators have examined this relationship and the problem of vitellogenesis in sponges (for reviews see Diaz et al., '75, or Diaz's dissertation, '79). Meiotic divisions in freshwater sponge oocyte development have not been adequately described, but in marine sponges maturational divisions and polar body formation generally occur following oocyte growth (Fell, '74a). In the freshwater sponges *S. lacustris, E. fluviatilis, Baicalospongia bacillifera,* and *Lubomirskia baicalensis* meiotic prophase is initiated prior to the period of yolk accumulation (Leveaux, '42; Gureeva, '72) in oocytes.

Spermatogenesis. Spermatogenesis in freshwater sponges has been described by Leaveaux ('42). Masses of spermatogonia are contained within cysts in *S. lacustris* and *E. fluviatilis.* The cysts are composed of follicle cells derived through the aggregation of mesohyl ameboid cells.

Following mitosis, spermatogonia of approximately 7–9-μm diameter become primary spermatocytes (Fig. 22). Secondary spermatocytes, of 3–5-μm diameter, divide to form spermatids. Spermiogenesis has not been adequately described. Leveaux ('42) reported that the spermatid nucleus becomes condensed, associates with two centrioles, and lacks an acrosome. A flagellum, approximately 130 μm long, develops and becomes oriented toward the periphery of the cyst.

Recent ultrastructural studies of spermatogenesis in the marine demosponges *Aplysilla rosea* (Tuzet et al., '70) and *Suberites massa* (Diaz and Connes, '80) provide additional insight into this process. In *A. rosea*

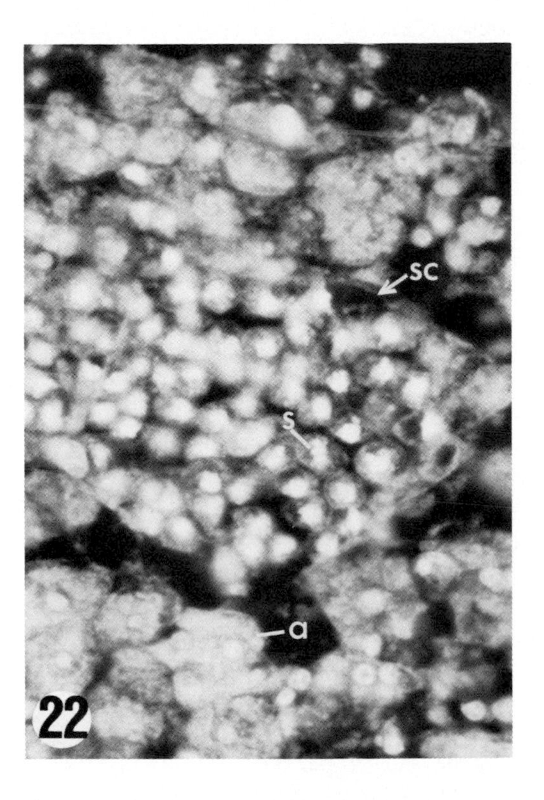

Fig. 22. A spermatic cyst (sc) contains dividing primary spermatocytes (s). Numerous archeocytes (a) surround the cyst; note their prominently fluorescent nucleoli. Acridine orange method for nucleic acids.

choanocytes are simultaneously transformed into spermatogonia by loss of microvillar collars and the posterior region of the choanocyte while the flagellum persists. The choanocyte flagellum is also present in primary spermatocytes, secondary spermatocytes, and spermatids. However, the spermatozoon flagellum develops from a centriole during spermiogenesis.

During spermiogenesis, centriole duplication in the spermatid is followed by the eccentric orientation of the nucleus in the cytoplasm, disappearance of the Golgi apparatus, and fusion of mitochondria into two large masses on either side of the oval-shaped sperm head, which lacks an acrosome.

Spermatogenesis in *S. massa* (Diaz and Connes, '80) differs slightly in that, although spermatogonia are derived from choanocytes as in *A. rosea*, early in spermatogonial development collars and choanocyte flagellae are lost. Both sponges exhibit synaptonemal complexes at the onset of the first meiotic prophase. The mature *S. massa* spermatozoon is bilaterally symmetrical with a cone-shaped head and a long flagellum implanted obliquely into the base of the intermediate piece, which contains three large mitochondrial masses. There is no acrosome but Golgi-derived vesicles occupy the apex of the sperm head.

There is considerable room for further investigation of virtually all aspects of spermatogenesis in freshwater sponges, including the basic morphological aspects of this process.

Fertilization and embryogenesis. Fertilization is apparently internal in all freshwater sponges. Sperm are released via the excurrent canals of the aquiferous system into the water, and are similarly transported into egg-bearing (female) sponges of the same species by the incurrent passages. Fertilization has never been observed in freshwater sponges. In those sponges in which fertilization has been described (for reviews see Brien, '73b, and Fell, '74a), a choanocyte-derived carrier cell is utilized. The spermatozoan enters a choanocyte that, after losing its collar and flagellum, migrates to the oocyte where the spermatozoan is deposited by direct transfer or by engulfment of the carrier-cell–spermatozoan complex. Further studies of this unusual mode of penetration and activation are warranted. Embryogenesis begins with the onset of cleavage. The oocyte, containing phagocytized trophocytes and vitelline platelets, is contained within a follicular envelope (Fig. 23). Proximally, the cytoplasm of trophocytes contains inclusions of the smaller size range seen in the oocyte, while trophocyte nuclear membranes and nucleoli persist. Unequal divisions (Figs. 24,25) produce a stereoblastula (Fig. 26) that, at least in some species, i.e., *E. fluviatilis* (Brien and Meewis, '38), consists of peripheral micromeres and internal macromeres. The micromeres differentiate into a peripheral flagellated epitheliocyte layer. Through separation the macromeres form an anterior larval cavity (Fig. 27). This anterior cavity is delimited by a layer of pinacocytes that is in turn

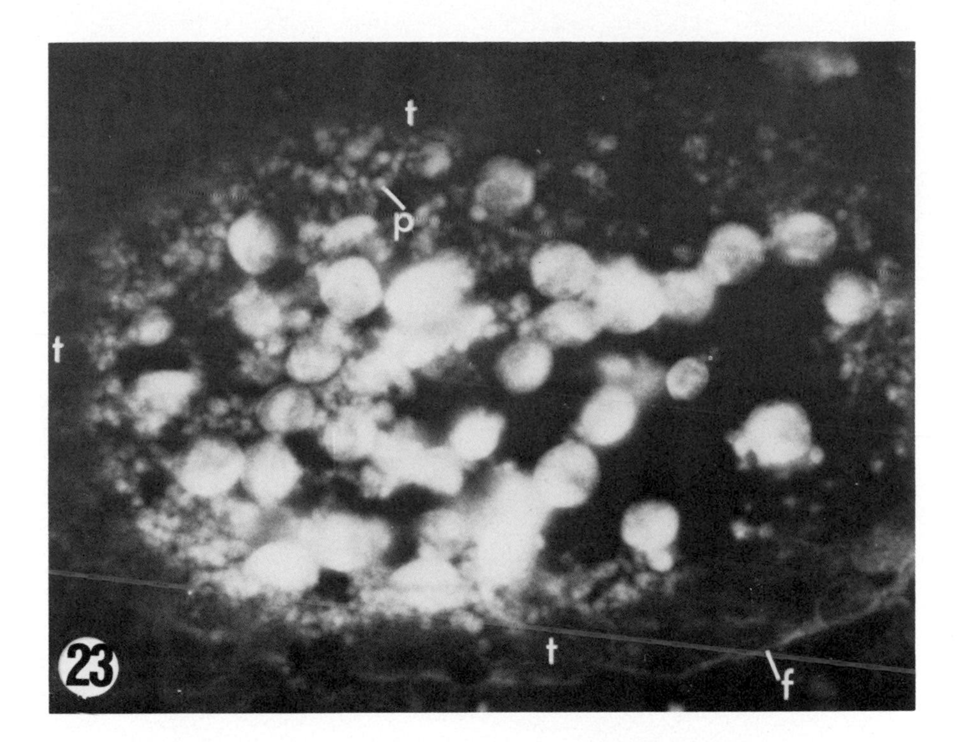

Fig. 23. Phagocytized trophocytes (t) are bounded externally by the sialic-acid-containing follicular layer (f) as embryogenesis is initiated. Numerous small inclusions (p) are associated with the internal surfaces of trophocytes, suggesting progressive degradation of phagocytized trophocytes. × 1,250. (Harrison and Cowden, '82.)

subjacent to an unorganized layer of ameboid collencytes. The collencytes are positioned immediately beneath the peripheral flagellated layer of epitheliocytes, having migrated into this region from the macromere-derived posterior archeocyte cell mass of the parenchymula larva.

The spongillid parenchymula larva is, as noted by Borojević ('70), characterized by morphogenetic precocity, a fact that influences the degradation and metabolism of cytoplasmic reserves, cytodifferentiation and the fate of individual cell types. With attainment of the escape stage, the larva is essentially a mobile adult sponge exhibiting choanocytic chambers (Fig. 28); canal systems; a well-defined connective tissue stroma, as well as a diverse cell population consisting of specialized elements and a population of totipotent archeocyte reserve cells; and a terminally differentiated epitheliocyte line (Brien and Meewis, '38; Harrison and Cowden, '75b, '75c).

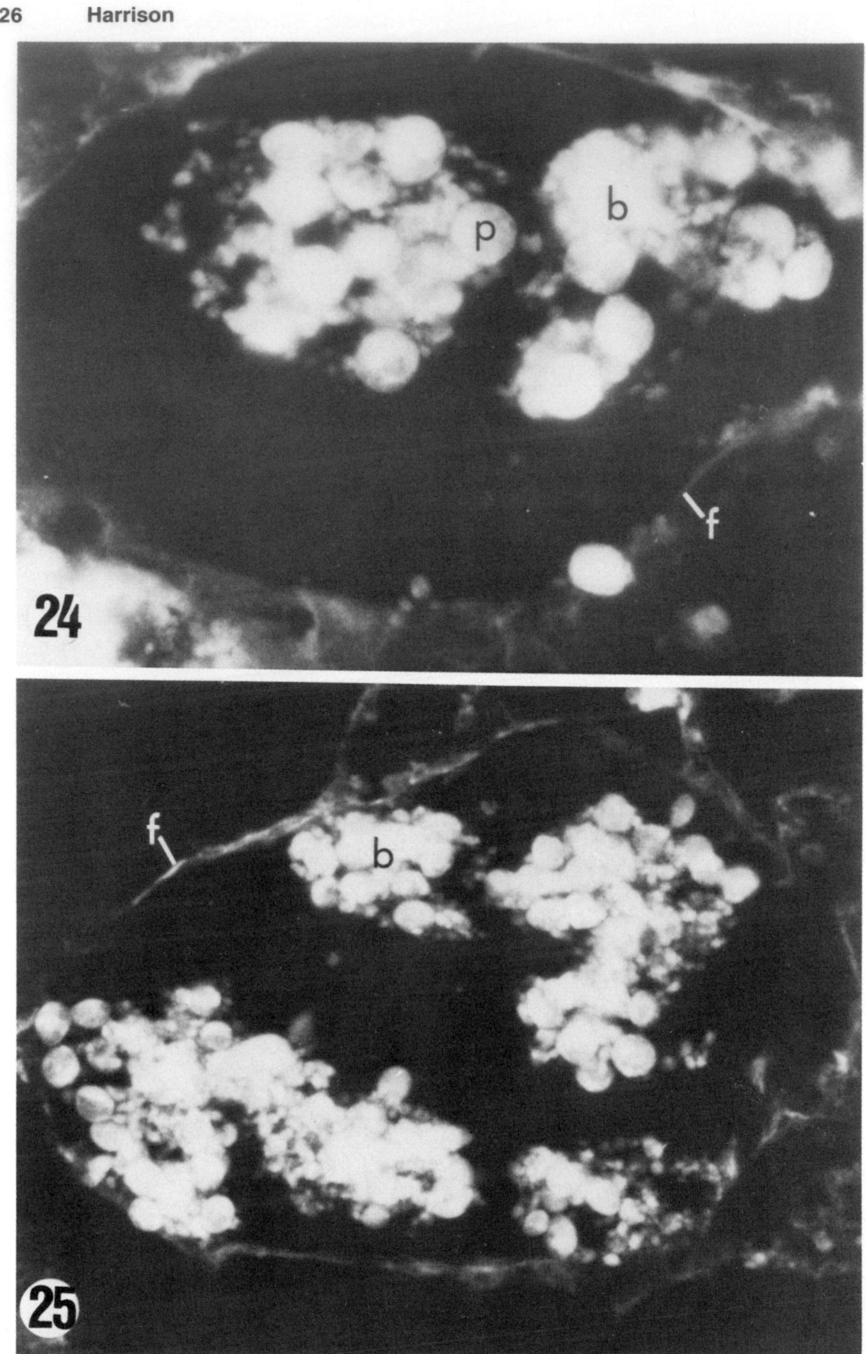

Figs. 24, 25. Two- and eight-celled embryos. b, blastomere; f, follicle; p, platelet × 1,250. (Harrison and Cowden, '82.)

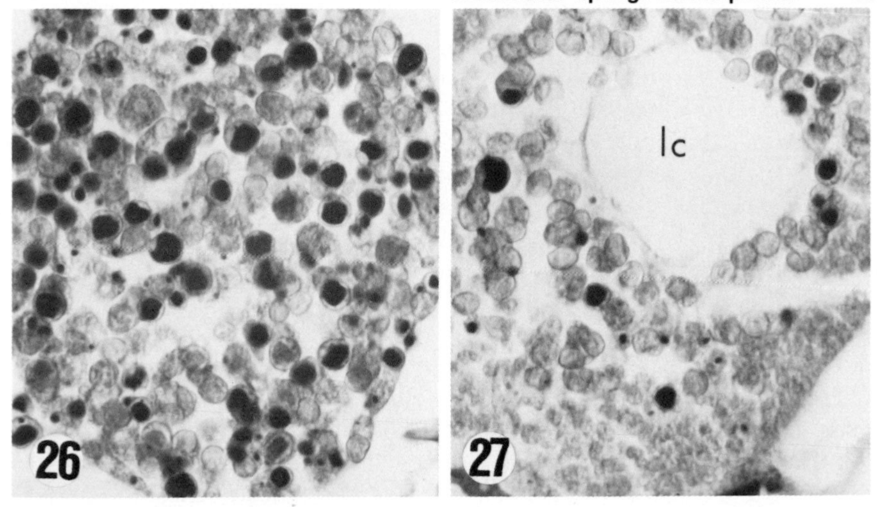

Figs. 26, 27. These figures demonstrate the mobilization of PAS-positive carbohydrates, principally neutral mucins during larval development. Figure 26 illustrates a blastula, characterized by considerable carbohydrate reserves within cytoplasmic vitelline platelets. Figure 27 represents a later phase in larval development. An anterior larval cavity (1c) is now present. Note the scarcity of PAS-positive reserves suggesting rapid mobilization of carbohydrates during this period of increased cytodifferentiation. Periodic-acid–Schiff method. × 220. (Harrrison and Cowden, '75c.)

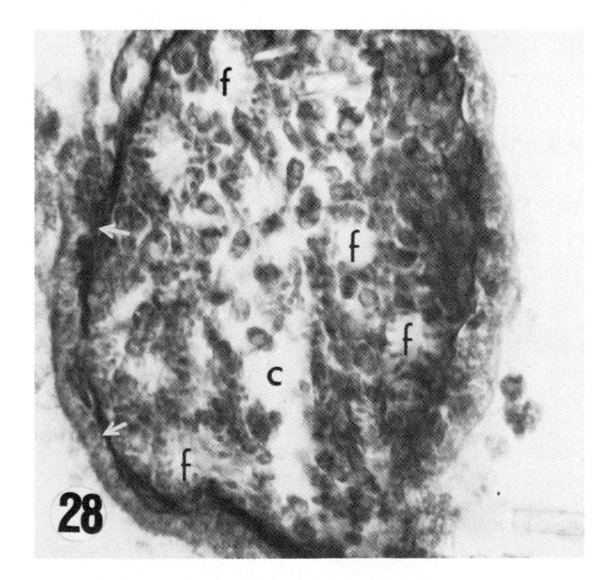

Fig. 28. A mature larva sectioned through the posterior cell mass. Note the prominent PAS-positive connective tissue layer (white arrows) underlying peripheral epitheliocytes. Choanocytic chambers (f) and a canal (c) are visible. Periodic-acid–Schiff method. × 220. (Harrison and Cowden, '75c.)

In contrast to all other poriferan larvae of the parenchymella class (Borojević, '70), the spongillid parenchymula exhibits no reversal of embryonic layers. Rather than migrating inward to form the definitive choanoderm of the adult as in *Mycale contarenii* and other species (Borojević, '66, '70), the peripheral epitheliocytes appear to have no role in morphogenesis of the spongillid parenchymula and are all eventually phagocytized by larval archeocytes. As Brien and Meewis ('38) and Brien ('67b, '73b) have reported, the parenchymula choanocytes arise from archeocytes, whereas in other poriferan larval forms choanocytes are typically differentiated from epitheliocytes—a cell type that is capable of differentiating into choanocytic chambers in tissue culture (Borojević, '66).

The progressive development of a well-defined connective tissue system in the parenchymula is another example of developmental precocity (Fig. 28). The connective tissue probably arises from two sources. In the earlier stages, when the framework is essentially a border of neutral mucin associated with the endopinacocytes peripheral to the anterior chamber, collencytes are the principal source of exoplasmic material. As the escape stage is approached, the presence of archeocytes containing cytoplasmic PAS-positive fibrillar strands suggests that these cells are functioning as spongocytes or lophocytes.

As would be expected, larval archeocytes present those morphological and cytochemical characteristics of an undifferentiated cell supporting a considerable level of RNA synthesis. Characteristically, these cells contain a vesicular nucleus with a prominent nucleolus, diffuse chromatin, and high levels of cytoplasmic RNA. The process of cytodifferentiation from larval archeocyte to terminal epitheliocyte is characterized by decreasing cytoplasmic basophilia and a progressively more condensed form of nuclear organization, leading finally to a high degree of chromatin condensation in the nuclei of terminal stage cells (Harrison and Cowden, '75b, '75c, '76b) (Figs. 29–33). Throughout this process, sialic-acid–enriched carbohydrates become increasingly localized at the distal surfaces of flagellated

Figs. 29–33. Cytodifferentiation during larval development. Figure 29 shows larval archeocyte demonstrating a vesicular form of nuclear organization with prominent nucleolus (n) and extended chromatin. Figure 30 shows maturing epitheliocytes exhibiting intermediate levels of chromatin condensation. Figure 31 shows epitheliocytes of the mature larva. These terminally differentiated cells exhibit pycnotic nuclei that are often tear-drop in shape. Flagella (f) are present. Collencytes (c) underlie the peripheral epitheliocytes, which will eventually be phagocytized. (Figs. 29–31, Feulgen reaction. × 1,500). Figure 32 shows larval blastomere fluorochromed to demonstrate basic protein. Note fluorescence associated with cytoplasmic reserve inclusions (i) and with vesicular nucleus (n). Brilliant sulfoflavine, pH 2.8 × 1,500. Figure 33 shows a larval archeocyte. Note vesicular nucleus (n) and high levels of cytoplasmic RNA. Methylene blue, pH 4, for nucleic acids. × 1,500. (Figs. 29–33 from Harrison and Cowden, '75c.)

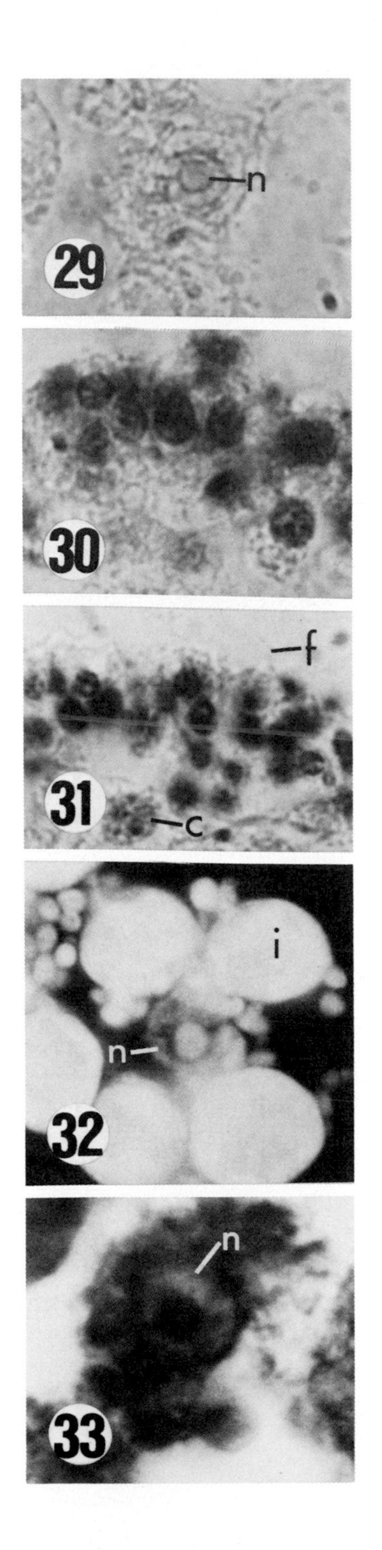
n
29
30
f
31
c
i
n
32
n
33

epitheliocytes (Harrison and Cowden, '82). The presence of sialic acid at this surface of the epitheliocytes, a cell type involved in swimming and in eventual contact with the substratum, is consistent with the role of sialic acid in interactions with the external environment of cells. It is not unreasonable to consider that larval settling and attachment to the substratum may be influenced to some degree by the probable overall negative charge generated by the carboxyl group of sialic acid.

There have been no studies of the mechanism of larval release or the factors influencing larval settling in freshwater sponges. These topics have been extensively reviewed for marine sponges by Sarà and Vacelet ('73) and Fell ('74a).

ASEXUAL REPRODUCTION

A number of asexual reproductive processes, often based upon the developmental potential of archeocytes, occur naturally in freshwater sponges. Wound healing and outgrowth formation are the most simple forms of asexual reproduction. In the former, basopinacocytes migrate to the wound area, initiate a "healing bridge" across the wound surface, and are followed by other ameboid cells (Harrison, '72a). The initial pinacocyte bridge is expanded to effect wound closure.

In outgrowth formation (Harrison, '74b), cellular migration from the parent sponge may be followed by cytodifferentiation in the outgrowth region, although fully differentiated cells may migrate directly into the region. This outgrowth area may become detached from the parent to form a separate "individual."

Gemmulation

Gemmule formation. Most freshwater sponges and a few marine sponges (Fell, '74b) evolved gemmulation as a means of species survival probably as a response to fluctuating water levels. The stimuli that actually initiate gemmule formation are not understood (Rasmont, '68; Harrison and Cowden, '75a; Simpson, '80). While a number of exogenous stimuli, e.g., pH and temperature, influence gemmulation, the process itself appears to be more complex. Rasmont ('74) demonstrated that theophylline, an inhibitor of cAMP-phosphodiesterase, strongly stimulated gemmulation. Simpson and Rodan ('76a) found that continuous exposure of *S. lacustris* hatching sponges to cAMP, cGMP, or aminophylline brought about gemmule formation. However, Rasmont was unable to induce gemmulation by direct addition of cyclic nucleotides. Allowing for species differences, the experimental data available suggest that cyclic nucleotides are involved in gemmulation, and our current knowledge of cyclic nucleotide function would lead us to infer that these changes are in response to some other control stimulus.

Although species variation exists (Evans, '01; Brien, '32; Leveaux, '39; Rasmont, '55a,b,c; De Vos, '71; Harrison and Cowden, '75a) the developmental patterns described during spongillid gemmule formation are quite uniform.

Aggregation of ameboid cells of the sponge mesohyl initiates gemmule formation (Fig. 34). In most cases, the aggregate consists of two cell types, archeocytes and trophocytes, although additional cell types, i.e., "granular cells" or spongocytes may be present in some species. By phagocytosis of either lysed or entire trophocytes, archeocytes (thesocytes) accumulate reserve materials. De Vos ('71) observed direct phagocytosis of entire trophocytes by thesocytes with subsequent transformation of secondary lysosomes into vitelline platelets, i.e., the transformation of portions of trophocyte cytoplasm into thesocyte vitelline platelets. In *E. fragilis* (Harrison and Cowden, '75a), the aggregrate polarized early into an internal region composed of trophocytes and thesocytes and a peripheral zone composed of several layers of rounded cells with vesicular nuclei—the future palisade spongocytes that will synthesize the gemmule coat (Fig. 35). Thesocytes at the margin of the central cell mass become flattened to form a transitory pinacodermlike simple squamous epithelium. A second squamous layer may also cover the exterior of the spongocyte layer. This layer also develops from cells with vesicular nuclei that originate from the spongocyte layer or from the mesohyl. The layer persists at least until the formation of the micropyle in the palisade spongocyte stage of coat formation, but it is also transient.

The appearance of the internal pinacodermlike layer reinforces the polarization of the gemmule into internal and peripheral regions. This epithelium is apparently present in some species but absent in others. Rasmont ('55c) observed it in *Ephydatia mülleri* but never was able to see it in *E. fluviatilis.* He did not observe the sclerification of this layer into a component of the internal gemmule membrane as reported by Leveaux ('39). The correspondence of staining patterns in the epithelium and the innermost stratum of the internal gemmule membrane indicates that in *E. fragilis,* as in *S. lacustris* and *E. mülleri,* the epithelium either synthesizes part of, or through sclerotization becomes incorporated into, the internal gemmule membrane.

Shortly after development of the internal epithelium, which separates the thesocyte mass from the region of the future gemmule coat, a striking reorganization occurs at the periphery of the aggregate. The spongocytes, which until this moment form a relatively unorganized layer, become arranged as an orderly palisade layer. This is a simple columnar epithelium positioned at right angles to the internal layer of squamous epithelial cells. Prior to attainment of the palisade form, spongocytes exhibit considerable cytoplasmic carbohydrate but do not show the marked polarization at the

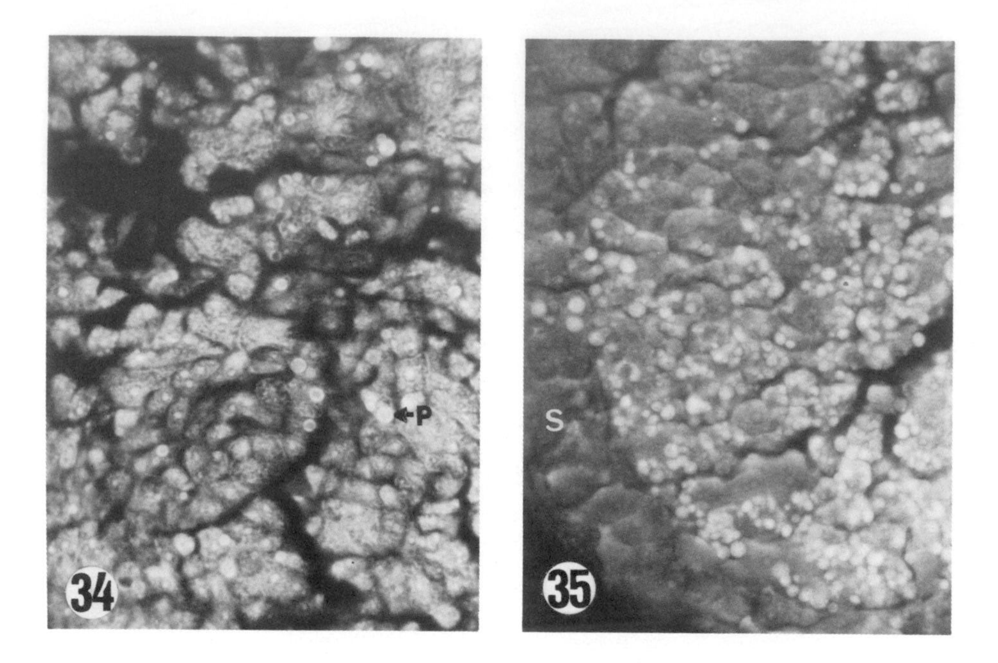

Fig. 34. Aggregation of ameboid cells initiating gemmule formation. Thesocytes and trophocytes show considerable cytoplasmic fluorescence demonstrating ionizable amino groups of basic proteins. Vitelline platelets (p) are recognizable in thesocytes. Brilliant sulfoflavine, pH 2.8 × 220. (Harrison and Cowden, '75a.)

Fig. 35. Early differentiation of gemmule aggregate into an internal thesocyte zone and a peripheral region composed primarily of spongocyte precursors (s). fluorescein mercuric acetate method for protein sulfhydryl groups. × 220. (Harrison and Cowden, '75a.)

proximal face of the cell or in the perinuclear region that develops when the palisade is organized (Fig. 36). These secretory cells form the outer regions of the internal gemmule membrane and then synthesize the pneumatic layer of the gemmule coat. Spongocytes exhibit morphological characteristics of actively secreting cells, i.e., vesicular nuclei with extended chromatin, high levels of cytoplasmic RNA, and prominent Golgi zones (Fig. 36). In an electron-microscopic study of *E. fluviatilis,* De Vos ('74) observed junctional sites consisting of both interdigitated and septate junctions between spongocytes. Most of the cytoplasm contained expanded cisternae of rough endoplasmic reticulum (200–300 mm in diameter), which was filled with a granulofibrillar secretory product, while the basal region of cells contained prominent vacuoles. Following secretion, these microfibrils eventually formed giant spongin fibers (De Vos, '72) (Fig. 37). As seen with the light micro-

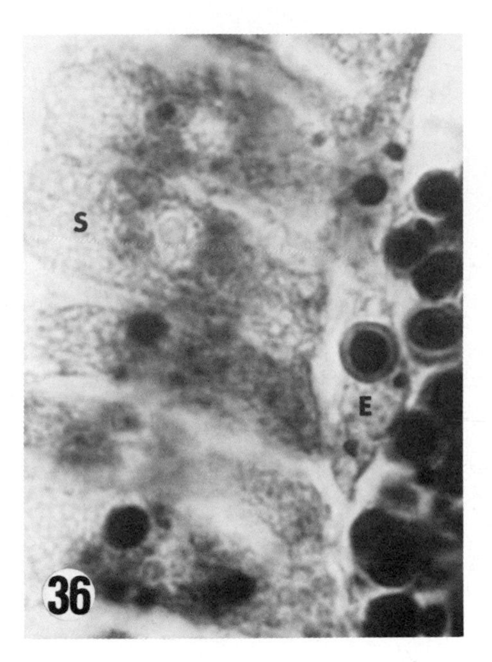

Fig. 36. Early stage in synthesis of internal gemmule membrane. Palisade spongocytes (S) contain PAS-positive Golgi areas. A portion of the transitory thesocyte-derived epithelium (E) underlies the spongocytes. Periodic-acid–Schiff. × 1,050. (Harrison and Cowden, '75a.)

scope (Harrison and Cowden, '75a), during deposition of the alveolar walls of the pneumatic layer, spongocytes are almost completely surrounded by secretory products that are quite rich in sulfated mucins (Fig. 38). In many cases in which spongocytes are completely enclosed, they become necrotic and appear to degenerate (Fig. 39). In other cases, the spongocyte apparently withdraws from the deposited alveolus, laying down the outer wall as it leaves. De Vos ('74) reported that during formation of the pneumatic layer the large basal vacuoles of the spongocytes are released. Their membranes line the cavities of the layer and possibly form a surface for spongin deposition as the pneumatic layer develops. The fully formed pneumatic layer consists of interconnected membrane-lined alveoli (Figs. 40–42). The framework of the layer consists of a chemically complex matrix in which giant spongin fibers are embedded (see Garrone, '78, for an extensive review). In *E. fragilis* and in several other species the final layer of the gemmule coat is the external gemmule membrane, which resembles the internal layer in structure and is synthesized by spongocytes in the same manner (Fig. 41). During pneumatic coat formation in many species, sclerocytes arriving from

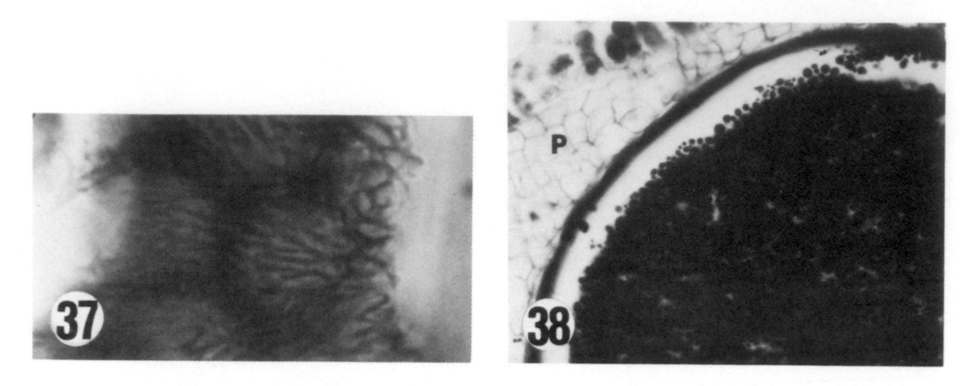

Fig. 37. Face view of fibrous layer of internal gemmule membrane. Alveoli of the preumatic layer are visible beneath the fibrous layer. Masson's connective tissue method. × 1,050. (Harrison and Cowden, '75a.)

Fig. 38. The pneumatic coat (p), in the process of being synthesized, is external to the strongly positive internal gemmule membrane. Vitelline platelets also contain considerable neutral mucin. Periodic-acid–Schiff. × 220. (Harrison and Cowden, '75a.)

the mesohyl deposit gemmoscleres in the matrix of the gemmule coat (Fig. 42). Sclerocytes degenerate following placement of the spicules.

Gemmules possess a differentiated area, the micropyle, thorough which cells emerge during gemmule hatching (Fig. 43). De Vos and Rozenfeld ('74) demonstrated that the micropyle membrane consists of both the inner and outer gemmule membrane layers and is similar in composition to them. A number of authors placed morphogenetic importance on the position of the micropyle (Brien, '32; Leveaux, '39; Rasmont, '55c; Rozenfeld, '70; De Vos and Rozenfeld, '74) and demonstrated that coat formation by spongocytes is initiated at the pole of the gemmule opposite the micropyle or, in some cases, at the pole of the gemmule adjacent to the basal pinacoderm of the sponge. However, Harrison and Cowden ('75a) found that in *E. fragilis* the only indication of a distinct polarization is the normal initiation of palisade spongocyte differentiation in an area at approximately right angles to the position of the developing micropyle. In development of a two-gemmule complex with a shared peripheral zone of contact, palisade spongocyte differentiation begins in the shared region. In no case was spongocyte differentiation initiated at the pole opposite the micropyle.

In addition to the three-layered gemmule coat described above, there are a number of morphological variations. Jorgensen ('46) described two types of gemmules in *S. lacustris,* a thin-walled green gemmule and a thick-walled

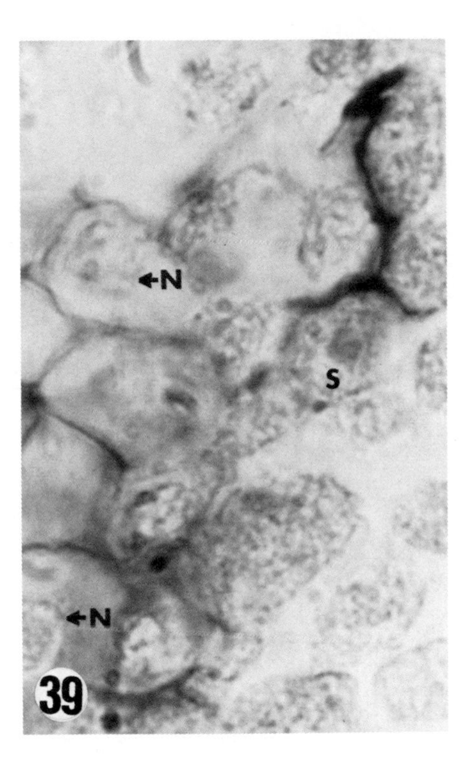

Fig. 39. Spongocytes (s) depositing alveolar walls of the pneumatic layer exhibit an extracellular zone of sulfomucins while lacking cytoplasmic staining. Note necrotic spongocytes (N) inside alveoli. Alcian blue 8GX, pH 0.5, for acid mucins. × 1,050. (Harrison and Cowden, '75a.)

brown gemmule. The coat of the thin-walled gemmule consists of two distinct layers, a thin, homogenous, possibly chitinous, inner layer and an outer layer containing collagen fibers (Simpson and Vaccaro, '73). Gilbert and Simpson ('76) conducted further studies on gemmule polymorphism in *S. lacustris.* Brien ('67a,c) described a number of African freshwater sponge species in which the internal gemmule layer was surrounded by a layer of spongin containing spicules. There was no third external layer in these gemmules. While gemmule polymorphism does exist in some species, e.g., *S. lacustris* (Gilbert and Simpson, '76), Poirrier ('76) demonstrated that the dimensions and even the presence of the pneumatic layer were influenced by environmental variables such as pH. In the field and through laboratory experimentation, he indicated that gemmule coat structure and gemmosclere formation

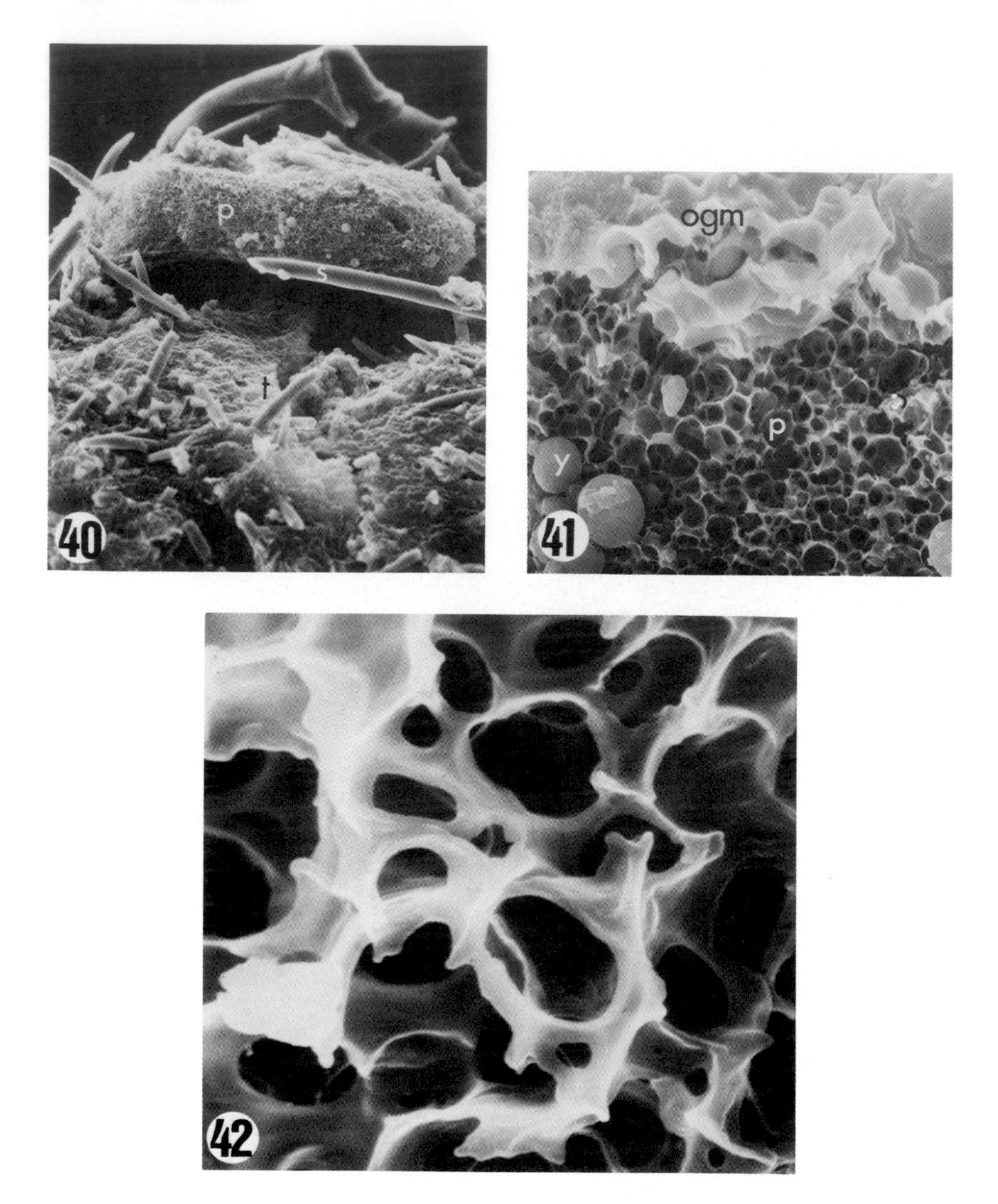

Figs. 40–42. Gemmule coat of *Stratospongilla penneyi.* Figure 40 shows the coat surrounding thesocytes (t) covered with spicular (s) debris. p, pneumatic layer. SEM × 250. Figure 41 shows a higher magnification of the coat demonstrating the pneumatic layer (p), the external gemmule membrane (ogm), and vitelline inclusions (y). SEM × 3,000. In Figure 42 the interconnected alveoli of the pneumatic layer are evident. SEM × 30,000. (Figs. 40–42 from Harrison, '79.)

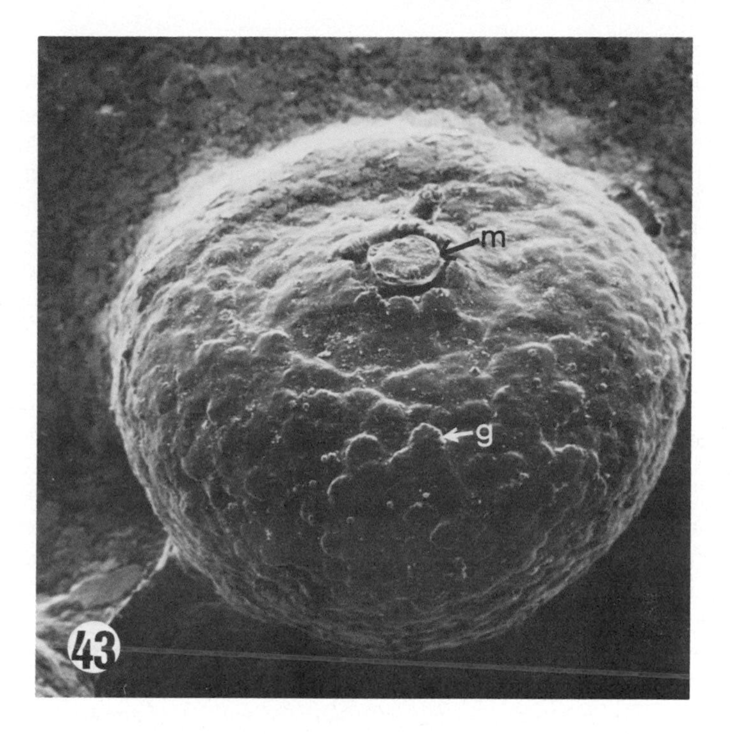

Fig. 43. Gemmule of *Ephydatia fluviatilis*. An armature of gemmoscleres (g) is positioned in the coat. The micropyle (m) is unopened in this unhatched gemmule. SEM × 280. (Courtesy of Dr. Louis De Vos.)

(Poirrier, '74) exhibited ecomorphic variation in a number of species. The development by Poirrier et al. ('81) of a system for a long-term laboratory culture of freshwater sponges should facilitate further studies of the significant problem of environmental influences upon morphogenetic expression during gemmule formation.

During development up to gemmule coat formation, thesocytes amass vitelline reserves, primarily through phagocytosis and modification of trophocytes (De Vos, '71). The chemical composition of vitelline platelets in thesocytes seems to vary considerably from species to species. Pourbaix ('34, '35) reported inclusions of *E. fluviatilis* and *S. lacustris* were composed of glycoprotein valves covering an internal lipid layer. Several authors (Kauffold and Spannhof, '63; Ruthmann, '65; Simons and Muller, '66) demonstrated that the lens-shaped platelets consist of peripheral valves of ribonucleoprotein, a central zone of basic protein, and lipid droplets in a zone surrounding the protein layer. Kauffold and Spannhof ('63) noted that the vitelline inclusions of *E. mülleri* contained no carbohydrate or mucopolysaccharide. Ruthmann ('65), however, found PAS-positive material in platelets in *S. lacustris* and *E. fluviatilis* as did Harrison and Cowden ('75a) in *E. fragilis*. We were unable to identify RNA positively in the outer valves of the platelets in that study, but after extracting RNA chemically instead of using an RNase extraction we have since been able to confirm that the valves contain RNA. Apparently, as we proposed at the time, the presence of multiple membrane layers surrounding the platelet hinders penetration of the enzyme. De Vos ('71) followed platelet formation in *E. fluviatilis* and observed the transformation of phagocytized trophocytes into vitelline platelets by a process of segregation and stratification. Initially, glycogen is located centrally, delimited by a mitochondrial zone. Peripherally, ribosomes and vesicles of rough endoplasmic reticulum form two plano-convex shells covering the mitochondrial layer. Lipid droplets form an equatorial ring. In later phases of vitellogenesis, the glycogen core disappears, being transformed into an isolated electron-dense mass. The mass is delimited by a ribonucleoprotein shell associated with each pole of the platelet. The entire platelet is surrounded by multiple layers of membranes.

As De Vos ('71) noted, following release from dormancy the thesocyte vitelline platelets are degraded. The considerable reservoir of trophocyte-derived rough endoplasmic reticulum sequestered in the outer valves of the platelets becomes available to thesocytes of the activated gemmule, enhancing their already considerable synthetic capabilities.

After completion of the gemmule coat and after all trophocytes have been phagocytized in vitellogenesis, the thesocytes divide mitotically but only undergo karyokinesis, not cytokinesis. The resulting binucleate cells (Fig. 44) persist throughout the period of gemmule dormancy (Brien, '32).

Gemmule dormancy. In most cases, there is a regular seasonal cycle of gemmule formation, dormancy, and subsequent hatching. In colder climates, gemmule formation usually occurs in autumn and hatching occurs in the spring. In warmer regions, gemmulation may coincide with the onset of hotter weather with hatching following in midautumn (Harrison, '74a). The period of dormancy may last for a number of years, particularly in arid environments (Harrison, '74c).

A number of freshwater sponges (Rasmont, '55a,b,c) and at least some marine sponges (Fell, '74a,b) require a period of cold temperature, a true diapause, in order to hatch at a later time. In other species, e.g., *E. fluviatilis*, a diffusible inhibitor, gemmulostasin, is produced by the tissues of the living sponge to prevent premature hatching of gemmules still contained within functional parent tissue (Rasmont, '63, '65; Rozenfeld, '70, '71, '74).

The nature and physiological role of gemmulostasin is not known. It has been shown to inhibit thymidine incorporation (Rozenfeld, '74), a cAMP-like effect (Simpson and Rodan, '76a,b). It has been suggested (Simpson and Rodan, '76a,b; Simpson, '80) that gemmulostasin may act through partial inhibition of phosphodiesterase activity and eventually by influencing osmotic relationships (Simpson et al., '73).

Gemmule germination. The initial events that occur in gemmule germination, the period prior to cell migration (hatching) as designated by Simpson and Fell ('74), have been described by Berthold ('69), Rozenfeld ('70, '71), De Vos and Rozenfeld ('74), Ruthmann ('65), Tessenow ('69), and Harrison et al. ('81) and reviewed by Simpson and Fell ('74). The first morphological event that can be seen at the light-microscope level in the process of release from dormancy and germination is the separation of nuclei in binucleate thesocytes (Berthold, '69). Binucleate thesocytes may undergo a second nuclear division that produces tetranucleate cells or may undergo complete mitosis to form four uninucleate daughter cells. Thesocyte progeny consist of an archeocyte reserve or a second cell population, histoblasts, which are destined to pass through the gemmule micropyle, possibly utilizing enzymatic digestion (Rozenfeld, '71; Garrone, '78).

Recent studies by Simpson and Rodan ('76a,b) correlated dormancy release in *S. lacustris* with a sharply decreased gemmular cyclic AMP content in the first 2 hours of gemmule germination and no significant increase in cAMP levels during the period of hatching. Harrison et al. ('81), using immunofluorescent methods, observed an early appearance of cGMP and cAMP in histoblast precursors approximately 12 hours before any morphological or cytochemically demonstrable events related to cytodifferentiation occurred (Fig. 45). Ostrom and Simpson ('78, '79) demonstrated that calcium ions were essential for germination of gemmules and suggested that stimulation of cell division and cell movement involves the interactions of calcium ions and cyclic nucleotides. Alterations in intragemmular osmotic pressure are also involved in the triggering of cell division and hatching (Zeuthen, '39; Schmidt, '70; Simpson et al., '73).

The first 48 hours following dormancy release is a period of accelerated DNA synthesis, digestion of vitelline platelets, and synthesis of RNA and protein as binucleate thesocytes divide to form histoblasts (Rozenfeld, '74). By 24 hours, cGMP levels increase significantly in thesocytes at the gemmule periphery prior to histoblast formation. This elevation occurs coincident with the onset of premitotic S-phase in these cells. Tritiated thymidine uptake begins to rise at 24 hours and peaks at 36 hours in germinating gemmules of *E. fluviatilis* (Rozenfeld, '74), a period that also coincides with the system's greatest sensitivity to puromycin, an inhibitor of protein synthesis acting on translation of mRNA (Rozenfeld, '80). Prior to hatching, there is considerable reduction in the osmotic pressure of gemmular fluids, from approximately 220 mOsm to approximately 40 mOsm (Zeuthen, '39). This is recognizable

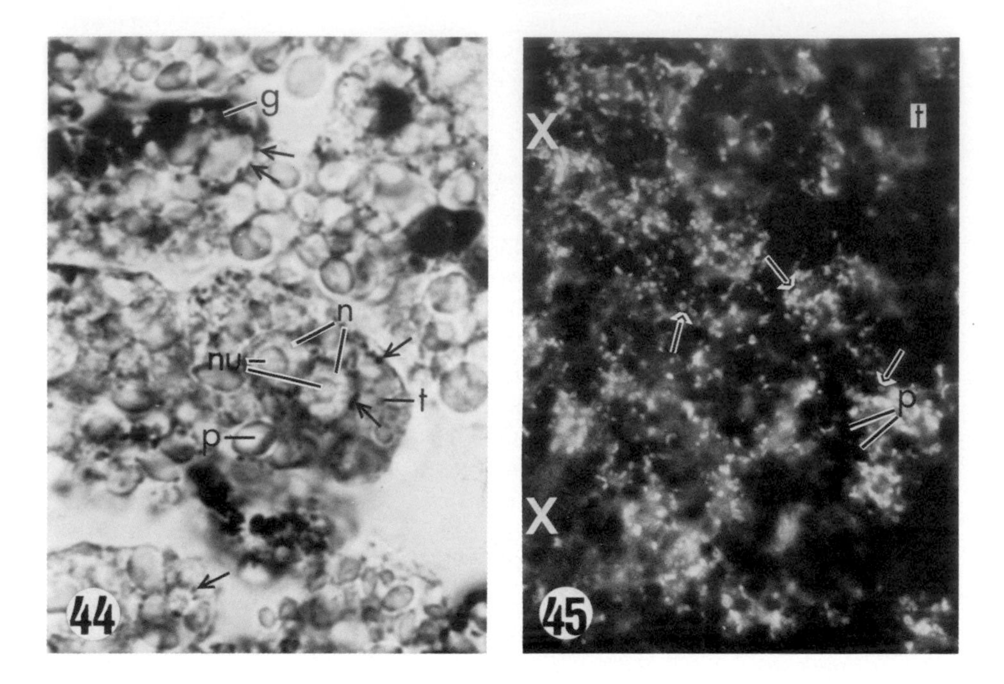

Fig. 44. Binucleate thesocytes (t) within an unhatched gemmule. As seen in the thesocyte located centrally in the figure, these cells contain two vesicular nuclei (n) with prominent nucleoli (nu) and large extrachromosomal volume. Vitelline platelets (p) are seen throughout the cytoplasm. Particulate glycogenic inclusions (arrows) or crescentric diffusion artifacts of stained glycogen (g) are present. Periodic-acid–Schiff. × 1,200. (Harrison et al., '81.)

Fig. 45. Immunofluorescent localization of cGMP in a frozen section of a gemmule incubated at room temperature for 24 hours. Cyclic GMP fluorescence levels are markedly elevated in thesocytes located toward the periphery of the unhatched gemmule. The peripheral direction in the gemmule is labeled by ×s. Dark areas (p) indicate vitelline platelets. Thesocytes (t) located more centrally in the gemmule show little fluorescence. × 1,200. (Harrison et al., '81.)

microscopically as thesocytes alter from the "tightly packed" to a more loosely organized relationship between 24 and 36 hours after dormancy release.

Binucleate thesocytes (Fig. 44) contain considerable cytoplasmic RNA and a pair of vesicular nuclei with prominent nucleoli, highly extended chromatin, and large extrachromosomal volume (Ruthmann, '65; Tessenow, '69; De Vos, '71; Harrison et al., '81). Nuclei supravitally fluorochromed with acridine orange, a probe of nucleic acid organization, exhibit almost

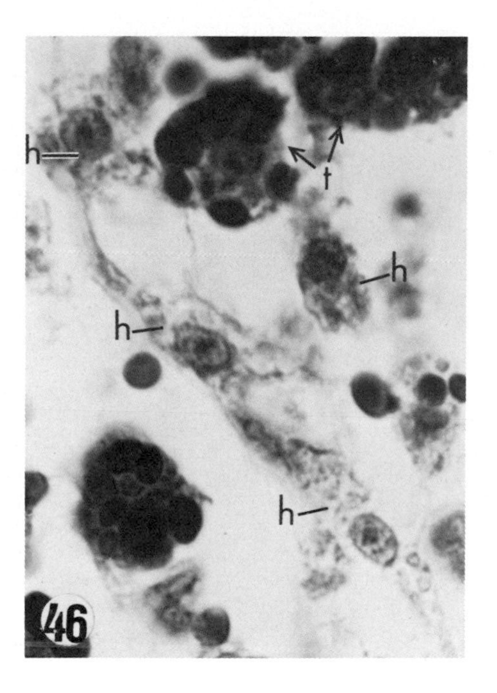

Fig. 46. In unhatched gemmules incubated for 48 hours at room temperature, cytoplasmic basophilia is considerably lower in histoblasts (h) than in thesocytes (t). Methylene blue, pH 4.0, method for nucleic acids. × 1,200. (Harrison et al., '81.)

transparent green nucleoplasm and prominent but pale green fluorescent nucleoli (Harrison and Cowden, in press). The cytoplasm is filled with vitelline platelets and contains diffusely distributed netural mucins. With differentiation into histoblasts, acid phosphatase levels increase coincident with the degradation of vitelline inclusions (Tessenow, '69). While still in the gemmule, histoblasts contain no yolk reserves and single nuclei (Fig. 46). Cytoplasmic carbohydrate increases throughout histoblast differentiation.

Gemmule hatching. Many studies of hatching have utilized the "sandwich" culture method (Ankel and Eigenbrodt, '50; Wintermann, '51) in which a gemmule is placed on a microscope slide, positioned between two cover slips, and allowed to hatch. This technique has also facilitated microcinematography of developmental events (see Rasmont, '79, for review).

Histoblasts migrate from the germinating gemmule and either differentiate into collencytes (Brien, '32) or become directly modified into pinacocytes which form the initial attachment to the substratum (Fig. 47). These cells at

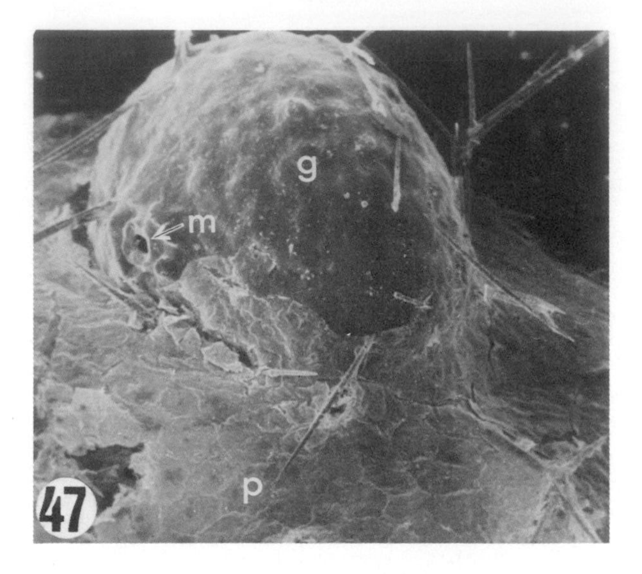

Fig. 47. Following hatching from the gemmule (g) the sponge attaches to the substratum. Note the open gemmule micropyle (m) through which hatching cells emerged. p, pinacocytes of the exopinacoderm of the hatched sponge, *Ephydatia fluviatilis*. SEM. × 270. (Courtesy of Dr. Louis De Vos.)

the outgrowth periphery are distinguished by strikingly elevated bright orange cytoplasmic fluorescence (Fig. 48) following application of acridine orange as a supravital dye (Harrison and Cowden, in press). These high levels of cytoplasmic RNA, lacking in quiescent pinacocytes deeper into the sponge interior, are probably associated with the synthesis and secretion of a basal attachment layer of collagen by these cells. The variation in pinacocyte cytoplasmic RNA levels seen in vital preparations explains the discrepancy between Garrone's observations ('78) that basopinacocytes contained extensive rough endoplasmic reticulum and Harrison's report ('72a) of low levels of cytoplasmic RNA and correspondingly low levels of secretory activity in these cells. The elevation of cytoplasmic RNA levels in cells at the periphery of the outgrowth exactly correlates with the pattern of cyclic nucleotide location, particularly cyclic GMP, as reported by Harrison et al. ('81) (Figs. 48,49). Cyclic GMP levels were elevated in cells of the basopinacoderm located at the outgrowth periphery, whereas other cells deeper into the sponge, including thesocytes, displayed low levels of cGMP as demonstrated by the indirect fluorescent antibody method. Although Harrison et al. ('81) associated cGMP localization with cytoplasmic organelles possibly involved

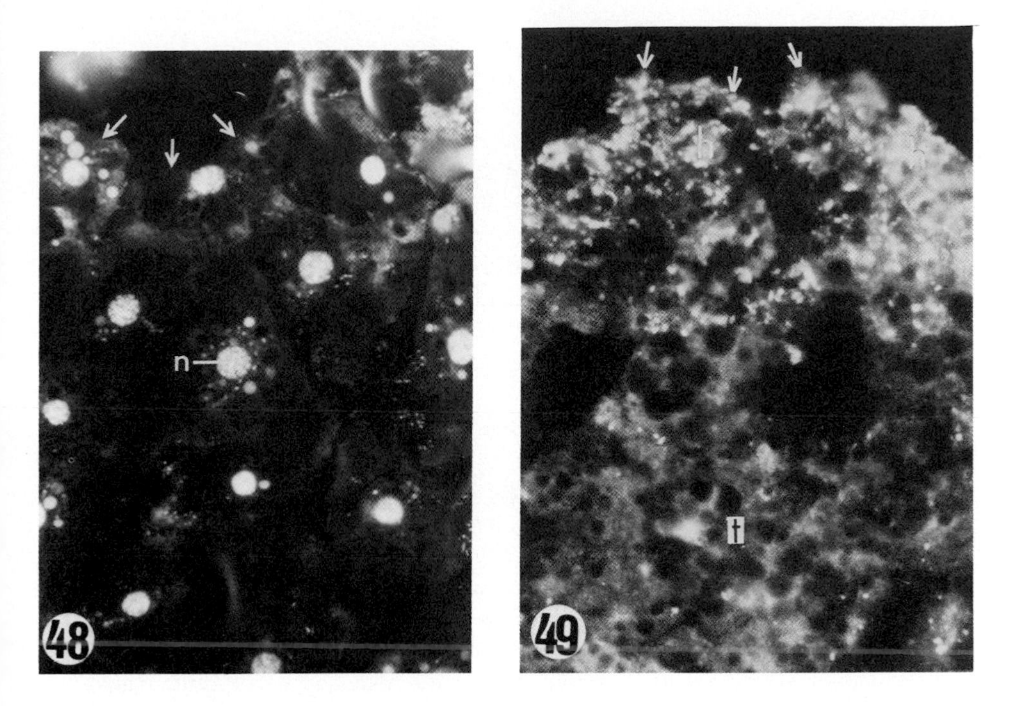

Fig. 48. As the basopinacoderm extends following hatching, cells at the outgrowth periphery (arrows) exhibit strikingly elevated RNA fluorescence levels after using acridine orange as a marker for nucleic acids. These high levels of cytoplasmic RNA are lacking in pinacocytes deeper into the sponge interior. (n) pinacocyte nucleus. Compare this figure with the immunofluorescent localization of cGMP demonstrated in Figure 49. × 1,200. Acridine orange, 10^{-4} M, used as a vital fluorochrome. (Harrison and Cowden, in press.)

Fig. 49. Immunofluorescent localization of cGMP in a frozen section of a gemmule incubated for 48 hours at room temperature. Cells at the periphery of the basopinacoderm of the hatched sponge (arrows) show markedly elevated cGMP levels. Cells (t) more centrally located in the outgrowth show little elevation in fluorescence. × 1,200. (Harrison et al., '81.)

in mobilization of nutrient reserves, it seems reasonable that some of the elevation in cyclic nucleotide levels could be associated with signals that initiate the elaboration of either collagen or a collagen-associated material, possibly fibronectin.

The cells at the periphery of the early outgrowth often exhibit gross irregularities in nuclear profiles (Fig. 50) (Harrison and Cowden, in press). Later in development, collencytes migrate to the sponge margin, overgrow the existing peripheral pinacocytes, exhibit a pulse of intensely brilliant cytoplasamic red fluorescence (Fig. 48), which seems to correlate with attachment, and are in turn overgrown as the sponge extends. However, these cells rarely, if ever, exhibit the irregular nuclear forms seen when the sponge initially attaches to the substrate. It is likely that the irregular nuclear shapes are indicative of a terminally differentiated condition. Having completed their attachment function, these cells will probably die. The absence of this pattern in later stages suggests that these terminally differentiated cells with irregular nuclei are derived through differentiation of histoblasts, whereas the peripheral cells observed in the extension of outgrowths developed from collencytes or from thesocytes that migrated from the gemmule after some mobilization of reserve material and differentiation.

By 72 hours after release from dormancy, a continuum of stages in the formation of choanocytic chambers can be seen throughout the growing sponge. Brien ('32) considered that chambers arose from yolk-laden monocucleate "amebocytes" that divided several times to form choanoblasts. Aggregates of choanoblasts, often joined by isolated migratory choanoblasts, formed the choanocytic chambers. Our supravital AO studies suggest the possibility of a slightly different ontogenetic sequence (Figs. 51–55). Division of the mononucleate cells is closely accompanied by the progressive development of a rather prominent cytoplasmic vacuole (Fig. 52). Development of the chamber involves successive divisions of choanoblasts that surround the periphery of the enlarging vacuole that will become the choanocytic chamber. Free choanoblasts may associate with the cells of the developing chamber. The high levels of red cytoplasmic fluorescence, which probably indicate a very high cytoplasmic RNA concentration, are seen in the cells after supravital fluorochroming with AO. This probably reflects the considerable level of synthetic activity that is associated with formation of the flagellum and microvillar collar during later stages of choanocyte differentiation.

Although choanocytes can develop via the differentiation of archeocytes in mature sponges, cells conforming to the classical description of this cell

---

Fig. 50. Cells at the periphery of early outgrowths of hatched sponges often exhibit irregular nuclear profiles (n) not seen later in development. × 720. Acridine orange, 10^{-4} M, used as a vital fluorochrome. (Harrison and Cowden, in press.)

Figs. 51–55. Formation of choanocytic chambers. Figure 51 shows mitosis in a choanoblast (c). In Figure 52 vacuoles (v) develop in choanoblasts with small nucleoli. A four-celled chamber is present. In Figure 53 a six-celled chamber is present. An isolated cell—possibly a migratory choanoblast (cb?)—is present. Figure 54 shows an eight-celled chamber. In Figure 55, in a 16-celled chamber, microvillar collars are beginning to grow into the chamber lumen. × 720. Acridine orange, 10^{-4} M, used as a vital fluorochrome. (Harrison and Cowden, in press.)

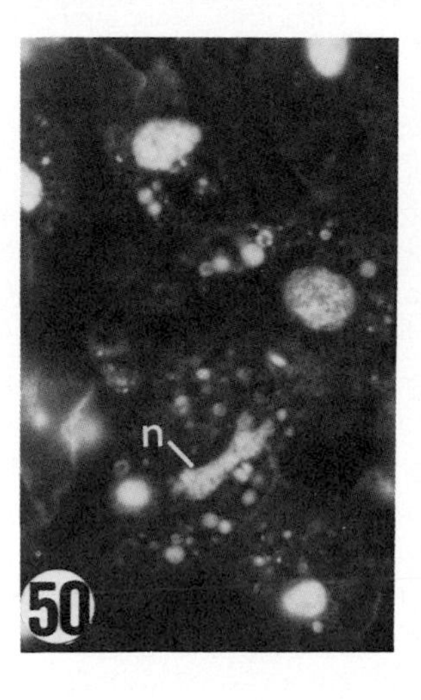
n
50

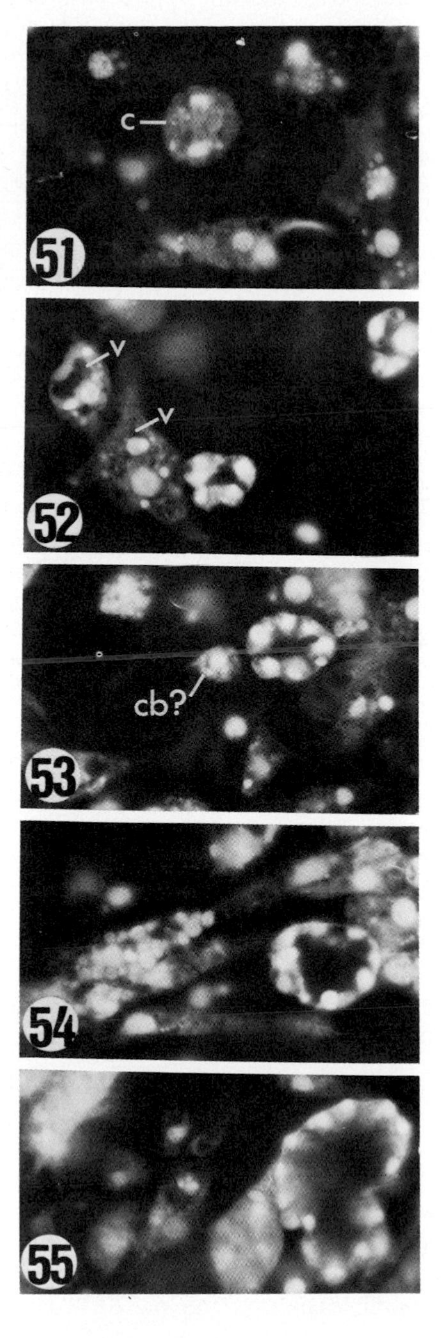
c
51
v
v
52
cb?
53
54
55

type (see above) did not appear in significant numbers in our study until about 90 hours following release from dormancy, a period relatively late in this developmental sequence. It is interesting, however, that Rozenfeld and Rasmont ('76) used hydroxyurea to inhibit differentiation of choanocytes while developing techniques for isolating archeocytes.

It is obvious from the studies of Brien ('32) and others that the pinacoderm layer, derived initially from histoblasts, separates early in development from the central thesocytes. In subsequent development these central thesocytes progressively metabolize vitelline material in order to support proliferation and differentiation. While this central mass of thesocytes mainly populates the mesohyl, their migration and morphological transformation appears to be accompanied by the expected increased synthesis of RNA, proliferation, enlargement of nucleoli, and ultimately decreases in cytoplasmic RNA content. Finally, nucleoli disappear and chromatin texture is altered, particularly in terminally differentiated choanocytes and pinacocytes (for a review of sponge cells and lineages refer to Lévi, '70).

The sequence of developmental events in choanocytic chambers is particularly interesting since it involves synthesis of cytoplasmic RNA, formation of a characteristic vacuole, and proliferation and synthesis of cytoskeletal and microvillar structures followed by loss of cytoplasmic RNA when these synthetic events are complete. The vacuole appears to have some special role in defining the choanocytic chamber, and probably defines both the axis of cell division and the symmetry of the structure. Although other cells may participate in the organization of the chamber, its special organization and symmetry appear to be established by this initial structure. At this time the precise details of cytoskeletal organization that are involved in this morphogenesis are not known, but it seems certain that cytoskeletal structures do have an important role in this developmental process.

Later events in the development of the aquiferous system, particularly the induction of oscula, have been treated in some detail by Mergner ('59, '64, '66, '70). For a review of gemmulation and development from gemmules, the reader is referred to Brien ('73b).

Reduction

A number of sponges, marine and freshwater, regress in periods of environmental stress and form reduction bodies or enter an "overwintering" phase (Bergquist and Sinclair, '73; Boreojević, '70; Fell, '74b; Hartman, '58; Müller, '11a,b; Penney, '33; Simpson, '68; Van de Vyver and Willenz, '75). Reduction, as seen in the freshwater sponges, differs from gemmule formation in that, while gemmule formation is a response to long-term environmental fluctuations, with likely modulation by humoral agents such as gemmulostasin (Rozenfeld, '70, '71) or by cAMP (Rasmont, '74), reduction is initiated in response to more short-lived deleterious fluctuations of environmental conditions. Freshwater sponge reduction displays similarities to overwintering in freshwater and marine species in that canal systems are obliterated, although in those freshwater sponges that overwinter, e.g., *E. fluviatilis,* overwintering is necessary for, and appears to be related to, the induction of sexual reproduction (Van de Vyver and Willenz, '75).

In freshwater sponge reduction body formation, the sponge tissue withdraws centrally from the sponge skeleton, forming a spherical structure composed of an inner cell mass of ameboid cells, principally archeocytes, within an investing simple squamous pinacocyte epithelium. Upon the return of more favorable environmental conditions, a complete sponge may be produced through differentiation of the internal somatic cell complex of the reduction body. If adverse conditions persist, the reduction body will undergo progressive degeneration leading to eventual death.

Müller ('11a,b) and Penney ('33) found that the reduction process could be initiated in the laboratory simply by placing freshwater sponges into an adverse environment, e.g., tap water. The process could be reversed by returning the degenerating sponge or the reduction body to pond water. The reduction system, therefore, offers a unique vehicle for the study of cytological aspects of degeneration and regeneration.

Recent studies in our laboratory (Harrison and Davis, '82) have demonstrated four trends during the first 72 hours leading toward reduction body formation. These are 1) a general withdrawal of the sponge into a central rounded mass; 2) the genesis of lysosomal activity concomitant with phagocytic activity, which increases noticeably throughout the period of observation; 3) cellular degeneration and decrease in the number of cell types in the sponge; and 4) initiation of collective tissue elaboration as cell mass decreases in proportion to an increasingly fibrous connective tissue matrix.

After 1 hour, there is definite withdrawal of the sponge mass (seen in outgrowth preparations). In the "withdrawn" area remnants of pinacocytes of the basal epithelium remain attached to the glass substratum. Cytochemically, there are very few differences noticeable from zero hour control preparations. There appears to be some slight evidence of increased cytoplasmic vacuolation in the archeocytes. Aside from these slight alterations, cellular organization appears normal and some mitotic figures may be seen.

By 6 hours, it is evident that the sponge cells have continued to withdraw toward the center of the tissue mass. In the withdrawal area remnants of basal pinacocytes, scattered nuclei, and patches of cytoplasm are present. A few cells migrate out over this area and remnants of connective tissue may be seen. Vacuolation is more apparent and PAS-positive inclusions that are not glycogen are present in the cytoplasm of some cells. The most noticeable event at 6 hours is the increase in deposition of connective tissue. Lophocytes, the collagen-secreting cells, are seen, and in a number of cases lophocyte precursor cells are present with numerous cytoplasmic inclusions, which display a green fluorescence emission when stained with acridine orange. Collagen-linked carbohydrates are fluorochromed in this instance, emitting a characteristic deep-green fluorescence, which is probably related to protein rather than nucleic acid of either class.

By 12 hours, the formation of lysosomal inclusions containing both netural and sulfated mucins is much more evident. In some sponges, the canals are filled with debris, such as algae, suggesting some degree of cytolysis. However, at this stage spiculogenesis is still occurring, mitotic figures are seen, and choanocytic chambers are still being formed.

After 24 hours, cytoplasmic RNA staining in archeocytes has become quite patchy, strongly suggesting dissolution of the normally prominent and compact rough endoplasmic reticulum characteristic of these cells. Choanocytic chambers are still present although choanocyte remnants and debris are present within canal systems. Lysosomes containing sulfomucin are increasingly evident.

At 48 hours, the trends established earlier are more advanced. Lysosome formation in archeocytes is accompanied by a breakdown of the homogeneity of cytoplasmic distribution of RNA. Algae are extruded and choanocytic debris fills the canal system. However, mitotic figures are still observed in this cell population.

After 72 hours, the sponges show considerable evidence of degeneration and are well into the reduction process although characteristic reduction bodies have not been formed at this stage. Cytoplasmic inclusions contain, in addition to the carbohydrates noted previously, basic proteins and nucleic acids. The sponge outgrowth is quite condensed and a substantial amount of cellular debris occupies the withdrawal area. Canal systems are loaded with debris. This elaboration of connective tissue, initiated at 6 hours, is now the most prominent aspect of the reduction process. The consistency of the sponge mesohyl is radically altered as sheets of connective tissue hang in festoons, draping spicules and other fixed elements, while great numbers of lophocytes migrate through the webbing, depositing even more collagen (Fig. 56).

Harrison et al. ('75) presented the initial observations of the histochemical and ultrastructural characteristics of tap-water–induced reduction bodies of a freshwater sponge. We observed that, when compared with normal adult freshwater sponges, reduction bodies exhibited considerable organizational degeneration. Subdermal spaces and elements of the aquiferous system were totally absent. There was also a considerable reduction in the diversity of cell types. Choanocytes were completely lacking in reduction bodies. The vast majority of cells of the reduction body interior were archeocytes (Fig. 57)

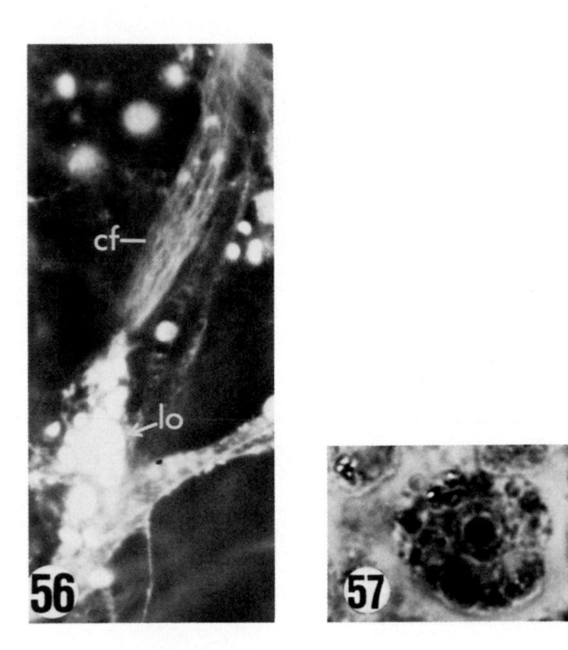

Fig. 56. Seventy-two hours after initiation of the reduction process, numerous lophocytes (lo) trailing tufts of collagen fibrils (cf) are seen throughout the degenerating sponge. × 720. Brilliant sulfoflavine for basic protein. (Harrison and Davis, '82).

Fig. 57. An archeocyte in the reduction body exhibits an enlarged nucleolus, extended chromatin, and peripherally displaced cytoplasmic inclusions. Hematoxylin and eosin. × 1,200. (Harrison et al., '75.)

exhibiting, in their cytoplasm, multiple spherical chemically complex inclusions, which by cytochemical criteria contained both RNA and protein (Fig. 58). Our ultrastructural observations of these archeocytes suggested that they were probably involved in both phagocytic and autophagic activities. The complete absence of choanocytes and the considerable reduction in the numbers of several other cell types normally present in adult sponges indicated that archeocytes phagocytized entire cells or cellular fragments. The phagocytic activities of archeocytes have been well documented by Müller ('11b) in reduction bodies; by De Vos ('71), who depicted somatic archeocyte phagosomes morphologically identical to those seen above; and by Harrison ('72a), who demonstrated high levels of acid phosphatase activity in somatic archeocytes following feeding. It seems notable that some fundamental alteration in self-recognition patterns occurs during the reduction process. Elements of the sponge once recognized as "self" by the archeocytes are in this

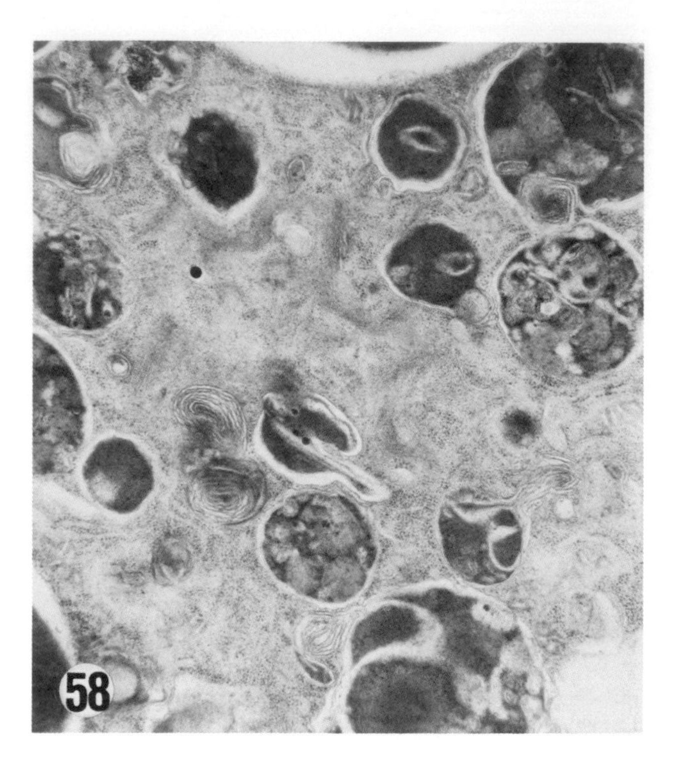

Fig. 58. The cytoplasm of an archeocyte within a reduction body exhibits lysosomes containing mitochondrial fragments, myelin figures, microtubules, and numerous free ribosomes. Note areas where smooth membranes of myelin figures are continuous with lysosomal membranes. Glutaraldehyde-osmium tetroxide fixation. Uranyl acetate–lead citrate. × 21,000. (Harrison et al., '75.)

process considered to be foreign or "nonself," which leads to subsequent phagocytosis of these cells by the archeocytes. An adverse environmental event initiates a radical alteration in the nature of cell-to-cell interaction in this animal, with a resulting increase in archeocyte phagocytic activity.

These events appear to be closely correlated with features of aging seen both in sponges and in higher animals (Pavans de Ceccatty, '79). A decrease in cell mass in proportion to an increasingly fibrous connective tissue matrix, "age involution" (Bellamy, '73), occurs during reduction. The phagocytic events described above are very similar to the "phagocytosis crisis" undergone by marine species, e.g., *Halisarca sp.* (Lévi, '56), in which a burst of phagocytic and autophagic cellular activity accompanies degeneration prior to dying (see Pavans de Ceccatty, '79, for further discussion).

It is apparent that the reduction body system of freshwater sponges is an extremely straightforward, if not simple, developmental system. Its developmental potential resides in the fact that the sponge is reduced primarily to reserve cells, the archeocytes, which possess virtual blastomeric capacities for differentiation. The reduction process can be initiated and manipulated experimentally. One can experimentally induce degeneration, with subsequent reduction body formation, or can initiate further degeneration from the reduction body toward death. One may also manipulate the obverse face of the system, observing regeneration from the reduction body to a healthy sponge containing all cellular elements. In both the degenerative and regenerative phases of the system, there are opportunities for cytological or biochemical examination of the process at selected intervals. The system presents an unusually straightforward vehicle for investigations of degeneration and regeneration as processes in developmental biology.

CELLULAR INTERACTIONS AND RECOGNITION

Wilson ('07), using the marine sponges *Microciona prolifera* and *Haliclona occulata,* demonstrated that a mixture of cells from two species would separate to form species-specific aggregates. Since Wilson's experiment, biologists have employed the technique of dissociation of sponges into free cells to examine the mechanisms of cellular interaction during the reaggregation process. As Van de Vyver ('79) notes, the absence of any circulatory system in sponges means that incompatibility reactions are expressed at cell-to-cell contact.

Wilson's ('07) experiments with marine sponges have been repeated and amplified upon using freshwater sponges (Brien, '73b; Rasmont, '61; Van de Vyver, '70, '71, '75, '79, '80; Curtis and Van de Vyver, '71; De Sutter and Van de Vyver, '77). For a number of years, species specificity during sponge cell aggregation was considered to be absolute. However, a number of investigators (Sarà, '68; Sarà et al., '66; Humphreys, '70) demonstrated the existence of at least transient bispecific aggregates and suggested that aggregation was only specific for some species pairs.

Specificity of aggregation involves a cell-surface glycoprotein aggregation factor. This factor, identified by Moscona ('63), Humphreys ('63), and McClay ('74), exhibits antigenic properties of the plasma membrane (MacLennan, '69) and interacts with "baseplate" receptors on the cell surface (Weinbaum and Burger, '73; Burger and Jumblatt, '77; Burger et al., '78; Müller et al., '78; Burkart et al., '79; Curtis, '79). When cells are washed in calcium–magnesium-free (CMF) seawater or CMF-"M" medium for freshwater sponges (Rasmont, '61) and then returned to the medium, they reaggregate quite slowly. However, addition of aggregation factor to the water initiates aggregation. McClay ('74) demonstrated that aggregation factors isolated from several marine sponges were species specific. They did not enhance aggregation in heterospecific cell-factor combinations. Aggregation factor from *Microciona parthena* was purified and characterized (Henkart et al., '73; Cauldwell et al., '73). It is a large, acidic 70S glycoprotein complex consisting of approximately equal parts of carbohydrate and protein. Electron microscopy of these large complex molecules shows a large fibrous molecule with an open, sunburst structure, i.e., 45-Å-diameter fibers arranged in an 800-Å-diameter circle with 11–15 arms 1,100 Å in length radiating from the circle. Following removal of calcium, the complex dissociates into glycoprotein subunit molecules of molecular weight 2×10^5 daltons plus its circular core. The structure of the *Geodia cydonium* aggregation factor is quite similar (Müller and Zahn, '73). (For further information on the aggregation process in marine sponges refer to Müller et al., '74, '76a,b,c, '79; Burkart and Burger, '77; Burkart et al., '78, '79; Jumblatt and Burger, '77; Jumblat et al., '78; Kuhns et al., '78; Mir-Lechaire and Burger, '78).

Recent studies by Buscema and Van de Vyver ('79), using electron microscopy, have added significantly to our knowledge of the morphogenetic events of reaggregation (Brien, '37) in the freshwater sponge *E. fluviatilis.* Following dissociation, pinacocytes and choanocytes regroup and establish the pinacoderm and some choanocytic chambers by 6 hours. Choanocytic chambers grow by the addition of free choanocytes. Archeocytes actively phagocytize debris and entire choanocytes throughout the aggregation process (refer to archeocyte activity during reduction, above).

Following density gradient centrifugation (De Sutter and Van de Vyver, '77, '79; De Sutter and Buscema, '77), cell suspensions of *E. fluviatilis* were separated into two main fractions. The first consisted of archeocytes whereas the second contained mostly choanocytes plus some pinacocytes and archeocytes. With this technique, De Sutter and Van de Vyver ('77) determined the developmental potential of specific cell lines. The archeocyte fraction alone builds abnormally flat but functional sponges. Although the aquiferous system differentiates, these sponges are deficient in the number of choanocytic chambers. The choanocyte fraction aggregates and forms a pinacoderm, but does not settle. Although choanocytes differentiate into chambers, the aggregates die within a few days.

Although heterospecific rejection is well known, Van de Vyver ('70, '71) demonstrated the presence of three physiological and genetic strains of the freshwater sponge species *E. fluviatilis.* When grown in close proximity to each other different strains exhibit infraspecific or "allogenic" recognition and form a nonmerging front between them (Fig. 59). This front consists of the juxtaposed but separate pinacoderm margins and a barrier of collagen, which is continuous with the basal attachment mat secreted by the basopina-

Fig. 59. SEM of nonmerging front (arrows) between two different strains of *Ephydatia fluviatilis* (S_1,S_2) brought into contact. The Xs indicate a region where the sponges have withdrawn away from each other, leaving the collagenous barrier that was deposited earlier. SEM. × 450. (Courtesy of Dr. Louis De Vos.)

coderm, positioned between them (Van de Vyver, '80; Van de Vyver and De Vos, '79). De Sutter and Van de Vyver ('79) demonstrated clearly that the property of allogenic recognition extended to individual cell lines in isolated sponge cell fractions. Allogenic non-coalescence also occurs naturally in a number of marine sponges (Curtis, '79).

In nature, allogenic recognition and xenogenic recognition, the recognition of foreign species, are important ecological adaptations for species with an encrusting growth habit and high levels of asexual reproduction (Van de Vyver, '79). In these situations, the development of a barrier, a "nonmerging front," or elaboration of a cytotoxic response is important in competition for the substrate. This recognition of "nonself" is also seen in the deposition of a collagen or spongin barrier through enhanced fibrogenesis following sponge invasion by parasitic or commensal organisms (Connes, '67; Connes et al., '71b). Another form of rejection is by phagocytosis. This may involve destruction of live or damaged cells or of cellular debris during reduction (Harrison et al., '75) or in bispecific aggregates (Van de Vyver and Buscema, '77; Buscema and Van de Vyver, '79), and in the "phagocytosis crisis" of aging (Lévi, '56; Diaz, '79). A third recognition response, cytotoxicity, in which a region of foreign tissue is killed, has been observed in some species (Van de Vyver, '79; Evans and Curtis, '79).

All the elements of cellular interaction and recognition—e.g., aggregation factors (aggregation-promoting factors [PAFs]) (Curtis and Van de Vyver, '71; Curtis, '79; Evans and Curtis, '79) and "baseplate" receptors (Weinbaum and Burger, '73) [see also aggregation inhibition factors, IAF's (Curtis, '79; Van de Vyver, '79)], enhanced fibrogenesis, and phagocytosis—suggest the evolutionary emergence of a cellular defense mechanism that bears some relationship to primitive immunity. Indeed, studies by Hildemann et al. ('79), Evans et al. ('80), and Evans and Curtis ('79) of the responses of marine sponges appear to have identified graft rejection characterized by specific and accelerated second-set rejection with an alloimmune memory component. These studies are provocative and, in fact, Van de Vyver ('79) found no suggestion of alloimmune memory following second-set allograft experiments on *E. fluviatilis* (which rejects by erecting a collagen barrier) or the marine sponge *Axinella polypoides* (which rejects through a cytotoxic response). Anemnesis has not been unequivocally demonstrated in this group, but a refined capacity for "self" and "nonself" recognition is present.

The degree of variability should not be unexpected and reflects a major hazard in extrapolating from one sponge species to another, and particularly from marine sponges to freshwater sponges. The considerable increase in genome size (Harrison and Cowden, '75b, '76b) in the freshwater sponges, the appearance of sialic acid (Harrison and Cowden, in press) in freshwater sponges but not in marine sponges, and the recent (Garrone et al., '81) demonstration of fibronectin production by freshwater sponge cells all suggest the considerable degree of diversity present within the phylum. While this degree of diversity creates a versatile and intriguing research vehicle, it can create hazards, and challenges, for both the novice and the experienced investigator.

CONCLUDING REMARKS

While both freshwater and marine sponges have commanded the attention of experimental biologists since Wilson's ('07) exciting findings, there has been an acceleration in the quantity and quality of information about this group since investigators working with these organisms began to use the considerable tools of contemporary biomedical technology. As in so many other cases, the very diversity and scope of specialization of various species within the group confounds attempts at facile generalization. Indeed, the progress of recent events in freshwater and marine sponge biology emphasizes the desirability of obtaining a broader scope of detailed information.

Freshwater sponges are a relatively late evolutionary modification within an ancient group, and while the Western European and North American species have received considerable attention, little is known about the biology of freshwater sponges in most of Asia and the Southern Hemisphere. It is known that some species from very deep lakes may not undergo gemmulation whereas species from arid environments may conversely suppress sexual reproduction. Tidal marine sponges frequently display forms of gemmulation and may exhibit reduction phenomena, but these simply have not received systematic study. Within this very broad continuum, it is not surprising that findings in other matters differ in detail: how the system handles foreign material, the conditions that trigger sexual expression and reproduction, and modes of asexual reproduction.

Perhaps the most striking attribute of sponges in general is the very complex adaptation of connective tissue elements to various specialized environments. It has been pointed out that sponge connective tissue can be extremely dense, elaborate, and regular, approaching the structure of vertebrate tendon. It may be extremely loose with large amounts of amorphous connective tissue material, or in special instances chondroid tissue may be formed. The point needs to be emphasized that virtually the entire repertoire of connective-tissue specializations that occur in multicellular animals can be encountered within this group.

However, the truly fascinating aspect of this group, and of freshwater sponges in particular, is the amazing plasticity of the system. Whereas it may be possible in other groups to talk about probable "totipotent" cells, it is only the sponge archeocyte that has been truly demonstrated to have this capacity. Thus, by withdrawal to an archeocyte population, the individual can survive a transient environmental insult or, through formation of gemmules, withstand dessication and other lengthy environmental alterations. The precise mechanism of dormancy and the fundamental changes in cytoplasmic organization that are involved in the induction or breaking of dormancy are not known.

Of the freshwater sponge species that have been studied, the diploid DNA content of their genomes is much greater than that of the marine sponge with the highest DNA content. Other than the token representation of some marine species in the Woods Hole survey article on phylogenetic aspects of DNA organization, nothing is known or understood about the details of these differences. Clearly there are large differences, differences on the order of the differentials between anuran amphibia and urodele amphibia, but the evolutionary and functional significance of this is elusive.

Since it is possible to culture cells from known species and from geographic variants within a species, it should be possible to determine in due course the degree of genetic variation within genotypes. This would require cell isolation, cell culture, and probably some biochemical analysis.

Virtually nothing is known about the cytogenetics of this group. However, since methods are known that allow short-term culture of dissociated cells and since the stages in development or response to some other developmental sequences in which mitosis occurs are known, it should be possible to apply modern karyotyping methods to the group and arrive at some generalizations about evolutionary patterns. Similarly microfluorometric or cytophotometric determinations of DNA amounts and of some subsets of DNA (rapidly renaturing DNA) might be informative.

Similarly, the plasticity of the cytoskeletal elements in the small number of definitive cell types could probably stand examination by contemporary technology, particulary during cytodifferentiation from stem archeocytes. The organisms are available throughout virtually all parts of the world. The techniques required to handle and maintain freshwater sponges and sponge cells are not entirely simple, but are far less complex and demanding of resources than vertebrate cell culture. Given reasonable laboratory support facilities, this group—particularly by virtue of its relatively low number of definitive cell types—offers an extremely useful and challenging model of differentiation, morphogenesis, and cell-to-cell communication. The system can particularly be recommended to those investigators with relatively limited funds who wish to study important problems in cytodifferentiation. Some investment in experience could have handsome rewards.

LITERATURE CITED

Ankel, W.E., and H. Eigenbrodt (1950) Über die Wuchsform von *Spongilla* in sehr flachen Raumen. Zool. Anz. *145:*195–204.

Bagby, R.M. (1970) The fine stucture of pinacocytes in the marine sponge *Microciona prolifera* (Ellis and Sollander). Z. Zellforsch. Mikrosk. Anat. *105:*579–594.
Bellamy, D. (1973) Aging as a process. In J. Lobue (ed): Humoral Control of Growth and Differentiation. Vol. II. Nonvertebrate Neuroendocrinology and Aging. New York: Academic Press, pp. 219–280.
Bergquist, P.R. (1978) Sponges. London: Hutchinson.
Bergquist, P.R., and M.E. Sinclair (1973) Seasonal variation in settlement and spiculation of sponge larvae. Mar. Biol. *20:*35–44.
Berthold, G. (1969) Untersuchengen über die histoblastendifferenzierung in der gemmule von *Ephydatia fluviatilis.* A. Wiss. Mikros. *69:*227–243.
Borojević, R. (1966) Etude experimentale de la différenciation des cellules de l'éponge au cours de son développement. Dev. Biol. *14:*130–153.
Borojević, R. (1970) Différenciation cellulaire dans l'embryogénèse et la morphogénèse chez les spongiaires. In W.G. Fry (ed): Biology of Porifera. Symp. Zool. Soc. London, No. 25, London: Academic Press, pp. 467–490.
Boury-Esnault, N. (1973) L'exopinacoderme des Spongiaires. Bull. Mus. Natl. Hist. Nat. (Paris) *117:*1193–1206.
Brauer, E.B. (1975) Osmoregulation in the fresh water sponge, *Spongilla lacustris.* J. Exp. Zool. *192:*181–192.
Bretting, H. (1979) Purification and characterization of sponge lectins and a study of their immunochemical and biological activities. In C. Levi and N. Boury-Esnault (eds): Biologie des Spongiaires. Coll. Internat. C.N.R.S., No. 291 (Paris) pp. 247–255.
Brien, P. (1932) Contribution à l'étude de la régénération naturelle chez les Spongillidae. Arch. Zool. Exp. Gen. *74:*462–506.
Brien, P. (1937) La réorganisation de l'éponge après dissociation par filtration et phénomènes d'involution chez *Ephydatia fluviatilis.* Arch. Biol. *48:*185–268.
Brien, P. (1943) La formation des orifices inhalants chez les Spongillidae (*Spongilla lacustris* L., *Ephydatia fluviatilis*). Bull. Mus. R. Hist. Nat. Belg. *19:*1–10.
Brien P. (1966a) Le polyphylétisme des éponges d'eau douce. L'embryogénèse et la larve chez *Potamolepis stendelli* (Jaffe). C.R. Acad. Sci. (Paris) *263:*725–728.
Brien, P. (1966b) Le polyphylétisme des éponges d'eau douce. Formation de statoblastes chez *Potamolepsis stendelli* (Jaffe). C.R. Acad. Sci. (Paris) *263:*725–728.
Brien, P. (1967a) L'embryogénèse d'une éponge d'eau douce africaine: *Potamolepis stendelli* (Jaffe). Larves des Potamolepides et des Spongillides. Polyphylétisme des Eponges d'eau douce. Bull. Acad. Belg. Cl. Sci. *53:*752–777.
Brien, P. (1967b) Les Eponges. Leur nature metazoaire, leur gastrulation, leur état colonial. Ann. Soc. R. Zool. Belg. *97:*197–235.
Brien, P. (1967c) Embryogénèse de *Potamolepis Stendelli* et *Spongilla Moori.* Polyphylétisme des Eponges d'eau douce. Bull. Acad. R. Belg. Cl. Sc. *53:*752–757.
Brien, P. (1969) Nouvelles éponges du Lac Moero. Res. Sci. Exp. Hydrob. Bassin Lac Bangweolo Luapula *11:*1–39.
Brien, P. (1970) Les Potamolepides africaines nouvelles du Luapula et du Lac Moero. In W.G. Fry (ed): Biology of the Porifera. Symp. Zool. Soc. Lond, No. 25. London: Academic Press, pp. 163–187.
Brien. P. (1972) Les feuillets embryonnaires des Eponges. Acad. R. Belg. Bull. Cl. Sci. *58:*715–732.
Brien, P. (1973a) *Malawispongia echinoides* Brien, études complémentaires, histologie, sexualité, embryologie, affinités systematiques. Rev. Zool. Bot. Afr. *87:*50–76.
Brien, P. (1973b) Les Demosponges. Morphologie et Reproduction. In P.-P. Grasse (ed): Traité de Zoologie, Vol. 3. Spongiaires. Paris: Masson, pp. 133–461.
Brien, P., and H. Meewis (1938) Contribution à l'étude de l'embryogénèse des Spongillidae. Arch. Biol. *49:*177–250.

Brill, B. (1973) Untersuchungen zur Ultrastruktur der Choanocyte von *Ephydatia fluviatilis* L. Z. Zellforsch. Mikrosk. Anat. *144:*231–246.

Burger, M.M., W. Burkart, G. Weinbaum, and J. Jumblatt (1978) Cell–cell recognition: Molecular aspects, recognition and its relation to morphogenetic processes in general. Symp. Soc. Exp. Biol. *32:*1–24.

Burger, M.M., and J. Jumblatt (1977) Membrane involvement in cell–cell interactions: A two component model system for cellular recognition that does not require live cells. In J.W. Lash and M.M. Burger (eds): Cell and Tissue Interactions. New York: Raven, pp. 155–172.

Burkhart, W., and M.M. Burger (1977) Studies on cell populations from *Microciona prolifera* separated by Ficoll gradients. Biol. Bull. *153:*417.

Burkhart, W., J. Jumblatt, T.L. Simpson, and M. Burger (1979) Macromolecules which mediate cell–cell recognition in *Microciona prolifera.* In C. Levi and N. Boury-Esnault (eds): Biologie des Spongiaires, Coll. Internatl. C.N.R.S., No. 291. (Paris) pp. 239–246.

Burkhart, W., T.L. Simpson, and M.M. Burger (1978) Cell surface properties of larval and adult cell types in the marine sponge *Microcioina prolifera.* Biol. Bull. *155:*430.

Buscema, M., and G. Van de Vyver (1979) Etude ultrastructurale de l'agrégation des cellules dissociées de l'éponge *Ephydatia fluviatilis.* In C. Lévi and N. Boury-Esnault (eds): Biologie des Spongiaires, Coll. Internatl. C.N.R.S., No. 291. (Paris) pp. 225–231.

Cauldwell, C., P. Henkart, and T. Humphreys (1973) Physical properties of sponge aggregation factor: A unique proteoglycan complex. Biochemistry *12:*3051–3055.

Connes, R. (1967) Réactions de défense de l'éponge *Tethya lyncurium* Lamarck, vis-à-vis des micro-organismes et de l'amphipode *Leucothoe spinicarpa* Abildg. Vie et Milieu, Sér. A. *18:*281–289.

Connes, R., J.P. Diaz, and J. Paris (1971a) Choanocytes et cellule centrale chez la Démosponge *Suberites massa* Nardo. C.R. Acad. Sci. (Paris) *273:*1590–1593.

Connes, R., J. Paris, and J. Sube (1971b) Réactions tissulaires de quelques démosponges vis-à-vis de leurs commensaux et parasites. Naturaliste Can. *98:*923–935.

Curtis, A.S.G. (1979) Recognition by sponge cells. In C. Lévi and N. Boury-Esnault (eds): Biologie des Spongiaires, Coll. Internatl. C.N.R.S., No. 291. (Paris) pp. 205–209.

Curtis, A.S.G., and G. Van de Vyver (1971) The control of cell adhesion in a morphogenetic system. J. Embryol. Exp. Morphol. *26:*295–312.

Dendy, A. (1914) Observations on the gametogenesis of *Grantia compressa.* Q. J. Microsc. Sci. *60:*313–376.

De Sutter, D., and M. Buscema (1977) Isolation of a highly pure archeocyte fraction from the freshwater sponge *Ephydatia fluviatilis*. Wilhelm Roux's Archiv. *181:*151–161.

De Sutter, D., and G. Van de Vyver (1977) Aggregative properties of different cell types of the fresh water sponge *Ephydatia fluviatilis* isolated on Ficoll gradients. Wilhelm Roux's Arch. *181:*151–161.

De Sutter, D., and G. Van de Vyver (1979) Cell recognition properties of isolated sponge cell fractions. In C. Levi and N. Boury-Esnault (eds): Biologie des Spongiaires, Coll. Internatl. C.N.R.S., No. 291. (Paris) pp. 217–224.

De Vos, L. (1971) Etude ultrastructurale de la gemmulogénèse chez *Ephydatia fluviatilis*. I. Le vitellus-formation, teneur en ARN et glycogène. J. Microscopie *10:*283–304.

De Vos, L. (1972) Fibres géantes de collagène chez l'Eponge *Ephydatia fluviatilis*. J. Microscopie *15:*247–252.

De Vos, L. (1974) Etude ultrastructurale de la formation et de l'éclosion des gemmules d'*Ephydatia fluviatilis*. Thèse doctoral, Bruxelles.

De Vos, L. (1977) Etude au microscope électronique à balayage des cellules de l'éponge *Ephydatia fluviatilis*. Arch. Biol. *88:*1–14.

De Vos, L. (1979) Structure tridimensionelle de l'éponge d'eau douce *Ephydatia fluviatilis*. In C. Lévi and N. Boury-Esnault (eds): Biologie des Spongaires, Coll. Internatl. C.N.R.S., No. 291. (Paris) pp. 159–164.

De Vos, L., and F. Rozenfeld (1974) Ultrastructure de la coque collagène des gemmules d'*Ephydatia fluviatilis* (Spongillides). J. Microscopie *20:*15–20.

Diaz, J.-P. (1974) De l'origine de certains endopinacocytes chez la Demosponge *Suberites massa* Nardo. Bull. Soc. Zool. Fr. *99:*687–693.

Diaz, J.-P. (1979) Variations, differenciations et fonctions des categories cellulaires de la Demosponge d'eaux saumatres, *Suberites massa* Nardo, au cours du cycle biologique annuel et dans des conditions expérimentales. Doctoral thesis, Montpellier, pp. 1–332.

Diaz, J.-P., and R. Connes (1980) Etude ultrastructurale de las spermatogénèse d'une Démosponge. Biol. Cell. *38:*225–230.

Diaz, J.-P., R. Connes, and J. Paris (1973) Origine de la lignée germinale chez une Démosponge de l'étang de Thau: *Suberites massa* Nardo. C.R. Seances Acad. Sci. (Paris) *277:*661–664.

Diaz, J.-P., R. Connes, and J. Paris (1975) Etude ultrastructurale de l'ovogénèse d'une Démosponge: *Suberites massa* Nardo. J. Microscopie Biol. Cell. *24:*105–116.

Drum, R.W. (1968) Electron microscopy of siliceous spicules from the freshwater sponge *Heteromeyenia*. J. Ultrastruct. Res. *22:*12–21.

Duboscq, O., and O. Tuzet (1944) L'ovogénèse, la fécondation et les premiers stades du développement de *Sycon elegans* Bow. Arch. Zool. Exp. Gen. *83:*445–459.

Efremova, S.M. (1970) Proliferation activity and synthesis of protein in the cells of freshwater sponges during development after dissociation. In W.G. Fry (ed): Biology of the Porifera. Symp. Zool. So. London, No. 25. London: Academic Press, pp. 399–413.

Efremova, S.M., and V.I. Efremova (1979) Proliferation cellulaire chez la larve nageante de l'éponge d'eau douce. In C. Lévi and N. Boury-Esnault (eds): Biologie des Spongiaires, Coll. Internatl. C.N.R.S., No. 241. (Paris) pp. 59–65.

Evans, R. (1901) The structure and metamorphosis of the larva of *Spongilla lacustris*. Q. J. Microcs. Sci. *42:*363–476.

Evans, C., and A.S.G. Curtis (1979) Graft rejection in sponges: Its relation to cell aggregation studies. In C. Lévi and N. Boury-Esnault (eds): Biologie des Spongiaires, Coll. Internatl. C.N.R.S., No. 291. (Paris) pp. 211–215.

Evans, C.W., J. Kerr, and A.S.G. Curtis (1980) Graft rejection and immune memory in sponges. In M.J. Manning (ed): Phylogeny of Immunological Memory. Amsterdam: Elsevier/North-Holland, pp. 27–34.

Feige, W. (1969) Die Feinstrukture der Epithelien von *Ephydatia fluviatilis*. Zool. Jb. *86:*177–237.

Fell, P.E. (1969) The involvement of nurse cells in oogenesis and embryonic development in the marine sponge, *Haliclona ecbasis*. J. Morphol. *127:*133–150.

Fell, P.E. (1974a) Porifera. In A.C. Giese and J.S. Pearse (eds): Reproduction of Marine Invertebrates, Vol. I. New York: Academic Press, pp. 51–132.

Fell, P.E. (1974b) Diapause in the gemmules of the marine sponge, *Haliclona loosanoffi* with a note on the gemmules of *Haliclona oculata*. Biol. Bull *147:*333–351.

Fincher, J.A. (1940) The origin of the germ cells in *Stylotella heliophila* Wilson (Tetraxonida). J. Morphol. *67:*175–197.

Fjerdingstad, E.J. (1961) The ultrastructure of choanocyte collars in *Spongilla lacustris* (L.). Zeit. Zellforsch. *53:*645–657.

Frost, T.M. (1978) Impact of the freshwater sponge *Spongilla lacustris* on a *Sphagnum* bog-pond. Verh. Int. Verein. Limnol. *20:*2368–2371.

Frost, T.M. (1980a) Selection in sponge feeding processes. In D.C. Smith and Y. Tiffon (eds): Nutrition in the Lower Metazoa. Oxford: Pergamon Press, pp. 33–44.

Frost, T.M. (1980b) Clearance rate determination for the freshwater sponge *Spongilla lacustris*: Effects of temperature, particle type and concentration, and sponge size. Arch. Hydrobiol. *90:*330–356.

Frost, T.M., and C.E. Williamson (1980) In situ determination of the effect of symbiotic algae on the growth of the freshwater sponge *Spongilla lacustris*. Ecology *61:*1361–1370.

Garrone, R. (1969) Collagène, spongine et squelette minéral chez l'éponge *Haliclona rosea* (O.S.). J. Microscopie *8:*581–598.

Garrone, R. (1978) Phylogenesis of Connective Tissue, Vol. 5. In L. Robert (ed): Frontiers of Matrix Biology. Basel: Karger, pp. 1–250.

Garrone, R., C. Lethias, and J. Escaig (1980) Freeze-fracture study of sponge cell membranes and extracellular matrix. Preliminary results. Biol. Cell. *38:*71–74.

Garrone, R., T.L. Simpson, and J. Pottu-Boumendil (1981) Ultrastructure and deposition of silica in sponges. In T.L. Simpson and B.E. Volcani (eds): Silicon and Siliceous Structures in Biological Systems. New York: Springer-Verlag, pp. 495–525.

Gatenby, J.B. (1920) The germ-cells, fertilization and early development of *Grantia (Sycon) compressa*. J. Linn. Soc. Zool. *34:*261–297.

Gilbert, J.J. (1974) Field experiments on sexuality in the freshwater sponge *Spongilla lacustris*. The control of oocyte production and the fate of unfertilized oocytes. J. Exp. Zool. *188:*165–178.

Gilbert, J.J., and H.L. Allen (1973a) Studies on the physiology of the green freshwater sponge, *Spongilla lacustris*: Primary productivity, organic matter, and chlorophyll content. Verh. Int. Verein. Limnol. *18:*1413–1420.

Gilbert, J.J., and H.L. Allen (1973b) Chlorophyll and primary productivity of some green, freshwater sponges. Int. Rev. Ges. Hydrobiol. *58:*633–658.

Gilbert, J.J., and S. Hadzisce (1977) Life cycle of the freshwater sponge *Ochridaspongia rotunda* Arndt. Arch. Hydrobiol. *79:*285–318.

Gilbert, J.J., and T.L. Simpson (1976) Sex reversal in a freshwater sponge. J. Exp. Zool. *195:*145–151.

Gureeva, M.S. (1972) Sorites and oogenesis in the Baikal endemic sponges. Histologie *14:*32–45.

Harrison, F.W. (1972a) The nature and role of the basal pinacoderm of *Corvomeyenia carolinensis* Harrison (Porifera: Spongillidae). Hydrobiologia *39:*495–508.

Harrison, F.W. (1972b) Phase contrast photomicrography of cellular behaviour in spongillid porocytes (Porifera: Spongillidae). Hydrobiologia *40:*513–517.

Harrison, F.W. (1974a) Sponges (Porifera: Spongillidae). In C.W. Hart, Jr., and S.L.H. Fuller (eds): Pollution Ecology of Freshwater Invertebrates. New York: Academic Press, pp. 29–66.

Harrison, F.W. (1974b) Histology and histochemistry of developing outgrowths of *Corvomeyenia carolinensis* Harrison (Porifera: Spongillidae). J. Morphol. *144:*185–194.

Harrison, F.W. (1974c) The localization of nuclease activity in spongillid gemmules by substrate film enzymology. Acta Histochem. *51:*157–163.

Harrison, F.W. (1977) The taxonomic and ecological status of the environmentally restricted spongillid species of North America. III. *Corvomeyenia carolinensis* Harrison 1971. Hydrobiologia *56:*187–190.

Harrison, F.W. (1979) The taxonomic and ecological status of the environmentally restricted spongillid species of North America. V. *Ephydatia subtilis* (Weltner) and *Stratospongilla penneyi* sp. nov. Hydrobiologia *65:*99–105.

Harrison, F.W., and R.R. Cowden (1975a) Cytochemical observations of gemmule development in *Eunapius fragilis* (Leidy): Porifera; Spongillidae. Differentiation *4:*99–109.

Harrison, F.W., and R.R. Cowden (1975b) Feulgen microspectrophotometric analysis of deoxyribonucleoprotein organization in larval and adult freshwater sponge nuclei. J. Exp. Zool. *193:*131–136.

Harrison, F.W., and R.R. Cowden (1975c) Cytochemical observations of larval development in *Eunapius fragilis* (Leidy): Porifera; Spongillidae. J. Morphol. *145:*125–142.

Harrison, F.W., and R.R. Cowden (1976a) Aspects of Sponge Biology. New York: Academic Press.

Harrison, F.W., and R.R. Cowden (1976b) Feulgen microspectrophotometric analysis of deoxyribonucleoprotein organization in sponge nuclei. In F.W. Harrison and R.R Cowden (eds): Aspects of Sponge Biology. New York: Academic Press, pp. 141–152.

Harrison, F.W., and R.R. Cowden (1982) Sialic acid patterns during oogenesis and larval development in the freshwater sponge *Eunapius fragilis* (Porifera: Spongillidae). Acta Embryol. Morphol. Exp. (In press).

Harrison, F.W., and R.R. Cowden (In press) Application of acridine orange as a vital fluorescent probe during dormancy release and development from gemmules in *Spongilla lascustris* L. Trans. Am. Microsc. Soc.

Harrison, F.W., and D.A. Davis (1982) Morphological and cytochemical patterns during early stages of reduction body formation in *Spongilla lacustris* (Porifera: Spongillidae). Trans. Am. Microsc. Soc. (In press).

Harrison, F.W., D. Dunkelberger, and N. Watabe (1974) Cytological definition of the proferan stylocyte: A cell type characterized by an intranuclear crystal. J. Morphol. *142:*265–276.

Harrison, F.W., D. Dunkelberger, and N. Watabe (1975) Cytological examination of reduction bodies of *Corvomeyenia carolinensis* Harrison (Porifera: Spongillidae). J. Morphol. *145:*483–492.

Harrison, F.W., and M.B. Harrison (1977) The taxonomic and ecological status of the environmentally restricted spongillid species of North America. II. *Anheteromeyenia biceps* (Lindenschmidt, 1950). Hydrobiologia *55:*167–169.

Harrison, F.W., and M.B. Harrison (1979) The taxonomic and ecological status of the environmentally restricted spongillid species of North America. IV. *Spongilla heterosclerifera* Smith 1918. Hydrobiologia *62:*107–111.

Harrison, F.W., E.M. Rosenberg, D.A. Davis, and T.L. Simpson (1981) Correlation of cyclic GMP and cyclic AMP immunofluorescence with cytochemical patterns during dormancy release and development from gemmules in *Spongilla lacustris* L. (Porifera: Spongillidae). J. Morphol. *167:*53–63.

Hartman, W.D. (1958) Natural history of the marine sponges of southern New England. Bull. Peabody Mus. *12:*1–155.

Hartman, W.D. (1981) Form and distribution of silica in sponges. In T.L. Simpson and B.E. Volcani (eds): Silicon and Siliceous Structures in Biological Systems. New York: Springer-Verlag, pp. 453–493.

Hartman, W.D., J.W. Wendt, and F. Wiedenmayer (1980) Living and Fossil Sponges. Sedimentia VIII. Comparative Sedimentology Laboratory, University of Miami, Florida.

Henkart, P.S., S. Humphreys, and T. Humphreys (1973) Characterization of sponge aggregation factor: A unique proteoglycan complex. Biochemistry *12:*3045–3050.

Hildemann, W.H., I.S. Johnson, and P.L. Jokiel (1979) Immunocompetence in the lowest Metazoan phylum: Transplantation immunity in sponges. Science *204:*420–422.

Humphreys, T.D. (1963) Chemical dissolution and in vitro reconstruction of sponge cell adhesions. Dev. Biol. *8:*27–47.

Humphreys, T.D. (1970) Species specific aggregation of dissociated sponge cells. Nature (London) *25:*685–686.

Hyman, L.H. (1940) The Invertebrates: Protozoa Through Ctenophora. New York: McGraw-

Hill, pp. 284–364.

Imlay, M., and M.L. Paige (1972) Laboratory growth of freshwater sponges, unionid mussels, and sphaerid clams. Prog. Fish Cult. *34:*210–215.

Jones, W.C. (1979) The microstructure and genesis of sponge biominerals. In C. Lévi and N. Boury-Esnault (eds): Biologie des Spongiaires, Coll. Internatl. C.N.R.S., No. 291. (Paris) pp. 425–447.

Jorgensen, C.B. (1946) On the gemmules of *Spongilla lacustris* auct. together with some remarks on the taxonomy of the species. Vidensk. Medd. Danks. Naturhist. Foren. *109:*69–79.

Jumblatt, J., and M.M. Burger (1977) Studies on the binding of *Microciona prolifera* aggregation factor to dissociated cells. Biol. Bull. *153:*431.

Jumblatt, J., V. Schlup, and M.M. Burger (1978) Segments of sponge aggregation factor which bind specifically to homotypic cells. Biol. Bull. *155:*448.

Kauffold, P., and L. Spannhof (1963) Histochemische Untersuchungen an den Reservestoffen der Archeocyten in Gemmulen von *Ephydatia mülleri* Lbk. Naturwiss. *50:*384–385.

Kilian, E.F. (1952) Wasserströmung und Nahrungsaufnahme beim Susswasserschwamm *Ephydatia fluviatilis.* Z. Vergl. Physiol. *34:*407–447.

Kuhns, W.J., S. Bramson, J. Jumblatt, J. Burkart, and M.M. Burger (1978) Fluorescent antibody labelling of *Microciona prolifera* aggregation factor and its baseplate component. Biol. Bull. *155:*449.

Ledger, P.W. (1975) Septate junctions in the calcareous sponge *Sycon ciliatum.* Tissue Cell *7:*13–18.

Ledger, P.W., and W.C. Jones (1977) Spicule formation in the calcareous sponge *Sycon ciliatum.* Cell Tiss. Res. *181:*553–567.

Leveaux, M. (1939) La formation des gemmules chez les Spongillidae. Ann. Soc. R. Zool. Belg. *70:*53–94.

Leveaux, M. (1942) Contribution à létude histologique de ovogénèse et de la spermatogénèse des Spongillidae. Ann. Soc. R. Zool. Belg. *72:*251–269.

Lévi, C. (1956) Etude des *Halisarca* de Roscoff. Embryologie et systématique des Démosponges. Arch. Zool. Exp. Gen. *93:*1–181.

Lévi, C. (1970) Les cellules des éponges. In W.G. Fry (ed): Biology of the Porifera, Symp. Zool. Soc. London, No. *25.* London: Academic Press, pp. 353–364.

MacLennan, A.P. (1969) An immunochemical study of the surfaces of sponge cells. J. Exp. Zool. *172:*253–266.

Mank, A., and E.F. Kilian (1979) The ingestion and digestion of food of the freshwater sponge *Spongilla lacustris.* In C. Lévi and N. Boury-Esnault (eds.): Biologie des Spongiaires, Coll. Internatl. C.N.R.S., No. 291. (Paris) pp. 353–360.

Marshall, W. (1883) On some new siliceous sponges collected by Mr. Pechuel-Loesche in the Congo. Ann. Mag. Nat. Hist. *12:*391–412.

McClay, D.R. (1974) Cell aggregation: Properties of cell surface factor from five species of sponge. J. Exp. Zool. *188:*89–102.

Mergner, H. (1959) Über die Induktion neuer Oscularrohre bei Susswasserschwämmen. Naturwissenschaften *46:*632.

Mergner, H. (1964) Über die Induktion neuer Oscularrohre bei *Ephydatia fluviatilis.* Wilhelm Roux Arch. Entw. Mech. Org. *155:*9–128.

Mergner, H. (1966) Zum Nachweis der Artspezifität des Induktionsstoffes bei Oscularrohr-Neubildungen von Spongilliden. Verh. It. Zool. Ges. *1966:*522–564.

Mergner, H. (1970) Ergebnisse der Entwicklungsphysiologie bei Spongilliden. In W.G. Fry (ed): Biology of the Porifera, Symp. Zool. Soc. London, No. 25. London: Academic Press, pp. 365–397.

Mir-Lechaire, F.J., and M.M. Burger (1978) Isolation of an inhibitor of sponge cell aggregation from membranes. Biol. Bull. *155:*456–457.

Moscona, A.A. (1963) Studies on cell aggregation: Demonstration of materials with selective cell binding activity. Proc. Natl. Acad. Sci. U.S.A. *49:*742–747.

Müller, K. (1911a) Das Regenerationsvermogen der Süsswasserschwämme. Arch. Entw. Mech. Org. *32:*397–446.

Müller, K. (1911b) Reduction bei Süsswasserschwämmen. Arch. Entw. Mech. Org. *32:*557–607.

Müller, W.E.G., B. Kurelec, R.K. Zahn, I. Müller, P. Vaith, and G. Uhlenbeck (1979) Aggregation of sponge cells. Function of a lectin in its homologous biological system. J. Biol. Chem. *254:*7479–7481.

Müller, W.E.G., I. Müller, B. Kurelec, and R.K. Zahn (1976a) Species-specific aggregation factor in sponges. IV. Inactivation of the aggregation factor by mucoid cells from another species. Exp. Cell. Res. *98:*31–40.

Müller, W.E.G., I. Müller, and R.K. Zahn (1974) Two different aggregation principles in reaggregation process of dissociated sponge cells (*Geodia cydonium*). Experentia *30:*899–902.

Müller, W.E.G., I. Müller, and R.K. Zahn (1976b) Species-specific aggregation factor in sponges. V. Influence on programmed synthesis. Biochim. Biophys. Acta *418:*217–225.

Müller, W.E.G., I. Müller, R.K. Zahn, and B. Kurelec (1976c) Species-specific aggregation factor in sponges. VI. Aggregation receptor from the cell surface. J. Cell Sci. *21:*227–241.

Müller, W.E.G., and R.K. Zahn (1973) Purification and characterization of a species-specific aggregation factor in sponges. Exp. Cell. Res. *80:*95–104.

Müller, W.E.G., R.K. Zahn, B. Kurelec, and I. Müller (1978) Species-specific aggregation factor in sponges. Differention *10:*55–60.

Okada, Y. (1928) On the development of a Hexactinellid sponge *Farrea sollasii.* J. Fac. Sci. Imp. Univ. Tokyo *2:*1–27.

Ostrom, K.M., and T.L. Simpson (1978) Calcium and the release from dormancy of freshwater sponge gemmules. Dev. Biol. *64:*332–338.

Ostrom, K.M., and T.L. Simpson (1979) A recent study of calcium and other divalent cations in the release from dormancy of freshwater sponge gemmules. In C. Lévi and N. Boury-Esnault (eds): Biologie des Spongiaires, Coll. Internatl. C.N.R.S., No. 291. (Paris) pp. 39–46.

Pavans de Ceccatty M. (1979) Cell correlations and integration in sponges. In C. Lévi and N. Boury-Esnault (eds): Biologie des Spongiaires. Coll. Internatl. C.N.R.S., No. 291. (Paris) pp. 123–135.

Pavans de Ceccatty, M., Y. Thiney, and R. Garrone (1970) Les bases ultrastructurales des communications intercellulaires dans les oscules de quelques éponges. In W.G. Fry (ed): Biology of the Porifera. Symp. Zool. Soc. London, No. 25. London: Academic Press, pp. 449–466.

Penney, J.T. (1932) A simple method for the study of living freshwater sponges. Science *75;*341.

Penney, J.T. (1933) Reduction and regeneration of the freshwater sponges (*Spongilla discoides*). J. Exp. Zool. *65:*475–492.

Penney, J.T., and A.A. Racek (1968) Comprehensive revision of a worldwide collection of freshwater sponges (Porifera: Spongillidae). U.S. Nat. Mus. Bull. *272:*1–184.

Poirrier, M.A. (1969) Louisiana freshwater sponges: Ecology, taxonomy, and distribution. Doctoral thesis, Louisiana State University. University Microfilms Inc., Ann Arbor,

Michigan, No. 70-9083.
Poirrier, M.A. (1974) Ecomorphic variation in gemmoscleres of *Ephydatia fluviatilis* Linnaeus (Porifera: Spongillidae) with comments upon its systematics and ecology. Hydrobiologia *44:*337–347.
Poirrier, M.A. (1976) A taxonomic study of the *Spongilla alba, S. cenota, S. wagneri* species group (Porifera: Spongillidae) with observations of *S. alba.* In F.W. Harrison and R.W. Cowden (eds): Aspects of Sponge Biology. New York: Academic Press, pp. 203–213.
Poirrier, M.A., J.C. Francis, and R.A. LaBiche (1981) A continuous-flow system for growing freshwater sponges in the laboratory. Hydrobiologia *79:*255–259.
Pottu-Boumendil, J. (1975) Ultrastructure, cytochimie et comportements morphogénétiques des cellules de l'Eponge *Ephydatia mülleri* (Lieb.) au cours de l'éclosion des gemmules. Doctoral thesis, Lyons.
Pourbaix, N. (1934) Etude histochimique des substances de reserve au cours de la reproduction asexuée. Ann. Soc. R. Zool. Belg. *65:*41–58.
Pourbaix, N. (1935) Formation histochemique des gemmules d'Eponges. Ann. Soc. R. Zool. Belg. *66:*33–37.
Racek, A.A. (1969) The freshwater sponges of Australia (Porifera: Spongillidae). Austral. J. Mar. Freshw. Res. *20:*267–310.
Racek, A.A., and F.W. Harrison (1975) The systematic and phylogenetic position of *Paleospongilla chubutensis* (Porifera: Spongillidae). Proc. Linn. Soc. New South Wales *99:*157–165.
Rasmont, R. (1955a) La gemmulation des Spongillides. II. Modalités de la diapause gemmulaire. Bull. Acad. R. Belg. Cl. Sci. *41:*214–223.
Rasmont, R. (1955b) La gemmulation des Spongillides. (III). Rupture de la diapause par des agents chimiques. Ann. Soc. R. Zool. Belg. *85:*173–181.
Rasmont, R. (1955c) La gemmulation des Spongillides. IV. Morphologie de la gemmulation chez *Ephydatia fluviatilis* et *Spongilla lacustris.* Ann. Soc. R. Zool. Belg. *86:*349–387.
Rasmont, R. (1959) L'ultrastructure des choanocytes d'éponges. Ann. Sci. Nat. Zool. *1:*253–262.
Rasmont, R. (1961) Une technique de culture des éponges d'eau douce en milieu controlé. Ann. Soc. R. Zool. Belg. *91:*147–156.
Rasmont, R. (1963) Le rôle de la taille et de la nutrition dans le déterminisme de la gemmulation chez les Spongillides. Dev. Biol. *8:*243–271.
Rasmont, R. (1965) Existence d'une régulation biochimique de l'éclosion des gemmules chez les spongillides. C.R. Acad. Sci. Paris *261:*845–847.
Rasmont, R. (1968) Chemical aspects of hibernation. In M. Florkin and B.T. Scheer (eds): Chemical Zoology, Vol. 2, Porifera, Coelenterata, and Platyhelminthes. New York: Academic Press, pp. 65–77.
Rasmont, R. (1974) Stimulation of cell aggregation by theophylline in the asexual reproduction of freshwater sponges (*Ephydatia fluviatilis*). Experentia *30:*792–794.
Rasmont, R. (1975) Freshwater sponges as a material for the study of cell differentiation. In A.A. Moscona (ed): Current Topics in Developmental Biology, Vol 10. New York: Academic Press, pp. 141–159.
Rasmont, R. (1979) Les éponges: Des metazoaires et des sociétés de cellules, In C. Lévi and N. Boury-Esnault (eds): Biologie des Spongiaires, Coll. Internatl., C.N.R.S., No. 291. (Paris) pp. 21–29.
Rozenfeld, F. (1970) Inhibition du développement des gemmules de Spongillides: Specificité et moment d'action de la gemmulostasine. Arch. Biol. *81:*193–214.
Rozenfeld, F. (1971) Effets de la perforation de la coque des gemmules d'*Ephydatia fluviatilis*

(Spongillides) sur leur développement ulterieur en présence de gemmulostasine. Arch. Biol. (Liège) *82:*103–113.

Rozenfeld F. (1974) Biochemical control of freshwater sponge development: Effect on DNA, RNA, and protein synthesis of an inhibitor, secreted by the sponge. J. Embryol. Exp. Morphol. *32:*287–295.

Rozenfeld, F. (1980) Effects of puromycin on the differentiation of the freshwater sponge: *Ephydatia fluviatilis*. Differentiation *17:*193–198.

Rozenfeld, F., and A.S.G. Curtis (1980) A technique for sterile culture of freshwater sponges. Experentia *36:*371–373.

Rozenfeld, F., and R. Rasmont (1976) Hydroxyurea: An inhibitor of the differentiation of choanocytes in freshwater sponges and a possible agent for the isolation of embryonic cells. Differentiation *7:*53–60.

Ruthmann, A. (1965) The fine structure of RNA-storing archaeocytes from gemmules of freshwater sponges. Q. J. Microsc. Sci.*106:*99–114.

Sarà, M. (1955) La nutrizione dell ovcita in Calcispongie Omoceli. Ann. 1st Museo. Zool. Univ. Napoli *7:*1–30.

Sarà, M. (1968) Bispecific cell aggregation of the sponges *Haliclona elegans* and *Tethya citrina*. Acta Embryol. Morphol. Exp. *10:*228–239.

Sarà, M. (1974) Sexuality in the Porifera. Boll. Zool. *41:*327–348.

Sarà, M., L. Liaci, and N. Melone (1966) Bispecific cell aggregation in sponges. Nature (London) *210:*1167–1168.

Sarà, M., and J. Vacelet (1973) Ecologie des Demosponges. In P.-P. Grasse (ed): Traité de Zoologie. Paris: Masson, pp. 462–576.

Schmidt, I. (1970) Phagocytose et pinocytose chez les Spongillidae. Z. Vergl. Physiol. *66:*398–420.

Schmidt, I. (1970) Etude préliminaire de la différenciation des thésocytes d'*Ephydatia fluviatilis* L. extraits méchaniquement de la gemmule. C.R. Acad. Sci. Paris *271:*924–927.

Simons, J.R., and L. Muller (1966) Ribonucleic acid-storage inclusions of freshwater sponge archaeocytes. Nature *210:*347–348.

Simpson, T.L. (1968) The biology of the marine sponge *Microciona prolifera* (Ellis and Solander). II. Temperature-related, annual changes in functional and reproductive elements with a description of larval metamorphosis. J. Exp. Mar. Biol. Ecol. *2:*252–277.

Simpson, T.L. (1980) Reproductive processes in sponges: a critical evaluation of current data and views. Int. J. Invert. Reprod. *2:*251–269.

Simpson, T.L. (1981) Effects of germanium on silica deposition in sponges. In T.L. Simpson and B.E. Volcani (eds): Silicon and Siliceous Structures in Biological Systems. New York: Springer-Verlag, pp. 527–550.

Simpson, T.L., and P.E. Fell (1974) Dormancy among the Porifera: Gemmule formation and germination in freshwater and marine sponges. Trans. Am. Microsc. Soc. *93:*544–577.

Simpson, T.L. and J.J. Gilbert (1973) Gemmulation, gemmule hatching, and sexual reproduction in freshwater sponges. I. Life cycle of *Spongilla lacustris* and *Tubella pennsylvanica*. Trans. Am. Microsc. Soc. *92:*422–433.

Simpson, T.L., L.M. Refolo, and M.E. Kaby (1979) Effects of germanium on the morphology of silica deposition in a freshwater sponge. J. Morphol. *159:*343–354.

Simpson, T.L., and G.A. Rodan (1976a) Role of cAMP in the release from dormancy of freshwater sponge gemmules. Dev. Biol. *49:*544–547.

Simpson, T.L., and G.A. Rodan (1976b) Recent investigations of the involvement of 3′,5′-cyclic AMP in the developmental physiology of sponge gemmules. In F.W. Harrison and R.R. Cowden (eds) Aspects of Sponge Biology. New York: Academic Press, pp.

83–97.
Simpson, T.L., and C.A. Vaccaro (1973) The role of intragemmular osmotic pressure in cell division and hatching of gemmules of the freshwater sponge *Spongilla lacustris* (Porifera). Z. Morphol. Tiere *76:*339–357.
Simpson, T.L., and C.A. Vaccaro (1974) An ultrastructural study of silica deposition in the freshwater sponge *Spongilla lacustris.* J. Ultrastruct. Res. *47:*296–309.
Simpson, T.L., C.A. Vaccaro, and R.I. Sha'afi (1973) The role of intragemmular osmotic pressure in cell division and hatching of gemmules of the freshwater sponge *Spongilla lacustris* (Porifera). Z. Morphol. Tiere *76:*339–357.
Tessenow, W. (1969) Lytic processes in development of freshwater sponges. In J.T. Dingle and H.B. Fell (eds): Lysosomes in Biology and Pathology. North-Holland Research Monographs. Vol. 14A. Amsterdam: North-Holland, pp. 392–405.
Tuzet, O. (1930) Spermatogénèse de *Reniera.* C.R. Soc. Biol. *103:*970–973.
Tuzet, O. (1947) L'ovogénèse et al fécondation de l'éponge calcaire *Leucosolenia (Clathrina) coriacea* Mont. et de l'éponge siliceuse *Reniera elegans* Bow. Arch. Zool. Exp. Gén. *85:*127–148.
Tuzet, O. (1974) L'origine de la lignée germinale et al gametogénèse chez les Spongiaires. In E. Wolff (ed): Origine de la Lignée Germinale. Paris: Herman, pp. 79–111.
Tuzet, O. (1973) Introduction et place des Spongiaires dans la classification. In P.-P. Grasse (ed): Traité de Zoologie, Vol. 3. Paris: Masson, pp. 1–26.
Tuzet, O., R. Garrone, and M. Pavans de Ceccatty (1970) Observations ultrastructurales sur la spermatogénèse chez la Demosponge *Aplysilla rosea* Schulze (Dendroceratide): Une metaplasie exemplaire. Ann. Sci. Nat. *12:*27–50.
Tuzet, O., and M. Pavans de Ceccatty (1958) La spermatogénèse, l'ovogénèse la fécondation et les premiers stades du développement d'*Hippospongia communis* LMK (= *H. equina* O.S.). Bull. Biol. Fr. Belg. *92:*331–348.
Van de Vyver, G. (1970) La non confluence intraspécifique chez les spongaires et la notion d'individu. Ann. Embryol. Morphogen. *3:*251–262.
Van de Vyver, G. (1971) Mise en évidence d'un facteur d'agrégation chez l'éponge d'eau douce *Ephydatia fluviatilis.* Ann. Embryol. Morphol. *4:*373–381.
Van de Vyver, G. (1975) Phenomena of cellular recognition in sponges. Curr. Top. Dev. Biol. *10:*123–140.
Van de Vyver, G. (1979) Cellular mechanisms of recognition and rejection among sponges. In C. Lévi and N. Boury-Esnault (eds): Biologie des Spongiaires, Coll. Internatl. C.N.R.S., No. 291. (Paris) pp. 195–204.
Van de Vyver, G. (1980) Second-set allograft rejection in two sponge species and the problem of an alloimmune memory. In M.J. Manning (ed): Phylogeny of Immunological Memory. Amsterdam: Elsevier/North-Holland, pp. 15–26.
Van de Vyver, G., and M. Bucsema (1977) Phagocytic phenomena in different types of freshwater sponge aggregates. In J.B. Soloman and J.D. Horton (eds): Developmental Immunobiology. Amsterdam: Elsevier/North-Holland Biomedical Press, pp. 3–8.
Van de Vyver, G., and L. De Vos (1979) Structure of a non-merging front between two freshwater sponges *Ephydatia fluviatilis* belonging to different strains. In C. Lévi and N. Boury-Esnault (eds): Biologie des Spongiaires, Coll. Internatl. C.N.R.S., No. 291. (Paris) pp. 233–237.
Van de Vyver, G., and P. Willenz (1975) An experimental study of the life-cycle of the freshwater sponge *Ephydatia fluviatilis* in its natural surroundings. Wilhelm Roux's Arch. *177:*41–52.
Weinbaum, G., and M.M. Burger (1973) A two-component system for surface guided reassociation of animal cells. Nature (London) *244:*510–512.

Weissenfels, N. (1976) Bau und Funktion des Süsswasserschwamms *Ephydatia fluviatilis* (Porifera). III. Nahrungsaufnahme, Verdauung und Defäkation. Zoomorphol. *85:*73–88.

Willenz, P. (1980) Kinetic and morphological aspects of particle ingestion by the freshwater sponge *Ephydatia fluviatilis* L. In D.C. Smith and Y. Tiffon (eds): Nutrition in the Lower Metazoa. Oxford: Pergamon Press, pp. 163–178.

Williamson, C.E. (1979) An ultrastructural investigation of algal symbiosis in white and green *Spongilla lacustris* (L.) In D.C. Smith and Y. Tiffon (eds): Nutrition in the Lower Metazoa. Oxford: Pergamon Press, pp. 163–178.

Williamson, C.E. (1979) An ultrastructural investigation of algal symbiosis in white and green *Spongilla lacustris* (L.) (Porifera: Spongillidae). Trans. Am. Microsc. Soc. *98:*59–77.

Wilson, H.V. (1907) On some phenomena of coalescence and regeneration in sponges. J. Exp. Zool. *5:*245–258.

Wintermann, G. (1951) Entwicklungsphysiologische Untersuchungen an Süsswasserschwamme. Zool. Jahrb. Abt. Anat. *71:*427–486.

Wintermann-Kilian, G., E.F. Kilian, and W.E. Ankel (1969) Musterbildung und Entwicklungsphysiologie der Epithelien beim Susswasserschwamm *Ephydatia fluviatilis.* Zool. Jb. Anat. *86:*459–492.

Zeuthem, E. (1939) On the hibernation of *Spongilla lacustris.* Z. Vergl. Physiol. *26:*537–547.

Developmental Biology of Freshwater Invertebrates, pages 69–127

Hydra

Georgia E. Lesh-Laurie

ABSTRACT The freshwater coelenterate *Hydra* is introduced from the perspective of its structural organization and natural history. Methods for maintaining and manipulating the animal in culture are examined, as are its sexual and asexual reproductive processes. The experimental literature on polarity and its establishment, maintenance, reestablishment, and reversal is reviewed. Experimental manipulations of cellular events, including cell-cycle studies, cell commitment, and cell differentiation, are explored. Investigations reporting the isolation, chemical characterization, and biological role of morphogenetic substances are discussed, as are the biochemical events accompanying normal regeneration. Finally, the potential of hydra as a system for pursuing questions of pattern formation is considered.

Hydra has been a favorite experimental system of developmental biologists since Abraham Trembley's pioneering experiments in 1744. Superficially, the widespread use of hydra as an organism in which to examine developmental phenomena might appear to be a curious occurrence. Experimental organisms usually manifest the events representative of their taxonomic group in a "classic" or "diagrammatic" fashion. Hydra, being a member of the predominately marine Phylum Coelenterata and Class Hydrozoa, would predictably exhibit a metagenic life cycle with a polypoid form reproducing asexually and a medusa being the vehicle for sexual reproduction. Embryogenesis would result in the formation of a definitive planula larva, which after settlement would form a colonial organism.

Hydra, however, is a solitary, freshwater animal devoid of a medusoid stage in its life cycle (Fig. 1). It reproduces both asexually and sexually in the polypoid form, passing its entire embryogenesis within a nearly impenetrable embryotheca. Why, then, one might inquire, the fascination with hydra as an experimental organism? In part, the answer lies in the developmental phenomena first explored by Trembley. His preliminary experiments, involv-

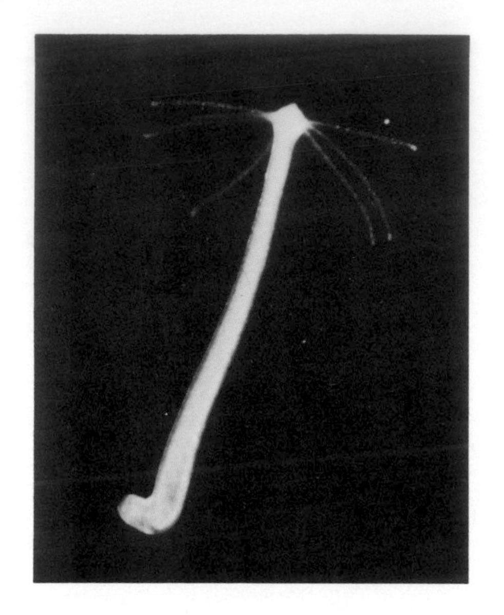

Fig. 1. Nonbudding *H. pirardi.* × 8.

ing simple amputations, revealed that hydra possessed an extensive capacity for repair and regeneration. As investigators directed their attention to a further understanding of this developmental event, they soon discovered within hydra a unique combination of structural and functional properties that enable the animal to serve as a model *in vivo* system for the study of several developmental processes:

1. The existence of a pluripotent cell population within the mature animal allows examination of the feedback controls and cell cycle parameters associated with cell commitment and differentiation.
2. The inductive properties exhibited by the organism's distal and proximal structures permit further clarification of this fundamental form of tissue interaction.
3. The animal's size and simple construction facilitate light-microscopic, histochemical, and ultrastructural study.
4. The ease with which the animal can be grafted and/or disaggregated and reassembled makes questions of positional information amenable to experimental analysis.
5. The organism's primitive nervous system coupled with the investigator's ability to eliminate it selectively affords an unparalleled opportunity to explore the neural control of developmental processes.

NATURAL HISTORY OF THE ADULT

Hydra unobtrusively populate most aerobic freshwater environments. Although in an experimental situation the animals express a preference for a rough substratum over a smooth one (Kanaev, '52), hydras' frequent occurrence attached to floating or submerged vegetation facilitates their collection by naturalists. The animals show no marked rheotropic response (Wagner, '05; Kanaev, '52), and their distribution may range over depths of several meters, depending upon the species (Welch and Loomis, '24). Irrespective of species-specific distributional variation, hydra are found in greatest frequency near the surface.

Initially this surface congregation of hydra was considered an adaptive response to their general food source, crustacean larvae. Later, other investigators argued that this surface location represented a positive phototropic behavior (see discussion in Kanaev, '52). Haug ('33), however, simply and elegantly distinguished between these alternatives by placing hydra in an aquarium in which its food (retained in a gauze bag) was placed in the darkest portion of the container. Observations over a 8- to 10-day period revealed hydra continually collected on the light side of the aquarium, despite a complete absence of food in this region.

In nature, once objects to which hydra may be attached have been obtained, one may ascertain the presence of the organisms by allowing these materials to stand undisturbed for several hours, preferably in a container that allows visualization of the material from all possible directions (e.g., open glass jar or fingerbowl). Hydra, if present, will relax and become visible suspended from or attached to the collected materials. After the animals have been located, they may be cleaned of adherent debris with fine watchmaker's forceps and placed in fresh pond water or an artificial culture solution.

Alternatively, if collection from a natural environment is inconvenient, hydra may be obtained from most commercial animal suppliers. As the animals are somewhat temperature-sensitive, it is recommended that organisms be purchased from a supplier who guarantees live delivery. Among suppliers known to provide healthy organisms: A.H. Forbes Laboratories, Verona, PA 15147; Ann Arbor Biological Center, Ann Arbor, MI 48103; Carolina Biological Supply Company, Burlington, NC 27215; Connecticut Valley Biological Supply Company, Southampton, MA 01073; Ward's Natural Science Establishment, Rochester, NY 14603.

Organisms obtained from commercial sources share with those from natural environments the characteristic of being somewhat undernourished and nonbudding. Freshly hatched brine shrimp nauplii provide an ample food source, and 1 week of daily feeding will normally result in the development of budding animals in a culture. In the spring and fall, periods of sexuality occur in hydra, and if sexual animals are desired they can usually be specified from commercial suppliers during these periods.

STRUCTURAL ORGANIZATION OF THE ORGANISM

Hydras extend 5–10 mm along their oral-aboral axis. Their cylindrical body column terminates distally in a conical hypostome with a central mouth encircled by tentacles. In spite of the animal's overt radial symmetry, this oral region is frequently referred to as the hydra's "head." At the proximal end of the body column, an attachment device or basal disc is found. This aboral region is called the animal's "foot." Histological characterization of the animal (see below) has resulted in the further demarcation of its body into five general regions extending distoproximally from the head to the foot (Fig. 2). The hypostome extends from the animal's distal mouth and includes the region from which hollow tentacles emerge. The gastric region begins immediately subjacent to the tentacles and continues proximally to the area where asexual reproductive products, or buds, are formed. The area of asexual reproduction is the budding region. In nonbudding animals one frequently considers the gastric region as extending proximally until the peduncle, or stalk, is reached. The peduncle or stalk is a generally narrowed area immediately distal to the terminal basal disc.

Histologically, the animal consists of two epithelial layers separated by an acellular mesoglea (Fig. 3). The outer epithelium—the epidermis or ectoderm—is constructed of autoreproductive epitheliomuscular cells. The interstices between these cells are occupied by a population of autoreproductive interstitial cells (I cells), or stem cells, and their derived cell types. Within the epidermis, I-cell–derived cell types include nerves, the animal's stinging cells or nematocytes, and gametes, if present. None of these cell types, when terminally differentiated, are capable of cell division. At least two types of nervous elements, ganglion cells and sensory cells, have been described from electron microscopic studies in hydra (Lentz and Barrnett, '65; Davis et al., '68; Davis, '68, '69, '70; Westfall, '73; Westfall and Kinnamon, '78; Westfall et al., '80; Kinnamon and Westfall, '81). Light microscopists, however, have provided evidence for a greater diversity in this cell population (Burnett and Diehl, '64a; Tardent and Weber, '76; Epp and Tardent, '78).

Four nematocyte types are generally recognized: large, pear-shaped stenoteles used in prey capture; small, comma-shaped desmonemes, which entangle prey; and ovoid atrichous and heterotrichous isorhizas, which assist hydra in attachment and translocation (Fig. 4). Interstitial-cell–derived cell types are not randomly distributed along the animal's oral-aboral axis. Rather, unique concentrations of cell types occupy particular positions along the body

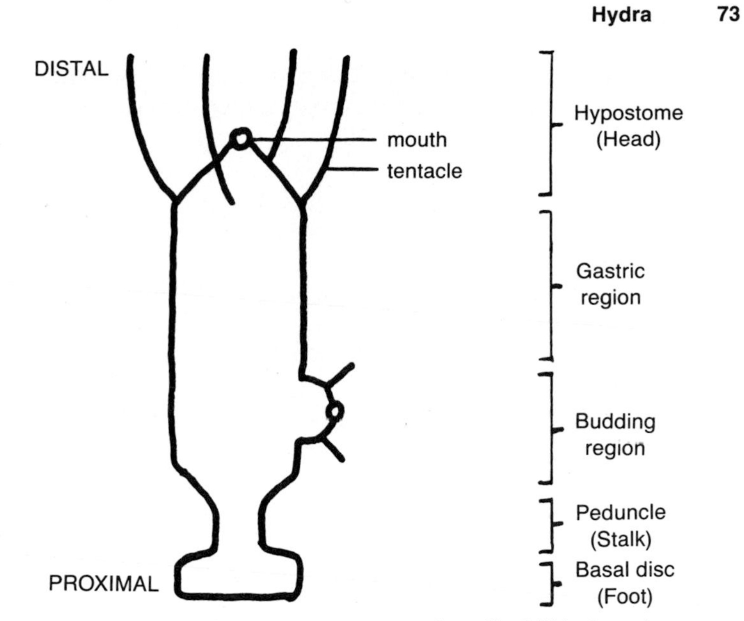

Fig. 2. Hydra morphology. Diagrammatic representation of hydrid body regions.

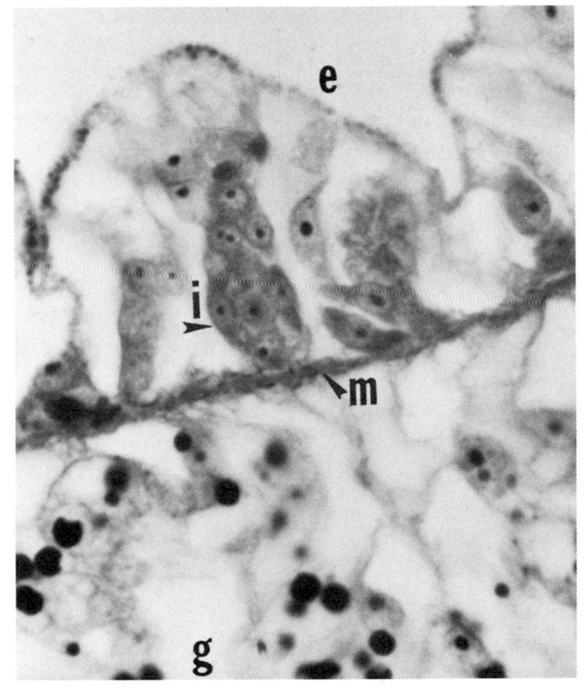

Fig. 3. Histological organization of hydra. The outer epidermis or ectoderm (e) contains epitheliomuscular cells and a nest of I cells (i). The inner gastrodermis or endoderm (g) is populated by digestive cells with dark, spherical inclusions. Separating the two layers is an acellular mesoglea (m). × 880.

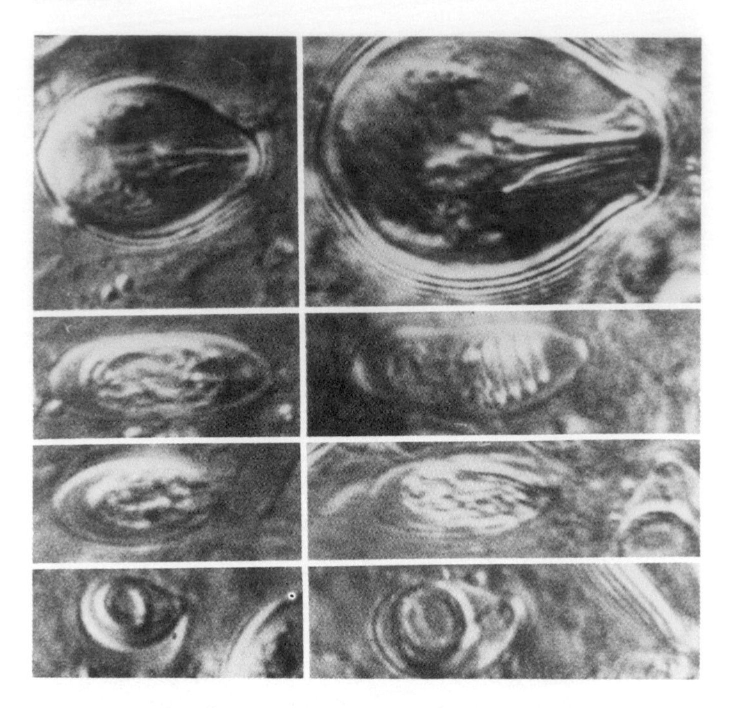

Fig. 4. Nematocysts of *P. oligactis* (left panel) and *H. attenuata* (right panel). From top to bottom these are the stenotele, holotrichous isorhiza, atrichous isorhiza, and desmoneme. These photographs were made from macerated tissue viewed with phase-contrast optics. × 2,500. Reprinted from Lee and Campbell ('79), with permission.

column, in part resulting in the distoproximal regionalization of the animal described above. For example, nerves are concentrated in the hypostome and basal disc regions. Desmonemes and atrichous isorhizas are found exclusively in the hypostome and tentacle regions, whereas stenoteles occur throughout the body column. Interstitial cells themselves are located predominately in the gastric region (Bode et al., '73, '76). Considerable attention has been directed toward clarifying the mechanisms regulating these restricted distributions, a topic that will be examined later in the discussion of cell commitment.

The inner epithelial layer—the gastrodermis or endoderm—consists of digestive cells, gland cells, and mucous cells, all of which are autoreproductive. Some evidence exists to suggest that gland and/or mucous cells may be I-cell derivatives (Haynes and Burnett, '63; Burnett et al., '66; Rose and Burnett, '68a; Znidaric and Lui, '69; Rose and Burnett, '70; Znidaric, '70). These cell types are also asymmetrically distributed along the distoproximal axis of the animal with mucous cells concentrated in the hypostomal region.

Gland cells, alternatively, are widely distributed within the remainder of the column. Gastrodermal nerve cells have also been observed, but no detailed morphological or developmental analysis of them has been completed.

In addition to the references already presented, numerous excellent histological, histochemical (Burnett, '59; Cowden and Glocker, '60; Sanyal and Mookerjee, '60; Lentz and Barrnett, '61), and ultrastructural (Lentz, '65b, '66; Davis et al., '66; Davis and Haynes, '68; Davis and Bursztajn, '73; Davis, '73, '74, '75; Westfall and Townsend, '77; West, '78a; Epp et al., '79; Wood, '77, '79a, b, '80; Wood and Kuda, '80a, b; Epp, '80; Holstein, '81) studies characterizing the tissue organization and normal cell types of hydra are available if further anatomical detail is desired.

SPECIATION

Species variation in hydra occurs primarily in characteristics such as tentacle length and pattern of emergence, nematocyte morphology, coloration and gross morphological form (Kanaev, '52; Lee and Campbell, '79). Of these, coloration, tentacle patterning, and gross morphological form have been of particular significance to developmental biologists. Coloration will reflect, to some extent, the nutritional state of the animal. However, the presence of symbiotic algae within the digestive cells of *Hydra virdis* (*Chlorohydra viridissima*) results in a distinct green color in this species, irrespective of nutrition. These algae provide a natural cell marker that can be exploited in studies of tissue movements or origins.

Tentacle patterning has been most frequently recorded from developing buds, with two generalized "patterns" being expressed. In *H. attenuata*, for example, all tentacles emerge in a nearly synchronous manner. Alternatively, in *H. oligactis* tentacles arise in a rigorous, predictable sequence. Initially, a midlateral pair forms, followed by a single dorsal tentacle, a single ventral tentacle and finally two intercalating lateral tentacles (Ito and Ohata,'63; Baird and Burnett, '67; Sanyal and Mookerjee, '67; Lesh-Laurie and Hang, '72; Otto and Campbell, '77a; Lee and Campbell, '79). Buds from chimeric hydra, constructed from *H. attenuata* epithelia and *H. oligactis* I cells, express the synchronous *H. attenuata* tentacle emergence pattern (Campbell, '79; Lee and Campbell, '79), providing evidence that this morphogenetic event in hydra is epithelially directed.

The most apparent gross morphological difference between hydra species concerns the presence or absence of a clearly defined peduncle or stalk. Several species (e.g., *H. oligactis, H. fusca, H. vulgaris*) possess an obvious constricted area immediately distal to the basal disc and are often grouped as the stalked hydra. Other species (e.g., *H. attenuata, H. littoralis*) possess a less distinct demarcation between the body column and the basal disc (Fig.

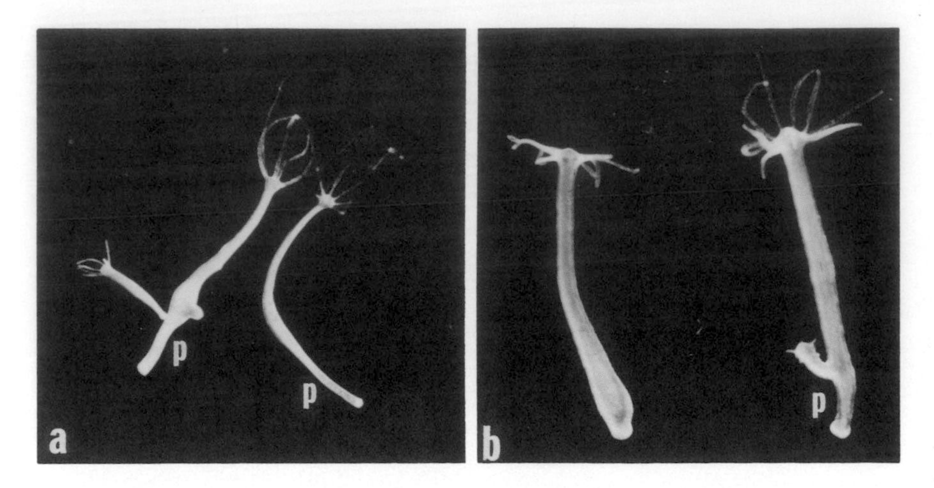

Fig. 5. Stalked (a) versus nonstalked (b) hydra. Note the prominent peduncle or stalk (p) in the *H. littoralis* shown in 5a, compared to the virtual absence of this region in *H. pirardi* (5b). × 5.

5). Developmentally, this difference is expressed, at least in part, in differences in the ability of the animal to regenerate in a proximal direction.

Hydra include members of what have been described historically as three genera, *Chlorohydra, Hydra* and *Pelmatohydra.* Conventionally, most investigators today place all animals in a general genus *Hydra.* Taxonomically, species within the genus remain somewhat ill-defined, with numerous synonyms in existence. A simple taxonomic guide to the most frequently encountered species has been compiled recently by Campbell ('82).

MAINTENANCE AND MANIPULATION OF ADULTS

Numerous methods are currently available for the mass culture of mature hydra. All of these involve the use of buffered salt solutions and a crustacean food source. As these methods have recently been published in collected form (Lenhoff, '82), only the two most frequently employed culture solutions will be presented.

Culturing Media

Historically, routine laboratory culture of hydra derived from the pioneering experiments of Loomis and Lenhoff on the environmental factors influencing growth in hydra (Loomis, '53, '54; Loomis and Lenhoff, '56; Lenhoff and Loomis, '57; Lenhoff and Bovaird, '60; Lenhoff, '66). As a result of their investigations, they developed an easily prepared culture solution consisting of 20 mg sodium bicarbonate and 10 mg disodium ethylenediaminetetraacetic acid (versene) in 1 liter of tap water. This culture solution had limited applicability in critical developmental and physiological studies, however, because versene-treated tap water is a solution of variable and unknown ionic composition. Furthermore, although the versene chelated toxic copper ions from the solution, it also removed essential calcium. Therefore, the original bicarbonate-versene–tap water medium is usually modified as:

Stock solution A:	10 gm (ethylenedinitrilo)-tetraacetic acid tetrasodium salt
	20 gm $NaHCO_3$
	1 liter distilled or deionized water
Stock solution B:	40 gm $CaCl_2 \cdot 2H_2O$
	1 liter distilled or deionized water

The working culture solution is prepared by adding 5 ml of stock solution A to 990 ml of distilled or deionized water, mixing, adding 5 ml of stock solution B, and mixing again. It is necessary in preparing this working culture medium always to be certain to allow the solution containing the chelating agent (stock solution A) to mix with the water, chelating it, *before* the calcium-containing solution (stock solution B) is added.

Currently, many investigators routinely culture hydra in modified M solution, a medium modified from that first reported by Muscatine and Lenhoff ('65). The advantages of this culture medium and its effects on the growth of several hydra species are superbly reviewed by Lenhoff and Brown ('70).

Modified M solution

Component	Concentration in working culture medium
$NaHCO_3$	10^{-3}M
KCl	10^{-4}M
$MgCl_2 \cdot 6H_2O$	10^{-4}M
$CaCl_2 \cdot 2H_2O$	10^{-3}M
Tris buffer (pH 7.6)	10^{-3}M

To facilitate preparation of the large quantities of solution necessary for mass culture, concentrated stock solutions of these media components are also frequently employed. A convenient method for preparing stock solutions for modified M solution is given below:

Stock solution A: Prepared in 4 liters distilled or deionized water		
$NaHCO_3$	67.2 gm	(2×10^{-1}M)
KCl	5.96 gm	(2×10^{-2}M)
$MgCl_2 \cdot 6H_2O$	16.24 gm	(2×10^{-2}M)
Stock solution B: Prepared in 3 liters distilled or deionized water		
$CaCl_2 \cdot 2H_2O$	88.2 gm	(2×10^{-1}M)
Tris buffer (pH 7.6)		(2×10^{-1}M)

Prepare Tris buffer by mixing 2,100 ml Tris-HCl (0.2 M) and 900 ml Tris base (0.2 M). Check pH and adjust as necessary. Keep buffer refrigerated.

The working culture solution is prepared by adding 5 ml of stock solution A to 990 ml of distilled or deionized water, mixing, adding 5 ml of stock solution B, and mixing again.

Spring water can be substituted for distilled or deionized water in either of the culture solutions described. Also, slight ionic modifications may be introduced to accommodate individual water sources. Finally, in order to reduce any bacterial contamination that may be contained in the animal's external mucus or introduced with its food, some investigators add extremely low concentrations of bacteriostatic agents such as rifampicin or kanamycin sulfate to their working culture solutions.

The most commonly employed culturing vessels include fingerbowls and shallow glass or Plexiglas baking dishes. Animals should be disbursed into the culture solution so that they exist at reasonable densities (e.g., ~1,000 animals/20-cm finger bowl). Few hydra cultured in large volumes of media will not exhibit the growth rate observed in somewhat denser cultures.

Feeding and Maintenance

Hydra's principal food source is crustacean larvae. In the laboratory these are most reasonably and inexpensively obtained by hatching brine shrimp. Brine shrimp "eggs" (actually encysted gastrulae) may be purchased from numerous suppliers. It is recommended that the cysts be purchased directly from a supplier rather than from a secondary source, as it is difficult to assess the condition of cysts under the latter circumstance. Once hatched, larvae may be collected and concentrated using 125 mesh/square inch bolting cloth, thoroughly washed free of sea salts, resuspended in a small volume of hydra culture water, and fed to hydra cultures using a pipette. Shrimp larvae should be applied to the cultures at a density sufficient to allow hydra to feed to repletion in ~15 minutes.

Two and one-half to 6 hours after feeding, undigested food must be removed from the cultures and fresh culture solution provided. This is most frequently done by either vortexing animals or by employing bolting cloth. In the vortexing method, one detaches animals from the bottom of the dish or finger bowl using a clean finger or pipette bulb. The animals are then gently swirled to the center of the dish and removed by pipette to clean culture medium. Several swirlings may be necessary to separate the hydra from unconsumed food and debris. Alternatively, hydra, after being detached from the culture vessel, may be poured gently through 38-mesh/square inch bolting cloth. This mesh size generally retains the hydra (some buds may be lost), and allows the unconsumed food and debris to pass. The hydra are then rinsed with hydra culture water and returned to a cleaned culture vessel. Both of these methods have the advantage of allowing a thorough cleaning of the culture vessel before returning hydra to it.

Vigorous cultures may be maintained using a 2–3 times per week feeding regime, although animals will survive with only periodic feeding so long as they are provided with a clean culture vessel and fresh culture medium. Lighting conditions are not critical; nor is temperature, so long as it is restricted to between 15 and 25°C. Low temperature extremes will retard the animals' growth and may encourage the development of sexual organs. High temperature extremes are lethal.

Again, one is referred to the reviews by Lenhoff and Brown ('70) or Lenhoff ('82) for additional considerations of current hydra culturing techniques.

Surgical Manipulation

Surgery is usually performed on hydra using either iridectomy scissors or a long-handled scalpel with a No. 15 blade. Watchmaker's No. 3 or 5 forceps are the best aid in positioning the animals. A thin, flat, glass or plastic surface (for example, a Petri dish) provides a convenient vessel in which to perform surgical manipulations. Although these operations may be completed without magnification, the use of a dissecting microscope is recommended.

Grafting Procedures

Hydra is a particularly amenable organism for studying cell and/or tissue interactions as the animal's gross morphological organization may be manipulated with ease using grafting techniques. Grafting involves the transfer of donor tissue (which has been surgically separated from a donor animal) to a host animal. Two Petri dishes are generally used for grafting experiments, one of which is modified for holding host and donor tissues in apposition during healing. Amputations are performed in the Petri dish that contains culture solution. The host animal is usually prepared first and transferred to the second dish. The second Petri dish has been layered with paraffin wax, with alcohol-cleaned human hairs embedded vertically in the wax. Culture solution is present over the wax to a level covering the projecting hairs.

Using watchmaker's No. 5 forceps, host and donor tissue may be pierced so they can be positioned on the vertical hairs as desired. Once placement is complete, donor and host tissues may be secured with an alcohol-cleaned hair loop. It is important to ensure that cut or abraided surfaces of donor and host tissue are in contact and that the axes of the two tissues are aligned. Healing will occur within 2–4 hours, after which the hair loop may be carefully removed and the grafted animal gently lifted from its position on the implanted hair. Removal of the grafted animal is most easily accomplished by placing forceps beneath the grafted animal and gradually lifting it off the hair. The graft should then be placed into a clean culture dish using an appropriate sized pipette. Grafting procedures are also most conveniently executed using dissecting microscopes (Sersig and Lesh-Laurie, '81).

Vital Marking of Cells

Developmental biologists have long sought effective mechanisms for marking living specimens to study tissue movements. Local applications of aqueous vital stains, such as methylene blue or neutral red, present several disadvantages, however. First, there is a progressive dilution of the marker as the dye diffuses, making boundary regions between stained and unstained tissue indistinct. Furthermore, the diffusibility of these dyes prohibits the marking of single cells.

Some hydra species, such as *H. viridis*, possess a natural vital marker in the algal symbionts of their gastrodermal digestive cells. *Hydra viridis* that contain symbiotic algae are green in color, whereas animals rendered aposymbiotic or alga-free are white. As both symbiotic and aposymbiotic animals are of the same species, they may be easily grafted, with host and donor gastrodermis unambiguously marked. Aposymbiotic hydra may be prepared by subjecting green animals to high light intensity in the presence of a photosynthesis inhibitor (Pardy, '76). For example, hydra may be maintained at 15°C in the presence of 10^{-6} M 3-(3,4-dichlorophenyl)-1,1-dimethylurea with a light irradiation at the surface of the culture container of 825 Wm^{-2}. Animals are not fed during this treatment, but the culture medium is changed daily. After 5 days of continuous exposure to these conditions, many animals are alga-free. These individuals may then be isolated and aposymbiotic clones grown from them.

If an epidermal marker is desired, Campbell ('73) has designed a colloidal carbon-marking technique that also allows individual cells to be tagged. Hydra are injected with India ink which apparently enters the epidermal cells by phagocytosis, becoming concentrated in what are presumed to be lysosomes. Two methods of application are suggested depending upon whether one wishes a localized or diffuse mark. To tag a small group of cells, the ink is injected into the epidermis by placing a mouth pipette (with a tip drawn to approximately 5 μm) nearly parallel to the animal's body surface. A small amount of ink is then very slowly released. Excessive quantities of ink will result in its penetration of the mesoglea and collection in the animal's gastrovascular cavity. If a more broadly marked area is desired, the pipette may be placed perpendicular against the body surface (the surface should be slightly depressed at the point of contact), and a large amount of ink forcibly released.

Interestingly, neither the algal symbiont nor the colloidal carbon method marks I cells. Likewise, vital dyes do not seem to stain I cells. Although the mechanism for this occurrence is not known, it has been suggested that epithelial cell functions may include scavenging the animal's intercellular spaces, thus restricting access of superficial markers to the I-cell population (Campbell, '73).

Cytological Studies

Cell separation and quantification in hydra are frequently accomplished using a modification of Béla Haller's tissue maceration technique (David, '73). Selected tissue samples are macerated in a small quantity of glycerin–acetic acid–distilled water (1:1:13) at room temperature for 10–20 minutes. Once tissue disaggregation is completed, the cells are postfixed in one-tenth volume of 20% formalin. Cell suspensions are then mixed with one-tenth volume of 2% sodium dodecyl sulfate, spread on an ethanol-cleaned 3″ × 1″ glass microscope slide, and allowed to evaporate to dryness in a dust-free place.

Preparations are examined as wet mounts using phase contrast microscopy. Cell counts may be made of random edge-to-edge sweeps across the short dimension of the slide. This methodology reduces the effects of selective cell accumulation at the slide's edges (David, '72).

Hydra are also readily prepared for histological study as either whole-mount or sectioned specimens. Prior to fixation animals may be relaxed 3–5 minutes in either 5% ethanol or 1–3% urethane (made up in culture solution). Whole-mount fixation is frequently completed in 100% ethanol for 1 hour. Following fixation the animals are transferred to distilled water and then stained, often in 0.05% aqueous toluidine blue (pH 8), for 1–2 minutes. Stained animals are rinsed in distilled water and destained, as desired, in 70% ethanol. After destaining, the animals are dehydrated, cleared in xylene or toluene, and mounted (Diehl and Burnett, '64).

Bouin's or Lavdowsky's fixatives are often employed if tissue sections are desired. Following fixation (2–24 hours), animals are dehydrated in ethanol, cleared in either xylene or toluene, and embedded in paraffin (or plastic) by conventional methods. Paraffin sections between 5 and 7 μm are easily obtained. Staining may be by any appropriate procedure.

Application of Exogenous Substances

Exogenous materials such as metabolic agonists or antagonists, radioactive substances, etc. are normally applied to hydra by immersing animals in a solution containing an appropriate concentration of the desired compound(s). Drawbacks exist to this type of exposure, however, as entry to the organism normally occurs via the epithelial cells of the gastrovascular cavity. Therefore, placing organisms in a fluid of a specific composition does not ensure that the concentration of the solution is the effective concentration reaching the cells of the organism. In addition, the animal only periodically flushes its gut cavity, with the result that immersion and effective exposure may not be simultaneous events. To counteract these difficulties, investigators often inject the desired material directly into the animal's gastrovascular cavity, as well as immersing them in it.

Because of these problems it is necessary to ascertain whether, and to what extent, the predicted effect is being obtained when exogenous compounds are presented to hydra. Although it is difficult to estimate biologically effective amounts of material, concentrations above 10^{-3} M may deleteriously affect the pH and/or the ionic balance of the culture medium. Thus, one frequently encounters concentrations between 10^{-4} and 10^{-8} M being employed.

SEXUAL REPRODUCTION AND EMBRYOGENESIS

Sexual reproduction is an episodic phenomenon in hydras. In nature, gonads are elaborated at times of environmental instability, such as the varying temperature and ionic conditions accompanying seasonal change. In spite of extensive study, the mechanism(s) regulating gonad and gamete formation in hydras remain imperfectly understood (Kanaev, '52; Loomis and Lenhoff, '56; Loomis, '64; Tardent, '66, '74; Stagni, '74). Several investigators have achieved considerable success in initiating sexual differentiation through environmental manipulation. For example, Loomis ('64) reported that diverse media enrichments released sexuality in *H. littoralis*, whereas lowering culturing temperatures allowed gonad and gamete formation in *H. pirardi, H. oligactis, H. pseudoligactis,* and *H. hymanae* (Burnett and Diehl, '64b; Davison, '76; West, '78b). These observations have led to the speculation that sexual differentiation results from feedback mechanisms sensitive to microenvironmental conditions. In other species, however, the spontaneous appearance of sexually mature organisms has permitted the serendipitous examination of this physiological process.

Hydra species are variously described as gonochoristic (*H. oligactis*), unstable gonochoristic (*H. attenuata, H. carnea*), or hermaphroditic (*H. circumcinta, H. viridis*) (Stagni, '66; Tardent, '66, '74; Vannini, '74). Even in stable gonochorists, however, sexual states can be altered if the two sexes are grafted together. Parabiosing of male and female *H. oligactis* results in the female's becoming male (Brien, '61; '63); similar observations have been reported in *H. attenuata* (Tardent, '68, '74). In those species exhibiting hermaphroditism, male gonads normally appear first and occupy the distal-most portion of the body column (Fig. 6). Female gonads develop later in the lower, proximal gastric region (Fig. 7). Hermaphroditic species may simultaneously display mature male and female gonads while continuing to reproduce asexually (Fig. 8).

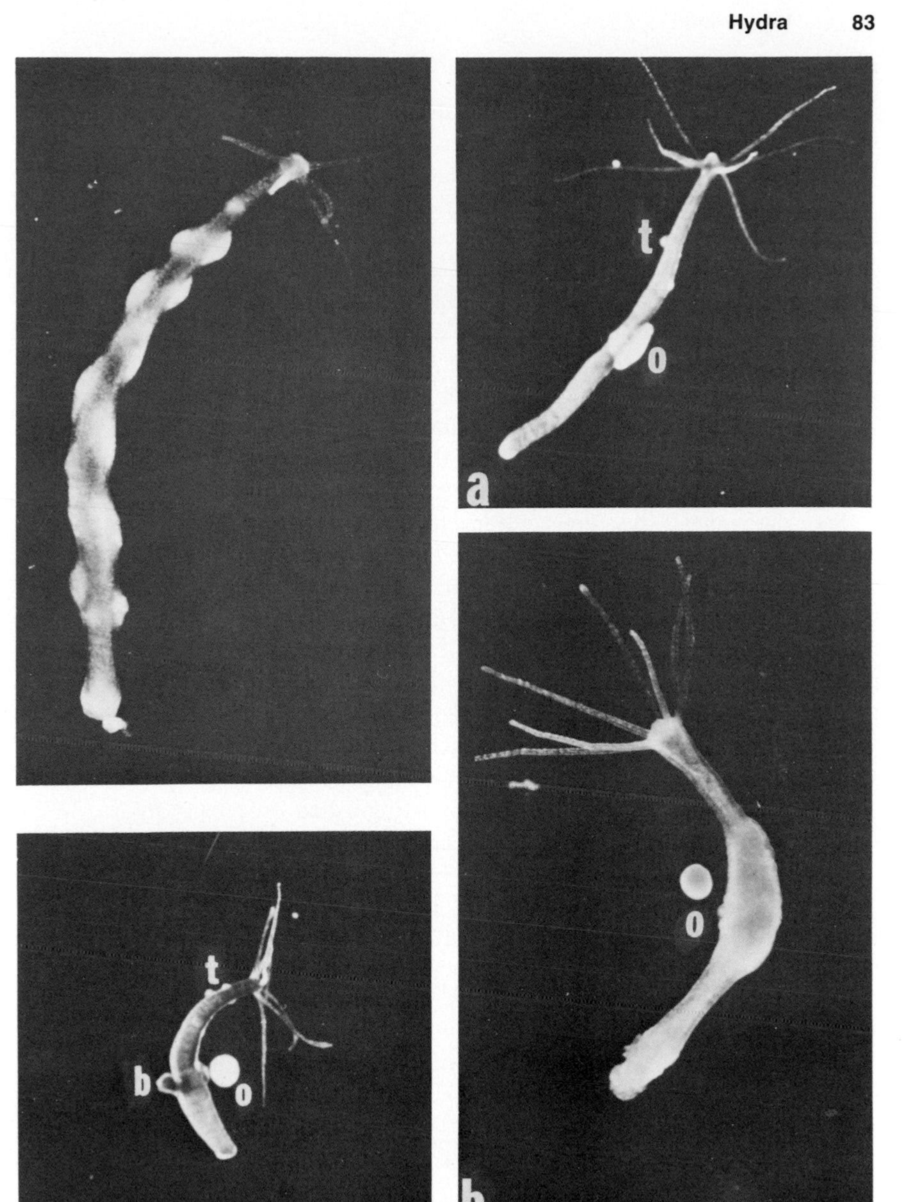

Fig. 8. Fig. 7a, b.

Fig. 6. *Hydra pirardi* with testes extending throughout the gastric region. × 10.

Fig. 7. Sexual hydra with ovaries. a) *H. viridis* with distal testes (t) and a proximal ovary (o). × 10. b) Female *H. pirardi* with an ovary (o) and recently released egg. × 6.

Fig. 8. Hermaphroditic *H. viridis* containing testes (t), an ovary with egg (o), and a bud (b). × 5.

Gametogenesis

Because of the extemporaneous nature of sexual reproduction in hydra, some controversy initially arose over the origin of the animal's gametes. Downing ('05, '08, '09) postulated the existence of a separate and continuous germinal line in hydra. However, in 1872 Kleinenberg had already suggested an alternative origin for these cells from the animal's I-cell population. The subsequent light microscopic investigations of Tannreuther ('09) and Brien and his colleagues (Brien, '50a, b, '66; Brien and Reniers-Decoen, '49, '51), the reconstitution studies of Burnett and colleagues (Haynes and Burnett, '63; Burnett et al., '66; Davis et al., '66), and the more recent electron microscopic observations (Schincariol et al., '67; Moore and Dixon, '72; Zihler, '72; West, '78b) leave little doubt, however, regarding the correctness of Kleinenberg's original postulation that gametes are derived from I cells, and not from a distinct germinal population.

Spermatogenesis has been extensively studied in hydra, and numerous excellent ultrastructural reviews of the process are available (Weissman et al., '69; Moore and Dixon, '72; Zihler, '72; West, '78b). The process begins with a local accumulation of I cells within the intercellular spaces between the epitheliomuscular cells (Brien and Reniers-Decoen, '51). It is believed that this congregation derives from the combined migratory and mitotic activity of these cells (Schincariol et al., '67; Moore and Dixon, '72; Tardent, '72; Stagni, '74; West, '78b). As these cells represent the progenitors of sperm, they are usually referred to as spermatogenic cells or spermatogonia.

Spermatogenesis continues as an undistinguished process in hydra. In the mature testes, epitheliomuscular cells are elongated with their bases flattened along the mesoglea. The interstices between the these cells define actual "compartments," which contain cells in all stages of gamete development. In general, the cells within a compartment are zoned apicobasally on the basis of their maturation with spermatogonia located near the mesoglea and more mature stages extending apically. The only peculiarity associated with spermatogenesis in hydra is the incomplete separation of meitoic division products. As a result, the occurrence of four spermatids joined to each other by cytoplasmic bridges and contained within a single membrane is commonplace (Tardent, '74; West, '78b).

The fully differentiated hydra sperm (Fig. 9) is considered to be of a primitive type, possessing a generally cylindrical head with unmodified mitochondria (West, '78b). No structurally apparent acrosome has been reported in hydra, nor have the apical vesicles seen in marine coelenterates been observed. However, the space between the nucleus and the cell membrane of a mature sperm contains what has been described as a "flocculent" material, which has been suggested to serve an acrosomelike function (West, '78b).

Mature sperm are released into the medium in a relatively continuous manner, thereby ensuring sperm availability over an extended time period (Zihler, '72; Tardent, '74; Honegger, '81). Although the exact mechanism of release has not been described, Honegger believes that the sperm, which are maturing within several compartments in a given testis, are probably released through openings within individual epitheliomuscular cells. Once released into the environment, mature sperm remain viable for approximately 3 hours (Zihler, '72).

Oogenesis begins, as did spermatogenesis, with an accumulation of I cells within the intercellular spaces between gastric region epitheliomuscular cells. Clusters of 32–64 cells have been observed (Stagni, '74; Honegger, '81). Eventually, one cell in each cluster becomes a primary oocyte and begins to ingest the other congregating cells by phagocytosis. Several of the developing oocytes then merge to form an initially multinucleate secondary oocyte. One of these nuclei will persist to become the germinal vesicle (Zihler, '72; Tardent, '74). Regulation of this oocyte selection process remains wholly unexplained in hydra. Zihler ('72) did successfully separate the developing oocyte into two parts at its multinucleate stage, and was able to fertilize and initiate embryogenesis in each fragment. Therefore, many, possibly any, of the nuclei in the transitory multinucleate oocyte can become the functional egg nucleus.

Many of the above-described events of oogenesis are difficult to observe with the light microscope owing to combined effects of the opacity of the egg and the presence of nematocysts within the overlying epidermal epithelial cells (Honegger, '81). Immediately following the release of the second polar body, however, the developing egg, which has been forming in the intercellular spaces between epitheliomuscular cells, is extruded from the epithelial sheet. The egg, which is surrounded by a clear jelly layer (McConnell, '38a–c; Tardent, '74; Honegger, '81), remains viable for about 12 hours.

Fertilization

No chemotactic mechanism has been described for freshwater coelenterates, but Honegger ('81) reports that the hydra egg appears attractive to sperm when it bursts through its epidermal covering. Sperm penetration occurs within a restricted area of the egg, where the female pronucleus lies close to the egg membrane. In spite of this seeming sperm-egg fusion site, no micropyle has been observed. Honegger ('81), however, using light and scanning electron microscopy, has shown a cup-shaped depression of the egg membrane and overlying jelly layer at the fusion and penetration site.

Fertilization in hydra may be experimentally prevented if the jelly layer surrounding the unfertilized egg is removed with trypsin. Similar treatment does not affect the development of an already fertilized egg (Honegger, '81).

This corresponds to the situation in, for example, sea urchins where sperm-egg interactions are presumably mediated, at least in part, by trypsin-sensitive glycoproteins within the egg coats.

With no acknowledged chemical attraction between the egg and sperm of hydra, one might question whether sufficient circumstances would exist to generate random collisions between the egg and sperm in a natural environment. Tardent in 1974 presented a cogent argument that the following conditions, when acting in concert, would be adequate to ensure successful natural fertilization without a chemotactic mechanism:

1. In natural populations, gametogenesis is synchronized between the sexes.
2. Male gametes are produced over a much longer time period than female gametes, thus providing an uninterrupted supply of sperm during egg development.
3. Sperm are released continuously, and the 3-hour viability period for sperm is long.

Embryogenesis

Following fertilization, cleavage is initiated at the site of polar body release. Hydra, consistent with other coelenterates, exhibits a unilateral cleavage pattern. Blastomere formation remains synchronous and easily observable up to the 16-cell stage. Continued cleavage yields various sized blastomeres and ultimately the formation of a typical hydrozoan coeloblastula (Honegger, '81).

Endoderm formation has been described as occurring via delamination from the cells of the coeloblastula and/or by a multipolar ingression of prospective endodermal cells. In either situation, the cavity of the blastula fills with cells, and, within a day of fertilization, the embryo begins to elaborate its protective, acellular embryotheca (Brauer, 1891; Kanaev, '52; Honegger, '81). Later events of embryogenesis are speculative, at best, as the embryo continues development within its nearly impenetrable embryotheca. Older literature reports that I cells form as a layer of cells beneath the ectoderm, presumably being of ectodermal origin (Brauer, 1891). Also, it is believed that the mesoglea is not deposited until shortly before hatching (McConnell, '38a–c). No recent confirmation of either of the latter observations has been recorded.

The embryo, protected by its embryotheca, remains attached to the polyp for a few days, then detaches. Hatching of a fully formed hydra occurs 12–70 days after fertilization (McConnell, '38a–c; Kanaev, '52; Honegger, '81).

ASEXUAL REPRODUCTION

Hydra also reproduces asexually by budding and stolonization. The morphological events accompanying the budding process, from the deformation of the parent body column to the elaboration of the mature bud before its detachment, have been extensively described (Chang et al., '52; Burnett, '66; Clarkson and Wolpert, '67; Otto and Campbell, '77a; Shostak, '77). Grafting experiments between symbiotic and aposymbiotic *H. viridis* have established that the cell source for buds is at least partly parental. Epithelial cells of the epidermal and gastrodermal layers move down the parental body column and out into the bud (Shostak and Kankel, '67). Nonepithelial cell types, such as nematocytes and I cells, most of which normally migrate distally, move quickly into any developing bud present (Herlands and Bode, '74a). The cellular composition of the developing bud also changes from an early homogeneous distribution of epithelial, interstitial, and nematoblast cells to the pattern of regional specialization characteristic of a nonbudding parent animal (Sanyal, '67; Bode et al., '73).

Mechanistically, budding is usually divided into three distinct stages: initiation, elongation, and separation (or individuation). Bud initiation appears to be related to a requisite quantity and localization of parental cells in the budding region (Shostak et al., '68; Webster and Hamilton, '72). Bud elongation, alternatively, has been described as a reorientation and repolarization of tissues. Inhibiting mitosis and DNA synthesis by radiation do not interrupt the process, suggesting that elongation may be a tissue movement phenomenon (Clarkson and Wolpert, '67). Some of the mechanisms that may be operating during the elongation process have been discussed by Campbell ('74) and Otto ('77). Bud separation (or individuation) was recognized as a distinct event by Rand in 1899. Nevertheless, the problem of detachment remains a mystery. Changing mesogleal fiber patterns have been postulated as possible forces directing these morphogenetic activities (Burnett and Hausman, '69; Hausman and Burnett, '70; Brooks, '76). Recently, Sersig and Lesh-Laurie ('81), through grafting techniques, identified the parental body column as instrumental in the detachment process.

In spite of extensive study, budding remains one of the most complex, unresolved developmental problems in hydra. The existence of nonbudding "mutants" (Fig. 10) and the capacity to produce them by sexual reproduction may provide fruitful avenues for further critical investigations (Lenhoff, '65; Lenhoff et al., '69; Lesh-Laurie, '71; Moore and Campbell, '73a, b; Brinkley, '74; Novak and Lenhoff, '80).

Stolonization is an infrequently described occurrence in hydra (Brien, '52,'56; Haynes et al., '64; Haynes and Burnett, '64). The process may be recognized initially by an increase in peduncular length. Eventually, the

Fig. 9. Longitudinal section through the head and middle piece of a mature hydra sperm. × 17,000. Reprinted from West ('78), with permission.

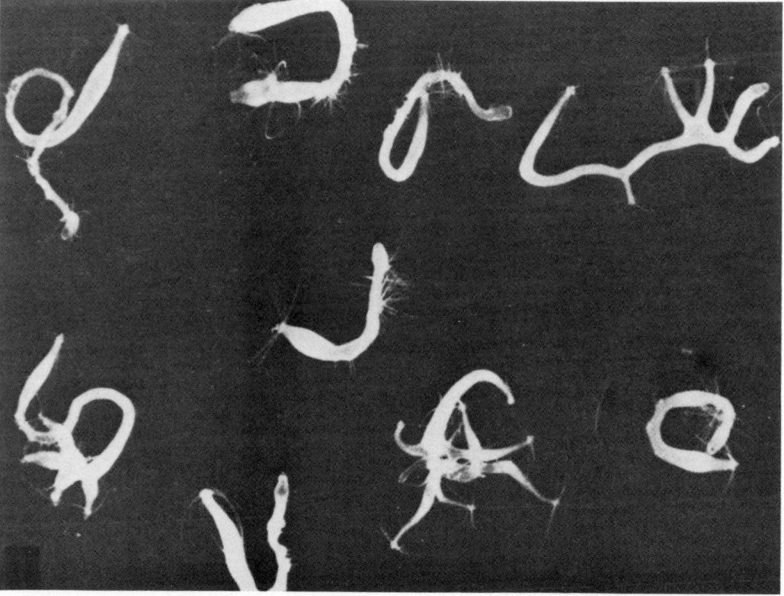

Fig. 10 (below). A clone of nine nonbudding mutant *C. viridissima.* Reprinted from Lenhoff et al. ('69), with permission.

portion of the peduncle that will become the stolon (Fig. 11a) exhibits an increased diameter so that the animal appears to consist of a hypostome-gastric region-budding region-peduncle-stolon-peduncle-basal disc (Fig. 11b). Two to three stolons may form simultaneously on a single animal (Fig. 11f). A basal disc finally forms at the original budding region-peduncle junction, freeing the stolon from its parent organism (Fig. 11c). Enzyme histochemistry reveals that the stolon is not a region of active growth (Haynes and Burnett, '64).

Within the central portion of the stolon a protrusion emerges that will develop a mouth and tentacles. Once the developing outgrowth obtains adult size, it may detach like a bud and/or an additional outgrowth may form (Fig. 11d). Most often, outgrowths do not detach, with the resultant individual evidencing a pseudocolonial appearance (Fig. 11e).

EXPERIMENTAL STUDIES: POLARITY

One of the developmental characteristics that first attracted investigators to hydra as an experimental system is the animal's deceptively simple polarity, and its capacity to maintain this asymmetry during normal growth and to restore it following an amputation. Curiosity regarding hydra's regulative abilities was piqued even further in 1909 when Browne demonstrated by grafting that tissue at the apex of hydra's asymmetry, the hypostome, exhibited inductive potency (Fig. 12). Browne's initial finding has been repeatedly confirmed (Koelitz, '11; Issajew, '26; Rand et al., '26; Tripp, '28; Mutz, '30; Li and Yao, '45; Yao, '45) and extended to demonstrate similar inductive capability within the animal's proximal terminus, the basal disc (Browne, '09; Yao, '45).

Further study revealed that the presence of either a hypostome or basal disc inhibited the formation of that structure (Rand et al., '26; Mookerjee and Sinha, '59; Webster, '66a; MacWilliams and Kafatos, '68; MacWilliams et al., '70). In addition, the hypostome and basal disc are the first regions to be regenerated and, once initiated, the formation of each appears to be an autonomous occurrence. Thus both the hypostome and basal disc satisfy Huxley and De Beer's ('34) criteria for a developmentally dominant region, and as a consequence considerable attention has been directed to determining their role(s) in the control of developmental processes in hydra.

With respect to hypostomal dominance, Webster, Wolpert, and Wilby, after an exhaustive series of grafting experiments, attributed this property to two interacting axial gradients. They envisioned a distoproximal gradient in inhibition to hypostome formation and in threshold for inhibition. Whenever the level of inhibition dropped below this threshold, hypostome formation occurred (Webster, '66a, b, '71; Webster and Wolpert, '66; Wilby and Webster, '70a).

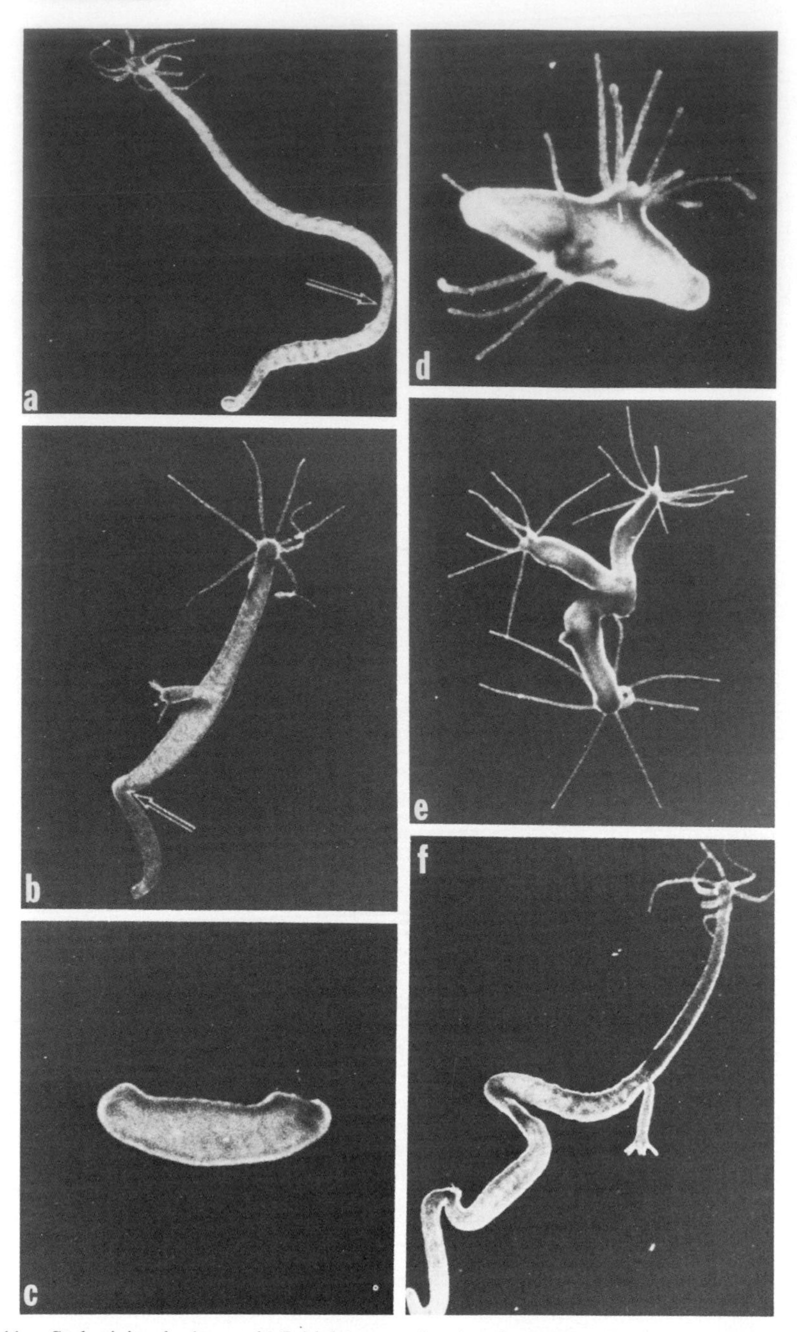

Fig. 11. Stolonizing hydra. a, b) Initial stages of stolon formation. The arrow indicates the position at which the basal disc would be found in a nonstolonizing animal. c) Isolated stolon. d) Two hypostomes developing simultaneously on an isolated stolon. e) Three individuals forming from a single stolon. f) Development of two stolons on a single animal. Reprinted from Haynes et al. ('64), with permission.

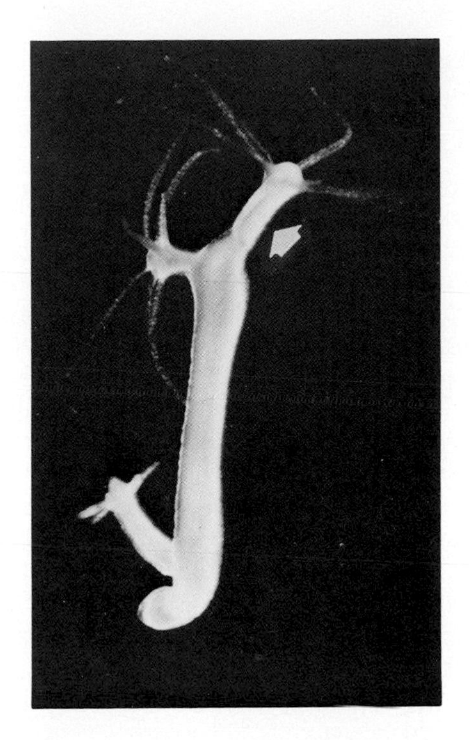

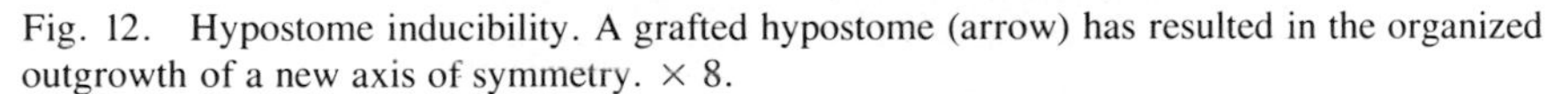

Fig. 12. Hypostome inducibility. A grafted hypostome (arrow) has resulted in the organized outgrowth of a new axis of symmetry. × 8.

Comparable studies with the basal disc are also consistent with its developmental dominance being explained as the result of gradients emanating from the basal disc (MacWilliams and Kafatos, '68, '74; MacWilliams et al., '70; Shostak, '72; Hicklin and Wolpert, '73a). In addition, Sacks and Davis ('80), using animals with experimentally altered cell numbers and cell-type densities, showed that the inhibitory properties of the basal disc were unaffected by changing these parameters.

Therefore, the developmentally dominant regions of hydra share the characteristic of controlling their own differentiation, predictably through the establishment of gradients. Both the hypostome and basal disc also possess epithelial cells and nerves as their principal cell populations.

Polarity Reestablishment: Regeneration

Following an amputation, hydra quickly reestablishes its asymmetrical organization. The morphological events accompanying normal regeneration have been extensively described (King, '01; Rowley, '02; Kanaev, '26; Ham and Eakin, '58; Spangenberg, '61; Spangenberg and Eakin, '61; Lentz and

Barrnett, '62; Lentz, 65a–c; Mookerjee and Bhattacherjee, '65; Lui and Znidaric, '68; Rose and Burnett, '68b; Znidaric, '70, '71a, b; Bhattacherjee, '72; Lesh-Laurie and Donaldson, '78; Wood and Kuda, '80a, b). Among the fascinating problems explored in regenerating hydra have been those addressing the participation of cell division in the regenerative process and the role of the nervous system in directing regeneration.

Distal regeneration in hydra has often been considered a morphallactic phenomenon. Morphallactic regeneration is characterized by a general absence of new cell formation until the existing cells have been organized to form the missing structures. It has been well documented in only a few experimental systems, most particularly postcephalic reorganizatin in annelids (Berrill and Mees, '36). A determination of the time and frequency of mitotic activity during hydrid regeneration would, therefore, provide insight into the mechanism(s) involved in the reestablishment of hydra's polarity.

Most of the data from the cytological examination of mitotic figures at discrete time intervals during regeneration (e.g., Rowley, '02; Park et al., '70) would argue that the events of distal regeneration occur in the absence of significant mitotic activity. Ham and Eakin ('58), however, reported a burst of mitosis within 1 hour of hypostomal amputation. Regenerating hydra exposed to agents that interfere with mitosis, such as colchicine, colcemid, nitrogen mustard, or ionizing radiation, although exhibiting a retarded rate and often a decreased capacity for complete regeneration, generally reform a normal hypostome and often initiate tentacle elaboration (Sturtevant et al., '51; Brien and Reniers-Decoen, '55; Park, '58; Diehl and Burnett, '64; Webster, '67; Corff and Burnett, '69; Corff, '73; Hicklin and Wolpert, '73b). It therefore appears that many of the events of distal regeneration take place without substantial mitotic involvement.

The coincident occurrence of developmental dominance in those regions of the hydra that possess the highest nerve cell density initially alerted investigators to the possible involvement of the nervous system in directing regenerative events in hydra. Bursztajn and Davis ('74) presented a detailed histological study of the changes in nerve-cell density associated with distal regeneration. They recorded a statistically significant increase in the concentration of nerve-cell droplets released at the wound surface within 1 hour of hypostomal amputation. This increase in the release of neurosecretory products and a corresponding increase in nerve-cell density continued for 24 hours. Finally, with the accumulation of nerve cells in the regenerating hypostome, tentacle formation was initiated. These observations led them to suggest that a threshold number of nerves might be necessary to establish the conditions requisite to tentacle elaboration. Their data revealed further that nerve-cell activity was not localized at the cut surface during regeneration. Nerves in the peduncle and basal disc regions also underwent marked histological changes during distal regeneration.

Burnett and his colleagues and Lentz also have presented evidence to substantiate a requirement for nervous involvement in the developmental processes of regeneration, asexual reproduction, and sexuality (Burnett and Diehl, '64a, b; Burnett et al., '64; Lentz, '65c).

Polarity Reestablishment: Isolated Tissue Layers, Reaggregation

Hydra is also capable of morphologic regulation following a rearrangement of its cell layers. Roudabush ('33) reported that everted hydra (i.e., hydra turned inside out) may regulate by a massive migration of cells across the mesoglea. Unfortunately, no ultrastructural confirmation of this observation exists (Macklin, '68a). Complete morphogenesis has been attained, however, following the separation of hydra's two tissue layers. The species selected for cell layer isolation studies is generally the symbiotic, green hydra *H. viridis.* In this species the algae-laden gastrodermal epithelial cells provide a convenient and natural intracellular marker to assist investigators in assessing the purity of their preparations. Several isolation procedures have been published (Cerame-Vivas, '61; Haynes and Burnett, '63; Lowell and Burnett, '69), which are reviewed and evaluated in Lenhoff ('82).

Few reports speak to the survival capacity of an isolated epidermal layer, although Lowell and Burnett achieved some success with epidermal explants (Lowell and Burnett, '69, '73; Burnett et al., '73). Gastrodermal isolates, alternatively, can develop an epidermis as a result of epithelial cell proliferation. The dedifferentiation of gastrodermal gland cells gives rise to epidermal I cells, which then form the specialized I-cell derivatives characteristic of the epidermis, nerves, and nematocytes (Normandin, '60; Haynes and Burnett, '63; Burnett et al., '66; Davis et al., '66; Davis, '68, '70a, b, '73).

More recently, Noda ('70a) and Gierer et al. ('72) have convincingly demonstrated that hydra may even be reconstructed from individual cells. After mechanically disrupting cells in a saline disaggregation medium, these investigators followed reaggregation histologically from an initial random collection of cells, through the reestablishment of tissue layers to the morphogenesis of head and foot structures (see also Bode, '74; Klimek et al., '80). The availability of a reaggregation technique presents unique opportunities for further study of hydrid morphogenesis by allowing the investigator to observe developmental regulation after combining specific cell types in defined proportions. Successful reconstruction of hydra from individual cells also permits the conclusion that organismal polarity in hydra is not due to the orientation or asymmetry of individual cells, but derives instead from some supracellular property or properties.

Polarity Reversal

Polarity can also be altered experimentally in hydra. Initially these experiments were undertaken with the hope that they would provide insights to both the nature of the axial gradients that presumably determine the asymmetrical organization of the organism, and to the mechanism by which the gradients are established and maintained during the life history of the animal. Literature references abound to attest to the fact that polarity in hydra may be modified by grafting (Wetzel, 1898; Peebles, 1900; King, '01, '03; Browne, '09; Goetsch, '29; Sinha, '65; Webster, '66a, b, '71; Mookerjee and Bhattacherjee, '67; Wilby and Webster, '70a; Lesh-Laurie, '73; Newman, '74). These experiments demonstrate unequivocally that both the hypostome and basal disc gradients influencing polarity in hydra can be displaced nearly anywhere along the body column, depending upon the nature of the tissue interaction constructed (Fig. 13). Such a translocation could most easily be explained if the gradients were substance gradients, derived from materials that could diffuse or be transported within hydra's tissues (Macklin, '68b; Shostak, '73, '74a, b, '75a).

Hydrid polarity can also be modified chemically using agents such as colchicine (Corff and Burnett, '69), colcemid (Webster, '67; Shostak and Tammariello, '69; Shostak, '75), cold (Corff and Burnett, '70), or dithiothreitol (Hicklin et al., '69). All of these compounds have disruptive effects on microtubule organization. However, the relevance of that observation to the mechanism(s) by which these agents influence organismal polarity will have to await further experimentation.

EXPERIMENTAL STUDIES: GROWTH

Hydra's ability to maintain a steady-state condition with respect to size and form has long intrigued students of growth. Initial experimentation employing vital dyes to mark tissue revealed general tissue movement away from a stationary zone located directly beneath the animal's hypostome (Tripp, '28; Kanaev, '30; Brien and Reniers-Decoen, '49; Semal van Gansen, '54; Burnett and Garofalo, '60; Burnett, '61, '62, '66; Mookerjee, '66). These observations resulted in the conclusion that the animal possessed a subhypostomal growth zone and to the incorrect belief that growth was restricted to this area. In spite of this generalization, which was popularized in the 1960s, histological evidence had always been consistent with the occurrence of mitotic figures throughout the animal's body column (Schneider, 1890; McConnell, '30, '33a, b, '36; Brien, '51).

Later, in an extensive and elegant study, Campbell ('67a–c) combined the techniques of vital dying and grafting with histology and autoradiography to examine simultaneously tissue movements and the distribution of mitotic activity in the animal. He detected mitotic activity in numerous cell types, and observed mitotic figures throughout hydra's body column irrespective of variations in culturing conditions, feeding, or circadian cycle. In fact, the mitotic index was virtually identical at all levels of the column, and little

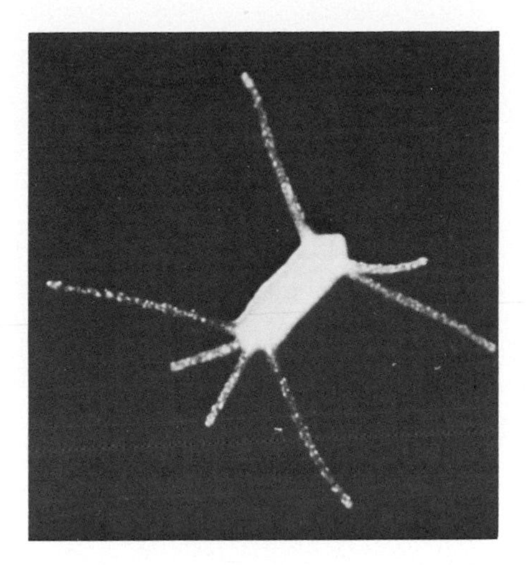

Fig. 13. Bipolar hydra produced by grafting. × 8.

difference was seen in the distribution of mitoses between budding and nonbudding animals (Campbell, '67a; Clarkson and Wolpert, '67). Thus Campbell argued that the pattern of tissue movement witnessed in hydra reflected a pattern of localized cell loss, rather than a pattern of localized proliferation, and he conjectured that hydra maintained its steady-state condition by balancing growth with cell loss through the processes of budding and tissue wear. Budding, therefore, becomes a vehicle to study growth dynamics as well as tissue morphogenesis in hydra (Shostak, '67, '68; Shostak et al., '78).

Budding

Although hydra achieve a steady-state condition with respect to size and form, all hydra of a given species are not the same size. Most particularly, hydra cultured at different temperatures (Stiven, '65; Park and Ortmeyer, '72; Bisbee, '73; Shostak, '81) or subjected to different feeding regimes (Kass-Simon and Potter, '71; Otto and Campbell, '77b; Shostak, '79; Gurkewitz et al., '80) attain different sizes and express different budding rates. Regulation, in these situations, may occur by balancing cell attrition at the tentacle tips and basal disc by wear with the cell-cycle times expressed by epithelial cells.

Unique ionic distributions also affect hydra's size and asexual reproductive process. Potassium and lead have been most critically examined, but there is as yet no clear understanding of the mechanism(s) operating to control budding under these conditions (Loomis, '54; Lenhoff and Bovaird, '60; Lenhoff, '65, '66; Schulz and Lesh, '70; Koblick and Epp, '75; Epp and Koblick, '77; Browne and Davis, '77).

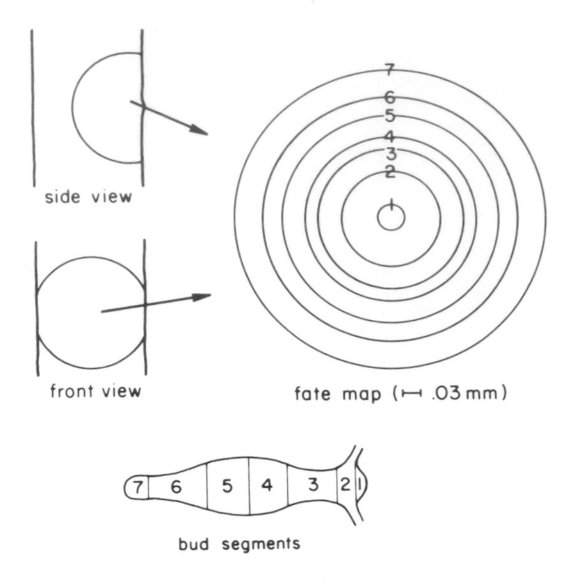

Fig. 14. Fate maps of *H. attenuata* bud. The side and front views of the hydra in the upper left reveal the outer limits of the bud fate map. The bull's-eye of concentric circles (upper right) shows the mean distance for marked cells entering each bud segment. The bud segments are shown in the bottom diagram. Reprinted from Otto and Campbell ('77), with permission.

With respect to bud morphogenesis, Webster ('71) reported that one could induce budding from the distal end of the body column if one grafted a hypostome to the proximal end of the digestive zone and removed the parent hypostome. Distal budding occurred before any reestablishment of axial polarity through regenerative events could be detected, and suggested to Webster that a budding region is organized quite independent of those factors influencing axiation in the animal (see also Tardent, '72).

Using vital intracellular markers (Campbell, '73) Otto and Campbell were able to define a fate map for the *H. attenuata* bud (Fig. 14). They observed that both epidermal and gastrodermal tissue are recruited from a circular region surrounding the developing bud tip, and that parental contribution to a bud terminates at approximately the time tentacle rudiments appear on the bud. These findings allowed the conclusions that bud tissue recruitment is a radially symmetrical process in hydra, and that it is an event distinct from bud hypostome morphogenesis as the two processes are chronologically separated (Sanyal, '66; Otto and Campbell, '77a).

Examining the morphogenetic events of bud initiation, Webster and Hamilton ('72) noticed an increase in the thickness of both the epidermal and gastrodermal layers concomitant with,or immediately antecedent to, the curvature of the tissue layers that initiates bud formation. They hypothesized that these tissue movements derived from an increase in the adhesive properties of the cells making up the budding region, in a manner analogous to that occurring during deuterostome gastrulation. Increased cell adhesion within a confined space will cause cells to pack more tightly, increasing the thickness of the cell layers and ultimately resulting in a curvature of tissue. With the aid of polarizing microscopy, Otto ('77) has further suggested a possible involvement for epidermal epithelial cell muscle processes in the morphogenetic events of budding. Her observations are consistent with an active reorientation of muscle-cell processes contributing to bud outgrowth.

Mesoglea

Any investigation of hydra's growth and morphogenesis cannot be undertaken, however, in the absence of a consideration of the substratum separating the animal's cell layers—the mesoglea. Mesoglea is basically an extracellular matrix whose chemical composition (Barzansky and Lenhoff, '74; Barzansky et al., '75; Brooks, '76) and ultrastructural organization (Davis and Haynes, '68; Haynes et al., '68; Hausman, '73) compare favorably with vertebrate basal lamina (Day and Lenhoff, '81). Steady-state hydra exhibit a characteristic mesogleal fiber pattern that is modified after feeding and during budding and regeneration (Burnett and Hausman, '69; Hausman and Burnett, '69, '70). Furthermore, if during distal regeneration one prevents collagen (and hence mesogleal) secretion using the proline analogue L-azetidine-2-carboxylic acid, the biosynthetic and differentiative events of regeneration proceed, but tentacle morphogenesis does not occur (Brooks and Lesh-Laurie, in preparation). Thus, by perturbing mesoglea formation one can seemingly uncouple morphogenesis from differentiation during hydra regeneration.

Morphologically, the mesoglea provides structural support for the hydra as attested to by the observation that mesoglea isolated from a hydra retains the perfect form of the hydra (Hausman, '73). It also serves as a place for cell insertion. No specific evidence exists for mesogleal involvement in directing epithelial tissue–mediated morphogenesis, although Shostak and coworkers (Shostak et al., '65; Shostak and Globus, '66) have shown that hydrid cell layers move independently of one another and of their underlying mesogleal surface. Campbell ('80) recently presented cinematographic evidence that epithelial cells may provide the force for morphogenesis in hydra by displacing their muscle processes along the mesogleal net.

EXPERIMENTAL STUDIES: CELLULAR EVENTS

Eventually, it becomes necessary to seek a cellular basis for the dynamicism of hydra's growth processes. Whole-mount techniques (Tardent, '54;

Burnett, '59; Zumstein and Tardent, '72) and detailed histological studies had already provided a qualitative basis for defining hydra cell types and assessing their distribution along the distoproximal axis of the animal. These methodologies, however, did not permit a quantitative analysis of the kinetics of cellular proliferation, commitment, or differentiation. To do this, methods had to be devised to examine and make quantitative measurements on individual cells. As *in vitro* studies in hydra had been less than satisfactory (Li et al., '63; Trenkner et al., '73), David in 1973 introduced a modification of Béla Haller's tissue maceration technique that allowed unambiguous cell classification and quantification.

A second requirement for effective cellular analysis was the capacity to study single hydrid cell types selectively. This objective could be attained either by examining individual cell types in isolation or by the selective elimination of specialized cell types. Strelin ('29), Zawazzin ('29), and Brien and Reniers-Decoen ('55) devised the first method for selective cell elimination by exposing hydra to X-rays (Mookerjee and Aditya, '66; Mookerjee and Mitra, '70). Later γ-irradiation was also employed (Clarkson and Wolpert, '67; Hicklin and Wolpert, '73b; Fradkin et al., '78). Diehl and Burnett ('64, '65a, b, '66) first explored chemical methods for destroying specific cell types, treating hydra with 0.01% nitrogen mustard (methyl-bis-(beta-chloroethyl) amine hydrochloride) for 10 minutes. Hydroxyurea exposure also affected selective cell populations when hydra were treated with 10^{-2} M hydroxyurea in three 24-hour cycles, each separated by 12 hours in normal culture solution (Bode et al., '76; Sacks and Davis, '79).

All of the above methods succeeded in significantly reducing hydra's I-cell population; all also eventually depleted I-cell–derived cell types such as nerves and nematocytes. All of these techniques also possessed drawbacks, however. Either the method failed to eliminate I cells completely (e.g., irradiation, hydroxyurea) or, when the I-cell population was totally destroyed, the animal also succumbed (e.g., irradiation, nitrogen mustard), leaving unresolved the question of possible deleterious effects of the technique on non-I-cell populations.

Recently, Campbell and his colleagues have achieved remarkable success in eliminating the I-cell population using the mold alkaloid colchicine. By subjecting *H. attenuata* to two successive 8-hour exposures to 0.4% colchicine (made up in culture solution), spaced approximately 10 days apart, between one-third and one-half of the treated hydra become completely I-cell-free (Campbell, '76, '79; Marcum and Campbell, '78a). I cells are eliminated, presumably by a phagocytic process, and within about 3 weeks no I cells nor cells of an I-cell lineage (i.e., nerves, nematocytes) may be detected histologically or by using cell maceration methods (Fig. 15) (Marcum and Campbell, '78a; Wanek et al., '80a).

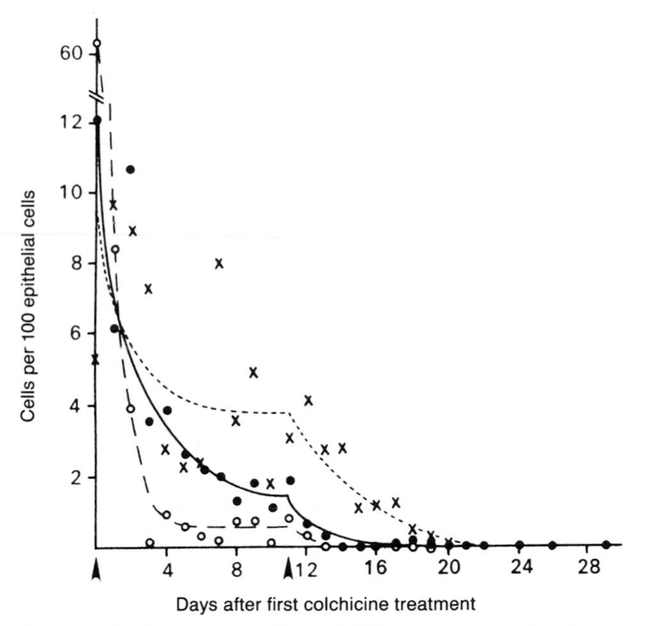

Fig. 15. Time of cell disappearance after colchicine treatment. The disappearance of nerves (●), I cells (○) and gland cells (X) is followed after exposing *H. attenuata* to 0.4% colchicine for 8 hours on days 0 and 11 (indicated by arrows). Population levels of these cell types are expressed relative to the epithelial cell number. Reprinted from Marcum and Campbell ('78), with permission.

The resulting "epithelial hydra" (Fig. 16) are fully viable, and the only inconvenience associated with them is that they must be hand-fed to be maintained. Epithelial hydra often display more tentacles than normal animals, but their tentacles are straight, motionless, and often not in a perfect whorl. Some distention is also evident in the hypostome and distal gastric regions of epithelial hydra. Finally, to date, success in developing epithelial hydra has only been obtained using *H. attenuata.*

Nevertheless, epithelial hydra express the full range of developmental potential evident in hydra. Epithelial hydra bud, thereby allowing clones of them to be formed. They regenerate in a nearly normal fashion, restoring the same percentage of original tentacles as do normal hydra. The hypostomes of epithelial hydra have inductive ability. Epithelial hydra also maintain their normal tissue polarity; however, this axiation may be reversed by grafting as occurs in normal animals. Therefore, nerve-free tissue may express some of the properties of developmental dominance in hydra, previously attributed to the nervous system. With respect to morphogenesis, epithelial cell cycles and tissue movements are normal in epithelial hydra. Thus, epithelial cells alone

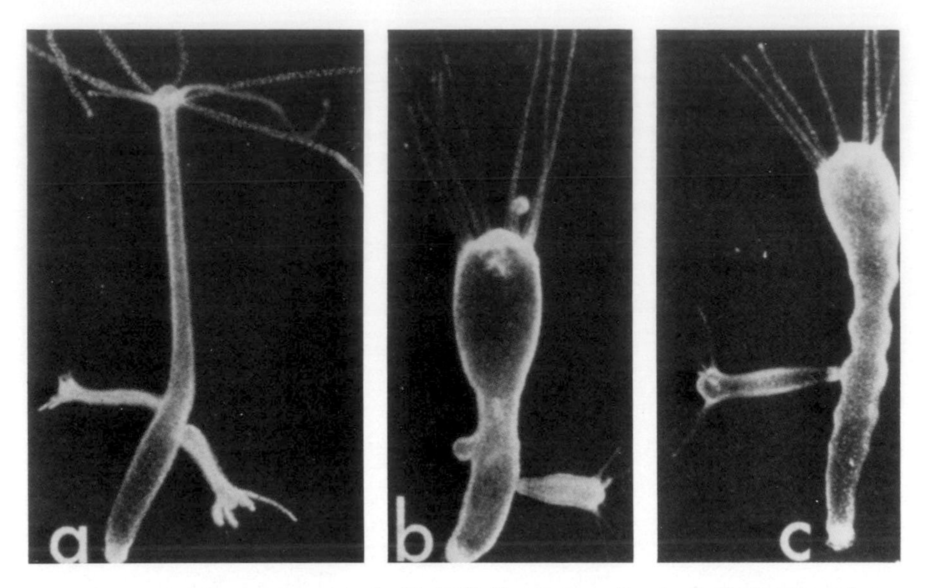

Fig. 16. Normal *H. attenuata* (a); "epithelial" *H. attenuata* (b, c). × 10. Reprinted from Campbell ('79), with permission.

are capable of supporting morphogenesis in hydra. Also, using epithelial hydra, hydrostatic pressure was identified as an important element in hydrid morphogenesis. If the body column of epithelial hydra is deflated and held in that condition, both budding and the histological organization of the animal become abnormal (Marcum et al., '77; Marcum and Campbell, '78a; Campbell, '79; Wanek et al., '80b).

Cell Cycle Studies

Using primarily the maceration technique, cell cycle parameters were determined for hydra. In their cell cycle studies of hydra's epithelial cell populations, David and Campbell ('72) confirmed the earlier conclusion that growth is distributed throughout a hydra. More than 90% of the epithelial cell population proliferated at a rate equal to the organism's growth rate. Their investigation also produced the interesting finding that hydra epithelial cells apparently possess a long and variable G_2 period, indicating that DNA synthesis and cell division may not be tightly coupled events in this cell population.

The maceration technique also allowed definition of the absolute number of cells constituting a hydra and the distribution of these cell types along the axis of the animal (Fig. 17) (Bode et al., '73). Similar quantification during budding and hypostome and basal disc regeneration revealed that the initial

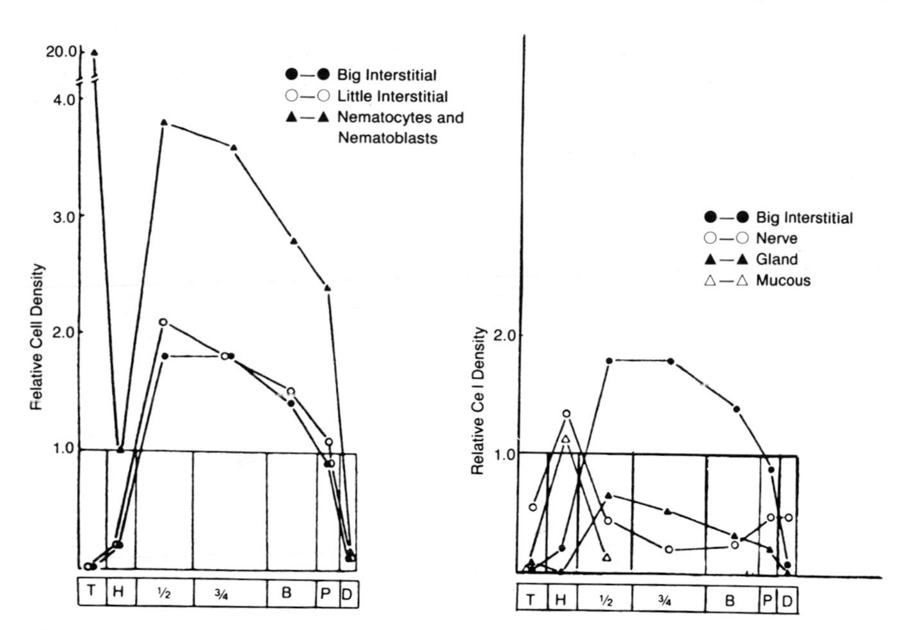

Fig. 17. Axial distribution of cell types in hydra. The relative density of each cell type is given as the ratio of the number of that cell type to the number of epitheliomuscular cells. The abscissa indicates the relative size of each axial region in terms of epitheliomuscular cells. Tentacles (T), hypostome (H), upper gastric region (1/2), lower gastric region (3/4), budding region (B), peduncle (P), basal disc (D). Reprinted from Bode et al. ('73), with permission.

cellular response associated with each of these events was an increase in the concentration of nerve cells at the site of the event (Bode et al., '73; Berking, '80a). This discovery confirms previous histological and ultrastructural observations and provides additional data in support of a role for the nervous system during normal morphogenetic processes.

Cell cycle studies were also completed for I cells and their derived cell types (Campbell and David, '74; David and Challoner, '74; David and Gierer, '74). These studies provided evidence for at least two populations of what had previously been called I cells in hydra, a proliferating stem-cell population and a population of committed cells destined to form either nerves or nematocytes.

I-Cell Commitment and Differentiation

With the cellular definition of a "standard" hydra, opportunity was provided to examine the regulation of these cell populations. Of particular interest and curiosity was the purported multipotent I-cell or stem-cell population. The I-cell system (stem cells, plus all differentiating intermediates and product cells) comprises approximately 75% of the cells in an asexually reproducing hydra (Bode and David, '78). Until the advent of techniques that permitted the quantitative study of individual cells and the unequivocal elimination of selected cell types, however, one could only provide indirect evidence that the stem-cell population was truly multipotent (Brien, '53, '61; Diehl and Burnett, '66; Saffitz et al., '72). Then in 1977 David and his colleagues announced the development of a procedure to clone I cells (David and Murphy, '77; Sproull and David, '79a, b; see also Venugopal and David, '81a, for small-scale modification). Essentially, the technique involves the introduction of I cells into a nitrogen mustard–treated host. Nitrogen mustard–treated hydra are I-cell–depleted, and although they will die in 3-4 weeks their epithelia remain fully viable in a morphogenetic sense during this interval. Viable, labeled I cells are isolated and introduced into the nitrogen mustard–treated host by reaggregation methods. Later, clones arising from the transplanted I cells can be examined; in these studies several were seen to contain both nerves and nematocytes, thus elegantly demonstrating the existence of multipotent stem cells within hydra's I-cell population.

Within the I-cell population nerves and nematocytes differentiate continuously, whereas the stem-cell population maintains itself by proliferation. Yet the relative proportion of each cell type within a hydra remains essentially unchanged over many generations and in spite of size differences that exist among animals. When Bode and his colleagues critically examined hydra's cell populations they discovered that although the absolute size of the epithelial population changed with, for example, feeding, the I-cell/epithelial-cell ratio remained unaltered. Thus, regulation in the two populations might be tightly coupled, and homeostatic mechanisms might be involved in the control of at least one of the populations (Bode and Flick, '76; Bode et al., '76, '77; Bode and Smith, '77; Bode and David, '78).

Of the two populations, the I-cell group is both the most amenable to manipulation and is also the population whose regulation assures hydra of the specialized cell types it requires to maintain itself. Therefore, Bode et al. ('76) reduced the I-cell population of *H. attenuata* to about 1% of its normal level using hydroxyurea and then studied the repopulation of this cell type and its derivatives. They hypothesized that regulation within the I-cell population would occur either through control of cell cycle time or through control of the fraction of cells within the population undergoing cell division at any given time versus the fraction of cells differentiating.

The next question Bode and his co-workers confronted was whether regulation within the stem-cell population was a local phenomenon influenced by the position of the stem cell within the axial organization of the animal, or whether stem cells somehow monitored their entire population. Two experimental observations argued for the former alternative. First, the type of stem-cell product differentiating in hydra varies along the body column of the animal. Second, in the I-cell depletion study cited above, if regulation were an "organismal" phenomenon one should see no evidence of nematocyte differentiation until recovery of the stem-cell population was completed. Yet, before the stem-cell population had attained a 10% recovery level, nematocyte differentiation was detected. Therefore, regulation within the stem-cell population is a local event, sensitive to the axial position of each particular cell in hydra's distoproximal polarity (Bode and Flick, '76; Bode et al., '76, '77; Bode and Smith, '77; Bode and David, '78; David and Plotnick, '80; Venugopal and David, '81a).

Directly examining stem-cell commitment to nerve cell differentiation, Yaross and Bode ('78a) showed that nerve cells feed back to the stem cell population by influencing additional stem cells to be committed to form nerves. In addition to being correlated with nerve-cell density, nerve-cell commitment is also position-dependent in both steady-state and regenerating animals (Yaross and Bode, '78a, b; Venugopal and David, '81a, c). Nerve-cell commitment occurs during the S phase of the cell cycle (Berking, '79; Venugopal and David, '81b), and committed nerves form at a distal regenerating surface within 6 hours after amputation (Venugopal and David, '81b). Thus, the pattern of nerve-cell differentiation in a regenerate is believed to reflect a pattern of nerve-cell commitment, rather than a recruitment of committed precursors to the wound site (Venugopal and David, '81c).

Nematocyte differentiation is also influenced by feedback mechanisms (Vögeli, '72; Herlands and Bode, '74b). Zumstein and Tardent (Tardent et al., '71; Zumstein and Tardent, '71; Zumstein, '73) and Bode and David ('78) reported that if stenotele nematocyte populations were selectively reduced 90%, increased numbers of stem cells become committed to stenotele differentiation. Bode and David ('78) doubt, however, that this type of negative feedback control plays a major role in regulating nematocyte populations in steady-state animals. If negative feedback loops prominently influenced nematocyte production, nematocyte population sizes would be expected to remain fairly constant in steady-state hydra; they do not (Bode et al., '77).

Recently, Yaross and Bode ('78a) were able to correlate nematocyte commitment with the size of the stem-cell population, leading to the suggestion that nematocyte differentiation may be regulated by feedback from the stem-cell population. Fujisawa and David ('81) have also shown that, although the number of cell divisions through which a stem cell will pass en route to nematocyte differentiation is fixed when the cell enters the nematocyte pathway, commitment occurs during the terminal cell cycle prior to differentiation. Seeking to extend their studies to the regenerate, Yaross and Bode ('78c) have discovered that the regenerating head region is unable to support nematocyte differentiation.

Evaluation of Chimeric Hydra

The existence of epithelial hydra allows an examination of which developmental processes are controlled by epithelial tissues and which are directed by I cells (stem cells) or I-cell–derived cell types such as nerves. To test these alternatives, chimeric hydra are constructed in which an epithelial hydra of one strain is repopulated with I cells from another strain or species that is developmentally distinct from the epithelial animal.

Chimeras are easily formed by grafting; one need only confine one's pairings to those species that graft readily to one another (Campbell and Bibb, '70; Noda, '70b; Bibb and Campbell, '73). Within a few days after parabiosing an epithelial hydra to a selected I-cell donor, I cells will invade the epithelial hydra (Brien and Reniers-Decoen, '55; Tardent and Morgenthaler, '66). After separating the graft, the transplanted I cells will repopulate the epithelial host to normal levels of both I cells and I-cell–derived cell types. The construction and evaluation of chimeric hydra has been aided immeasurably by the recent detailed analyses of genetic and developmental mutant hydra strains by Sugiyama and Fujisawa ('77a, b, '78a, b, '79a; Fujisawa and Sugiyama, '78; Marcum et al., '80).

Essentially, morphological characteristics follow the epithelial lineage, whereas cell characteristics are expressed consistent with the I-cell source in the chimera. Among the developmental processes directed by epithelial tissues are size, dry weight, growth rate, tentacle number, tentacle patterning on a bud, and budding rate. Nematocyst morphology and nerve- and I-cell densities reflect the I-cell source (Marcum and Campbell, '78b; Sugiyama and Fujisawa, '78b, '79a; Campbell, '79; Lee and Campbell, '79). Sugiyama and Fujisawa ('79b) have now begun a quantitative statistical analysis of all possible correlations between the characteristics combined in chimeric hydra to provide a basis for determining the mechanisms involved in these developmental manifestations.

EXPERIMENTAL STUDIES: BIOCHEMICAL ANALYSES

Morphogens

In the early 1960s Burnett presented evidence that the factors or morphogens influencing growth and developmental processes in hydra were diffusible materials. His data were derived from a series of grafting experiments in which he employed agar blocks in ways similar to their utilization in the discovery of plant growth substances (Burnett, '61, '62, '66). Later, Lesh and Burnett ('64, '66) and Lentz ('65a) isolated and partially characterized a material functionally analogous to one of Burnett's proposed growth factors, the "growth-stimulatory material." The overt response of tissue exposed to concentrated amounts of this low molecular weight, protease-sensitive com-

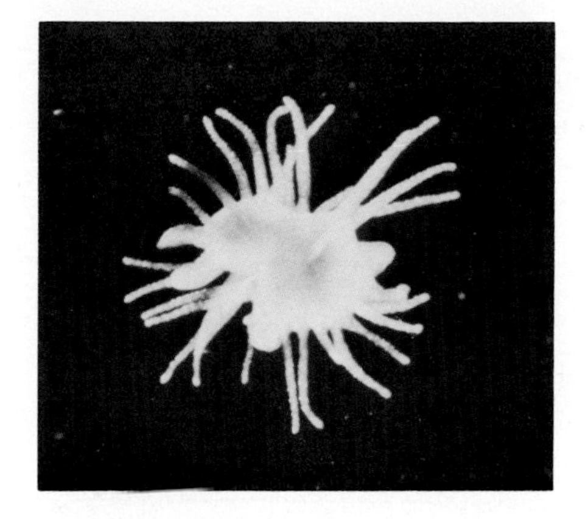

Fig. 18. *Hydra pirardi* with supernumerary tentacles, 47 days after exposure to concentrated amounts of a crude morphogen preparation. × 5.

pound was a stimulation of hypostome and tentacle development (Fig. 18) (see also Shostak et al., '78). Because supernumerary hypostome and tentacle formation are accompanied by increased I-cell differentiation to form nerves and nematocytes, Lesh and Burnett hypothesized (and then provided quantitative evidence to support their contention) that differences in the concentration of the isolated peptidyl component affected qualitative differences in the direction of I-cell differentiation (Lesh and Burnett, '66; Lesh, '70).

Simultaneously with Burnett's pronouncements, Lenique and Lundblad ('66) also announced the isolation of growth substances from hydra hypostomal extracts. Their studies, using either electrophoretic or gel-filtration preparative procedures, yielded two populations of antagonistic factors. The "growth promoter" compounds were electronegative proteins, presumably nucleoproteins, whereas the inhibitory substances were electropositive proteins, and were predicted to be histones. Both groups of compounds had molecular weights > 40,000, acted in low concentrations, and exerted their most profound effect for a limited period of time at the beginning of regeneration (see also Macklin, '71).

Coincident with both of these investigations, Davis ('66, '67) reported the isolation of an inhibitory material from hydra tissue homogenates. Davis's studies, however, showed that the inhibition affected by his materials was due to the presence of quaternary ammonium compounds in the animal's nematocysts, thus raising a cautionary consideration for subsequent biochemical investigations.

In 1971 Müller and Spindler also validly criticized the original isolation and characterization studies, as the bioassays were generally performed using concentrated amounts (pharmacological dosages) of the isolated compo-

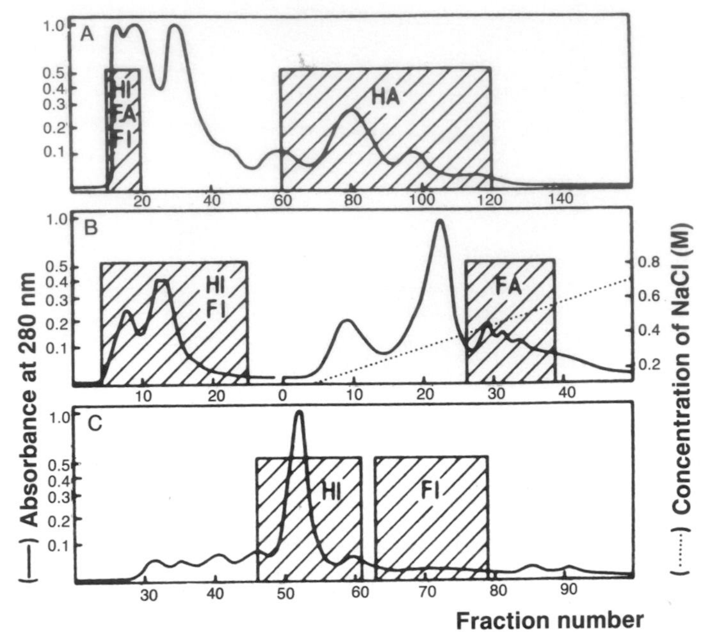
A
HI
FA
FI
HA
B
HI
FI
FA
C
HI
FI
(—) Absorbance at 280 nm
(......) Concentration of NaCl (M)
Fraction number

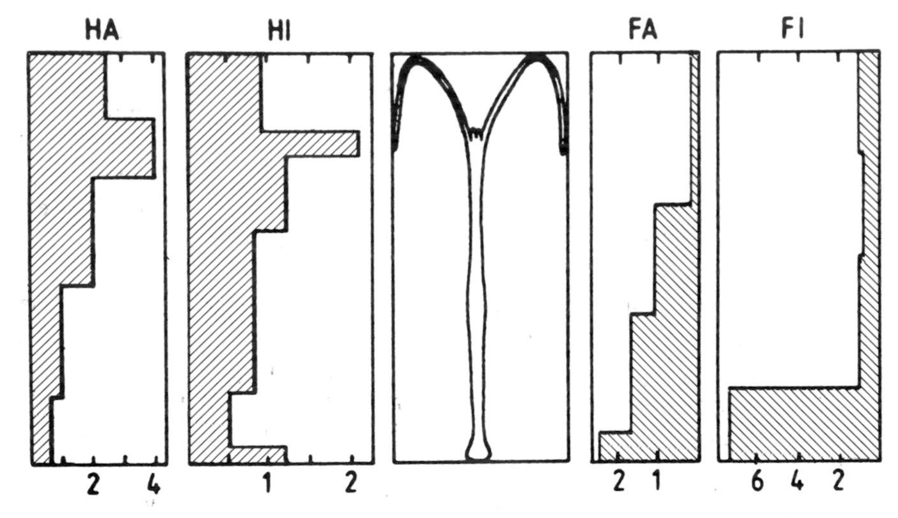
HA
HI
FA
FI
Specific Activity (B.U./O.D.280)

nent(s). Müller and Spindler argued that if the isolated substances were true morphogens, they should, after sufficient purification, be capable of eliciting their biological activity at extremely low concentrations (physiological dosages).

These criticisms have been completely resolved by the work of Schaller and her colleagues, who have confirmed and significantly extended the earlier morphogen studies. These investigators have purified and assayed four morphogenetic substances from hydra tissues, all presumably derived primarily from the nervous system. They are a head activator (HA), a head inhibitor (III), a foot activator (FA), and a foot inhibitor (FI) (Fig. 19). Their data argue that these components are unequally distributed along the animal's body column, functioning interactively to influence the formation of hydra's two organizational centers, the hypostome and basal disc (Schaller, '73, '75, '76c, '78; Schaller and Gierer, '73; Schmidt and Schaller, '76, '80; Grimmelikhuijzen and Schaller, '77; Grimmelikhuijzen, '79; Schaller et al., '79; Bodenmüller et al., '80; Schmidt et al., '80; Bodenmüller and Schaller, '81; Schaller and Bodenmüller, '81; Birr et al., '81).

Throughout these investigations attention has been focused most prominently on a biochemical and developmental analysis of the HA compound. Recently the application of reverse-phase high pressure liquid chromatography (HPLC) has allowed purification of the HA to homogeneity. Chemically, the molecule is an undecapeptide with the amino acid sequence pGlu-Pro-Pro-Gly-Gly-Ser-Lys-Val-Ile-Leu-Phe (MW 1,142). This sequence has been confirmed by synthesis, and Schaller and Bodenmüller have reported a widespread occurrence for this peptide within the animal kingdom (Schaller, '75; Schaller et al., '77; Bodenmüller et al., '80; Bodenmüller and Schaller, '81; Schaller and Bodenmüller, '81; Birr et al., '81).

Head activator is necessary for head regeneration to occur, and also serves to stimulate bud outgrowth. Its mode of action is as a mitogen, stimulating cells in the G_2 period to proceed through mitosis. In the presence of HA,

---

Fig. 19. Hydra morphogens. Top: Separation of morphogens. A) Separation of the head activator (HA) from the other three substances by chromatography on Sephadex G-10 of a hydra methanol extract. Distilled water is the eluent. B) Separation of the foot activator (FA) from the head inhibitor (HI) and foot inhibitor (FI) by chromatography of fractions 10–20 from (A). Chromatography is performed on DEAE Sephadex A-25 with 35 mM Tris HCl (pH 7.6) as eluent. The NaCl gradient from 0.0 to 0.7 M is shown on the right side of the figure. C) Separation of HI from FI by chromatography of fractions 5–25 from (B). Chromatography is performed on Sephadex LH-20 with methanol as an eluent. Bottom: Distribution of morphogens along the body axis of hydra. Data were obtained by cutting hydra into pieces and assaying these pieces for their morphogen content. Reprinted from Schmidt et al. ('80), with permission.

stem-cell commitment to nerve cells is stimulated, whereas the commitment to nematocyst formation is inhibited (Schaller, '75, '76a–c).

In contrast to the HA, the HI inhibits head regeneration and bud outgrowth, and prevents the induction of new buds. It is further predicted to be nonpeptidyl in character and of low molecular weight (< 500) (Schaller, '79). The FA activates foot formation, but is without influence on bud outgrowth or hypostome formation. It is a low molecular weight (500–1,000) peptide (Grimmelikhuijzen and Schaller, '77; Grimmelikhuijzen, '79). At low concentrations (10^{-7} M) the FI inhibits foot formation in regenerating hydra. However, at higher concentrations it also retards head regeneration, bud induction, and bud outgrowth. Like the HI, the FI is purported to be of low molecular weight (< 500) and to contain no peptide bonds (Schmidt and Schaller, '76, '80; Schmidt et al., '80).

Berking ('77, '79, '80b; Berking and Gierer, '77) has also recently reported the isolation and purification of a low molecular weight (300–1,000) endogenous inhibitory substance that acts at low concentrations ($< 10^{-3}$ M). The principal effect of this material is the prevention or retardation of asexual reproduction, although it also has inhibitory influences on head and foot formation. This component, like those of Schaller, is believed to be stored within nervous elements. Using this material Berking has begun to investigate questions of prepatterning and positioning in the budding process.

The current availability of epithelial hydra and morphogenetic mutants has added new dimensions to the study of morphogenetic substances. Schaller's discovery of normal or even higher than normal concentrations of all four hydra morphogens in nerve-depleted animals has contributed to a general reassessment of the chemical role of the nervous system in developmental regulation. Morphogenetic mutants, alternatively, may provide a more definitive means of evaluating the effects of endogenous substances on morphogenesis and patterning (Schaller et al., '77a–c, '79, '80).

Biochemical Events of Normal Regeneration

In addition to a biochemical analysis of growth factors in hydra, investigators have also been attempting to define the biochemical events occurring during normal regeneration to provide a basis for meaningful experimental manipulation of the system. It has been shown, for example, that during distal regeneration DNA synthesis precedes tentacle outgrowth (Lesh-Laurie et al., '76). Hypostome determination and tentacle initiation, however, take place in the virtual absence of DNA synthesis (Clarkson, '69a; Lesh-Laurie et al., '76). These observations are consistent with suggestions that hypostome formation in hydra is a morphallactic occurrence, whereas tentacle elaboration involves both morphallactic and epimorphic mechanisms. Interference with DNA synthesis using, for example, 5-fluorouracil, hydroxyurea, or bromodeoxyuridine, allows hypostome development but variously interrupts the differentiative and morphogenetic events associated with tentacle formation (Clarkson, '69b; Goyns and Stanisstreet, '74; Lesh-Laurie et al., '76; Lesh-Laurie and Frank, '77).

Distal regeneration is also accompanied by major peaks of RNA and protein synthesis. Ribonucleic acid synthetic activity has been recorded concomitant with hypostome formation and preceding the elaboration of tentacles (Clarkson, '69a; Brooks et al., '77) and is considered a requirement for normal distal regeneration (Kass-Simon, '69; Clarkson, '69b; Lesh-Laurie and Hang, '72; Lesh-Laurie, '74; Kolenkine, '73). Ribonucleic acid species have been isolated and characterized in hydra (Voland et al., '77), and the specific manipulation of RNA synthetic patterns with the antibiotic actinomycin D has yielded some dramatic results. Predictably, high concentrations of actinomycin D (e.g., 10-60 μg/ml) both suppressed RNA synthesis and retarded or inhibited distal regeneration (Clarkson, '69b; Datta and Chakrabarty, '70; Venugopal and Mookerjee, '81). These dosages were most effective, however, when applied several hours prior to hypostomal amputation. Lower concentrations of actinomycin D (3 μg/ml), which suppressed RNA synthesis only in the hypostome region of the animal, resulted in regeneration progressing only to the two-tentacled stage. Regenerates formed normal-appearing hypostomes and two tentacles, 180° separated from each other (Lesh-Laurie and Hang, '72; Lesh-Laurie, '74; Voland and Lesh-Laurie, '80).

Actinomycin D–treated hypostomes exhibit no inductive ability (Clarkson, '69b), and at the cellular level actinomycin D exposure depletes the animal's I-cell population (Kolenkine, '73; Lesh-Laurie, '74). Replacing an actinomycin D–treated hypostome with a normal one will reverse the I-cell effect of actinomycin D exposure (Lesh-Laurie, '74). Basal disc regeneration appears unaffected by actinomycin D treatment (Kolenkine, '73; Venugopal and Mookerjee, '77).

In the literature these RNA data have been discussed in terms of an mRNA involvement in hypostome development. It is very difficult, however, to confidently draw these types of conclusions based primarily on inhibitor studies performed using whole organisms. Additional detailed molecular measurements must be made before specific significance can be attributed to the actinomycin D observations.

Very few investigations assessing the involvement of specific protein syntheses in regeneration have been conducted. Clarkson ('69a) initially demonstrated a requirement for protein synthesis during distal regeneration, and some reports exist in which inhibitors have been used to reveal insights into this requirement (Clarkson, '69b; Brooks, '76). Brooks ('76), in a general review of protein synthetic patterns during distal regeneration, found peaks of synthesis associated with the major morphogenetic events of regeneration. The only specific protein examined in his study was collagenase-sensitive material. He discovered that a major increase in collagen secretion coincided precisely with the time of tentacle initiation. Using the proline analog L-azetidine carboxylic acid, Brooks confirmed that the cellular and synthetic activities of regeneration proceeded in the absence of collagen secretion. Morphogenesis, however, did not occur. This led Brooks to hypothesize an involvement for collagen secretion in the mechanical rather than the inductive aspects of morphogenesis.

The recent refinements in polyacrylamide gel electrophoresis techniques present new opportunities for investigators to examine specific classes of proteins. Regeneration should be a process amenable to meaningful investigation with this methodology (Hood, '81).

The discovery of peptide morphogens in hydra has also aroused interest in the possible role of cyclic nucleotides in regulating developmental phenomena in the organism. Cyclic AMP and GMP levels have been measured in hydra, and distribution studies have found concentrated amounts of cAMP in the hypostomal region of the animal; cGMP alternatively displays its highest concentration in the peduncle-basal disc area. Levels of both cyclic nucleotides fluctuate in predictable patterns during distal regeneration (Hill and Lesh-Laurie, '81; Berkowitz and Lesh-Laurie, in preparation).

The discovery of neural-derived peptide morphogens in hydra has also led investigators to explore the possible existence of known neuropeptides in hydra. Recently, Grimmelikhuijzen et al. ('80, '81a, b), using immunocytochemistry, have detected gastrin/CCK-like immunoreactivity in the sensory nerves of hydra's mouth region, substance P-like immunoreactivity in the nerves of the tentacles and basal disc (Taban and Cathieni, '78, '79), and neurotensinlike immunoreactivity in nerve fibers throughout the organism. These findings are of particular interest when evaluated in light of hydra's primitive nervous system and questions of the evolution of peptide hormones.

Recently, high concentrations of free glutamate have been recorded in hydra. This observation, coupled with the interest in the developmental role played by hydra's nervous system, has led investigators to begin an evaluation of glutamate as a possible neural-active agent in the animal (Lesh-Laurie et al., '80; Flechtner et al., '81) and to examine the enzymes associated with the γ-glutamyl moiety and its relation to hydra's glutathione feeding response (Danner et al., '76, '78; Cobb et al., '80).

Application of Exogenous Materials

In addition to these investigations, the hydra literature is replete with studies probing the effects of exogeneous chemical manipulations on developmental processes. These range from the application of neuropharmacological agents (Lentz and Barrnett, '63) and thalidomide (Lenique, '67), which variously inhibited regeneration and gave rise to aberrant forms, to osmotic and respiratory perturbations (Datta, '68; Datta and Mitra, '72; Yasugi, '74; Mulherkar and Therwath, '78). Among the most provocative, yet unconfirmed, observations is that reporting the commitment of the entire I-cell population to nematocyte differentiation and the transformation of all gland cells to mucous cells following exposure to 0.2% $NaHCO_3$ + 0.2% $CaCl_2$ (Macklin and Burnett, '66).

PATTERNING

Because hydra manifests many developmental phenomena in a relatively simple manner, the animal has also served as a focus for numerous theoretical inquiries. Within the past decade, for example, developmental biologists have begun to confront problems of patterning, and the subsequent expression of differentiation in regulating systems. Patterning was initially explored, in hydra, from the perspective of positional information and the role of positional signaling in the generation of biological fields (Wolpert, '69, '71). The facility with which specific hydrid tissue interactions may be constructed through grafting permitted experimental evaluation of these models (Wolpert et al., '71, '74; Hicklin et al., '73).

Later, Gierer and Meinhardt developed a theory of biological pattern formation based on short-range autocatalytic activation and longer-range inhibition (Gierer and Meinhardt, '72, '74; Meinhardt and Gierer, '74). This theory is consistent with the emerging analyses of morphogenetic substances in hydra, and has been applied in an attempt to comprehend the animal's polarity gradients. Computers have allowed, and will continue to allow, the almost unrestricted theoretical extension of patterning models (Gierer, '73, '77a, b, '80, '81). Experimental verification to date, however, has been limited to grafting (Newman, '74; Cohen and MacWilliams, '75) and regeneration (Bode and Bode, '80) studies. In the future, increased attention will undoubtedly be directed to developmental problems such as patterning, which are amenable to mathematical analysis.

Acknowledgments. The author extends her sincerest thanks to M. Berkowitz, K. Boehm, S. Hill, and P. Suchy for their critical reading of the manuscript. The patience of G. Shuga and T. Allen in the preparation of the manuscript is also gratefully acknowledged.

LITERATURE CITED

Baird, R., and A.L. Burnett (1967) Observations on the discovery of a dorso-ventral axis in hydra. J. Embryol. Exp. Morphol. *17*:35–81.

Barzansky, B., and H.M. Lenhoff (1974) On the chemical composition and developmental role of the mesoglea of hydra. Am. Zool. *14*:575–581.

Barzansky, B., H.M. Lenhoff, and H.R. Bode (1975) Hydra mesoglea: Similarity of its amino acid and neutral sugar composition to that of vertebrate basal lamina. Comp. Biochem. Physiol. *50B:*419–424.

Berking, S. (1977) Bud formation in hydra: Inhibition by an endogenous morphogen. Wilhelm Roux' Arch. *181:*215–225.

Berking, S. (1979) Control of nerve cell formation from multipotent stem cells in hydra. J. Cell Sci. *40:*193–205.

Berking, S. (1980a) Commitment of stem cells to nerve cells and migration of nerve cell precursors in preparatory bud development in hydra. J. Embryol. Exp. Morphol. *60:*373–387.

Berking, S. (1980b) Analysis of morphogenetic processes in hydra by means of an endogenous inhibitor. In P. Tardent and R. Tardent (eds): Developmental and Cellular Biology of Coelenterates. Amsterdam; Elsevier/North-Holland, pp. 377–381.

Berking, S., and A. Gierer (1977) Analysis of early stages of budding in *Hydra* by means of an endogenous inhibitor. Wilhelm Roux' Arch. *182:*117–129.

Berrill, N.J., and P. Mees (1936) Reorganization and regeneration in *Sabella.* I. Nature of gradient, summation and posterior reorganization. J. Exp. Zool. *73:*67–83.

Bhattacherjee, S. (1972) Cell interaction during regeneration of normal and radiated hydras. Indian J. Exp. Biol. *10:*8–12.

Bibb, C., and R.D. Campbell (1973) Cell affinity determining heterospecific graft intolerance in hydra. Tissue Cell *5:*199–208.

Birr, C., B. Zachmann, H. Bodenmüller , and H.C. Schaller (1981) Synthesis of a new neuropeptide, the head activator from hydra. FEBS Letts *131:*317–321.

Bisbee, J.W. (1973) Size determination in *Hydra*: The roles of growth and budding. J. Embryol. Exp. Morphol. *30:*1–19.

Bode, H.R. (1974) Activity of hydra cells *in vitro* and in regenerating cell reaggregates. Am. Zool. *14:*543–550.

Bode, H.R., S. Berking, C.N. David, A. Gierer, H. Schaller, and E. Trenkner (1973) Quantitative analysis of cell types during growth and morphogenesis in *Hydra.* Wilhelm Roux' Arch. *171:*269–285.

Bode, H.R., and C.N. David (1978) Regulation of a multipotent stem cell, the interstitial cell of hydra. Prog. Biophys. Mol. Biol. *33:*189–206.

Bode, H.R., and K.M. Flick (1976) Distribution and dynamics of nematocyte populations in *Hydra attenuata.* J. Cell Sci. *21:*15–34.

Bode, H.R., K.M. Flick, and P.M. Bode (1977) Constraints on the relative sizes of the cell populations in *Hydra attenuata.* J. Cell Sci. *24:*31–50.

Bode, H.R., K.M. Flick, and G.S. Smith (1976) Regulation of interstitial cell differentiation in *Hydra attenuata.* I. Homeostatic control of interstitial cell population size. J. Cell Sci. *20:*29–46.

Bode, H.R., and G.S. Smith (1977) Regulation of interstitial cell differentiation in *Hydra attenuata.* II. Correlation of the axial position of the interstitial cell with nematocyte differentiation. Wilhelm Roux' Arch. *181:*203–213.

Bode, P.M., and H.R. Bode (1980) Formation of pattern in regenerating tissue pieces of *Hydra attenuata.* Dev. Biol. *78:*484–496.

Bodenmüller, H., and H.C. Schaller (1981) Conserved amino acid sequence of a neuropeptide, the head activator, from coelenterates to humans. Nature *293:*579–580.

Bodenmüller, H., H.C. Schaller, and G. Darai (1980) Human hypothalamus and intestine contain a hydra-neuropeptide. Neurosci. Lett. *16:*71–74.

Brauer, A. (1891) Uber die Entwicklung von Hydra. Z. Wiss. Zool. *52:*169–216.

Brien, P. (1950a) Etude d'*Hydra viridis.* Ann. Soc. R. Zool. Belg. *81:*33–110.

Brien, P. (1950b) La vesicule germinative de l'oocyte des Hydres. Bull. Cl. Sci. Acad. R. Belg. *36:*561–573.

Brien, P. (1951) Contribution à l'étude des hydres d'eau douce (*Hydra fusca, H. viridis, H. attenuata*). Croissance et reproduction. Bull. Soc. Zool. Fr. *76:*277–296.

Brien, P. (1952) Apparition de la stolonisation chez l'hydre vert et sa transmissibilité. Bull. Biol. Fr. Belg. *86:*1–30.

Brien, P. (1953) La perennite somatique. Biol. Rev. *28:*308–349.

Brien, P. (1956) La mutation somatique chez l'hydre vert. Bull. Acad. R. Belg. 5th Ser. *42:*906–920.

Brien, P. (1961) Induction sexuelle et interservalité chez une Hydra gonochorique (*Hydra fusca*) par la méthode des greffes. Comptes Rend. *253:*1997–1999.

Brien, P. (1963) Contribution à l'étude de la biologie sexuelle chez les hydres d'eau douce. Induction gamétique et sexuelle par la méthode des greffes en parabiose. Bull. Biol. Fr. Belg. *97:*213–283.

Brien, P. (1966) Biologie de la Reproduction Animale. Paris: Masson.

Brien, P., and M. Reniers-Decoen (1949) La croissance, la blastogénèse, l'ovogénèse chez *Hydra fusca* (Pallas). Bull. Biol. Fr. Belg. *23:*293–386.

Brien, P., and M. Reniers-Decoen (1951) La gamétogénèse et l'intersexualité chez *Hydra attenuata.* Ann. Soc. R. Zool. Belg. *82:*285–327.

Brien, P., and M. Reniers-Decoen (1955) La signification des cellules interstitielles des hydres d'eau douce et le problème de la réserve embryonnaire. Bull. Biol. Fr. Belg. *89:*258–325.

Brinkley, L.L. (1974) Non-budding hydra: Form regulation and bud induction. Am. Zool. *14:*603–618.

Brooks, D. (1976) An analysis of some biosynthetic events of hydrid tentacle morphogenesis. Ph.D. thesis, Case Western Reserve University, Cleveland, Ohio, 171 pp.

Brooks, D., J.R. Voland, and G.E. Lesh-Laurie (1977) Biosynthetic events of hydrid regeneration. II. Patterns and profiles of RNA synthesis during distal morphogenesis. J. Embryol. Exp. Morphol. *37:*149–161.

Browne, E.N. (1909) The production of new hydranths in *Hydra* by the insertion of small grafts. J. Exp. Zool. *7:*1–23.

Browne, C.L., and L.E. Davis (1977) Cellular mechanisms of stimulation of bud production in hydra by low levels of inorganic lead compounds. Cell Tissue Res. *177:*555–570.

Burnett, A.L. (1959) Histophysiology of growth in hydra. J. Exp. Zool. *140:*281–342.

Burnett, A.L. (1961) The growth process in hydra. J. Exp. Zool. *146:*21–84.

Burnett, A.L. (1962) The maintenance of form in *Hydra.* Symp. Soc. Dev. Biol. *20:*27–52.

Burnett, A.L. (1966) A model of growth and cell differentiation in hydra. Am. Nat. *100:*165–190.

Burnett, A.L., L.E. Davis, and F.E. Ruffing (1966) A histological and ultrastructural study of germinal differentiation of interstitial cells arising from gland cells in *Hydra viridis.* J. Morphol. *120:*1–8.

Burnett, A.L., and N.A. Diehl (1964a) The nervous system of hydra. I. Types, distribution and origin of nerve elements. J. Exp. Zool. *157:*217–226.

Burnett, A.L., and N. Diehl (1964b) The nervous system of hydra. III. The initiation of sexuality with special reference to the nervous system. J. Exp. Zool. *157:*237–250.

Burnett, A.L., N.A. Diehl, and F. Diehl (1964) The nervous system of hydra. II. Control of growth and regeneration by neurosecretory cells. J. Exp. Zool. *157:*227–236.

Burnett, A.L., and M. Garofalo (1960) Growth pattern in the green hydra, *Chlorohydra viridissima.* Science *131:*160–161.

Burnett, A., and R. Hausman (1969) The mesoglea of hydra. II. Possible role in morphogenesis. J. Exp. Zool. *171:*15–24.

Burnett, A.L., R. Lowell, and M. Cyrlin (1973) Regeneration of a complete *Hydra* from a single differentiated somatic cell type. In A.L. Burnett (ed): Biology of Hydra. New York; Academic Press, pp. 255–270.

Bursztajn, S., and L.E. Davis (1974) The role of the nervous system in regeneration, growth and cell differentiation in Hydra. I. Distribution of nerve elements during hypostomal regeneration. Cell Tissue Res. *150:*213–220.

Campbell, R.D. (1967a) Tissue dynamics of steady state growth in *Hydra littoralis.* I. Patterns of cell division. Dev. Biol. *15:*487–502.

Campbell, R.D. (1967b) Tissue dynamics of steady state growth in *Hydra littoralis.* II. Patterns of tissue movement. J. Morphol. *121:*19–28.

Campbell, R.D. (1967c) Tissue dynamics of steady state growth in *Hydra littoralis.* III. Behavior of specific cell types during tissue movements. J. Exp. Zool. *164:*379–392.

Campbell, R.D. (1973) Vital marking of single cells in developing tissues: India ink injection to trace tissue movements in hydra. J. Cell Sci. *13:*651–661.

Campbell, R.D. (1974) Cell movements in hydra. Am. Zool. *14:*523–535.

Campbell, R.D. (1976) Elimination of hydra interstitial and nerve cells by means of colchicine. J. Cell Sci. *21:*1–13.

Campbell, R.D. (1979) Development of hydra lacking interstitial and nerve cells ("epithelial hydra"). Symp. Soc. Dev. Biol. *37:*267–293.

Campbell, R.D. (1980) Role of muscle processes in hydra morphogenesis. In P. Tardent and R. Tardent (eds): Developmental and Cellular Biology of Coelenterates. Amsterdam; Elsevier/North-Holland, pp. 421–428.

Campbell, R.D. (1982) Identifying hydra species. In H.M. Lenhoff (ed): Methods in Hydra Research. New York: Plenum, (in press).

Campbell, R.D., and C. Bibb (1970) Transplantation in coelenterates. Transplant Proc. *2:*202–211.

Campbell, R.D., and C.N. David (1974) Cell cycle kinetics and development of *Hydra attenuata.* J. Cell Sci. *16:*349–358.

Cerame-Vivas, M.J. (1961) Separation of cell layers in hydra. Wilhelm Roux' Arch. *153:*213–216.

Chang, J.T., H.H. Hsieh, and D.D. Liu (1952) Observations on hydra, with special reference to abnormal forms and bud formation. Physiol. Zool. *25:*1–10.

Clarkson, S.G. (1969a) Nucleic acid and protein synthesis and pattern regulation in hydra. I. Regional patterns of synthesis and changes in synthesis during hypostome formation. J. Embryol. Exp. Morphol. *21:*33–54.

Clarkson, S.G. (1969b) Nucleic acid and protein synthesis and pattern regulation in hydra. II. Effect of inhibition of nucleic acid and protein synthesis on hypostome formation. J. Embryol. Exp. Morphol. *21:*55–70.

Clarkson, S., and L. Wolpert (1967) Bud morphogenesis in hydra. Nature *214:*780–783.

Cobb, M.H., W. Heagy, J. Danner, H.M. Lenhoff, and G.R. Marshall (1980) Effect of glutathione on cyclic nucleotide levels in *Hydra attenuata.* Comp. Biochem. Physiol. *65C:*111–115.

Cohen, J.E., and H.K. MacWilliams (1975) The control of foot formation in transplantation experiments with *Hydra viridis.* J. Theor. Biol. *50:*87–105.

Corff, S. (1973) Organismal growth and the contribution of cell proliferation to net growth and maintenance of form. In A. Burnett (ed): Biology of Hydra. New York, Academic Press, pp. 345–389.

Corff, S., and A. Burnett (1969) Morphogenesis in hydra. I. Peduncle and basal disc formation at the distal end of regenerating hydra after exposure to colchicine. J. Embryol. Exp. Morphol. *21:*417–443.

Corff, S., and A. Burnett (1970) Morphogenesis in hydra. II. Peduncle and basal disc formation at the distal end of regenerating hydra after exposure to low temperatures. J. Embryol. Exp. Morphol. *24:*21–32.

Cowden, R.R., and J. Glocker (1960) A topological histochemical study of *Pelmatohydra oligactis.* Trans. Am. Microsc. Soc. *79:*180–190.

Danner, J., H.M. Lenhoff, M. Houston-Cobb, W. Heagy, and G.R. Marshall (1976) γ-Glutamyl transpeptidase in *Hydra.* Biochem. Biophys. Res. Commun. *73:*180–186.

Danner, J., M.H. Cobb, W. Heagy, H.M. Lenhoff, and G.R. Marshall (1978) Interaction of glutathione analogues with *Hydra attenuata* γ-glutamyltransferase. Biochem. J. *175:*547–553.

Datta, S. (1968) Effects of antimetabolites and morphostatic substances on regeneration in hydra. Indian J. Exp. Biol. *6:*190–192.

Datta, S., and A. Chakrabarty (1970) Effects of actinomycin D on the distal end regeneration in *Hydra vulgaris* Pallas. Experientia *26:*855.

Datta, S., and J. Mitra (1972) Biochemical events of hydroid regeneration. Indian J. Exp. Biol. *10:*463–464.

David, C.N. (1973) A quantitative method for maceration of hydra tissue. Wilhelm Roux' Arch. *171:*259–268.

David, C.N., and R.D. Campbell (1972) Cell cycle kinetics and development of *Hydra attenuata.* I. Epithelial cells. J. Cell Sci. *11:*557–568.

David, C.N., and D. Challoner (1974) Distribution of interstitial cells and differentiating nematocytes in nests in *Hydra attenuata.* Am. Zool. *14:*537–542.

David, C.N., and A. Gierer (1974) Cell cycle kinetics and development of *Hydra attenuata.* III. Nerve and nematocyte differentiation. J. Cell Sci. *16:*359–375.

David, C.N., and S. Murphy (1977) Characterization of interstitial stem cells in hydra by cloning. Dev. Biol. *58:*372–383.

David, C.N., and I. Plotnick (1980) Distribution of interstitial stem cells in *Hydra.* Dev. Biol. *76:*175–184.

Davis, L.V. (1966) Inhibition of growth and regeneration in hydra by crowded culture water.Nature *212:*1215–1217.

Davis, L.V. (1967) The source and identity of a regeneration-inhibiting factor in hydroid polyps. J. Exp. Zool. *164:*187–194.

Davis, L.E. (1968) Ultrastructural evidence for division of cnidoblasts in hydra. Exp. Cell Res. *52:*602–607.

Davis, L.E. (1969) Differentiation of neurosensory cells in *Hydra.* J. Cell Sci. *5:*699–726.

Davis, L.E. (1970a) Cell division during dedifferentiation and redifferentiation in the regenerating isolated gastrodermis of hydra. Exp. Cell Res. *60:*127–132.

Davis, L.E. (1970b) Further observations on dividing and nondividing cnidoblasts in the regenerating isolated gastrodermis of hydra. Z. Zellforsch. *105:*526–537.

Davis, L.E. (1971) Differentiation of ganglionic cells in *Hydra.* J. Exp. Zool. *176:*107–128.

Davis, L.E. (1973) Histological and ultrastructural studies of the basal disk of hydra. I. The glandulomuscular cell. Z. Zellforsch. *139:*1–27.

Davis, L.E. (1974) Ultrastructural studies of the development of nerves in hydra. Am. Zool. *14:*551–573.

Davis, L.E. (1975) Histological and ultrastructural studies of the basal disk of hydra. III. The gastrodermis and the mesoglea. Cell Tissue Res. *162:*107–118.

Davis, L.E., A.L. Burnett, and J.F. Haynes (1968) Histological and ultrastructural study of the muscular and nervous systems in *Hydra.* II. Nervous system. J. Exp. Zool. *167:*295–332.

Davis, L.E., A.L. Burnett, J.F. Haynes, and V.R. Mumaw (1966) A histological and ultrastructural study of dedifferentiation and redifferentiation of digestive and gland cells in *Hydra viridis.* Dev. Biol. *14:*307–329.

Davis, L.E., and S. Bursztajn (1973) Histological and ultrastructural studies of the basal disk of hydra. II. Nerve cells and other epithelial cells. Z. Zellforsch. *130:*29–45.

Davis, L.E., and J.F. Haynes (1968) An ultrastructural examination of the mesoglea of hydra. Z. Zellforsch. *92:*149–158.

Davison, J. (1976) *Hydra hymanae:* Regulation of the life cycle by time and temperature. Science *194:*618–620.

Day, R.M., and H.M. Lenhoff (1981) *Hydra* mesoglea: A model for investigating epithelial cell–basement membrane interactions. Science *211:*291–294.

Diehl, F.A., and A.L. Burnett (1964) The role of interstitial cells in the maintenance of hydra. I. Specific destruction of interstitial cells in normal, asexual, non-budding animals. J. Exp. Zool. *155:*253–260.

Diehl, F.A., and A.L. Burnett (1965a) The role of interstitial cells in the maintenance of hydra. II. Budding. J. Exp. Zool. *158:*283–298.

Diehl, F.A., and A.L. Burnett (1965b) The role of interstitial cells in the maintenance of hydra. III. Regeneration of hypostome and tentacles. J. Exp. Zool. *158:*299–318.

Diehl, F.A., and A.L. Burnett (1966) The role of interstitial cells in the maintenance of hydra. IV. Migration of interstitial cells in homografts and heterografts. J. Exp. Zool. *163:*125–140.

Downing, E.R. (1905) The spermatogenesis of *Hydra*. Zool. Jahrb. Abt. Anat. *21:*379–426.

Downing, E.R. (1908) The ovogenesis of *Hydra fusca.* A preliminary paper. Biol. Bull. *15:*63–66.

Downing, E.R. (1909) The ovogenesis of *Hydra*. Zool. Jahrb. Abt. Anat. *28:*295–322.

Epp, L.G. (1980) An SEM portrait of hydra. In P. Tardent and R. Tardent (eds): Developmental and Cellular Biology of Coelenterates. Amsterdam: Elsevier/North-Holland, pp. 295–300.

Epp, L.G., and D.C. Koblick (1977) Relationship of intracellular potassium to asexual reproduction in *Hydra*. J. Exp. Biol. *69:*45–51.

Epp, L.G., and P. Tardent (1978) The distribution of nerve cells in *Hydra attenuata* Pall. Wilhelm Roux' Arch *185:*185–193.

Epp, L.G., P. Tardent, and R. Banninger (1979) Isolation and observation of tissue layers in *Hydra attenuata* Pall. (Cnidaria, Hydrozoa). Trans. Am. Microsc. Soc. *98:*392–400.

Flechtner, V.R., G. E. Lesh-Laurie, and M.K. Abbott (1981) Evaluation of tentacle regeneration as a biological assay in hydra. Wilhelm Roux' Arch. *190:*67–72.

Fradkin, M., H. Kakis, and R.D. Campbell (1978) Effects of γ-irradiation of hydra: Elimination of interstitial cells from viable hydra. Radiat. Res. *76:*187–197.

Fujisawa, T., and T. Sugiyama (1978) Genetic analysis of developmental mechanisms in hydra. IV. Characterization of a nematocyst-deficient strain. J. Cell Sci. *30:*1975–1985.

Fujisawa, T., and C.N. David (1981) Commitment during nematocyte differentiation in hydra. J. Cell. Sci. *48:*207–222.

Gierer, A. (1973) Molecular models and combinatorial principles in cell differentiation and morphogenesis. Cold Spring Harb. Symp. Quant. Biol. *38:*951–961.

Gierer, A. (1977a) Biological features and physical concepts of pattern formation exemplified by hydra. Curr. Top. Dev. Biol. *11:*17–59.

Gierer, A. (1977b) Physical aspects of tissue evagination and biological form. Q. Rev. Biophys. *10:*529–593.

Gierer, A. (1980) Hydra as a model for physical concepts of biological pattern formation. In P. Tardent and R. Tardent (eds): Developmental and Cellular Biology of Coelenterates. Amsterdam: Elsevier/North-Holland, pp. 363–371.

Gierer, A. (1981) Generation of biological patterns and form: Some physical, mathematical and logical aspects. Prog. Biophys. Mol. Biol. *37:*1–47.

Gierer, A., S. Berking, H.R. Bode, C.N. David, K. Flick, G. Hansmann, H.C. Schaller, and E. Trenkner (1972) Regeneration of hydra from reaggregated cells. Nature New Biol. *239:*98–101.

Gierer, A., and H. Meinhardt (1972) A theory of biological pattern formation. Kybernetik *12:*30–39.

Gierer, A., and H. Meinhardt (1974) Biological pattern formation involving lateral inhibition. Lect. Math. Life Sci. *7:*163–182.

Goetsch, W. (1929) Das Regenerationsmaterial und seine experimentelle Beeinflusung. Wilhelm Roux' Arch. *117:*211–311.

Goyns, M.H., and M. Stanisstreet (1974) The effects of bromodeoxyuridine and bromouracil on regeneration in hydra. Wilhelm Roux' Arch. *175:*87–90.

Grimmelikhuijzen, C.J.P. (1979) Properties of the foot activator from hydra. Cell Diff. *8:*267–273.

Grimmelikhuijzen, C.J.P., A. Balfe, and P.C. Emson (1981a) Substance P-like immunoreactivity in the nervous system of hydra. Histochemistry *71:*325–333.

Grimmelikhuijzen, C.J.P., R.E. Carraway, A. Rokaeus, and F. Sundler (1981b) Neurotensin-like immunoreactivity in the nervous system of hydra. Histochemistry *72:*199–209.

Grimmelikhuijzen, C.J.P., and H.C. Schaller (1977) Isolation of a substance activating foot formation in hydra. Cell Diff. *6:*297–305.

Grimmelikhuijzen, C.J.P., F. Sundler, and J.F. Rehfeld (1980) Gastrin/CCK-like immunoreactivity in the nervous system of coelenterates. Histochemistry *69:*61–68.

Gurkewitz, S., M. Chow, and R.D. Campbell (1980) *Hydra* size and budding rate: Influence of feeding. Int. J. Invert. Reprod. *2:*199–201.

Ham, R.G., and R.E. Eakin (1958) Time sequence of certain physiological events during regeneration in *Hydra*. J. Exp. Zool. *139:*33–53.

Haug, G. (1933) Die Lichtreaktionen der *Hydra*. Z. Vergl. Physiol. *19:*246–303.

Hausman, R.E. (1973) The mesoglea. In A Burnett (ed): Biology of Hydra. New York: Academic Press, pp. 393–453.

Hausman, R.E., and A.L. Burnett (1969) The mesoglea of *Hydra*. I. Physical and histochemical properties. J. Exp. Zool. *171:*7–14.

Hausman, R.E., and A.L. Burnett (1970) The mesoglea of *Hydra*. III. Fiber system changes in morphogenesis. J. Exp. Zool. *173:*175–186.

Haynes, J.F., and A.L. Burnett (1963) Dedifferentiation and redifferentiation of cells in *Hydra viridis*. Science *142:*1481–1483.

Haynes, J.F., and A.L. Burnett (1964) A study of a stolonizing mutant of the European green hydra, *Hydra viridissima*. II. An analysis of the form regulating processes in the stolonizing animal. J. Morphol. *115:*193–206.

Haynes, J.F., A.L. Burnett, and L.E. Davis (1968) Histological and ultrastructural study of the muscular and nervous systems in hydra. J. Exp. Zool. *167:*283–294.

Haynes, J.F., A.L. Burnett, and W. Deutschman (1964) A study of a stolonizing mutant of the European green hydra, *Hydra viridissima*. I. The process of stolonization and some characteristics of the stolonizing animals. J. Morphol. *115:*185–192.

Herlands, R.L., and H.R. Bode (1974a) Oriented migration of interstitial cells and nematocytes in *Hydra attenuata*. Wilhelm Roux' Arch. *176:*67–88.

Herlands, R.L., and H.R. Bode (1974b) The influence of tissue polarity on nematocyte migration in *Hydra attenuata*. Dev. Biol. *40:*323–339.

Hicklin, J., A. Hornbruch, and L. Wolpert (1969) Inhibition of hypostome formation and polarity reversal in *Hydra*. Nature *221:*1268–1271.

Hicklin, J., A. Hornbruch, L. Wolpert, and M. Clarke (1973) Positional information and pattern regulation in hydra: The formation of boundary regions following axial grafts. J. Embryol. Exp. Morphol. *30:*701–725.

Hicklin, J., and L. Wolpert (1973a) Positional information and pattern regulation in hydra: Formation of the foot end. J. Embryol. Exp. Morphol. *30:*727–740.

Hicklin, J., and L. Wolpert (1973b) Positional information and pattern regulation in hydra: The effect of γ-irradiation. J. Embryol. Exp. Morphol. *30:*741–752.

Hill, S.K., and G.E. Lesh-Laurie (1981) Cyclic AMP concentrations during distal regeneration in *Hydra oligactis*. J. Exp. Zool. *218:*233–238.

Holstein, T. (1981) The morphogenesis of nematocytes in *Hydra* and *Forskalia*: An ultrastructural study. J. Ultrastruc. Res. *75:*276–290.

Honegger, R. (1981) Light and scanning electron microscopic investigations of sexual reproduction in *Hydra carnea*. Int. J. Invert. Reprod. *3:*245–255.

Hood, R. (1981) Biological and chemical assessment of growth factors and an electrophoretic analysis of proteins from the freshwater coelenterate, *Hydra*. Ph.D. thesis, Cleveland State University, 81 pp.

Huxley, J.S., and G. De Beer (1934) The Elements of Experimental Embryology. Cambridge: Cambridge University Press.

Issajew, W. (1926) Studien an organischen Regulationen. Wilhelm Roux' Arch. *108:*1–67.

Ito, T., and I. Ohata (1963) Experimental study on the pattern of tentacle formation in Hydra. Mem. Ehime Univ. *4:*525–537.

Kanaev, I.I. (1926) Uber die histologischen Vorgänge bei der Regeneration von *Pelmatohydra oligactis* Pall. (Reference in Kanaev, 1952.)

Kanaev, I.I. (1930) Zur Frage der Bedeutung der interstitiellen Zellen bei Hydra. Wilhelm Roux' Arch. *122:*736–759.

Kanaev, I.I. (1952) Hydra: Essays on the Biology of Fresh Water Polyps. Moscow: Soviet Academy of Sciences. (Edited and published from the Russian edition by H. Lenhoff.)

Kass-Simon, G. (1969) The regeneration gradients and the effects of budding, feeding, actinomycin and RNase on reconstitution in *Hydra attenuata* Pall. Rev. Suisse Zool. *76:*565–599.

Kass-Simon, G., and M. Potter (1971) Arrested regeneration in the budding region of hydra as a result of abundant feeding. Dev. Biol. *24:*363–378.

King, H.D. (1901) Observations and experiments on regeneration in *Hydra viridis*. Wilhelm Roux' Arch. *13:*135–178.

King, H.D. (1903) Further studies on regeneration in *Hydra viridis*. Wilhelm Roux' Arch. *16:*200–242.

Kinnamon, J.C., and J.A. Westfall (1981) A three dimensional serial reconstruction of neuronal distributions in the hypostome of a hydra. J. Morphol. *168:*321–329.

Kleinenberg, N. (1872) "Hydra." Eine anatomisch entwicklungsgeschichtliche Untersuchung. Leipzig: Engelmann.

Klimek, F., L. Graf, G. Hansmann, and A. Gierer (1980) Cell interaction and form in regenerating and budding hydra. In P. Tardent and R. Tardent (eds): Developmental and Cellular Biology of Coelenterates. Amsterdam: Elsevier/North-Holland, pp. 441–446.

Koblick, D.C., and L.G. Epp (1975) Control of growth of *Hydra* cultures by tissue potassium. Comp. Biochem. Physiol. *50A:*387–389.

Koelitz, W. (1911) Morphologische und experimentelle Untersuchungen an Hydra. Wilhelm Roux' Arch. *31:*423–455.

Kolenkine, X. (1973) Action de l'actinomycine D sur la régénération et le bourgeonnement de *Pelmatohydra oligactis*. Ext. Ann. d'Embryol. Morphol. *6:*109–123.

Lee, H., and R.D. Campbell (1979) Development and behavior of an intergeneric chimera of hydra (*Pelmatohydra oligactis* interstitial cells: *Hydra attenuata* epithelial cells). Biol. Bull. *157:*288–296.

Lenhoff, H.M. (1965) Cellular segregation and heterocytic dominance in hydra. Science *148:*1105–1107.

Lenhoff, H.M. (1966) Influence of monovalent cations on growth of *Hydra littoralis*. J. Exp. Zool. *163:*151–156.

Lenhoff, H.M. (1982) Methods in Hydra Research. New York: Plenum.

Lenhoff, H.M., and J. Bovaird (1960) The requirement of trace amounts of environmental sodium for growth and development of *Hydra*. Exp. Cell Res. *20:*384–394.

Lenhoff, H.M., and R.D. Brown (1970) Mass culture of hydra: An improved method and its application to other aquatic invertebrates. Lab. Anim. *4:*139–154.

Lenhoff, H.M., and W.F. Loomis (1957) Environmental factors controlling respiration in hydra. J. Exp. Zool. *134:*171–182.

Lenhoff, H.M., C. Rutherford, and H.D. Heath (1969) Anomalies of growth and form in hydra: Polarity, gradients, and a neoplasia analog. Natl. Cancer Inst. Monogr. *31:*709–737.

Lenicque, R.M. (1967) Action of thalidomide on the induction of tentacles in regenerating *Hydra littoralis*. Acta Zool. *48:*1–13.

Lenicque, R.M., and M. Lundblad (1966) Promoters and inhibitors of development during regeneration of the hypostome and tentacles of *Hydra littoralis*. Acta Zool. *47:*1–11.

Lentz, T.L. (1965a) Hydra: Induction of supernumerary heads by isolated neurosecretory granules. Science *150:*633–635.

Lentz, T.L. (1965b) The fine structure of differentiating interstitial cells in hydra. Z. Zellforsch. *67:*547–560.

Lentz, T.L. (1965c) Fine structural changes in the nervous system of the regenerating hydra. J. Exp. Zool. *159:*181–194.

Lentz, T.L. (1966) The Cell Biology of Hydra. Amsterdam: North-Holland.

Lentz, T.L., and R.J. Barrnett (1961) Enzyme histochemistry of hydra. J. Exp. Zool. *147:*135–147.

Lentz, T.L., and R.J. Barrnett (1962) Changes in the distribution of enzyme activity in the regenerating hydra. J. Exp. Zool. *150:*103–117.

Lentz, T.L., and R.J. Barrnett (1963) The role of the nervous system in regenerating hydra: The effect of neuropharmacological agents. J. Exp. Zool. *154:*305–328.

Lentz, T.L., and R.J. Barrnett (1965) Fine structure of the nervous system of hydra. Am. Zool. *5:*341–356.

Lesh, G.E. (1970) A role of inductive factors in interstitial cell differentiation in hydra. J. Exp. Zool. *173:*371–382.

Lesh-Laurie, G.E. (1971) Observations on pseudocolonial growth in hydra. Biol. Bull. *141:*278–298.

Lesh-Laurie, G.E. (1973) Expression and maintenance of organismic polarity. In A.L. Burnett (ed): Biology of Hydra. New York: Academic Press, pp. 143–167.

Lesh-Laurie, G.E. (1974) Tentacle morphogenesis in hydra: A morphological and biochemical analysis of the effect of actinomycin D. Am. Zool. *14:*591–602.

Lesh-Laurie, G.E., D.C. Brooks, and E.R. Kaplan (1976) Biosynthetic events of hydrid regeneration. I. The role of DNA synthesis during tentacle elaboration. Wilhelm Roux' Arch. *180:*157–174.

Lesh, G.E., and A.L. Burnett (1964) Some biological and biochemical properties of the polarizing factor in hydra. Nature *204:*492–493.

Lesh, G.E., and A.L. Burnett (1966) An analysis of the chemical control of polarized form in hydra. J. Exp. Zool. *163:*55–78.

Lesh-Laurie, G.E., and D.D. Donaldson (1978) Cellular events during early distal regeneration in *Hydra oligactis*. Acta Embryol. Exp. *3:*319–331.

Lesh-Laurie, G.E., V.R. Flechtner, and R.L. Hood (1980) Purification and analysis of hydra morphogen. In P. Tardent and R. Tardent (eds): Developmental and Cellular Biology of Coelenterates. Amsterdam: Elsevier/North-Holland, pp. 401–406.

Lesh-Laurie, G.E., and W.N. Frank (1977) The influence of BrdU on interstitial cell differentiation in hydra. Experientia *33:*909–910.

Lesh-Laurie, G.E., and L. Hang (1972) Tentacle morphogenesis in hydra. I. The morphological effect of actinomycin D. Wilhelm Roux' Arch. *169:*314–334.

Li, H.P., and T. Yao (1945) Studies on the organizer problem in *Pelmatohydra oligactis*. III. Bud induction by the developing hypostome. J. Exp. Biol. *21:*155–160.

Li, Y.Y.F., F.D. Baker, and W. Andrew (1963) A method of tissue culture of *Hydra* cells. Proc. Soc. Exp. Biol. *113:*259–262.

Loomis, W.F. (1953) The cultivation of hydra under controlled conditions. Science *117:*565–566.

Loomis, W.F. (1954) Environmental factors controlling growth in hydra. J. Exp. Zool. *126:*223–234.

Loomis, W.F. (1964) Microenvironmental control of sexual differentiation in hydra. J. Exp. Zool. *156:*289–306.

Loomis, W.F., and H.M. Lenhoff (1956) Growth and sexual differentiation of hydra in mass culture. J. Exp. Zool. *132:*555–573.

Lowell, R.D., and A.L. Burnett (1969) Regeneration of complete hydra from isolated epidermal explants. Biol. Bull. *137:*312–320.

Lowell, R.D., and A.L. Burnett (1973) Regeneration from isolated epidermal explants. In A.L. Burnett (ed): Biology of Hydra. New York: Academic Press, pp. 223–232.

Lui, A., and D. Znidaric (1968) Das Gastroderm in Prozess der Regeneration der *Hydra*. Wilhelm Roux' Arch. *160:*1–8.

Macklin, M. (1968a) Reversal of cell layers in hydra: A critical reappraisal. Biol. Bull. *134:*465–472.

Macklin, M. (1968b) Analysis of growth factor gradients in hydra. J. Cell. Physiol. *72:*1–8.

Macklin, M. (1971) The absence of electrically demonstrable polarizing factors in hydra. J. Cell. Physiol. *77:*83–92.

Macklin, M., and A.L. Burnett (1966) Control of differentiation by calcium and sodium ions in *Hydra pseudoligactis*. Exp. Cell Res. *44:*665–668.

MacWilliams, H., and F.C. Kafatos (1968) *Hydra viridis*: Inhibition by the basal disk of basal disk differentiation. Science *159:*1246–1247.

MacWilliams, H., F.C. Kafatos, and W.H. Bossert (1970) The feedback inhibition of basal disk regeneration in hydra has a continuously variable intensity. Dev. Biol. *23:*380–398.

MacWilliams, H., and F.C. Kafatos (1974) The basal inhibition in hydra may be mediated by a diffusing substance. Am. Zool. *14:*633–645.

Marcum, B.A., R.D. Campbell, and J. Romero (1977) Polarity reversal in nerve-free hydra. Science *197:*771–773.

Marcum, B.A., and R.D. Campbell (1978a) Development of hydra lacking nerve and interstitial cells. J. Cell Sci. *29:*17–33.

Marcum, B.A., and R.D. Campbell (1978b) Developmental roles of epithelial and interstitial cell lineages in hydra: Analysis of chimeras. J. Cell Sci. *32:*233–247.

Marcum, B., T. Fujisawa, and T. Sugiyama (1980) A mutant hydra strain (SF-1) containing temperature sensitive interstitial cells. In P. Tardent and R. Tardent (eds): Developmental and Cellular Biology of Coelenterates. Amsterdam: Elsevier/North-Holland, pp. 429–434.

McConnell, C.H. (1930) The mitosis found in *Hydra*. Science *72:*170.

McConnell, C.H. (1933a) Mitosis in *Hydra*: Mitosis in the ectodermal epithelio-muscular cells of *Hydra*. Biol. Bull. *64:*86–95.

McConnell, C.H. (1933b) Mitosis in *Hydra*. Mitosis of the secretory cells of the endoderm of *Hydra*. Biol. Bull. *64:*96–102.

McConnell, C.H. (1936) Mitosis in *Hydra*. Mitosis in the indifferent interstitial cells of *Hydra*. Wilhelm Roux' Arch. *135:*202–210.

McConnell, C.H. (1938a) The hatching of *Pelmatohydra oligactis* eggs. Zool. Anz. *123:*161–174.

McConnell, C.H. (1938b) The fate of the unfertilized eggs of *Hydra attenuata.* Zool. Anz. *124:*321–324.

McConnell, C.H. (1938c) Transportation of the embryos of *Hydra attenuata.* Zool. Anz. *124:*324–333.

Meinhardt, H., and A. Gierer (1974) Applications of a theory of biological pattern formation based on lateral inhibition. J. Cell Sci. *15:*321–346.

Mookerjee, S. (1966) Descending column of cells in hydra. Indian J. Exp. Biol. *4:*239–241.

Mookerjee, S., and A. Aditya (1966) Regeneration time in irradiated hydra. Indian J. Exp. Biol. *4:*201–205.

Mookerjee, S., and A. Bhattacherjee (1966) Cellular mechanics in hydroid regeneration. Wilhelm Roux' Arch. *157:*1–20.

Mookerjee, S., and S. Bhattacharjee (1967) Regeneration time at the different levels of hydra. Wilhelm Roux' Arch. *158:*301–413.

Mookerjee, S., and J. Mitra (1970) Morphogenetic abnormality in radiated hydra. Z. Biol. 441–451.

Mookerjee, S., and A. Sinha (1959) Regionality in the inductive power of hydra tentacle. J. Exp. Zool. *141:*379–388.

Mookerjee, S., and A. Sinha (1967) System of stability and lability in hydra. Wilhelm Roux' Arch. *158:*331–340.

Moore, G.P.M., and K.E. Dixon (1972) A light and electron microscopical study of spermatogenesis in *Hydra cauliculata.* J. Morphol. *137:*483–502.

Moore, L.B., and R.D. Campbell (1973a) Bud initiation in a non-budding strain of hydra: Role of interstitial cells. J. Exp. Zool. *183:*397–408.

Moore, L.B., and R.D. Campbell (1973b) Non-budding strains of hydra: Isolation from sexual crosses and developmental regulation of form. J. Exp. Zool. *185:*73–82.

Mulherkar, L., and A.M. Therwath (1978) Induction of hypostome and tentacles in the supra basal region of *Pelmatohydra oligactis* by cysteine hydrochloride. Indian J. Exp. Biol. *16:*1253–1255.

Müller, W.A., and K. Spindler (1971) The "polarizing inducer" in hydra. A reexamination of its properties and its origin. Wilhelm Roux' Arch. *167:*325–335.

Muscatine, L., and H.M. Lenhoff (1965) Symbiosis of hydra and algae. I. Effects of some environmental cations on growth of symbiotic and aposymbiotic hydra. Biol. Bull. *128:*415–424.

Mutz, E. (1930) Transplantations versuche an *Hydra* mit besonderer Berücksichtigung der Induktion Regionalität und Polarität. Wilhelm Roux' Arch. *121:*210–271.

Newman, S.A. (1974) The interaction of the organizing regions in hydra and its possible relation to the role of the cut end in regeneration. J. Embryol. Exp. Morphol. *31:*541–555.

Noda, K. (1970a) The fate of aggregates formed by two species of hydra (*Hydra magnipapillata* and *Pelmatohydra oligactis*). J. Fac. Sci. Hokkaido Univ. *17:*432–439.

Noda, K. (1970b) On the incompatibility of two species of hydra, *Hydra magnipapillata* and *Pelmatohydra robusta,* in a mixed culture. J. Fac. Sci. Hokkaido Univ. *17:*440–445.

Normandin, D.K. (1960) Regeneration of hydra from the endoderm. Science *131:*678.

Novak, P., and H.M. Lenhoff (1980) Regulation of bud induction and site of tentacle sprouting in a non-budding strain of *Hydra viridis.* In P. Tardent and R. Tardent (eds): Developmental and Cellular Biology of Coelenterates. Amsterdam: Elsevier/North-Holland, pp. 237–242.

Otto, J.J. (1977) Orientation and behavior of epithelial cell muscle processes during hydra budding. J. Exp. Zool. *202:*307–322.

Otto, J.J., and R.D. Campbell (1977a) Budding in *Hydra attenuata:* Bud stages and fate map. J. Exp. Zool. *200:*417–428.

Otto, J.J., and R.D. Campbell (1977b) Tissue economics of hydra: Regulation of cell cycle, animal size and development by controlled feeding rates. J. Cell Sci. *28:*117–132.

Pardy, R.L. (1976) The morphology of green *Hydra* endosymbionts as influenced by host strain and host environment. J. Cell Sci. *20:*655–669.

Park, H.D. (1958) Sensitivity of hydra tissues to X-rays. Physiol. Zool. *31:*188–193.

Park, H.D., and A.B. Ortmeyer (1972) Growth and differentiation in *Hydra*. II. The effect of temperature on budding in *Hydra littoralis.* J. Exp. Zool. *179:*283–288.

Park, H.D., A.B. Ortmeyer, and D.P. Blankenbaker (1970) Cell division during regeneration in hydra. Nature *227:*617–619.

Peebles, F. (1900) Experiments in regeneration and in grafting of hydrozoa. Wilhelm Roux' Arch. *10:*435–488.

Rand, H.W. (1899) The regulation of graft abnormalities in hydra. Wilhelm Roux' Arch. *9:*161–214.

Rand, H.W., J.F. Bovard, and D.E. Minnich (1926) Localization of formative agencies in Hydra. Proc. Natl. Acad. Sci. USA *12:*565–570.

Rose, P.G., and A.L. Burnett (1968a) An electron microscopic and histochemical study of the secretory cells in *Hydra viridis.* Wilhelm Roux' Arch. *161:*281–297.

Rose, P.G., and A.L. Burnett (1968b) An electron microscopic and radioautographic study of hypostomal regeneration in *Hydra viridis.* Wilhelm Roux' Arch. *161:*298–318.

Rose, P.G., and A.L. Burnett (1970) The origin of mucous cells in *Hydra viridis.* II. Mid-gastric regeneration and budding. Wilhelm Roux' Arch. *165:*177–191.

Roudabush, R.L. (1933) Phenomenon of regeneration in everted hydra. Biol. Bull. *64:*253–258.

Rowley, H.T. (1902) Histological changes in *Hydra viridis* during regeneration. Am. Nat. *35:*579–581.

Sacks, P.G., and L.E. Davis (1979) Production of nerveless *Hydra attenuata* by hydroxyurea treatments. J. Cell Sci. *37:*189–203.

Sacks, P.G., and L.E. Davis (1980) Developmental dominance in *Hydra*. I. The basal disk. Dev. Biol. *80:*454–465.

Saffitz, J.E., A.L. Burnett, and G.E. Lesh (1972) Nervous system transplantation in hydra. J. Exp. Zool. *179:*215–226.

Sanyal, S. (1966) Bud determination in hydra. Indian J. Exp. Biol. *4:*88–92.

Sanyal, S. (1967) Cellular dynamics in the morphogenesis of hydra during bud development and regeneration. Anat. Anz. *120:*1–13.

Sanyal, S., and S. Mookerjee (1960) Cytochemistry of the cell types in hydra and their functional significance. Proc. Natl. Inst. Sci. India *26:*119–125.

Sanyal, S., and S. Mookerjee (1967) Emergence of tentacular pattern in the hypostome fragments of hydra after successive extirpations. Folia Biol. *15:*237–244.

Schaller, H.C. (1973) Isolation and characterization of a low-molecular-weight substance activating head and bud formation in hydra. J. Embryol. Exp. Morphol. *29:*27–38.

Schaller, H.C. (1975) Head activator controls head formation in reaggregated cells of hydra. Cell Diff. *4:*265–272.

Schaller, H.C. (1976a) Action of the head activator as a growth hormone in hydra. Cell Diff. *5:*1–11.

Schaller, H.C. (1976b) Action of the head activator on the determination of interstitial cells in hydra. Cell Diff. *5:*13–20.

Schaller, H.C. (1976c) Head regeneration in hydra is initiated by the release of head activator and inhibitor. Wilhelm Roux' Arch. *180:*287–295.

Schaller, H.C. (1978) Action of a morphogenetic substance from hydra. Symp. Soc. Dev. Biol. *35:*231–241.
Schaller, H.C. (1979) Neuropeptides in hydra. Trends Neurosci. *1:*120–122.
Schaller, H.C., and H. Bodenmüller (1981) Isolation and amino acid sequence of a morphogenetic peptide from hydra. Proc. Natl. Acad. Sci. USA *78:*7000–7004.
Schaller, H.C., K. Flick, and G. Darai (1977) A neurohormone from hydra is present in brain and intestine of rat embryos. J. Neurochem. *29:*393–394.
Schaller, H.C., and A. Gierer (1973) Distribution of the head-activating substance in hydra and its localization in membranous particles in nerve cells. J. Embryol. Exp. Morphol. *29:*39–52.
Schaller, H.C., C.J.P. Grimmelikhuijzen, T. Schmidt, and H. Bode (1979) Morphogenetic substances in nerve-depleted hydra. Wilhelm Roux' Arch. *187:*323–328.
Schaller, H.C., T. Rau, and H. Bode (1980) Epithelial cells in nerve-free hydra produce morphogenetic substances. Nature *283:*589–591.
Schaller, H.C., T. Schmidt, K. Flick, and C.J.P. Grimmelikhuijzen (1977a) Analysis of morphogenetic mutants of hydra. I. The aberrant. Wilhelm Roux' Arch. *183:*193–206.
Schaller, H.C., T. Schmidt, K. Flick, and C.J.P. Grimmelikhuijzen (1977b) Analysis of morphogenetic mutants of hydra. II. The non-budding mutant. Wilhelm Roux' Arch. *183:*207–214.
Schaller, H.C., T. Schmidt, K. Flick, and C.J.P. Grimmelikhuijzen (1977c) Analysis of morphogenetic mutants of hydra. III. Maxi and mini. Wilhelm Roux' Arch. *183:*215–222.
Schaller, H.C., T. Schmidt, and C.J.P. Grimmelikhuijzen (1979) Separation and specificity of action of four morphogens from hydra. Wilhelm Roux' Arch. *186:*139–149.
Schaller, H.C., T. Schmidt, C.J.P. Grimmelikhuijzen, T. Rau, and H. Bode (1980) Morphogenetic substances in nerve-free *Hydra.* In P. Tardent and R. Tardent (eds): Developmental and Cellular Biology of Coelenterates. Amsterdam: Elsevier/North-Holland, pp. 395–399.
Schincariol, A.L., J. Habowsky, and G. Winner (1967) Cytology and ultrastructure of differentiating interstitial cells in spermatogenesis in *Hydra fusca*. Can. J. Zool. *45:*591–593.
Schneider, K.C. (1890) Histologie von *Hydra fusca* mitbesonderer Berücksichtigung des nervensystems des Hydropolypen. Arch. Mikrosk. Anat. *35:*321–388.
Schmidt, T., C.J.P. Grimmelikhuijzen, and H.C. Schaller (1980) Morphogenetic substances in hydra. In P. Tardent and R. Tardent (eds): Developmental and Cellular Biology of Coelenterates. Amsterdam: Elsevier/North-Holland, pp. 495–499.
Schmidt, T., and H.C. Schaller (1976) Evidence for a foot-inhibiting substance in hydra. Cell Diff. *5:*151–159.
Schmidt, T., and H.C. Schaller (1980) Properties of the foot inhibitor from hydra. Wilhelm Roux' Arch. *188:*133–139.
Schulz, J.K.R., and G.E. Lesh (1970) Evidence for a temperature and ionic control of growth in *Hydra viridis.* Growth *34:*31–55.
Semal van Gansen, P. (1954) L'histophysiologie de l'endoderme de l'hydre d'eau douce. Ann. Soc. Zool. Belg. *85:*217–278.
Sersig, B.L., and G.E. Lesh-Laurie (1981) Bud separation in *Hydra oligactis.* Biol. Bull. *160:*431–437.
Shostak, S. (1967) Bud movement in hydra. Science *155:*1567–1568.
Shostak, S. (1968) Growth in *Hydra viridis*. J. Exp. Zool. *169:*431–446.
Shostak, S. (1972) Inhibitory gradients of head and foot regeneration in *Hydra viridis*. Dev. Biol. *28:*620–635.

Shostak, S. (1973) Evidence of morphogenetically significant diffusion gradients in *Hydra viridis* lengthened by grafting. J. Embryol. Exp. Morphol. *29:*311–330.

Shostak, S. (1974a) The complexity of hydra: Homeostasis, morphogenesis, controls and integration. Q. Rev. Biol. *49:*287–310.

Shostak, S. (1974b) Bipolar inhibitory gradients influence on the budding region of *Hydra viridis*. Am. Zool. *14:*619–632.

Shostak, S. (1975a) The budding region as source of diffusible inhibitors of head and foot regeneration in *Hydra viridis*. Wilhelm Roux' Arch. *176:*241–251.

Shostak, S. (1975b) Effects of colcemid on *Hydra viridis* with multiple peduncle grafts. Dev. Growth Diff. *17:*323–333.

Shostak, S. (1977) Vegetative reproduction by budding in hydra: A perspective on tumors. Perspect. Biol. Med. *20:*545–568.

Shostak, S. (1979) Digestive cell and tentacle number in freshly detached buds of *Hydra viridis*. Int. J. Invert. Reprod. *1:*167–178.

Shostak, S. (1981) Variation in hydra's tentacle numbers as a function of temperature. Int. J. Invert. Reprod. *3:*321–331.

Shostak, S., J.W. Bisbee, C. Ashkin, and R.V. Tammariello (1968) Budding in *Hydra viridis*. J. Exp. Zool. *169:*423–430.

Shostak, S., and M. Globus (1966) Migration of epithelio-muscular cells in hydra. Nature *210:*218–219.

Shostak, S., and D.R. Kankel (1967) Morphogenetic movements during budding in hydra. Dev. Biol. *15:*451–463.

Shostak, S., D. Medic Jr., F.A. Sproull, and C.C. Jones (1978) Tentacle number in cultured *Hydra viridis*. Biol. Bull. *155:*220–234.

Shostak, S., N.G. Patel, and A.L Burnett (1965) The role of mesoglea in mass cell movement in hydra. Dev. Biol. *12:*434–450.

Shostak, S., and R.V. Tammariello (1969) Supernumerary heads in *Hydra viridis*. Natl. Can. Inst. Monogr. *31:*739–750.

Sinha, A.K. (1965) Experimental determination of polarity in hydra. Wilhelm Roux' Arch. *157:*101–116.

Spangenberg, D.B. (1961) A study of normal and abnormal regeneration of hydra. In H.M. Lenhoff and W.F. Loomis (eds): The Biology of Hydra and of Some Other Coelenterates. Coral Gables, Florida: University of Miami Press, pp. 413–423.

Spangenberg, D.B., and R. Eakin (1961) A study of the variation in the regeneration capacity of hydra. J. Exp. Zool. *147:*259–270.

Sproull, F., and C.N. David (1979a) Stem cell growth and differentiation in *Hydra attenuata*. I. Regulation of the self renewal probability in multiclone aggregates. J. Cell. Sci. *38:*155–169.

Sproull, F., and C.N. David (1979b) Stem cell growth and differentiation in *Hydra attenuata*. II. Regulation of nerve and nematocyte differentiation in multiclone aggregates. J. Cell Sci. *38:*171–179.

Stagni, A. (1966) Some aspects of sexual polymorphism in *Chlorohydra viridissima*. J. Exp. Zool. *163:*87–92.

Stagni, A. (1974) Some aspects of sexuality in fresh-water hydras. Bull. Zool. *41:*349–358.

Stiven, A.E. (1965) The relationship between size, budding rate and growth efficiency in three species of hydra. Res. Popul. Ecol. Kyoto *7:*1–15.

Strelin, G.S. (1929) Roentgenologische Untersuchungen an Hydren. II. Die histologischen Veränderungen im Körperbau von *Pelmatohydra oligactis*. Wilhelm Roux' Arch. *115:*27–51.

Sturtevant, F.M., R.P. Sturtevant, and C.L. Turner (1951) Effect of colchicine on regeneration in *Pelmatohydra oligactis*. Science *114:*241–242.

Sugiyama, T., and T. Fujisawa (1977a) Genetic analysis and developmental mechanisms in hydra. I. Sexual reproduction of *Hydra magnipapillata* and isolation of mutants. Dev. Growth Diff. *19:*187–200.

Sugiyama, T., and T. Fujisawa (1977b) Genetic analysis of developmental mechanisms in hydra. III. Characterization of a regeneration deficient strain. J. Embryol. Exp. Morphol. *42:*65–77.

Sugiyama, T., and T. Fujisawa (1978a) Genetic analysis of developmental mechanisms in hydra. II. Isolation and characterization of an interstitial cell-deficient strain. J. Cell Sci. *29:*35–52.

Sugiyama, T., and T. Fujisawa (1978b) Genetic analysis of developmental mechanisms in hydra. V. Cell lineage and development of chimera hydra. J. Cell Sci. *32:*215–232.

Sugiyama, T., and T. Fujisawa (1979a) Genetic analysis of developmental mechanisms in hydra. VI. Cellular composition of chimera hydra. J. Cell Sci. *35:*1–15.

Sugiyama, T., and T. Fujisawa (1979b) Genetic analysis of developmental mechanisms in hydra. VII. Statistical analyses of developmental-morphological characters and cellular compositions. Dev. Growth Diff. *21:*361–375.

Taban, C.H., and M. Cathieni (1978) Stimulation of head regeneration by substance P. Experientia *34:*958.

Taban, C.H., and M. Cathieni (1979) Localization of substance P-like immunoreactivity in hydra. Experientia *35:*811–812.

Tannreuther, G.W. (1909) Observations on the germ cells of hydra. Biol. Bull. *16:*205–209.

Tardent, P. (1954) Axiale Verteilungs-Gradienten der interstitellen Zellen bei *Hydra* und *Tubularia* und ihre Bedeutung für die Regeneration. Wilhelm Roux' Arch. *146:*593–649.

Tardent, P. (1966) Zur sexualbiologic von *Hydra attenuata* (Pall.). Rev. Suisse Zool. *73:*357–381.

Tardent, P. (1968) Experiments about sex determination in *Hydra attenuata* Pall. Dev. Biol. *17:*483–511.

Tardent, P. (1972) Experimente zum Knospungsprozeb von *Hydra attenuata* Pall. Rev. Suisse Zool. *79:*355–375.

Tardent, P. (1974) Gametogenesis in the genus *Hydra.* Am. Zool. *14:*447–456.

Tardent, P., and U. Morgenthaler (1966) Autoradiographische Untersuchungen zum Problem der Zellwanderungen bei *Hydra attenuata* (Pall.). Rev. Suisse Zool. *73:*468–480.

Tardent, P., F. Rich, and V. Schneider (1971) The polarity of stenothele differentiation in *Hydra attenuata* Pall. Dev. Biol. *24:*596–608.

Tardent, P., and C. Weber (1976) A qualitative and quantitative inventory of nervous cells in *Hydra attenuata* Pall. In G.O. Mackie (ed): Coelenterate Ecology and Behavior. New York: Plenum, pp. 501–512.

Trembley, A. (1744) Memoires pour servir a l'histoire d'un genre de polypes d'eau douce à bras en forme de cornes. Leyden: J. UH. Verbeek.

Trenkner, E., K. Flick, G. Hansmann, H. Bode, and P. Bode (1973) Studies on hydra cells *in vitro.* J. Exp. Zool. *185:*317–326.

Tripp, K. (1928) Die Regenerationsfahigkeit von Hydren in den verschildenen Korperregionen nach Regenerations und Transplantationsversuchen. Z. Wiss. Zool. *132:*476–525.

Vannini, E. (1974) Introduction to some aspects of sex differentiation in pluricellular animals at a lower order of organization: Porifera, fresh-water hydras and planarians. Boll. Zool. *41:*291–326.

Venugopal, G., and C.N. David (1981a) Nerve commitment in hydra. I. Role of morphogenetic signals. Dev. Biol. *83:*353–360.

Venugopal, G., and C.N. David (1981b) Nerve commitment in hydra. II. Localization of commitment in S phase. Dev. Biol. *83:*361–365.

Venugopal, G., and C.N. David (1981c) Spatial pattern of nerve differentiation in hydra is due to a pattern of nerve commitment. Dev. Biol. *83:*366–369.

Venugopal, G., and S. Mookerjee (1977) Requirement of RNA synthesis for bud morphogenesis in hydra. Indian J. Exp. Biol. *15:*154–155.

Venugopal, G., and S. Mookerjee (1981) Selective inhibition of regeneration in *Hydra vulgaris* by actinomycin D. Int. J. Invert. Reprod. *3:*49–55.

Vögeli, G. (1972) Autoradiographische Untersuchungen zur kurzzeitigen Zellmarkierung sowie Polarität und Ausmass der individuellen Zellwanderung bei *Hydra attenuata* (Pall.). Rev. Suisse Zool. *79:*649–674.

Voland, J.R., and G.E. Lesh-Laurie (1980) The effect of actinomycin D on RNA biosynthesis and morphogenesis in hydra. In P. Tardent and R. Tardent (eds): Developmental and Cellular Biology of Coelenterates. Amsterdam: Elsevier/North-Holland, pp. 287–292.

Voland, J.R., G.E. Lesh-Laurie, and S.S. MacIntyre (1977) A procedure for the extraction and characterization of RNA from the fresh water cnidarian hydra. Comp. Biochem. Physiol. *57B:*203–208.

Wagner, G. (1905) On some movements and reactions of hydra. Q. J. Microsc. Sci. *48:*585–622.

Wanek, N., B.A. Marcum, and R.D. Campbell (1980a) Histological structure of epithelial hydra and evidence for the complete absence of interstitial and nerve cells. J. Exp. Zool. *212:*1–11.

Wanek, N., B.A. Marcum, H. Lee, M. Chow, and R.D. Campbell (1980b) Effect of hydrostatic pressure on morphogenesis in nerve-free hydra. J. Exp. Zool. *211:*275–280.

Webster, G. (1966a) Studies on pattern regulation in hydra. II. Factors controlling hypostome formation. J. Embryol. Exp. Morphol. *16:*105–122.

Webster, G. (1966b) Studies on pattern regulation in hydra. III. Dynamic aspects of factors controlling hypostome formation. J. Embryol. Exp. Morphol. *16:*123–141.

Webster, G. (1967) Studies on pattern regulation in hydra. IV. The effect of colcemid and puromycin on polarity and regulation. J. Embryol. Exp. Morphol. *18:*181–197.

Webster, G. (1971) Morphogenesis and pattern formation in hydroids. Biol. Rev. *46:*1–46.

Webster, G., and S. Hamilton (1972) Budding in hydra: The role of cell multiplication and cell movement in bud initiation. J. Embryol. Exp. Morphol. *27:*301–316.

Webster, G., and L. Wolpert (1966) Studies on pattern regulation in hydra. I. Regional differences in the time required for hypostome determination. J. Embryol. Exp. Morphol. *16:*91–104.

Weissman, A., T.L. Lentz, and R.J. Barrnett (1969) Fine structural observations on nuclear maturation during spermiogenesis in *Hydra littoralis.* J. Morphol. *128:*229–240.

Welch, P., and H.A. Loomis ((1924) A limnological study of *Hydra oligactis* in Douglas Lake, Michigan. Trans. Am. Micros. Soc. *43:*203–235.

West, D.L. (1978a) The epitheliomuscular cell of hydra: Its fine structure, three-dimensional architecture and relation to morphogenesis. Tissue Cell *10:*629–646.

West, D.L. (1978b) Ultrastructural and cytochemical aspects of spermiogenesis in *Hydra hymanae* with references to factors involved in sperm head shaping. Dev. Biol. *69:*139–154.

Westfall, J.A. (1973) Ultrastructural evidence for a granule-containing sensory-motor-interneuron in *Hydra littoralis.* J. Ultrastruct. Res. *42:*268–282.

Westfall, J.A., and J.C. Kinnamon (1978) A second sensory-motor interneuron with neurosecretory granules in hydra. J. Neurocytol. *7:*365–379.

Westfall, J.A., J.C. Kinnamon, and D.E. Sims (1980) Neuroepitheliomuscular cell and neuroneuronal gap junctions in hydra. J. Neurocytol. *9:*725–732.

Westfall, J.A., and J.W. Townsend (1977) Scanning electron stereomicroscopy of the gastrodermis of hydra. Scan. Electron Microsc. *2:*623–629.

Wetzel, G. (1898) Tranplantationversuche mit hydra. Arch. Mikrosk. Anat. *52:*70–96.

Wilby, O.K., and G. Webster (1970a) Studies on the transmission of hypostome inhibition in hydra. J. Embryol. Exp. Morphol. *24:*583–593.

Wolpert, L. (1969) Positional information and the spatial pattern of cellular differentiation. J. Theor. Biol. *25:*1–47.

Wolpert, L. (1971) Positional information and pattern formation. Curr. Top. Dev. Biol. *6:*183–244.

Wolpert, L., J. Hicklin, and A. Hornbruch (1971) Positional information and pattern regulation in regeneration of hydra. Symp. Soc. Exp. Biol. *25:*391–415.

Wolpert, L., A. Hornbruch, and M.R.B. Clarke (1974) Positional information and positional signalling in hydra. Am. Zool. *14:*647–663.

Wood, R.L. (1977) The cell junctions of hydra as viewed by freeze-fracture replication. J. Ultrastruct. Res. *58:*299–315.

Wood, R.L. (1979a) The fine structure of the hypostome and mouth of hydra. I. Scanning electron microscopy. Cell Tissue Res. *199:*307–317.

Wood, R.L. (1979b) The fine structure of the hypostome and mouth of hydra. II. Transmission electron microscopy. Cell Tissue Res. *199:*319–338.

Wood, R.L. (1980) Freeze fracture studies on cell junction formation in regenerating hydra. In P. Tardent and R. Tardent (eds): Developmental and Cellular Biology of Coelenterates. Amsterdam: Elsevier/North-Holland, pp. 447–452.

Wood, R.L., and A.M. Kuda (1980a) Formation of junctions in regenerating hydra: Septate junctions. J. Ultrastruct. Res. *70:*104–117.

Wood, R.L., and A.M. Kuda (1980b) Formation of junctions in regenerating hydra: Gap junctions. J. Ultrastruct. Res. *73:*350–360.

Yao, T. (1945) Studies on the organizer problem in *Pelmatohydra oligactis.* J. Exp. Zool. *21:*147–160.

Yaross, M.S., and H.R. Bode (1978a) Regulation of interstitial cell differentiation in *Hydra attenuata.* III. Effects of I cell and nerve cell densities. J. Cell Sci. *34:*1–26.

Yaross, M.S., and H.R. Bode (1978a) Regulation of interstitial cell differentiation in *Hydra attenuata.* IV. Nerve cell commitment in head regeneration is position-dependent. J. Cell. Sci. *34:*27–38.

Yaross, M.S., and H.R. Bode (1978c) Regulation of interstitial cell differentiation in *Hydra attenuata.* V. Inability of regenerating head to support nematocyte differentiation. J. Cell Sci. *34:*39–52.

Yasugi, S. (1974) Observations on supernumerary head formation induced by lithium chloride treatment in the regenerating hydra, *Pelmatohydra robusta.* Dev. Growth Diff. *16:*171–180.

Zawazzin, A.A. (1929) Röntgenologische Untersuchungen an Hydren. I. Die Wirkung der Röntgenstrahlen auf die Vermehrung und Regeneration von *Pelmatohydra oligactis.* Wilhelm Roux' Arch. *115:*1–26.

Zihler, J. (1972) Zur Gametogenese und Befruchtungsbiologie von Hydra. Wilhelm Roux' Arch. *169:*239–267.

Znidaric, D. (1970) Comparison of the regeneration of the hypostome with the budding process in *Hydra littoralis.* Wilhelm Roux' Arch. *166:*45–53.

Znidaric, D. (1971a) Regeneration of the foot. I. *Hydra littoralis.* Z. Mikroskanat. Forsch. *84:*503–510.

Znidaric, D. (1971b) Regeneration of the foot. II. *Hydra pseudoligactis.* Z. Mikroskanat. Forsch. *84:*511–519.

Znidaric, D., and A. Lui (1969) Dedifferentiation of gland cells in hydra and further development of interstitial cells arising from them. Wilhelm Roux' Arch. *162:*374–383.

Zumstein, A. (1973) Regulation der Nematocyten—Production bei *Hydra attenuata.* Wilhelm Roux' Arch. *173:*294–308.

Zumstein, A., and P. Tardent (1971) Beitrag zum Problem der Regulation der Nematocyten Production bei *Hydra attenuata* Pall. Rev. Suisse Zool. *78:*505–714.

Zumstein, A., and P. Tardent (1972) Autoradiography of whole-mounts of isolated ectoderm of the body column of hydra. A new method. Experientia *28:*1124-1125.

Developmental Biology of Freshwater Invertebrates, pages 129–150

Development of the Freshwater Medusa *Craspedacusta sowerbii*

Charles F. Lytle

ABSTRACT The freshwater medusa *Craspedacusta sowerbii* exhibits a complex life cycle with both polyp and medusa stages and two distinct larval types. The polyp stage produces three different types of buds: 1) polyp buds, which form new polyps that remain attached to the parent to form a colony; 2) frustule buds, a form of asexual larva which separates from the parent and forms a new polyp; and 3) medusa buds, which detach to become the free-swimming medusae. The major features of embryonic development are described, and methods for the collection and culturing of polyps and medusae are presented.

INTRODUCTION

The freshwater medusa *Craspedacusta sowerbii* (Fig. 1) provides interesting material for the study of several important aspects of invertebrate development. Like other cnidarians (coelenterates) *C. sowerbii* exhibits a simple diploblastic structure and a simple level of metazoan organization. Despite its simple organization, *C. sowerbii* demonstrates a complex pattern of sexual and asexual reproduction and development. As pointed out by Campbell ('74), cnidarians exhibit a high degree of reversibility of morphogenetic processes and plasticity of development. This plasticity of developmental processes is clearly evident in *C. sowerbii.*

C. sowerbii is unusual in that it is the only widely distributed freshwater medusa in the world and one of the few members of the phylum Cnidaria to colonize fresh water successfully. A few other species of *Craspedacusta* occur in the Far East (Kramp, '50) and in Africa and India (Jones, '51; Bouillon, '58; Thiel, '73), but several closely related forms (*Gonionemus, Vallentinia*) are marine. All of these genera are presently considered to be members of the order Limnomedusae, class Hydrozoa, of the phylum Cnidaria (Coelenterata).

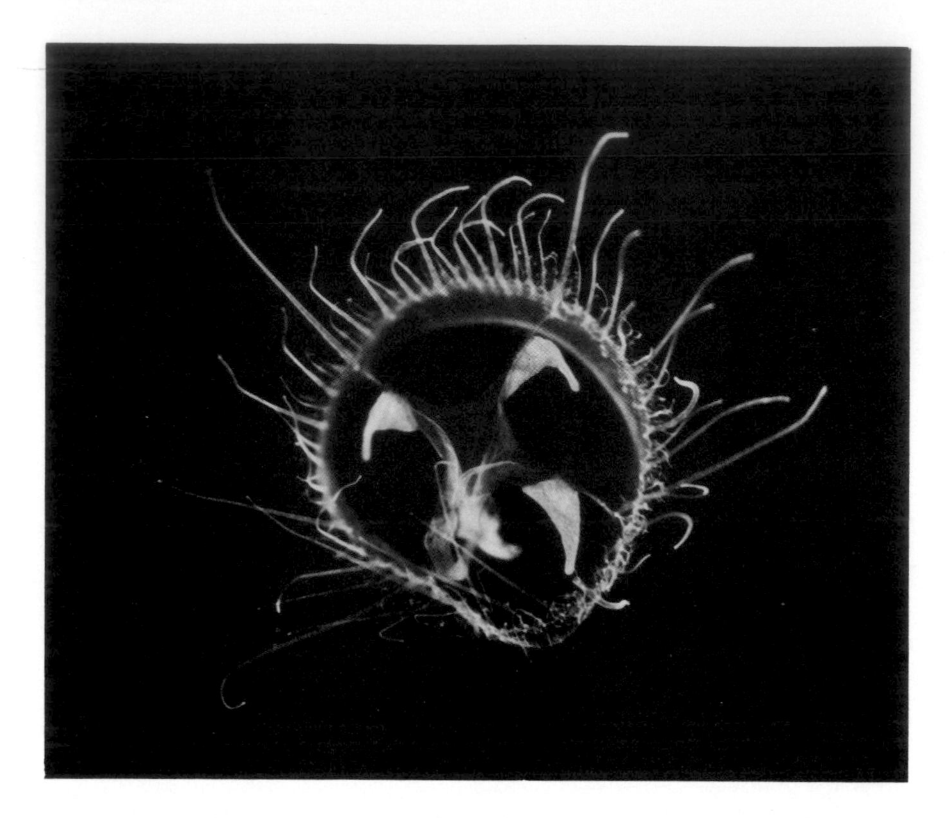

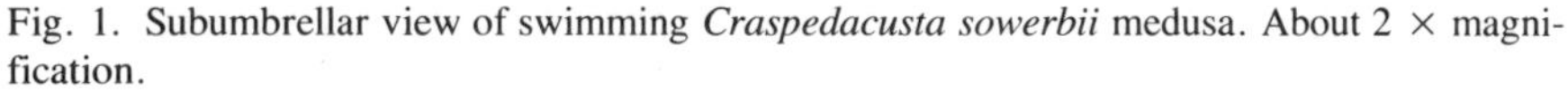

Fig. 1. Subumbrellar view of swimming *Craspedacusta sowerbii* medusa. About 2 × magnification.

Unlike the freshwater *Hydra, Craspedacusta* has both a medusa stage and a polyp stage in its life cycle, thus resembling many of the marine hydrozoans. The polyps exhibit several morphogenetic processes associated with asexual reproduction and the formation of the free-swimming medusa. Fertilized eggs and embryos are difficult to obtain, as explained later, but for the fortunate investigator embryological studies offer many opportunities for the extension of present knowledge.

LIFE CYCLE AND NATURAL HISTORY

C. sowerbii has a complex life cycle including a free-swimming medusa stage, a sessile polyp stage, and two distinct larval stages (Fig. 2). The organism thus demonstrates several different types of development. The

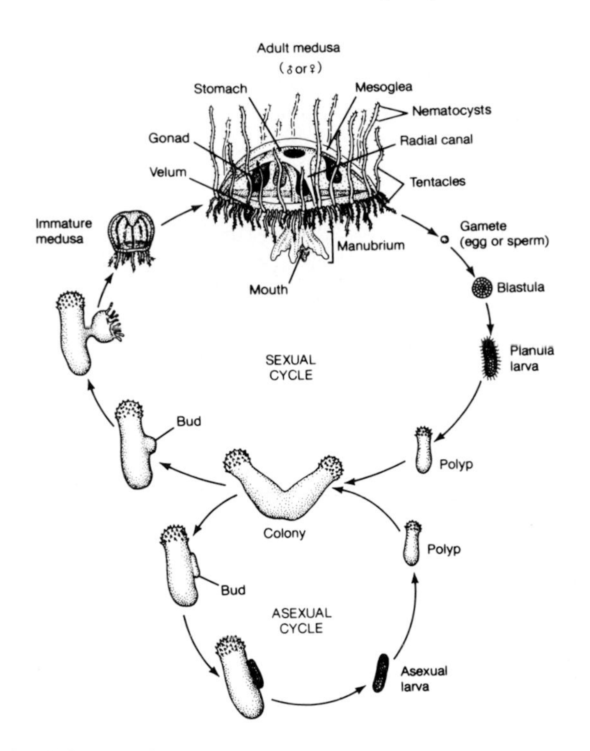

Fig. 2. Life cycle of *C. sowerbii*. From "Five Kingdoms: An Illustrated Guide to the Phyla of the Earth," by Lynn Margulis and Karlene V. Schwartz. W.H. Freeman and Company. Copyright© 1982. Drawn by L. Meszoly with information from C.F. Lytle.

medusae bear four gonads suspended from the radial canals into the subumbrellar cavity. Gametes develop on the outer surface of the gonads. Since there are no apparent secondary sexual characteristics, sex of the medusae can be determined only by microscopic examination of gonadal tissue.

In North America, as in most parts of the world, populations of *C. sowerbii* often appear to be monosexual (Rice, '58). For this reason sexual reproduction and normal embryological development have rarely been observed and little studied. Most populations appear to be maintained through asexual reproduction of the polyp stage.

When both male and female medusae are present, however, the pattern of early embryological development is similar to that reported for many other hydrozoans. After the eggs are shed from the adults, the eggs are fertilized and undergo several equal cleavages to form a blastula, gastrula, and later transform into a nonciliated planula larva. The planula settles to the substrate, moves about for a short time, attaches at one end, and develops into a polyp.

The polyp grows in size and can produce three types of buds: polyp buds, frustules, and medusa buds (Reisinger, '34; Lytle, '61). Polyp buds normally remain attached to the parent to form a small colony; frustules are asexual larvae that detach and metamorphose into polyps, and medusa buds grow and differentiate into free-swimming medusae.

C.sowerbii is widely distributed in most parts of the world except India and Africa. Despite its widespread distribution, however, the species is sporadic in appearance. It appears in some years but seems to be absent in others. The medusae, commonly but erroneously called "jellyfish," appear seasonally during the warm months of the year. At other times the species survives as a microscopic, nearly transparent, sessile polyp.

The polyps of *C. sowerbii* are small, transparent, and lack tentacles (Fig. 3). They live attached to submerged sticks, stones, and other objects. Polyps may consist of one to several hydranths joined at the base, usually with a continuous gastrovascular cavity. In nature colonies usually have from two to five hydranths, but colonies up to 24 hydranths (Fig. 4) have been produced under experimental conditions (Lytle, '60).

Polyps are permanently attached to the substrate and do not reattach if dislodged. The polyps feed on small worms, rotifers, crustaceans, and small insect larvae associated with the substrate community (see Payne, '24; Dejdar, '34; Bushnell and Porter, '67). Since they lack tentacles, the polyps depend chiefly on chance contact of motile prey for their food supply. A cluster of nematocysts surrounding the mouth serves to stun and hold the prey, which is then engulfed through the mouth by a series of contractions of the oral region.

The medusae (Fig. 1) are nearly transparent and are most commonly observed near the surface of a lake or pond. Adult medusae usually range from 10 to 20 mm in diameter. Many tentacles arise from the margin of the bell. In the center of the lower (subumbrellar) surface of the bell is the stomach region of the gastrovascular cavity. This chamber narrows below and terminates in a smaller manubrium, which bears a mouth surrounded by four oral lobes. On the inner margin of the bell is a thin, circular flap of tissue, the velum.

Extending outward from the stomach are four radial canals. These radial canals connect with a circular canal which extends around the circumference of the bell. Together the cavities of the manubrium, stomach, radial canals, and the circular canal comprise the gastrovascular cavity. The radial canals bear four gonads which hang downward into the subumbrellar cavity. The shape of the gonads varies widely in different populations of medusae. In large specimens the gonads may be quite elongate and extend several millimeters below the margin of the bell.

The activity of the medusae is influenced by light and temperature as reported by Milne ('38) and Deacon and Haskell ('67). Swimming move-

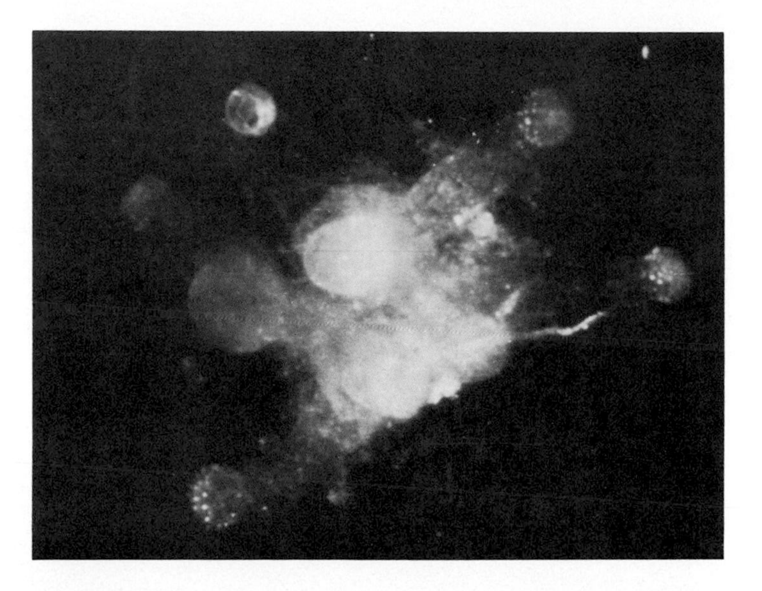

Fig. 3. Polyp stage of *C. sowerbii* with five hydranths (individuals) and four medusa buds (white spherical areas).

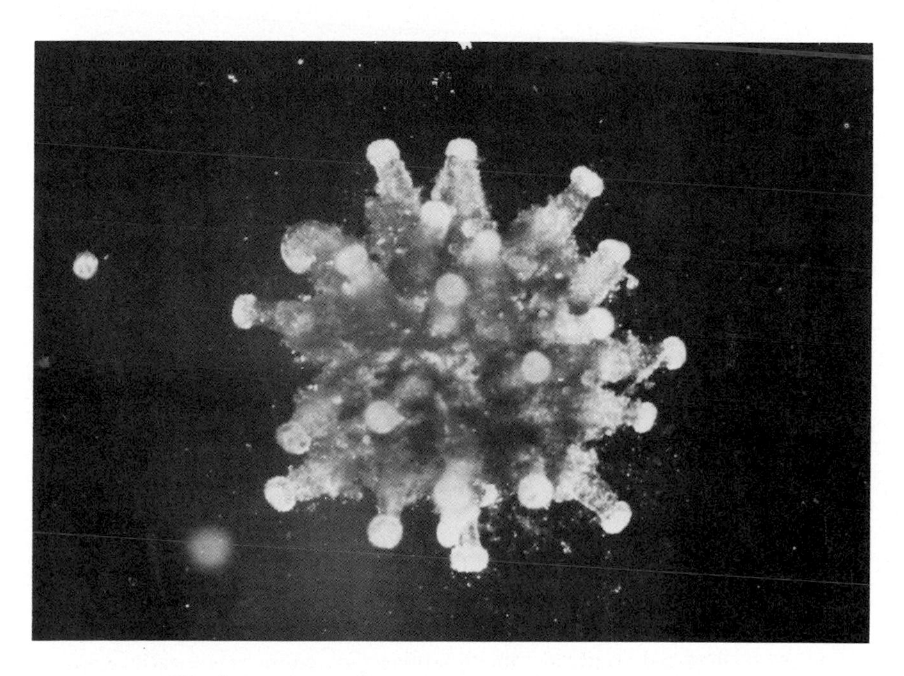

Fig. 4. Large experimental polyp colony with 22 members.

ments are stimulated by increasing light intensity and higher temperatures. When active, the medusae swim vertically by a series of rhythmic contractions of the bell until they reach the surface. Upon reaching the surface they cease active swimming and float slowly downward with the larger tentacles extending upward or downward (Fig. 5). These larger tentacles appear to aid in stabilizing the bell so that the downward movement is very nearly vertical. The descent may occur either with the manubrium (subumbrella) directed downward or with the manubrium directed upward; the tentacles extend rigidly in the opposite direction from the manubrium in either case.

The chief food of the medusae appears to be rotifers, copepods, cladocerans and other small zooplankters (Davis, '55).

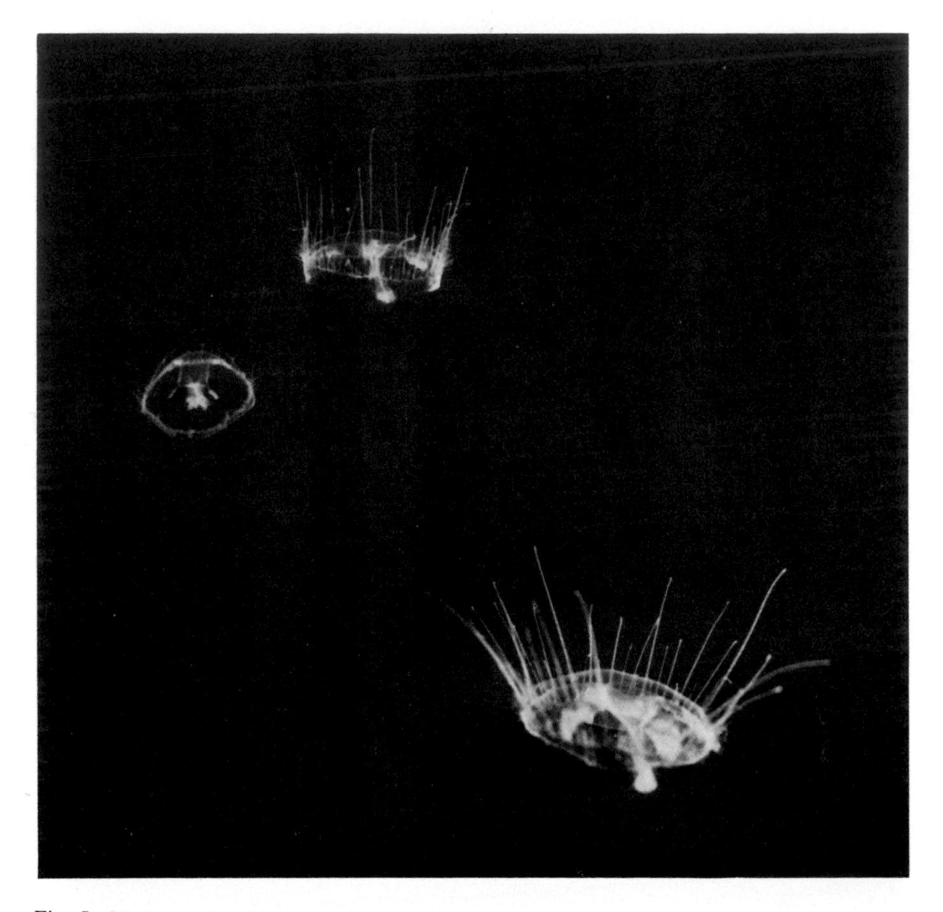

Fig. 5. Medusae showing the typical downward floating posture with large tentacles erect and short tentacles arched around margin of bell.

The medusae of *C. sowerbii* are distinctly seasonal in appearance. In most regions of North America they appear only during the warmer months of the year. Most recorded collections have been between July and October (Dexter et al., '49; Lytle, '60, '62). In warm regions, however, medusae may appear at other times. Deacon and Haskell ('67) observed medusae between October and January in Lake Mead, Nevada.

C. sowerbii is unusual in that in most parts of the world it has been found principally in relatively recent impoundments. More rarely has it been found in natural lakes or in streams and rivers (Payne, '24; Pennak, '56; Lytle, '60, '62; Bushnell and Porter, '67; Acker and Muscat, '76). Kramp ('50) provided convincing evidence that *C. sowerbii* is native to the Yangtze River basin of China, whence it has been transported nearly world wide in recent times in connection with human activities.

SOURCES OF MATERIAL

Locating a source of living *C. sowerbii* is a significant problem for many investigators. There are no commercial sources for this species, so cultures must be obtained from current investigators or collected from nature.

Collection from nature can be difficult because of the comparative rarity of the species and its sporadic appearances and disappearances. In some locations *C. sowerbii* has been known to persist for many years, but in other locations it has been found only during one or a few years. The species has been found in most parts of the United States and several places in southern Canada, and in South America, Europe, Australia, China, and Japan. Compilations of locations in the United States have been provided by Schmitt ('39), Dexter et al. ('49), Pennak ('56), Lytle ('60, '62), and Acker and Muscat ('76). Some more recent reports of *C. sowerbii* in various areas of the United States and Canada are provided in Table I.

Collection Techniques for Medusae

Medusae can be collected easily with a dip net from surface waters when they are active. They are most active during the day when the water is warmed by solar radiation. They react positively to light and warm temperatures. Medusae tend to be less active at night and during cloudy or overcast periods. During inactive periods they tend to settle at the bottom. Active stirring with a dip net or pole during such periods will sometimes bring them to the surface, where they can be collected more easily.

The medusae are rarely found in water shallower than 1–2 meters, so it is usually necessary to collect them from a dock or a boat. Care must be taken to avoid damaging the fragile medusae against the mesh of the net. They should be transferred to a jar, bucket, or other container with water from the collection site for transport.

TABLE I. Some Additional Locations Where *C. sowerbii* Has Been Collected

Area	Reference
United States	
Arizona	Kynard and Tasch ('74)
California	Arnold ('51, '68)
Hawaii	Matthews ('63, '66)
Illinois	Walley ('72)
	Lipsey and Chimney ('78)
Indiana	Eberly ('72)
Michigan	Bushnell and Porter, ('67)
	Smrchek ('70)
Nevada	Deacon and Haskell ('67)
Ohio	Hubschman and Kishler ('72)
	Acker and Muscat ('76)
	Beckett and Turanchik ('80)
Oklahoma	Kimmel et al. ('80)
Tennessee	Pennington and Fletcher ('80)
Wisconsin	Howmiller and Ludwig ('72)
Canada	
Ontario	Wiggins et al. ('57)

Medusae have also been collected from greater depths by scuba divers or by towing a plankton net beneath the surface. Specimens collected in the latter fashion, however, are often badly damaged.

Culture of Medusae

Most investigators have found the medusae difficult to culture or to maintain for extended periods in the laboratory. Most experimental work has therefore been done with recently collected medusae. Medusae may be held for several days in aquaria but become progressively inactive and slowly degenerate. Pennak ('56) briefly describes some aspects of the progressive degeneration.

Most workers have had limited success in feeding medusae in aquaria, but Reisinger ('57) reported that he was able to maintain sexually mature medusae for more than 3 months in the laboratory. He fed them chopped *Tubifex,* using a pipette to deliver food to the manubrium of a quiescent medusa. Reisinger also describes a method for rearing newly liberated medusae to sexual maturity. He fed the tiny young medusae illoricate rotifers (*Synchaeta pestinata* worked best) on the first day, and subsequently fed them copepod nauplii. He later switched to a diet of chopped *Tubifex* as the medusae approached maturity.

Several workers have found infestations of *Hydrameba* on the medusae collected in nature (see Rice, '60). Hausmann ('70) reported another ectoparasite, *Trichodina,* on *C. sowerbii* medusae. Specimens should be examined for such infestations before attempts at laboratory culture.

Collection Techniques for Polyps

Polyps of *C. sowerbii* are most easily collected by the submerged slide technique used to collect diatoms and sessile invertebrates (Woodhead, '43; Welch, '48).

Glass microscope slides are placed 2–3 slots apart in a wooden microscope slide box or similar rack and secured with rubber bands and/or a cord. If a microscope slide box is used, the solid bottom should be removed and replaced by one or two wooden strips to facilitate water flow through the rack. The rack is then weighted and suspended 1–2 m below the surface for one to several weeks. Larval settlement on the slides (frustules or planulae) will result in the development of polyps which are permanently attached to the glass slides. Welch ('48) gives a method for constructing a similar sampling device using wire mesh covers.

The polyps are located after removal of the slides from the sampling device by examining the slides under a dissecting microscope using transmitted light. Side illumination with the light source carefully adjusted to give a combination of transmitted and reflected light on the slide greatly facilitates location of the transparent polyps.

Polyps can also be collected by removing from the habitat submerged sticks, stones, or vegetation covered with a thin algal mat. In limestone quarries with vertical walls, I have had some success in finding polyps among algal scrapings from the side walls. When the algal mat is gently teased apart under a dissecting microscope, polyps can sometimes be found. Once polyps are located they can be used as a source of larvae for the initiation of laboratory cultures.

Culture Techniques for Polyps

Subcultures can be initiated by transferring frustule larvae into a small Petri dish or a watch glass containing pond water at about 18–20°C. After transfer the frustules settle to the bottom, attach by a mucous secretion, and begin to move about slowly on the bottom of the dish. These asexual larvae lack cilia and move along the substrate in a peculiar wormlike creeping fashion described by Kuhl ('47) and Crowell and Lytle ('55). After a few days the frustule ceases movement, attaches at the anterior end, and differentiates a mouth at the posterior end. Further differentiation results in the formation of a primary polyp (Fig. 6).

Daily water changes in the cultures should be initiated after the larvae have attached to the substrate. Pond water or dechlorinated tap water can be used for the cultures. Slowly running water produces the best results. Culture

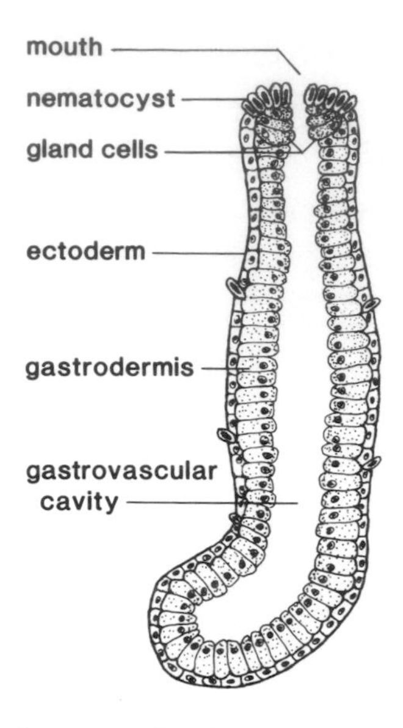

Fig. 6. Longitudinal section of a young polyp showing histological organization. After Payne ('24).

dishes can be placed in a shallow glass tray through which charcoal-filtered tap water slowly flows. Care must be taken to avoid strong flow, which may dislodge the larvae or wash away newly formed larvae produced in the cultures.

Polyps may be fed small crustacea, crustacean larvae, rhabdocoel flatworms, or oligochaetes. I have had the greatest success using the oligochaete *Aeolosoma,* which is easily cultured on rice-agar plates (Lytle, '59). Other workers have used other types of food. Reisinger ('57) and Dejdar ('34) used rotifers and chopped *Tubifex,* and McClary ('59) used *Artemia* nauplii.

Feeding on alternate days produces active growth and reproduction. Culture dishes should be removed from the flowing water, and food organisms added to each dish. After 20–30 min the cultures should be checked under a stereomicroscope. Those polyps that have not succeeded in capturing a food organism should be aided by pushing the food into contact with the mouth region with a pair of fine forceps or a glass rod. Simply providing food in the culture dish does not necessarily insure feeding.

NORMAL PATTERNS OF DEVELOPMENT

Embryological Development

Very little research has been done on the embryological development of *C. sowerbii* since few workers have had access to fertilized eggs and em-

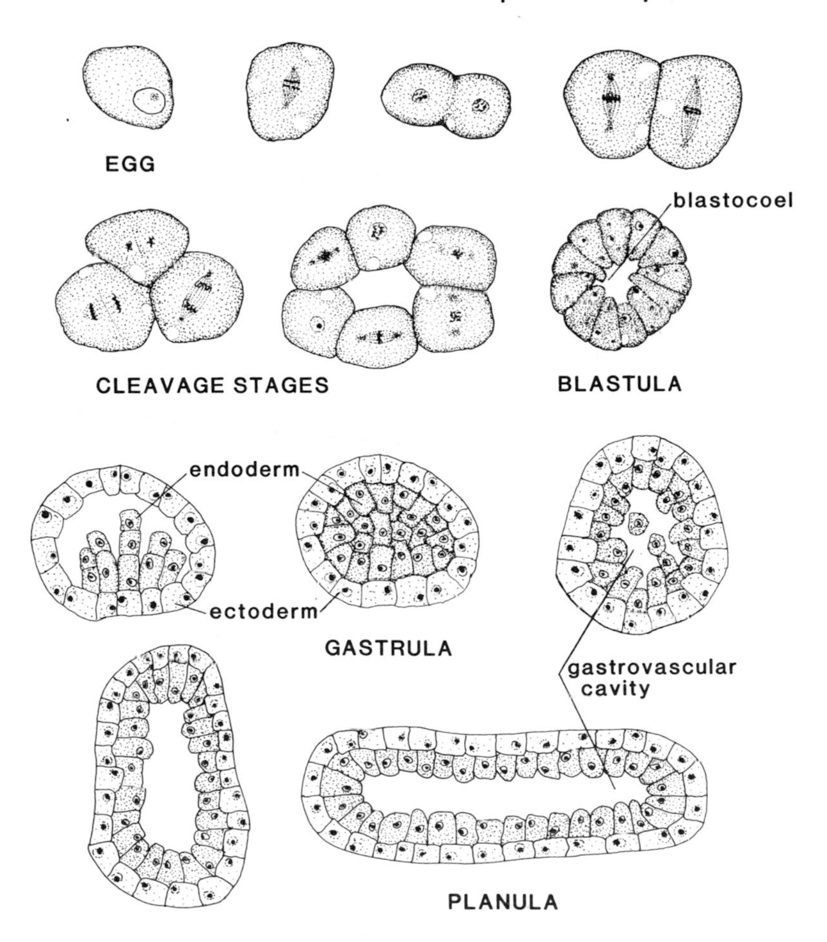

Fig. 7. Embryological development of *Craspedacusta.* Partly after Payne ('26).

bryos. Payne ('26) has provided the only published account of embryological development. The following account and Figure 7 are based on Payne's report and my own limited observations on material collected in Swift Creek, Monroe Country, Indiana. Development is generally similar to that in *Gonionemus* as described by Joseph ('25) and Kumé and Dan ('68).

Mature eggs released into the water have relatively little yolk. Sperm released from male medusae apparently fertilize the newly liberated eggs. Ovarian eggs appear to be arrested in the first meiotic prophase until fertilization. Nothing is known about spawning behavior in this species.

The first embryonic cleavage is vertical, total, and equal. I have also observed that, as noted by Payne, there is little synchrony of division among the blastomeres after the first division. Three cell stages are occasionally observed, and later stages often show nuclei of the cells in various stages of the cell cycle.

Cleavages subsequent to the first continue to be equal and produce blastomeres of equal size. A central cavity appears early in cleavage. Further divisions produce a blastula with a small central cavity. The cells of the blastula are broad at the surface and taper toward the central blastocoel. The blastula is ciliated and free-swimming.

The method of endoderm formation in this species is not clear. Payne describes it as a form of polar ingression in which cells proliferate on one side (presumably the vegetal pole) and migrate into the blastocoel. His Figure 5, however, and my observations of similar stages seem more to suggest of the delamination of endoderm from one side of the gastrula. Subsequently the blastocoel becomes filled with endoderm cells and forms a stereogastrula.

Later the gastrovascular cavity appears as a new space within the endoderm, and further division results in elongation of the embryo, thus transforming it into a planula larva. Cilia appear to be lost from the embryo at the onset of elongation. The planulae settle to the substrate and move about in the creeping fashion described for frustules prior to differentiating into a polyp. The planula ceases creeping, attaches at the anterior end (formerly the animal pole), and forms a mouth at the posterior end (former vegetal pole). The new oral end then grows upward so the mouth and capitulum surrounding the mouth are raised above the substrate. Nematocysts appear in a circle around the mouth. No tentacles are normally formed, and nematocysts are borne directly on the capitulum.

Asexual Reproduction and Development

Most studies of development in *C. sowerbii* have centered around asexual reproduction and associated morphogenesis (Reisinger, '57; McClary, '59; Lytle, '61).

The polyps of *C. sowerbii* offer an interesting opportunity to study the control of morphogenesis in asexual development because they produce three morphologically distinct types of buds: 1) hydranth or polyp buds, 2) frustules, and 3) medusa buds (Fig. 8). All three bud types originate as evaginations of the polyp wall and subsequently exhibit varying degrees of cell and tissue differentiation. The frustule bud exhibits the lowest degree of differentiation, and the medusa bud exhibits the greatest differentiation.

The formation of polyp buds was briefly described by Payne ('24), Persch ('33), and Dejdar ('34). Bouillon ('58) has provided a similar description of the process in the African freshwater medusa *Limnocnida*.

The typical development of a polyp bud is shown in the left column of Figure 8. The first clear indication of a new hydranth bud is a small bulge

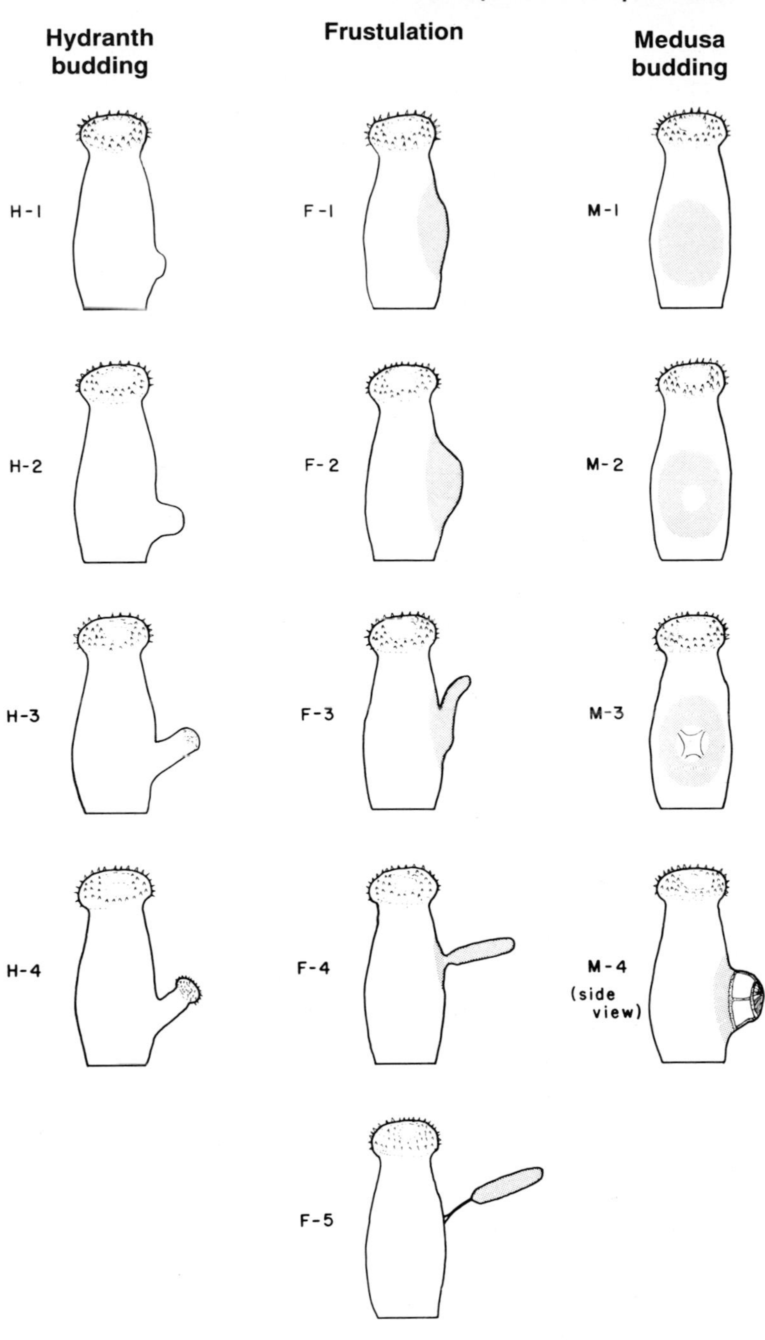

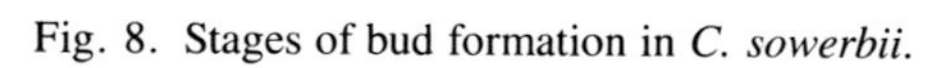
Fig. 8. Stages of bud formation in *C. sowerbii*.

on the side of the parent, usually near its base (stage H-1). I have observed a slight accumulation of optically dense granules in and around the young bud in some cases. The density of these granules in young hydranth buds is much less than in the other two types of buds, and often new hydranths are formed without any evident concentration of them. The granules will be discussed in more detail later.

The young bud elongates and enlarges slightly (stage H-2), after which nematocysts appear at the distal end (stage H-3). The final steps in the process involve the formation of the mouth, further development of nematocysts, and the expansion of the capitulum (stage H-4). The new hydranth becomes capable of feeding and begins to assume the typical flask shape at this time. It continues to increase in size, and within a few days (about 10 days at 27°C) it reaches full size and becomes indistinguishable from the other members of the colony.

The formation of frustules in *C. sowerbii* has been described by several workers, notably Payne ('24), Persch ('33), Moser ('30), Dejdar ('34), and Kuhl ('47). Frustules or nonciliated planuloid buds formed by vegetative processes are known in several other genera of hydroids including *Obelia, Campanularia, Corymorpha, Limnocnida, Vallentinia,* and *Gonionemus.*

The typical formation of a frustule is first indicated by an accumulation of opalescent granules within the wall of the parent hydranth (stage F-1); the distal portion of this region then bulges out to form an elongated ridge on the side of the hydranth (stage F-2). The distal end of the frustule becomes free and bends away from the axis of the hydranth early in the process (stage F-3). This separation continues, and nearly the entire frustule is free before the proximal invagination can be observed (stage F-4).

The newly formed larva is sausage-shaped with a solid endodermal core and a single outer layer of ectodermal cells (Fig. 9). Finally, the frustule separates from the parent by breaking the thin terminal connection (stage F-5). Later it attaches at one end to form a new polyp.

In the most extreme cases, the frustule actually seems to begin as a dark bulge on the side of the parent hydranth, and the distal end (future anterior end of the creeping frustule and the base of the subsequent polyp) then appears to be molded directly into cylindrical form as the bud increases in length. This molding process continues until nearly all of the frustule-forming material is shaped into the small sausage-shaped larva extending at an angle from the side of the hydranth. Only then is there any evidence of a second "invagination" or constriction to mark the proximal end of the frustule. Regardless of the variations in the earlier steps of frustule formation, the frustule ultimately has only a terminal attachment prior to release.

Payne ('24) gave a slightly different account of the formation of a frustule. According to his description, the first indication of a frustule is the develop-

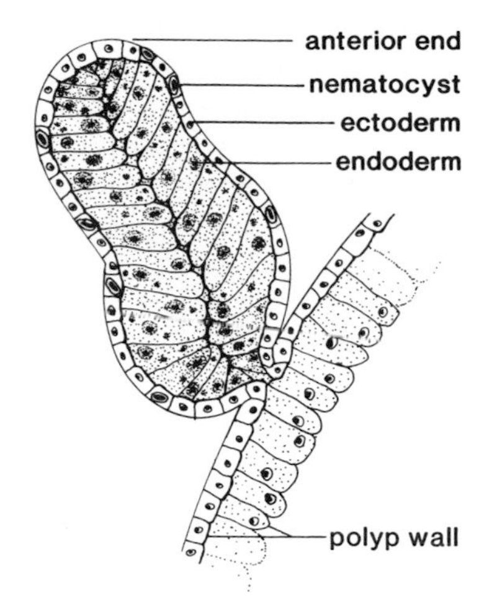

Fig. 9. Longitudinal section of a frustule prior to release from polyp. After Payne ('24).

ment of an opaque area along the side of the hydranth. By studying fixed and stained preparations, he found that this opacity is due to the accumulation of small, deeply staining basophilic granules within the cells of the endoderm. The further formation of a frustule was described largely as the result of two invaginations. The first marks the distal end of the frustule and pushes inward and proximally. Later a second invagination is initiated, marking the proximal end of the frustule. These two invaginations gradually approach each other and eventually pinch off the bud.

Payne stated that the distal invagination "does not push the bud out much, if any, beyond the periphery of the polyp," and indicated this in his illustrations. The illustrations of Persch and Kuhl, however, clearly show extensions of the bud beyond the periphery of the polyp wall. Most of the frustules formed in my cultures have closely resembled those illustrated and described by Persch and Kuhl; however, a few were also formed by the process described by Payne.

The formation of medusa buds (gonophores) by the polyps has been described by several workers, including Fowler ('90), Potts ('06, '08), Payne ('24), and Reisinger ('57). The first indication of medusa bud formation in a living animal that can be observed under the binocular microscope (Fig. 8) is the accumulation of opalescent granules in a region near the middle of the hydranth (stage M-1).

The accumulation of granules in the presumptive medusa bud is less localized than in the case of the early frustule bud, and the density of the granules within the presumptive medusa bud is somewhat less. The granules in the early frustule bud are very densely packed into a small area on the side of the hydranth, whereas the granules of an early medusa bud appear as a broad "plaque" sometimes distributed over half the side of a hydranth. This difference is more apparent in animals grown at 20–23°C than in animals grown at 26–27°C. The accumulation of granules in the early hydranth bud is much less than in either of the other two types, and is usually noticed only after some outgrowth from the polyp wall has occurred.

The next stage of the medusa bud is marked by the appearance of a clear area in the center of the medusa "plaque" which is devoid or nearly devoid of the granules (stage M-2). This stage is followed by the first appearance of tetraradial symmetry indicated by rudimentary radial canals at the edges of this clear area (stage M-3). The medusa bud begins to enlarge rapidly at this point, and subsequent stages are based on the diameter of the growing bud. The medusa bud at stage M-4 has a diameter of 0.17 mm; at stage M-5 it has a diameter of 0.33 mm, etc.

The buds at the time of release are larger at lower temperatures (19–23 and 20°C) than at higher temperatures. After release and the opening of the velum, the young medusa flattens out and increases considerably in diameter, usually measuring 0.85–1.0 mm in diameter shortly after liberation.

In well-nourished animals I have always found medusa buds to be formed on the lateral wall of the polyps. Payne and Reisinger both describe and illustrate medusa buds forming at the distal end of the polyp. I have observed such terminal medusa buds and subsequent regression of the polyp only in cases where the animals were inadequately fed. Properly nourished polyps are capable of forming several successive medusa buds.

Payne ('24) and Reisinger ('57) provide detailed accounts of the histological development of a medusa bud. The principal features are illustrated in Figure 10.

The medusa bud first appears as an ectodermal thickening on the lateral wall of the polyp (Fig. 10-1). Continued division of both ectodermal and endodermal cells in the medusa primordium results in an invagination in the polyp wall (Fig. 10-2). A cavity forms within the invaginated ectodermal cells (Fig. 10-2); this cavity is the future subumbrellar cavity of the medusa and the ectodermal cells surrounding it constitute the entocodon (Glockenkern).

The entocodon enlarges as the medusa bud grows out from the side of the polyp. Four radial canals form between the endoderm lateral to the entocodon and the adjacent endoderm of the polyp wall (Fig. 10-4a,b).

Reisinger describes two types of medusa buds: 1) one in which the ectoderm of the entocodon is completely separated by endoderm from the outer ectoderm of the medusa bud (Agassiz-type, Fig. 10-4a, 5a); and 2) one

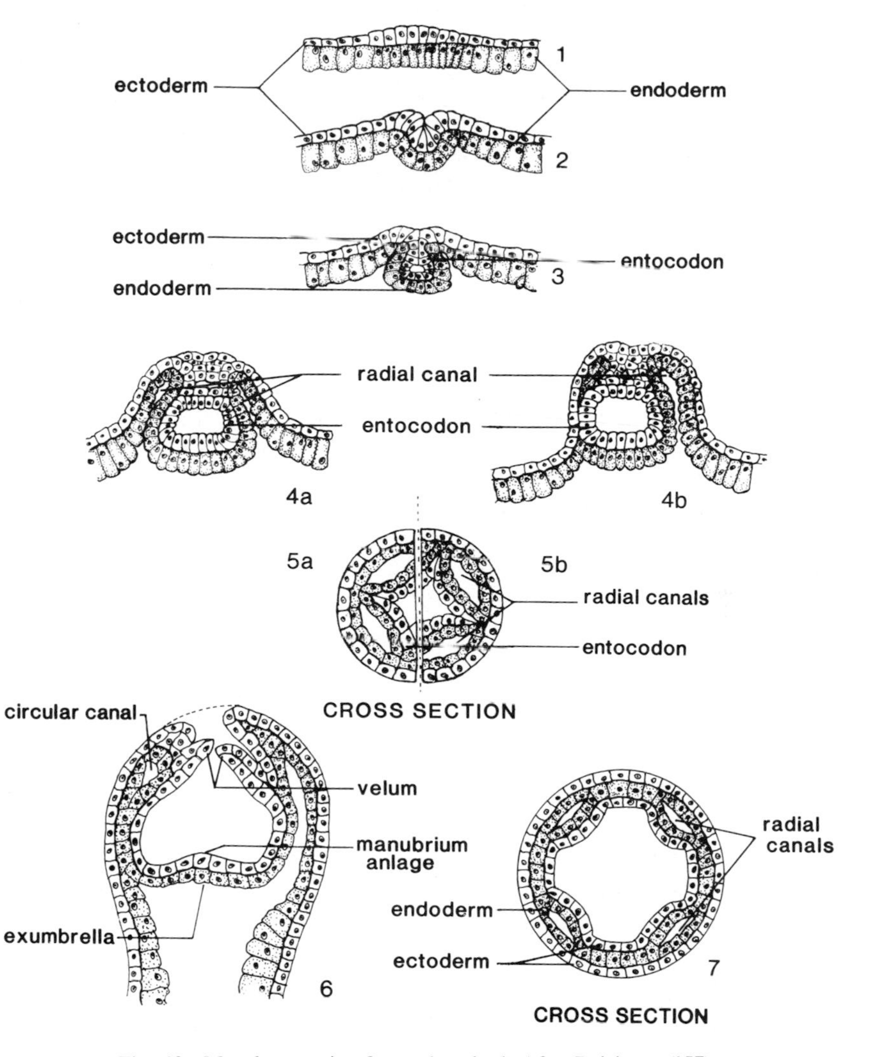

Fig. 10. Morphogenesis of a medusa bud. After Reisinger ('57).

in which the ectoderm of the entocodon is in contact with the outer ectoderm between adjacent radial canals (Goette-type, Fig. 10-4b, 5b).

A later stage is illustrated in Figures 10-6 and 10-7, which shows the formation of the circular canal, the separation of the distal end of the entocodon to form the velum, and the initiation of the manubrium anlage on the proximal side of the entocodon.

REGULATION OF ASEXUAL REPRODUCTION AND DEVELOPMENT

Several investigators have studied factors which influence asexual reproduction and development. Reisinger ('34, '57) found that medusa budding could be induced in *C. sowerbii* by a sudden temperature shift from 20°C to 25–27°C. Well-fed polyps cultured at 20°C began to produce medusa buds when raised to 25–27°C within a few days and reached a peak of medusa production after about 2–3 weeks at the higher temperature.

McClary ('59, '60) was able to induce medusa budding in regularly fed polyps cultured at constant temperatures of 26 and 33°C but obtained no medusa buds at 12, 20, or 25°C. He found that frustules were the most common type of bud produced at all temperatures and polyp buds the least common type.

Lytle ('59, '61) found that regularly fed polyps cultured in a flow system produced medusa buds at temperatures of 20, 19–23, and 27°C. The production of buds tended to follow a sequence of three phases: an initial phase of polyp budding, followed by a phase of maximum frustule budding, and finally a phase of medusa budding. The length of the cycle and the relative number of buds of each type were found to be temperature-dependent. Also the relative number of buds of the three types varied independently with temperature; i.e., different ratios of the three bud types were obtained. Experiments with altered feeding rates provided further evidence of a definite hierarchy among these three types of morphogenetic processes. The evidence suggests that polyp budding when in progress takes precedence and limits (inhibits?) medusa budding, and that medusa budding when in progress limits (inhibits?) frustule budding. All three reproductive processes are energy-intensive and require substantial food intake. Polyps can survive long periods with little food, but reproduce little. Food made suddenly available to such a starved colony results in budding activity. In well-nourished polyps energy derived from the food is channeled in specific ways into the formation of the three bud types.

It should be reported also that some workers have reported difficulty in stimulating the production of medusa buds in culture by temperature shifts (Matthews, '66) and by regular feeding at constant temperatures (Acker and Muscat, '76). The reasons for the apparent difficulty in repeating the earlier work are not understood, and further experiments by additional workers are needed.

REGENERATION

A few workers have studied regeneration in *C. sowerbii*. Reisinger ('57) studied the regeneration of gonadal tissue in female medusae. He found that gonads removed from young female medusae did not inhibit development of the medusae and that castrated females were able to partially regenerate the gonadal tissue and to produce mature eggs within 3 weeks.

Reisinger ('57) also performed a series of transplantation experiments on the polyps of *C. sowerbii* to investigate the role of the entocodon in the formation of a medusa bud. His experiments provide evidence that the entocodon serves as the organizer for the medusa bud.

McClary ('61) studied regeneration in polyps, frustules, and young medusae of *C. sowerbii.* He found that amputated or wounded polyps tended to dedifferentiate and to form globular masses of tissue termed "resting bodies." These resting bodies later transformed into frustules and then into new polyps. Frustules cut in various ways also formed globular masses of tissue and later became frustules.

Experiments with amputation of medusa buds showed that buds removed during early stages reverted to a resting body, which produced frustules and reverted to polyp form. More advanced medusa buds when excised continued to develop into medusoid form, thus indicating that determination of medusoid form had taken place.

In a subsequent report McClary ('64) described experiments in which he separated proximal and distal halves of polyps. He found that amputated distal halves (containing the capitulum and mouth) first dedifferentiated to form ovoid regression bodies which later formed a frustule and ultimately into a new polyp. Proximal halves regenerated in a manner similar to the formation of a polyp bud on the lateral wall of an intact polyp.

DEVELOPMENTAL ANOMALIES

Two anomalous developmental forms related to *C. sowerbii* have been reported by previous workers. Reisinger ('57) described a halammohydra stage produced from *C. sowerbii* medusae as a result of cold shock. Medusae shifted from optimal temperatures of 25–28°C to 6–10°C invert and lose much of their medusoid form, exhibiting gonads, stomach, manubrium, and some tentacles. This stage resembles the aberrant *Hyalammohydra* described by Remane ('27) as a neotenic larva of a narcomedusa from the North and East Seas which lacked most medusoid features. The hyalammohydra stage of *C. sowerbii* appears simply to be a moribund stage.

Another small freshwater hydroid, *Calpasoma dactyloptera,* was first described from Switzerland by Fuhrmann ('39) from an aquarium that also contained polyps of *C. sowerbii. Calpasoma* is smaller in size than *C. sowerbii* and has tentacles. Several workers have subsequently found *Calpasoma* in various parts of the world—always in aquaria or laboratory cultures—and usually in association with *C. sowerbii* (Buchert, '60; Kuhl, '60; Rahat, '61; Froelich, '63; Matthews, '66; Rahat and Campbell, '74). The relationship of *Calpasoma* to *C. sowerbii* is not yet understood, but Buchert ('60) believed that it was an alternate form of *C. sowerbii.* Other investigators, including Rahat and Campbell ('74), believe that they are distinct but related forms, perhaps distinct species. Further evidence is needed to establish the true relationship between these interesting forms.

LITERATURE CITED

Acker, T.S., and A.M. Muscat (1976) The ecology of *Craspedacusta sowerbii* Lankester, a freshwater Hydrozoan. Am. Midl. Nat. *95(2):*323–336.

Arnold, J.R. (1951) Fresh-water jellyfish (*Craspedacusta sowerbii*) found in California. Wasmann J. Biol. *9(1):*81–82.

Arnold, J.R. (1968) Freshwater jellyfish records in California—1929–1967. Wasmann J. Biol. *26(2):*255–261.

Beckett, D.C., and E.J. Turanchik (1980) Occurrence of the freshwater jellyfish *Craspedacusta sowerbyi* in the Ohio River USA. Ohio J. Sci. *80(2):*95–96.

Bouillon, J. (1958) Etude monographique du genre *Limnocnida* (Limnomedusae). Ann. Soc. R. Zool. Belg. 1956-1957 *87:*254–500.

Buchert, A. (1960) *Craspedacusta sowerbyi* Lank., eine Süsswasser Meduse und ihre beiden Polypentypen in der ungarischen Fauna. Acta Zool. Hung. *6:*29–55.

Bushnell, J.H., and T.W. Porter (1967) The occurrence, habitat, and prey of *Craspedacusta sowerbyi* (particularly polyp stage) in Michigan. Trans. Am. Microsc. Soc. *86(1):*22–27.

Campbell, R.D. (1974) Cnidaria. In A.C. Giese and J.S. Pearse (eds): Reproduction of Marine Invertebrates. Vol. 1, Acoelomate and Pseudocoelomate Metazoans. New York: Academic Press, pp. 133–199.

Crowell, S., and C.F. Lytle (1955) The locomotion of frustules of *Craspedacusta.* Proc. Indiana Acad. Sci. *64:*255.

Davis, C.C. (1955) Notes on the food of *Craspedacusta sowerbii* in Crystal Lake, Ravenna, Ohio. Ecology *36(2):*364–366.

Deacon, J.E., and E.L. Haskell (1967) Observations on the ecology of the freshwater jellyfish in Lake Mead, Nevada. Am. Midl. Nat. *78(1):*155–166.

Dejdar, E. (1934) Die Süsswassermeduse *Craspedacusta sowerbii* Lankester in monographische Darstellung. Z. Morph. Ökol. Tiere *28:*595–691.

Dexter, R.W., T.C. Surrarrer, and C.W. Davis (1949) Some recent records of the freshwater jellyfish *Craspedacusta sowerbii* from Ohio and Pennsylvania. Ohio J. Sci. *49(6):*235–241.

Eberly, W.E. (1972) New records for *Craspedacusta* in Indiana. Proc. Indiana Acad. Sci. *80:*178–179.

Fowler, G.H. (1890) Notes on the hydroid phase of *Limnocodium sowerbyi.* Q. J. Microsc. Sci. *30:*507–513.

Froelich, C.G. (1963) Ocorrência de forma polipoide de *Craspedacusta sowerbyi* Lank. (Limnomedusae) em São Paulo. An. Acad. Bras. Cienc. *35:*421–422.

Fuhrmann, O. (1939) Sur *Craspedacusta sowerbyi* Lank. et un noveau coelentere d'eau douce, *Calpasoma dactyloptera,* n.g.n.sp. Rev. Suisse Zool. *46:*363–368.

Hausmann, K. (1970) Trichodinae on the tentacles of the fresh water medusa *Craspedacusta sowerbii.* Mikrokosmos *68(1):*1–9.

Howmiller, R.P., and G.M. Ludwig (1972) A record of *Craspedacusta sowerbyi* in Wisconsin. Trans. Wisconsin Acad. Sci. Arts Lett. *60:*181–182.

Hubschman, J.H., and W.J. Kishler (1972) *Craspedacusta sowerbyi* and *Cordylophora lacustris* in western Lake Erie Coelenterata. Ohio J. Sci. *72(6):*318–321.

Jones, S. (1951) On the occurrence of the freshwater medusa, *Limnocnida indica* Annandale, in the western drainage of the Sahyadris. J. Bombay Nat. Hist. Soc. *49:*799–801.

Joseph, H. (1925) Zur Morphologie und Entwicklungsgeschichte von *Haleremita* und *Gonionemus*. Z. Wiss. Zool. *125:*374–434.

Kimmel, B.L., M.M. White, S.R. McComas, and B.B. Looney (1980) Recurrence of the freshwater jellyfish *Craspedacusta sowerbii* Cnidaria Hydrozoa in the Little River system. Southwestern Nat. *25(3):*426–428.

Kramp, P. (1950) Freshwater medusae in China. Proc. Zool. Soc. Lond. *120(1):*165–184.

Kuhl, G. (1947) Zeitrafferfilm—Untersuchungen über den Polypen von *Craspedacusta sowerbii* (Ungeschlechtliche Fortpflantzung, Ökologie, und Regeneration). Abhandl. Senckenbergischen Naturforschenden Ges. *473:*1–72.

Kuhl, G. (1960) Über die Umbildung einer "Meduse" von *Craspedacusta sowerbyi* Lank. in eine Frustel. Zeits. Morphol. Okol. Tiere *48:*439–446.

Kumé, M., and K. Dan (1968) Invertebrate Embryology. Washington: National Library of Medicine, Public Health Service, U.S. Department of Health, Education and Welfare, and the National Science Foundation. Translated by J.C. Dan. 605 pp.

Kynard, B.E., and J.C. Tash (1974) Freshwater jellyfish (*Craspedacusta sowerbyi*) in Lake Pategonia, Southern Arizona. J. Arizona Acad. Sci. *11:*76–77.

Lipsey, L.L. Jr., and M.J. Chimney (1978) New distribution records of *Cordylophora lacustris* new record and *Craspedacusta sowerbyi* new record Coelenterata in southern Illinois USA. Ohio J. Sci. *78(5):*280–281.

Lytle, C.F. (1959) Studies on the developmental biology of *Craspedacusta*. Doctoral dissertation, Indiana University (Diss. Abstr. 20:1491), 116 pp.

Lytle, C.F. (1960) A note on distribution patterns in *Craspedacusta*. Trans. Am. Microsc. Soc. *79:*461–469.

Lytle, C.F. (1961) Patterns of budding in the freshwater hydroid *Craspedacusta*. In H.M. Lenhoff and W.F. Loomis (eds): The Biology of Hydra and Some Other Coelenterates. Coral Gables, Florida: University of Miami Press, pp. 317–336.

Lytle, C.F. (1962) *Craspedacusta* in the southeastern United States. Tulane Studies Zool. *9:*309–314.

Matthews, D.C. (1963) Freshwater jellyfish *Craspedacusta sowerbyi* Lank. in Hawaii. Trans. Am. Microsc. Soc. *82(1):*18–22.

Matthews, D.C. (1966) A comparative study of *Craspedacusta sowerbyi* and *Calpasoma dactyloptera* life cycles. Pacific Sci. *20(2):*246–259.

McClary, A. (1959) The effect of temperature on growth and reproduction in *Craspedacusta sowerbii*. Ecology *40:*158–162.

McClary, A. (1960) Growth and differentiation in *Craspedacusta sowerbii*. Doctoral dissertation, University of Michigan.

McClary, A. (1961) Experimental studies of bud development in *Craspedacusta sowerbii*. Trans. Am. Microsc. Soc. *80:*343–353.

McClary, A. (1964) Histological changes during regeneration of *Craspedacusta sowerbii*. Trans. Am. Microsc. Soc. *83(3):*349–357.

Milne, L.J. (1938) Some aspects of the behavior of the freshwater jellyfish, *Craspedacusta* sp. Am. Nat. *72(742):*464–472.

Moser, J. (1930) *Microhydra* E. Potts. Sitzber. Ges. naturf. Freunde, pp. 238–303, 4 pl.

Payne, F. (1924) A study of the freshwater medusa, *Craspedacusta ryderi*. J. Morphol. *38:*387–430.

Payne, F. (1926) Further studies on the life history of *Craspeducusta ryderi,* a freshwater hydromedusan. Biol. Bull. *50(6):*433–443.

Pennak, R.W. (1956) The fresh-water jellyfish *Craspedacusta* in Colorado with some remarks on its ecology and morphological degeneration. Trans. Am. Microsc. Soc. *75:*324–331.

Pennington, W.L., and J.W. Fletcher (1980) Two additional records of *Craspedacusta sowerbyi* in the Tennessee River system USA. J. Tenn. Acad. Sci. *55(1):*31–34.

Persch, H. (1933) Untersuchungen über *Microhydra germanica* Roch. Z. Wiss. Zool. *144:*163–210.

Potts, E. (1906) On the medusa of *Microhydra ryderi* and on the known forms of medusae inhabiting fresh water. Q. J. Microsc. Sci. *50(N.S.):*623–633.

Potts, E. (1908) *Microhydra ryderi* during 1907. Proc. Delaware County Inst. Sci. *3:*89–108.

Rahat, M. (1961) Two polyps of limnotrachylina from Israel. Bull. Res. Council Isr. *10B:*171–172.

Rahat, M., and R.C. Campbell (1974) Three forms of the tentacled and non-tentacled freshwater coelenterate polyp genera *Craspedacusta* and *Calpasoma.* Trans. Am. Microsc. Soc. *93(2):*235–241.

Reisinger, E. (1934) Die Süsswassermeduse *Craspedacusta sowerbii* Lankester und ihr Vorkommen in Flussgebiet von Rhein und Maas. Natur Niederrhein *10:*33–43.

Reisinger, E. (1957) Zur Entwicklungsgeschichte und Entwicklungsmechanik von *Craspedacusta* (Hydrozoa, Limnotrachylina). Z. Morph. Okol. Tiere *45:*656–698.

Remane, A. (1927) *Halammohydra,* ein eigenartiges Hydrozoon der Nord- und Ostsee. Zeits. Morphol. Okol. Tiere *7:*643–677.

Rice, N.E. (1958) Occurrence of both sexes of the fresh-water medusa *Craspedacusta sowerbii* Lankester in the same body of water. Am. Midl. Nat. *59:*525–526.

Rice, N.E. (1960) *Hydramoeba hydroxena* (Entz), a parasite on the freshwater medusa *Craspedacusta sowerbii* Lankester, and its pathenogenicity for *Hydra cauliculata* Hyman. J. Protozool. *7(2):*151–156.

Schmitt, W. (1939) Fresh-water jellyfish records since 1932. Am. Nat. *73:*83–89.

Smrchek, J.C. (1970) *Craspedacusta sowerbyi* northern extension of range in Michigan. Trans. Am. Microsc. Soc. *89(2):*325–327.

Thiel, H. (1973) *Limnocnida indica* in Africa. Publ. Seto. Mar. Biol. Lab. *20:*73–79.

Walley, H.D. (1972) The fresh-water jellyfish *Craspedacusta sowerbyi* in Illinois. Trans. Ill. State Acad. Sci. *63(1–2):*80–81.

Welch, P.S. (1948) Limnological Methods. Philadelphia: Blakiston, pp. 260–262.

Wiggins, G.B., R.E. Whitfield, and F.A. Walden (1957) Notes on fresh-water jellyfish in Ontario. Contrib. R. Ontario Museum Div. Zool. Paleontol. No. 45, 6 pp.

Woodhead, A.E. (1943) Around the calendar with *Craspedacusta sowerbyi.* Trans. Am. Microsc. Soc. *62:*379–381.

Developmental Biology of Freshwater Invertebrates, pages 151–211

Developmental Biology of Triclad Turbellarians (Planaria)

Mario Benazzi and Vittorio Gremigni

ABSTRACT Triclads belong to the phylum Platyhelminthes, class Turbellaria, are simple Metazoa, and represent interesting and useful material for research in developmental biology. They have a high regeneration capacity correlated with the ability to reproduce asexually. Experimental research in this field has yielded fruitful results leading to a better understanding of cellular problems in planarian developmental biology. On the contrary, the embryonic development of triclads has been thoroughly investigated from a morphological point of view, but only a few papers have been published dealing with experimental investigations. This is principally due to technical difficulties relating to internal fertilization and embryonic development that occurs within a mass of vitelline cells inside a hard, nontransparent capsule. The very irregular spiral cleavage of planarian is often described as "anarchic" since it is impossible to follow the fate of blastomeres, or to identify primary embryonic layers and a gastrula stage. The basic biogeographical, ecological, and developmental features of triclads are described. Lastly, gaps in our current understanding of embryonic development are pointed out along with possible future avenues of research.

This chapter is dedicated to Dr. Giuseppina Benazzi-Lentati, who for over forty years has devoted her scientific activity to the cytology and cytogenetics of freshwater planarians. One of us has known her as a research companion and loving wife. The other began his scientific career with her, and fondly recalls the time spent with her. We both are grateful for our long and fruitful association with Giuseppina, who has always given and continues to give us scholarly advice and keen insight into the complexities of nature and the intricacies of life.

GENERAL INTRODUCTION

The term planarian is derived from the latin word *planus* (flat) and refers to worms having a dorso-ventrally flattened body that belong to the order Tricladida (class Turbellaria, phylum Platyhelminthes). The term Tricladida (or Triclads) refers to the three main branches (one anterior and two posterior) of the digestive system. Triclads inhabit fresh or salt water or humid terrestrial zones and their taxonomic classification in three suborders is mainly based on this ecological distinction: Terricola (or land planarians), Maricola (or marine planarians), and Paludicola (or freshwater planarians). This classification is still retained even though the ecological criteria separating the suborders seem inadequate and are of low phylogenetic value (Ball, '81).

Triclads belong to the Neoophora level of organization, which along with the Archoophora level defines the primary division of the class Turbellaria (Karling, '67, '74). This subdivision is based on the morphology of the female gonad and the structure of eggs, features that play an important role in determining the type of embryonic development. The Archoophora level of organization is the more primitive and includes organisms with simple female gonads, i.e., ovaries provided with entolecithal eggs and no vitellaria. The Neoophora level of organization refers to organisms provided with complex female gonads, i.e., ovaries containing alecithal (or nearly alecithal) eggs and vitellaria consisting of cells rich in yolk.

The Tricladida Paludicola, which are the subject of this article, are the best known planarians and are by far the most widely used in experimental research because they are easily cultured under laboratory conditions. According to the classification recently proposed by Ball ('74a, '77, '81) freshwater triclads are divided into three families: Dugesiidae, Planariidae, and Dendrocoelidae.

Planarians, by virtue of their ability for sexual and asexual reproduction, have provided researchers of developmental biology with an excellent organism for study. However, until now, most research has centered around asexual propagation and regeneration, and fewer studies have dealt with embryonic development. This is principally because planarians have an extraordinarily high regenerative capacity, which is easily studied. On the other hand, several technical and morphological difficulties hamper embryological studies. These include internal fertilization and development occurring inside a hard, red–brown capsule (usually called the cocoon), which does not allow one to observe directly embryonic development, and the presence of several embryos within each cocoon. A solution to this problem would be to investigate embryonic development *in vitro*, but, unfortunately, few attempts have been made in this line of research and these have not been successful. Moreover, Triclads are Neoophora, which have numerous vitelline cells entering the cocoon together with eggs. These cells become greatly involved in the process of embryonic development, making it very irregular, and strongly modify it with respect to the primitive type of development of this class where spiral cleavage occurs. Last, the irregularity of cleavage and of the subsequent stages of development makes it nearly impossible to identify and follow the fate of various cell lines.

Features of Development of the Organism

The most primitive type of development occurring in Turbellaria is unanimously considered to be that of Polyclads, which have entolecithal eggs and therefore belong to the Archoophora level of organization. In this order the cleavage is of the "quartet spiral type" and resembles that of annelids and mollusks (Kato, '68). However, this primitive type of cleavage does not occur in all Archoophora (for instance the Acoela, which have a modified "duet spiral cleavage") and is completely altered and undefinable in the Neoophora, which have ectolecithal eggs and vitelline cells supplying yolk for the embryonic development (for a review see Galleni and Gremigni, '82).

Triclads produce cocoons containing several eggs and thousands of vitelline cells and are thus typical representatives of Neoophora. The embryonic development of planarians is consequently highly irregular and no features of the primitive spiral cleavage can be identified. The fate of eggs in the cocoon varies: most of them form embryonic "spheres" after completing some segmental divisions; however, several eggs do not develop and instead degenerate more or less precociously (Benazzi and Benazzi-Lentati, '45, '48). Each developing "sphere" is composed of dividing blastomeres immersed in a yolk syncytium which is formed from the fusion of several vitelline cells. Other vitelline cells form a yolk mass destined to be ingested later in the embryonic digestive system.

A few blastomeres differentiate to give rise to a thin envelope, or primary epidermis, surrounding the "sphere," while other blastomeres differentiate into an embryonic pharynx and an embryonic unbranched intestine, which are dorsally located and ultimately engulf the yolk mass. Later on, the embryonic digestive system progressively disappears while definitive structures (epidermis, pharynx, intestine, nervous system, etc.) arise from a germinal cord consisting of a mass of embryonic cells proliferating in the ventral area. All the structures seem to arise from a single kind of undifferentiated embryonic cell. No trace of an archenteron or a coelom is evident during development of Triclads. In addition, there is no larval stage; instead development is direct and several complete, minute worms hatch from the cocoon. The developmental process takes 2 to 4 weeks, or even more, to complete, depending on species variation and ambient temperature.

Natural History of the Adults

Anatomy. A brief description of the general morphology of freshwater triclads will be presented here to help readers who are using these organisms in developmental biology for the first time. Readers are invited to consult the monographs by Bresslau ('33), Hyman ('51), de Beauchamp ('61), and Ball and Reynoldson ('81) or the original articles and reviews for more detailed systematic and morphological descriptions of triclads.

Planarians are free-living worms lacking a coelom, anus, skeleton, circulatory and respiratory system. Their length usually ranges from a few millimeters up to 3–4 cm and sometimes even more in some forms from Lake Baikal. In some species the anterior head has two small lateral projections called auricles that are provided with sensory cells. The head is nearly always provided with two or more eyes (up to a hundred in the genus *Polycelis*) and in some species a necklike region is present behind the head. Planarians are often uniformly colored brown, gray, or black by a parenchymal pigment, but they can also be streaked, spotted, or striped. Some cave forms are depigmented and therefore white (Fig. 1).

The planarian body is surrounded by a one-cell-layered epidermis that is ciliated especially on the ventral surface, where the cilia serve for locomotion. The epidermis is also provided with gland and sensory cells. Mucous cells are numerous near the surface of the animal and the mucus is abundantly discharged in water. Characteristic rod-shaped bodies known as rhabdites are produced by cells located in the subepidermal parenchyma and seem to have a defensive function. In fact, they are discharged in water when the animal is disturbed.

The parenchyma is a loose connective tissue located beneath the epidermis and separated from it by a thin basement membrane. It is considered to be a filling tissue that binds the epidermis with the gastrodermis and supports the internal organs. It is composed of two main types of cells: fixed parenchymal cells and the so-called neoblasts, which are considered to be free undifferentiated cells involved in regeneration (see "Experimental Procedures," below).

The muscular system consists of a subepidermal, parenchymal, and organ-specific musculature, and is particularly thick and strong in the pharynx and copulatory apparatus. The subepidermal and parenchymal muscles allow the planarian to lengthen or shorten its body. Glandulo-muscular adhesive organs are present in some Dendrocoelids where they are used for locomotion and for capturing prey.

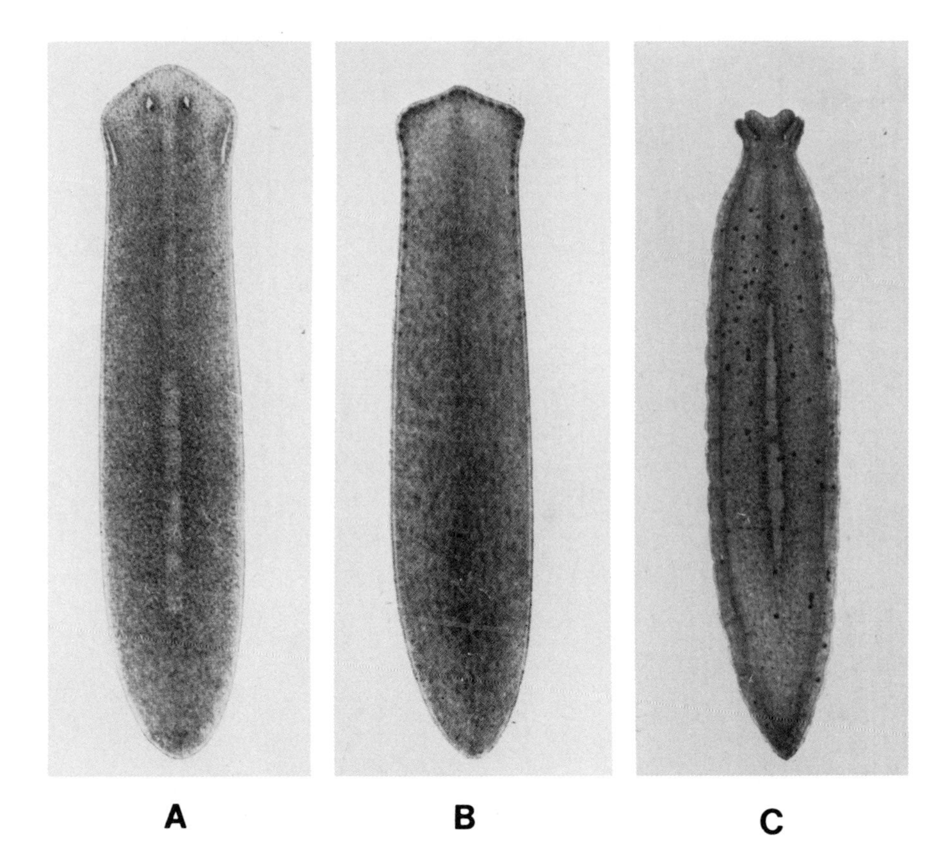

Fig. 1. General view of A) *Dugesia lugubris* s.l. (Dugesiidae), B) *Polycelis nigra* (Planariidae), C) *Bdellocephala punctata* (Dendrocoelidae). (From Steinmann and Bresslau, '13.)

The digestive system, which should be considered a gastrovascular cavity, is composed of a mouth, a cylindric and protrusible muscular pharynx, and a branched intestine (Fig. 2a). The mouth and pharynx are generally ventrally located in the middle of the body and the pharynx, when fully protruded, appears as a white cylinder. It is mostly muscular, but also has epithelial, glandular, nervous, and free parenchymal components. Some species of the genera *Phagocata* and *Crenobia* are polypharyngeal.

The pharynx empties into the intestine, which consists of one anterior and two posterior branches. The three branches are themselves highly ramified and their diverticula reach nearly every region of the body thus transporting food everywhere. The wall of the intestine is composed of gastrodermal, phagocytic, and eosinophilic gland cells.

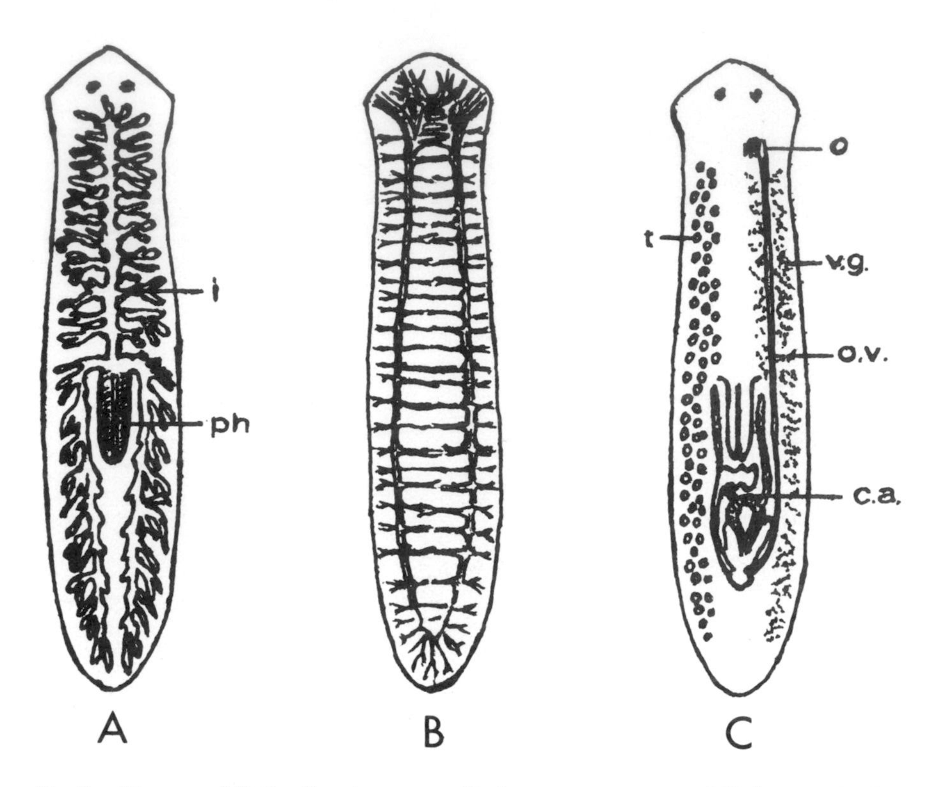

Fig. 2. Diagram of A) the digestive system, B) the nervous system, and C) the reproductive system of a planarian. ca, copulatory apparatus; i, intestine; o, ovary; ov, oviduct; ph, pharynx; t, testis; vg, vitelline gland.

The excretory system consists of protonephridial cells provided with terminal flame bulbs which communicate with either the dorsal or ventral surface of the body.

The nervous system is composed of two anterior interconnected cerebral ganglia (the brain) from which two longitudinal ventral cords arise (Fig. 2b). Peripherally, numerous nerves radiate from the cords. They can form plexuses around organs and contain neurosecretory cells.

Besides the eyes, which consist of pigment-cup ocelli with retinal and pigment cells, tactile and chemoreceptor sensory cells are scattered throughout the body, especially in the highly sensitive auricular grooves.

All freshwater planarians are hermaphroditic and their reproductive system is composed of very complex male and female apparatuses (Fig. 2c). The male apparatus in most genera consists of numerous follicular testes, which form two long lateral strands starting from behind the head and almost reaching the caudal end of the body. Inside the testes, maturing germ cells have centripetal distribution and the spermatozoa, which are located in the inner lumen of the gonad, each have two flagella with typical "9+1" organization of their axoneme. Short efferent ducts unite to form ciliated sperm ducts, which open via the vas deferens into a seminal vesicle that stores sperm. The copulatory organ is made up of a very muscular protrusible penis consisting of a bulb and a papilla, which are located in the male antrum. The latter opens into the common sexual antrum, which in turn communicates with the external surface by a genital pore situated posteriorly to the mouth.

The female apparatus consists of two small ovaries, usually located behind the head between the second and third ramifications of the anterior branch of the intestine, and numerous vitelline follicles that are scattered bilaterally in an area between the ovaries and the genital pore. Two short oviducts, each forming a seminal receptacle, start from the ovaries and join with two laterally lying ovovitelline ducts. These latter ducts receive yolk cells that come from vitelline follicles through short vitelline ducts. A copulatory bursa is situated in front of the penis and is connected to the female antrum by a bursal canal.

Ecology. Triclads typically inhabit streams, spring ponds, and lakes in temperate regions and also the tropics. Some species are rheophilous, which means that they only inhabit flowing waters, whereas others are limnadophilous, meaning that they inhabit still waters. Some are stenothermous, being limited to mountain springs and streams, whereas others are eurythermous and can easily adapt to wide changes in temperature. Several species of the family Dendrocoelidae inhabit only cave waters, are generally white, and lack eyes.

An interesting aspect of the ecological flexibility of freshwater planarians concerns the ability of some species to become permanent inhabitants of temporary waters. This is an unexpected feature because most of the species do not have terrestrial stages and adults are susceptible to dessication. This flexibility is seen in some species of the genus *Phagocata*, which are frequently found in ground waters that exist for only a few months of the year. Two well-studied North American species, *Phagocata velata* and *Hymanella retenuova*, are able to withstand dessication. The mechanisms involved are different in each species and relate to their different reproductive modalities. *P. velata*, studied by Castle ('27, '28) and Castle and Hyman ('34), is fissiparous. As reported by Hyman ('51, p. 159),

> In the spring the worms grow to 12 to 15 mm in length, then gradually fragment into small pieces, each of which encysts in a mucous cyst. Within this cyst the pieces reorganize into minute worms that emerge in a few weeks if the cysts remain submersed. In habitats that dry up in summer, emergence may not occur until the following spring, and such an asexual cycle may continue indefinitely. In permanent waters, mostly spring-fed marshes, worms hatched from cysts in the fall become sexually mature and breed in winter.

H. retenuova only reproduces sexually and is rarely found in permanent bodies of water (Castle, '41). It is the only species that can produce a large thick-shelled cocoon capable of withstanding prolonged periods of drought. Ball et al. ('81), based on morphological, cytological, and biological studies, showed that *Phagocata vernalis* Kenk, which was known as a typical vernal pond form, in reality is conspecific with *H. retenuova.* This species is also unique for its marked protandry, and the authors advance the hypothesis that the development of sequential hermaphroditism can avoid the potentially deleterious effects of self-fertilization. This phenomenon also would reduce the probability of inbreeding among siblings, which in the case of *H. retenuova* could easily occur in a dense and well-circumscribed population where possibilities for dispersal away from the parental zone are limited.

The classic division of planarians into freshwater and marine species, as previously mentioned, is not absolute. Marine species can survive and reproduce in freshwater and *vice versa* (Ball, '77). For example, species of the marine genus *Procerodes* can live indefinitely in highly diluted sea water. Conversely, *Dendrocoelum lacteum* was found in a stream opening into the Danish Wiek where it lives in water having a salt gradient from 0.01% to 0.5% (Gresens, '28). In the Gulf of Finland, freshwater planarians such as *Polycelis tenuis* live and reproduce. In this context it is interesting to mention that planarians showing morphological characters similar to Maricola have recently been found in fresh waters. *Opisthobursa mexicana* Benazzi and *O. josephinae* Benazzi, both troglodytic species from Mexico, are typical examples of this situation. Also *Balliania thetisae* Gourbault from Tahiti, *Eviella hinesae* Ball from Australia, and *Debeauchampia anatolica* Benazzi from Turkey exhibit many similarities with marine forms. The hypothesis that they represent marine relics seems reasonable.

Geographic distribution. We think it opportune to present a brief list of the more common genera and species present in various continents, taking into account the species that are most suitable for laboratory culture. Representatives of the three planarian families include, as was mentioned previously: 1) Dugesiidae, for the most part having spherical and stalked cocoons. *Cura* and *Dugesia* are the more important genera. The genus *Dugesia* comprises the subgenera *Girardia, Dugesia,* and *Schmidtea.* 2) Planariidae, having spherical or ovoid cocoons without a stalk. *Planaria, Phagocata, Polycelis, Crenobia* and a few other genera belong to this family. 3) Dendrocoelidae, which have unstalked cocoons, are comprised of two subfamilies: the Kenkinae, represented by the genera *Kenkia* and *Sphalloplana*, which are usually blind and unpigmented cave dwellers, and the Dendrocoelinae, usually unpigmented but with eyes, including various genera, such as *Dendrocoelum, Bdellocephala,* and *Dendrocoelopsis.* According to some authors these two subfamilies should be considered as two distinct families.

For North America, Kenk ('72) lists 12 genera. *Dugesia (Girardia) tigrina* (Girard) and *D. (Girardia) dorotocephala* (Woodsworth) are widely distributed in the United States and Southern Canada and are the best-known species. They are frequently used for experimental research and in classroom demonstrations and are capable of both sexual and asexual reproduction. *Cura foremanii* (Girard) is distributed throughout the eastern half of North America and only reproduces sexually after self-fertilization. *Polycelis coronata* (Girard), with sexual and asexual reproduction, inhabits springs and cold creeks. The species belonging to the genus *Phagocata* are numerous and include *morgani* (Stevens and Boring) (sexual and fissiparous), *gracilis* (Haldeman) (only sexual), *velata* Castle, and *vernalis* Kenk (=*Hymanella retenuova* Castle) found in vernal pools. The last is peculiar since the cocoon is carried in the genital atrium of the parent for up to 4 weeks before being deposited. A great number of species belonging to the genus *Sphalloplana* have been found in over 200 caves in at least 16 states. The Dendrocoelinae are represented by some *Dendrocoelopsis* species and by *Procotyla fluviatilis* Leidy, a planarian that was confused with the European *Dendrocoelum lacteum* for a long time.

In Central and South America one can find several species, particularly those belonging to the genera *Cura* and *Dugesia*. It should be noted that all *Dugesia* species of the New World belong to the subgenus *Girardia*.

The European planarians have long been the subject of study. The genus *Dugesia* includes two subgenera: *Dugesia* and *Schmidtea*. *Dugesia (Dugesia)gonocephala* s.l. represents a superspecies largely diffused throughout the Old World, with various species reproducing sexually or by fission. The species of the subgenus *Schmidtea*, namely *Dugesia lugubris* (O. Schmidt), *D. polychroa* (O. Schmidt), and *D. mediterranea* Benazzi et al., are indigenous to Europe, although *D. polychroa* has also been found as an immigrant in North America. Conversely, *D. tigrina*, an American species, is now widely diffused in Europe. Widely distributed in the European limnetic habitats are *Planaria torva* (Müller), *Polycelis nigra* (Müller), and *P. tenuis* Iijima. All these species are exclusively sexual whereas *Polycelis felina* (Dalyell) is a cold-running-water dweller and is frequently asexual. A typical inhabitant of European cold-running waters is *Crenobia alpina* (Dana). Many species of *Phagocata* and of the closely related genus *Atrioplanaria* are known; they frequently reproduce by fission and inhabit caves or underground waters. The Dendrocoelidae are represented by the lentic *Dendrocoelum lacteum* (Müller) and by some subterranean species; the genera *Dendrocoelopsis* and *Bdellocephala* are also found in Europe.

With regard to the Asiatic planarians, Japanese species have been the most extensively studied, and Kawakatsu ('74) stated that 19 species belonging to six genera are known. These are *Dugesia, Phagocata, Polycelis, Sphallo-plana, Bdellocephala,* and *Dendrocoelopsis,* plus one uncertain genus, *Mon-ocotylus*. The most widely distributed species is *Dugesia (Dugesia) japonica,* which belongs to the *D. gonocephala* group and exhibits various karyologically and reproductively different races. *Phagocata vivida* and the *Polycelis* species *auriculata, sapporo,* and *schmidti* are also widely distributed.

In Australia many species belonging to the family Dugesiidae are known and have been the object of a critical review by Ball ('74b). From a taxonomic point of view, Australia is interesting in that all of the freshwater planarians from this country described hereto belong to the most primitive family—the Dugesiidae—and Ball ('74b, '77) believes that these planarians originated in the Southern Hemisphere, possibly in what is now Antarctica. The genus *Cura* with the species *pinguis* (Weiss) is the most widely distributed throughout Australia. Other genera are *Neppia* Ball, *Spathula* Nurse, and *Reynoldsonia* Ball. The genus *Romankenkius* Ball has been described from Tasmania.

Life cycles. Planarian life cycles differ greatly in relation to their reproductive modalities. We may schematically distinguish three life-cycle types: 1) sexual, 2) asexual, and 3) mixed sexual and asexual. Planarians of the first and third type often show seasonal sexual cycles, which are related to their habitat and especially to the ambient temperature.

Sexual cycles. In sexual populations worms pass through a juvenile, a sexual, and a resting phase. The first phase lasts from the time of hatching until the attainment of sexual maturity, or second phase, during which individuals reciprocally copulate and deposit cocoons. The third phase begins with an involution of the reproductive system. The regression of the sexual reproductive system usually begins with the vitellaria and extends to testes, ovaries, and finally the copulatory apparatus (Stephan-Dubois and Gusse, '74). But Teshirogi and Fujiwara ('70) stated that in *Bdellocephala brunnea* the copulatory apparatus disappears before male and female gonads. Ovaries and testes are usually the first structures to reappear followed by the copulatory apparatus and finally by the vitellaria. Several sexual phases can occur in the same year or in the subsequent one, with variation in their frequency according to individual species.

Many species such as *Dugesia dorotocephala, D. gonocephala, D. lugub-ris-polychroa,* and *Polycelis nigra-tenuis* do not die after breeding and produce new sexual phases, usually in the subsequent years. On the contrary, other species such as *Polycelis felina, Planaria torva, Dendrocoelum lac-teum,* and *Bdellocephala punctata* are basically annuals that die after breeding (Vandel, '21; Young and Reynoldson, '66; Reynoldson and Sefton, '72).

Almost all planarians are hermaphroditic with distinct sex organs, a condition known as monoecism. Only two marine triclads, *Sabussowia dioica* and *Cercyra teissieri*, are gonochoric, a condition probably secondarily derived by the reduction of one set of sex organs.

Male gonads usually develop seasonally before the female ones; thus most planarians are slightly protandrous. Sexual reproduction usually involves copulation of two individuals and only a few exceptions to this rule are known, although self-fertilization is anatomically possible in most species. Self-fertilization has been observed in marine triclads and among the paludicola in *Cura foremanii* by Anderson ('52) and Anderson and Johann ('59) and in *Cura pinguis* by Gourbault and Benazzi ('75). In these two species self-fertilization probably represents their only reproductive modality. It also occurs sporadically in *Polycelis nigra* (Benazzi, '52; Lanfranchi, '65).

Copulation may last a rather long time (up to several hours) during which the sperm of one partner are deposited in the copulatory bursa of the other. In some species of the *Dugesia gonocephala* group, *Phagocata* and *Planaria,* the sperm are contained in spermatophores that may be seen inside the copulatory bursa. From the bursa the sperm are conducted through the bursal canal, enter the oviduct, and are stored in the seminal receptacles. Each worm may copulate repeatedly and lay cocoons in a space of a few days. Fertile cocoon deposition can occur even if the last copulation occurred some months before, because sperm retained in the seminal receptaculum maintain their fertility (Nobili, '58). Copulation and cocoon deposition can be inhibited by prolonged starvation, and also temporarily by excessive nutrition. The oocytes are fertilized in the oviducts as they pass through receptacles on their way to the genital atrium where they accumulate along with vitelline cells. The oocytes leave the ovaries in a late prometaphase I stage when the nuclear envelope is disrupted, and the chromosomes in most cases are represented by bivalents that appear scattered along the spindle. Only after sperm penetration do the further stages of oogenesis, namely the metaphase congression and the emission of the polocytes, take place. To study these events it is necessary to remove the cocoon during its formation in the genital atrium, where these events of oogenesis occur. In freshly deposited cocoons the fertilized oocytes are already in the interphase preceding the first cleavage division, whereas the unfertilized oocytes in the cocoon are stopped at the first meiotic metaphase stage and then degenerate.

During a seasonal cycle a certain percentage of cocoons (25% in *P. nigra*, according to Reynoldson, '60) are sterile. Although Reynoldson found no relation between fertility and the season, the authors' long experience in the field demonstrated that usually the first, and particularly the last, cocoons laid in each cycle are sterile, since they usually contain only vitelline cells.

In the genital atrium eggs accumulate along with thousands of vitelline cells and here cocoon-envelope formation takes place (Fig. 3). Peripheral granules (the so-called cocoon-shell globules) of the vitelline cells containing

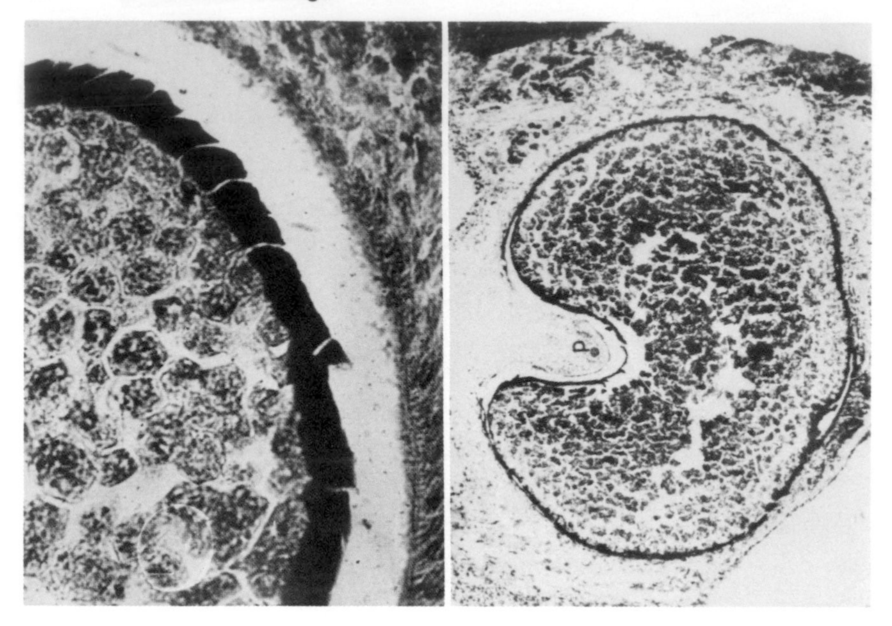

Fig. 3. Micrographs showing, on the right, a newly formed cocoon within the genital antrium of *Dugesia lugubris* s.l.; on the left, the thick black line represents the cocoon envelope. (From Marinelli, '72.)

basic proteins, polyphenols, and the phenol-oxidase system (Gerzeli and Gerzeli-Pedrazzi, '65; Marinelli, '72; Gremigni and Domenici, '74) are released from the cells. The granular contents undergo a process of quinone tanning and produce the sclerotin of the cocoon envelope, which becomes progressively harder and browner after deposition. Immediately before the cocoon is laid the parent adheres to the substratum by the margin of its body and remains in this position while the cocoon is formed in the genital atrium and subsequently laid. This entire process lasts from 2 to 4 hours and enables the researcher to identify the approximate time of cocoon deposition.

The cocoon, which can be detected as a bulge inside the parental atrium before deposition, is spherical or ellipsoidal and is 0.5 to over 3 mm in diameter depending on species. It can be attached to the substrate by a gelatinous secretion as in the families Planariidae and Dendrocoelidae, or by a thin stalk as in the genus *Dugesia* and some other genera of the family Dugesiidae. Each cocoon contains a small number of eggs (2 to 20, rarely more) and thousands of yolk cells (10,000 to 12,000 in *Planaria torva* according to Mattiesen, '04) produced by the vitelline glands. The development of the embryos in the cocoon lasts from about 10 days to several weeks according to the species and the ambient temperature. Le Moigne (unpublished data) on the basis of 66 observations stated that the development lasts in *P. nigra* from 14 to 38 days in relation to temperature variations. Gourbault ('72) compiled a table of the available information about the length of embryonic development. The shortest periods listed for surface species were 10 to 20 days for *D. lugubris-polychroa*, 14 to 28 days for *D. gonocephala*, and 21 to 28 days for *D. dorotocephala*. The longest periods were 70 days for *Crenobia alpina* and 90 days for *Dendrocoelum lacteum*. However, it should be pointed out that the culture temperatures were not standardized. Development in cave-dwelling species generally takes longer. Developmental times were about 60 days for *Atrioplanaria delamarei* at 12°–13°C and 75–120 days at 8°–9°C; about 60–120 days and 150 days for two different populations of *Phagocata vitta*; and 90–150 days for *Dendrocoelum regnardi*.

When the young worms hatch by the rupture of the cocoon they have the general morphology of adults, but lack any trace of reproductive apparatus, which will develop soon after birth.

Some data on the juvenile phase have been collected in laboratory cultures. Voigt ('28) found that in *P. nigra* the animals were 10 months old before the first cocoon was deposited, while Le Moigne (personal communication) observed that specimens of the same species begin to lay cocoons 3 months after birth. Abeloos ('30) found that sexual maturity occurred at 2 months in *D. gonocephala*. In *D. lugubris*, according to Bàlasz and Burg ('62) sexual maturity occurs before the worms attain their maximum volume. Haranghi and Bàlasz ('64) noted an initial but unsustained rise in cocoon production in this last species, which they attributed to asynchronous sexual maturation. Maximum cocoon production was reached when the animals were 10 months old, after which production gradually decreased and continued at a slower rate for 3 or 4 years. Jenkins ('74) has investigated a sexual race of *D. dorotocephala* four times a year (13-week intervals) beginning from the emergence of the juvenile from the cocoon. She found that nearly two thirds of the animals reached maturity during the second or third 13-week interval. A few became sexual before the end of the first interval, but only two of them produced cocoons. During the fourth and fifth interval another 27% became mature, and the remaining 6% began sexual reproduction at various times during the seventh interval. Jenkins also noted that sexual maturity was not a function of size, and at the time the first cocoons were deposited most individuals had reached lengths of 20 to 25 mm. Growth continued after maturity until a length of 30 to 35 mm or more was attained. The number of cocoons deposited increased rapidly during the first year and thereafter the rise was slight. The overall peak of cocoon production occurred near the end of the third year.

Jenkins and Brown ('63, '64) investigated several other variables, such as fertility of inbred and crossbred groups, sterility, juvenile mortality, and senescence, in the same species. These authors maintained that the reproductive characteristics of planarians are under genetic control, since whatever factors are present are likely to be manifested to a greater degree in inbred lines. Parental longevity, fertility, and juvenile mortality may be heritable; but the latter two characteristics are attributable to separate factors, because high juvenile mortality and low parental fertility do not necessarily coexist. They also concluded that the sexual race of *D. dorotocephala* undergoes a true senescence, since continuous sexual activity during maturity is followed by a postreproductive period characterized by the decline and eventual cessation of cocoon production. Quantitative data on age-dependent reproductive abilities (cocoon production, fertility, and number of embryos) were also collected in *D. lugubris* by Bàlasz and Burg ('62) and Haranghi and Bàlasz ('64). Lange ('68) has proposed a possible explanation of the physiological aging in planarians on a cellular basis, in terms of the relationship between chronological age, size, number, and tissue density of neoblasts.

Various observations on the sexual behavior of freshwater planarians maintained for long periods of time in laboratory culture have been conducted by Benazzi and colleagues. The life-span of certain species of the *D. gonocephala* group is very long, in some cases exceeding 20 years. However, the cessation of cocoon production after a more or less prolonged culture period frequently occurs. There are specific or racial differences, which suggest some kind of control, but the culture medium is also an important factor because it is possible to restore the fecundity of certain planarians by using various types of fresh water.

Asexual reproduction. Asexual reproduction occurs in some groups of Turbellaria by transverse fission, which may be of the architomic or paratomic type. In architomy, fission is not preceded by differentiation of the new individuals, whereas in paratomy, the animal undergoes division into a chain of zooids that become well differentiated before breaking off from the chain. In planarians architomy represents by far the more prevalent modality. The process is very simple and rapid: the posterior end of the animal suddenly adheres firmly to a substrate, while the anterior part of the body elongates, splits longitudinally, and divides the worm in half. The missing half is quickly regenerated. The division plane begins, in most cases, slightly behind the pharynx but is sometimes prepharyngeal. Binary division is the most common mechanism, but in some species of *Phagocata* the animal divides into a number of pieces that first encyst and then regenerate within the cyst.

Only a few cases of paratomy have been described in planarians. Kennel (1888), who studied *Planaria fissipara* from Central and South America (a "species inquirenda" since the reproductive system is unknown), observed that the brain, together with the eyes and mouth, is regenerated before fission takes place. Paratomical division also may occur in *Dugesia paramensis* from

Columbia (Fuhrmann, '14) and perhaps in *D. mertoni* from the Kay Islands (Indonesia) (Steinmann, '14), although paratomy in these two species was only studied in preserved specimens.

There are also cases that may represent an intermediate situation between architomy and paratomy. In *Rhodax evelinae* from Brazil the posterior third to fifth part of the body begins to separate by a constriction on both sides about 2 days before division takes place. This partial division along with an accumulation of formative cells (neoblasts) in the fission zone are the only preparatory stages occurring before division (Marcus, '46).

A peculiar mechanism has also been observed in species of the European genus *Atrioplanaria*: the individuals undergo transverse constrictions into two, three, or occasionally four segments, which remain united, however, for a long time without any differentiation into new organisms. Some segments finally divide and regenerate into new worms with a frequent concurrent regression of other constrictions (de Beauchamp, '37; Benazzi, '38; Benazzi and Gourbault, '77).

Some authors have even found indications of preparatory steps for division in typical architomic species. Stagni and Grasso ('65) found an accumulation of neoblasts at the presumptive plane of division in specimens of *D. tigrina* showing signs of impending fission. Lender and Zgahl ('73a) observed high rates of RNA synthesis in an asexual strain of *D. gonocephala* during the second half of the interval between two successive divisions. Zaccanti and Tognato ('79) also describe the numerical increase in the population of a fissiparous race of *D. gonocephala* as characterized by three different multiplication rhythms, the first concerning animals derived from anterior pieces, and the second and third concerning animals derived from posterior pieces. According to these authors the evidence suggests the presence of predetermined fission centers (cryptic paratomy).

All these data could be regarded as a confirmation of the existence of physiologically isolated zooids in planarians in agreement with Child's well-known hypothesis ('20, '41). However, we think that this hypothesis has yet to be definitively confirmed.

Mixed sexual and asexual cycles. Many investigators have studied the relationship between asexual and sexual reproduction and their effects on the life history of the individuals. In some populations fissioning is the sole mode of reproduction, but in others it alternates with sexuality. In the first case no different reproductive phases take place during the year, the only variation being the arrest or the slowing of divisions when the ambient temperature falls below a certain level. In the second case various situations may be found, with different life cycles controlled by both environmental and genetic factors.

These different reproductive patterns were first illustrated by Curtis ('02) in the American species *Planaria maculata (=Dugesia tigrina).* This species breeds from early spring through summer laying many cocoons. Then the copulatory apparatus disappears and the worm becomes asexual, reproducing by fission until the following spring.

South European planarians of the genera *Dugesia* and *Phagocata* also undergo a sexual cycle during which they can breed and produce cocoons. It usually lasts from the beginning of winter up to the onset of summer. Then the genital system shrinks and disappears while the animals reproduce again by fission.

These species possess a high regenerative power; however, this is not an absolute prerequisite, because there are species that never show asexual reproduction even though they display a remarkable regenerative capacity (for instance *Polycelis nigra-tenuis*).

Fissioning depends primarily on genetic factors even though it is also influenced by external conditions, especially temperature. This fact was brought to light by investigations carried out by Benazzi (see especially '38, '74) on species in the *Dugesia gonocephala* group in which there are races with only sexual individuals, races with only (or almost only) asexual individuals, and races with individuals of both types. Benazzi demonstrated that individuals destined to fission do not form a reproductive apparatus, whereas those destined to sexuality have an early formation of this apparatus. The asexual state of the fissiparous specimens is determined by the same genetic factors that determine the power of division (fission-controlling genes). Their primary effect is probably to prevent the transformation of neoblasts into germ cells, which are absent in fissiparous individuals. The lack of development of gonads is associated with the absence of the other reproductive organs. On the other hand, some specimens of fissiparous lines may eventually attain sexual maturity and most of them simultaneously lose their ability to divide. The behavior of these "ex-fissiparous" specimens varies greatly. In some races several individuals become sexual and lay fertile cocoons from which both sexual and fissiparous offspring originate. Frequently the reproductive cycle is related to a seasonal cycle: most commonly, fissioning occurs during the summer and sexual reproduction occurs in winter. In other races, however, reproduction is exclusively asexual and, even though a few individuals become morphologically "sexual," they are sterile in the sense that they lay either infertile cocoons or none at all. Morphological examination of these sterile ex-fissiparous individuals has revealed that the copulatory system as well as vitellaria may develop completely, while the gonads are highly abnormal in structure and function (Benazzi, '69, '74; Benazzi and Ball, '72; Sakurai, '81). The ovaries are very large and distributed in a wide area of the anterior part of the animal. There are a great number of oocytes present, which suggests that a hyperplastic mechanism is involved. Gremigni and Banchetti ('72b) have shown that ovarian hyperplasia involves an extensive transformation of neoblasts into oogonia, but the oocytes formed from these oogonia are only rarely able to complete their growth. Electron-microscopic investigations (Gremigni and Banchetti, '72a) have shown that larger oocytes undergo a progressive nuclear vacuolization, an increase of the lysosomal system at the beginning of vitellogenesis, and eventually cell lysis. With regard to the male gonads, Benazzi and Deri ('80) have found that, while the number of testes varies greatly in the ex-fissiparous specimens of different populations, they usually do not reach complete maturity, and spermatozoa are scant or absent.

All these facts show that these ex-fissiparous specimens do not represent a sexually reproducing stage alternating with fission. They must instead be considered an abnormal expression of sexuality, which we regard as a partial expression of the fission-controlling genes that are repressed in the ex-fissiparous specimens.

It is interesting to remember that sexuality in fissiparous individuals may be experimentally induced by grafting a fragment of a sexual individual (Kenk, '41; Okugawa, '57; Lender and Briançon, '74, '75) or by feeding the fissiparous individuals on crushed tissues of sexual planarians for many weeks (Grasso, '71; Grasso and Benazzi, '73; Benazzi and Grasso, '77; Sakurai, '81). In any case, the experimentally sexualized specimens share both the morphological anomalies and the sterility that characterize the spontaneously occurring ex-fissiparous individials. It is therefore reasonable to assume that sexualizing substances (neurosecretions) might exert their action by interfering with the fission-controlling genes.

Summing up, it seems clear that fissioning in planarians is a multifactorial phenomenon that may represent either a permanent or transient condition depending on whether individuals have true alternating sexual stages or other irreversible alternations of their sexual potentialities. Benazzi ('74) has advanced the hypothesis that the number, penetrance, or expressivity of the fission-controlling genes varies a great deal among populations, causing their different reproductive behaviors. The assumption that fissioning is basically determined by genetic factors does not undervalue the importance of environmental conditions in controlling its expression. Physiological mechanisms may also be able to influence this phenomenon. Indeed, Child ('10) showed that decapitation can induce fission, a result later confirmed by Vandel ('21), Okugawa and Kawakatsu ('56), and Kanatani ('57). On the other hand, crowding has been shown to have an inhibiting effect on fission (Kanatani, '57). Other investigators have shown the importance of neurosecretory substances in this process (Lender, '70, '74; Lender and Zgahl, '68, '69, 73b; Best et al., '69).

Collection Technique and Maintenance of Adults

Kenk ('72) has described the collection, transport, and laboratory culture of freshwater planarians. Planarians may be collected by examining the underside of flat stones, other objects that have fallen in the water, or stems of water plants, and then removed from the substrate with a soft paintbrush or a moistened finger tip. They must then be placed immediately in a jar containing water taken from the same location. Where there is an accumulation of leaf litter or other debris, a sample may be taken in a glass jar filled with water and kept in a cool place. Planarians, if present, will tend to accumulate in the upper layer. Many planarian species can be attracted by bait (a piece of liver or other meat) placed under a stone and examined after several hours. According to Kenk, a very effective method is that of placing the bait in a glass or plastic jar with a lid bearing many small round perforations, 3–5 mm in diameter. The jar is then submerged in a shady location in a stream, pond, or lake and left for some time. This method is also applicable for lakes of greater depths.

Live planarians should be transported to the laboratory as quickly as possible. This is particularly important for species living in cold springs or ground water because they are very sensitive to temperature fluctuations. A thermos is quite useful to transport specimens to the laboratory and may be also used successfully to ship specimens by air. In this regard, it is advisable to mail recently laid cocoons, when they are available, along with adults, since the former are less sensitive to temperature fluctuations and could still hatch some juveniles even if the adults arrive dead.

Culturing is usually carried out in shallow glass aquaria or jars kept either in the dark or in dim light; the temperature must be in accordance with the tolerance limits of the population. Planarians are also very sensitive to the chemical composition of the culture water, therefore chlorinated tap water should not be used unless the chlorine is removed by appropriate chemical methods, by letting the tap water stand for at least 1 day in an open container, or by bubbling air through it. Filtered spring or pond water may also be used.

All planarians are carnivorous and can feed on small pieces of liver, other meat, earthworms, or clotted blood. In addition, some living animals such as the annelid *Tubifex* (obtainable in pet stores), the larvae of the dipterus *Chironomus*, or the small crustaceans *Asellus* and *Gammarus* are quite suitable for cultures containing a limited number of planarians. The food may be given once or twice a week and left in the aquaria for several hours.

Then it should be removed to avoid organic contamination of the culture water, which should be changed after feeding, especially if worms were fed on liver or other meat.

It must be pointed out that sick or recently sectioned planarians should not be fed since they do not eat even if they still have a pharynx. Similarly, food must not be put in vials, jars, or thermos bottles containing adults that are to be shipped. Planarians may survive starvation for a long time, even up to some months, during which they undergo a gradual diminution of various organs. Through starvation, adults may be reduced to the size of very young worms; and evidence has been obtained that they actually become "young" again. Therefore, starvation may act like fissioning or artificial division of the body, which is a very effective method of increasing specimen populations.

Obtaining and Rearing Embryos

Planarian embryos can only be obtained from cocoons laid by fertilized organisms and they must be reared inside the cocoon. Attempts have been made to obtain normal and complete development by culturing artificially hatched embryos, but not always successfully. This topic will be treated more fully in the section below on "Experimental Procedures." At present, the best way to study the complete series of embryological events is to collect freshly laid cocoons, isolate them, rear them in small vials labeled with date of deposition, and open the cocoons according to a timed sequence.

GENERAL NORMAL EMBRYOLOGY

The development of triclads is usually amphimictic. A single spermatozoon (polyspermy has been rarely observed) penetrates each oocyte in the late prometaphase stage and activates the maturation division. In the freshly laid cocoon the fertilized eggs are in the prefusion zygote stage during which the male and female pronuclei still appear to be separate. The fusion of the two pronuclei occurs some hours later. Acconci ('19) observed that the fusion takes place in *D. lacteum* 1–2 hours after deposition of the cocoon. Melander ('63) claimed that in *P. nigra* the second meiotic division of the fertilized egg is usually completed in the late afternoon and segmentation divisions start early in the morning, independently of the time of cocoon deposition. Benazzi-Lentati ('60, '70) observed that the duration of the stages preceding the first cleavage division varies greatly in relation to temperature, size of the eggs, and probably other undetermined factors. In fact, there may even be asynchronous development in eggs of the same cocoon. In species of the genus *Dugesia*, Benazzi-Lentati recognized the following events: 1) after deposition, the pronuclei rapidly lose their stainability; 2) 3–6 hours later they again appear stainable, showing a fine network with evident nucleoli; and 3) 10–12 hours later the chromosomes are well organized but frequently still separate in two pronuclear areas (Fig. 4). In this last case, amphimixis takes place at the metaphase of the first cleavage.

In some races or biotypes, however, amphimixis does not occur. Parthenogenesis in triclads has not been absolutely demonstrated although it has been assumed to occur by Dahm ('64) in *Atrioplanaria racovitzai* based on the anomalies of spermatogenesis and a peculiar modality of oogenesis. On the contrary, there is clear documentation of pseudogamy (or gynogenesis), which is known to occur in the following species group: *D. gonocephala sensu lato, D. lugubris-polychroa, P. nigra-tenuis* (for a general review see Benazzi and Benazzi-Lentati, '76). In these species, diploid and polyploid biotypes may be present. The polyploid biotypes develop pseudogamically in most cases; the sperm head penetrates the oocyte, activates the maturation divisions, and then degenerates (Fig. 5a) or is expelled along with a cytoplasmic bud or a polar body (Fig. 5b). The sperm head is actively involved in the expulsion process, because after intense irradiation (15,000–20,000 r) it degenerates in the egg and is not extruded (Fig. 6).

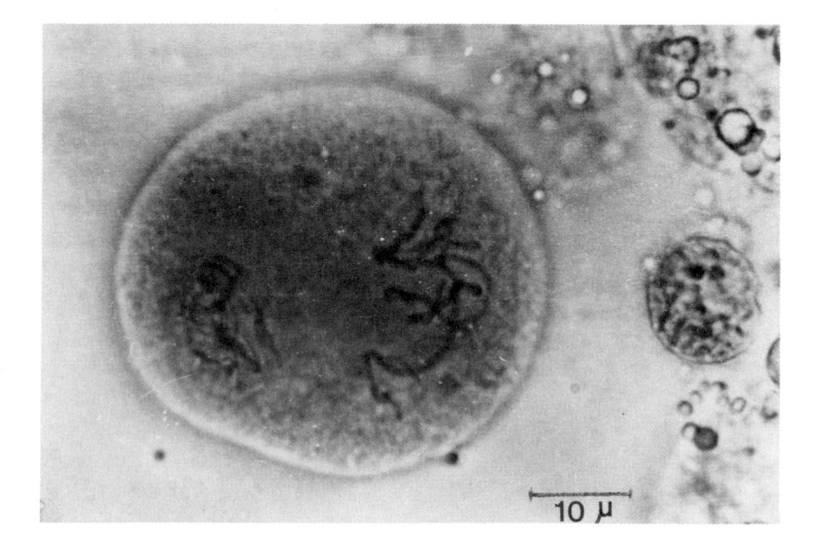

Fig. 4. Prophase of the first cleavage division in a zygote of the amphimictic biotype of *Dugesia benazzii*. The maternal and paternal chromosomes are still separated. (From Benazzi-Lentati, '70.)

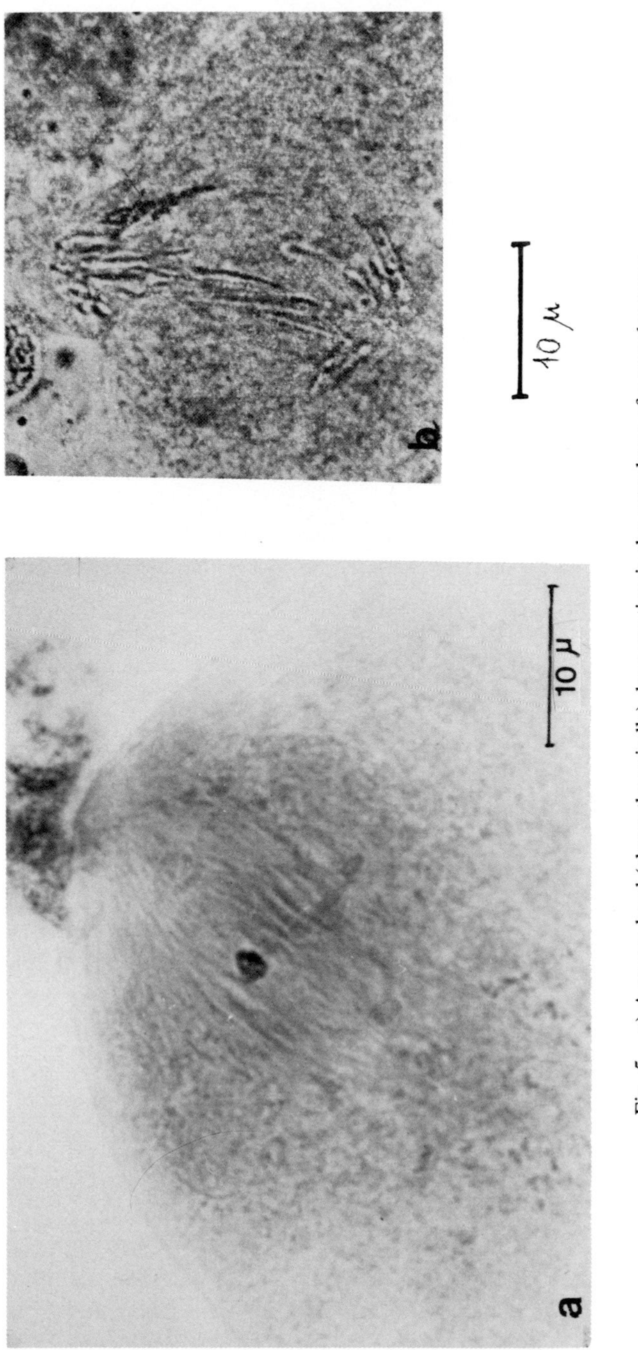

Fig. 5. a) A sperm head (above the spindle) degenerating in the cytoplasm of a pseudozygote of *Dugesia polychroa.* (From Benazzi-Lentati, '70.) b) A spermatozoon being expelled along with a polar body from an activated egg of *D. polychroa.* (From Benazzi-Lentati, '76.)

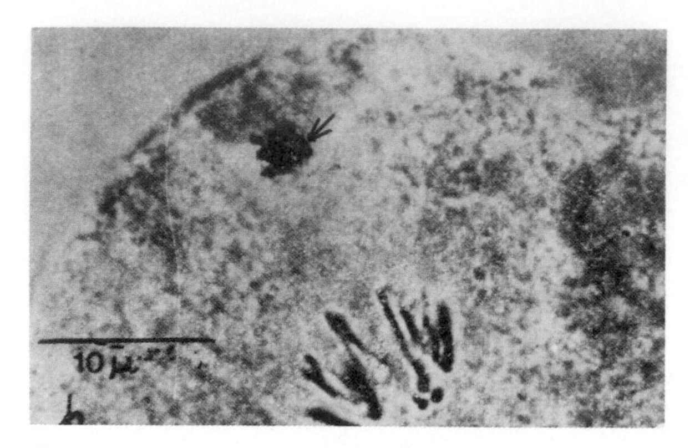

Fig. 6. An irradiated spermatozoon degenerating in the cytoplasm (arrow) of an activated egg of *D. polychroa*. (From Benazzi-Lentati, '76.)

Oocyte dimensions vary greatly according to species; moreover, in species showing different biotypes, the dimensions are correlated with the ploidy level (Mirolli, '56).

The interphase nuclei have an irregular contour and there are areas of cytoplasm that lie in infoldings of the nuclear envelope. A nucleolus with distinct fibrillar and granular components is evident. The cytoplasm is predominantly basophilic because of the large number of ribosomes and has several mitochondria. The primary oocyte often shows significant and characteristic species-specific differences. In *Polycelis nigra*, Gremigni and Domenici ('75) demonstrated that cortical granules that were distributed in a monolayer under the oolemma in ovarian oocytes are no longer present in the fertilized eggs (Fig. 7). Most probably a similar feature characterizes the eggs of *Planaria torva* and *Dendrocoelum lacteum*, which also contain cortical granules before fertilization (Gremigni, '69c, '74; Achtelik, '72). No conclusive data have been obtained so far regarding the fate of the autosynthetic yolk globules that accumulate in the ooplasm of Dugesiids (Fig. 8) (Gremigni, '69a,b; '76; '79) or of the "Balbiani body" observed in the ooplasm of *Dugesia dorotocephala* (Fig. 9) (Gremigni, '76).

The vitelline cells inside the cocoon have been divided into two types in relation to their morphological, chemical, and functional characteristics (Le Moigne, '63; Koscielski, '66; Marinelli and Vagnetti, '73). The first and more prevalent type consists of large round to oval cells (up to 40 μm in diameter) with a very low nuclear:cytoplasmic ratio. The cytoplasm is slightly basophilic and contains a great amount of reserve material, including yolk globules, lipid droplets, and glycogen (Fig. 10) (Domenici and Gremigni,

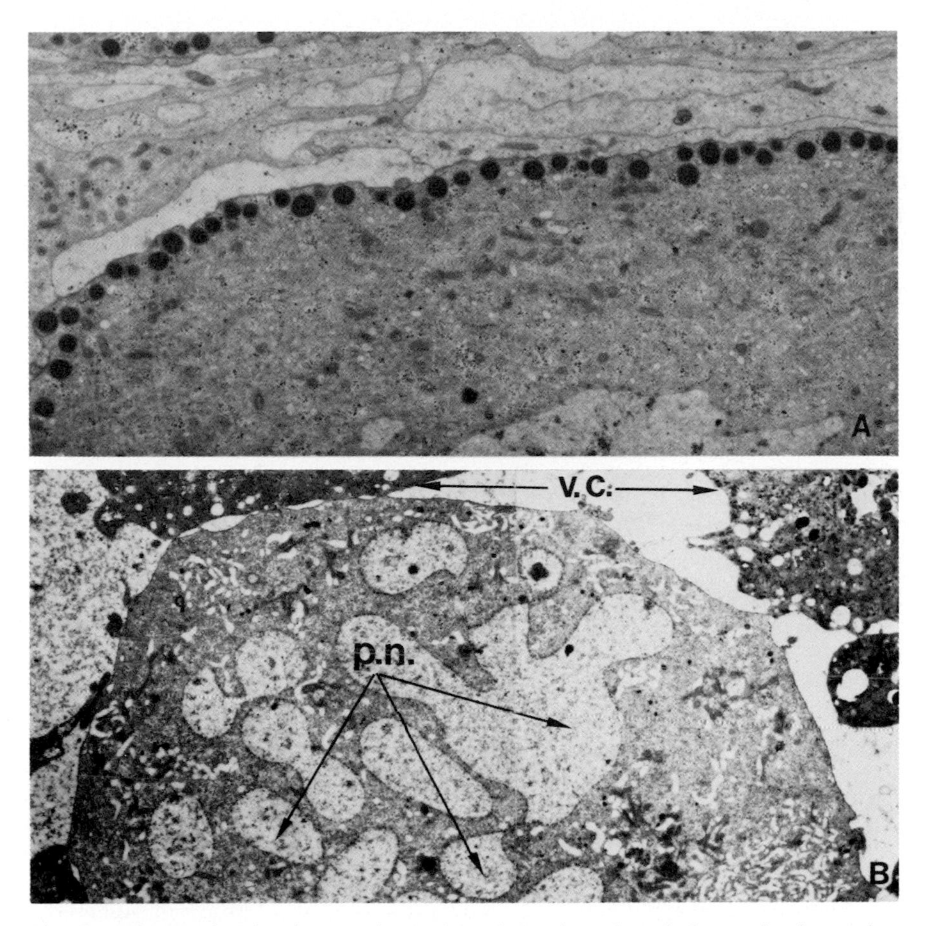

Fig. 7. Micrographs showing A) the peripheral location of cortical granules in ovarian oocytes (× 7,000) and B) the absence of cortical granules in a zygote within the cocoon of *Polycelis nigra*. (× 4,000) v.c., vitelline cells; p.n., polylobate nucleus. (From Gremigni and Domenici, '75.)

'74). There are only a few remnants of cocoon-shell globules, since most of them were extruded from the cell to form the cocoon envelope (Marinelli and Vagnetti, '73). These large vitelline cells will eventually be engulfed and digested by the growing embryo.

The second type of vitelline cell is smaller than that previously described and has an oblong shape. It is characterized by a nearly homogeneous basophilic cytoplasm with short projections. Lipid or protein inclusions

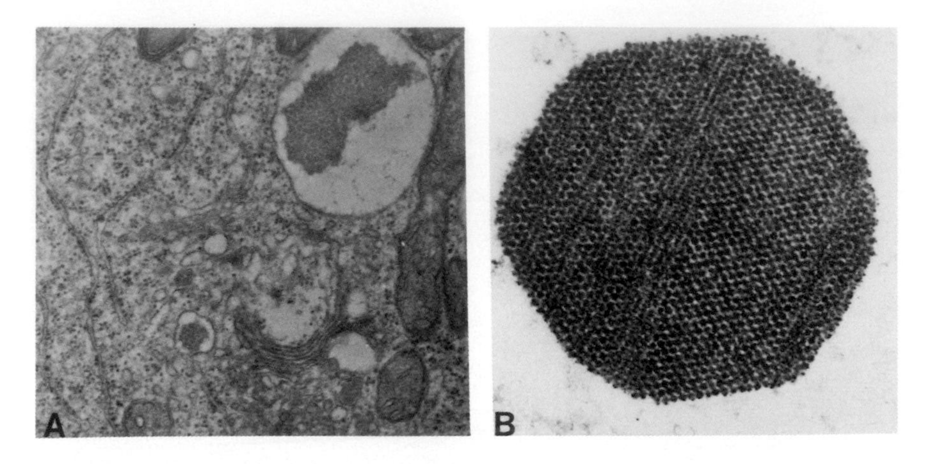

Fig. 8. A) forming and B) mature yolk globules in ovarian oocytes of *D. lugubris* s.l. × 30,000. (From Gremigni, '69b.)

similar in shape to those of type I vitelline cells are present (Marinelli and Vagnetti, '73). This type of vitelline cell is destined to form the yolk syncytium where blastomeric divisions occur.

Features of embryonic development were thoroughly described by authors over 50 years ago (see "Review of the Literature," below) and have been reinvestigated by Le Moigne ('63, '66a) in *P. nigra* and by Koscielski ('66) in *D. lacteum*. The following description (when not cited otherwise) is taken mainly from those careful studies. Le Moigne ('63) distinguished seven stages of development, which are characterized as follows:

Stage 1: Cleavage division of the egg and multiplication of blastomeres within a syncytium of fused vitelline cells (Figs. 11a, 12c,d).

Stage 2: Formation of an embryonic pharynx and intestine; delineation of a primary epidermis (Fig. 11b,c,d,e,f).

Stage 3: Engulfing of external vitelline cells within the embryonic intestine and multiplication of embryonic cells.

Stage 4: Delineation (stage 4A) and differentiation (stage 4B) of a temporary pharynx (Fig. 11g,l).

Stage 5: Appearance of the definitive organs: nervous ganglia or brain, pharynx, branched intestine; lengthening of the embryo.

Stage 6: Appearance of nervous fibers in the pharyngeal area.

Stage 7: Appearance of eyes, lengthening of nerve cords up to the tail.

The complete development takes about 24–28 days at 18°C in *P. nigra* (Le Moigne, '62).

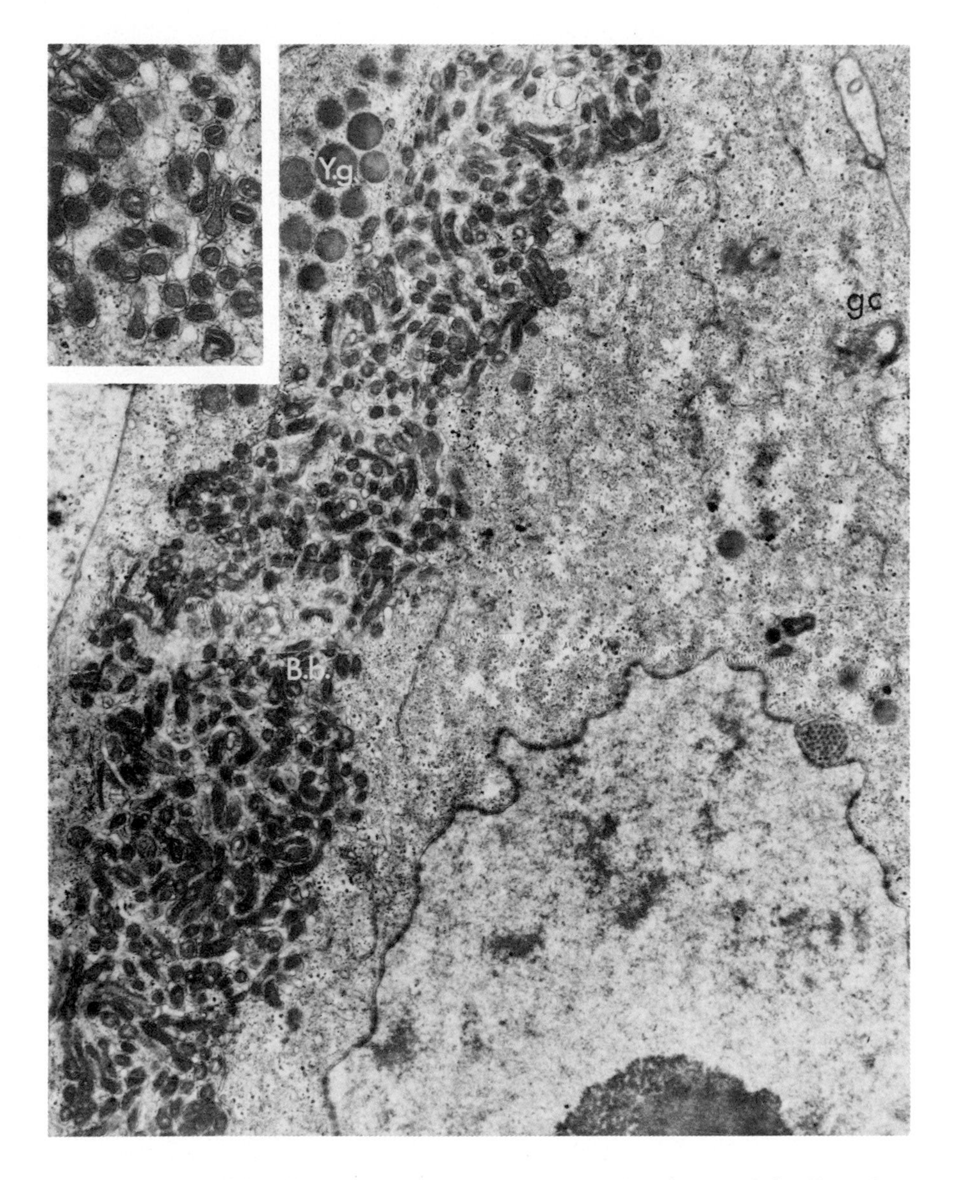

Fig. 9. Mitochondrial "Balbiani body" in an ovarian oocyte of *D. dorotocephala*. (From Gremigni, '76.) B.b., Balbiani body; g.c., Golgi complex; Y.g., yolk globules. × 12,000.

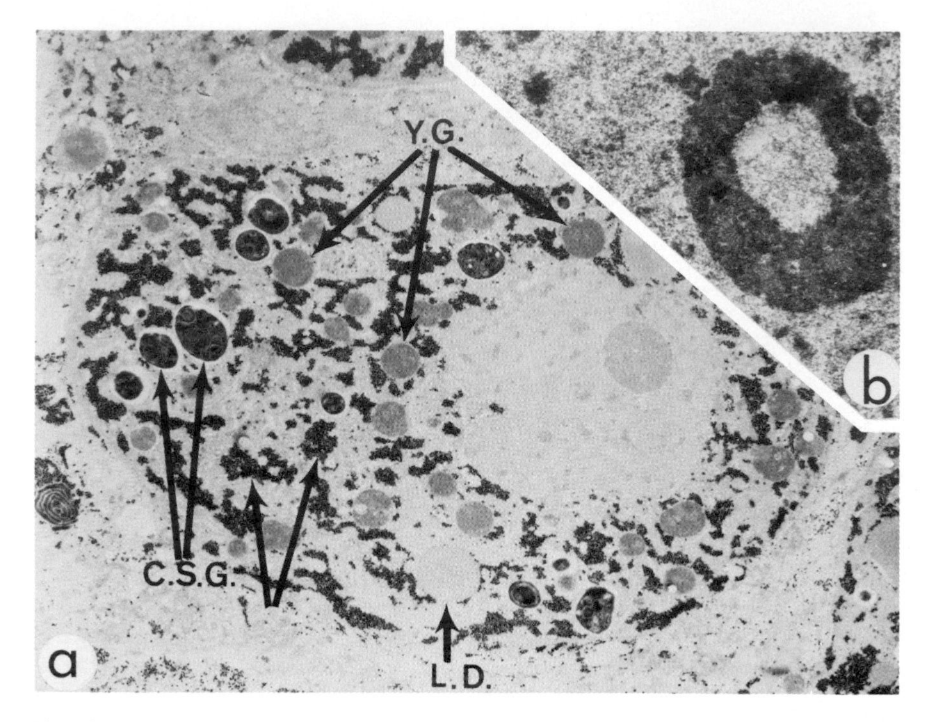

Fig. 10. a) General morphology of a mature type I vitelline cell in the vitelline glands of *D. lugubris* s.l.. Thin section stained with Thiery's method for mucopolysaccharides. × 3,850. b) Ring-shaped nucleolus from a similar cell. × 12,800. L.D., lipid droplet; Y.G., yolk globules; C.S.G., cocoon-shell globules. (From Domenici and Gremigni, '74.)

Fig. 11. Stages 1 to 4 of *P. nigra* embryonic development. a) Vitelline syncytium formation (stage 1); b–f) formation of embryonic pharynx and intestine (stages 2 and 3); g,i) formation of the temporary pharynx (stage 4); h) embryonic cells with cytoplasmic projections (stage 3). B, blastomeres; Cb, embryonic mouth cells; Cc, external cells of the pharynx; Ci, intermediate cells between pharynx and intestine; Cp, internal cells of the pharynx; E, epidermis; In, intestine; Le, lumen of the embryonic pharynx; Lt, lumen of the temporary pharynx; M, muscle cells of the temporary pharynx; Pe, embryonic pharynx; Pt, temporary pharynx; R, reticular tissue of the pharynx; Sy, vitelline syncytium; V1, type I vitelline cells; V2, type II vitelline cells. (From Le Moigne, '63.)

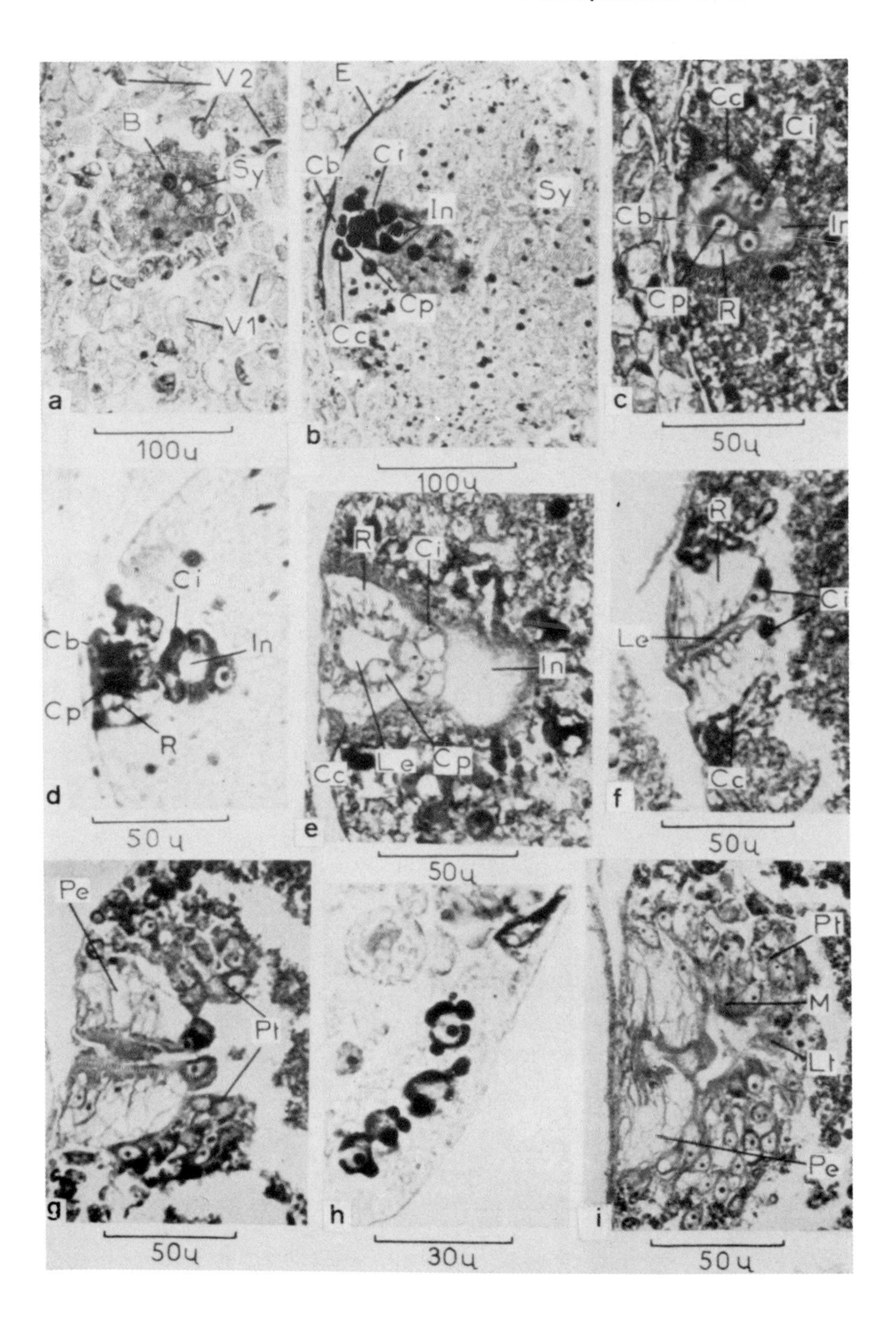
V2
B
Sy
V1
a
100u
E
Cb
Ci
In
Sy
Cp
Cc
b
100u
Cc
Ci
Cb
In
Cp
R
c
50u
Ci
Cb
In
Cp
R
d
50u
R
Ci
In
Cc
Le
Cp
e
50u
R
Ci
Le
Cc
f
50u
Pe
Pt
g
50u
h
30u
Pt
M
Lt
Pe
i
50u

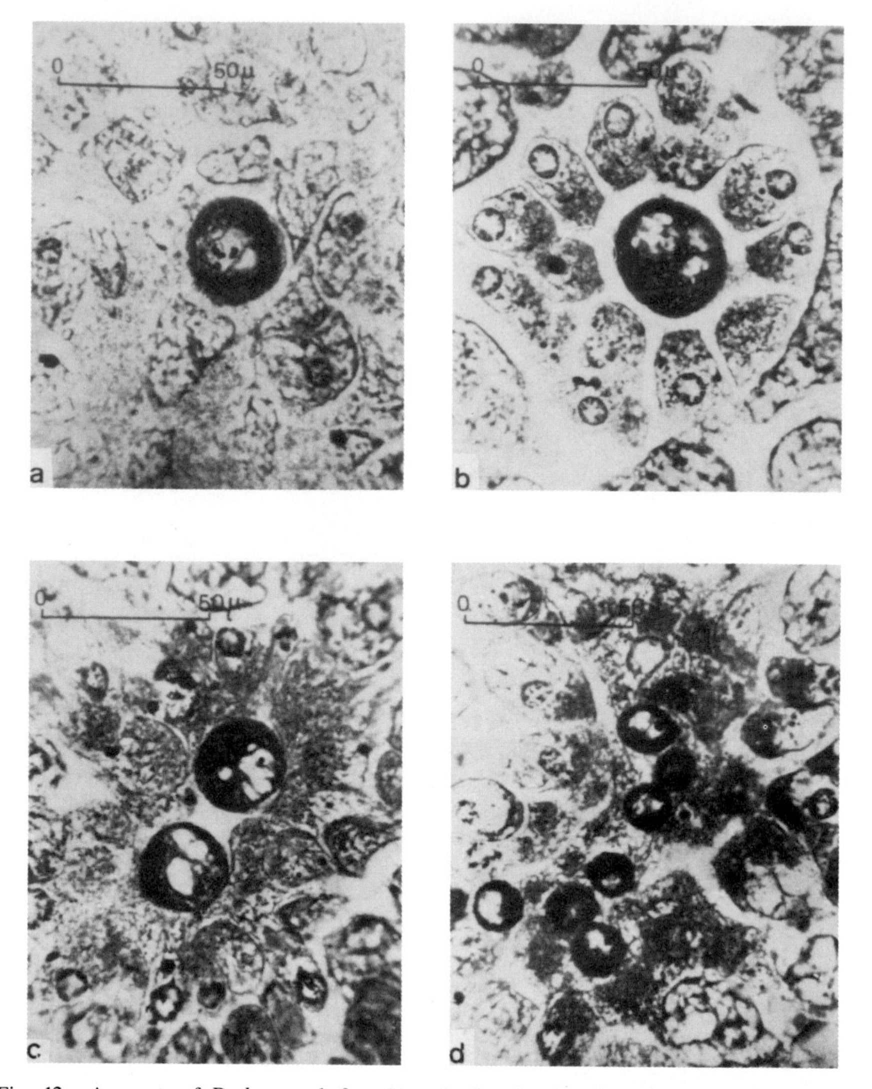

Fig. 12. A zygote of *D. lacteum* before (a) and after (b) clumping of vitelline cells; c) two-cell stage; d) a later stage of development showing several blastomeres in the forming vitelline syncytium. In b) vitelline cells show a proximal pole rich in RNA and a distal pole containing the nucleus and a highly vacuolar cytoplasm. (From Koscielski, '66.)

Stage 1

Just before the first cleavage division some vitelline cells approach and surround each egg (Fig. 12a,b) and by fusion of plasma membranes form a nutritive syncytium of an irregularly spherical or ovoid shape that progressively increases in size by the apposition of new vitelline cells at its periphery (Figs. 11a, 12c,d). According to Seilern-Aspang ('58) and Koscielski ('66), migration of vitelline cells towards the egg begins some hours after cocoon deposition and is due to so-called "grouping excretory substances" produced initially by the egg and later by blastomeres.

Ultrastructural investigations carried out on vitellaria would suggest that only one type of vitelline cell is produced in the vitelline glands. Later on, inside the cocoon, some vitelline cells might shrink and elongate, so that two poles become evident in them. The proximal pole is directed toward the egg and has a loose reticular cytoplasm apparently rich in RNA, while the distal pole contains the nucleus and a highly vacuolized cytoplasm (Fig. 12b). These cells have now assumed the characteristics of type II vitelline cells. According to Seilern-Aspang ('58) and Koscielski ('66) these are exposed to the action of syncytial substances produced by blastomeres and fuse together to form the yolk syncytium. This process occurs at the 8–20 cell stage and gives rise to the so-called "monocentric embryonic area" (Seilern-Aspang, '58). The internal part of this area is rich in RNA and mitochondria and corresponds to the proximal part of the type II vitelline cells. In contrast, the external part contains glycogen granules, lipid droplets, and vitelline cell nuclei. These last frequently degenerate into spherical pycnotic bodies, which will later migrate toward and penetrate the internal part of the syncytium.

The first cleavage division occurs through a regular mitosis and gives rise to two regular round-to-ovoid blastomeres, which separate from one another at telophase (Fig. 12c). They are a little smaller than the egg and their nuclei are highly lobulated with numerous deep infoldings of cytoplasm. The chromatin is dispersed, and one large nucleolus with segregated fibrillar and granular components is usually present. The nuclear envelope has numerous pores. The cytoplasm is highly basophilic and is poorly distinguished from the nucleus by light microscope. The cytoplasm is mainly filled with free ribosomes, which are often distributed in clusters or rosettes. There are also some mitochondria, and a limited amount of endoplasmic reticulum in the form of small vesicles. Other organelles are absent.

Subsequent cleavage divisions soon become asynchronous (Acconci, '19) so that one often sees stages with 5–7 blastomeres distributed in rows that are loosely dispersed in the syncytium (Le Moigne, '63; Koscielski, '66).

Melander ('63) studying *P. nigra*, showed that there is no synchronization of the division cycle during the first hours of development and that blastomeres exist in very different stages of the mitotic cycle. Melander also claimed that "after the second to third day of embryonic development (at 23°C) differences in size and morphology of blastomeres appear through unequal mitotic divisions. . ." and that ". . . there is a relation between shape, nuclear morphology and functions of blastomeres." However, due to irregular and asynchronous divisions, the fate of blastomeres is impossible to ascertain and early divisions do not yield a morula with determined polarity; localization of blastomeres is, rather, highly variable.

In the 24–30 cell stage blastomeres migrate to the periphery of the internal part of the syncytium, which forms a regular blastula-shaped structure whose blastocoel is filled with the internal mass of the syncytium (Fig. 13). Koscielski ('66, p. 91) states, "Despite the so-called 'anarchy' of cleavage stated in *Dendrocoelum lacteum*, there exists a stage where the regular arrangement of blastomeres corresponds to the blastula stage." This phenomenon, according to the author, represents a phylogenetic remnant.

During the cleavage period, the overall mass of blastomeres progressively increases as the vitelline syncytium supplies essential substrates for nucleic acid and protein synthesis, as well as sources of cellular energy, especially glycogen and fat droplets. Some degenerating blastomeres with pycnotic nuclei begin to appear between 10 and 34 hours after deposition (Melander, '63); their significance is not known.

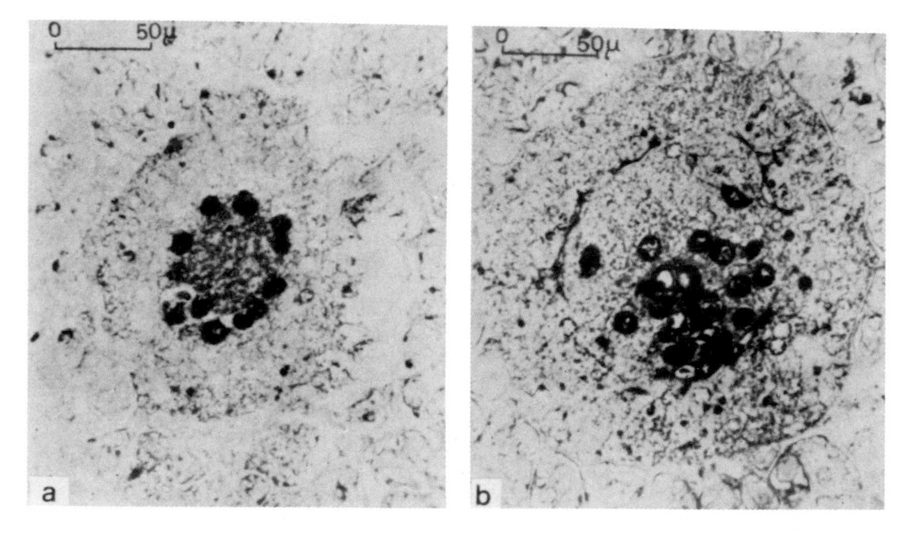

Fig. 13. a) Blastula-shape stage and b) stage of formation of the primary epidermis and embryonic pharynx in *D. lacteum*. (From Koscielski, '66.)

Stage 2

Five or six blastomeres migrate to the periphery of the syncytium after the embryo is composed of about 50 cells and has a 0.2 mm diameter. At the periphery the blastomeres flatten and form a thin, monolayered cap in the area where the future embryonic pharynx will develop. This represents the first delineation of the embryonic or primary epidermis (Fig. 11b).

The central blastomeres then approach the cap and some of them will form the embryonic pharynx and intestine. In *P. nigra* the onset of differentiation of the embryonic digestive system is detectable in cocoons laid 2 to 4 days previously and corresponds to embryos composed of about 60 cells, a dozen of which become epidermal cells. From the careful and detailed descriptions of different species given by various authors (Mattiesen, '04, in *D. lacteum;* Acconci, '19, in *D. lacteum* and *D. polychroa*; Fulinski, '38, in *P. torva*; Le Moigne, '63, in *P. nigra*), one can conclude that the origin of the embryonic digestive system is similar in all planarians.

During delineation of the embryonic pharynx, blastomeres differentiate and take up the following locations: a) four small cells at the level of the future mouth in continuity with the primary epidermis; b) four large cells that will form the internal cells of the pharynx; c) four external cells that will form the pharyngeal envelope; d) some small branched cells located between the last two groups of cells that will form the contractile tissue of the embryonic pharynx; e) four intermediate cells located between the internal pharyngeal cells and intestinal cells; and f) four cells that will delimit the embryonic intestine. The embryonic digestive system is initially solid and a lumen will form at a later time (Fig. 11b–f).

In its final form the embryonic digestive system consists of: 1) a mouth in contact with the epidermis; 2) a pharynx delimited by four external cells and formed by a reticular contractile tissue consisting of ramified cells from which muscle fibers originate; and 3) a hollow intestine separated from the pharynx by four intermediate cells.

During this stage the epidermal envelope becomes complete and is composed of extremely flat cells with small nuclei and a nonbasophilic strip of cytoplasm. Twenty to 30 free blastomeres rich in RNA are present within the syncytium, which is now acidophilic.

According to Seilern-Aspang ('58) and Koscielski ('66) the formation of the embryonic pharynx causes a transformation of the monocentric area into an area characterized by radial symmetry, the axis of which passes through the embryonic pharynx. According to these authors the area has a morphological gradient caused by differential displacement of syncytial RNA and mitochondria. This gradient will determine the dorsoventral axis of the future animal.

Stage 3

This stage is mainly characterized by the absorption of the type I vitelline cells that remained free in the cocoons. It occurs in 3–7-day-old *P. nigra* embryos. The vitelline cells are "swallowed" through the mouth by contractions of the reticular tissue of the pharynx and then go into the dilated lumen of the intestine. Thus the embryo assumes the aspect of an ovoid wineflask about 0.8 mm in length (Fig. 14).

The nutritive syncytium is confined to the periphery of the embryo and becomes transformed into a thin acidophilic layer in which some vitelline cell nuclei are still visible. The embryonic cells located within it actively divide, maintain the cytological characters of blastomeres, and mainly lie in the ventral hemisphere, which is occupied by the embryonic pharynx. During development some of the embryonic cells have cytoplasmic out-pocketings rich in RNA (Fig. 11h).

The epidermal cells become progressively thinner as the embryo enlarges and form a 1-μm-thick envelope around the embryo. Free embryonic cells migrate under this envelope and flatten to become secondary, ciliated epidermal cells (Skaer, '65). In the meantime, other free cells begin to differentiate into intestinal cells with phagocytic activity that sends cytoplasmic projections into the mass of vitelline cells located in the intestinal lumen. Muscle cells also differentiate at this stage.

Stage 4

This stage is characterized by an initial delineation (stage 4A) (Fig. 11g) and later complete formation (stage 4B) (Fig. 11c) of a temporary pharynx. It seems to occur exclusively in *P. nigra* and begins around the 7th–8th day after cocoon deposition when a migration of free embryonic cells forms a halo around the embryonic pharynx. The halo increases in size while the embryonic pharynx decreases in size, becomes confined to the external wall of the embryo, and then disappears. The halo is composed of pyriform cells rich in RNA that originate from embryonic cells. These divide actively and differentiate into muscular cells that will cover the pharynx with a thick layer of circular fibers. Undifferentiated cells containing vacuoles in a basophilic cytoplasm are present at the periphery of the temporary pharynx.

In the syncytium, some embryonic cells become flat and differentiate into epidermal cells, which are particularly numerous in the ventral region, while other free embryonic cells divide actively. Some of the latter produce cytoplasmic buds that are released into the syncytium.

According to Le Moigne ('66b, '68) free embryonic cells (blastomeres) maintain constant general features from stage 2 to stage 4A. They are round, are about 10–20 μm in diameter and have a nuclear:cytoplasmic ratio of 0.7.

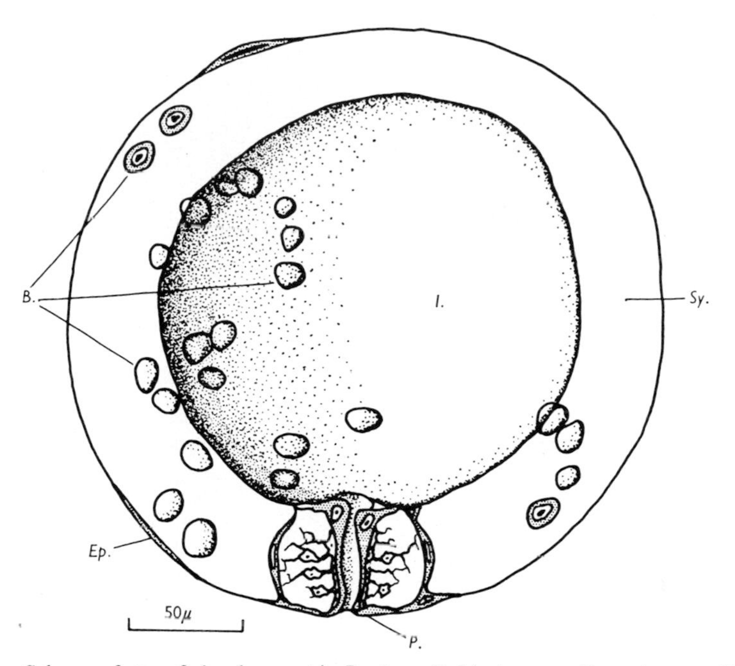

Fig. 14. Scheme of stage-3 development in *P. nigra.* B, blastomeres; Ep, primary epidermis; I, embryonic intestine; P, embryonic pharynx; Sy, vitelline syncytium. (From Le Moigne, '66a.)

They frequently appear surrounded by numerous mitochondria that lie free in the vitelline syncytium (Fig. 15). The nucleus is polylobulated and thus has deep infoldings occupied by cytoplasm. The chromatin is mainly diffuse and numerous interchromatin granules are present in the nucleoplasm. The nucleolus is 2–3 μm in diameter and has clearly separated fibrillar (central) and granular (peripheral) components. It is usually located near the nuclear envelope, which is rich in pores (Fig. 16).

The cytoplasm is particularly rich in free and clustered ribosomes but the endoplasmic reticulum and the Golgi complex are absent or very scarce; when present they take the form of vesicles. Mitochondria, round or elongated in shape, have slightly differentiated cristae and a poorly electron dense matrix. They are often associated with masses of fibrillar or finely granular material of uncertain origin and nature, usually known as "nuclear extrusions" or "chromatoid bodies" (Fig. 17). According to some authors these are characteristic of undifferentiated or differentiating somatic or germ cells (Sauzin, '66, '68; Le Moigne, '67a; Morita et al., '69; Pederson, '72;

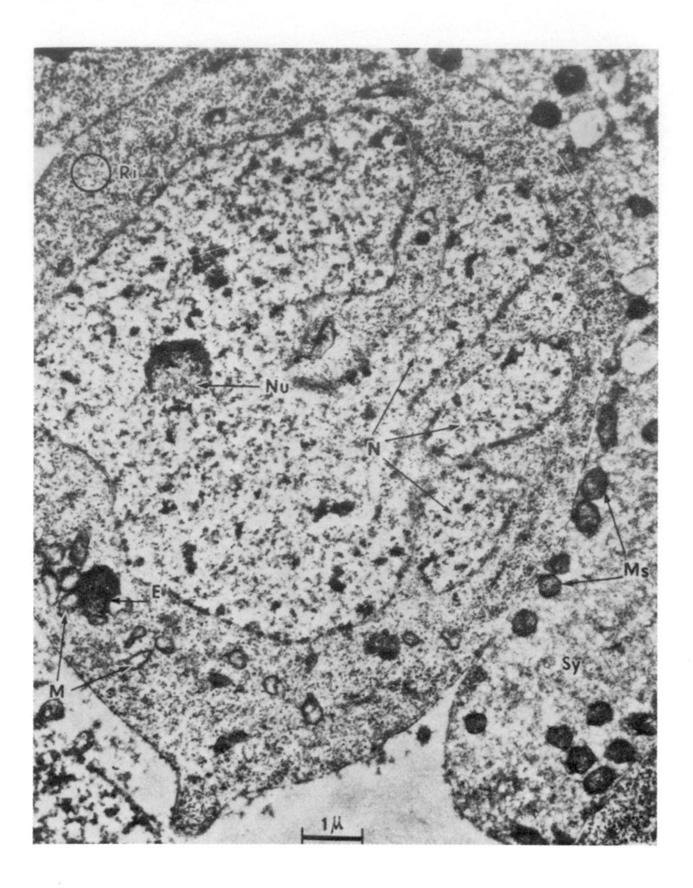

Fig. 15. Electron micrograph of an embryonic undifferentiated cell from a *P nigra* embryo, lying in the vitelline syncytium. The polylobate nucleus (N), the nucleolus (Nu), a nuclear extrusion (E) surrounded by mitochondria (M) and free or clustered ribosomes (ri) are seen. Several mitochondria (Ms) are also evident in the cortical area of the vitelline synoytium (Sy) facing the blastomeres. (From Le Moigne. '69.)

Franquinet and Lender, '73; Gremigni, '76); other authors find them characteristic of dedifferentiating cells (Coward et al., '74). Some microtubules, which probably represent remnants of the mitotic spindle, are visible in nondividing cells.

The role of the temporary pharynx is unknown; Le Moigne, who first described it, suggested that it could represent an ancestral remnant of a once-functioning embryonic pharynx.

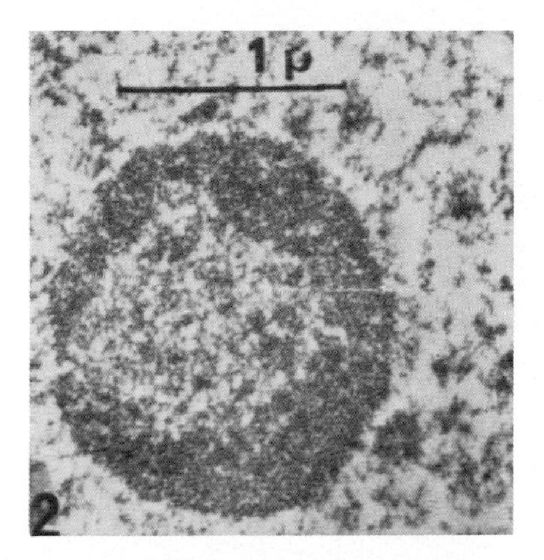

Fig. 16. Electron micrograph of a nucleolus from a stage-4A embryonic cell of *P. nigra* with the fibrillar and granular components. (From Le Moigne, '66b.)

Stage 5

This stage, occurring in *P. nigra* between the 9th and 14th day after cocoon deposition, is characterized by the delineation of the cerebral ganglia and the organization of the definitive pharynx and intestine. At the beginning of this stage a ventral anlage is clearly separated from the epidermal cells (Fig. 18a).

The first traces of the nervous system appear in the anterior area of the anlage just below the epidermis (Fig. 18b), where roundish cells actively divide and differentiate into a compact mass that will give rise to the cerebral ganglia.

The definitive pharynx (initially solid) appears in the posterior area of the ventral anlage in the same regions where the early embryonic and later temporary pharynx (now disappearing) were previously formed in *P. nigra*. In fact, the definitive pharynx results from a transformation of the temporary pharynx tissues. The pharyngeal atrium is the result of a tissue delamination between the ventral epidermis and the temporary pharynx. In species devoid of a temporary pharynx, the definitive pharynx can form either posteriorly (*P. torva, D. lacterum*) or anteriorly (*D. tigrina, Cura foremanii*) to the embryonic pharynx.

The intestine, which is initially formed as a simple epithelial tube, becomes progressively elongated in the posterior pharyngeal area, and forms one anterior and two posterior branches in which several smaller ramifications or caeca also appear.

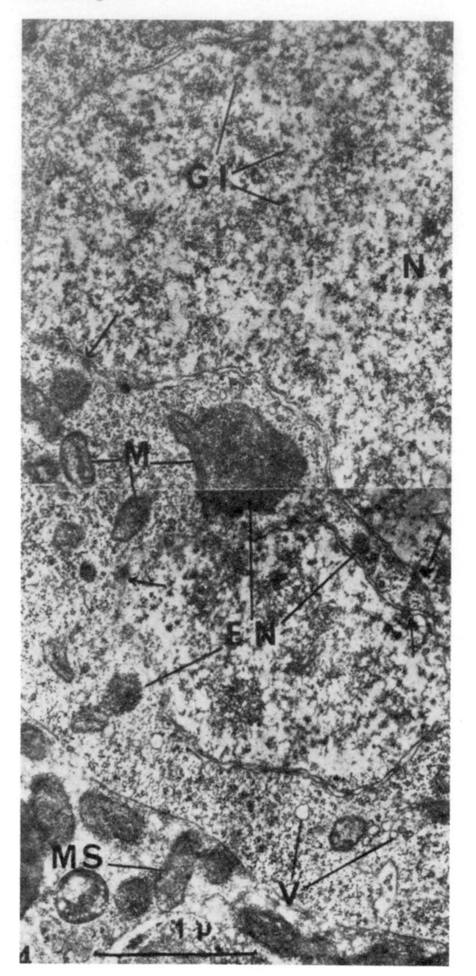

Fig. 17. Electron micrograph of an embryonic cell in the same stage as Figure 16, showing interchromatin granules (GI) and pores of the nuclear envelope (arrows) of the nucleus (N). Some nuclear extrusions (EN) surrounded by mitochondria (M) are seen in the cytoplasm. MS, mitochondria; V, vacuoles. (From Le Moigne, '66b).

In this stage the nutritive syncytium disappears and the organization of the definitive parenchyma, along with differentiation of several cell types, becomes evident. At the ultrastructural level cell differentiation begins with the development of the cisternae of the endoplasmic reticulum and Golgi complex, along with an increasing number and structural complexity of mitochondria. These differentiating cells continue to develop along different pathways and give rise to secondary epidermal cells, muscular fibers,

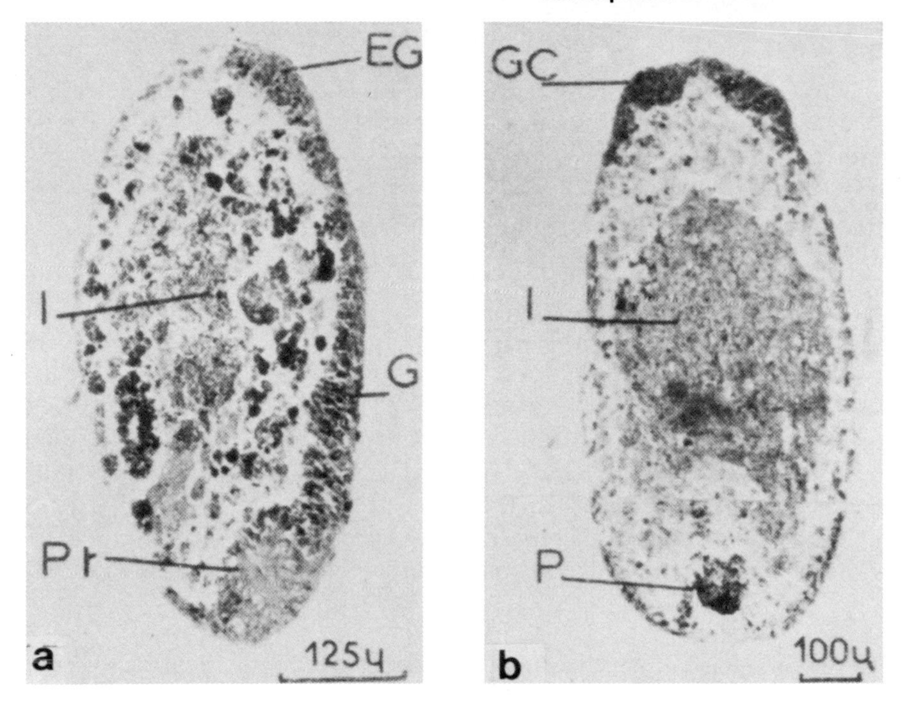

Fig. 18. *P. nigra* embryo at the onset a) and end b) of stage 5. EG, delineation of cerebral ganglia; G, ventral germinative cord; I, intestine; GC, cerebral ganglia; P, definitive pharynx; Pt, temporary pharynx. (From Le Moigne, '63.)

gastrodermal, protonephridial, nervous, and fixed parenchymal cells. (See Le Moigne, '67b, and Sauzin, '67a,b, for ultrastructural features of cell differentiation.)

The embryonic cells that do not undergo specialization or take part in organogenesis maintain the characteristic features of undifferentiated cells until hatching. They remain free in the parenchyma and strongly resemble the neoblasts described in adult worms (Le Moigne et al., '66). These embryonic cells differ slightly from blastomeres of the previous stages and show an increased nuclear:cytoplasmic ratio (1.7) and a more regular nuclear contour. Their nuclei still maintain a predominantly diffuse chromatin pattern and a nucleolus with segregation of fibrillar and granular components. Their cytoplasm is poorly differentiated and is rich in free or clustered ribosomes and chromatoid bodies surrounded by mitochondria. It is still poor in endoplasmic reticulum, Golgi complex, and other organelles.

In this stage the embryo progressively elongates to a length of 1.0–1.2 mm and assumes the basic shape of the hatching animal (Fig. 18b).

Stage 6

This stage is characterized by an increase in the size of the cerebral ganglia and by the appearance of nervous fibers, which occur between the 12th and 18th days after cocoon deposition in *P. nigra*. The nervous cells become spindle shaped and their axons become elongated and form the nerve cords while the brain commissure increases by differentiation of new fibers. The nerve cords elongate antero-posteriorly in the ventral area in front of the pharyngeal region.

In the other organs morphological and functional differentiation continues. The pharynx moves anteriorly and within it the first circular muscular fibers appear. Also, the intestinal caeca become deeper and more numerous and sensory cells appear in the apical area. The parenchyma is filled with undifferentiated embryonic cells rich in RNA and devoid of cytoplasmic buds. The eyes begin to appear about 14 days after deposition and signal the demarcation between stages 6 and 7.

The embryo now has the general appearance of a juvenile worm and is visible and mobile inside the cocoon. It can survive if experimentally hatched.

Stage 7

The final stage of embryogenesis is characterized by a complete differentiation and growth of the nervous system and eyes. The brain assumes its final shape with a prominent transverse commissure. The longitudinal nerve cords extend to the caudal area and lateral branching nerves derive from them to innervate the pharynx, intestine, and other organs. Other nerves appear anteriorly to innervate the eyes and the anterior sensorial area. The mouth will open into the external surface and the pharynx will acquire its final morphology with circular, longitudinal, and radial muscles. The epithelium becomes rich in cilia and gland cells, while the spaces between intestinal rami are filled with parenchymal tissue that assumes the characteristics of that seen in adult worms. These include both the presence of fixed cells and free undifferentiated cells, which are particularly abundant in prepharyngeal and postpharyngeal areas where the gonads and copulatory apparatus will form several weeks after hatching.

In this final stage the embryo becomes flat dorso-ventrally, is 2–5 mm long, and can successfully survive in a tap-water culture.

REVIEW OF THE LITERATURE

The first mention of embryological studies on planarians is usually attributed to Kölliker (1846), who described the contractile capacity of primary epidermal cells. However, the first extensive paper on general planarian embryology was published by Knappert (1865). He studied *Planaria fusca* (a

European planarian now interpreted variously taxonomically) and *Polycelis nigra* and found that each newly deposited cocoon contained four to six eggs surrounded by many vitelline cells. Each egg was described as being bounded by an envelope, the existence of which has never been confirmed by other authors. Knappert also observed the presence of two to three small globules (most likely polar globules) within the perivitelline space that forms around the egg. Moreover, he noted that a solid mass of blastomeres originated from total and equal cleavage divisions. This mass supposedly became enlarged through osmotic absorption of nutritive material, and at its periphery a membrane (the primary epidermis) was formed from blastomeres. In addition, a mass of yolk accumulated in the inner part of the developing embryo. The author also described the formation of a structure now known as the embryonic pharynx, which developed from thickening of the embryonic wall. Later on, the pharynx elongated, became hollow and very muscular, and was thus able to engulf vitelline cells. The author observed the development of the definitive pharynx as occurring simultaneously with a division of the embryo into a peripheral portion differentiating into epidermis and muscle and an internal portion differentiating into the intestinal wall. He also observed that embryos are initially round and only after intestinal branching do the worms assume the definitive, elongated shape. One is impressed by the fact that, although Knappert could only use rudimentary experimental methods, he was still able to individuate many of the main features of embryonic development in planarians.

After Knappert's paper those of the Russian zoologist Metschnikoff (1883), who was working at that time in Messina (Sicily) on *Planaria (=Dugesia) polychroa*, and of the Japanese researcher Iijima (1883, 1884), working at Leuckart's school in Leipzig on *Dendrocoelum lacteum*, were published. The results described by these authors showed that embryonic development in the two species was very similar, although some of their interpretations differed. Each assumed the existence of the three classic embryonic sheets: ecto, meso, and entoderm. However, Iijima suggested that the ectoderm originates from a peripheral layer of blastomeres that fuse together forming a syncytium. The syncytium would be able to absorb nutritive materials from the yolk cells by osmosis. In contrast, Metschnikoff suggested that the peripheral syncytium is formed from vitelline cell fusion, whereas the ectoderm originates from blastomeres migrating to the periphery. Moreover, the Russian scientist believed that the entoderm originated from fusion of vitelline cells ingested by the embryo, while the Japanese scientist claimed that the entoderm consists of a central mass of blastomeres that do not fuse together and that give rise to the embryonic pharynx. The mesoderm originates from free embryonic cells scattered throughout the synctium according to both authors. Metschnikoff and Iijima also described the formation of the embryonic and definitive pharynx and suggested that all organs in the adult originated from what they supposed to be the mesoderm.

A few years later the French researcher Hallez (1887) published a very important monograph on the embryonic development of *D. lacteum* and *Planaria (=Dugesia) polychroa*. He studied all stages of embryogenesis from fertilization to the development of all definitive organs. He first stated that blastomeres are undifferentiated cells, all of which are able to give rise to each organ. Hallez also underlined the difficulty in distinguishing blastomeres into embryonic sheets. In fact, from the very beginning of development the ectoderm is formed from a variable number of blastomeres that continuously become more numerous; therefore original ectodermal and entodermal cells would not both exist. Moreover, Hallez suggested that migratory cells are not to be interpreted as true mesoderm; rather they are similar to cells present in the interstitial tissue of Coelenterates.

The papers published in the final years of the last century laid the groundwork of our present knowledge about planarian embryogenesis. However, at the beginning of this century many important contributions were also published. The work of Bardeen merits mention, since his paper ('02) was the first to contain an experimental approach to planarian embryology. In fact, the author attempted to study the development of hatched embryos in *Planaria maculata* (=*Dugesia tigrina*) and also investigated the regenerative capacity of transected embryos. During the same period Curtis ('02) published the results of a descriptive embryological study on the same American species, and Stevens ('04) made similar descriptive studies on the other North American species *Planaria simplissima* (=*Cura foremanii*). In particular, Curtis reported on the time of appearance and on relative locations of the embryonic and definitive pharynx, giving rise to a long and lively debate ('05) with Mattiesen ('04), who worked on *Planaria torva* and criticized Curtis' data. Mattiesen's fine paper clarified several uncertain points concerning planarian development with conclusions that are now widely accepted. These include the mechanism of segmentation, the formation of the yolk syncytium and the embryonic pharynx, and, particularly, that early differentiation into embryonic sheets does not occur, the whole animal instead originating from embryonic mesenchyme.

Later on, some fine studies on the descriptive and experimental embryology of *D. lacteum* were published by Fulinski ('14, '16) who did, however, claim that embryonic sheet differentiation existed. About 20 years later the same author (Fulinski, '38) reinvestigated the significance of the genesis and evolutionary meaning of the embryonic pharynx trying to show the correlation between this structure and the "blastopore" of Polyclads.

Several fine observations on the fertilization, segmentation, and early stages of embryonic development of *D. lacteum* and *D. polychroa* were made by Acconci ('19). Later on Carlé ('35) investigated several aspects of development with special emphasis on the genesis and function of the embryonic pharynx in the terrestrial species *Geoplana notocelis* and *Rhynchodemus terrestris*.

More recently, important studies on planarian developmental biology include those of Seilern-Aspang ('56, '57a,b, '58) in freshwater and marine planarians, Koscielski ('64, '66, '67) in *D. lacteum*; the karyological study of Melander ('63) in *P. nigra* and *D. polychroa*, and the fine investigations of Le Moigne ('62, '63, '65a,b,c, '66a,b, '67a,b,c, '69) on *P. nigra*. All the papers by these authors are frequently referred to in this chapter.

Finally the review articles by Böhmig ('12–'17), Bresslau ('33), Hyman ('51), de Beauchamp ('61), Kato ('68), Skaer ('71), and Galleni and Gremigni ('82) should be mentioned; although they do not contain original data they provide the reader with interesting perspectives and overall views of planarian developmental biology.

EXPERIMENTAL PROCEDURES

Experimental studies of planarian development have received much less attention than descriptive studies. Le Moigne ('63, p 403) stressed this point: "Les études descriptives du développement embryonnaire des Triclades sont assez nombreuses et déjà anciennes. Les travaux expérimentaux sont par contre assez rares." Little had changed when, 8 years later, Skaer ('71, p.104), in a review paper on experimental embryology of marine and fresh water planarians, wrote, "Since the power of regeneration of adult planarians is so great and has been so much worked on, the experimental embryology of planarians is of considerable interest. Very little experimental embryology of planarians has, however, been carried out."

And now, more than 10 years later, we cannot unfortunately contest those comments even if some advances have been made by the Parisian school of Le Moigne, Franquinet, and Martelli to develop an experimental system permitting an easier approach to a molecular study of planarian embryonic development.

Development of Artificially Hatched Embryos

The most basic method used in the experimental embryology of planarians concerns the capacity of embryos to continue their development after being artificially hatched. Bardeen ('02) found that precociously hatched embryos of *Planaria maculata* (=*Dugesia tigrina*) can survive only if they are cultured from a certain stage, which is believed to correspond approximately to stage 6 of Le Moigne ('63). Le Moigne (personal communication) stated that *P. nigra* embryos that are artificially hatched even as soon as stage 4 are generally able to develop normally to form animals provided with a brain, eyes, pharynx, etc., even if grown in water with 50 μg/ml of actinomicin D. Moreover, he stated (Le Moigne, '63) of 38 embryos of *P. nigra* extracted from cocoons 10–15 days after deposition and cultured in tap water, 10 survived for only 24–48 hours, while 13 showed normal development, since they began to eat after 24 days. After eye appearance (stage 7) embryo survival was complete.

Seilern-Aspang ('57b) studied the first developmental steps in *P. torva* by spreading the contents of freshly laid cocoons on a cover-slide previously placed in a penicillin-streptomycin solution for 5–10 minutes. Although protected against drying, the life-span of such cultures was limited and embryos died quickly at early stages when the yolk syncytium was surrounded by the cells forming the primary epidermis. The author also suggested that a flattening of the "embryonic monocentric area" resulted in polyembryony.

Further investigations were undertaken on *D. lacteum* by Koscielski ('64, '66, '67) who, through a very simple culture method, occasionally succeeded in obtaining a nearly regular development of embryos and juveniles. Koscielski also confirmed the results obtained by Seilern-Aspang on polyembryony and ascertained that the thickness of the vitelline syncytia is a critical factor controlling embryonic development. In fact, if the cellular mass spread on the coverslip is sufficiently thick, a regular development occurs that gives rise to a normal juvenile (Fig. 19). On the contrary, if the cellular mass is too thin, the very flattened syncytia divide into two, three, or sometimes four areas, giving rise to several embryos that do not continue to develop.

Methods of culturing dissociated tissues and isolated cells have been improved by the experiments of Sengel ('60, '63), Ansevin and Buchsbaum ('61), Betchaku ('67), Chandebois ('68), and particularly Franquinet ('73, '76, '81); and it is now possible to obtain regular and complete embryonic development *in vitro*.

Centrifugation Experiments

In an attempt to obtain information about the type of developmental regulation (mosaic or regulatory eggs) in planarians, Seilern-Aspang ('58) performed several centrifugation experiments, which have been thoroughly analyzed and commented on by Skaer ('71). Seilern-Aspang found that, when freshly laid cocoons are centrifuged even for long periods of time, embryos develop regularly, whereas centrifugation carried out on 2–6-hour-old cocoons (stage of clumping of vitelline cells) may cause abnormal development and formation of symmetrical twins caused by displacement of several eggs in the same area. Prolonged centrifugation at this stage results in total nonhatching of juveniles.

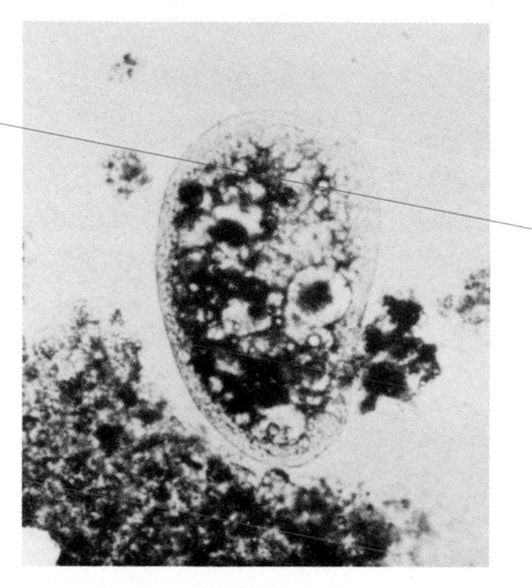

Fig. 19. A 10-day-old juvenile worm (embryo) of *D. lacteum* developed on a cover slip. (From Koscielski, '67.)

The period of sensitivity to centrifugation was identified to be that corresponding to Le Moigne's stage 1, when vitelline cell migration towards the egg occurs. Seilern-Aspang suggests that even after centrifugation the egg can still influence the surrounding vitelline cells to clump, thus showing that they possess a regulatory capacity. However, he does not discuss the effect of centrifugation on hypothetical "grouping substances."

When freshly laid cocoons are centrifuged continuously for 3–4 days, no embryos develop because embryonic pharynx formation is inhibited. When cocoons laid 7–10 days previously are centrifuged, displacement of some structures from the periphery to the central part of the embryo and *vice versa* is observed. However, after a short while regulatory forces are able to correct these abnormalities and the embryo again assumes normal morphology.

Based on Seilern-Aspang's experiments one could conclude that the antero-posterior polarity of embryos is not destroyed or modified by centrifugation. In fact such treatment can produce well-oriented twins.

Experiments With Ionizing Radiation

Planarians have long been the subject of experiments in radiation biology. Bardeen and Baetjer ('04) showed that regenerative capacity is lost following a strong dose of X-rays, and Shaper ('04) obtained the same effect with radium. Similar results were then obtained by several authors and the failure to regenerate was attributed to the selective death of the so-called "neoblasts," which seem to be highly radio-sensitive (see Brøndsted, '69, for a review).

The influence of ionizing radiation on embryos and adults of planarians has been studied by Benazzi ('62). He observed that adults of *D. lugubris* s.l. that had received 3,000 to 6,000 r of X-rays behaved quite normally for about one month. They were also able to copulate and lay fertile cocoons. However, 3–5 weeks after radiation, signs of damage appeared and led progressively to the animal's death. This effect was dose-related, with highly irradiated animals being affected earlier.

Benazzi also irradiated cocoons at different times from deposition with X- and γ-rays. The results clearly showed that the effect of radiation varies greatly in relation to the particular stage of embryonic development. Doses of 1,000 to 1,300 r may not be lethal for eggs, even before the first cleavage begins. In fact, some cocoons hatched and the juvenile worms reached sexual maturity and produced eggs with a normal karyotype. In contrast, doses from 1,500 to 2,000 r are consistently lethal for both eggs and early spherical embryos. The cocoons irradiated a few hours after deposition and examined after many days sometimes contain early spherical embryos with necrotic blastomeres. This shows that irradiated eggs are able to divide and organize early embryonic stages but then development stops. The necrotic blastomeres are characterized by the loss of nuclear chromatin and by a "lamellar pattern" in their cytoplasm. The same dose of 1,500–2,000 r applied to fully developed embryos (stage 7), with flattened body shape, definitive pharynx, and eyes, does not kill the embryos; thus the cocoons hatch and normally shaped juveniles are produced. However, in the next few days pathologic changes such as swelling, depigmentation, and parenchymal cell rarefaction occur (Fig. 20) followed in most cases by the death of the young planarian. These alterations are most probably due to the destruction of neoblasts. These results were confirmed and extended by Lanfranchi ('64).

The use of ionizing radiation also enabled Benazzi ('63) to provide a new direct proof of the independence of the two events involved in amphygony, namely egg activation and karyogamy. Benazzi demonstrated that some species have both diploid and polyploid biotypes, the first being amphimictic and the latter pseudogamous. Research on the radiation effect on these biotypes was conducted on *D. polychroa* using the diploid biotype (biotype A according to Benazzi, '57) and the polyploid biotypes B and C, which have synaptic and asynaptic oogenesis, respectively. Some adults were whole-body X-irradiated with 1,000 to 7,000 r and then mated to untreated, unfertilized specimens acting as females. In cases of amphimictic oocytes, the development could be quite normal as long as the spermatozoa received an X-ray dose lower than 2,000 r. When 2,000 to 3,000 r were employed, the

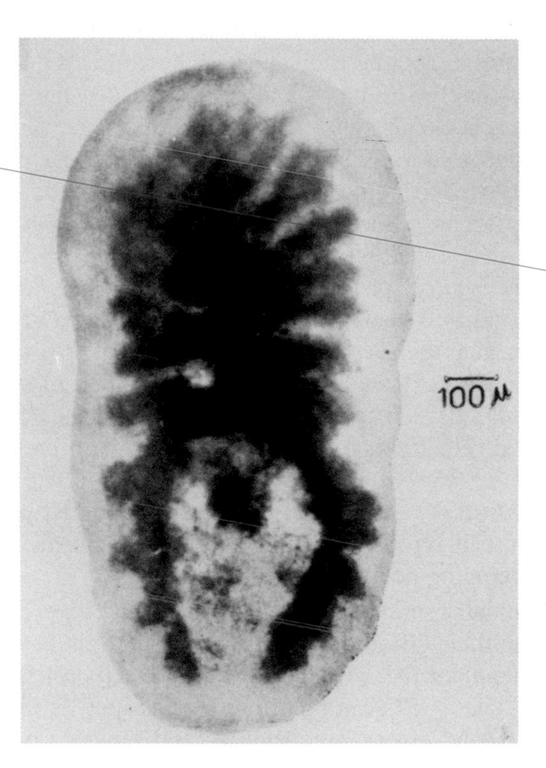

Fig. 20. A young *D. lugubris* s.l. specimen born from an irradiated embryo showing specific regressive features (swelling, depigmentation, and parenchymal cell rarefaction) probably due to destruction of neoblasts. (From Benazzi, '63.)

cleavage was regular and spheric embryos would sometimes form. However, development in most cases did not continue; therefore the highly irradiated sperm gave rise to a lethal constitution, which produced an effect similar to the one caused by the direct irradiation of fertilized eggs.

On the contrary, the development of pseudogamous eggs is normal even following more intense irradiation of the sperm. This proves that X-rays do not diminish the sperm-activating power, at least within the limits of the experimental doses. Thus, Benazzi's research on amphimictic and pseudogamous biotypes of planarians confirms the distinction between karyogamy and sperm-activating power that was first evidenced in the classic research of Hertwig ('11) and then by other authors in the normally amphimictic species of Amphibia.

It is also interesting that the irradiated sperm maintain their activating power for several weeks within the seminal receptacle of the partner; thus again one sees that X-rays act only upon nuclear material. Benazzi-Lentati ('64) observed normal spindle regulation of the egg even when the spermatozoa had been strongly irradiated and concluded that the regulation of the maturation process is exclusively under the control of the egg.

Le Moigne also ('65a, '66a, '68) studied the effects of irradiation on developing embryos. He found that in *P. nigra* moderate doses of X-rays (1,000 r) given prior to stage 4 cause the death of the embryo. In contrast, starting from stage 4, even higher doses (3,000 r) do not prevent differentiation and hatching in most specimens. However, stage-7 embryos and hatchlings that were irradiated could not regenerate after being sectioned. Indeed, they could only heal at the site of the cut but not form a blastema. The author suggests the following explanation for this: Starting from stage 4, embryos have two types of cells: a) differentiating (or determined) cells (type I) and b) undifferentiated cells (type II) both of which show different behavior after X-ray treatment. In fact, starting from stage 4 the blastomeres can meet either of two fates: they can a) undergo determination and begin to differentiate into various cell types, or b) maintain an embryonic undifferentiated condition thus becoming neoblastlike cells.

Type I cells initially show some radiolesions, which are repaired, and the cells can continue their differentiation, giving rise to tissues and organs that become smaller than normal, presumably because cell proliferation decreases after irradiation. These cells can therefore be classified as radioresistant.

Type II cells show lesions precociously and rapidly degenerate. A few days after irradiation the nucleolus shows obvious alterations and subsequently dissociates completely; the nucleus shows rarefaction and aggregation of chromatin clumps and later undergoes fragmentation. In the cytoplasm, ribosomes decrease in number and eventually disappear, and large vacuoles appear. In addition, mitochondria and endoplasmic reticulum degenerate (Fig. 21). Consequently, type II cells are destroyed and can be classified as being radiosensitive. Thus, at hatching juvenile planarians would be devoid of neoblasts and incapable of regenerating.

Regeneration Experiments on Embryos

Bardeen ('02) studied the regenrative capacity of *D. tigrina* embryos following a prepharyngeal section made in a developmental stage when nervous cords were forming. Some of the anterior fragments succeed in regenerating both the pharynx and the tail; but, while the posterior fragments survived and became pigmented, they never regenerated a head. The latter could be regenerated only in older embryos where well-formed nerve cords reached the posterior part of the body. Thus, the author suggested that the lack of head regeneration was due to insufficient development of the nerve cords.

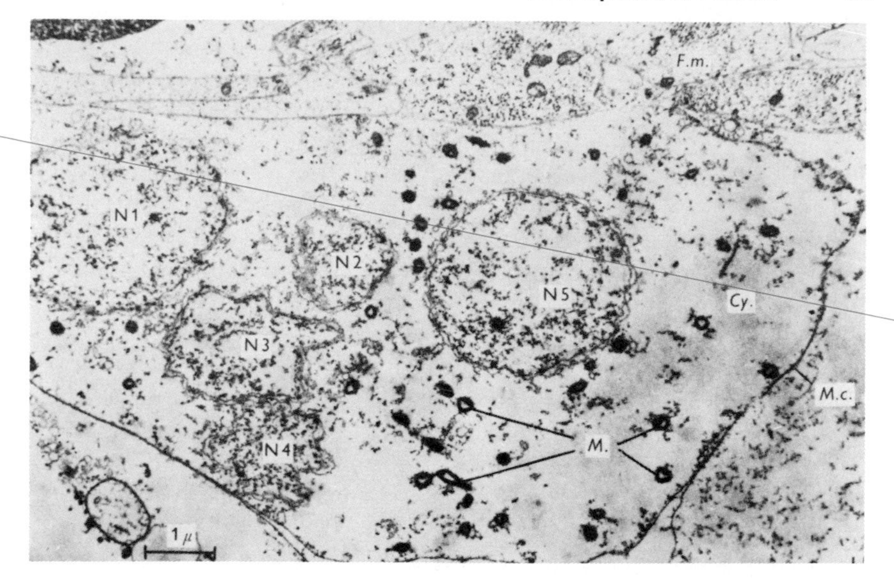

Fig. 21. Electron micrograph of an undifferentiated cell from a stage-4B embryo of *P. nigra* 72 hours after X-ray treatment (3,000 r). The cell is completely lysed. N1–N5, nuclear fragments; M, mitochondria; Cy, rarefied cytoplasm; Mc, fragmented plasma membrane; Fm, muscle fibers of a surrounding differentiated cell. (From Le Moigne, '68.)

Similar results were obtained by Liotti and Bruschelli ('64) in *D. lugubris*. According to the authors, the lack of head regeneration was due to the absence of well-differentiated nerve cells in the amputated fragment.

Le Moigne ('65a,b,c, '66a, '69) repeated these experiments in *P. nigra* embryos. He stated that regeneration of both the anterior and posterior fragments following a prepharyngeal section is possible after embryos reached stage 5 when nerve cords are not yet differentiated. However, at stage 5 a high mortality was observed and regeneration occurred in only a few cases. The author also noticed that starting with this stage the remaining, still developing tissues and organs, along with the nervous system, continue to differentiate before the blastema is formed. Thus, the blastema is formed more slowly in stage-5 fragments than in stage-6 or -7 fragments because in stage 5, the differentiation of the nerve cords must precede the normal process of regeneration. Anterior fragments showed a higher regenerative power (85% at stage 6 and 90.6% at stage 7) than posterior fragments (58.7% and 58.5% at stages 6 and 7, respectively). Moreover Le Moigne ('66a) stated that in some cases regeneration is improved by a rapid cicatrisation in embryos which were not differentiated enough. In fact, after a second cut was performed at the level of the first healing, a higher percentage of stage 7 posterior fragments (84% versus 58.5% after the first cut) regenerated. By contrast, stage 6 embyros are too young to bear a second cut so that the percentage of regenerates decreases (26% versus 58.7%).

In conclusion, Le Moigne suggests that morphologically undifferentiated cells are available for regeneration at stage 5, a stage when determined nerve cells are already present along the ventral part of the embryo. After the cut the determined nerve cells differentiate, thus forming the nerve cords, and then the posterior fragment of the embryo is able to build a blastema. A second cut will in some cases improve the regeneration process.

Regeneration Experiments in Adults

It is a widely known fact that most freshwater planarians have an extraordinarily high regenerative capacity, which has always interested and stimulated scientists to investigate this complex phenomenon; an enormous number of papers have been published on this area (for reviews see books by Brøndsted, '69, and by Chandebois, '76).

In the introduction to his excellent book, *Planarian Regeneration*, Brøndsted succinctly enumerated the many reasons why researchers are enthusiastic about studying regeneration. One of the most important reasons is that, "One of the foremost goals in elucidating the factors involved in regeneration is to reconcile the principles in regeneration processes with those in ontogenesis from the egg." The author continues, "In fact, ...so many similarities are found between the ontogenetic process and that of regeneration that it is permissible as a working hypothesis to postulate the same kind of major morphogenetic principle in ontogenesis and regeneration." However, Brøndsted suggests that regeneration is even more complex than embryonic development, because tissue and organ regeneration occur in a system that is already organized and full of stimulatory influences.

When cut or wounded, a planarian first forms a "healing" and then a "blastema" at the wound level. From the blastema the regeneration of missing or damaged parts of the body begins. One of the major open questions in regeneration research is identifying the source of blastema cells, as well as their fate during regeneration. At present it is almost unanimously believed that the blastema consists of undifferentiated cells that are able to differentiate into any functional cell type, depending on the morphogenetic field influences they encounter during regeneration (Child, '20, '41; Wolff et al., '64; Vannini, '65). Thus blastema cells, in a sense, behave during regeneration as embryonic cells do during ontogenesis. A particularly strict correlation between the differentiation on events occurring in regeneration and ontogenesis has been proposed by several authors (see Wolff, '62; Lender, '62; Brøndsted, '69; Pedersen, '72; Gremigni, '74, '81 for reviews), who claim that the blastema is formed by embryonic reserve cells (the frequently cited neoblasts) that remain freely scattered in the parenchyma of the adult worm throughout its life. Neoblasts are considered to be undifferentiated cells that maintain all the original potentialities of blastomeres even if the neoblasts are still present and functioning in adults.

The "neoblast theory," although supported by a great deal of experimental evidence, has always had and still has numerous opponents. In recent years Hay ('68) and Coward ('69) in particular have criticized this theory and proposed the so-called "dedifferentiation theory" that was previously suggested by Steinmann ('08, '25) and Lang ('12) and accepted, among others, by Hyman ('51), Woodruff and Burnett ('65), and Rose and Shostak ('68). According to this theory blastema cells arise from functionally differentiated cells that, following an injury, can dedifferentiate and reacquire multiple potentialities.

The possibility that cell dedifferentiation actually occurs in planarians during both traumatic and physiological regeneration seems quite likely. Results obtained from both regeneration and starvation experiments indicate that dedifferentiation is perhaps a more common phenomenon than was previously thought. Ultrastructural observations support this belief (Coward et al '74; Hay and Coward, '75; Gremigni and Puccinelli, '77). In addition, experimental evidence supporting the existence of dedifferentiation–redifferentiation in germ cells during regeneration has recently been obtained by Gremigni and co-workers (Banchetti and Gremigni, '73; Gremigni and Puccinelli, '77; Puccinelli and Gremigni, '78; Gremigni and Miceli, '80; Gremigni et al., '80a,b). These studies employed a polyploid biotype of *D. lugubris* s.l. that is provided with a natural karyological marker. Moreover karyological and histological investigations (Baguña, '76; Gremigni and Puccinelli, '79) demonstrate that specialized cells progressively decrease in number in starved planarians while the number of undifferentiated cells remains relatively high and the percentage of mitoses increases. Moreover, Gremigni and Puccinelli ('79) demonstrated that a small percentage (3%) of dividing undifferentiated cells were clearly derived from male or female germ cells. The latter investigation suggests that during starvation some specialized cells degenerate and supply the surviving cells with elementary substances while other differentiated cells dedifferentiate and help to maintain a sufficiently large pool of undifferentated cells required for the continuous turn-over of cells and tissues that is indispensable for survival.

Based on the results of regeneration and starvation studies Gremigni and co-workers proposed that blastema formation and regeneration in planarians involve both undifferentiated reserve cells and dedifferentiated cells that have not reached an irreversible specialized state.

Martelly and Le Moigne recently observed that immature young planarians form a blastema and regenerate even if maintained in an actinomicin D solution after the cut, while in adults this drug determines a complete inhibition of regeneration (Martelly and Lender, '79). Furthermore, the increase in RNA and protein synthesis is very different in transected young and adult planarians. The authors suggest that regeneration in young planarians involves predetermined cells which may be morphologically undifferentiated, but are provided with stable mRNA (Martelly, Franquinet, and Le Moigne, '81).

SUPPLEMENT

No special equipment and reagents other than those usually present in an average zoological or embryological laboratory are needed to undertake descriptive or experimental study of planarian embryology. Collection of planarians is usually done by hand, sometimes a pierced jar or fine net is used. A paintbrush or finger tip can be used to gently handle planarians. Also Pasteur pipettes with a small aperture can be used to transfer cocoons, growing embryos, juveniles, and adults. Glass jars of various size, thermoses, bottles, vials, or small tubes can be used to maintain grouped, coupled, or single planarians alive; the last type of container is particularly useful for developing embryos within the cocoons. Tungsten-tipped needles, micropipettes and fine razor blades are needed to gently pierce the cocoon shell. Oxygenators and aerators are not necessary, unless one wishes to culture and maintain reophilous worms. However, controlling the ambient temperature with thermostatic chambers and controlling light intensities can be very useful since most planarians avoid light and are better maintained in the dark.

To culture separate cells and embryos *in vitro,* the usual equipment is necessary along with various substances to make up specific culture media, as reported in the specific literature.

CONCLUDING REMARKS

The normal embryonic development of freshwater planarians has been widely investigated since the last century through classic histological and, more recently, also by electron-microscope techniques. All the investigations show that ontogenesis of triclads is unique in the animal kingdom. It does not resemble the primitive type of development—"the spiral type"—which is characteristic of some Archoophora groups of Turbellaria. Indeed, the large number of vitelline cells surrounding the developing eggs within the cocoon greatly interfere with embryonic development so that no trace of primitive spiral cleavage is identifiable. The segmentation is usually defined as "anarchic," which means that it is so irregular that the fate of blastomeres, the precise origin of cell lines and organs, and a gastrula stage cannot be identified.

According to most recent studies of planarian embryology, all definitive structures arise from a single cell type—undifferentiated and undetermined free blastomeres. Thus the theory of embryonic sheets (ecto-, meso-, and entoderm) does not seem to be applicable to these animals, and early segregation of cell lines is not believed to occur. However, Melander ('63) expressed a contrary point of view for the germ lines. On the basis of a cytogenetic investigation carried out mainly on *P. nigra*, this author claims that a precocious differentiation or determination of male and female germ cells can be identified in the early stages of embryonic development. He stated that, "...there is a relation between shape, nuclear morphology (different constellation of large and small chromocenters) and function of the blastomeres," and, in particular, that, "...it seems probable that the two dark chromosome segments seen in the oocytes are descendants of the chromocenters present in certain blastomeres. Similarly the spermatocytes which lack conspicuous chromocenters might be descendants of blastomeres without such structures." Moreover, Melander surprisingly suggests that, "The two different chromocenters behave differently in oocytes and blastomeres as regards formation of nucleoli. The chromosome segments which form nucleoli in the oocytes, form no nucleoli in the blastomeres. Nucleolus in the blastomeres and somatic cells is organized from minute terminal satellites." Finally, Melander concludes that, "A cellular sex determination within these hermaphroditic worms would occur, governing certain cells to produce epithelium of the ovaries and others that of testes."

Melander's view can be criticized on the basis that there is no mechanism to explain nucleolus formation according to which different cells in the same organism would have a different number and location of nucleolar organizers. In addition, Melander's theory has never received further support from any investigations into either ontogenesis or regeneration of gonads. In particular, recent findings of Stephen-Dubois ('64, '65), Ghirardelli ('65), Fedecka-Bruner ('67), and Gremigni's group ('81, '82) on germ cell "transdifferentiation" seem to refute the theory of early differentiation of germ lines.

The question of an early differentiation of the female and male lines has been discussed also by Benazzi ('66) on the basis of karyological data supporting the vote of territorial influence in the evolution of the female or male gonocytes. This influence is certainly in contrast with Melander's claims; however, Benazzi does not exclude a premature sexual differentiation of neoblasts. He suggests the possibility that neoblasts with male or female determination migrate preferentially into the territories of respective competence. Another possibility is that elements already directed toward female or male differentiation would find further induction in the competent territories.

Another area of uncertainty regarding possible early differentiation of germ lines concerns the meaning and fate of the mitochondrial "Balbiani body" found in *D. dorotocephala* oocytes (Gremigni, '76), which could represent a "germ determinant" as proposed for other organisms (see Beams and Kessel, '74, for a review).

A problem closely related to that of early cell line segregation is that of the egg type characterizing planarians: most evidence obtained up to now favors the "regulatory theory," but the possibility of mosaicism in the eggs of planarians cannot be excluded at present.

Other unresolved questions that could be the object of worthwhile research include, what are the mediators of the fine mechanism that regulate and induce the differentiation into types I and II vitelline cells in the early stages of embryogenesis? Are the hypothetical "grouping substances" really produced by the egg and blastomeres and what is their chemical nature? Do the vitelline cells specifically differentiate within the yolk follicles? What kind of factors regulate the appearance and maintenance of the antero-posterior polarity? What are the fate and meaning of yolk globules present in ovarian oocytes of Dugesiids, which Gremigni ('79) proposed might supply reserve material during the very early stages of embryonic development and be remnants inherited from an ancestral Archoophoran progenitor? Is it possible that cell transdifferentiation also occurs during ontogenesis besides its very likely occurrence in adult regeneration?

These and many other problems could be better understood if more experimental investigations were conducted. As was stated previously, very few studies have been made up to now, particularly those combining biochemical, autoradiographic, ultrastructural, and other modern laboratory methods, because of technical difficulties. We are encouraged by the interesting and important research now being conducted by the Parisian school of Le Moigne, Franquinet, and Martelli (see Martelli et al, '81). These authors recently obtained very good baseline data on the fine mechanisms of planarian regeneration by studying cultured cells. Such studies will give investigators more opportunities to conduct direct biochemical research on embryonic cells, although one must remember that the natural developmental environment, particularly the important role played by the vitelline cells, is hard to reproduce under *in vitro* conditions.

Acknowledgments. We appreciate Dr. Irving Kaufman's advice and assistance in preparing the English translation of this chapter.

LITERATURE CITED

Abeloos, M. (1930) Recherches expérimentales sur la croissance et la régénération chez les Planaires. Bull. Biol. Fr. Belg. *64:*1–140.

Acconci, C. (1919) Osservazioni sullo sviluppo delle planarie d'acqua dolce. Boll. Ist. Zool. R. Univ. Palermo *1:*49–76.

Achtelik, W. (1972) Electron microscopic studies on the genesis of cortical granules in the oocytes of *Dendrocoelum lacteum* (Müller). Zool. Polon. *21:*59–66.

Anderson, J.M. (1952) Sexual Reproduction without cross-copulation in the fresh-water Triclad Turbellarian, *Curtisia foremanii.* Biol. Bull. *102:*1–8.

Anderson, J.M., and S.C. Johann (1959) Some aspects of reproductive biology in the fresh-water Triclad Turbellarian, *Cura foremanii.* Biol. Bull. *115:*375–383.

Ansevin, K.D., and P. Buchsbaum (1961) Observations on planarian cells cultivated in solid and liquid media. J. Exp. Zool. *146:*153–161.

Baguña, J. (1976) Mitosis in the intact and regenerating planarian *Dugesia mediterranea* n.sp. Mitotic studies during growth, feeding and starvation. J. Exp. Zool. *195:*53–64.

Bàlasz, A., and M. Burg (1962) Quantitative data to the changes of propagation according to age. I. Cocoon production of *Dugesia lugubris.* Acta Biol. Hung. *12:*297–304.

Ball, I.R. (1974a) A contribution to the phylogeny and biogeography of the fresh-water triclads (Platyhelminthes:Turbellaria). In N.W. Riser and M.P. Morse (eds): "Biology of the Turbellaria." New York: McGraw-Hill, pp. 339–401.

Ball, I.R. (1974b) A new genus and species of fresh-water planarian from Australia (Platyhelminthes: Turbellaria). J. Zool. London *174:*149–158.

Ball, I.R. (1977) On the phylogenetic classification of aquatic planarians. Acta Zool. Fenn. *154:*21–35.

Ball, I.R. (1981) The phyletic status of the Paludicola. Hydrobiologia *84:*7–12.

Ball, I.R., N. Gourbault, and R. Kenk (1981) The planarians (Turbellaria) of temporary waters in Eastern North America. Royal Ontario Museum, Life Sciences Contributions *127:*1–27.

Ball, I.R., and T.B. Reynoldson (1981) British Planarians D.M. Kermack and R.S.K. Barnes, Cambridge Univ. Press, pp 1–141.

Banchetti, R., and V. Gremigni (1973) Indirect evidence for neoblast migration and for gametogonia dedifferentiation in ex-fissiparous specimens of *Dugesia gonocephala* s.l. Accad. Naz. Lincei *55:*107–115.

Bardeen, C.R. (1902) Embryonic and regenerative development in planarians. Biol. Bull. *3:*262–288.

Bardeen, C.R., and F.H. Baetjer (1904) The inhibiting action of roentgen rays on regeneration in planarians. J. Exp. Zool. *1:*191–195.

Beams, H.W., and R.G. Kessel (1974) The problem of germ cell determinants. Int. Rev. Cytol. *39:*413–479.

Beauchamp, P. de (1937) Nouvelles diagnoses de Triclades oscuricoles. Bull. Soc. Zool. Fr. *62:*265–272.

Beauchamp, P. de (1961) Classe des Turbellariés. In P.P. Grassé (ed): "Traité de Zoologie." Paris: Masson, *4:*69–123.

Benazzi, M. (1938) Ricerche sulla riproduzione delle planarie. Tricladi paludicoli con particolare riguardo alla moltiplicazione asessuale. Accad. Naz. Lincei *7:*31–89.

Benazzi, M. (1952) Sulla possibilità di riproduzione sessuata senza accoppiamento in Tricladi di acqua dolce. Atti Soc. Tosc. Sc. Nat. Ser. B *59:*107–110.

Benazzi, M. (1957) Cariologia di *Dugesia lugubris* (O. Schmidt) (Tricladida Paludicola). Caryologia *10:*276–303.

Benazzi, M. (1962) L'azione delle radiazioni ionizzanti sullo sviluppo delle planarie. Accad. Naz. Lincei *32:*26–29.

Benazzi, M. (1963) L'azione dello spermio irradiato nello sviluppo anfimittico e pseudogamico delle planarie. Atti Ass. Gen. It. *8:*255–256.

Benazzi, M. (1966) Considerations on the neoblasts of planarians on the basis of certain karyological evidence. Chromosoma *19:*14–27.

Benazzi, M. (1969) Régénération et fonctionnalité de l'-appareil génital chez les planaires scissipares. Ann. Embryol. Morphol. [Suppl.] *1:*244–245.

Benazzi, M. (1974) Fissioning in planarians from a genetic standpoint. In N.W. Riser and M.P. Morse (eds): "Biology of the Turbellaria." New York: McGraw-Hill, pp. 476–492.

Benazzi, M., and I.R. Ball (1972) The reproductive apparatus of sexual specimens from fissiparous populations of *Fonticola morgani* (Tricladida, Paludicola). Can. J. Zool. *50:*703–704.

Benazzi, M., and G. Benazzi-Lentati (1945) Su di un fattore letale in *Dugesia* (*Euplanaria*) *gonocephala.* Boll. Soc. It. Biol. Sper. *20:*514–515.

Benazzi, M., and G. Benazzi-Lentati (1948) Sulla mortalità prenatale nei Tricladi. Boll. Soc. It. Biol. Sper. *24:*807–808.

Benazzi, M., and G. Benazzi-Lentati (1976) Platyhelminthes. In B. John (ed): "Animal Cytogenetics." Berlin-Stuttgart: Gebrüder Borntraeger, pp. 1–77.

Benazzi, M., and P. Deri (1980) Histo-cytological study of ex-fissiparous planarian testicles (Tricladida, Paludicola). Monitore Zool. It. (N.S.) *14:*151–163.

Benazzi, M., and N. Gourbault (1977) *Atrioplanaria morisii* n.sp., a new cave planarian from Italy. Boll. Zool. *44:*327–335.

Benazzi, M., and M. Grasso (1977) Comparative research on the sexualization of fissiparous planarians treated with substances contained in sexual planarians. Monitore Zool. It. (N.S.) *11:*9–19.

Benazzi-Lentati, G. (1960) Precoce separazione dei due cromatidi in cromosomi di *Dugesia lugubris.* Caryologia *12:*482–496.

Benazzi-Lentati, G. (1964) Sulla regolazione dei processi maturativi di ovociti di planarie attivati da spermi normali ed irradiati. Boll. Zool. *31:*963–971.

Benazzi-Lentati, G. (1970) Gametogenesis and egg fertilization in planarians. Int. Rev. Cytol. *27:*101–179.

Benazzi-Lentati, G. (1976) Different relationship between egg and sperm in the pseudogamy of planarians according to the type of oogenesis, synaptic or asynaptic. Accad. Naz. Lincei *59:*287–289.

Best, J.B., A.B. Goodman, and A. Pigon (1969) Fissioning in planarians: control by the brain. Science *164:*565–566.

Betchaku, T. (1967) Isolation of planarian neoblasts and their behaviour *in vitro* with some aspects of the mechanism of the formation of regeneration blastema. J. Exp. Zool. *164:*407–434.

Böhmig, L. (1912-17) Entwicklungsgeschichte der Tricladen. In Dr. H.G. Bronn's Klassen U. Ordnungen d. Tierreichs. Leipzig.

Bresslau, E. (1933) Turbellaria. In Küenthal and Krumbach (eds): "Handbuch der Zoologie." Berlin, pp. 52–304.

Brøndsted, H.V. (1969) "Planarian Regeneration." Oxford: Pergamon Press, pp. 1–276.

Carlé, R. (1935) Beiträge zur Embryologie der Landplanarien. I. Frühentwicklung, Bau und Funktion des embryonal Pharynx. Zeit. Morph. Ökol. Tiere *29:*527–558.

Castle, W.A. (1927) The life history of *Planaria velata.* Biol. Bull. *53:*139–144.

Castle, W.A. (1928) An experimental and histological study of the life cycle of *Planaria velata.* J. Exp. Zool. *51:*417–483.

Castle, W.E. (1941) The morphology and life-history of *Hymanella retenuova*, a new species of triclad from England. Am. Mid. Nat. *26:*85–97.

Castle, W.E., and L.H. Hyman (1934) Observations on *Fonticola velata* (Stringer), including a description of the anatomy of the reproductive system. Trans. Am. Microsc. Soc. *53:*154–171.

Chandebois, R. (1968) Action d'un milieu synthétique convenant pour des coltures histiotypique sur l'activité du tissu indefférencié de fragments de Planaires. 2nd Int. Coll. Invertebrate Tissue Colture, Como 1967. Pavia: Fusi, pp. 32–62.

Chandebois, R. (1976) In A. Wolsky (ed): "Histogenesis and Morphogenesis in Planarian Regeneration." Basel: Karger, pp. 1–182.

Child, C.M. (1910) Physiological isolation of parts and fission in *Planaria*. Arch. Entw. Mech. Org. *30:*159–205.

Child, C.M. (1920) Some considerations concerning the nature and origin of physiological gradients. Biol. Bull. *39:*147–187.

Child, C.M. (1941) "Patterns and Problems of Development." Chicago: University of Chicago Press.

Coward, S.J. (1969) Regeneration in planarians: Some unresolved problems and questions. J. Biol. Psychol. *11:*15–19.

Coward, S.J. (1974) Chromatoid bodies in somatic cells of the planarian: Observations on their behaviour during mitosis. Anat. Rec. *180:*533–546.

Coward, S.J., C.E. Bennett, and B.L. Hazlehurst (1974) Lysosome and lysosomal activity in the regenerating planarian: Evidence in support of dedifferentiation. J. Exp. Zool. *189:*133–146.

Curtis, W.C. (1902) The life story, the normal fission and the reproductive organs of *Planaria maculata.* Proc. Boston Soc. Nat. Hist. *30:*515–559.

Curtis, W.C. (1905) The location of the permanent pharynx in the planarian embryo. Zool. Anz. *29:*169–175.

Dahm, A.G. (1964) The taxonomic relationship of the European species of *Phagocata* (=*Fonticola*) based on karyological evidence. Arkiv Zool. *16:*481–509.

Domenici, L, and V. Gremigni (1974) Electron microscopical and cytochemical study of vitelline cells in the fresh-water triclad *Dugesia lugubris* s.l. II. Origin and Distribution of reserve materials. Cell Tiss. Res. *152:*219–228.

Fedecka-Bruner, B. (1967) Études sur la régénération des organs génitaux chez la planaire *Dugesia lugubris.* I. Régénération des testicules après destruction. Bull. Biol. Fr. Belg. *101:*255–319.

Franquinet, R. (1973) Cultures *in vitro* de cellules de la Planaire d'eau douce *Polycelis tenuis* Iijima. C.R. Acad. Sci. Paris *276:*1733–1736.

Franquinet, R. (1976) Étude comparative de l'évolution des cellules de planaire d'eau douce *Polycelis tenuis* (Iijima) dans des fragments dissociés en culture *in vitro*: Aspects ultrastructuraux, incorporation de leucine et d'uridine tritiée. J. Embryol. Exp. Morphol. *36:*41–54.

Franquinet, R. (1981) Synthèse d'ADN dans les cellules de Planaires cultivées *in vitro*. Rôle de la sérotonine. Biol. Cell. *40:*41–46.

Franquinet, R., and T. Lender (1973) Etude ultrastructurale des testicules de *Polycelis tenuis* et *Polycelis nigra* (Planaires). Evolution des cellules germinales mâles avant la spermiogenèse. Z. Mikorsk. Anat. Forsch. *87:*4–22.

Fuhrmann, O. (1914) Turbellariés d'eau douce de Colombie. Mem. Soc. Neuchâteloise Sci. Nat. *5:*793–804.

Fulinski, B. (1914) Die Entwiklungesgichte von *Dendrocoelum lacteum* Oerst. I. Teil: Die erste Entwiklungsphase vom Ei bis zur Embryonalpharynx bildung. Bull. Acad. Sci. Cracovie, 147–190.

Fulinski, B. (1916) Die keimblätterbildung bei *Dendrocoelum lacteum* Oerst. Zool. Anz. *47:*380–400.

Fulinski, B. (1938) Zur Embryonalpharynxfrage der Trikladiden. Zool. Polon. *2:*185–207.

Galleni, L., and V. Gremigni (1982) Fertilization, Development and Parental care in Turbellaria. In K.G. Adiyodi and R.G. Adiyodi (eds): "Reproductive Biology of Invertebrates." Vol. IV. Chichester, England: J. Wiley, in press.

Gerzeli, G., and G. Gerzeli-Pedrazzi (1965) Aspetti istomorfologici e istochimici della differenziazione dei vitellogeni e della formazione del bozzolo nelle Planarie. Arch. Zool. It. *50:*1–19.

Ghirardelli, E. (1965) Differentiation of the germ cells and regeneration of the gonads in planarians. In V. Kiortsis and H.A.L. Trampush (eds): "Regeneration in Animals" Amsterdam: North-Holland, pp. 177–184.

Gourbault, N. (1972) Recherches sur les Triclades Paludicoles hypogés. Mém. Mus. Nat. Hist. Nat. Paris, Ser. A *73:*1–249.

Gourbault, N., and M. Benazzi (1975) Karyological data on some species of the genus *Cura* (Tricladida, Paludicola). Can. J. Genet. Cytol. *17:*345–354.

Grasso, M. (1971) Esperimenti sul controllo della maturazione sessuale in ceppi agami di planarie. Boll. Zool. *38:*532.

Grasso, M., and M. Benazzi (1973) Genetic and physiologic control of fissioning and sexuality in Planarians. J. Embryol. Exp. Morphol. *30:*317–328.

Gremigni, V. (1969a) Ricerche istochimiche e ultrastrutturali sull' ovogenesi dei Tricladi. I. Inclusi deutoplasmatici in *Dugesia lugubris* e *Dugesia gonocephala.* Accad. Naz. Lincei *47:*101–108.

Gremigni, V. (1969b) Origine del vitello negli ovociti di *Dugesia lugubris* e *D. benazzii* (Tricladida, Paludicola). Atti VII Congr. It. Micr. Elettr., Modena, pp. 68–70.

Gremigni, V. (1969c) Ricerche istochimiche e ultrastrutturali sull' ovogenesi dei Tricladi. II. Inclusi citoplasmatici in *Planaria torva, Dendrocoelum lacteum* e *Polycelis nigra.* Accad. Naz. Lincei *47:*397–404.

Gremigni, V. (1974) The origin and cytodifferentiation of germ cells in the planarians. Boll. Zool. *41:*359–377.

Gremigni, V. (1976) Genesis and structure of the so-called "Balbiani body" or "yolk nucleus" in the oocyte of *Dugesia dorotocephala* (Turbellaria, Tricladida). J. Morphol. *149:*265–278.

Gremigni, V. (1979) An ultrastructural approach to planarian taxonomy. Syst. Zool. *28:*345–355.

Gremigni, V. (1981) The problem of cell totipotency, dedifferentiation and transdifferentiation in Turbellaria. Hydrobiologia *84:*171–179.

Gremigni, V. (1982) 3. Platyhelminthes–Turbellaria. In K.G. Adiyodi and R.G. Adiyodi (eds): Reproductive Biology of Invertebrates. Vol. I: Oogenesis, Oviposition, and Oosorption. Chicester, England: J. Wiley and Sons Ltd., pp. 67–107.

Gremigni, V., and R. Banchetti (1972a) Submicroscopic morphology of hyperplasic ovaries of ex-fissiparous individuals in *Dugesia gonocephala* s.l. Accad. Naz. Lincei *52:*539–543.

Gremigni, V., and R. Banchetti (1972b) The origin of hyperplasia in the ovaries of ex-fissiparous specimens of *Dugesia gonocephala* s.l. Accad. Naz. Lincei *53:*477–485.

Gremigni, V., and L. Domenici (1974) Electron microscopical and cytochemical study of vitelline cells in the fresh-water Triclad *Dugesia lugubris* s.l. I. Origin and morphogenesis of cocoon-shell globules. Cell Tiss. Res. *150:*261–270.

Gremigni, V., and L. Domenici (1975) Genesis, composition and fate of cortical granules in the eggs of *Polycelis nigra* (Turbellaria, Tricladida). J. Ultrastr. Res. *50:*277–283.

Gremigni, V., and C. Miceli (1980) Cytophotometric evidence for cell "Transdifferentiation" in planarian regeneration. Wilhelm Roux' Arch. *188:*107–113.

Gremigni, V., C. Miceli, and E. Picano (1980b) On the role of germ cells in planarian regeneration. II. Cytophotometric analysis of the nuclear Feulgen-DNA content in somatic regenerated tissues. J. Embryol. Exp. Morphol. *55:*65–76.

Gremigni, V., C. Miceli, and I. Puccinelli (1980a) On the role of germ cells in planarian regeneration. I. A karyological investigation. J. Embryol. Exp. Morphol. *55:*53–63.

Gremigni, V., M. Nigro, and I. Puccinelli (1982) Evidence of male germ cell redifferentiation into female germ cells in planarian regeneration. J. Embryol. Exp. Morphol. *70:*29–36.

Gremigni, V., and I. Puccinelli (1977) A contribution to the problem of the origin of the blastema cells in planarians: A karyological and ultrastructural investigation. J. Exp. Zool. *199:*57–72.

Gremigni, V., and I. Puccinelli (1979) Sdifferenziamento e migrazione di cellule germinali in planarie tenute a digiuno: indagine cariologica. Ric. Sci. Educ. Perm. Atti XLVII Convegno U.Z.I., Bergamo-Milano, suppl. *6:*115–116.

Gresens, J. (1928) Versu he über die Widerstandsfähigkeit einiger Süsswassertiere gegenüber Salzlösungen. Z. Morphol. Ökol. Tiere *12:*706–800.

Hallez, P. (1887) Embryogénie des Dendrocoeles d'eau douce. Mem. Soc. Sci. Lille *16:*1–107.

Haranghi, L., and A. Bàlasz (1964) Ageing and rejuvenation in planarians. Exp. Gerontol. *1:*77–91.

Hay, E.D. (1968) Dedifferentiation and metaplasia in vertebrate and invertebrate regeneration. In Ursprung (ed): "The Stability of Differentiated State." New York: pp. 85–108.

Hay, E.D., and S.J. Coward (1975) Fine structure studies on the planarian *Dugesia.* I. Nature of the "Neoblast" and other cell types in non injured worms. J. Ultrastr. Res. *50:*1–21.

Hertwig, O. (1911) Die Radiumkrankheit tierischer Keimzellen. Ein Beitrag zur experimentellen Zeugungs-und Verebungslehre. Arch. Mikr. Anat. *77:*1–164.

Hyman, L. (1951) The Invertebrates: Platyhelminthes and Rhyncocoela. Vol. 2. New York: McGraw-Hill, pp. 1–458.

Iijima, I. (1883) Über die Embryologie von *Dendrocoeleum lacteum.* Zool. Anz. *605:*610.

Iijima, I. (1884) Untersuchungen über den Bau und die Entwiklungesgichte der Süsswasser-Dendrocoelen (Tricladen). Z. Wiss. Zool. *40:*359–464.

Jenkins, M.M. (1974) Relationship between reproductive activity and parental age in a sexual race of *Dugesia dorotocephala.* In N.W. Riser and M.P. Morse (eds): "Biology of the Turbellaria." New York: McGraw-Hill, pp. 493–516.

Jenkins, M.M., and H.P. Brown (1963) Cocoon production in *Dugesia dorotocephala* (Woodworth) 1897. Trans. Am. Microsc. Soc. *82:*167–177.

Jenkins, M.M., and H.P. Brown (1964) Copulatory activity and behaviour in the planarian *Dugesia dorotocephala* (Woodworth) 1897. Trans. Am. Microsc. Soc. *83:*32–40.

Kanatani, H. (1957) Further studies on the effect of crowding on supplementary eye-formation and fission in the planarian, *Dugesia gonocephala.* J. Fac. Sci. Univ. Tokyo *8:*23–39.

Karling, T.G. (1967) Zur Frage von dem systematischen Wert der Kategorien Archoophora und Neoophora (Turbellaria). Commentati Biol. Soc. Sci. Fenn. *30:*1–11.

Karling, T.G. (1974) On the anatomy and affinities of the turbellarian orders. In N.W. Riser and M.P. Morse (eds): "Biology of the Turbellaria." McGraw-Hill, New York, pp. 1–16.

Kato, K. (1968) Platyhelminthes (Class Turbellaria). In M. Kume and K. Dan (eds): "Invertebrate Embryology." Belgrade: Nolit, pp. 125–143.

Kawakatsu, M. (1974) Further studies on the vertical distribution of fresh-water planarians in the Japanese Islands. In N.W. Riser and M.P. Morse (eds): "Biology of the Turbellaria." New York: McGraw-Hill, pp. 291–338.

Kenk, R. (1941) Induction of sexuality in the asexual form of *Dugesia tigrina* (Girard). J. Exp. Zool. *87:*55–69.

Kenk, R. (1972) Freshwater Planarians (Turbellaria) of North America. In Biota of Freshwater Ecosystem, Identification Manual. Washington, D.C. U.S. Government Printing Office, pp. 1–81.

Kennel, J. (1888) Untersuchungen an neuen Turbellarien. Zool. Jahr. Abth. Anat. Ont. Tiere *3:*447–486.

Knappert, B. (1865) Bijdragen tot de Ontwikkelings-Geschiedenis der zoetwater Planarien. Nat. Verh., Utrecht; quoted by Iijima (1884).

Kölliker, A. (1846) Ueber die contractilen Zellen der Planarienembryonen. Arch. Naturgesch. *12:*291–295.

Koscielski, B. (1964) Polyembryony in *Dendrocoelum lacteum* O.F. Müller. J. Embryol. Exp. Morphol. *12:*633–636.

Koscielski, B. (1966) Cytological and cytochemical investigations on the embryonic development of *Dendrocoelum lacteum* O.F. Müller. Zool. Polon. *16:*83–96.

Koscielski, B. (1967) Experimental studies on embryonic development of *Dendrocoelum lacteum* O.F. Müller. Experientia *23:*212–215.

Lanfranchi, A. (1964) Aspetti citomorfologici in embrioni irradiati della planaria *Dugesia lugubris* (O. Schmidt). Boll. Zool. *31:*555–565.

Lanfranchi, A. (1965) Peculiare caso di autofecondazione nella planaria *Polycelis nigra* Ehrenberg. Accad. Naz. Lincei *38:*960–961.

Lang, A. (1912) Über Regeneration bei Planarien. Arch. Mikr. Anat. *79:*361–426.

Lange, C.S. (1968) A possible explanation in cellular terms of the physiological ageing of the planarians. Exp. Gerontol. *3:*219–230.

Le Moigne, A. (1962) Etude de formules chromosomiques de quelques *Polycellis* (Turbellarié, Triclades) de la région parisienne. Bull. Soc. Zool. Fr. *87:*259–270.

Le Moigne, A. (1963) Etude du développement embryonnaire de *Polycelis nigra* (Turbellarié, Triclade). Bull. Soc. Zool. Fr. *88:*403–422.

Le Moigne, A. (1965a) Effet des irradiations aux rayons x sur le développement embryonnaire et le pouvoir de régénération à l'éclosion, de *Polycelis nigra* (Turbellarié, Triclade). C.R. Acad. Sci. Paris *260:*4627–4629.

Le Moigne, A. (1965b) Sur la régénération et la différenciation de fragments d'embryons de *Polycelis nigra-tenuis* (Turbellarié, Triclade). C.R. Seanc. Soc. Biol. *159:*54–57.

Le Moigne, A. (1965c) Mise en évidence d'un pouvoir de régénération chez l'embryon de *Polycelis nigra* (Turbellarié, Triclade). Bull. Soc. Zool. Fr. *90:*355–361.

Le Moigne, A. (1966a) Etude du développement embryonnaire et recherches sur les cellules de régénération chez l'embryon de la Planaire *Polycelis nigra* (Turbellarié, Triclade). J. Embryol. Exp. Morphol. *15:*39–60.

Le Moigne, A. (1966b) Etude au microscope électronique de cellules d'embryons de *Polycelis* (Turbellarié, Triclade), au début de leur développement. C.R. Acad. Sc. Paris *263:*550–553.

Le Moigne, A. (1967a) Présence d'émissions nucléaires fréquemment associées à des mitochondries, dans les cellules embryonnaires de Planaires. C.R. Seanc. Soc. Biol. *161:*508–511.

Le Moigne, A. (1967b) Etude au microscope électronique de la différenciation des principaux types cellulaires chez l'embryon de la Planaire *Polycelis nigra.* Bull. Soc. Zool. Fr. *92:*617–628.

Le Moigne, A. (1967c) Mise en évidence au microscope électronique de la persistence de cellules indifférenciées au cours du développement embryonnaire de la Planaire *Polycelis nigra.* C.R. Acad. Sci. Paris *265:*242–244.

Le Moigne, A. (1968) Etude au microscope éléctronique de l'évolution des structures embryonnaires de Planaires aprés irradiation aux rayons x. J. Embryol. Exp. Morphol. *19:*181–192.

Le Moigne, A. (1969) Étude du développement et de la régénération embryonnaires de *Polycelis nigra* (Ehr.) et *Polycelis tenuis* (Iijima), Turbellariés Triclades. Ann. Embryol. *2:*51–69.

Le Moigne, A., M.J. Sauzin, and T. Lender (1966) Comparaison de l'ultrastructure du néoblaste et de la cellule embryonnaire des Planaires d'eau douce. C.R. Acad. Sci. Paris *263:*627–629.

Lender, T. (1962) Factors in morphogenesis of regenerating fresh-water Planaria. Adv. Morphogen. *2:*305–331.

Lender, T. (1964) Mise en évidence et role de la neurosécretion chez les planaires d'eau douce. Ann. Biol. *13:*165–172.

Lender, T. (1974) The role of neurosecretion in freshwater planarians. In N.W. Riser and M.P. Morse (eds): "Biology of the Turbellaria." New York: McGraw-Hill, pp. 460–475.

Lender, T., and C. Briançon (1974) Induction de la maturation sexuelle chez la planaire scissipare *Dugesia gonocephala* (Turbellarié, Triclade). J. Embryol. Exp. Morphol. *32:*159–168.

Lender, T., and C. Briançon (1975) Sexual differentiation in the fissiparous strain of *Dugesia gonocephala.* In R. Reinboth (ed): "Intersexuality in the Animal Kingdom." Berlin: Springer-Verlag, pp. 20–29.

Lender, T., and F. Zgahl (1968) Influence du cerveau et de la neurosécrétion sur la scissiparité de la Planaire *Dugesia gonocephala.* C.R. Acad. Sci. Paris *267:*2008–2009.

Lender, T., and F. Zgahl (1969) Influences des conditions d'élevage et de la neurosécrétion sur le rythmes de scissiparité de la race asexuée de *Dugesia gonocephala.* Ann. Embryol. Morphol. *2:*379–385.

Lender, T., and F. Zgahl (1973a) Les synthèses des acides ribonucléiques au cours du cycle de scissiparité de la planaire *Dugesia gonocephala.* C.R. Acad. Sci. Paris *276:*1859:1862.

Lender, T., and F. Zgahl (1973b) Mode d'action de la neurosécrétion sur la scissiparité de la planaire *Dugesia gonocephala.* C.R. Acad. Sci. Paris *277:*525–528.

Liotti, F.S., and G. Bruschelli (1964) Rapporti tra sistema nervoso e rigenerazione nelle planarie. Prime osservazioni sulle capacità rigenerative di esemplari di *Dugesia lugubris* prima e dopo l'apertura dei cocon. Riv. Biol. *57:*121–149.

Marcus, E. (1946) Sobre Turbellaria brasileiros. Bol. Fac. Cienc. Letr. Zool. S. Paulo. *11:*5–254.

Marinelli, M. (1972) Observations on the shell formation in the cocoon of *Dugesia lugubris* s.l. Boll. Zool. *39:*337–341.

Marinelli, M., and D. Vagnetti (1973) Electron microscopic investigations on the yolk cells in the cocoon of *Dugesia lugubris* s.l. Boll. Zool. *40:*367–369.

Martelly, I., R. Franquinet, and A. Le Moigne (1981) Relationship between variations of cAMP, neuromediators and stimulation of nucleic acid synthesis during planarian (*Polycelis tenuis*) regeneration. Hydrobiologia *84:*195–201.

Martelly, I. and A. Le Moigne (1979). Comparison des effets de l'actinomycin D sur les synthèses d'ARN des planaires jeunes et adultes en régénération. C.R. Soc Biol. *173:*1023–1030.

Mattiesen, E. (1904) Ein Beitrag zur Embryologie der Süsswasserdendrocoelen. Z. Wiss. Zool. *77:*274–361.

Melander, Y. (1963) Cytogenetic aspects of embryogenesis in Paludicola Tricladida. Hereditas

*49:*119–166.

Metschnikoff, E. (1883) Die Embryologie von *Planaria polychroa.* Z. Wiss. Zool. *38:*331–354.

Mirolli, M. (1956) Le dimensioni degli ovociti di *Dugesia lugubris* (O. Schmidt) Triclade paludicolo, in relazione al corredo cromosomico. Atti Soc. Tosc. Sc. Nat. *62:*156–172.

Morita, M., J.B. Best, and J. Noel (1969) Electron microscopic studies of Planarian regeneration. I. Fine structure of neoblasts in *Dugesia dorotocephala.* J. Ultrastr. Res. *27:*7–23.

Nobili, R. (1958) La durata della funzionalità degli spermi nei ricettacoli seminali di tricladi paludicoli. Arch. Zool. It. *43:*157–171.

Okugawa, K.I. (1957) An experimental study of sexual induction in the asexual form of Japanese freshwater planarian *Dugesia gonocephala* (Dugès). Bull. Kyoto Gakugei Univ. Ser. B *11:*8–27.

Okugawa, K.I., and M. Kawakatsu (1956) Studies on the fission of Japanese freshwater planaria *Dugesia gouocephala* (Dugès). V. On the influence of fission frequencies of the animals of sexual and asexual races by means of head removal operations. Bull. Kyoto Gakugei Univ. ser B. *8:*43–59.

Pedersen, K.J. (1972) Studies on regeneration blastemas of the planarian *Dugesia tigrina* with special reference to differentiation of the muscle-connective tissue filament system. Wilhelm Roux' Arch. *169:*134–169.

Puccinelli, I., and V. Gremigni (1978) Sulla presenza nei blastemi di planarie di cellule provenienti dal territorio gonadico femminile. Accad. Naz. Lincei *63:*588–592.

Reynoldson, T.B. (1960) A quantitative study of the population biology of *Polycelis tenuis* Iijima (Turbellaria, Tricladida). Oikos *11:*125–141.

Reynoldson, T.B., and A.D. Sefton (1972) The population biology of *Planaria torva* (Müller) (Turbellaria, Tricladida). Oecologia *10:*1–16.

Rose, C., and S. Shostak (1968) The transformation of gastrodermal cells to neoblasts in regenerating *Phagocata gracilis* (Leidy). Exp. Cell Res. *50:*553–561.

Sakurai, T. (1981) Sexual induction by feeding in an asexual strain of the freshwater planarian *Dugesia japonica japonica.* Zool. Soc. Japan *54:*103–112.

Sauzin, M.J. (1966) Etude au microscope éléctronique des néoblastes de la Planaire *Dugesia gonocephala* (Turbellarié, Triclade) et des ses changements ultrastructuraux au cours des premiers stades de la régénération. C.R. Acad. Sci. Paris *263:*605–608.

Sauzin, M.J. (1967a) Etude ultrastructurale de la différenciation du néoblaste au course de la régénération de la planaire *Dugesia gonocephala.* I. Différenciation en cellule nerveuse. Bull. Soc. Zool. Fr. *92:*313–318.

Sauzin, M.J. (1967b) Etude ultrastructurale de la différenciation du néoblaste au cours de la régénération de la planaire *Dugesia gonocephala.* I. Différenciation musculaire. Bull. Soc. Zool. Fr. *92:*613–616.

Sauzin, M.J. (1968) Présence d'émissions nucléaires dans les cellules différenciées et en différenciation de la Planaire adulte *Dugesia gonocephala.* C.R. Acad. Sci. Paris *267:*1146–1148.

Seilern-Aspang, F. (1956) Frühentwicklung einer mariner Triclade (*Procerodes lobata* O. Schmidt). Wilhelm Roux' Arch. *148:*589–595.

Seilern-Aspang, F. (1957a) Polyembryonie als abnorme Entwicklung bei *Procerodes lobata* O. Schmidt (Turbellaria). Zool Anz. *159:*187–193.

Seilern-Aspang, F. (1957b) Polyembryonie in der Entwicklung von *Planaria torva* (M. Schultz) auf Deckglaskultur. Zool. Anz. *159:*193–202.

Seilern-Aspang, F. (1958) Entwicklungeschichliche Studien an Paludicolen Tricladen. Wilhelm Roux' Arch. *150:*425–480.

Sengel, C. (1960) Culture *in vitro* de blastème de régénération des Planaires. J. Embryol. Exp. Morphol. *8:*468–476.

Sengel, C. (1963) Culture *in vitro* de blastème de régénération de la planaire *Dugesia lugubris.* Ann. Epiphyties *14:*173–183.

Shaper, A. (1904) Experimentelle untersuchungen über den Einfluss der Radiumstrahlen und der Radiumemanation auf embryonale und regenerative Entwicklungsvorgänge. Anat. Anz. *25:*298–314.

Skaer, R.J. (1965) The origin and continuous replacement of epidermal cells in the planarian *Polycelis tenuis* (Iijima). J. Embryol. Exp. Morphol. *13:*129–139.

Skaer, R.J. (1971) Planarians. In G. Reverberi (ed): "Experimental Embryology of Marine and Freshwater Invertebrates." Amsterdam: North-Holland, pp. 104–125.

Stagni, A., and M. Grasso (1965) Osservazioni preliminari sulla schizogenesi in *Dugesia tigrina.* Accad. Naz. Lincei *38:*905–910.

Steinmann, P. (1908) Untersuchungen über das Verhalten des Verdau ungssystem bei der Regeneration der Tricladen. Arch. Entw. Mech. Org. *25:*523–568.

Steinmann, P. (1914) Beschreibung einer neuen Süsswassertriclade von den Kei-Insel nebst einigen allgemeinen Bemerkungen über Tricladen. Anat. Abh. Jenck. Naturf. Gezel. *35:*111–121.

Steinmann, P. (1925) Das Verhalten der Zellen und Gewebe im regenerierenden Tricladenkörper. Verh. Naturforsch. Gesellsch. *36:*133–162.

Steinmann, P., and E. Bresslau (1913) Die Strudelwürmer (Turbellarien). Leipzig, pp. 1–380.

Stephan-Dubois, F. (1964) La lignée germinale des Turbellariés et des Annélides dans l'évolution normale et la régénération. In E. Wolff (ed): "L'Origine de la Lignée Germinale Chez les Vertébrés et Chez Quelques Groupes d'Invertébrés." Paris: Hermann, pp. 115–136.

Stephan-Dubois, F. (1965) Les néoblastes dans la régénération chez les planaires. In V. Kiortsis and H.A.L. Trampush (eds): "Regeneration in Animals and Related Problems." Amsterdam: North-Holland, pp. 112–130.

Stephan-Dubois, F., and M. Gusse (1974) Origine et différenciation des cellules vitellines lors de la régénération saisonnière des vitellogènes chez la planaire *Dendrocoelum lacteum.* Wilhelm Roux' Arch. *174:*181–194.

Stevens, M. (1904) On the germ cells and the embryology of *Planaria simplissima.* Proc. Acad. Nat. Sci. Philadelphia *56:*208–220.

Teshirogi, W., and H. Fujiwara (1970) Some experiments on regression and differentiation of genital organs in a freshwater planarian, *Bdellocephala brunnea.* Sci. Rep. Hirosaki Univ. *17:*38–49.

Vandel, A. (1921) Recherches expérimentales sur les modes de reproduction des Planaires Triclades Paludicoles. Bull. Biol. Fr. Belg. *55:*343–518.

Vannini, E. (1965) Regeneration and sex-gradient in hermaphroditic animals. In V. Kiortsis and H.A.L. Trampush (eds): "Regeneration in Animals and Related Problems." Amsterdam: North-Holland, pp. 160–176.

Voigt, W. (1928) Verschwinden des Pigmentes bei *Planaria polychroa* und *Polycelis nigra* unter dem Einfluss ungünstiger Existenzbedingungen. Zool. Jahrb. *45:*293–316.

Wolff, E. (1962) Recent researches on the regeneration of Planaria. In Rudnick (ed): "Regeneration." New York: Ronald Press, pp. 53–84.

Wolff, E., T. Lender, and C. Ziller-Sengel (1964) Le rôle des facteurs autoinhibiteurs dans la régénération des planaires. Rev. Suisse Zool. *71:*75–98.

Woodruff, L., and A.I. Burnett (1965) The origin of blastema cells in *Dugesia tigrina.* Exp. Cell Res. *38:*295–305.

Young, J.O., and T.B. Reynoldson (1966) The quantitative study of the population biology of *Dendrocoelum lacteum* (Müller) (Turbellaria, Tricladida). Oikos *15:*237–264.

Zaccanti, F., and G. Tognato (1979) Osservazioni sui ritmi di riporoduzione agama in un ceppo scissiparo di *Dugesia gonocephala* s.l. Rend. Accad. Sc. Ist. Bologna *267:*275–284.

Developmental Biology of Freshwater Invertebrates, pages 213–220

Supplement: Collection, Maintenance, and Manipulation of Planarians

Ronald R. Cowden

Planarians may be found in streams, ponds, lakes, and marshes. Virtually any reasonably permanent freshwater body may contain planarians. The methods of collection differ from various habitats. Collections from small shallow streams are most easily made by turning over flat rocks and removing the worms by finger or a fine paintbrush into a collecting container. The same method might be used around the margins of lakes, or of small ponds, but collections from deeper bodies of water or marshes require other methods. The most successful approach has been to lower a jar, a fine mesh wire basket, or a fine mesh net to the bottom baited with a piece of fish flesh or other meat (e.g. raw liver), and allow it to stand for a few hours or days. Depending on the depth and general weather conditions, the free end of the suspending cord may be secured to a float-moored raft, stake, or stout vegetation in marshy areas. The methods of removal from these nets are the same. It is desirable to carry a number of bottles or flasks to transport the collections. In the cases of planarians collected from very cold mountain streams, it is necessary to maintain these animals in thermos flasks, and may be desirable to transfer these to containers in a portable ice chest for transport to the laboratory. If collections from temperate habitats extend over hours, water should be changed in the jars every few hours, and it may even be a reasonable precaution—if the site is distant from the laboratory—to take a jug of water from the environment, cool it to the environmental temperature, and change the water one or more times on the trip back from the field. Temperature sensitivity varies with species, and the investigator should be guided by prevailing temperature conditions in the collection environment. This also holds for maintaining the animals in the laboratory.

Land planarians may be found in greenhouses in the cold months, but in the spring and summer they are most frequently found under flagstones, boards that been have lying flat on the ground for some time, or old logs. They are probably most abundant deeper in the ground, but this is the most general way of collecting them. These animals are extremely sensitive to desiccation, and it is wise to put several inches of moist earth in a collecting jar, as well as some leaves or moss to prevent desiccation. If they are to be maintained in a terrarium, some attention must be given to providing a suitably moist environment, both by introduction of water and by providing a moisture-preserving cover. A sensible examination of where and how the animals live should be a useful guide to their maintenance. Since they feed on earthworms, these should be provided as a diet. These animals regenerate well, and have been been used in regeneration studies.

Returning to the aquatic forms that are more central to our concern, finger bowls of various sizes, Stender dishes, or Pyrex utility baking dishes are suitable vessels in which to culture free-living flatworms. The 2-quart flat or low-utility dish is particularly recommended since, depending on the size of the worms, they can conveniently hold 50–200 animals. These vessels are usually filled half to two-thirds full with spring, lake, or artificial pond water, and 50–100 animals are added depending on their size. Whereas North American *Dugesia* seem to thrive on hard boiled egg yolk, the white planarians are best fed with shreds of beef liver. Actually, both *Dugesia* and the white planarians can be fed with beef liver. After the animals have fed for several hours, the food should be removed and the water changed. In practice, the worms are generally moved to clean vessels, and the population is split if the density requires it. Then the vessel can be thoroughly washed out, and scoured out with a sponge or brush. Though there may be some detergent-rinse sequence that would be harmless to the planarians, we have not found detergent necessary, and do not use it with our live material glassware. Feeding every 2 days is sufficient.

Although some of the earlier experimental embryologists have proposed special knives and other equipment for experimental mainpulation of planarians in regeneration experiments, this does not seem really necessary. Most of the cuts an experimenter might wish to make can easily be made with a single-edged razor blade or a No. 11 Bard-Parker surgical blade. It is useful to have the lower half of a Petri dish filled with paraffin so the worms can be placed on the paraffin surface for cutting, and it is frequently convenient to cut the worms under a stereomicroscope with incident illumination. Anterior or posterior parts or smaller fragments can then be transferred to labeled vessels with pipettes or brushes for regeneration. The planarians are generally starved for about a week before cutting experiments are undertaken; this precaution reduces infection and increases viability.

The results of over half a century of detailed morphogenetic experiments have been reviewed by Bronstead ('69) and will not be discussed here. It has already been noted in the main chapter that it is necessary to collect and segregate freshly deposited cocoons that contain multiple embryos. Unfortunately, the two commonest and most readily available North American species of *Dugesia* are seldom encountered in sexual phase, certainly not from most commercial sources.[1] *Dugesia trigrinum* goes through a sexual reproductive phase in the early spring—March, in Louisiana. In contrast, many, if not most, of the Eurasian species display prominent sexual phases in the spring or during the warm months of the year. Although there are sexual North American species, they are not the most common forms. Most of the information and references leading to the literature and methods that are used in manipulating turbellarian embryos was reviewed in the body of this chapter. However, there have been some developments that relate to the biology of turbellarians that seem to be worthy of special comment.

There were a number of earlier attempts at culturing planarian cells, but the most successful method was developed by Franquient ('73). He cultured fragments of *Polycoelis tenuis* Iijima by the following method. A basal medium of this composition was prepared:

Double-distilled water	1 liter
NaCl	0.7 gm
KCl	0.5 gm
$CaCl_2 \cdot 2H_2O$	0.4 gm
$MgSO_4 \cdot 7H_2O$	0.16 gm
$KH2PO_4$	0.07 gm
$NaHCO_3$	0.05 gm
Lactalbumin hydrolysate	5.0 gm
Yeast extract	2.5 gm
Glucose	20.0
1,500 I.U. penicillin/ml	

The planarians are cut into pieces of 1–2 mm^3, washed several times with sterile basal medium to remove adherent microorganisms, then homogenized in a loose-fitting homogenizer or mechanically dissociated by other means. Use of stainless steel wire cloth for this purpose could be recommended. The resulting cell suspension is then repeatedly washed with basal medium, and transferred to appropriate culture vessels. All glass- or plasticware should be sterile.

Although potentially mitotic cells are found in virtually all parts of the worms, the highest concentration is found in the prepharyngeal region. The main function of this medium is apparently to retard differentiation and

[1]Specimens of *Dugesia tigrinum* in sexual phase which will produce cocoons may be obtained from Carolina Biological Supply Co., Burlington, NC 27215, through their south Louisiana facility in March of each year. However, it is wise to check with the supplier well in advance of this requirement.

maintain mitotic proliferation of "neoblasts" (Franquient, '76). By use of this medium the cells can be maintained in division for up to 12 days, but they reach the highest mitotic index on days 2–4. Though the cells will grow in basal medium, the mitotic index increases substantially if 5% horse serum is added.

Franquient ('79) also discovered that serotonin and some other catecholamines can promote regeneration and the growth of cultured cells, and that biochemical determinations of biogenic amines are correlated with the distribution of nervous tissue, which is greatest in the cephalic region. For some time it has been believed that inervation or products of nerve cells were controlling factors in regeneration, and Fanquient's experiments have given strong support to this hypothesis.

There has been a substantial amount of fairly recent cytogenetic work on planarians. Most of the groups currently working on planarian cytogenetics are represented in the Third Turbellarian Symposium (Schockaert and Ball, '81). Benazzi and Benazzi-Lentati ('76, summary) were not the first to work cytogenetically with planarians, but they were the first to vigorously apply standards of population cytogenetics to the group, and to discover polyploid and anuploid populations. They described a diploid (n=4), *Dugesia lugubris*, and races with triploid male sex cells and hexaploid oocytes. Since then a number of innvestigators have discovered polyploidy and supernumerary choromosomes in polyclads (Galleni and Puccinelli, '81) or triclads (Gourbault, '81; Oki et al., '81; and Teshirogi and Ishida, '81). In the latter case a urea-banding method was successfully employed on mitotic and meiotic chromosomes. In spite of the fact that planarian material was looked upon as "difficult," there have been a substantial number of studies of planarian chromosomes over the past decade. Unfortunately, it seems difficult to come by straightforward descriptions of preparative techniques: Either the problem is passed off as "routine," or a confusing variety of technical proposals are encountered. Dutrallaux and Linicque ('71) recommended the use of diluted serum to cause the typical hypotonic swelling obtained in higher animal cells by treatment with solutions of low-salt concentrations. The maximum rate of mitosis is usually encountered in anterior regeneration on the third day. However, they did not recommend use of colchicine or colcemid to accumulate metaphase cells. Other workers suggest the use of colchicine or colcemid for 3–18 hours to accumulate metaphase cells, then dissociate the cells after fixation and dissociation in acetic acid (2%) and prepare rapid squash preparations by squashing in 45% acetic acid containing a chromosome dye such as orcein. Presumably fragments squashed in 45% acetic acid could be stained by other chromosome stains. Coverslips are removed by the dry-ice method, and air-dried after brief exposure to ethanol or methanol–acetic acid (3:1) or ethanol. Teshirogi and Ishida ('81) have used the preparation methods of Imai et al. ('77), and the urea-banding method of Kato and Yoshida ('72).

The preparation method of Imai et al. ('77) involved dissecting the tissue in 0.005% colchicine plus 1% sodium citrate (1:1) in a cavity slide, then moving the tissue into this mixture for another 20 minutes. This fluid was drained off, and the tissue was moved to a freshly cleaned slide. All the adherent fluid was removed by inclining and draining the slide, and then by removal with dissecting needles or forceps. A few drops of fixative I (3 parts glacial acetic acid, 3 parts absolute ethanol, and 4 parts distilled water) are added, and the slide is inclined so that the fixative flows around the tissue fragments. This is drained off and a few additional drops of fixative I are added. The preparation is then placed under a stereomicroscope, and the tissue is macerated with dissecting needles or watchmaker's forceps. Immediately after the tissue is dispersed into single cells or small clumps, a few drops of fixative II (glacial acetic acid, absolute ethanol, 1:1) are added. After 15–30 seconds the fluid is drained off by inclining the slide laterally. Then a few drops of fixative III (glacial acetic acid) are added for about 15 seconds, and again drained by inclining it along the long side. The slide is placed horizontally to dry. These preparations may be stained by the standard pH 6.8 Giemsa method.

The urea-banding method of Kato and Yosida ('72) consisted of exposing the fixed and dried preparations to 0.07% 2-mercaptoethanol, 2 M urea, and 0.05% sodium lauryl sulfate in pH 8.0 phosphate buffer. This was preferably performed on preparations that had been allowed to dry overnight at room temperature. After briefly washing these preparations with tap water, they may be stained by the standard Giemsa method.

Obviously, it has been possible to make conventional aceto-orcein, lacto-acetic-orcein, or Feulgen preparations from spreads or squashes of planarian tissue. Since it is now possible to culture planarian cells, it should also be possible to shock them osmotically after treatment with colcemid to accumulate metaphase cells, fix them with methanol-acetic acid (or some variant), conventionally prepare air-dried chromosome spreads, insert one of the banding methods (if desired), and stain or flurochrome the chromosomes with any of a number of DNA-specific dyes. If chromosome preparations from tissues are desired, it should be possible, using 3-day balstema preparations, to treat with a spindle inhibitor (colchicine, colcemid, etc.) for 3–4 hours to accumulate metaphase cells, insert the method of Dutrallaux and Linicque ('71) or some other modification to obtain osmotic swelling, mechanically dissociate the cells, fix them in some variant of methanol-acetic acid (3:1), and prepare dried spreads. The "standard method" of preparing rapid squashes of planarian tissue is to place the regenerating fragments in 0.03% colchicine for 3–4 hours, transfer small anterior fragments to 2% acetic acid for 5 minutes, then squash the fragments in lacto-acetic orcein (equal parts of 45% acetic acid and 45% lactic acid to which 2 gm/100 ml orcein has been added). Species that live in cold habitats may require a longer time in colchicine. Gourbrault and Benazzi ('74) placed fragments of *Dendroceolopsis chattoni* in 0.03% colchicine for 18 hours. Obviously, the methods of Imai et al. ('77), which were developed for preparation of chromosome spreads from small insect organs, are more complicated and fastidious. The extent to which these methods, or modifications of these methods, are helpful or necessary in the preparation of planarian chromosomes is not clear.

Frequient's ('73) method of culturing "neoblasts" clearly works well for *Polycoelis tenuis*; however, the extent to which it might be used on other aquatic triclads has not been formally introduced into the literature. Furthermore, since these animals live in a low osmotic environment, the conventional approach to swelling cells with hypotonic salt may not work even on cells in suspension; dilute protein (serum) solutions appear to be preferable for these cells. However, the responses in detail, and thus optimal conditions, may differ from species to species. The foregoing should introduce the reader to the literature on this subject and offer a reasonable starting point.

Gremigni and Miceli ('80) and Gremigni et al. ('80a) have made elegant use of the race of *Dugesia lugubris* with triploid male sex cells, hexaploid oocytes, and diploid somatic tissue to demonstrate that *both* differentiated germ cells and somatic cells can participate in regeneration. This was demonstrated first cytogenetically (Gremigni et al., '80b), then by use of Feulgen microspectrophotometry. With the demonstrated variations in ploidy within populations, it seems certain that microspectrophotometry, microfluorometry, or "flow" fluoresence cytometry should offer some interesting experimental options within the group.

Baguñá and Romera ('81) have used a cell-maceration procedure for obtaining dispersed cells of planarians. This approach is similar to that described for hydra. Planarians, or, more properly, planarian pieces, are placed in a mixture of methanol, glacial acetic acid, glycerol, and distilled water (3:1:2:14) at 8–10°C. for 24–48 hours. The pieces are then gently shaken until a suspension of single cells is obtained. The suspension is then fixed by 0.1 volume of 20% formaldehyde and/or 0.1 volume of 1% osmium tetroxide and spread on microscope slides to dry on a level surface. They recommend that a coverslip be mounted with a drop or so of water and examined with 40× phase-contrast optics. If the investigator has access to a Shandon Cytocentrifuge or Cyto/Buckets, the suspension can be fixed, spun down, and dried (Cytocentrifuge), or spun down without further fixation then fixed with 10% formalin, 3.5% gluteraldehyde on 1% OsO4. These cells may be stained with toluidine blue, azure A, or any of a number of procedures that are compatible with these fixatives. This includes most of the cytochemical and general morphological methods. These preparations are rated according to percentage of incidence in several hundreds of cells.

Thirteen types of cells, and some unclassifiable cells—excluding gonadal cells—may be recognized. It should also be possible to make use of flow analysis and sorting to classify these cells. The method should also be useful for studies of regeneration in juveniles.

Finally, as in most other groups, it has been established that the analysis of constitutive proteins by variations of acrylamid gel electropohoresis offers a suitable basis for comparisons among populations and of screening for genetic variability, and can be of value in making taxonomic distinctions (Teshirogi and Ishida, '81). Though the membrane lipoproteins extracted with detergents seem to be relatively uniform within the group that was studied, proteins obtained by extraction in a low ionic strength buffer seem to offer more diversity.

This Supplement has been intended as an introduction to the collection, maintenance, and manipulation of planarians, and to indicate the progress and directions that investigative work related to development in this group has taken. As an aside, probably because of their lesser availability, recent studies of regenerative capacities in other groups of free-living flatworms have not appeared, and cytogenetic information concerning karyotype organization is at best episodic. In addition to the obvious areas of interest within the triclads proper, there is a real need for similar or comparative information about other Turbellarians.

LITERATURE CITED

Benazzi, M., and G. Benazzi-Lintati (1976) Plathyheminthes. In B. John (ed): Animal Cytogenetics. Berlin: Gebrüder Borntraeger, pp. 1–182.

Baguñá, J., and R. Romero (1981) Quantitative analysis of cell types during growth, degrowth and regeneration in the planarians, *Dugesia mediterranea* and *Dugesia tigrina.* In E.R. Schockaert and I.M. Ball (eds): The Biology of Turbellaria, The Hague: W. Junk, pp. 181–194.

Bronstead, H.V. (1969) Planarian Regeneration. London: Pergamon.

Dutrillaux, B., and P. Linicque (1971) Analyse du caryotype de cineq espèces de Planaires par la méthode du choc hypotonique. Acta Zool. *52:*241–248.

Franquinet, R. (1973) Cultures in vitro de cellules de la Planaire d'eau douce *Polycelis tenuis* Iijima. C.R. Acad. Sci. Paris *276,*Ser. D:1733–1736.

Franquinet, R. (1976) Etude comparative de l'évolution des cellules de planaires d'eau douce, *Polycelis tenuis* (Iijima), dans des fragments dissociés en coulture in vitro. Aspects ultrastructuraux, incorporations de lucine et d'uridine tritiée. J Embryol Exp. Morphol. *36:*41–54.

Franquinet, R. (1979) Rôle de la sérotonine et des catecholamines dans la regeneration de la planaire *Polycelis tenuis*. J. Embryol. Exp. Morphol. *51:*85–95.

Gallini, L., and I. Puccinelli (1981) Karylogical observations on polyclads. In E.R. Schockaert and I.R. Ball (eds): The Biology of Turbellaria. The Hague: W. Junk, pp. 31–44.

Gourbault, N. (1981) The karyotypes of Dugesia species from Spain (Turbellaria, Tricladida). In E.R. Schockaert and I.M. Ball (eds): The Biology of Turbellaria. The Hague: W. Junk, pp. 45–52.

Gourbault, N., and M. Benazzi (1974) Etude caryologique du triclade hypogé *Dendrocoelopsis chattoni* (De Beauchamp). Ann. Speleol. *29:*621–626.

Grimigni, V., and C. Miceli (1980) Cytophotometric evidence for cell "transdifferentiation" in planarian regeneration. W. Roux Arch. *188:*107—113.

Grimigni, V., C, Miceli, and E. Picano (1980a) On the role of germ cells in planarian regeneration. II. Cytophotometric analysis of the nuclear Feulgen-DNA content in cells of regenerated somatic tissues. J. Embryol. Exp. Morphol. *55:*65–76.

Grimigni, V., C. Miceli, and I. Puccinelli (1980b) On the role of germ cells in planarian regeneration. I. A karyological investigation. J. Embryol. Exp. Morphol. *55:*53–63.

Imai, H.T., R.H. Crozier, and R.N. Taylor (1977) Karyotype evolution in Australian ants. Chromosoma *59:*347—393.

Kato, H., and T.H. Yosida (1972) Banding patterns of Chinese hamster chromosomes revealed by new techniques. Chromosoma *36:*272–280.

Oki, I., S. Tamura, T. Yamayoshi, and M. Kawakatsu (1981) Karyological and taxonomic studies of *Dugesia japonica* Ichikawa et Kawakatsu in the Far East. In E.R. Schockaert and I.M. Ball (eds): The Biology of Turbellaria. The Hague: W. Junk, pp. 53–68.

Schockaert, E.R., and I.M. Ball (eds) (1981) The Biology of Turbellaria. The Hague: W. Junk (also published in Hydrobiologia *84:*1-301).

Teshirogi, W., and S. Ishida (1981) Studies on the speciation of Japanese freshwater planarian *Polycoelis auricula* based on the analysis of its karyotypes and constitutive proteins. In E.R. Schockaert and I.M. Ball (eds): The Biology of Turbellaria. The Hague: W. Junk, pp. 69–77.

NOTE ADDED IN PROOF

A "standard" method of preparing chromosome preparations was given in a recent article by Redi, C.R., S. Garagna, and C. Pellicciari (Chromosome preparation from planarian blastemas: A procedure suitable for cytogenetic and cytochemical studies, in Stain Tech. *57:*190–192). They also note that chromosome banding studies have been undertaken by Bennazzi et al. (Caryologia *34:*129–139). These should be useful sources.

Developmental Biology of Freshwater Invertebrates, pages 221–248

Developmental Biology of *Schistosoma mansoni*, With Emphasis on the Ultrastructure and Enzyme Histochemistry of Intramolluscan Stages

Burton J. Bogitsh and O. Stephen Carter

ABSTRACT The life cycle and development of *Schistosoma mansoni* are presented as the most convenient model for experimental studies of digenetic trematodes which cause schistosomiasis in man. Literature related to the development of the free-swimming larva, the miracidium, is presented; but the main focus of this presentation is on the intramolluscan developmental stages. This covers penetration of the molluscan (snail) intermediate host by the miracidium, transformation of the miracidium into a mother sporocyst, and the propagation and migration of daughter sporocysts. Particular attention has been given to the ultrastructure and enzyme histochemistry of these developmental stages. Methods for the experimental management and propagation in the laboratory of *Schistosoma mansoni* have been given, including methods and experimental results obtained from studies of the development of sporocysts in vitro.

INTRODUCTION

The schistosomes are digenetic trematodes that belong to the family Schistosomatidae Poche, 1907. Members of this family are generally dioecious Digenea, but some members are hermaphroditic. The life cycles of schistosomatids are broadly similar although they are found in a wide variety of definitive vertebrate hosts, including mammals, birds, amphibians, and

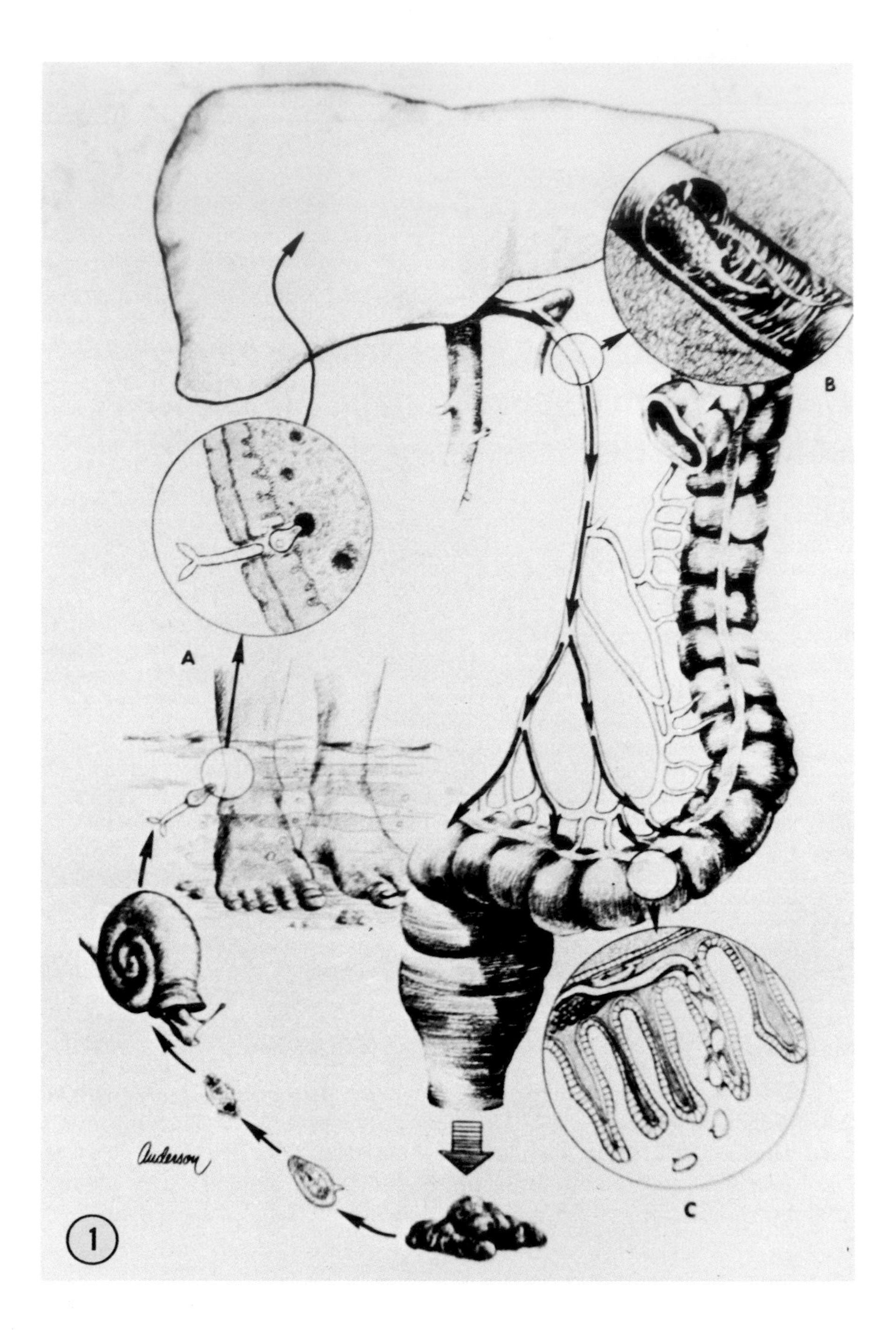
A
B
C
Anderson
1

reptiles, depending on the species in question. The intermediate host is always a mollusc.

This chapter is primarily directed toward studies of the organism *Schistosoma mansoni*, one of the tetrad of organisms causing human schistosomiasis (or bilharziasis). *S. mansoni* and the other human schistosomes (*S. haematobium, S. japonicum,* and *S. mekongi*) have been most extensively studied because of their direct physical and economic impact upon tropical and subtropical man.

Adults of the genus *Schistosoma* live in blood vessels of the definitive host, the female worms producing shelled embryos which are fertilized *in utero* (Fig. 1). Patterns of oviposition vary among the four species of Schistosoma. For instance, *S. mansoni* eggs are deposited singly whereas the other three species deposit large aggregates, or "rafts," of eggs. Interestingly, the eggs of all species when deposited contain immature embryos that are rarely developed beyond the morula stage. The length of time required for development of the embryo to a viable miracidium capable of swimming and infecting a snail intermediate host varies from 6–8 days for *S. mansoni* (Gönnert, '55) to 9–10 days for *S. japonicum* (Vogel, '42). Other than the fact that it always occurs in the tissues of the mammalian host, few investigative data are available regarding embryonic development of the miracidium during this period. Among the few reports available are those by Pellegrino et al. ('62) and Michaels and Prata ('68), who have investigated rates of development of the miracidium of *S. mansoni in vivo* and *in vitro*, respectively. In assessing the developmental state, Pellegrino et al. ('62) employ morphological criteria to divide development into five stages. Stage 1 is exemplified by the presence of an embryonic disc that occupies one-third the width of the egg. Stage 2 is reached when this embryo occupies one-half of the egg. At stage 3, the embryo fills the entire egg. The final stage is the miracidial stage (stage 5), at which time the miracidium displays such signs of vitality as the beating of the flagella of the protonephridia and general muscular movement within the eggshell.

Although no description of the cell lineage is available for the development of the schistosome miracidia specifically, there is no reason to believe that

---

Fig. 1. Life cycle of *Schistosoma mansoni*. Eggs passed in feces, hatch in freshwater streams and yield miracidia, which penetrate appropiate species of snails. Cercariae, produced by asexual multiplication, leave the snail host and penetrate the skin of humans (A) exposed in freshwater streams. Schistosomula, by way of the bloodstream, eventually reach the liver where development into adulthood continues. Mature male and female adults, in copulation, migrate against the portal flow, chiefly by way of the inferior mesenteric vein (B) to the small venules in the rectosigmoid area of the large intestine. Many eggs deposited like beads in a chain in the small venule (C) break out into the lumen of the intestine and pass out in the feces (from Beck, J.W. and J.E. Davies (1976) *Medical Parasitology*. 2nd Ed. 244pp).

they vary greatly from the basic pattern of embryogenesis of *Paraorchis acanthus* as described by Rees ('40). In this digenetic trematode, differentiation of cells occurs at the first cleavage division in which a propagatory cell and a somatic cell are formed from the zygote. The somatic cell gives rise to the vitelline membrane surrounding the embryo and located just beneath the shell and to the remaining tissues of the embryo or miracidium. The propagatory cell undergoes repeated divisions with each division producing one somatic cell, and another which retains the cytological characteristics of the mother propagatory cell. After a number of these divisions, the propagatory cell migrates to the rear of the developing miracidium where it divides into numerous propagatory cells and forms the "germ ball." The germ ball is destined to give rise to the next generation of larvae; in the case of the schistosomes, the secondary or daughter sporocysts. It is of interest that during this development, no germ layers form (Rees, '40). These "eggs" reach fresh water in the excretions of the host and hatch in the water, releasing a free-swimming larva, the miracidium. The larva then must penetrate a snail, the intermediate host, within a period of 24–48 hours. Following direct entry into the tissues of the molluscan host, the miracidium undergoes a series of developmental transformations that result in the first intramolluscan larval stage known as the "mother" or "primary" sporocyst. This tubular, sacular, unbranched larva containing germinal tissue remains at the original locus of entry, giving rise, asexually, to the second generation of sporocysts which emerge from the "mother" and are disseminated throughout the snail host. This second generation of sporocysts comprises the "daughter" or "secondary" sporocysts, which proliferate extensively within the snail host giving rise to myriads of cercariae, furcocercous trematode larvae. The cercariae, when mature, leave the sporocyst and emerge from the snail into water, then encounter and penetrate the vertebrate host to begin the vertebrate stage of the life cycle.

In this chapter, the ultrastructure and some aspects of the histochemistry of the two intramolluscan stages are considered, and information is included pertaining to *in vitro* culture methods for these stages.

MOTHER SPOROCYST

The free-swimming, ciliated miracidium that escapes from the "egg" is the direct precursor of the mother sporocyst. The miracidium must locate and penetrate its specific snail host (in the case of *Schistosoma mansoni*, a snail of the genus *Biomphalaria*). Once the miracidium has found such a snail, it penetrates the epithelium by chemical and mechanical means.

In order to understand the morphology of the mother sporocyst, it is essential to first understand the morphology of the miracidium. One may think of the miracidium as a packet of germinal tissue enclosed within a special living transport vessel equipped with water-resistant plates, active ciliary "oars," and sensory receptors, and armed with penetrating apparatus to deliver the payload of larval parasite material.

The outer covering of the miracidium is composed of tiers of large, ciliated epithelial cells between which appear thin, cytoplasmic ridges. The ridges intercalate between the epithelial plates (Figs. 2, and 3), and, as the plates are shed, the cytoplasmic ridges spread and eventually fuse to form the surface of the primary sporocyst (Fig. 4). The method by which the plates are shed remains obscure at present; however, Wikel and Bogitsh ('74) have described vacuolations occurring in the basal lamina of the ciliated cells that may contain hydrolytic enzymes which may have a role in this phenomenon. Preliminary histochemical observations confirming the presence of acid phosphatase in these vacuoles support this assumption. The final sloughing of ciliated epithelial cells varies from one trematode group to another; however, in the schistosome, the final shedding occurs within the snail, following completion of penetration by the miracidium.

Meuleman et al. ('78) attribute miracidial plate shedding to at least three phenomena—namely, the ciliary beat, muscle contraction, and expansion of the ridges against and underneath the plates. Osmolarity changes resulting from a change from freshwater to snail hemolymph may also be a factor. Another theory holds that basal cavitations or vacuoles found beneath the plates (Wikel and Bogitsh, '74; Basch and DiConza, '74) are responsible for the shedding. Meuleman et al ('78) observe only occasional vacuoles at this site, and suggest that they are a *result* of plate shedding rather than a *cause*.

Maldonado and Acosta-Matienzo ('47) meticulously describe the miracidium of *S. mansoni*, stating that migration of miracidia may occur after penetration and that sporocysts localize in the kidney and digestive gland in small snails (a point that is inconclusive in later studies). They describe the elongation and folding of the mother sporocyst and show that the appearance of multiple chambers in sections of mother sporocysts may have misled Gordon et al. ('34) into believing that miracidia undergo fragmentation into numerous sporocysts. They describe development of daughter sporocysts, their breaking free from the mother sporocyst, the subsequent migration to the digestive gland and ovotestis of the snail, and the development of cercariae within these organs. Damage to snail tissue is evident in late stages of infection.

Considerable study has been dedicated to miracidial structure and changes that occur after penetration of snails (Wilson, '69; Basch and Di Conza, '74; Smith and Chernin, '74; Wikel and Bogitsh, '74; LoVerde, '75; Koie and Frandsen, '76; Meuleman et al., '78; Rivera and Liang, '80). The ciliated plates have been extensively described earlier, and the probability of sunken

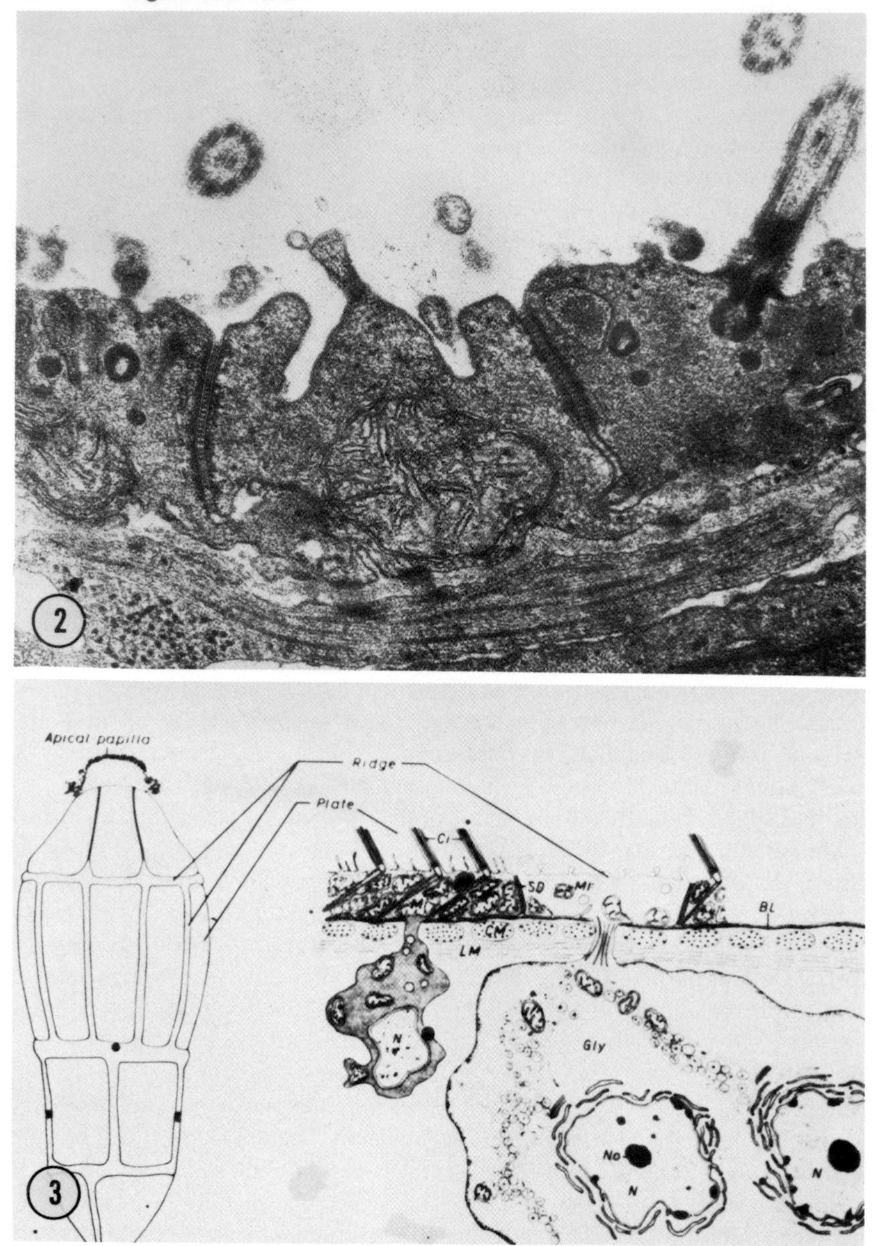

Fig. 2. Cytoplasmic ridge contains a prominent mitochondrion and is connected to adjacent epidermal cells by septated desmosomes. × 38,000.

Fig. 3. Drawing of the miracidium of *Schistosoma mansoni* showing the outer structure with apical papilla (left) and the ultrastructure of the body wall (right). Note the plates and ridges that cover the body, both connected to sunken nuclei. BL, basal lamina, Ci, cilia, CM, circular muscle, Gly, glycogen, Mi, mitochondrion, N, nucleus, No, nucleolus, SD, septate desmosome. (from Meuleman, 1978).

nucleated cytons suggested (Wikel and Bogitsh, '74); the presence and position of inconspicuous sunken cell bodies of the epidermal plates are clearly defined by Meuleman et al. ('78) (Fig. 3), who, in a more recent study, have also determined the ontogeny of the cytons (Meuleman et al., '80). Ridge structure between the plates has been described and discussed in detail by Wikel and Bogitsh ('74), Meuleman et al. ('78), and Rivera and Liang ('80). All investigators agree that the loss of miracidial epidermal plates and the spreading and expansion of the cytoplasmic ridges with the subsequent appearance of an outer microvillar surface herald the beginning of the mother sporocyst.

Meuleman et al. ('78) show that the syncytial tegument of the mother sporocysts develops to its full extent by 48 hours after penetration into the intermediate host, displaying at that time a more uniform thickness with a highly microvillar outer surface (Fig. 5). The outer microvillar surface is covered, in addition, by a coat of electron-dense material. Meuleman et al. ('78) report the presence of endocytotic vesicles extending from the base of folds of the tegument; however, Smith and Chernin ('74) and the present authors (Bogitsh and Carter) fail to confirm this observation. The cytoplasm of the outer layer contains, in addition to these vesicles, mitochondria, polyribosomes, rough endoplasmic reticulum, glycogen, and lipids (Fig. 5). An irregular basal plasma membrane, under which is found a thin basement membrane, develops infoldings which penetrate deep into the cytoplasm. After 48 hours, the tegumental surfaces of the mother sporocyst and the daughter sporocyst in the area of the digestive gland are very similar in morphology (Meuleman, '72; Meuleman and Holzmann, '75).

Meuleman et al. ('78) report aggregation of snail amoebocytes in the area of miracidial penetration which scavenge the remnants of shed ciliated epidermal plates. By the time the tegument of the mother sporocyst is completely formed, a few amoebocytes in the vicinity contain remnants of almost completely digested plates (Fig. 6).

DAUGHTER SPOROCYSTS

Young mother sporocysts retain some parenchymal tissues from the miracidial stage, but as their germinal centers develop, the parenchymal tissues begin to degenerate (Fig. 7). Meuleman et al. ('80) elaborate further upon the development of the mother sporocyst and its relationships to the enclosed daughter sporocysts via sequential studies of the developmental process in which they examine sections of sporocysts obtained daily at intervals from the first to the eleventh day postexposure.

Several hours after snail penetration, aggregates of cells within the mother sporocyst begin to proliferate as embryonic daughter sporocysts. It is appar-

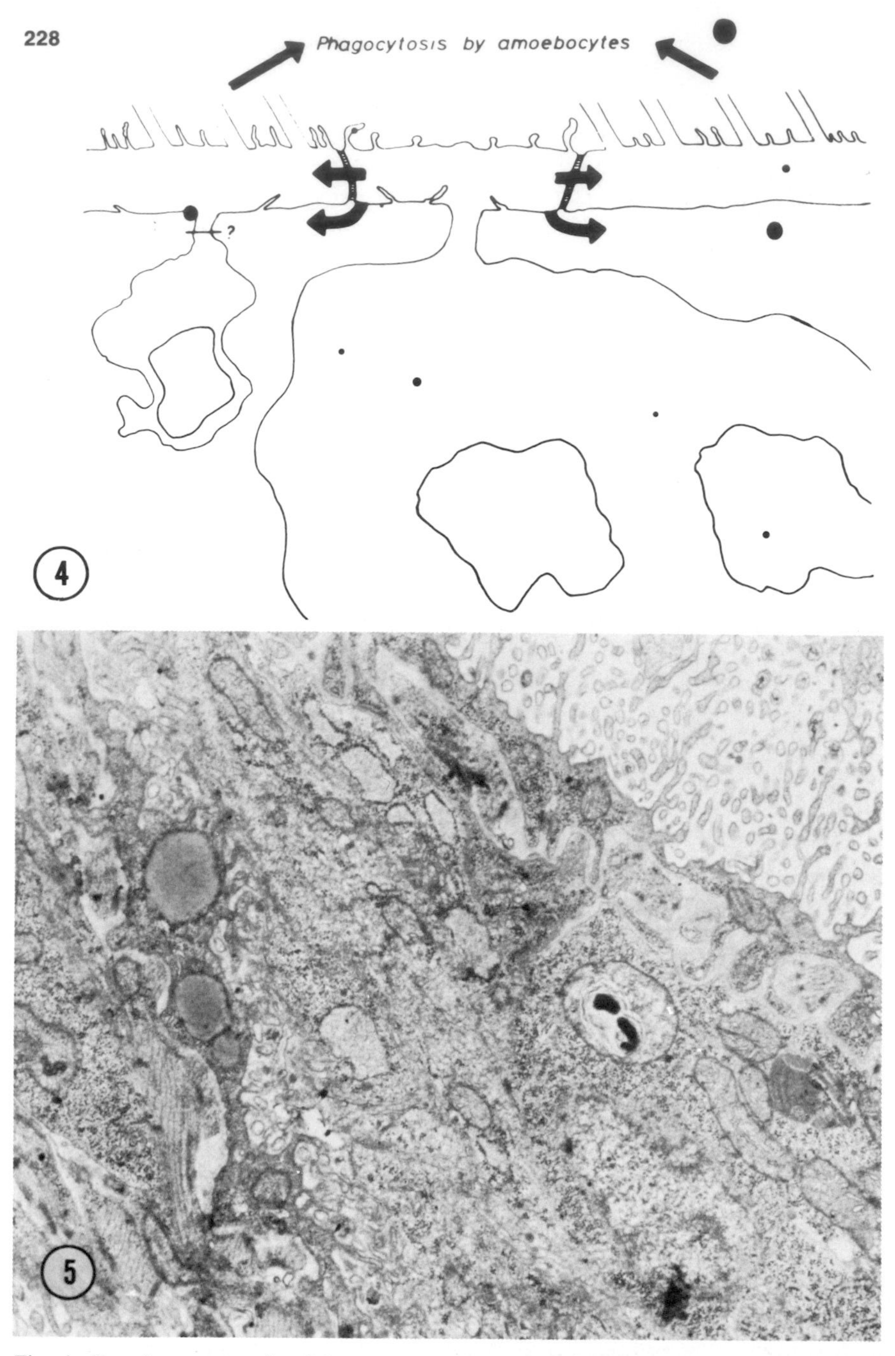

Fig. 4. Drawing representing 2-hr post-penetration changes in the body wall of *S. mansoni* miracidium. Plates are shed and subsequently phagocytized by host amoebocytes. The fates of the cell bodies and their connections are unknown. The ridges themselves expand, forming the outer layer of the mother sporocyst (from Meuleman, 1978).

Fig. 5. Extensive microvillar outer surface of mother sporocyst. ×35,000.

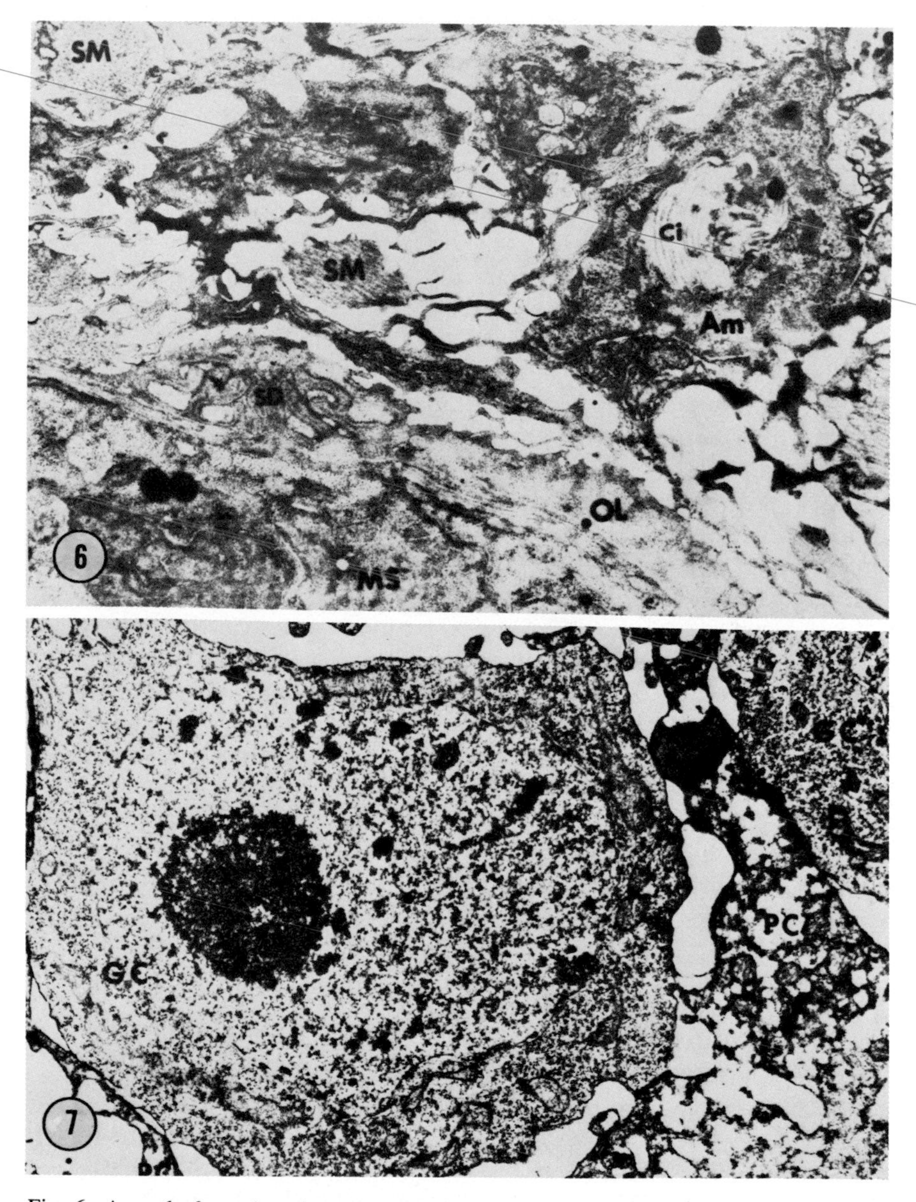

Fig. 6. A newly formed mother sporocyst (MS), 2-hr post-penetration showing snail amoebocytes (Am), which phagocytize shed plates and cilia (Ci). The outer layer (OL) of the sporocyst is now a continuous tegument. SD, remnants of septate desmosomes, SM, snail muscle. × 10,800 (from Meuleman, 1978).

Fig. 7. Photomicrograph showing the almost total degeneration of parenchyma cells (PC) surrounding germinal cells (GC) within a four-day old sporocyst. × 7,250 (from Meuleman et al., 1980).

ent that the development and migration of daughter sporocysts is an extended process, since the first daughter sporocysts appear in various parts of the snail 8 days to 2 weeks postpenetration.

Intramaternal daughter sporocysts are differentiated from extramaternal daughter sporocysts by virtue of their location and developmental state. It is almost certain that in some instances late daughter sporocysts that remain at their site of origin within the mother sporocyst may have cercarial bodies developing within them, but the vast majority of the daughter sporocysts leave the mother sporocyst and migrate through the snail hemolymph system seeking the digestive gland and ovotestis of the snail in which to mature and produce cercariae.

Meuleman et al. ('80) provide an excellent study for broader understanding of the intramaternal development of the daughter sporocysts. They demonstrate that the basic developmental pattern of the body wall of the daughter sporocyst within the mother sporocyst parallels that of the cercaria inside the daughter sporocyst (Meuleman and Holzmann, '75).

The major events in the development of the daughter sporocyst are 1) the increase in the number of germinal cells and their emergence as the most prominent feature of the mother sporocyst (Fig. 8A); 2) the occupation of the space around the germinal cells by larger irregularly shaped parenchymal cells, which are a prominent feature of the miracidium; 3) the elongation of the mother sporocyst into an increasingly tortuous shape after about 2 days postexposure; 4) the expansion of the nucleated region of the tegument into the inner side of the outer layer and formation of thin extensions that enwrap the germinal cells 4 days postexposure (the parenchyma cells disappear after a few more days (Fig. 8B)); 5) the budding development of daughter sporocyst embryos from germinal cells in the brood chamber of the mother sporocyst, 5 or 6 days postexposure (Fig. 8C); 6) the initial envelopment of the daughter sporocysts are by a primitive epithelium, formed by fusion of the extensions of the tegument of the mother sporocyst (Fig. 8C); 7) the expansion beneath the primitive epithelium of peripherally located somatic cells in the daughter embryos and their coalescence into a continuous syncytial envelope 8 days postexposure (Fig. 8D); 8) several rapid changes, including degeneration of the primitive epithelium 9–10 days postexposure; and 9) the development of the syncytial envelope, which formed beneath the primitive epithelium into the final tegumental outer layer with subsequent nuclear pycnosis and disappearance, and the linking of the tegumental layer to internally situated nucleated cell bodies (Fig. 8E).

Closely packed germinal cells appear in the medial part of the fully developed daughter sporocyst, and, 10 or 11 days after infection, the daughter sporocysts begin to escape from the mother sporocyst.

Smith and Chernin ('74) have studied the ultrastructure of mother and daughter sporocysts of *S. mansoni* 6 and 13 days after infection of the snail

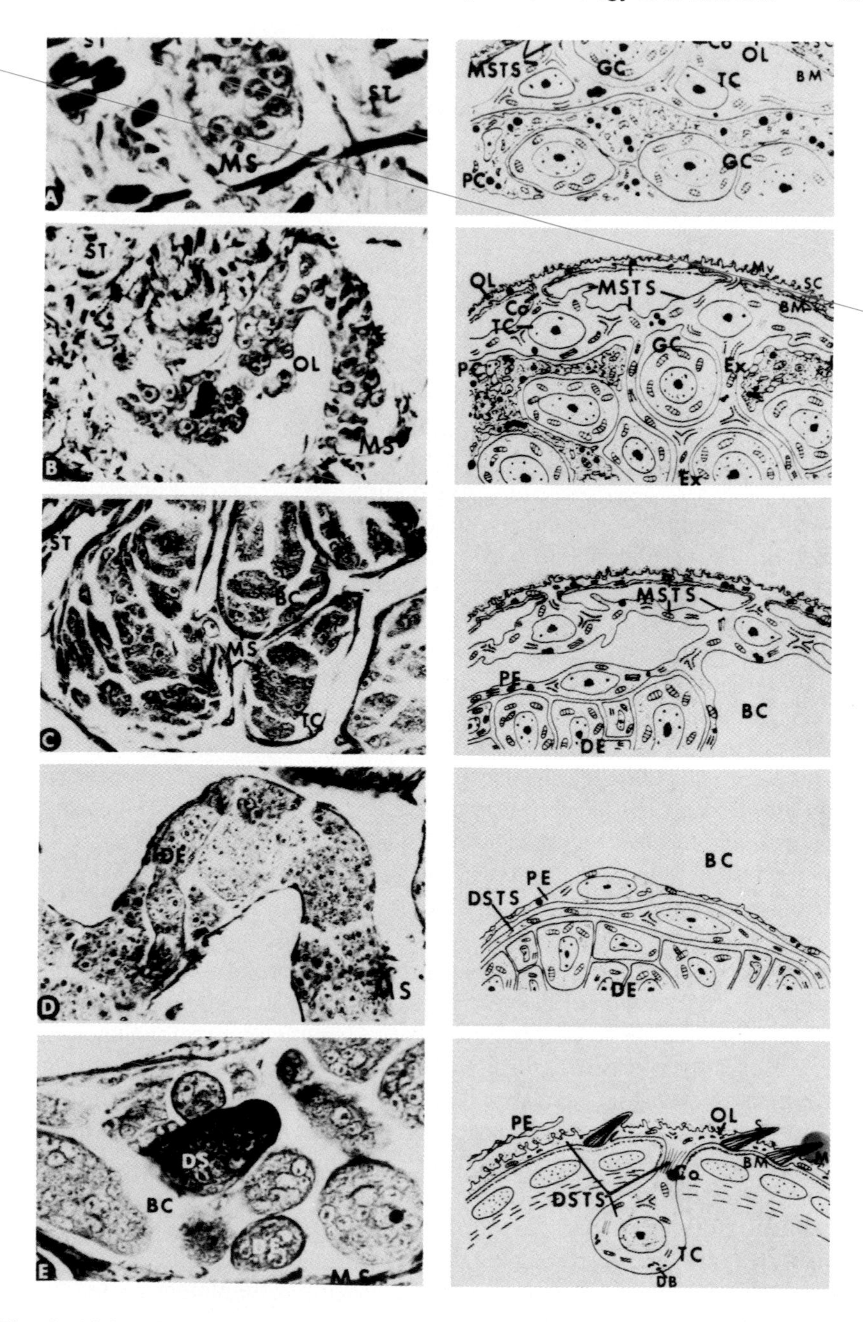

Fig. 8. Light micrographs (left) and drawings from electron micrographs (right) showing developing daughter sporocysts inside a sporocyst: A=2 days, B=4 days, C=6 days, D=9 days, and E=11 days after infection. (See text for explanation. (from Meuleman et al., 1980).

intermediate host. They note several cytological differences between the teguments of mother and daughter sporocysts. The major distinctions are the more uniform depth of the tegument and the far shorter surface amplifications in young daughter sporocysts. In addition, the tegument of the daughter sporocysts possesses spines in the anterior region (Fig. 9).

HISTOCHEMISTRY

There have been comparatively few histochemical studies concerning the intramolluscan stages of the schistosomes, and those have been of a general nature since relatively little is known of the physiology of trematode sporocysts (Erasmus, '72).

An investigation by Etges et al. ('75) deals with the nervous system of several schistosomes—specifically, that of the sporocysts and intrasporocyst cercariae of *Schistosoma mansoni.* In the study, substrates are employed for the localization of nonspecific esterases as well as cholinesterase. Nonspecific esterase activity is evident in a node in the anterior region of young daughter sporocysts and in germ-cell masses. These results led the investigators to hypothesize that nervous elements may be present in the anterior region of the daughter sporocyst. However, the examination by Kinoti et al. ('71) of daughter sporocysts of *S. mattheii* and *S. bovis* in *Bulinus africanus* for the presence of nonspecific esterases yields no reaction, and the negative findings are interpreted as an indication that lipids play little part, if any, in the metabolism of the daughter sporocysts in these organisms. Likewise, DiConza and Basch ('75) detect no nonspecific esterases or lipases in *in vitro* cultured *S. mansoni* sporocysts.

On the basis of ultrastructural observations, Meuleman ('72) suggests that the tegument of the daughter sporocyst of *S. mansoni* is involved in the uptake of nutrients by absorption and possibly endocytosis. She also notes the presence of lipid deposits in the tegument and, contrary to the views of Kinoti et al. ('71), hypothesizes that they may serve as energy stores for the organism. Finding no glycogen stores in the tegument, she concludes that, since the organism is bathed in a low molecular weight carbohydrate-rich hemolymph, there is no need for it to store carbohydrates in the form of glycogen. Parasitized snails experience rapid depletion of glucose stores (Christie et al., '74; Carter and Bogitsh, '75), and there is strong indication that the developing sporocysts divert glucose from the snail intermediate host and that this diversion probably occurs via the tegument (see Carter and Bogitsh, '75; and Becker, '80; for reviews).

The lack of a digestive tract and the presence of a tegument replete with surface amplifications (Fig. 10) provide persuasive evidence that the epithelium is the primary means by which nutrients enter the sporocysts. Several

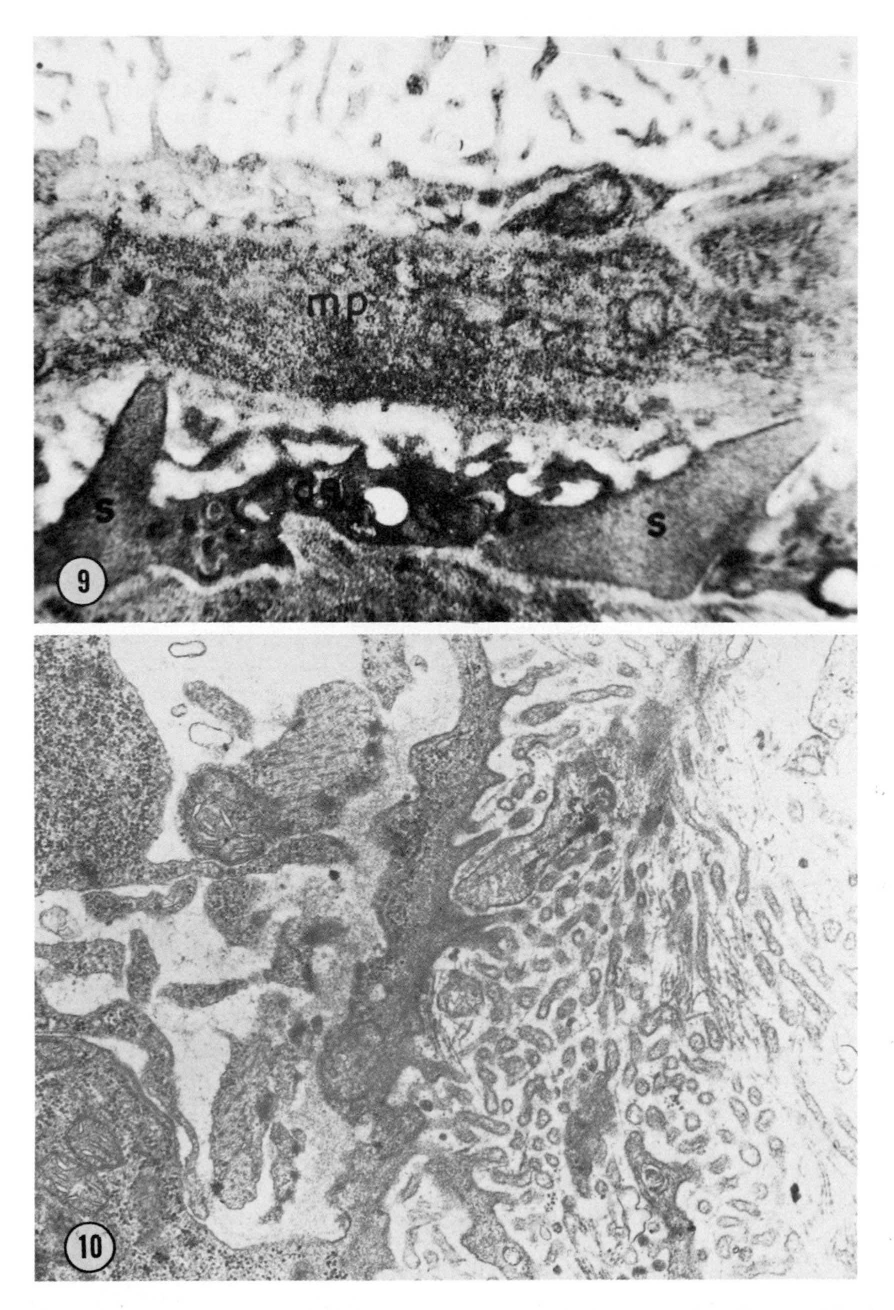

Fig. 9. Mother and daughter sporocysts of *S. mansoni*. Note tegumental perikaryon (Mp) in a 13-day mother sporocyst, and tegument (de) with spines (S) in the young daughter sporocyst ×28,500 (from Smith and Chernin, 1974).

Fig. 10. Mother sporocyst with enclosed daughter sporocyst. Note the high degree of surface amplifications. × 21,000.

cytochemical enzyme studies dealing with this surface utilize reactions designed to illustrate the mechanisms available for the uptake of molecules. To this end, Krupa et al. ('71) and Krupa and Bogitsh ('75) demonstrate cytochrome *c*-oxidase in the tegumental mitochondria of *S. mansoni* daughter sporocysts (Fig. 11). Visualization of the activity is accomplished by oxidation of 3,3′-diaminobenzidine (DAB) with a number of variables, such as inhibitors and changes in pH, as controls. It is of interest that the mitochondria of younger intrasporocyst cercariae (i.e., those without a noticeable surface coat) likewise exhibit cytochrome *c*-oxidase activity. In older cercariae, however, the tegumental mitochondria are unable to oxidize DAB (Fig. 12). Krupa and Bogitsh ('75) suggest that enzyme activity in the mitochondria reflects the metabolic status of the tissue; consequently, the tegument of the older intrasporocyst cercariae is considered to be quiescent metabolically. Conversely, the active mitochondria in daughter sporocyst and young intrasporocyst cercarial teguments indicate metabolically active tissues. In the same report, Krupa and Bogitsh ('75) note the presence of beta-glycogen in the sporocyst tegument, whereas Meuleman ('72), as noted above, fails to observe any glycogen stores in the tegument of the daughter sporocyst.

Another indicator of the absorptive properties of a tissue is the presence of phosphatase activities. Indirect evidence exists of a relationship between membrane-nonspecific phosphatase activity and membrane transport. Perhaps the most positive statement that can be made is that tissues active in the uptake of glucose also demonstrate nonspecific phosphatase activity. On the other hand, most biochemical and histochemical studies of solute transport across membranes and employing concentration gradients have been linked to specific adenosine triphosphate phosphohydrolase systems in various animal tissues (see Skou, '65; Bonting, '70, for reviews). Krupa and Bogitsh ('72) depict ultrastructural localization of phosphohydrolase activities in the daughter sporocysts and intrasporocyst cercariae of *S. mansoni,* using adenosine triphosphate (ATP) as a substrate. Reaction product from the hydrolysis of ATP is evident on the outer surface of the apical plasma membrane of the daughter sporocyst tegument (Fig. 13), in mitochondria (Fig. 14), and in membranes of the brood chamber. The judicious use of inhibitors shows that the reaction in the tegument is due to the activity of a phosphatase of rather low specificity, probably nonspecific alkaline phosphatase. The reaction produced associated with the mitochondria, however, is classified as an ATPase due to its sensitivity to fixation, inhibition by parachloromercurobenzoate, and activation by cysteine. In the intrasporocyst cercariae, nonspecific alkaline phosphatase activity also occurs on the tegumental surface. In cercarial mitochondria, however, phosphatase activity is heterogeneous with only the smaller, condensed mitochondria displaying reaction product (Fig. 15). The difference in mitochondrial activity probably reflects a difference in

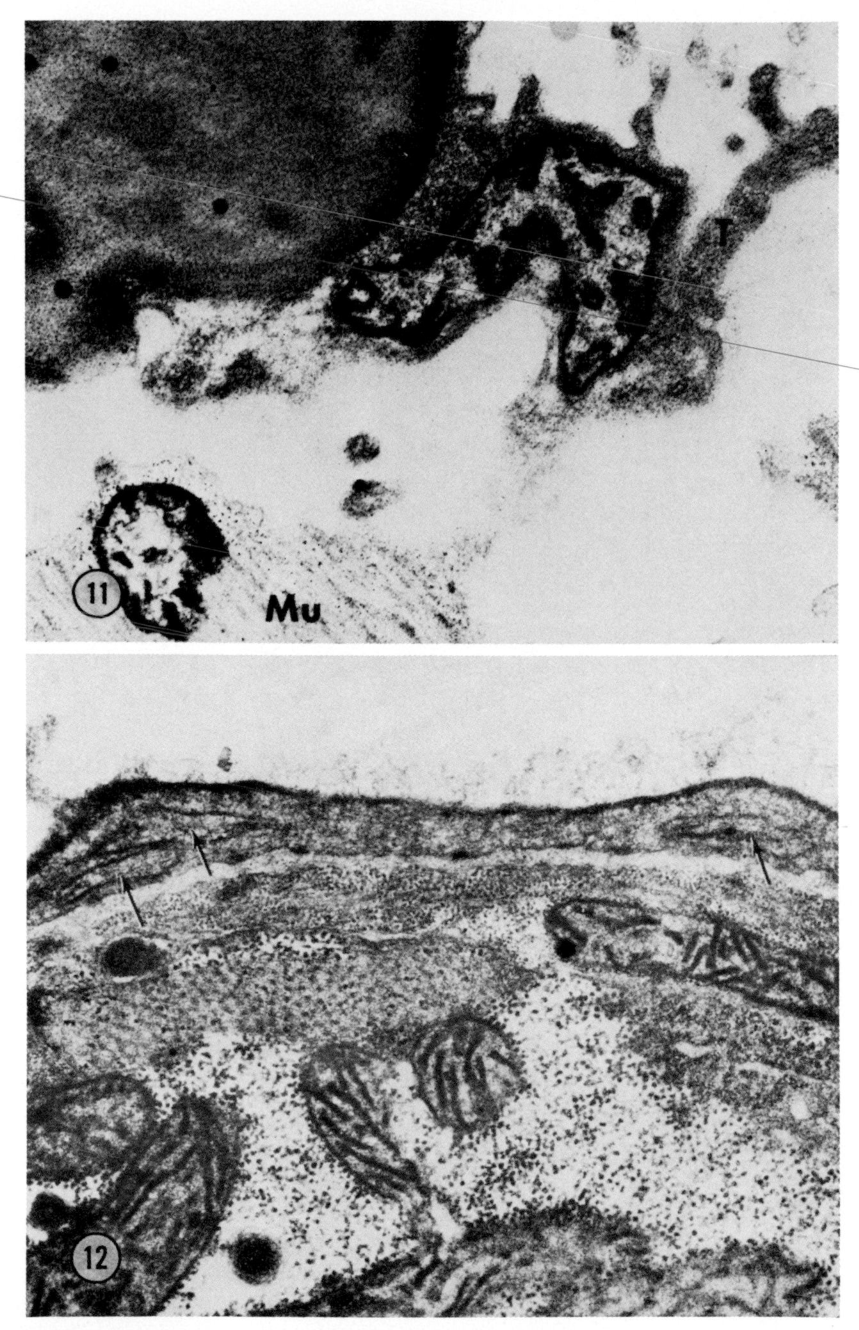

Fig. 11. Apical portion of *Schistosoma mansoni* daughter sporocyst, showing DAB-positive mitochondria in tegument (T) and muscle (Mu, at lower left). (DAB, pH 7.4, 60 min., Uranyl acetate and lead citrate) ×50,000 (from Krupa and Bogitsh, 1975).

Fig. 12. Apical portion of an intrasporocyst cercaria. Note the two DAB- negative mitochondria (arrows) in tegument covered by glycocalyx, and compare with subtegumental DAB-positive mitochondria encompassed by glycogen granules. (DAB, pH 7.4, 45 min.) No stain ×48,000 (from Krupa and Bogitsh, 1975).

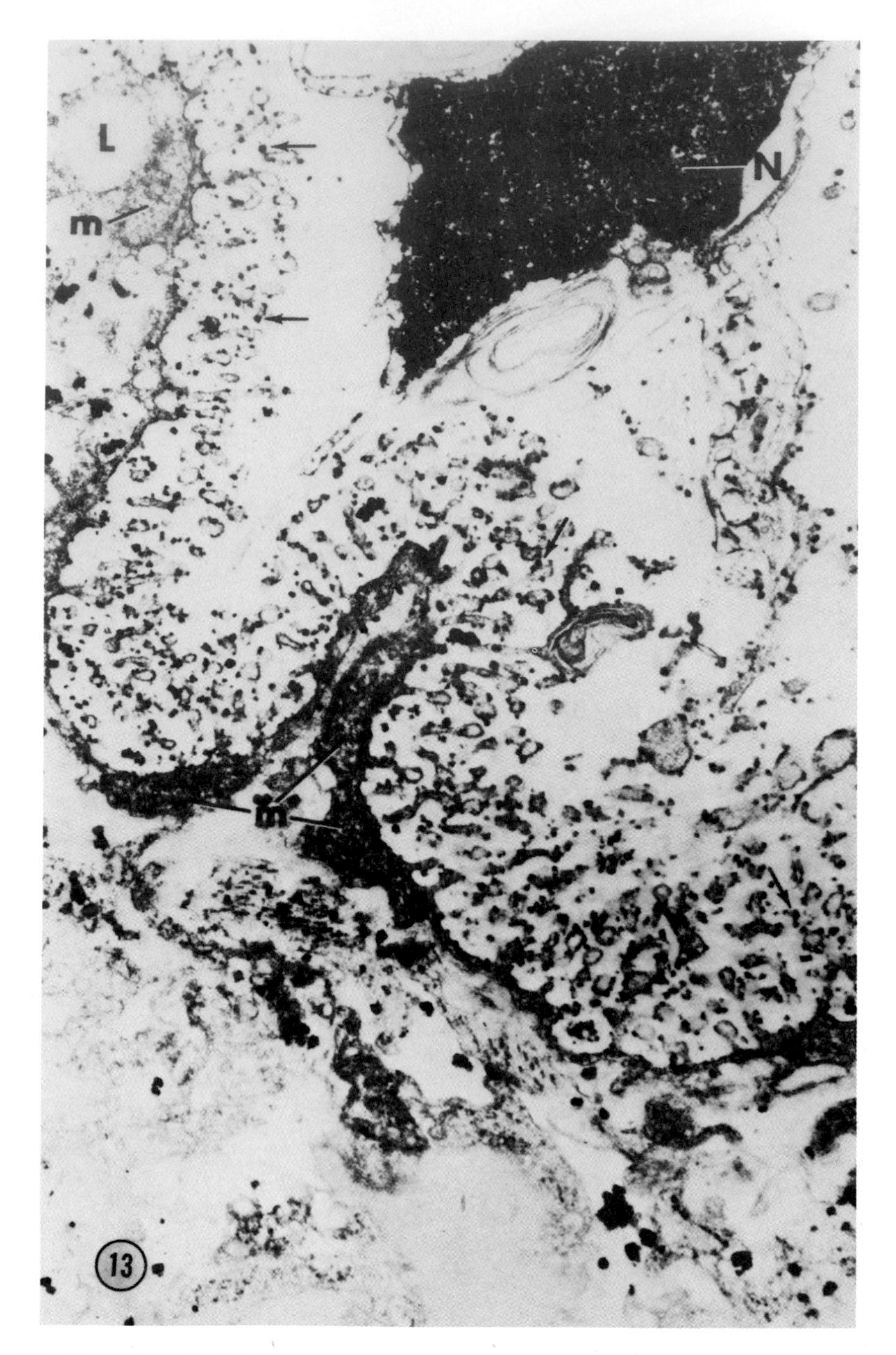

Fig. 13. Body wall of *Schistosoma mansoni* sporocyst surface, fixed in formaldehyde, shows reaction product indicative of phosphohydrolase activity aligned along outer surface of microvilli (arrows). Significance of larger dense clumps within the body wall is uncertain. The mitochondria (M) contain no demonstrable enzyme activity. The pycnotic nucleus (N) is of snail host origin. L = lipid droplets. (Substrate: ATP) ×28,000 (from Krupa and Bogitsh, 1972).

metabolic status existing between the mitochondria, although no similar heterogeneity exists relative to cytochrome *c*-oxidase activity.

In order to assess further the distribution of tegumental transport enzymes of schistosome daughter sporocysts, Krupa et al. ('75) use nitrophenyl phosphate as a substrate for the localization of Na-K-ATPase. Using ouabain as a specific inhibitor of transport ATPase, they document a ouabain-sensitive, potassium-activated ATPase system in *S. mansoni* and *S. haematobium* daughter sporocysts. Reaction product was observed on the cytoplasmic side of the apical plasma membranes of the teguments of both species (Fig. 16).

In summary, there have been relatively few histochemical studies on the intramolluscan stages of the schistosomes, and those have centered primarily on the properties of the tegument of the daughter sporocysts. As in vertebrate tissues and other helminths, phosphatase activity in the tegument at this stage occurs in association with the apical plasma membrane. Krupa and Bogitsh ('72), using ATP as a substrate, show a nonspecific alkaline phosphatase to be associated with the outer aspect of the apical plasma membrane of the daughter sporocyst. Another study, using nitrophenyl phosphate as substrate (Krupa et al., '75), reveals a transport ATPase on the cytoplasmic side of the apical plasma membrane. Although the particular relationship of the two enzymes is not addressed, it is reasonable to assume that the nonspecific phosphatase on the outer surface is involved with the hydrolysis of phosphate esters of glucose and that the transport ATPase on the cytoplasmic surface is a membrane-associated enzyme. Such an orientation may indicate that enzymatic cleavage and membrane transport are closely integrated (Crane, '68; Bogitsh, '75).

IN VITRO STUDIES

Recent investigations concerning *in vitro* cultivation of intramolluscan schistosomes can be conveniently divided into two phases.

The first phase deals with the transformation of newly hatched miracidia into mother sporocysts. Voge and Seidel ('72) identify the most favorable conditions for transformation of *S. mansoni* and *S. hematobium* miracidia. According to their report, an increase in osmolarity approximating that of snail hemolymph elicits the shedding of epidermal cells from newly hatched miracidia within 9 hours. They also show that the transformed organisms can tolerate for relatively long periods of time a wide variety of *in vitro* environmental conditions, including changes in osmolarity, concentrations of various amino acids, sera, pH, vitamins, and glucose. In the most satisfactory monoaxenic medium, mother sporocysts of *S. mansoni* can survive and grow up to 3 weeks; those of *S. japonicum*, up to 5 weeks.

Utilizing a previously devised daughter sporocyst culture medium (see DiConza and Basch, '74) for *S. mansoni* miracidial transformation, Basch

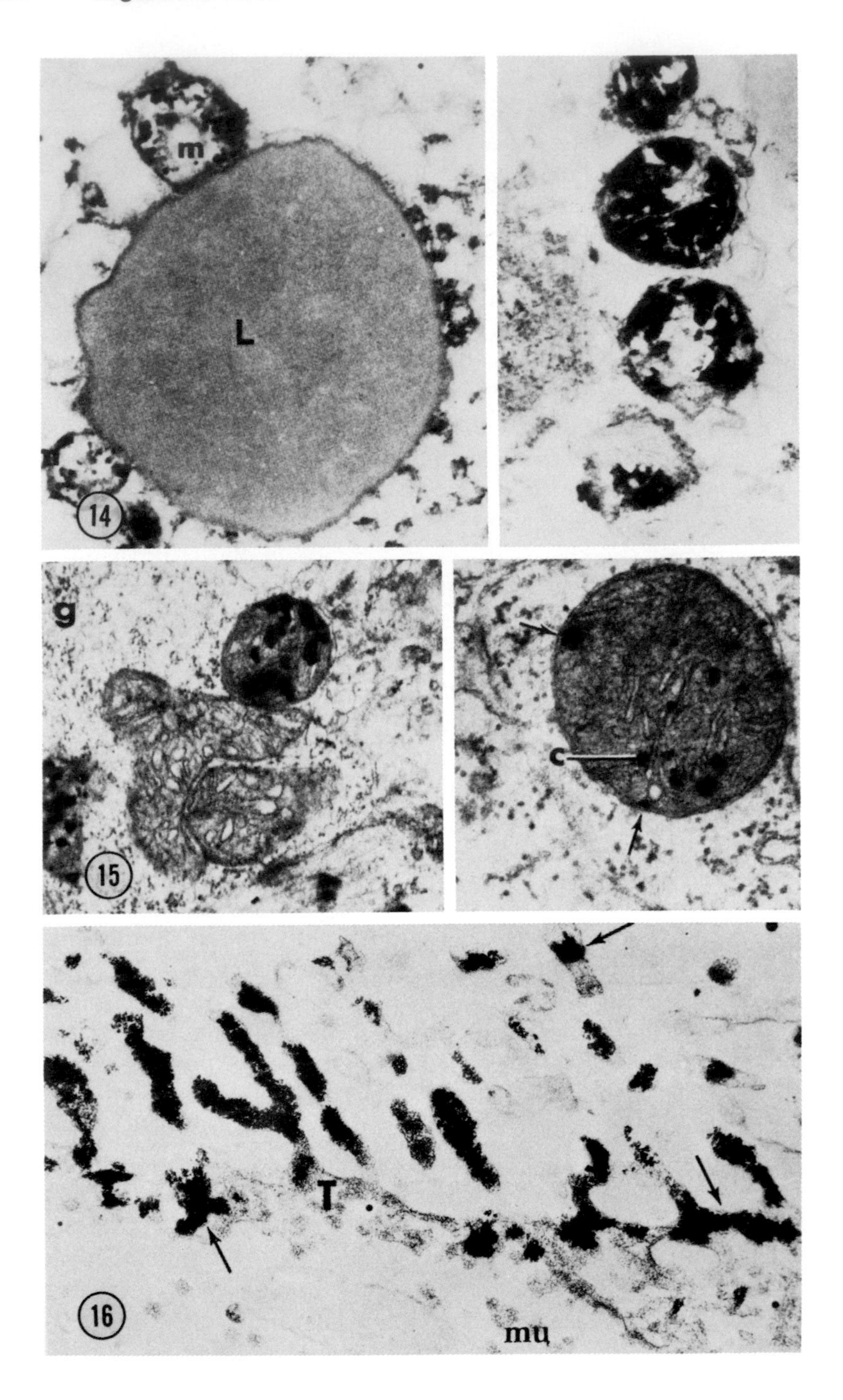
m
L
14
g
15
c
T
16
mu

and DiConza ('74) determined that the ciliated epidermis is shed within 3 hours, provided human serum is included in the medium; horse or fetal calf sera are less satisfactory for the maintenance and development of the organisms. After 200 minutes in culture, electron microscopic analysis reveals a tegument morphologically identical to that in mother sporocysts produced *in vivo* (Fig. 17).

Koie and Frandsen ('76), in studies of transformed miracidia of *S. mansoni via* scanning electron microscopy, accomplish *in vitro* transformation of miracidia using hemolymph from the snail, *Planorbarius corneus*. They observe that miracidial, posteriormost, ciliated epidermal cells are shed first and that subsequent shedding progresses in an anterior direction. These investigators assign no precise interval for completion of shedding but note that the cilia of the miracidia cease beating within an hour after exposure to the snail hemolymph medium.

The second phase of schistosome *in vitro* culture investigations deals with circumstances favorable to the growth and development of sporocysts. Hansen et al. ('73) and DiConza and Hansen ('73) have produced monoaxenically cultured daughter sporocysts of *S. mansoni* in a medium containing arthropod cells from an established mosquito line. Subsequently, Hansen et al. ('74a) have utilized a culture consisting, in part, of a *Drosophila* medium preconditioned with *Aedes* tissue culture, to show that daughter sporocysts dissected from mother sporocysts can survive up to 13 days in the medium and produce small embryos. Up to this point, results are comparable to those obtained when daughter sporocysts are grown in the earlier medium (Hansen et al., '73) in which conditioning of the medium with mosquito cells is continuous. In the latter instance, however, development proceeds to the liberation of

---

Fig. 14. Body wall of *Schistosoma mansoni* sporocyst—at least three mitochondria (M) of a body wall incubated for ATPase activity prior to fixation show intense reaction due to stimulation of cysteine. L = lipid droplets. (Substrate: ATP + cysteine) ×41,000 (from Krupa and Bogitsh, 1972).

Fig. 15. Intrasporocyst cercariae of *Schistosoma mansoni*. Glycogen particles (g) surround muscle mitochondria, some with reaction product next to other, presumably unreactive, mitochondria (top and lower left). Lower right, mitochondrion with reaction product in cristal (c) and in outer compartment (arrows). (Substrate: ATP) ×29,000 (from Krupa and Bogitsh, 1972).

Fig. 16. *S. haematobium* mother sporocyst incubated with NPP, showing K^+-NPPase reaction product at arrows. The active sites are seen to be restricted to the cytoplasmic side of the apical plasma membrane that covers the microvilli and to the cytoplasmic side of the apical and basal plasma membranes of the tegument (T); mu = muscle. ×50,000 (from Krupa et al., 1975).

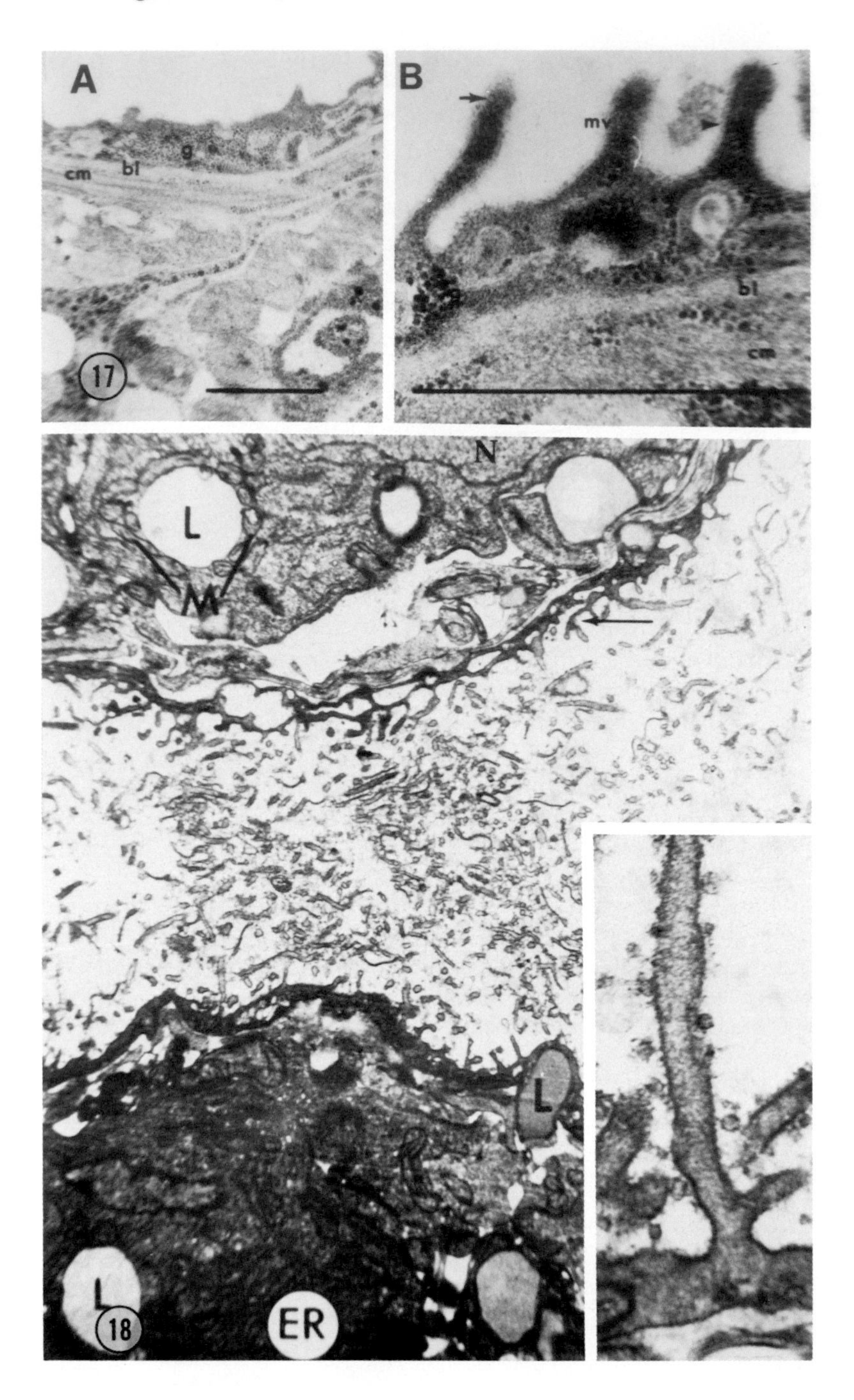
A
B
cm
bl
g
mv
bl
cm
17
N
L
M
L
L
18
ER

progeny daughter sporocysts. The nutritive contribution of the mosquito culture, though apparently necessary for the development of the progeny daughter sporocysts, remains vague.

In still later experimentation, Hansen et al. ('74b) circumvent the preconditioning requirement by adding sulfhydryl-containing compounds under a controlled gas environment of 0.5% carbon dioxide. Although there is no increase in the number of progeny daughter sporocysts produced, the new medium eliminates the influence of other animal cells (e.g., arthropod cells). DiConza and Basch ('74a) employ a culture medium derived in part from their analysis of *Biomphalaria glabrata* hemolymph. The basic medium, which consists of a combination of salts, amino acids, vitamins, organic acids, and glucose in HEPES buffer at pH 7.0–7.2, is supplemented with human serum. The final osmolality of the complete medium is 126 mOs. Daughter sporocysts placed in this medium survive up to 13 days and remain capable of producing cercariae when transplanted in *B. glabrata.* Electron microscopic studies on these sporocysts show morphologically normal tegument (Fig. 18). Using daughter sporocysts from these *in vitro* cultures, DiConza and Basch ('74b) also demonstrate that, when (^{3}H)-thymidine is integrated into the culture medium, the label is visible in the nuclei of cells in the germinative region of the organisms. Likewise, the cultured daughter sporocysts absorb labeled glucose through their teguments. Continuing studies on the cultured organisms, DiConza and Basch ('75) note the accumulation of neutral lipids in both mother and daughter sporocysts maintained *in vitro.* This accumulation is a function of time, with the amount of lipid increasing in direct proportion to the length of time in the cultured medium.

In summary, artificially cultured intramolluscan stages of the schistosomes depend primarily on empirically based media. However, attempts to simulate the natural molluscan environment in which these forms normally develop have been relatively successful. Although the organisms derived from these media are comparable in morphology and physiology to their *in vivo* counter-

---

Fig. 17 A & B. *Schistosoma mansoni* cultured mother sporocysts after 200 min. in culture. Epidermal plates are gone and new tegument is present. Note granules (g) disintegrating at bases of microvilli (mv) in which electronlucent bodies (arrow) are seen. The arrowheads show the microvillar surface unit membrane. Basal lamina (bl) and circular muscles (cm) are seen. Scale lines are 1 μm (from Basch and DiConza, 1974).

Fig. 18. Two 5-day old *Schistosoma mansoni* daughter sporocysts showing microvilli on their surfaces (arrow). Also note lipid droplets (1), mitochondria (M) endoplasmic reticulum (ER) and nucleus (N). $\times$12,500. Note: The enlarged view of a branched microvillus at the surface of the tegument ($\times$72,500) (from DiConza and Basch, 1974).

parts, it is apparent that much additional work is necessary before a truly defined medium is available. It is equally apparent that *in vitro*–maintained sporocysts are suitable for histochemical, physiological, and developmental studies in the absence of the influence of host tissues.

LABORATORY PROCEDURES

Schistosoma mansoni, of the four schistosome species, is by far the most readily and easily maintained in the laboratory. This is due, primarily, to the ease with which the intermediate molluscan host, *Biomphalaria glabrata*, is grown under laboratory conditions. It is a hardy snail, which grows and reproduces well under laboratory conditions.

The primary host in the laboratory is generally an inbred strain of albino mice although any strain of laboratory mouse is adequate. Other laboratory animals are not recommended for *Schistosoma mansoni,* although hamsters and gerbils have been used to a limited extent.

Procurement of Miracidia and Snail Infection Techniques

Infected mice (60 days post–cercarial exposure and older) may be used for obtaining miracidia. Following death by cervical dislocation, the livers are removed and placed in chilled 0.9% saline solution. The organs are then triturated in a Waring blender for 30–60 seconds, and the suspension is spun down in a Sorvall centrifuge at approximately 125 g for 5 minutes.

The supernatant is disposed of, and the pellet containing schistosome eggs is resuspended in spring water (aged tap water can be substituted). This suspension is placed into a 500-ml Erlenmeyer sidearm flask and illuminated 5 min with a 60- to 100-watt incandescent light. The sidearm flask is subsequently covered with a double thickness of aluminum foil, leaving the sidearm uncovered. The flask is filled with spring water, and the top covered with aluminum foil. A pipette is used to introduce a small amount of water into the uncovered sidearm, and the incandescent bulb is then situated 15–20 cm from it. The miracidia, which are negatively geotactic and positively phototactic, will aggregate in the illuminated uncovered sidearm and can be removed at intervals with a pipette. After each removal, more water is added to the sidearm, bringing the water level up to the top again.

Using a watch glass or other small container (such as a Petri dish), the miracidia can be counted, if desired. *En masse* infection of snails is best carried out in a finger bowl; individual vials are used when individual exposure is the chosen method of infection. Immersion of the exposure vessels in a water bath heated at 32°C for the duration of exposure time yields optimal percentages of infected snails. The exposure time should be 2–4 hours. Snails are removed from the exposure vessels and placed in covered pans with or without aeration for the incubation period.

After a 30-day period, snails may be checked for infection by isolating each in a glass vial, immersing in 10–20 ml of spring water, and illuminating with a 60-watt bulb placed approximately 30 cm from the vial. A stereoscan microscope is used to observe the water for the presence of emerging cercariae approximately 30 minutes to 1 hour following illumination.

Infected snails should be maintained in separate pans and shielded from light, and may be used for obtaining cercariae at weekly intervals. Too frequent cercerial sheds lead to early death of the snails; however, if shed on a weekly basis, some will live and produce cercariae for 4 or more months.

Infection Method for Laboratory Mice

Adult schistosomes are easily maintained in the laboratory in inbred laboratory mice. It is preferable to use female mice since they are less aggressive and not so likely to be killed by a cage mate during the course of infection.

To obtain cercariae, 15–20 infected snails are placed in a small bowl in approximately 100 ml spring water, and the bowl is illuminated with a 60-watt bulb set about 15 cm from the bowl. The light stimulates cercarial emergence, and within an hour cercarial density will be sufficient for exposing mice. The cercariae should be used within 4–6 hours after emergence to ensure maximum infectivity.

When cercariae have emerged, a count of 100–300 cercariae is made for mouse exposure. To do this, three 0.1-ml samples are pipetted into three Petri dishes. Beginning at the center of the dish, small drops are arranged in a spiral pattern from the center outward (Fig. 19). Cercariae are counted

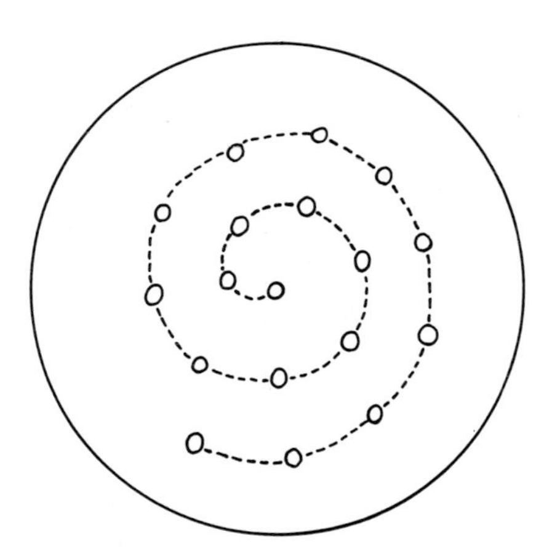

Fig. 19. See text for explanation.

from the center out by means of a stereoscan microscope. The average of three 0.1-ml samples produces a valid estimate of numbers of cercariae/0.1 ml from which larger multiples may be calculated.

The volume containing the number of cercariae to be used for exposing the mice is pipetted into small test tubes or vials that have been lined up on a small test tube rack and placed at the side of a mouse cage (Fig. 20). Mice are anesthetized *via* intraperitoneal injection of sodium nembutal (01 mg/kg body weight) and the tail of each anesthetized mouse is immersed in a vial (Fig. 21).

Covered cages are left undisturbed overnight. *Caution:* Cages should be placed in a quiet area for 24 hours, since the awakening mice jerk their tails splattering cercaria-containing water in the vicinity. The mice should be tagged noting the date of exposure, and adult worms or livers from infected mice may be used beginning about 60 days postexposure.

Other methods may be used for introducing cercariae into mice: Intraperitoneal injection of a counted cercarial suspension is one such method. Also, the belly may be shaved, a metal ring placed thereon, and the sample containing cercariae placed directly on the skin, allowing cercarial penetration.

For further details on the maintenance of *S. mansoni*, see Lee and Lewert ('56).

CONCLUDING REMARKS

There are monogenetic trematodes that would probably offer better models of early development of trematodes than schistosomes. Descriptions of early development in various species are given in the standard texts and reference works on parasitology and invertebrate zoology. The larval miracidium consists of a ciliated protective covering and cells that construct the next stage of the developmental progression, the daughter sporocyst. The main emphasis of this presentation has been on the intramolluscan stages of schistosome development since these stages represent the developmental phases in which the principal amplification in numbers of potentially infective units occurs. This step is an adaptation that is considered important to the success of a parasitic life style. Although there are speculations and even some experimental work on this topic, the basis of trophic selection by the larvae of an intermediate host, and the tolerance by that intermediate host of helminth intermediate forms, have not been satisfactorily explained.

The shedding of miracidial plates at the point of penetration clearly provokes a cellular reaction to these structures, but their germinal progeny are tolerated. It has been suggested that the selection of intermediate hosts is a matter of accidental convergent evolution, and that somehow the cell-surface proteins and carbohydrates of mutually tolerant forms match. An explanation of this tolerance is a significant and practical problem in comparative immunology. Methods are now available that should allow more direct experimental examination of this hypothesis, or the development of other hypotheses to explain this class of adaptation.

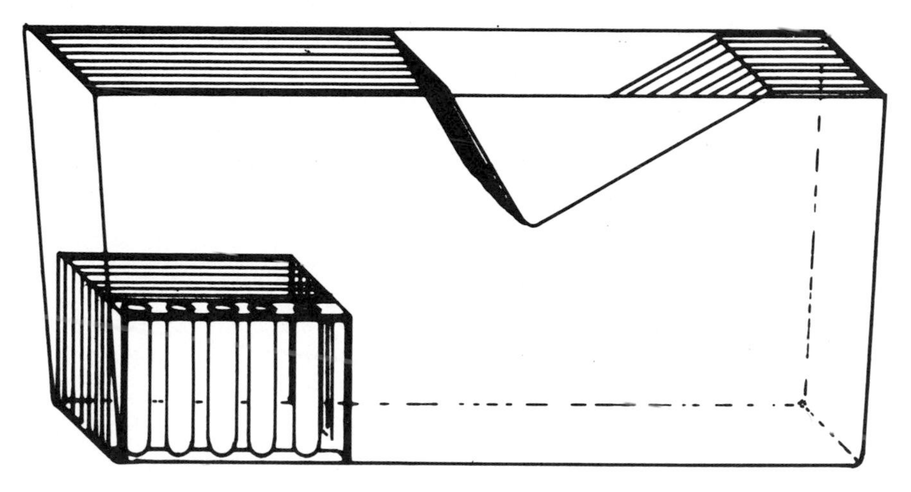

Fig. 20. See text for explanation.

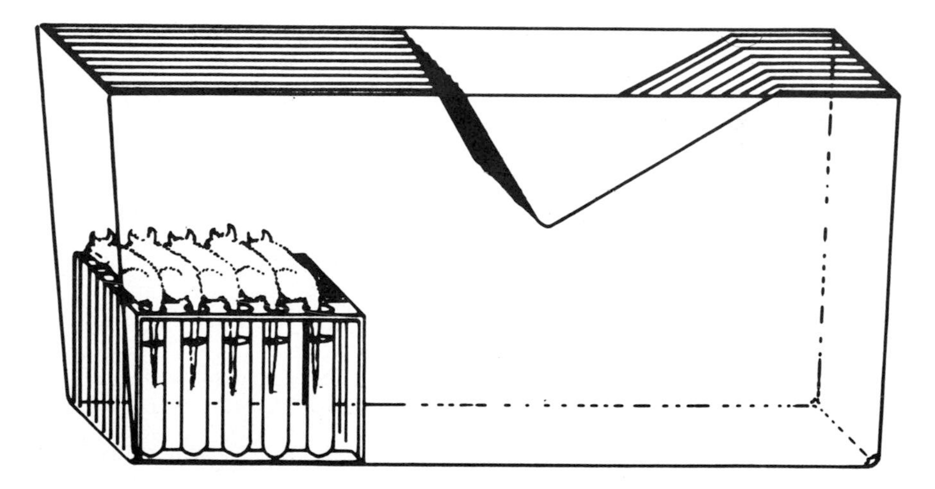

Fig. 21. See text for explanation.

When the trematode life cycle is considered from the prospective of a developmental biologist, development of the intramolluscan stages constitutes an alternation of generations from a sexually reproduced miracidium to an asexually reproducing mother sporocyst and the daughter sporocysts. The extent to which developmental restrictions occur in the germinal portions of larvae and sporocysts other than those tissues forming the integuments has not been vigorously examined, and there are grounds for believing that these intermediate stages in which proliferation of true germinal tissue—i.e., that tissue that will participate in the organization of cercariae and the definitive trematode—should offer some particular opportunities for the experimental modification of subsequent development and for studies of regulative capacities. A great deal of concern has not been dedicated to these issues, but it would seem probable that more interest in these problems and the developmental chemistry of sporocysts could emerge now that it is clearly possible to manipulate transformation of miracidia into mother sporocysts and the development of daughter sporocysts *in vitro*. This technical advance opens many experimental opportunities.

Although parasitic infections in general are not major public health problems in the industrial temperate countries of the world, schistosomiasis represents one of the major health problems of the tropics, which contain most of the world's underdeveloped nations. Indeed, the charge has been raised in the World Health Organization and other international bodies that the relatively feeble effort devoted to research on schistosomiasis and parasitology in general by the technologically advanced nations constitutes a cynical disregard of the well-being of the Asians, Africans, and Latin American Indians who are the principal victims of this set of diseases. There are specific practical humanitarian and geopolitical grounds for fostering an enhanced interest in research on schistosomes and other parasitic forms.

LITERATURE CITED

Basch, P.F., and J.J. DiConza (1974) The miracidium-sporocyst transition in *Schistosoma mansoni:* Surface changes *in vitro* with ultrastructural correlation. J. Parasitol. *60:*935–941.

Becker, W. (1980) Metabolic interrelationship of parasitic trematodes and molluscs, especially *Schistosoma mansoni* in *Biomphalaria glabrata.* Z. Parasitenkd. *63:*101–111.

Bogitsh, B.J. (1975) Cytochemical observations on the gastrodermis of digenetic trematodes. Trans. Am. Micros. Soc. *94:*524–528.

Bonting, S.L. (1970) Sodium-potassium activated adenosinetriphosphatase and cation transport. In E.E. Bittar (ed): Membranes and Ion Transport. New York: Interscience Publishers, Vol. 1, pp. 257–263.

Carter, O.S., and B.J. Bogitsh (1975) Histologic and cytochemical observations of the effects of *Schistosoma mansoni* on *Biomphalaria glabrata.* Ann. N.Y. Acad. Sci. *266:*380–393.

Christie, J.D., W.B. Foster, and L.A. Stauber (1974) ^{14}C uptake by *Schistosoma mansoni* from *Biomphalaria glabrata* exposed to ^{14}C-glucose. J. Invert Pathol. *23:*297–302.

Crane, R.K. (1968) A concept of the digestive-absorptive physiology. Am. Phys. Soc. *5:*2535–2542.

DiConza, J.J., and P.F. Basch (1974a) Axenic cultivation of *Schistosoma mansoni* daughter sporocysts. J. Parasitol. *60:*757–763.

DiConza, J.J., and P.F. Basch (1974b) Incorporation of ^{3}H-thymidine and ^{14}C-glucose by *Schistosoma mansoni* daughter sporocysts *in vitro*. J. Parasitol. *60:*1045–1046.

DiConza, J.J., and P.F. Basch (1975) Accumulation of lipids in *Schistosoma mansoni* sporocysts cultured *in vitro*. J. Invert. Pathol. *28:*337–340.

DiConza, J.J., and E.L. Hansen (1973) Cultivation of *Schistosoma mansoni* daughter sporocysts in arthropod tissue cultures. J. Parasitol. *59:*211–212.

Erasmus, D.A. (1977) The host-parasite interface of trematodes. Adv. Parasitol. *15:*201–242.

Etges, F.J., O.S. Carter, and G. Webbe (1975) Behavioral and developmental physiology of schistosome larvae as related to their molluscan hosts. Ann. N.Y. Acad. Sci. *266:*480–496.

Gönnert, R. (1955) Schistosomiasis Studien. II. Über die Eibildung bei *S. mansoni* und das Schiksal der Eier in Wirtsorganismus. Z. Tropenmed. Parasitol. *6:*33–52.

Gordon, R.M., T.H. Davey, and H. Peaston (1934) The transmission of human bilhaiziasis in Sierra Leone, with an account of the life-cycle of the schistosomes concerned, *S. mansoni* and *S. haematobium.* Ann. Trop. Med. Parasitol. *28:*323–418.

Hansen, E.L. (1975) Secondary daughter sporocysts of *Schistosoma mansoni:* Their occurrence and cultivation. Ann. N.Y. Acad. Sci. *266:*426–436.

Hansen, E., and G. Perez-Mendez (1972) Scanning electron microscopy of *Schistosoma mansoni* daughter sporocysts. Int. J. Parasitol. *2:*174.

Hansen, E.L., G. Perez-Mendez, S. Long, and E. Yarwood (1973) Emergence of progeny-daughter sporocysts in monoxenic culture. Exp. Parasitol. *33:*486–494.

Hansen, E.L., G. Perez-Mendez, and E. Yarwood (1974a) *Schistosoma mansoni:* Axenic culture of daughter sporocysts. Exp. Parasitol. *36:*40–44.

Hansen, E.L., G. Perez-Mendez, E. Yarwood, and E.J. Buecher (1974b) Second generation daughter sporocysts of *Schistosoma mansoni* in axenic culture. J. Parasitol. *60:*371–372.

Kinoti, G.K., R.G. Bird, and M. Barker (1971) Electron microscope and histochemical observations on the daughter sporocysts of *Schistosoma mattheei* and *Schistosoma bovis.* J. Helm. *45:*237–244.

Koie, M., and F. Frandsen (1976) Stereoscan observation of the miracidium and early sporocyst of *Schistosoma mansoni.* Z. Parasitol. *50:*335–344.

Krupa, P.L., and B.J. Bogitsh (1972) Ultrastructural phosphohydrolase activities in *Schistosoma mansoni* sporocysts and cercariae. J. Parasitol. *58:*495–514.

Krupa, P.L., and B.J. Bogitsh (1975) *Schistosoma mansoni* and *Biomphalaria glabrata:* Ultrastructural localization of enzyme with deaminobenzedine in larvae and host digestive glands. Exp. Parasitol. *37:*147–156.

Krupa, P.L., B.J. Bogitsh, and G.H. Couseneau (1971) Ultrastructural demonstration of cytochrome oxidase activity in mitochondria of digenetic trematode larvae. Biol. Bull. *131:*393.

Krupa, P.L., L.M. Lewis, and P. DelVecchio (1975) Electron microscopy of ouabain-inhibited, potassium-dependent transport adenosine triphosphatase activity in schistosome sporocysts. Ann. N.Y. Acad. Sci. *266:*465–479.

Lee, C.R., and R.M. Lewert (1956) The maintenance of *Schistosoma mansoni* in the laboratory. J. Infect. Dis. *99:*15–20.

LoVerde, P.T. (1975) Scanning electron microscope observations on the miracidium of *Schistosoma.* Int. J.Parasitol. *5:*95–97.

Maldonado, J.F., and J. Acosta-Matienzo (1947) The development of *Schistosoma mansoni* in the snail intermediate host, *Australorbis glabratus.* Puerto Rico J. Public Health Trop. Med. *22:*331–371.

Meuleman, E.A. (1972) Host-parasite inter-relationships between the freshwater pulmonate *Biomphalaria pfeifferi* and the trematode *Schistosoma mansoni.* Neth. J. Zool. *22:*355–427.

Meuleman, E.A., and P.J. Holzmann (1975) The development of the primitive epithelium and true tegument in the cercaria of *Schistosoma mansoni.* Z. Parasitenk. *45:*307–318.

Meuleman, E.A., P.J. Holzmann, and R.C. Peet (1980) The development of daughter sporocysts inside the mother sporocyst of *Schistosoma mansoni* with special reference to the ultrastructure of the body wall. Z. Parasitenk. *61:*201–212.

Meuleman, E.A., D.M. Lyarun, M.A. Khan, P.J. Holzman, and T. Sminia (1978) Ultrastructural changes in the body wall of *Schistosoma mansoni* during the transformation of the miracidium into the mother sporocyst in the snail host *Biomphalaria pfeifferi.* Z. Parasitenk. *56:*227–242.

Michaels, R.M., and A. Prata (1968) Evolution and characteristics of *Schistosoma mansoni* eggs laid *in vitro.* J. Parasitol. *54:*921–930.

Pellegrino, J., C.A. Oliverira, J. Faria, and A.S. Cunha (1962) New approach to the screening of drugs in experimental schistosomiasis mansoni in mice. Am. J. Trop. Med. Hyg. *11:*201–215.

Rees, G. (1940) Studies on the germ cell cycle of the digenetic trematode *Parorchis acanthus* Nicoll. II. Structure of the miracidium and germinal development in the larval stages. Parasitology *32:*372–391.

Rivera, E.R., and Y.S. Liang (1980) Intramolluscan ultrastructural aspects of miracidial-sporocyst-cercarial metamorphosis of *Schistosoma mekongi.* In J.I. Bruce and S. Sornmani (eds): The Mekong Schistosome. Malacol. Rev. (Suppl. 2) pp. 67–91.

Skou, J.C. (1965) Enzymatic basis for active transport of Na+ and K+ across cell membrane. Physiol. Rev. *45:*596–617.

Smith, J.H., and E. Chernin (1974) Ultrastructure of young mother and daughter sporocysts of *Schistosoma mansoni.* J. Parasitol. *60(1):*85–89.

Voge, M., and J.S. Seidel (1972) Transformation *in vitro* of miracidia of *Schistosoma mansoni* and *S. japonicum* into young sporocysts. J. Parasitol. *58:*699–704.

Vogel, H. (1942) Über Entwicklung, Lebensdauer und Tod der Eier von *Bilharzia japonica* im Wirtsgewebe. Deutsche Tropenmed. Z. *46:*57–91.

Wilson, R.A. (1969) Fine structure of the tegument of the miracidium of *Fasciola hepatica* L. J. Parasitol. *55:*124–133.

Wikel, S.K., and B.J. Bogitsh (1974) *Schistosoma mansoni:* Penetration apparatus and epidermis of the miracidium. Exp. Parasitol. *36:*342–354.

Developmental Biology of Freshwater Invertebrates, pages 249–281

Development of the Nematode *Caenorhabditis elegans*

Einhard Schierenberg

ABSTRACT *Caenorhabditis elegans* is a free-living, nonparasitic nematode. It is a self-fertilizing hermaphrodite. Males arise spontaneously by nondisjunction of X-chromosomes. Of all eukaryotic organisms *C. elegans* has probably been most extensively studied on the cellular level. Within 12 hours the fertilized egg develops into a young larva with 558 nuclei (560 in the male). During postembryonic development the animal proceeds through four larval stages increasing its number of nuclei to 958 (1,031 in the male) plus some 2,000 germ cells (about 1,000 in the male). The cell lineages from fertilization to adulthood have been completely analyzed in living embryos and animals. This and its well-established genetics (more than 300 genes have been mapped on the six linkage groups) make it a suitable model organism to study problems of gene action and development.

Various techniques have been used to interfere with normal development (including laser-induced cell ablations) and to analyze development on the subcellular level (including recombinant DNA technology).

The characteristic features of rigidly determined development, the low cell number, and the knowledge of cellular events should make it possible to identify molecular action *in situ* and relate it to the structure and function of cells and tissues.

INTRODUCTION

Historical Background

For more than a hundred years nematodes have been a subject for developmental research. A unique combination of favorable properties, including a small number of cells, small embryo size, and rapid, reproducible development, made nematodes classic models for studying embryogenesis. The analysis of egg maturation, meiosis, fertilization, zygote formation, and early cleavage led to several fundamental discoveries. These include the individual character of chromosomes, their reduction during meiosis, the early separation of somatic tissues from the germ line, determinate cleavage, and constant number (for reviews see Chitwood and Chitwood, '50; von Ehrenstein and Schierenberg, '80).

C. elegans was first collected in Algeria by Maupas (1900) and named *Rhabditis elegans*. Staniland reisolated it in 1946 from mushroom compost in Bristol, England. It was first proposed as a model for genetic studies by Dougherty and co-workers (Dougherty and Calhoun, '48; Dougherty et al., '59; Fatt and Dougherty, '63). Sexual development and differentiation were studied by Honda ('25) and in Nigon's laboratory (Nigon, '49, '65; Nigon and Brun, '55).

Brenner began to work on *C. elegans* in 1965 after an extensive search for a model organism to study the genetic basis of development and behavior (Brenner, '73). From a stock of the Bristol strain he isolated one hermaphrodite and established two lines: one, a line of hermaphrodites propagating by self-fertilization and the other, a line with males, which arise spontaneously on occasion and can be maintained by mating with hermaphrodites. These are the founder stocks and carry the code name N_2. Descendants of these founder stocks are used today in nearly all *C. elegans* laboratories. A different strain has been collected in Bergerac, France and is studied in the Laboratory of Physiological Genetics and Nematology in Lyon. The temperature optimum for growth and reproduction is 22°–24°C for the Bristol strain (Bergerac, 17°C) and the maximum temperature for reproduction 27°C (Bergerac, 20°C). The observations, experiments, and data reported in this chapter refer to the Bristol strain unless indicated otherwise.

Caenorhabditis elegans as a Model for Developmental Biology

Caenorhabditis elegans is a self-fertilizing, free-living nonparasitic soil nematode. The adult hermaphrodite is about 1 mm long and contains some 800 nongonadal cells (Sulston and Horvitz, '77). Occasionally slightly smaller males arise, which can be crossed with hermaphrodites (see "Genetics," below). *C. elegans* has the typical anatomy of a rhabditid nematode (Chitwood and Chitwood, '50). An outer layer of hypoderm cells covers the body muscles, which are arranged into four longitudinal rows. The contracting pharynx pumps food through the mouth cavity into the intestine, which ends in a rectum and anus. Most of the approximately 300 nerve cells are situated in the head region. The nerve ring, a bundle of nerve fibers, runs circumferentially around the pharynx. It contains the endings of the sensory receptors in the head, interneurons, motor neurons, and processes, coming from the tail (White, '74). The ventral cord of the adult contains 57 motor neurons, which innervate the subventral body muscles. These neurons send commissures to the dorsal side and form a dorsal nerve cord innervating the subdorsal body muscles (White et al., '76).

The reproductive system of the hermaphrodite is made of two reflexed gonadal tubes. Each consists of an ovary, oviduct, spermatheca, and uterus. It ends in a midventral vulva, through which the eggs are laid (Fig. 1). The male gonad consists of a single testis. The sperm is passed into a vas deferens, which is connected to the rectum forming a cloaca. The feature most clearly distinguishing the male from the hermaphrodite is the male's copulatory bursa at the tail (Fig. 1). Because of its favorable features, and as a result of Brenner's ('74) pioneering work on the genetics of *C. elegans*, an increasing number of scientists are concentrating on this organism to investigate problems of gene function during development (for a review see Zuckerman, '80a,b).

Sources, Availability, and Maintenance in the Laboratory

The *C. elegans* Genetics Center (CGC, Dr. Donald Riddle, Division of Biological Sciences, Tucker Hall, University of Missouri, Columbia, Missouri 65211) sponsored by the National Institute of Aging is responsible for strain acquisition, banking, and distribution. Strains are available without cost to all qualified investigators pursuing genetics-related studies with *C. elegans*. Other CGC functions include maintenance of the genetic map, coordination of genetic nomenclature, and bibliographic information (e.g., a complete list of *C. elegans* literature). In Europe many strains may be obtained from the Cambridge Laboratory (Dr. John Sulston, MRC Cambridge, Laboratory of Molecular Biology, Hills Road, Cambridge CB2 2QH, England). *C. elegans* wild type are available from any laboratory working with this organism (addresses are available from CGC).

C. elegans is easily and inexpensively grown under controllable laboratory conditions. The worms are maintained on Petri dishes filled with an agar layer containing a limited amount of uracil. The plates are seeded as a food source with a uracil-requiring mutant of *Escherichia coli* (see next paragraph). This prevents an overgrowth of the bacterial lawn, which would obscure the worms (Brenner, '74). The animals grow well in the temperature range between about 15°C–25°C. To have fresh cultures continuously available, some individuals are transferred every few days to a new plate with a sterile paper strip (Brenner, '74), with a toothpick, or on a small agar chunk. If the plates are prevented from drying out, living individuals can be still found as dauerlarvae after several months (see "Life Cycle," below).

The reagents and special equipment requirements for nematode growth medium plates are from the handbook of a *C. elegans* course held in Cold

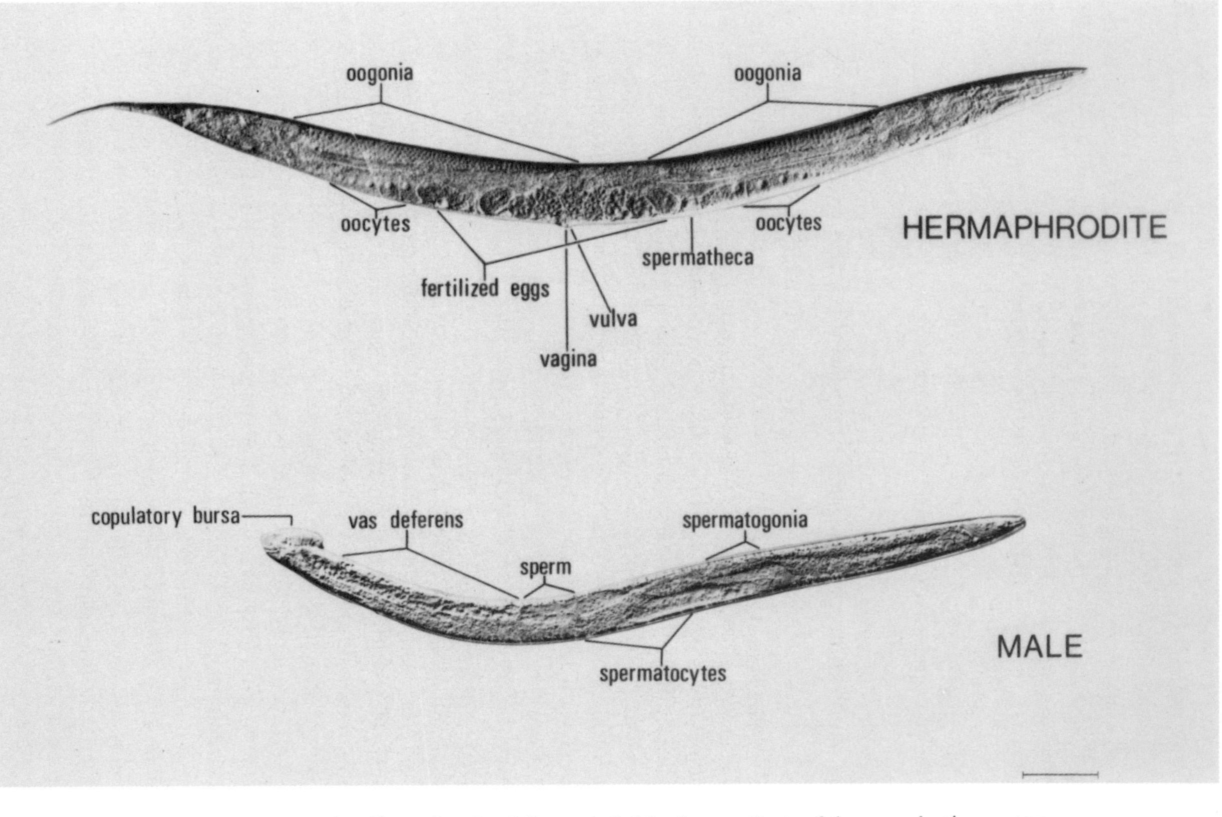

Fig. 1. Hermaphrodite and male of *Caenorhabditis elegans*. Parts of the reproductive system labeled (after Schierenberg, '78). Bar 100 μm.

Spring Harbor in 1972 by R. Russell and R. Pertel. There are four stock solutions: A) 0.5 gm cholesterol in 100 ml ethanol (95%); B) 18.38 gm $CaCl_2$, H_2O to make 250 ml; C) 61.62 gm $MgSO_4 \bullet 7\ H_2O$, H_2O to make 250 ml; D) 27.08 gm KH_2PO_4, 8.9 gm K_2HPO_4 (anhydrous), H_2O to make 250 ml (adjust to pH 6 with 1.0 M potassium phosphate buffer). Solutions B, C, and D are autoclaved for 30 minutes at 120°C.

One and one-half grams of NaCl, 8.5 gm Bacto-Agar (Difco), 1.25 gm Peptone (Difco), and H_2O to make 500 ml are autoclaved for 30 minutes at 120°C. Then add 0.5 ml solution A, 1.0 ml solution B, 0.5 ml solution C, and 12.5 ml solution D. Mix well, allow to cool to about 50°C, and pour into Petri dishes.

To make *E. coli* growth medium mix 1.0 mg Uracil, 2.0 gm Tryptone (Difco), 0.5 gm NaCl, and H_2O to make 100 ml, adjust to pH 7.4 with 0.1 N NaOH, and autoclave for 30 minutes at 120°C. Allow to cool, add some *E. coli* of the uracil-requiring strain (Cambridge strain collection number OP 50), and grow overnight at 37°C. Deposit 0.1 ml of an overnight bacterial culture on each agar plate. After a day's incubation at room temperature the plate is ready to use.

Obtaining and Rearing Embryos

Embryos are laid by young mothers from about the 50-cell stage onwards. With increasing age of the mother, and depending on environmental conditions, a major part of embryonic development occurs *in utero*. Young worms may even hatch inside the mother.

Advanced embryos can be collected from the plate. The worms and a portion of the bacteria are washed away from the plate by gently rinsing with tap water or M9 buffer.[1] The eggs stick to the agar surface. They can be sucked off with a drawn-out Pasteur pipette or glass capillary connected to a rubber tube.

To dissect young embryos out of the mother, some gravid adults are transferred with a sharpened toothpick or a platinum wire (which can be sterilized in a flame) into a small drop of tap water or medium (as the lipid layer of the egg shell is hardly penetrable at all, the eggs develop in nearly any kind of liquid medium). The worms are cut open with a scalpel under a dissecting microscope. Because of the high internal pressure the eggs pop out of the mother. Under the dissecting microscope eggs with a size of approximately $50 \times 30\ \mu m$ can be roughly categorized according to developmental stage. Embryos from the late 1-cell stage onwards develop normally outside the mother and hatch.

[1]To prepare M9 buffer (Brenner, '74), mix 6 gm Na_2HPO_4, 3 gm KH_2PO_4, 5 gm NaCl, and 0.25 gm $MgSO_4 \bullet 7\ H_2O$, H_2O to make 1,000 ml, then autoclave for 30 minutes at 120°C.

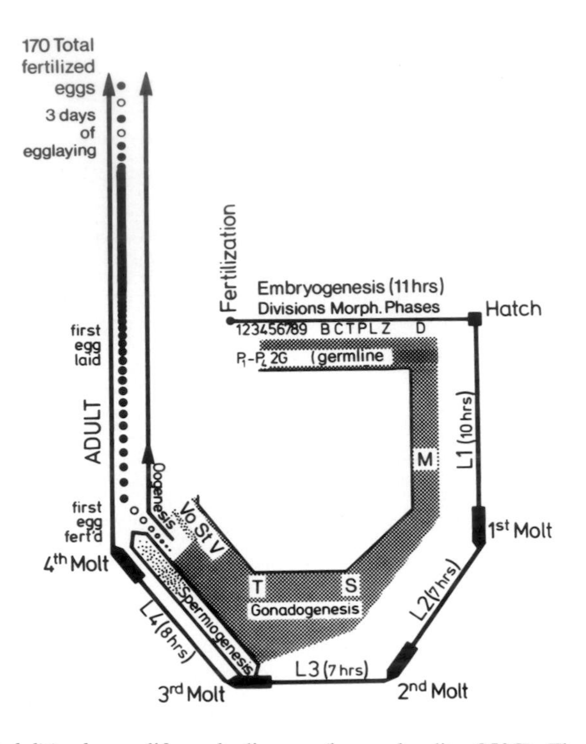

Fig. 2. *Caenorhabditis elegans* life-cycle diagram (hermaphrodite, 25°C). The lengths of the lines are proportional to time. The egg-to-egg generation time is 52 hours. L_1 is first larval stage and so forth and the times after names of stages are duration of each stage. The thickenings of the line indicate the duration of the molts, including lethargus. In embryogenesis, the numbers indicate the time of initiation of each round of cell division and the letters the time of appearance of each morphogenetic stage: B, lima bean; C, comma; T, tadpole; P, plum; L, loop; Z, pretzel; D, late pretzel; 2G, division of P_4 to form two germ cells; two nongerm line gonad cells are also present at hatching (Deppe et al., '78; Sulston and Schierenberg, unpublished results). In stippled area, visible events at each postembryonic stage in gonadogenesis (Hirsh et al., '76; Kimble and Hirsh, '79) are indicated as follows: M, mitogenesis begins; S, second wave of somatic gonad cell divisions begins; T, 180° turn; V, vagina forms; St, spermatheca formed; V_0, vulval opening formed; ●, fertilized eggs, ○, oocytes. (From Cassada et al., '81, with permission.)

DEVELOPMENT OF THE WILD TYPE

Life Cycle

The life cycle of *C. elegans* is rapid. It takes about 2.5 days at 25°C (Fig. 2). Within 12 hours (Schierenberg, '78) a fertilized egg develops into a juvenile with about 550 cells (Sulston and Horvitz, '77) and a length of about

1 mm. After hatching, the animal passes through four larval stages (L_1–L_4) within 32 hours (Fig. 2) before reaching adulthood.

At the end of its growing period, the adult is about six times longer than at the L_1 stage (Cassada and Russell, '75). Spermiogenesis starts in the early L_4 stage and is finished before the first mature oocytes are produced. Egg laying begins about 50 hours after hatching. The majority of fertilized eggs are laid within 24 hours. Growth and reproduction have been quantitatively analyzed at 16°, 20°, and 25°C (Cassada and Russell, '75; Byerly et al., '76). Between L_3 stage and adulthood the differentiation of secondary sexual structures (e.g., the hermaphrodite vulva and the male tail) takes place. When adverse environmental conditions interfere with normal development, *C. elegans* can arrest development after the second molt and enter a nongrowing stage called the dauerlarva. In this state animals have a thicker cuticular structure and resist destruction by a variety of chemicals (Cassada and Russell, '74). Dauerlarvae may survive for months without food (Klass and Hirsh, '76) and can resume normal development if suitable conditions are restored.

Synopsis of Embryogenesis

Embryogenesis can be separated into proliferation and morphogenesis phases. Within the first 5 hours the constant amount of cytoplasm is partitioned into about 500 cells. By that time nearly all cells of the first larval stage are produced prior to attainment of a vermiform structure. In the second half little division occurs. With cell elongation, association, and rearrangement the embryo stretches into a worm.

Twelve embryonic stages have been defined, based on easily identifiable landmarks such as the number and position of the intestinal precursor cells ("E-cells") and the outer contours of the developing embryo (Fig. 3) (Schierenberg, '78). The duration of embryonic development is temperature dependent, as is indicated by measurements at three different temperatures in Figure 4 (Schierenberg, '78). The sequence of visible developmental events seems to be temperature independent.

EXPERIMENTAL PROCEDURES

Microscopy

Light microscopical cell lineage studies of the wild type. Methods of mounting eggs on a microscope slide are described in Deppe et al. ('78) and Schierenberg et al. ('80). An even better image is obtained using the following procedure: a small piece of thin agar layer (3%–5% dissolved in H_2O) is placed on the microscope slide (Sulston, '76). Eggs dissected out of a young mother are transferred to it with a drawn-out Pasteur pipette in a small drop

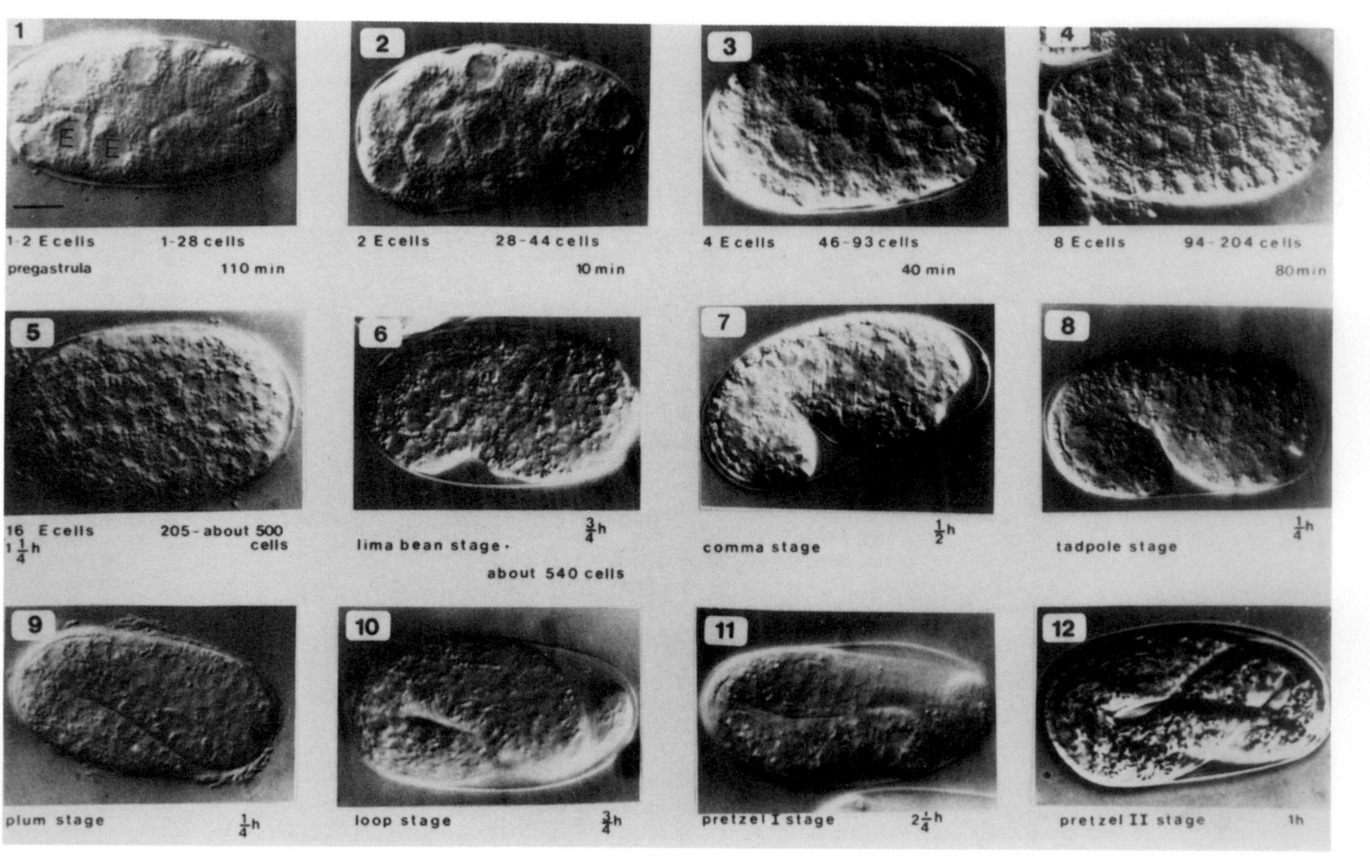
1
E
E
1-2 E cells 1-28 cells
pregastrula 110 min
2
2 E cells 28-44 cells
10 min
3
4 E cells 46-93 cells
40 min
4
8 E cells 94-204 cells
80 min
5
16 E cells 205-about 500 cells
$1\frac{1}{4}$h
6
lima bean stage - $\frac{3}{4}$h
about 540 cells
7
comma stage $\frac{1}{2}$h
8
tadpole stage $\frac{1}{4}$h
9
plum stage $\frac{1}{4}$h
10
loop stage $\frac{3}{4}$h
11
pretzel I stage $2\frac{1}{4}$h
12
pretzel II stage 1h

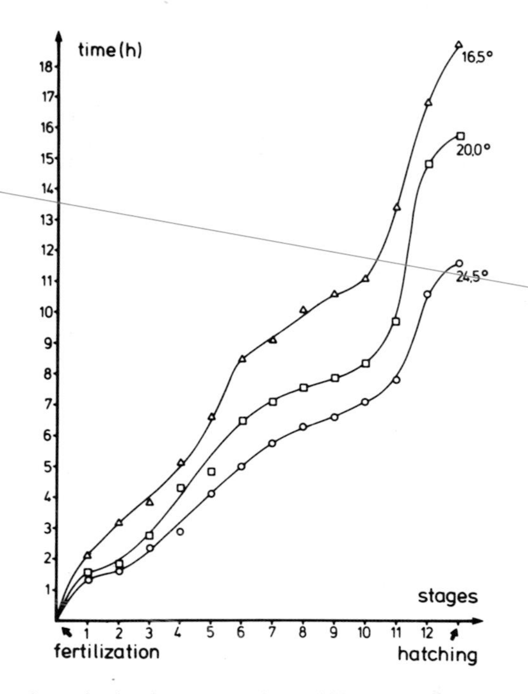

Fig. 4. Times of embryonic development at three different temperatures. For each temperature 10 individuals were tested. The time between fertilization and hatching is subdivided into 12 stages, as in Figure 3. (From Schierenberg, '78).

of liquid. A cover glass (bigger than the piece of agar), sealed on the edges with petrolatum, is placed on top. By gently pressing the cover glass, the eggs can be flattened to a certain extent to allow better viewing without interfering with normal development.

Depending on which cell line one wants to follow, eggs with left, right, dorsal, or ventral view are selected for observation. For an overall observation from a single view, the ventral orientation has proved optimal. The

---

Fig. 3. Events in embryogenesis of *Caenorhabditis elegans*. Embryogenesis is divided into 12 periods or stages. In stages 1–5, number and position of the intestinal precursor cells (E-cell, labeled in stage 1, before immigration) and in stages 6–12, the outer contours of the developing embryo are taken as identification landmarks. In stages 4 and 5 not all E-cells are visible in one focal plane. After the lima bean stage the embryo has approximately 550 living cells as the freshly hatched juvenile. The times given represent the approximate durations of each stage at 25°C. Orientation: anterior right, dorsal top. Bar 10 μm. (After Schierenberg, '78.)

microscope is fitted with Nomarski differential interference contrast optics, and optionally with a video camera and recorder. A servo motor can be attached to the micrometer screw to move the focus continuously up and down.[2] The shallow optical image breaks up the egg into different layers visible in a continuous sequence with little loss of information. For late lineage studies we favor direct observation through the oculars (with manual focusing), continuously observing a few specific cells.

The egg of *C. elegans*, as in other nematodes, is fertilized in the spermatheca by a single sperm. Rare cases of polyspermic fertilization lead to abnormal early cleavage patterns (Boveri, '10; Schierenberg et al., '80). The germinal vesicle breaks down, a refractive egg shell forms around the egg, and meiosis is completed with the extrusion of two polar bodies (Nigon, '49; Nigon and Brun, '55). After about 30 minutes the pronuclei reform at the two opposite poles, migrate towards each other, and fuse to form a zygote (Deppe et al., '78; Schierenberg, '78; for a review of all embryonic events see von Ehrenstein and Schierenberg, '80). Immediately afterwards, the first cleavage starts, taking about 5 minutes to complete. As in other nematodes five somatic founder cells (AB, MSt, E, C, D) are generated (Fig. 5) by the asymmetric division of the germ-line cell (P_0–P_4). The five somatic founder cells and the germ cell P_4 are determined to go through a fixed division and differentiation program. The cell cycles are rapid, the fastest requiring only 10 minutes. Each cell line has a characteristic synchronous division rhythm. Towards the end of the division program this synchrony becomes less pronounced. Many cell lines form a clear bilateral cell pattern, whereby the precursor cells for the left and right structures are already separated with the first division of a founder cell.

The determination of ancestry and function of each cell from fertilization to hatching has recently been completed (Deppe et al., '78; Schierenberg et al., '80; von Ehrenstein et al., '81; Sulston et al., unpublished results). These studies (Fig. 5) revealed the following. The founder cell AB generates nearly 70% of all cells in the young juvenile. These are mainly ectodermal (hypodermis, nervous system, part of the pharynx) but include the four rectal and intestinal muscles, some pharyngeal muscles, and a single body muscle.

[2]We use a Zeiss Universal Microscope at 1,600 × magnification, a high resolution video camera (Ikegami Tsushinki Corporation, New York, model CTC 6000, with Pasekon-tube), a 1-inch video recorder (model 711 P), or a ½-inch video recorder (model TVR-322) both from International Video Corporation, Sunnyvale, California and a video monitor (Chuomusen LTD, Tokyo, model QVM 310 B). A plexiglass pane is affixed to the screen to allow us to make direct drawings of the video image on transparent plastic sheets. The attachment changing the focal plane was built in our workshop. A motor drives an eccentric arm converting rotational into reciprocal motion.

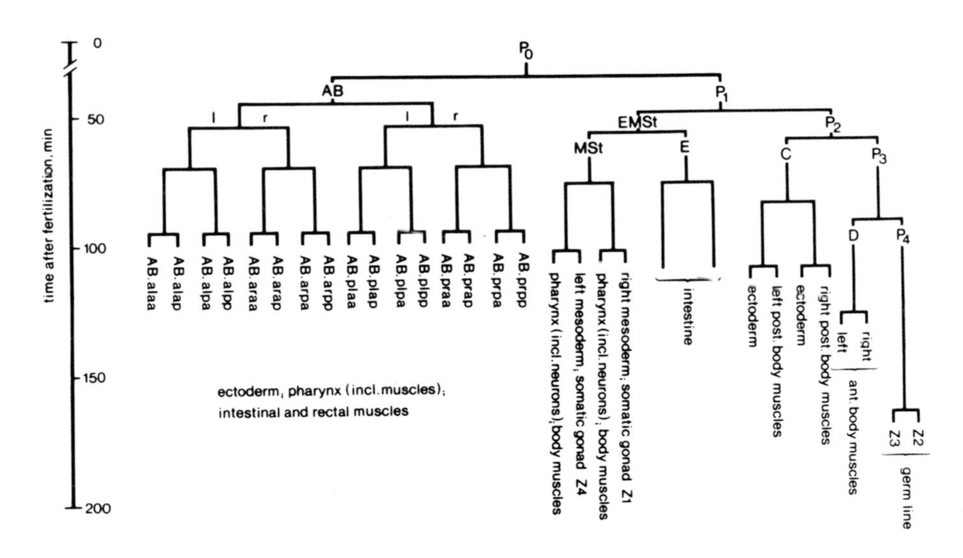

Fig. 5. Early cleavage pattern of the egg of *C. elegans* generating the five somatic founder cells AB, MSt, E, C, and D and the germ line precursor P_4. The first division of P_0, the zygote, occurs about 35 minutes (21°C) after fertilization. In the juvenile animal, the precursors give rise to the differentiated structures shown. Only the initial divisions of the founder cells are indicated. The descendents of cells are named by adding a period and one lowercase letter per division. This letter represents the position of a cell relative to its sister cell immediately after division. Anterior-posterior sisters are designated by "a" and "p," dorsal–ventral by "d" and "v," and left–right by "l" and "r." In oblique divisions only the predominant axis is indicated. Lineage tree branches are labeled by the same letters; the "a," "d," or "l" sister is always put on the left branch; two letters are allowed for oblique divisions. No labeling on the branches indicates an a–p division. (From von Ehrenstein et al., '81, with permission.)

The founder cell MSt generates 80 surviving cells. These include part of the pharynx (with some nerve cells), body muscles, the four coelomocytes, the single postembryonic myoblast cell, and the two precursor cells for the somatic gonad. The founder cell E generates the 20 intestinal cells. The founder cell C generates 47 surviving cells. From C.aa and C.pa ectodermal elements (mainly dorsal hypoderm) arise; from C.ap and C.pp exclusively, body muscles arise. The germ cell P_4 divides only once during embryogenesis. More than 110 embryonic cell deaths occur during embryogenesis. Nearly 90% arise in the AB lineage, the rest in the C and MSt lineages.

The traditional classification that each somatic founder cell gives rise to a distinctive germ layer must be modified. AB yields some muscle cells and MSt yields some neurons. In both cases the minor tissue type separates at a late stage from the major at multiple points. Thus the classic notion of absolute separation of tissue type at an early stage cannot be maintained.

To observe the development of hatched animals, single specimens are transferred to a microscope slide covered with a thin agar layer. The center of a 12 × 12 mm cover slip is very thinly coated with *E. coli* scraped from a culture plate, gently placed onto the agar, and sealed on the edges. After a short period of quiescence, the worm usually moves into the bacterial lawn, starts feeding, and continues development (Sulston, '76). If an animal must be followed for a long period, it can be arrested by refrigerating overnight (6°C–8°C) during a suitable quiescent period (lethargus before molts). After returning to 20°C the worm resumes normal development. The divison of 51 nongonadal postembryonic blast cells in the hermaphrodite (55 in the male) gives rise to a total of 816 nongonadal nuclei (976 in the male) in the adult (Sulston and Horvitz, '77; Sulston et al., '80). The two precursor cells for the somatic gonads divide into 143 cells in the hermaphrodite and 56 in the male (Kimble and Hirsch, '79). All this gives a virtually complete picture of *C. elegans* development on the cellular level from fertilization to adulthood.

Cellular analysis of embryonic arrest mutants. With the knowledge of wild-type development as a reference system, embryogenesis-defective mutants can be analyzed on the cellular level. Cleavage pattern, division axes, and timing of cell division of 11 temperature-sensitive embryonic arrest mutants (which are also genetically well characterized, Miwa et al., '80) have been followed at the nonpermissive temperature of 25°C to at least the 50-cell stage (Schierenberg et al., '80).

L_4 juveniles or very young adults (grown at the permissive temperature of 16°C) are shifted to 25°C; 7–12 hours later the eggs are excised and mounted on a microscope with a temperature-controlled stage connected to a constant temperature circulator. The temperature-controlled stage is manufactured by Zeiss. The hole in the center was enlarged to allow Köhler illumination with oil immersion of the condenser front lens at 100 × magnification. The constant temperature circulator is from Wobser KG, Lauda, Federal Republic of Germany (model K4R-CZ). The temperature between cover slip and slide can be monitored with a microthermocouple (copper and constantan wire 40 and 15 μm in diameter, respectively).

All but one mutant (the only one in this set in which gene expression during embryogenesis is necessary for normal development) show visible defects before the 50-cell stage. Nevertheless, most of these continue development for several hours until development is arrested and they die.

The cleavage patterns are essentially normal, but the timing of divisions is altered in nearly all mutants. Neither those with a slower overall division rate than the wild type nor those that are quicker are able to enter a normal morphogenesis. The abnormal division rhythm leads to positional defects of

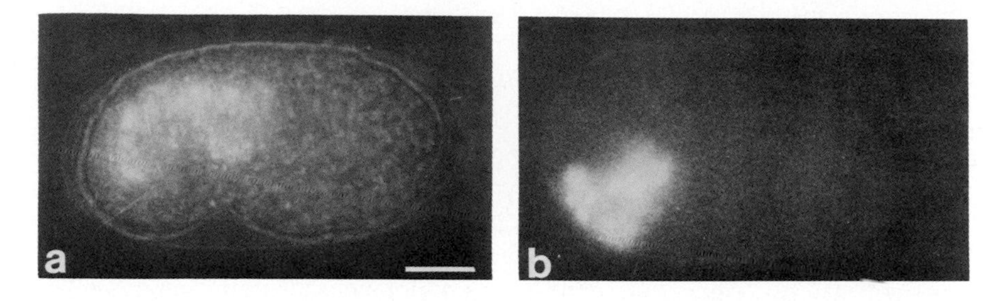

Fig. 6. Autofluorescent intestinal precursor cells ("E-cells") of a) lima bean stage of the wild type and b) terminal phenotype of the temperature-sensitive embryonic arrest mutant *emb-5* (allele hc61) at the nonpermissive temperature of 25°C. Bar 10 μm. (From Schierenberg et al., '80, with permission.)

cells. Thus correct positioning seems to be a prerequisite for normal development.

The programming of the egg by maternal genes involves the timing of cell division rhythms. Particularly obvious examples for abnormal cell behavior are the intestinal precursor cells ("E-cells"). In two mutants of this set (two alleles of one gene) the division of two to four E cells occurs prematurely, leading to an abnormal cell pattern after their immigration (Fig. 6). Several cases of similar reversals in the E-cell development have been found in a more recently obtained larger set of "emb" mutants (Cassada et al., '81; Denich et al., unpublished results). Thus, mutations in different genes can lead to the same kind of visible defects. Phenocopies of these defects have been obtained by laser-induced cell-fusion experiments (see "Laser Microbeam," below). The observations are consistent with the view that timing is controlled by cytoplasmic components and is involved in establishing regional differences in the embryo of *C. elegans*.

In a third set of ts-embryonic arrest mutants, where lineage was not followed in detail, 6 of 21 (for all of them maternal gene expression is sufficient) showed an abnormal first cleavage (Wood et al., '80).

Mutants affecting postembryonic development. The phenotypes of postembryonic cell lineage mutants, defining 14 different genes (Horvitz and Sulston, '80), have been described recently (Sulston and Horvitz, '81; Chalfie et al., '81). Besides mutants with general abnormalities in cell division and those where only specific hypodermal cells or vulval development was affected, mutants in two genes produce reiterations in certain cell lineages. In this last class, specific cells assume fates normally associated with certain of their own ancestors. Normal development of *C. elegans* may be thought of as proceeding by a series of developmental decisions that accompany each cell division. These mutants appear to fail to execute certain of these hypothetical switches. It has been suggested (Chalfie et al., '81) that they may reveal underlying reiterative aspects of normal development, masked in the wild type by the action of modifying genes inactivated in the mutants.

Two postembryonic cell cycle defective mutants have been found. In one of these DNA replication is blocked. An explanation of why this mutant shows normal locomotory behavior as a young larva but is later uncoordinated was suggested by White et al. ('78). Certain motoneurons of the ventral cord apparently displace their neuromuscular junctions after hatching from the ventral to the dorsal side in the wild type as well as in this mutant. But, as the block in DNA replication prevents the generation of additional motoneurons in the mutant, the ventral muscles are left without synaptic input.

In the second mutant, postembryonic cell and nuclear division are absent, but normal cell cycling including DNA replication continues (see also "Biochemistry" below), producing cells with large polyploid nuclei (Albertson et al., '78).

Many mutants affecting gamete development and sex determination have been isolated. These include gonadogenesis defectives (Abdulkader and Brun, '76, in the Bergerac strain; Hirsh and Vanderslice, '76), spermatogenesis and fertilization defectives (Hirsch and Vanderslice, '76; Ward and Miwa, '78), mutants that transform genotypic hermaphrodites into phenotypic males (Hodgkin and Brenner, '77; Klass et al., '76, '79), and intersex mutants (Beguet and Gilbert, '78, in the Bergerac strain; Nelson et al., '78).

Electron Microscopy

Detailed electron-microscopical data are available for the structure of the adult animal (Ward et al., '75; Ware et al., '75; White et al., '76; Albertson and Thomson, '76; Hall, '77; Chalfie and Thomson, '79) and sperm morphogenesis (Ward et al., '81). Methods for fixing, embedding, and serially sectioning embryos encased in the egg shell have been described (Krieg et al., '78). Fixation is normally carried out at an elevated temperature of 40°C in order to penetrate the surrounding impermeable lipid layer. As a fixative, 2% osium tetroxide is used for good cell membrane visualization. For better intracellular ultrastructure visualization 4% glutaraldehyde followed by 2% osmium tetroxide is used.

Early embryos have been matched to the same stages in the light microscope, and each cell can be assigned its lineage name (Fig. 7a). In cross-sections of stages in the morphogenesis phase, different cell layers and cell types have been identified (Fig 7b). It has been found that membranes

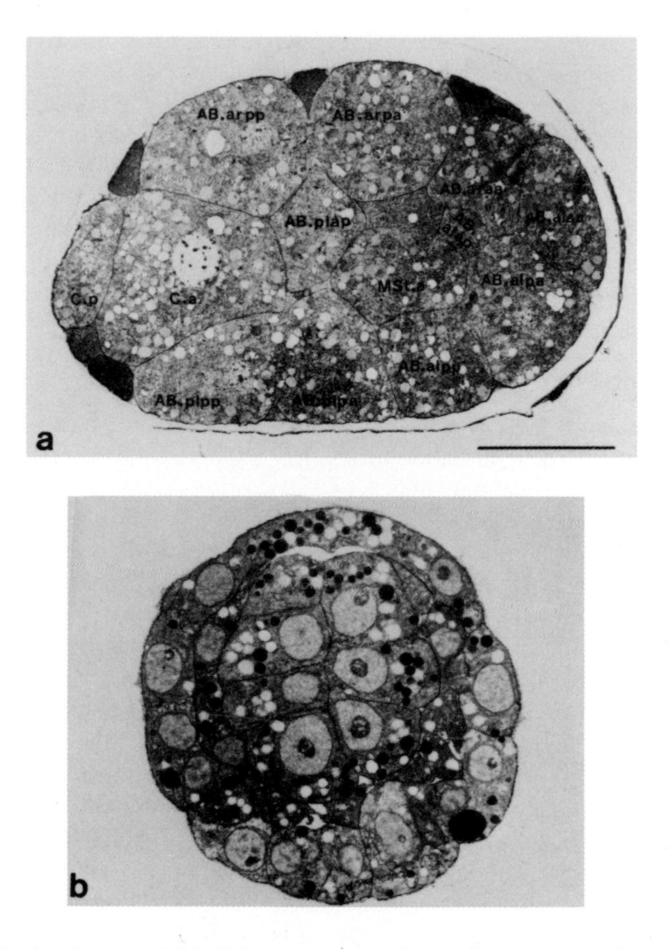

Fig. 7. a) Section from the left half from a complete longitudinal electron-micrograph series of a 24-cell embryo. The cells were identified by matching the reconstructed series to a living embryo. For nomenclature, see Figure 5. Note open membranes between sister pairs. Fixation with osmium tetroxide; orientation: anterior, right; right, top. Bar 10 μm. b) Section from the complete transverse of a lima bean stage embryo with 540 cells, posterior region. This section shows the relatively simple organization in this stage, even though cell division is virtually complete. The cells are arranged in tissue layers. In this region an inner layer of endoderm cells (future gut) is surrounded by a layer of mesoderm cells (future body muscles) enclosed by a basal membrane. The outer layer of ectoderm consists dorsally of future hypoderm cells and ventrally of future nerve cells. Fixation with osmium tetroxide. Orientation: dorsal, top. Bar 10 μm. (From Krieg et al., '78, with permission.)

between cell sisters stay open for up to 15 minutes after division. The characteristic chromatin pattern of nuclei allows one to define their position in the cell cycle (Krieg et al., '78; Schierenberg, '78). To avoid elevated temperature, the egg shell can be enzymatically digested with chitinase (Bazzicalupo et al., personal communication). Using chitinase treatment, followed by osmium tetroxide fixation, J. Sulston (personal communication) has identified most cells of a sectioned embryo in the tadpole stage (see Fig. 3).

A method developed in our laboratory using a laser microbeam to shoot a hole in the egg shell to allow penetration of the fixative (von Ehrenstein et al., '81) gives good ultrastructural details after glutaraldehyde and osmium tetroxide fixation.

Laser Microbeam

Besides being a tool for fixation, a laser microbeam attached to a microscope is used in several laboratories for experimental interference with normal *C. elegans* development.

Several versions are in use. J. White has created a simple setup where a flash-light-pumped dye laser enters the microscope directly above the objective (Planachromat 100 only). A modified version has been constructed by R. Horvitz in which the beam enters the microscope from above the objective revolver, so that all objectives can be used. A more sophisticated attachment using an N_2 pumped dye laser (Lambda Physics, Göttingen, Federal Republic of Germany) is used in our laboratory (Biotechnik GmbH, Mönchengladbach, Federal Republic of Germany). It allows the quick change of dyes with different wave lengths and a high repetition rate of laser pulses.

Three types of laser experiments have been carried out so far.

a) During embryonic (Sulston and Schierenberg, unpublished results) and postembryonic development (Sulston and Horvitz, '77; Kimble et al., '79; Kimble and White, '81; Sulston and White, '80), single nuclei were ablated and the effect on the remaining cells was studied. In the embryonic and postembryonic lineages several cases of regulation have been found. This demonstrates that even in nematodes, the best classical example of mosaic development, this developmental strategy is not exclusively utilized. Four types of cell interaction have been distinguished: induction, lineage regulation, functional regulation, and form regulation (Sulston and White, '80). During embryogenesis only two cases of functional regulation have been detected so far (J. Sulston, personal communication). When a specific cell was ablated it was replaced by a homologous cell from the contralateral lineage branch, which normally would have a different function. It must be emphasized that most lineages seem to be autonomous. Blastomere isolation experiments have confirmed the absence of lineage regulation in young embryos of *C. elegans* (Laufer et al., 1980).

b) In contrast, functional regulation can occur on an intracellular level, as has also been observed in uncleaved *Ascaris* eggs (Boveri, '10; for a review see von Ehrenstein and Schierenberg, '80). Considerable amounts of cytoplasm and the surrounding cell membrane were extruded through a laser-induced hole at the posterior end of the egg during the first cleavage, including all the cytoplasm normally destined for the germ cell P_4 (Fig. 8).[3] Nevertheless, normal embryos could develop and produce fertile adults (Laufer and von Ehrenstein, '81). This and similar experiments suggest that neither a prelocalization of (unbound) cytoplasmic determinants nor a particular amount of cytoplasm is responsible for the normal segregation of developmental potential.

c) The third kind of laser experiments on developing embryos are cell fusions (Schierenberg, unpublished results). With a single laser pulse the membrane between two adjacent cells can be disrupted. The cells fuse and the cytoplasm mixes. If this is done, for example, at the 2-cell stage at a time when the nuclei of both cells are in mitotic prophase, these divide into four nuclei within the refused cell, forming directly into a 4-cell stage. The subsequent division rhythms of the E- and P-cell lines are always considerably changed. Alterations in the sequence of events lead to positional defects of cells, which in turn do not allow normal morphogenesis. These results suggest that cytoplasmic components influence the cell cycle rhythm.

Computer-Aided Three-Dimensional Reconstructions

A computer can be used for storage and three-dimensional reconstruction of embryonic data. One major aim is to combine the light-microscopical lineage data with the ultrastructural details given by the electron microscope. A 24-cell stage of *C. elegans* has been reconstructed from electron micrographs and matched to the same stage observed in the light microscope, so that each cell can be assigned its lineage name (Krieg et al., '78). Recently a new method of three-dimensional reconstruction has been developed in our

[3]The eggs are mounted in extrusion medium containing 120 mM NaCl, 10 mM Hepes at pH 7.2, 10 μm/ml streptomycin sulfite, 100 units/ml penicillin, 0.045 mg/ml each of 18 of the 20 standard amino acids (omitting asparagine and glutamine), and 0.025 mg/ml of trypan blue. The eggshell absorbs the blue dye and is thereby sensitized to a Rhodamine 6 G dye laser-microbeam (wave length 590 nm). For fixing eggs the extrusion medium contains 0.3–2.0% glutaraldehyde. (For further details, see Laufer and von Ehrenstein, '81, and von Ehrenstein et al., '81.) For cell fusion the laser dye BibuQ (Lambda, Physics, Göttingen, wave length 386 nm) was used.

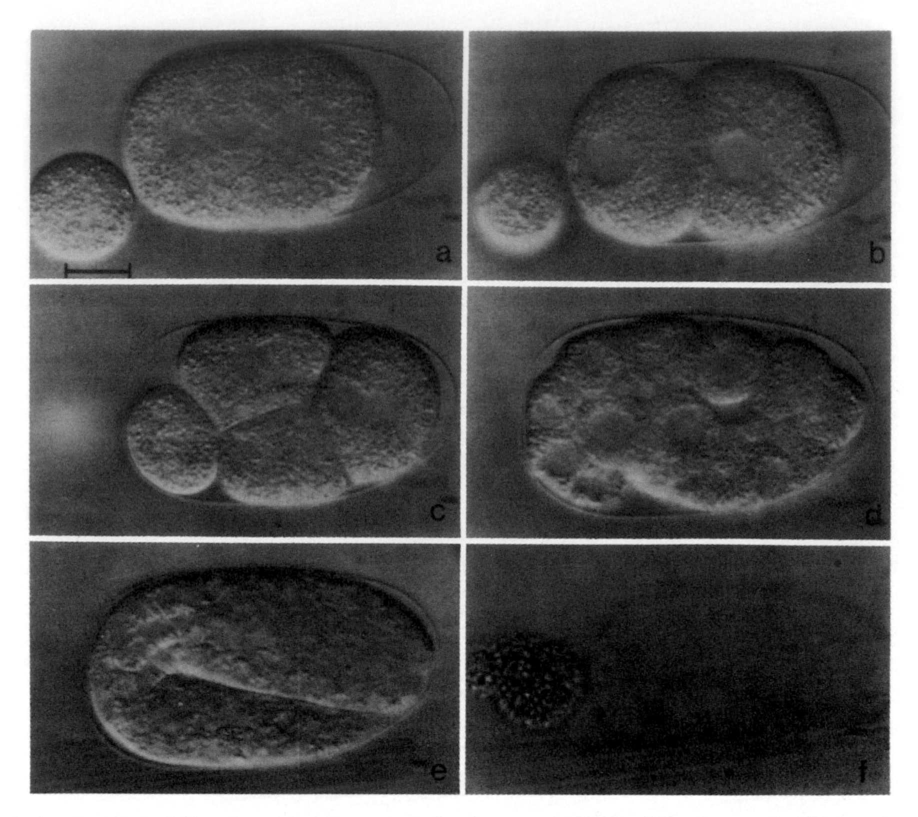

Fig. 8. Normal development after removal of egg cytoplasm with a laser microbeam. A posterior fragment containing about 10% of the total egg cytoplasm was extruded at the 1-cell stage (P_0) after the fusion of the pronuclei. a) Egg during first division showing extruded cell fragments. b) 2-cell stage. c) 4-cell stage. d) 28-cell stage. e) Plum-stage. f) Empty eggshell after hatching of the larva, which subsequently developed into a fertile adult. Bar 10 μm. (From Laufer and von Ehrenstein, '81; with permission. Copyright 1981 by the American Association for the Advancement of Science.)

laboratory.[4] As a first step, the three-dimensional coordinates of living embryos stored on video tapes are traced into the computer. As was described above under "Microscopy," a motor drives the micrometer screw continuously up and down, dissecting the embryo into several optical layers. In

[4]The graphics display system we use is an EYECOM II picture digitizer, display terminal, and color monitor from Spatial Data Systems, Goleta, California. It is connected to an image-processing system, which consists of a computer and its mass storage peripheral controlled by the operating system RT 11 (Digital Equipment Corporation, Maynard, Massachusetts). The system was assembled by W. Sidio; the software was written by C. Carlson.

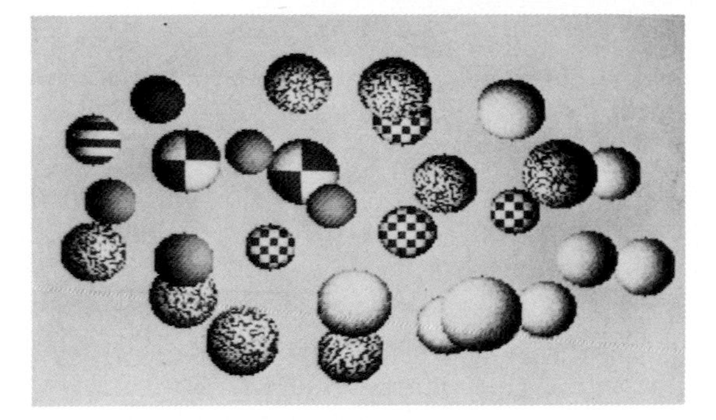

Fig. 9. Computer reconstruction of a 28-cell stage. Each nucleus is represented by a ball. Various codes in color or black and white are possible to distinguish different cell types: white, descendents of AB.a; dotted, descendants of AB.p; checkered, descendents of MSt; quartered, descendents of E; gray, descendants of C; striped, D, black, P_4. For nomenclature see figure 5. Orientation: anterior, right; left, top.

addition a video marker is coupled to the micrometer screw generating a black line on the video screen, which is a direct measure for the third dimension. The maximal diameter of each nucleus is traced in, and the computer reconstructs a model in which each nucleus is represented by a ball. The nuclei are assigned different color patterns according to lineage membership (Fig. 9). In a similar way, the electron microscopical series of the same embryo, which has been fixed after light-microscopical observation, is reconstructed. Here the third dimension is given by the number of sections. By matching the two reconstructions of several embryos with more than 100 cells we have begun to relate lineage information to ultrastructural details.

Genetics

One of the most favorable properties of *C. elegans* is its suitability for genetic analysis. Since Brenner's fundamental studies ('74), many genetic aspects of development have been worked out (for a review see Herman and Horvitz, '80). A uniform genetic nomenclature has been developed (Horvitz et al., '79, Herman et al., '80).

The hermaphrodites (with five pairs of autosomes and two X-chromosomes) are advantageous for genetic analysis because the animals are driven to homozygosity by self-fertilization. Males (with five pairs of autosomes and one X-chromosome), arising spontaneously 1 in 700 (Hodgkin, '74), can be crossed with the hermaphrodites that permit the transfer of genetic markers from one hermaphrodite to another. Their sperm are injected into the hermaphrodite and preferentially used to fertilize the oocytes (Ward and Carrel, '79). More than 300 genes have been mapped on the six linkage groups. Brenner ('74) estimated that a total of 2,000 genes are essential for the development of *C. elegans*. Most of the identified genes concern behavior and/or structure of the hatched worm. Nearly 50 of the identified genes (Fig. 10) are necessary for normal embryogenesis (Miwa et al., '80; Wood et al., '80; Cassada et al., '81). Cassada et al. ('81) estimated that the number of genes necessary for embryogenesis may be between 200 and 560.

Isolation of developmental mutants. A current strategy employed in searching for "embryogenesis genes" is the isolation of temperature-sensitive embryonic arrest mutants (Hirsch and Vanderslice, '76; Miwa et al., '80; Cassada et al., '81). These develop normally at the permissive temperature of 16°C and express the mutant phenotype at the nonpermissive temperature of 25°C. Mutations are induced in the worms with ethyl methane sulfonate as described by Brenner ('74). Single worms are incubated at 16°C until the F2 generation appears. Individual F2 worms are transferred into wells of a microtiter plate containing bacteria in liquid medium. After the F3 generation appears, the animals are replica plated onto microtiter plates at 25°C. After 3 days worms that have not reproduced (in particular those with accumulated eggs) are tested further, going back to the corresponding well of the 16°C master plate. The remaining candidates are cloned and backcrossed twice against the wild type. Nearly all emb-mutations turned out to be recessive, a few expressing semidominant behavior.

Mapping of emb genes. To assign a mutated gene to one of the six linkage groups, the new mutant is crossed at 16°C with easily identifiable tester mutants, each representing one linkage group. In the F2 generation one looks for the number of double mutants. Since the embryonic arrest phenotype is not expressed at 16°C, the progeny that show the tester mutant phenotype are shifted to 25°C and scored for embryonic arrest. If the two genes are linked, the fraction of F2 progeny with the tester mutant phenotype turning out to be homozygous for an embryonic arrest gene should be significantly less than one quarter.

The positioning of a mutation on its linkage group involves determining map distances from two-factor crosses and ordering mutants with respect to each other with three-factor crosses.

A complementation test is used to test for allelism of two mutations closely related in position. For autosomal recessive mutations males homozygous for one mutation are crossed at 16°C with hermaphrodites homozygous for the other mutation. When F1 heterozygotic double mutants are shifted to 25°C, no F2 eggs hatch if the genes do not complement each other. All F2 eggs are expected to hatch if the genes complement each other and the parental gene expression (see below) is sufficient. If for one (or both) of the two complementing genes embryonic gene expression is necessary, one quarter (or one half) of the F2 eggs are not expected to hatch.

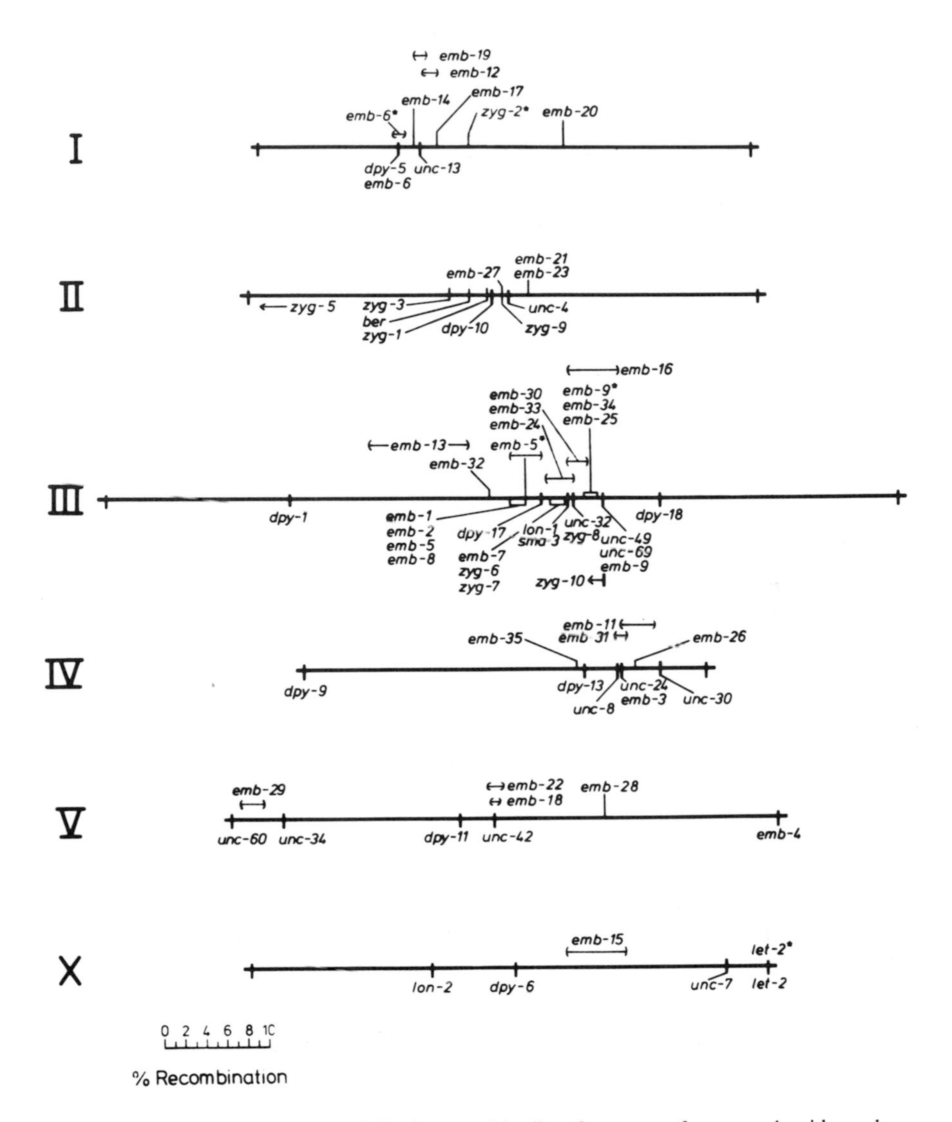

Fig. 10. Simplified genetic map of *C. elegans* with all *emb* genes so far mapped, with marker genes below the line. Bars indicate the range of locations for the genes. The relative order of *emb* genes with overlapping bars is unknown. An asterisk indicates data from allele(s) of already published genes (shown below the linkage group line), for *emb* genes, Miwa et al. ('80); for zyg genes, Wood et al. ('80); for let-2, Meneely and Herman ('79); and from the *C. elegans* standard map, Herman et al. ('80). (From Cassada et al., '81, with permission.)

Parental tests. Three genetic tests have been used for *C. elegans* by Hirsh et al. ('77) to determine where (in the mother, father, or embryo) the tested gene is expressed. Each of these tests answers a specific question. In the R-test (rescue by wild-type male): Is maternal gene expression necessary? In the S-test (self-fertilization): Is parental gene expression sufficient? In the H-test (rescue by heterozygous male): Is paternal gene expression sufficient? For most mutants tested so far (Hirsch et al., '77; Miwa et al., '80; Wood et al., '80; Isnenghi et al., unpublished results) maternal gene expression turned out to be sufficient for normal development.

Temperature shift experiments. To define the temperature-sensitive period (TSP) of these mutants, temperature shifts have been performed, either with worms containing eggs and oocytes of variable age (Hirsh and Vanderslice, '76; Vanderslice and Hirsh, '76; Wood et al., '80) or with embryos of selected cell stages (Miwa et al., '80).

Consistent with the parental tests, in many cases the TSP was found to be during oogenesis. Those in which the TSP is during embryogenesis often show the first visible defects shortly after the end of the TSP. Depending on when the embryos are shifted during the TSP, the early visible defects are more or less severe (Schierenberg et al., '80).

As defined by Hirsh and Vanderslice ('76), the latest time of down-shift (25°C to 16°C.) and earliest time of up-shift (16°C to 25°C) when the mutant phenotype is not observed define the beginning and end, respectively of the TSP. Miwa et al. ('80) instead defined "normal" and "defective" execution stages, allowing a more satisfactory explanation for those cases where the end of the TSP would otherwise be before the beginning.

Recombinant DNA Techniques

It has been postulated that rearrangements within DNA are part of the mechanism of cellular differentiation. In order to test the occurrence of rearrangements in the DNA of *C. elegans* during development, Emmons et al. ('79) have compared restriction fragments derived from DNA of sperm and somatic tissue. Sperm and somatic cells were isolated from a strain that produces an increased number of males. Compressing males between two plexiglass plates causes release of sperm cells into the medium, which can then be separated by filtration and pelleting (Klass and Hirsh, '81). Sperm nuclei were released by mild homogenization. DNA was prepared by conventional methods. As a source for somatic DNA, animals of the first larval stage (L_1), containing 550 somatic and only two germ cells, were used. Recombinant DNA molecules consisting of a plasmid with inserts of *Bam* H1 restriction fragments of *C. elegans* DNA were randomly cloned in an *E. coli* strain. The restriction digests were fractionated by agarose gel electrophoresis and transferred to Millipore filters by the method of Southern ('75). The cloned larval *Bam* H1 restriction fragments were radioactively labeled and hybridized to filters containing sperm or L_1 DNA. Comparing the patterns of molecular weights or relative intensities of the hybridization bands between somatic and germ line DNA, they could not detect any signs for rearrangements of the DNA during germ and somatic differentiation.

With similar recombinant DNA techniques Certa ('81) has cloned histone genes of *C. elegans* wild type. Using a sea-urchin H4 probe *C. elegans* histone gene sequences have been detected (Scharfenberg, personal communication).

K. Edwards has carried out *in situ* hybridization experiments with homologous recombinant DNA probes (personal communication). She detected abundant RNA using cloned, nick-translated DNA sequences containing ribosomal, histone, actin, and myosin genes. Following the squashing procedure of Gossett and Hecht ('80), differential hybridization was found between young (1–50-cell) and old ("comma" stage to hatching) embryos and between frozen sections of adult gonad and intestinal tissue. This indicates differential gene expression.

Biochemistry

Extensive biochemical studies have been carried out on *C. elegans* and closely related species dealing with nutrition and/or aging (for a review see Zuckerman, '80b). Several investigators have focused on muscle organization (for a review see Zengel and Epstein, '80) and on the study of enzymatic defects in *C. elegans* (for a review see Siddiqui and von Ehrenstein, '80).

Sulston and Brenner ('74) determined the content of DNA by chemical analysis and renaturation kinetics to be about 20 times the genome of *E. coli*; 83% of the DNA sequences are unique. No differences were detected between larval and adult DNA.

Two-dimensional gel electrophoresis has been used to analyze turnover and modification of proteins synthesized during postembryonic development (Johnson and Hirsh, '79). Comparing the patterns of pulse-labeled protein of the four larval stages and the adult, 113 of 800 total proteins were found to undergo modulation at one or more of the developmental stages.

Recently a biochemical analysis of the major sperm protein has been performed by Klass and Hirsh ('81). Using immunofluorescence with a specific antibody and pulse-labelling, they showed that this protein is synthesized prior to the visible differentiation of the primary spermatocytes in a specific region of the male gonad.

Only a few aspects of embryonic development have been investigated so far using biochemical procedures, probably for technical reasons. It is difficult to collect large quantities of eggs and separate different embryonic stages from each other.

For many biochemical studies it is necessary to start with gram quantities of worms. Vanfleteren ('76) describes an inexpensive axenic (sterile) large-scale cultivation. Sulston and Brenner ('74) use a suspension of bacteria in a simple medium yielding about 5 gm worms/liter (wet weight). For biochemical studies the harvest is cleaned from bacteria by centrifugation in a 35% sucrose solution. Using a modified culture and cleaning technique, Certa (personal communication) measured the changes of protein, lipid, and enzyme concentrations during postembryonic development. He separated the five developmental stages with nylon nets of different-size mesh.

Histones of different developmental stages can be separated by HPLC (Certa, '81; Certa and von Ehrenstein, '81). A variant of histone H2A is only found in embryos of *C. elegans*. This variant has a different amino acid composition from the postembryonic H2A. This has been shown by peptide mapping and *in vitro* translation of the histone messenger RNAs. Certa and Isnenghi (personal communication) found a mutant which contains only the embryonic variant of Histone H2A. Certa ('81) reported also mutants with Histone H2A variants that are absent from the wildtype (Fig. 11). He postulates specific division repressors for different cell types. A correct interaction between such hypothetic repressors and the altered chromatin may be disturbed in the mutants, leading to the observed abnormal cell division patterns (Schierenberg et al., '80; Denich et al., unpublished results).

Various procedures are used to stain specific structures in the hatched animal or the embryo. An easy method to stain chromosomes in intact worms with a fluorescent dye (Hoechst 33258) suitable for microfluorimetry has been described (Albertson et al., '78). They showed that, in a lineage defective mutant in which the ventral cord cells do not divide during postembryonic development, the normal number of DNA duplications still occurs. These polyploid cells display characteristics of the cells they would have ordinarily produced.

Gossett and Hecht ('80) used the same dye but a different procedure, including RNase treatment, to stain squashes of *C. elegans* embryos. This method allows an exact count of nuclei in all embryonic stages. By spreading out the embryonic cells in two dimensions one can observe them as discrete entities. Using this technique, appropriate histochemical and immunocytochemical probes may help to identify macromolecules specific to different cell typés or lineages. Recently Hecht et al. ('81) have examined the titer of poly (A) in squashes prepared from oocytes and variously staged embryos by *in situ* hybridization with a radiolabeled poly (U) probe. Individual grains in the autoradiograms were counted directly by light microscopy. The results suggest that transcription of the embryonic genome begins to be detectable around the 100-cell stage. After that time, the titer of nuclear and total embryonic poly (A) increases at a linear rate up to hatching.

For the histochemical analysis of acetylcholinesterase-deficient mutants, Culotti et al. ('81) stained whole mounts of worms with acetylthiocoline as

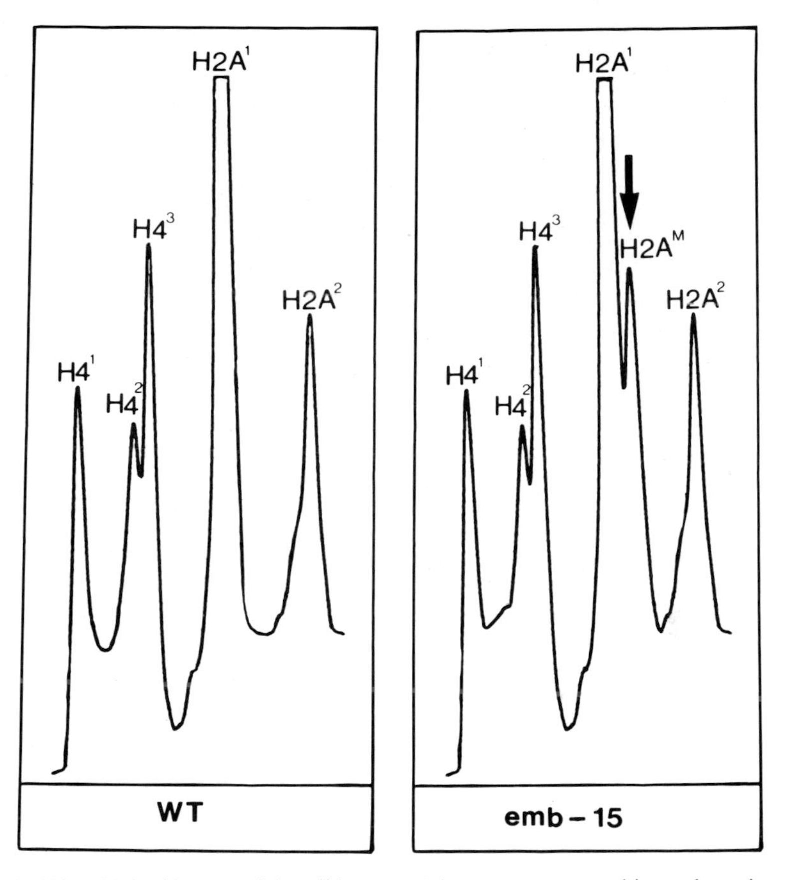

Fig. 11. H4 and H2A histones of the wild type and the temperature-sensitive embryonic arrest mutant *emb-15*. The histones were separated by high-performance liquid chromatography (HPLC) using a Hypersil C-18 column. Note the additional peak for a H2A variant in the mutant (arrow). (After Certa, '81, with permission).

substrate following a modified procedure of Karnovsky and Roots ('64). Analyzing *C. elegans* embryos with this staining method, the enzyme has not been detected before the late morphogenesis phase (Siddiqui et al., unpublished results).

Toward Tissue Culture

The tough and impermeable shell of the *C. elegans* egg is an obstacle to any kind of manipulation. As a first step toward growing embryonic cells in a culture medium, Laufer et al. ('80) have developed a medium in which the blastomeres of burst eggs continue to develop. Their embryonic culture medium combines 90 mM NaCl, 4 mM KCl, 18 mM $CaCl_2$, 18 mM $HGSO_4$, 5 mM HEPES (pH 7.2), 10 KI u/ml aprotonin (Sigma), and 100 μl/ml streptomycin sulfate (Sigma). To block cleavage 50 μg/ml colchicine and 25 μg/ml cytochalasin B (Sigma) were added.

As a special additive, this medium contains 25% *Ascaris* coelomic fluid. Similar results can be obtained when the coelomic fluid is replaced by 10% fetal calf serum (J. Laufer, personal communication). Under these conditions 2-cell eggs can cleave into several hundred cells. Many of the resulting embryos twitch demonstrating muscle contraction. But as the microenvironment inside the intact egg shell seems to be crucial, little morphogenesis has been observed to take place, resulting in irregularly shaped cell masses. Observation of isolated early blastomeres has confirmed Boveri's assumption ('10) that nematode embryos follow a mosaic type of development, i.e., differentiation potentials are due to intrinsic determinants. In contrast, potential for topogenesis (cell positioning) does not seem to be a cell autonomous property.

As the burst-egg embryos are permeable to drugs, Laufer et al. ('80) blocked further cleavage of 2-cell stages with colchicine and cytochalasin B. Fluorescent rhabditin granules, used as a characteristic for identifying the gut cells (fluorescence starting normally in the late proliferation phase), were detected several hours later in P_1, the precursor cell of E (for nomenclature, see Fig. 6). This demonstrates that at least some biochemical differentiation can take place appropriately in time without cleavage or nuclear division.

SUMMARY AND CONCLUSIONS

Caenorhabditis elegans illustrates the benefits of many scientists studying various aspects of a single organism. Recent progress initiated by Brenner's work ('73, '74), established *C. elegans* as a suitable model organism to study problems of genetics and development.

Of all eukaryotic organisms, cellular aspects of development have probably been most extensively studied in this animal. All cell lineages are now known from fertilization to adulthood and they are invariant. This is a favorable prerequisite for posing questions about general principles such as developmental strategies, and for investigating the underlying molecular machinery.

One strategy may be that an omnipotent cell (the germ line cell) throws off with each division a somatic cell (see Fig. 5) with different developmental options and thereby passes itself through a stepwise, progressive restriction of cell potency. If there are such developmental restrictions they must be reversible, at least in the germ cells, which express their totipotency again in the next generation. Such a mechanism has been also suggested for higher organisms (Papaioannou et al., '78; Monk, '81).

Another strategy may be the differential use of symmetrical (proliferative) and asymmetrical (stem cell-like) divisions to generate the kind and number of cells necessary to achieve development of the specific definitive structures. All lineages may be viewed as specific combinations or modifications of these two types of cell division (von Ehrenstein and Schierenberg, '80).

If a cell is only bipotent with respect to the developmental potential of its immediate daughter cells (see below), many asymmetrical division steps are necessary for extensive cell diversification. Many cells may then be unnecessary in themselves as sisters of other cells that carry specific developmental potential expressable only after cell division. This may explain the high number of embryonic cell deaths.

A third strategy may be timing as a tool for pattern formation. The correct control of the specific cell cycle rhythms may in itself be sufficient (and necessary) in many cases to position cells for correct cell–cell interaction. Cell lineage studies on temperature-sensitive embryonic arrest mutants showed that early alterations in the order of developmental events like division and migration often lead to abnormal cell positioning. Although timing and cell fate are normally closely related, they are controlled independently, as has been shown with mutants and by cell fusion experiments (Schierenberg et al., '80, and unpublished results).

Roux (1888) proposed two alternative mechanisms as to how the function of a cell could be determined—either by ancestry or by position. Nematodes are thought to be the best examples for the first of these developmental strategies, called mosaic development. But a few cases of natural variation during postembryonic development in which a cell can assume one of two alternative positions and later follow one of two alternative lineages (Sulston and Horvitz, '77; Kimble and Hirsh, '79; Sulston et al., '80) and some examples of limited regulation after laser-induced cell ablation (Kimble et al., '79; Sulston and White, '80; Kimble and White, '81) demonstrate that even in nematodes extrinsic factors can influence the fate of certain cells. The rare examples of external influence found so far are consistent with the concept that a cell is only bipotent and that the action of extrinsic factors is limited to preventing or allowing it to enter one of these alternative pathways.

It has been suggested that all developmental decisions in *C. elegans* are binary and depend on cell division (von Ehrenstein and Schierenberg, '80). Cell lineages based on binary decisions have been proposed earlier for higher organisms (Holtzer, '78); and a binary code has been proposed for imaginal disc determination and transdetermination in *Drosophila* (Kauffman, '73).

The cell lineage studies have given us a comprehensive picture of development in this organism on the cellular level. Our ideas of the molecular mechanisms are, however, rudimentary. Conventional biochemical assays are available and have been used to study subcellular structures, but with methods of tissue homogenization and separation into different components, we cannot expect to come close to an understanding of how gene action leads to cell specificity and how cell–cell interaction results in a complex organism. The characteristic feature of rigidly determined development, the low cell number, and our knowledge of cellular development make it most advantageous to try to identify molecular action *in situ* and relate it to the structure and function of cells and tissues. Burst eggs that continue development (Laufer et al., '80) or squashed embryos (Gossett and Hecht, '80) that may still allow the identification at least of cell types are promising areas for further study.

Procedures for staining eggs and worms with monoclonal antibodies have been developed (D. Albertson et al., personal communication). By using fluorescent antibody staining Strome and Wood ('82) have followed cyto plasmic granules unique to germ-line cells throughout the life cycle of *C. elegans.* These granules are segregated exclusively to germ cell precursors during early embryogenesis.

To investigate the ontogeny of muscle proteins during embryogenesis of *C. elegans*, Gossett et al. (personal communication) use antibodies against myosin and paramyosin. They identified myosin in 1–5 cells of the 400–500-cell stage. Half an hour later in the 450–500-cell stage myosin and paramyosin were found in 30–80 blastomeres.

The small size of *C. elegans* complicates external manipulation. A powerful technique has been worked out by J. Kimble (personal communication) for introducing macromolecules into living worms. With the help of a micromanipulator, substances can be injected into a narcotized worm under a dissecting microscope. She showed that the progeny could be radiolabeled by injection of radioactive liquid into the pseudocoelom or gonad of the parent. This method may allow one to rescue mutants by injecting the missing gene product or genomic clones of worm DNA in order to achieve transformation. A method for microinjection into living embryos has been developed by B. Masters (personal communication). She uses a water-immersion objective that gives enough working distance to insert a fine capillary from the side with a micromanipulator.

The implicit assumption in turning to a simple organism in order to study development is that there are fundamental mechanisms common to all metazoa. Although *C. elegans* seems to show more similarities to higher organisms than one might have expected some years ago, it is not clear to what degreee it can serve as a paradigm. Nevertheless, we expect that *C. elegans*, which offers us unique chances to investigate details of the relationship between gene action and development, will also give us new clues about the logical structure of the developmental network in higher organisms.

Acknowledgments. I thank Randy Cassada for critical reading of the manuscript, Mechthild Ziemer for expert photography, and Brigitte Knoke for skillful typing.

LITERATURE CITED

Abdulkader, N., and J. Brun (1976) Isolation of sterile or lethal temperature-sensitive mutants in *Caenorhabditis elegans* var. Bergevac. Nematologica *22:*222–223.

Albertson, D.G., J.E. Sulston, and J.G. White (1978) Cell cycling and DNA replication in a mutant blocked in cell division in the nematode *Caenorhabditis elegans*. Dev. Biol. *63:*165–178.

Albertson, D.G., and J.N. Thomson (1976) The pharynx of *Caenorhabditis elegans*. Phil. Trans. R. Soc. Lond. B. *275:*299–325.

Beguet, B., and M.A. Gilbert (1978) Obtaining a self-fertilizing hermaphrodite mutant with a male copulatory bursa in the free-living nematode *Caenorhabditis elegans*. C.R. Acad. Sci. Paris *286:*989–992.

Boveri, T. (1910) Dic Potenzen der *Ascaris*-Blastomeren bei abgeänderter Furchung. Festschr. R. Hertwig. Vol. 3, Jena: Gustav Fishcr, pp 133–214.

Brenner, S. (1973) The genetics of behavior. Br. Med. Bull. *29:*269–271.

Brenner, S. (1974) The genetics of *Caenorhabditis elegans*. Genetics *77:*71–94.

Byerly, L., R.C. Cassada, and R.L. Russell (1976) Life cycle of the nematode *Caenorhabditis elegans*. I. Wild type growth and reproduction. Dev. Biol. *51:*23–33.

Cassada, R., E. Isnenghi, M. Culotti and G. von Ehrenstein (1981) Genetic analysis of temperature-sensitive embryogenesis mutants in *Caenorhabditis elegans*. Dev. Biol. *84:*193–205.

Cassada, R.C., and R.L. Russell (1974) A positive selection for behavioral and developmental mutants of a nematode. Fed. Proc. *33:*1476.

Cassada, R.C., and R.L. Russell (1975) The dauerlarva, a postembryonic developmental variant of the nematode *Caenorhabditis elegans*. Dev. Biol. *46:*326–342.

Certa, U. (1981) Die Funktion der Histone bei der Entwicklung des Nematoden *Caenorhabditis elegans*, Ph.D. Thesis, University of Göttingen.

Certa, U. and G. von Ehrenstein (1981) Reversed-phase high performance liquid chromatography of histones. Anal. Biochem. *118:*147–154.

Chalfie, M., H.R. Horvitz, and J.E. Sulston (1981) Mutations that lead to reiterations in the cell lineages of *Caenorhabditis elegans*. Cell *24:*59–69.

Chalfie, M., and J.N. Thomson (1979) Organization of neuronal microtubules in the nematode *Caenorhabditis elegans*. J. Cell Biol. *82:*278–289.

Chitwood, B.G., and M.B. Chitwood (1950) Introduction to Nematology. Baltimore: University Park Press.

Culotti, J.G., G. von Ehrenstein, M.R. Culotti, and R.L. Russell (1981) A second class of acetylcholinesterase-deficient mutants of the nematode *Caenorhabditis elegans*. Genetics *97:*281–305.

Deppe, U., E. Schierenberg, T. Cole, C. Krieg, D. Schmitt, B. Yoder, and G. von Ehrenstein (1978) Cell lineages of the embryo of the nematode *Caenorhabditis elegans*. Proc. Natl. Acad. Sci. U.S.A. *75:*376–380.

Dougherty, E.C., and H.G. Calhoun (1948) Possible significance of free-living nematodes in genetic research. Nature *161:*29.

Dougherty, E.C., E.L. Hansen, W.L. Nicholas, J.A. Mollett, and E.A. Yarwood (1959) Axenic cultivation of *Caenorhabditis briggsae* with unsupplemented and supplemented chemically defined media. Ann. N.Y. Acad. Sci. *77:*176–217.

von Ehrenstein, G., and E. Schierenberg (1980) Cell lineages and development of *Caenorhabditis elegans* and other nematodes. In B. Zuckerman (ed): Nematodes as Biological Models, Vol. 1. New York: Academic Press, pp. 1–71.

von Ehrenstein, G., J.E. Sulston, E. Schierenberg, J.S. Laufer, and T. Cole (1981) Embryonic cell lineages and segregation of developmental potential in *Caenorhabditis elegans*. In H.G. Schweiger (ed): International Cell Biology 1980–1981. Berlin: Springer, pp. 519–525.

Emmons, S.W., M.R. Klass, and D. Hirsh (1979) Analysis of the constancy of DNA sequences during development and evolution of the nematode *Caenorhabitis elegans.* Proc. Natl. Acad. Sci. U.S.A. *76:*1333–1337.

Fatt, H.V., and E.C. Dougherty (1963) Genetic control of differential heat tolerance in two strains of the nematode *Caenorhabditis elegans.* Science *141:*266–267.

Gossett, L.A., and R. Hecht (1980) A squash technique demonstrating embryonic nuclear cleavage of the nematode *Caenorhabditis elegans.* J. Histochem. Cytochem. *28:*507–510.

Hall, D.H. (1977) The posterior nervous system of the nematode *Caenorhabditis elegans,* Ph.D. Thesis, California Institute of Technology, Pasadena.

Hecht, R.M., L.A. Gossett, and W.R. Jeffery (1981) Ontogeny of maternal and newly transcribed mRNA analyzed by *in situ* hybridization during development of *Caenorhabditis elegans.* Dev. Biol. *83:*374–379.

Herman, R.K., and H.R. Horvitz (1980) Genetic analysis of *Caenorhabditis elegans.* In B. Zuckerman (ed): Nematodes as Biological Models, Vol. 1. New York: Academic Press, pp. 227–262.

Herman, R.K., H.R. Horvitz, and D.L. Riddle (1980) The nematode *Caenorhabditis elegans.* In S.J.O. Brien (ed): Genetic Maps, Vol. 1. Bethesda: National Cancer Institute NIH, pp. 183–193.

Hirsh, D., D. Oppenheim, and M. Klass (1976) Development of the reproductive system of *Caenorhabditis elegans.* Dev. Biol. *49:*200–219.

Hirsh, D., and R. Vanderslice (1976) Temperature-sensitive developmental mutants of *Caenorhabditis elegans.* Dev. Biol. *49:*220–235.

Hirsh, D., W.B. Wood, R. Hecht, S. Carr, and R. Vanderslice (1977) Expression of genes essential for early development in the nematode *Caenorhabditis elegans.* In G. Wilcox, J.N. Abelson, and C.F. Fox (eds): Molecular Approaches to Eukaryotic Genetic Systems. New York: Academic Press, pp. 347–356.

Hodgkin, J. (1974) Genetic and anatomical aspects of the *Caenorhabditis elegans* male, Ph.D. Thesis, University of Cambridge, England.

Hodgkin, J.A., and S. Brenner (1977) Mutations causing transformation of sexual phenotype in the nematode *Caenorhabditis elegans.* Genetics *86:* 275–288.

Holtzer, H. (1978) Cell lineages, stem cells and the "quantal" cell cycle concept. In B.J. Lord, C.S. Potten, and R.J. Cole (eds): Stem Cells and Tissue Homeostasis. Cambridge, England: Cambridge University Press, pp. 1–22.

Honda, H. (1925) Experimental and cytological studies on bisexual and hermaphrodite free-living nematodes with special reference to problems of sex. J. Morphol. Physiol. *40:* 191–233.

Horvitz, H.R., S. Brenner, J. Hodgkin, and R.K. Herman (1979) A uniform genetic nomenclature for the nematode *Caenorhabditis elegans.* Mol. Gen. Genet. *175:*129–133.

Horvitz, H.R., and Sulston, J.E. (1980) Isolation and genetic characterization of cell-lineage mutants of the nematode *Caenorhabditis elegans.* Genetics *96:*435–454.

Johnson, K., and D. Hirsh (1979) Patterns of proteins synthesized during development of *Caenorhabditis elegans*. Dev. Biol. *70*:241–248.

Kauffman, S.A. (1973) Control circuits for determination and transdetermination. Science *181*:310–318.

Karnovsky, M.J., and L. Roots (1964) A "direct-coloring" thiocholine method for cholinesterases. J. Histochem. Cytochem. *12*:219–221.

Kimble, J., and D. Hirsh (1979) The postembryonic cell lineages of the hermaphrodite and male gonads in *Caenorhabditis elegans*. Dev. Biol. *70*:396–417.

Kimble, J.E., J.E. Sulston, and J.G. White (1979) Regulative development in the postembryonic lineages of *Caenorhabditis elegans*. In N. Le Douarin and A. Monroy (eds): Stem Cells, Cell Lineages and Cell Determination. New York: Elsevier/North-Holland, pp. 59–68.

Kimble, J., and J.G. White (1981) On the control of germ cell development in *Caenorhabditis elegans*. Dev. Biol. *81*:208–219.

Klass, M.R., and D. Hirsh (1976) Nonaging developmental variant of *Caenorhabditis elegans*. Nature *260*:523–525.

Klass, M.R., and D. Hirsh (1981) Sperm isolation and biochemical analysis of the major sperm protein from *Caenorhabditis elegans*. Dev. Biol. *84*:299–312.

Klass, M.R., N. Wolf, and D. Hirsh (1976) Development of the male reproductive system and sexual transformation in the nematode *Caenorhabditis elegans*. Dev. Biol. *52*:1–18.

Klass, M.R., N. Wolf, and D. Hirsh (1979) Further characterization of a temperature sensitive transformation mutant in *Caenorhabditis elegans*. Dev. Biol. *69*:329–335.

Krieg, C., T. Cole, U. Deppe, E. Schierenberg, D. Schmitt, B. Yoder, and G. von Ehrenstein (1978) The cellular anatomy of embryos of the nematode *Caenorhabditis elegans*, analysis and reconstruction of serial section electron micrographs. Dev. Biol. *65*:193–215.

Laufer, J.S., P. Bazzicalupo, and W.B. Wood (1980) Segregation of developmental potential in early embryos of *Caenorhabditis elegans*. Cell *19*:569–577.

Laufer, J.S., and G. von Ehrenstein (1981) Nematode development after removal of egg cytoplasm: Absence of localized unbound determinants. Science *211*:402–405.

Maupas, E. (1900) Modes et formes de reproduction des nematodes. Arch. Zool. Exp. Gen. *8*:463–624.

Meneely, P.M., and R.K. Herman (1979) Lethals, steriles, and deficiencies in a region of the X chromosome of *Caenorhabditis elegans*. Genetics *92*:99–115.

Miwa, J., E. Schierenberg, S. Miwa, and G. von Ehrenstein (1980) Genetics and mode of expression of temperature-sensitive mutations arresting embryonic development in *Caenorhabditis elegans*. Dev. Biol. *76*:160–174.

Monk, M. (1981) A stem-line model for cellular and chromosomal differentiation in early mouse development. Differentiation *19*:71–76.

Nelson, G.A., K.K. Lew, and S. Ward (1978) Intersex, a temperature sensitive mutant of the nematode *Caenorhabditis elegans*. Dev. Biol. *66*:386–409.

Nigon, V. (1949) Les modalités de la reproduction et le déterminisme de sexe chez quelques nématodes libres. Ann. Sci. Nat. Zool. *11*:1–132.

Nigon, V. (1965) Développement et reproduction des nématodes. In P.P. Grassé (ed): Traité de Zoologie. Vol. 4(3). Paris: Masson, pp. 218–316.

Nigon, V., and J. Brun (1955) L'évolution des structures nucléaires dans l'ovogénèse de *Caenorhabditis élégans*. Chromosoma *7*:129–169.

Papaioannou, V.E., J. Rossant, and R.L. Gardner (1978) Stem cells in early mammalian development. In B.J. Lord, C.S. Potten, and R.J. Cole (eds): Stem Cells and Tissue Homeostasis. Cambridge, England: Cambridge University Press, pp. 49–69.

Roux, W. (1888) Beiträge zur Entwicklungsmechanik des Embryo. Virchows Arch. Path. Anat. *114:*113–153, 246–291.

Schierenberg, E. (1978) Die Embryonalentwicklung des Nematoden *Caenorhabditis elegans* als Modell. Ph.D. Thesis, University of Göttingen.

Schierenberg, E., J. Miwa, and G. von Ehrenstein (1980) Cell lineages and developmental defects of temperature-sensitive embryonic arrest mutants in *Caenorhabditis elegans*. Dev. Biol. *76:*141–159.

Siddiqui, S.S., and G. von Ehrenstein (1980) Biochemical genetics of *Caenorhabditis elegans*. In B. Zuckerman (ed.): Nematodes as Biological Models. New York: Academic Press, pp. 285–304.

Southern, E.M. (1975) Detection of specific sequences among DNA fragments separated by gel electrophoresis. J. Mol. Biol. *98:*503–517.

Strome, S. and W.B. Wood (1982) Immunofluorescence visualization of germ-line-specific cytoplasmic granules in embryos, larvae, and adults of *Caenorhabditis elegans*. Proc. Natl. Acad. Sci. U.S.A. *79:*1558–1562.

Sulston, J.E. (1976) Post-embryonic development in the ventral cord of *Caenorhabditis elegans*. Phil. Trans. R. Soc. Lond. B. *275:*287–297.

Sulston, J.E., D.G. Albertson, and J.N. Thomson (1980) The *Caenorhabditis orhabditis elegans* male: Postembryonic development of nongonadal structures. Dev. Biol. *78:*542–576.

Sulston, J.E., and S. Brenner (1974) The DNA of *Caenorhabditis elegans*. Genetics *77:*95–104.

Sulston, J.E., and Horvitz (1977) Post-embryonic cell lineages of the nematode *Caenorhabditis elegans*. Dev. Biol. *56:*110–156.

Sulston, J.E., and R. Horvitz (1981) Abnormal cell lineages in mutants of the nematode *Caenorhabditis elegans*. Dev. Biol. *82:*41–55.

Sulston, J.E., and J.G. White (1980) Regulation and cell autonomy during postembryonic development in *Caenorhabditis elegans*. Dev. Biol. *78:*577–598.

Vanderslice, R., and D. Hirsh (1976) Temperature-sensitive zygote defective mutants of *Caenorhabditis elegans*. Dev. Biol. *49:*236–249.

Vanfleteren, J.R. (1976) Large-scale cultivation of a free living nematode *Caenorhabditis elegans*. Experientia *32:*1087–1088.

Ward, S., Y. Argon, and G.A. Nelson (1981) Sperm morphogenesis in wild-type and fertilization defective mutants of *Caenorhabditis elegans*. J. Cell Biol. *91:*26–44.

Ward, S., and J.S. Carrel (1979) Fertilization and sperm competition in the nematode *Caenorhabditis elegans*. Dev. Biol. *73:*304–321.

Ward, S., and J. Miwa (1978) Characterization of temperature sensitive fertilization defective mutants of the nematode *Caenorhabditis elegans*. Genetics *88:*285–304.

Ward, S., J.N. Thomson, J.C. White, and S. Brenner (1975) Electron microscopical reconstruction of the anterior sensory anatomy of the nematode *Caenorhabditis elegans*. J. Comp. Neurol. *160:*313–338.

Ware, R.W., D. Clark, K. Grossland, and R.L. Russell (1975) The nerve ring of the nematode *Caenorhabditis elegans:* Sensory input and motor output. J. Comp. Neurol. *162:*71–110.

White, J.G. (1974) Computer-aided reconstruction of the nervous system of *Caenorhabditis elegans*. Ph.D. Thesis, University of Cambridge, England.

White, J.G., D.G. Albertson, and M.A.R. Anness (1978) Connectivity changes in a class of motorneurons during the development of a nematode. Nature *271:*764–766.

White, J.G., E. Southgate, J.N. Thomson, and S. Brenner (1976) The structure of the ventral nerve cord of *Caenorhabditis elegans*. Phil. Trans. R. Soc. Lond. B *275:*327–348.

Wood, W.B., R. Hecht, S. Carr, R. Vanderslice, N. Wolf, and D. Hirsh (1980) Parental effects and phenotypic characterization of mutations that affect early development in *Caenorhabditis elegans*. Dev. Biol. *74:*446–469.

Zengel, J.M., and Epstein, H.F. (1980) Muscle development in *Caenorhabditis elegans*: A molecular genetic approach. In B. Zuckerman (ed.): Nematodes as Biological Models, Vol. 1. New York: Academic Press, pp. 73–126.

Zuckerman, B.M. (ed) (1980a) Nematodes as Biological Models. Vol. 1, Behavioral and Developmental Models. New York: Academic Press.

Zuckerman, B.M. (ed) (1980b) Nematodes as Biological Models. Vol. 2, Aging and Other Models. New York: Academic Press.

Developmental Biology of Freshwater Invertebrates, pages 283–316

Development in the Freshwater Oligochaete *Tubifex*

Takashi Shimizu

ABSTRACT The *Tubifex* egg is fertilized at metaphase of the first meiosis. After extrusion of polar bodies, pole plasms comprising endoplasmic reticulum and mitochondria accumulate around the animal and vegetal poles. The developmental stages of polar body formation are characterized by a dynamic shape change called deformation movement. The first cleavage produces the smaller AB- and larger CD-cell. The second cleavage gives rise to four cells: A, B, C and D; the D-cell is larger than the other three cells. Thereafter, these four cells divide in a spiral cleavage pattern, producing micromeres and yolky macromeres. The pole plasms are segregated into 2d- and 4d-cells resulting from divisions of the D-cell. The descendant cell ($2d^{111}$) of 2d-cell and 4d-cell exclusively participate in teloblastogenesis. Four ectoblasts and a mesoblast are located on either side of the embryo, and bud off small cells forming germ bands. Gastrulation movement consists of two events: 1) ventral shift and ensuing coalescence of germ bands, and 2) epibolic expansion of a micromere-derived epithelial sheet over the endodermal cells. Organogenesis begins during the gastrula stage. The ectodermal germ bands are responsible for the ventral nerve cord and circular muscle layer. The mesodermal organs are exclusively derived from the stem cells produced by the mesoblasts. Gastrulation is followed by elongation of the embryo. It finally changes its shape into a vermicular form. The embryonic period lasts for two weeks at 18°C, and is divided into 19 stages.

The central problem in *Tubifex* development is the relationship between the pole plasm localization and cell determination. To our regret, however, we have no data at present to look into this problem, except for the undisputed fact that the pole plasms are segregated first to D-cell, subsequently to 2d- and 4d-cell, and finally to germ band cells.

INTRODUCTION

Tubifex (Annelida; Oligochaeta; Tubificidae) is a cosmopolitan freshwater oligochaete distributed on all six continents, excluding Antarctica (for details, see Timm, '80). This animal is readily available to most investigators and can be collected in abundance. Furthermore, fertilized eggs are easily obtained in the laboratory. For these reasons, *Tubifex* has long been used as material for developmental studies.

The *Tubifex* egg undergoes a series of spiral, unequal divisions, and shows a determinate development. One of the outstanding features of the *Tubifex* egg is the pole plasm accumulation (PPA), which takes place shortly before the first cleavage. The pole plasms are subsequently segregated into determined blastomeres. Since the pioneering cell lineage study by Penners ('22, '24a), many students have investigated the relationship between the pole plasms and the determinate development in the *Tubifex* egg. Although this problem still remains unsettled, this freshwater oligochaete is expected to provide a good system to study the mosaic nature of the Spiralian eggs.

In the monograph "The Oligochaeta" by Stephenson ('30), eight species of *Tubifex* are cited. Among them, a European species *Tubifex tubifex* Müller (*T. rivulorum* Lamarck) and a Japanese species (*T. hattai* Nomura) have mainly been used for developmental studies. In this chapter, I describe the normal development in *Tubifex*, reviewing works on these two species. Results so far obtained from experimental studies are also reviewed.

Reproduction and Life Cycle of *Tubifex*

Tubifex is hermaphroditic and has a pair of testes and ovaries in segments 10 and 11, respectively (for the anatomy of *Tubifex*, see Dixon, '15; Nomura, '26; Stephenson, '30). In the natural condition, there are two breeding periods, spring and early fall (Matsumoto and Yamamoto, '66; Poddubnaya, '80). However, mature worms that possess a well-developed clitellum and eggs in the ovisac (Fig. 1A) are found throughout the year.

Fertilized eggs packed in a cocoon are laid by mature worms. The ovipository behavior of *Tubifex* has been described by Hirao ('65b); it can be divided into two successive steps: 1) cocoon formation around the clitellum (Figs. 1B and 2A–D), and 2) ensuing release of the formed cocoon from the worm by the backward movement of the worm (Fig. 2E-H). Mature eggs are released through the female pores into the interstice between the cocoon membrane and the clitellar epithelium (Figs. 1B and 2D). At the same time, the cocoon fluid, which is known to play an important role in embryonic development (Jaana et al., '80), is also secreted into this interstice. The cellular basis of cocoon formation in *Tubifex* has been established by Hirao ('65b), Suzutani ('77), and Suzutani-Shiota ('80).

The embryonic period lasts for about 15 days at 18°–20°C, and then juveniles hatch out. Worms attain sexual maturity within 2 months: the gonads are formed by the end of the first month of life; during the next 2 weeks formation of genital organs reaches completion (Poddubnaya, '80). Copulation occurs between mature worms, and exchanged sperm bundles

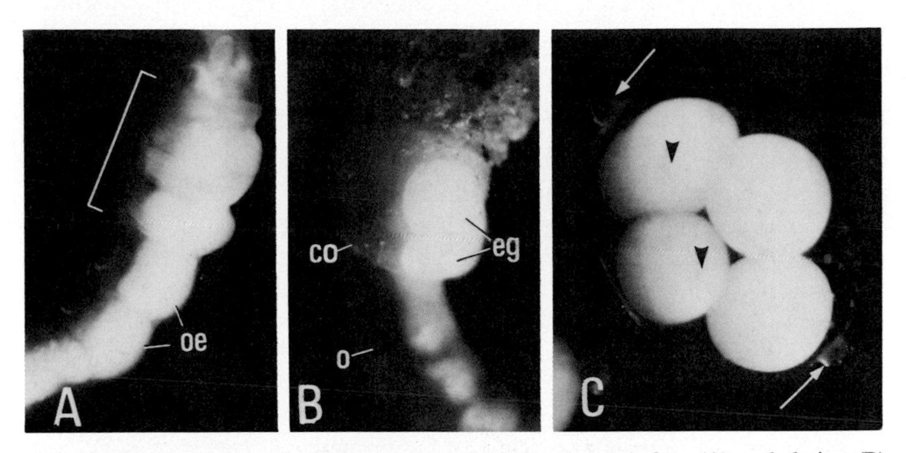

Fig. 1. A,B. Side view of clitellar region of *Tubifex hattai* before (A) and during (B) oviposition. During oviposition, ovisac eggs (oe) are released into the interstice between the clitellar epithelium (bracket) and the cocoon membrane (co) is formed around it. Released eggs (eg) are not yet fertilized. Note the absence of eggs in the ovisac (o). × 20. C. A cocoon 1 hour following deposition. Four eggs are seen through the transparent cocoon membrane. Arrowheads point to the animal pole where the first meiotic apparatus is located. White arrows indicate processes at both ends of the cocoon. × 40.

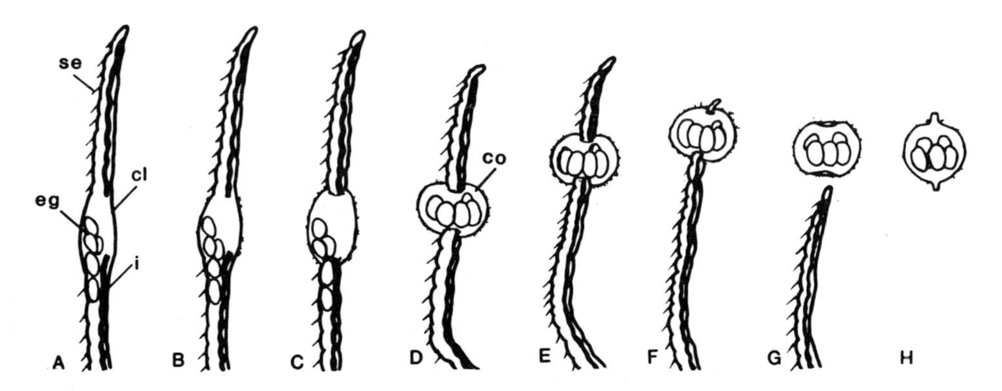

Fig. 2. Sequence of cocoon deposition in *Tubifex hattai*. cl, clitellum; co, cocoon membrane; eg, eggs; i, intestine; se, seta. (Redrawn from Hirao, '65b.)

(spermatozeugmata) are stored in the spermatheca (see Braidotti et al., '80). The life cycle of *Tubifex* has been estimated at 2–4 years (Matsumoto and Yamamoto, '66; Poddubnaya, '80).

The development of genital organs has been investigated by Gatenby ('16). Morphological studies of gametogenesis have been reported by several authors (oogenesis: Gathy, 1900; Specht, '61; Hirao, '64; spermiogenesis:Ferraguti and Lanzavecchia, '71; Braidotti et al., '80; Fleming, '81; Jaana, '82a,b).

Methods for Obtaining and Handling Animals and Eggs

Animals. *Tubifex* is common in the mud of ponds and streams with a poor water flow, especially where natural organic waste is concentrated. Usually *Tubifex* makes a clump with other oligochaete species (*Limnodrilus, Branchiura, Rhizodrilus,* etc.). To collect mature worms, sludge containing these clumps is put into a relatively tall cylinder and allowed to stand overnight. Worms move upwards; they finally assemble near the upper surface of the sludge. The assemblage of worms is taken out of the cylinder and washed repeatedly to remove sludge.

The separation of *Tubifex* from other species can be easily accomplished in two steps. By a gentle stirring of environmental water, *Tubifex* and *Limnodrilus* are separated from other species, since these worms rapidly change their bodies into a spiral in response to the artificial agitation. Furthermore, hair setae are present in *Tubifex*, but not in *Limnodrilus*; therefore, these two species are distinguished by the inspection of setal morphology under the dissecting microscope. In addition to these characteristics, cocoons of *Tubifex* and *Limnodrilus* are different in appearance (see Penners, '33).

For rearing, separated worms are placed in vats containing well water. The bottom of the vat is covered with a layer of soft mud 30–40 mm thick. It is very important that water in the vat be constantly supplied with running well water. The worms are reared at 12° ± 1°C and nourished with yeast, squash of boiled potatoes, and corn-flour. Under these rearing conditions, worms breed constantly. When mature worms are transferred to higher temperature (~ 20°C), they deposit cocoons within 1–4 days at any season of the year.

Cocoons. To obtain newly laid eggs, mature worms from which mud has been washed away are placed in finger bowls (10 cm) containing well water and a layer of sterilized sand 15 mm thick (Lehmann, '41c; Hirao, '66). Worms (about 200 per bowl) are placed under red light or in a dark room at 18°–20°C (Hirao, '65a). Water in the bowl should be changed twice a day. Deposited cocoons are usually buried in the layer of sand in the bowl. Since cocoons are much lighter than sand particles, they are flung up by puffing a jet stream into the layer of sand and allowing it to fall slowly. In this way, by inspection of the contents of the bowl at 3–4-hour intervals, cocoons are collected and transferred into a petri dish containing a culture medium (Table I). Worms that have already deposited cocoons are returned to vats described above.

Eggs. Usually a cocoon contains 4–8 eggs. To observe the developmental process in detail, eggs are freed from the cocoon in the culture medium. After cocoons are washed several times in the culture medium, the cocoon membrane is torn off piecemeal with a pair of watchmaker's forceps. At this time, since the cocoon membrane is elastic, great care must be taken not to compress the eggs with the membrane. Freed "intact" eggs are washed twice in newly prepared culture medium, and kept at temperatures below 20°C. Petri dishes and the culture medium are renewed every day. Under these culture conditions, especially when the eggs are within the intact vitelline membrane, it is not necessary to add antibiotics to the solution; however, if needed, Kanamycine sulfate (0.001%) may be used (Inase, '68b). When eggs are damaged, the vitelline space becomes cloudy with egg contents exuding through the wound made on the egg surface. Such eggs should be discarded even if the wound heals and eggs appear healthy, because the outer surface of the vitelline membrane of these eggs is liable to be covered with fungi within a few days, which more or less affect the development of other eggs in the same dish.

Abnormality. When eggs are freed from the cocoon, some eggs which develop abnormally are found. The abnormality most frequently emerges at the time of the first cleavage: the cleavage furrow is perpendicular to the egg axis and formed near the animal pole; therefore, it looks as if a large "polar body" is formed. To avoid such abnormal cleavage as much as possible, eggs should not only be kept at temperatures below 20°C, but also be allowed to remain still during the time of formation of the mitotic apparatus for the first cleavage (see below).

"Unfertilized" eggs. At present, "unfertilized" eggs cannot be taken out of the ovisac without activation. A part (segments 14–16) of the mature worm is removed in the culture medium; after a thorough wash in the medium,

TABLE I. Composition of Culture Medium for *Tubifex* Embryos

	Grams per liter				
Solution[a]	NaCl	KCl	$CaCl_2$	$MgSO_4$	(Addition)
Lehmann's[b]	0.09	0.03	1.44	0.06	—
Inase's[c]	0.28	0.09	0.89	0.24	—
CMF[d]	0.93	0.29	—	—	(5 mM EDTA)

[a]Final adjustment to pH 7.4 with Tris-HCl.
[b]For European species (Lehmann, '48).
[c]For Japanese species (Inase, '60b).
[d]Calcium-magnesium free solution (Shimizu, '78a).

eggs with the spindle of the first meiosis are taken out of the ovisac. These ovisac eggs (see Fig. 1A), after exposure to the exterior, continue development at least up to the early gastrula stage (Shimizu, unpublished). On the other hand, oocytes with germinal vesicles do not show any sign of activation; they degenerate within an hour. It is not known whether spermatozoa are involved in this activation.

Fixation of eggs for light and electron microscopy. For light microscopy, Zenker's fluid is recommended as fixative. Bouin's fluid may also be used for embryos older than the gastrula stage; however, it cannot be used for uncleaved eggs, because the endoplasm comprising yolk granules is not well preserved. For electron microscopy, eggs and embryos are fixed in 5% glutaraldehyde (6 hours) and postfixed in 1% osmium tetroxide (2 hours); fixatives are buffered with 0.05 M cacodylate-HCl (pH 7.4) (for details of procedures, see Shimizu, '81b).

NORMAL DEVELOPMENT

Early cleavages, teloblastogenesis, and germ band formation in *Tubifex tubifex* have been thoroughly described by Penners ('22, '24a). Later, Inase ('67) studied the cell lineage in *T. hattai*, and showed that there is no difference in the developmental pattern between these two species. Figure 3 depicts representative embryos of developmental stages compiled in Table II.

Fertilization (Stage 1a)

The entrance of the spermatozoon into the *Tubifex* egg occurs within the cocoon during or shortly after its deposition. Fertilizing spermatozoa are discharged from the spermatheca into the cocoon when the cocoon passes over the opening of the spermatheca at the time of cocoon deposition (see Fig. 2D,E; Hirao, '68).

The fertilization cone is invariably formed on the vegetal hemisphere, especially near the vegetal pole (Hirao, '68). The cone is resorbed into the egg within 30 minutes. For a short time after the regression of the cone, the spermatozoon can be recognized beneath the egg surface near the vegetal pole (Fig. 4A,B).

Fig. 3. Diagrammatic illustration of developmental stages in *Tubifex*. Stage numbers (1–18) correspond to those in Table II. Stages 1a–12c, animal pole view; stages 13–15, side and ventral view; stages 16–18, side view. N, O, P, and Q denote ectoblasts; NOPQ, OPQ, and OP denote precursor cells of ectoblasts. Mesoblasts (M) are located posteriorly behind ectoblasts (stage 12). Anteroposterior (Ar–Pr) and dorsoventral (Dl–Vl) axes are indicated for stages 13–15. Double arrowheads (stages 16–18) point to the anterior end of the embryo. E, endodermal cell; GB, ectodermal germ band; PB, Polar body; PP, pole plasm; S, seta; Sg, segment.

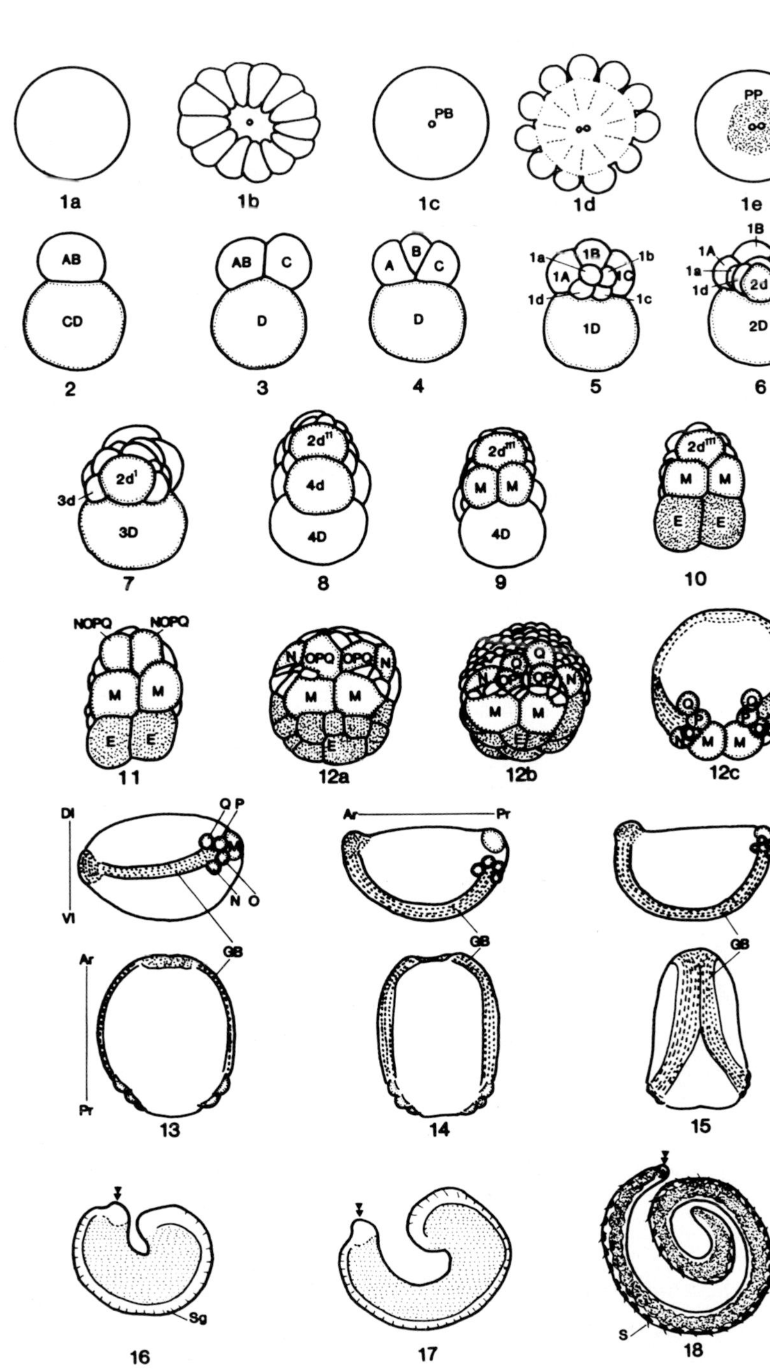
1a
1b
PB
1c
1d
PP
1e
AB
CD
2
AB
C
D
3
A
B
C
D
4
1a
1B
1b
1A
1C
1d
1c
1D
5
1B
1A
1C
1a
1b
1d
2d
1c
2D
6
2d
3d
3D
7
2d
4d
4D
8
2d
M
M
4D
9
2d
M
M
E
E
10
NOPQ
NOPQ
M
M
E
E
11
N
OPQ
OPQ
N
M
M
E
12a
Q
N
N
M
M
12b
Q
Q
P
P
N
N
M
M
12c
Dl
Q P
N O
Vl
GB
Ar
Pr
GB
GB
SIDE
Ar
GB
Pr
13
GB
14
GB
15
VENTRAL
Sg
16
17
S
18

TABLE II. Developmental Stages of *Tubiflex hattai* Embryos at 18°C[a]

Stage	Description	Hours
1	Uncleaved egg	
	a) Metaphase I (fertilization)	0
	b) First polar body formation	2
	c) Metaphase II	3
	d) Second polar body formation	4
	e) Pole plasm accumulation	6
2	First cleavage (formation of cells AB and CD); two cells	9.5
3	Second cleavage (formation of cells C and D); three cells	14
4	Second cleavage (formation of cells A and B); four cells	15
5	Third cleavage (formation of first quartet of micromeres)	
	a) Formation of cells 1d and 1c; six cells	17
	b) Formation of cells 1a and 1b; eight cells	18
6	Formation of 2d-cell (fourth cleavage); nine cells	20.5
7	Formation of 3d-cell (fifth cleavage); 17 cells	24
8	Formation of 4d-cell (sixth cleavage); 22 cells	28.5
9	Formation of mesoblast pair	33
10	Division of 4D-cell	34
11	Formation of NOPQ-cell pair	36
12	Formation of ectoblast pairs	
	a) N-teloblast pair	48
	b) Q-teloblast pair	64
	c) O- and P-teloblast pairs	72
13	Initial elongation of germ bands (early gastrula)	96
14	Completion of epiboly (mid gastrula)	120
15	Coalescence of germ bands (late gastrula)	
	a) Anterior end of the embryo	144
	b) Two thirds of the embryo length	168
16	Appearance of peristaltic constriction	192
17	Elongation of the body	240
18	Completion of a vermicular form	288
19	Hatching out	360

[a]This schedule is applicable to the European species (see Woker, '44).

Surface changes. Following sperm penetration, the vitelline membrane is separated from the egg surface. The vitelline membrane elevation starts at the site of sperm penetration (Hirao, '68). The separation of the vitelline membrane from the egg results from severance of each microvillus from the egg surface (Fig. 4C,D; Shimizu, '76b). The microvilli are first constricted near their bases and then severed from the egg (Fig. 4D). The tips of villi become thinner and are embedded in the vitelline membrane.

Polar Body Formation and Deformation Movement (Stages 1b–1d)

The meiotic apparatus of the first meiosis is located away from the egg surface at the time of oviposition. It then moves toward the surface and

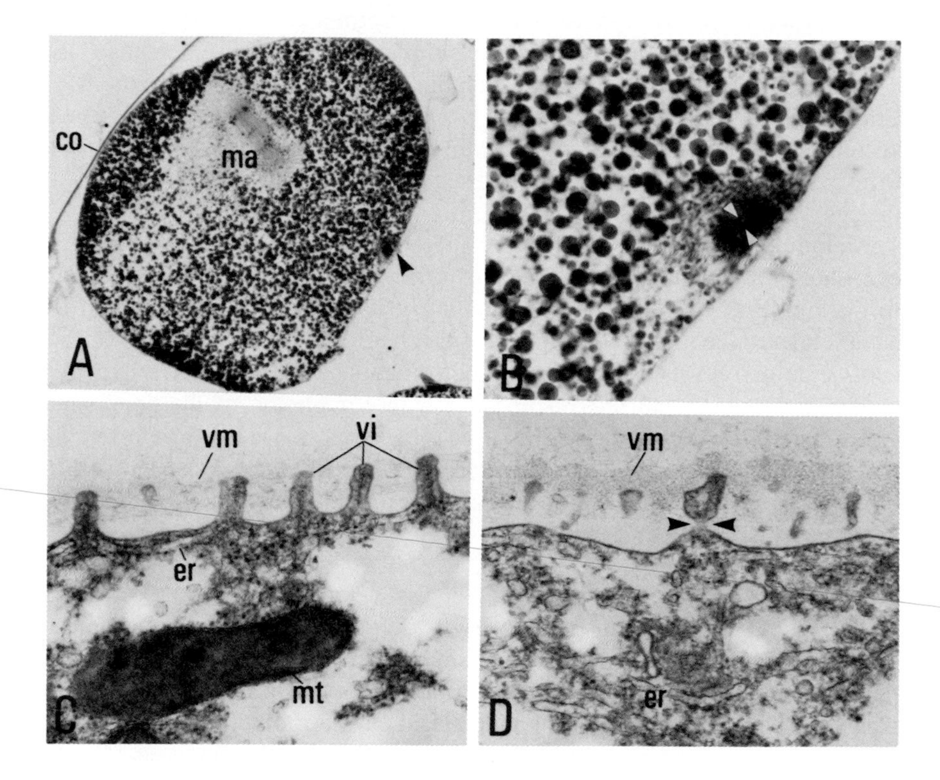

Fig. 4. A. Photomicrograph of an egg about 30 minutes after oviposition, showing a site of sperm penetration (arrowhead). co, cocoon membrane; ma, meiotic apparatus. × 125. B. Higher magnification of A. Note granular ooplasm surrounding sperm nucleus (arrowheads). × 250. C,D. Transmission electron microscope (TEM) micrographs showing the surface of eggs before (C) and after (D) fertilization. Note constriction (arrowheads in D) of a microvillus. er, endoplasmic reticulum; mt, mitochondria; vi, microvilli; vm, vitelline membrane. × 30,000.

attaches to the egg cortex. This attachment site is easily recognized under the dissecting microscope; it appears as a bright spot resulting from exclusion of yolk granules by the peripheral aster (Fig. 1C). This spot is the first landmark of the animal pole of the *Tubifex* egg.

At 15–20 minutes following the appearance of the animal pole spot, fertilized eggs, which have been flaccid, become round with the increase in surface rigidity. After another 10 minutes, a small bulge is formed in the bright spot. This bulge is subsequently given off and becomes the first polar body. Shortly after the extrusion of the first polar body, the second meiotic apparatus is formed at the site where the first polar body was given off. The meiotic apparatus is tethered to the surface by the microfilamentous cortical layer, which is localized at the animal pole (Shimizu, '81b). Eggs remain at metaphase of the second meiosis for 90 minutes and then enter anaphase. The second polar body is extruded within 70–80 minutes (Shimizu, '81c). Chromosomes left behind in the egg proper are gradually decondensed and become karyomeres with some nucleoli. Karyomeres fuse to one another to form a female pronucleus, which subsequently moves toward the center of the egg.

Meiotic phases in the *Tubifex* egg can be easily recognized, because the egg shows a dynamic shape change during the time of the polar body formation (Fig. 5). Meridionally running grooves are formed first on the equatorial surface then on the animal hemisphere (Shimizu, '79). This surface contractile activity is designated the deformation movement.[1] Grooves disappear and eggs round up concurrently with completion of the polar body extrusion. In relation to the deformation movement, Hess ('59) reported that unsaturated lipids are concentrated in protuberances formed by the movement (see Fig. 5). The mechanism of the deformation movement will be discussed later.

Pole Plasm Accumulation (Stage 1e)

Following the cessation of the second deformation movement, ooplasm devoid of yolk granules gradually accumulates toward the animal and vegetal pole (Fig. 6A; Shimizu, '76a). This ooplasmic mass has been called "pole plasm." The accumulation of the pole plasm lasts for about 90 minutes, i.e., during the time of karyogamy and formation of the mitotic apparatus for the first cleavage. The pole plasms show high activities of mitochondrial enzymes (Lehmann, '41a, '48; Carrano and Palazzo, '55); in fact, as Figure 6B depicts, the pole plasms consist of mitochondria and endoplasmic reticulum (Lehmann and Wahli, '54; Weber, '56, '58; Lehmann and Mancuso, '58; Lehmann and Henzen, '63; Henzen, '66). Inase ('68a) observed that the pole plasms are positively stained by Feulgen's stain, suggesting the presence of deoxyribonucleic acid (DNA) in the pole plasms. The mechanism underlying this ooplasmic segregation is discussed later.

First Cleavage (Stage 2)

The mitotic apparatus for the first cleavage is formed in the center of the egg about 7 hours after fertilization. According to Huber ('46), during its formation the apparatus orients itself parallel to the egg axis; however, during

[1]This surface activity has been called variously: "amöboid Bewegung" (Penners, '22), "Protuberanzenbildung" (Lehmann, '56), and "deformation movement" (Matsumoto and Kusa, '66; Shimizu, '75).

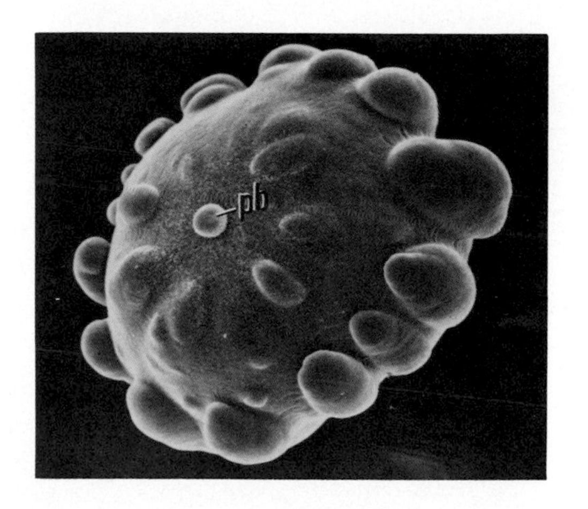

Fig. 5. Scanning electron microscope (SEM) micrograph of an egg undergoing the second deformation movement (stage 1d). Note protuberances on the animal hemisphere and the equatorial region. pb, the second polar body at the animal pole. × 120.

anaphase, it rotates to orient itself perpendicularly to the egg axis. Asters at both poles of the spindle are extremely different in size (Penners, '22).

At the inception of the first cleavage, the egg surface, especially at the equatorial region, shows a slight distortion. Later, some meridionally running grooves are formed. Within an hour, grooves other than the cleavage furrows disappear; the cleavage furrows become evident. The first cleavage is unequal and meridional. The egg divides into smaller AB-cell and larger CD-cell. Both pole plasms are segregated in CD-cell (Fig. 7).

Second Cleavage (Stages 3–4)

The mitotic apparatus is first formed in CD-cell then in AB-cell (Penners, '22; Woker, '44). Unlike the apparatus of the first cleavage, this apparatus possesses asters of approximately similar size. The division also occurs precociously in CD-cell. Initially, CD-cell forms a protuberance on the right side of AB-cell when viewed from the animal pole. The cleavage furrow is formed around the proximal region of this protuberance. CD-cell divides into smaller C-cell and larger D-cell. About one hour later, AB-cell divides into A- and B-cell; usually A-cell is slightly larger than B-cell. The cleavage planes are meridional. The pole plasms are segregated into D-cell (Fig. 7) and sink deep in the interior of the cell (Penners, '22).

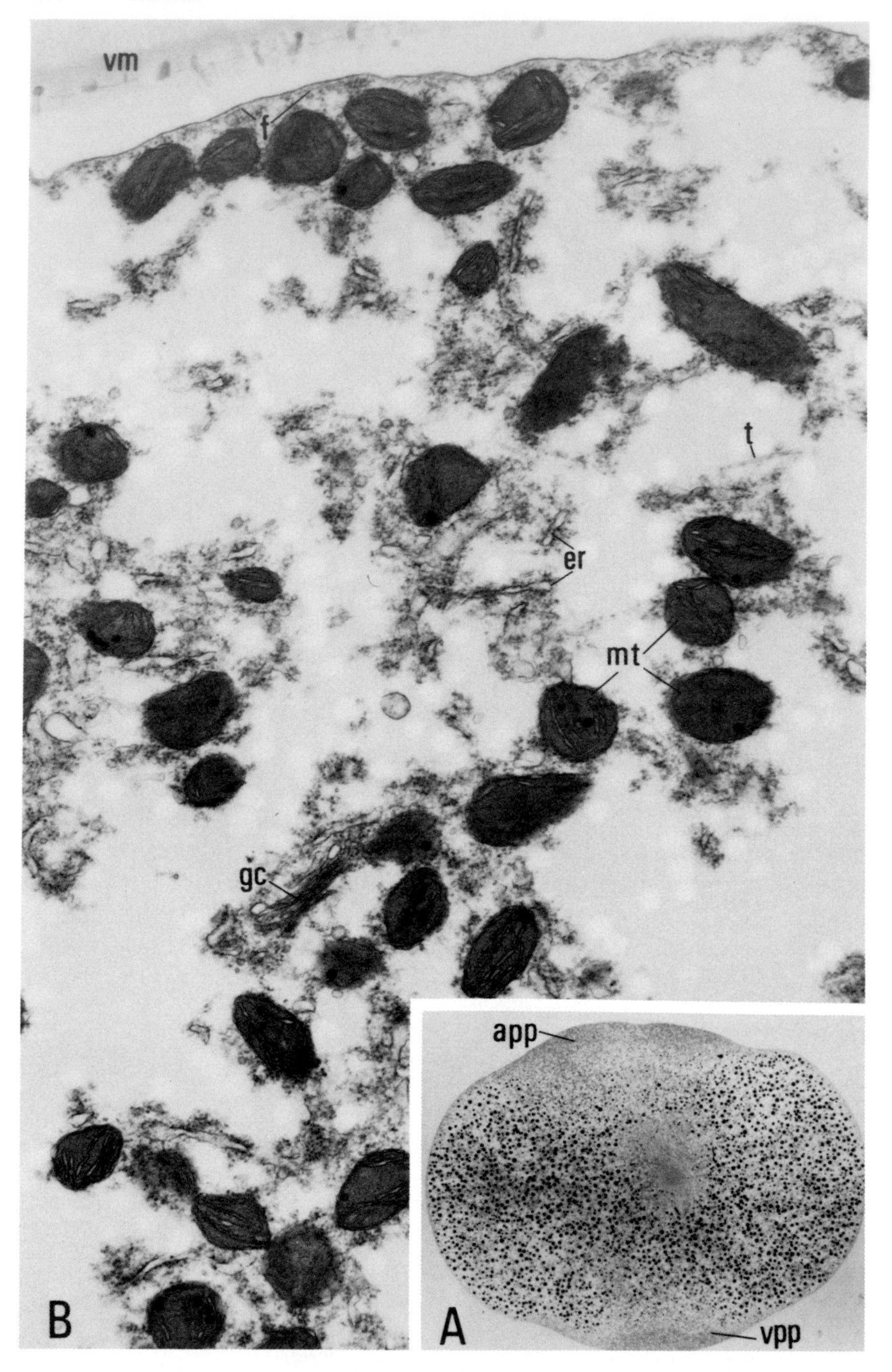

Fig. 6. A. An egg with the animal (app) and vegetal (vpp) pole plasm (stage 1e). × 100. B. TEM micrograph of the animal pole plasm. er, endoplasmic reticulum; f, microfilament; gc, Golgi complex; mt, mitochondria; t, microtubule; vm, vitelline membrane. × 20,000.

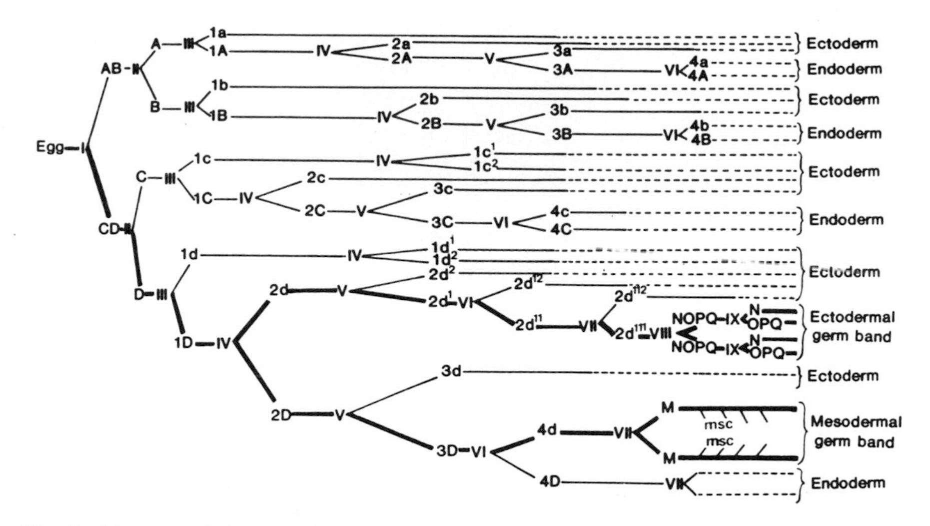

Fig. 7. Diagram of the cell lineage of *Tubifex*. Roman numerals denote the number of cleavages. The designations represent names of individual cells (see Fig. 3); msc stands for mesodermal stem cell produced by the mesoblasts (M). The traces of the pole plasm segregation are indicated by the heavy lines. Details of cell division cycles are omitted in the portions indicated by dashed lines. (Redrawn from Penners, '22.)

Third Cleavage (Stage 5)

The asynchrony of the division cycle established at the second cleavage is retained in the third one: C- and D-cell enter the mitosis about one hour earlier than A- and B-cell. C- and D-cell each form a small protuberance on the animal pole side (Fig. 8A); then the cleavage furrow is formed around the proximal region of the protuberance giving rise to highly unequal division that produces macromeres 1C and 1D and micromeres 1c and 1d. Since the cleavage plane is rather equatorial, 1c- and 1d-cell straddle 1C- and 1D-cell and 1D- and A-cell, respectively, on the animal pole side. Similarly, micromeres 1a and 1b are separated from A- and B-cell, respectively. The pole plasms remain in 1D-cell.

Fourth Cleavage: Formation of the Precursor Cell of Ectoblasts (Stage 6)

The fourth cleavage begins with 1D-cell dividing into larger 2D-cell and smaller 2d-cell. Following this, the division of other macromeres takes place sequentially, rather than concomitantly, with 1C-cell dividing first, 1A-cell dividing next, and 1B-cell dividing finally: 1A-, 1B-, and 1C-cell separate off second micromeres 2a, 2b, and 2c, respectively (Fig. 7).

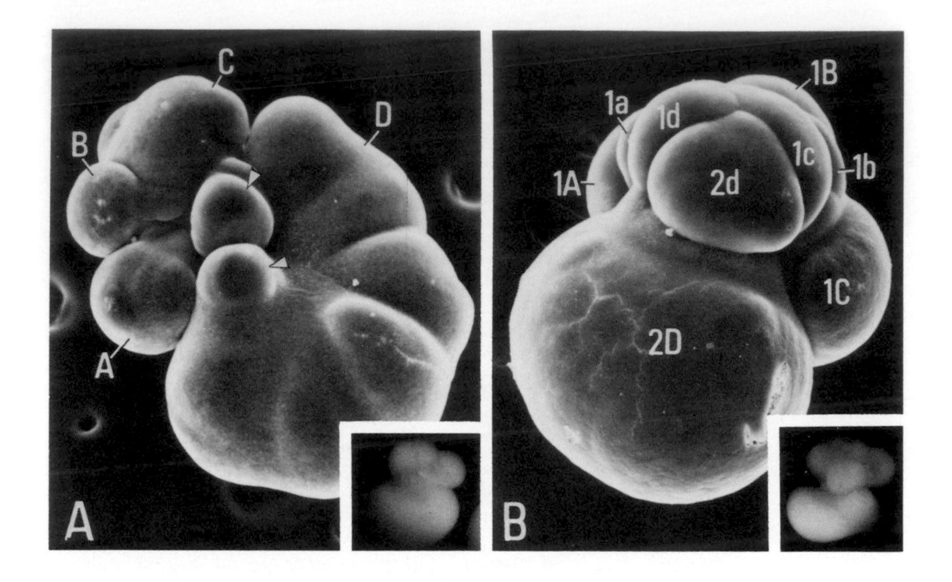

Fig. 8. A. SEM micrograph of an embryo at stage 5. Arrowheads point to protuberances of C- and D-cell, which will become 1c- and 1d-cell, respectively. × 90. Inset, the live embryo 20 minutes before fixation for SEM. × 20. B. SEM micrograph of an embryo at stage 6. Note an arc of micromeres around 2d-cell. × 90. Inset, the live embryo 20 minutes before fixation for SEM. × 20.

The 2d-cell is much larger than any other cell of the first and second quartet of micromeres. At the beginning of the division of 1D-cell, a large protuberance is formed toward the first quartet of micromeres, so that these micromeres are arranged as an arc around the distal surface of the protuberance, which later becomes 2d-cell (Fig. 8B). The pole plasms are shared between 2d- and 2D-cell (Fig. 7). A bulk of yolk granules is left behind in 2D-cell; 2d-cell is a precursor cell of ectoblasts (ectodermal teloblasts).

An asynchrony in division timing becomes more pronounced in the subsequent cleavage divisions (Fig. 7). Since a tendency to precocious division of D-quadrant cells is maintained in the subsequent divisions, it is convenient to characterize the following developmental stages by the cleavage divisions and differentiation in these cells.

Sixth Cleavage: Formation of the Precursor Cell of Mesoblasts (Stage 8)

About 4 hours after formation of 2d-cell, 2D-cell divides into larger 3D-cell and smaller 3d-cell (fifth cleavage, stage 7). The latter lies on the left side of $2d^1$-cell, which is derived from the division of 2d-cell. Meanwhile,

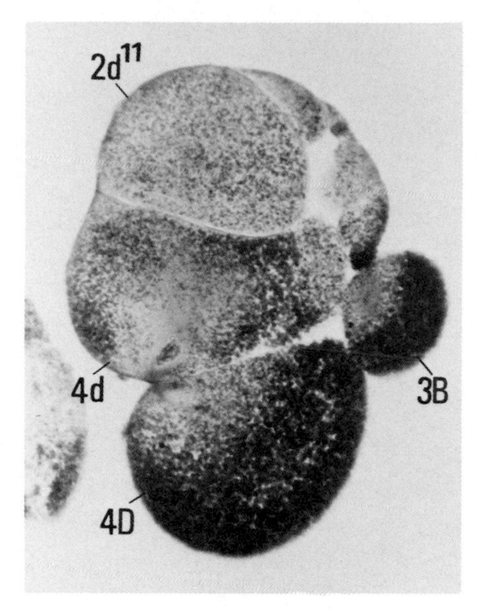

Fig. 9. Section of an embryo at stage 8 through the midline. × 100.

$2d^1$-cell divides into larger $2d^{11}$-cell and smaller $2d^{12}$-cell; the pole plasms remaining in $2d^1$-cell are segregated into $2d^{11}$-cell (Fig. 7). After another 4½ hours, 3D-cell divides into 4D-cell and 4d-cell(stage 8); 4d-cell is a little larger than $2d^{11}$-cell (Fig. 9). At this developmental stage, $2d^{11}$-, 4d-, and 4D-cell all come to lie in the future midline of the embryo.

The pole plasms remaining in 3D-cell are finally segregated into 4d-cell (Fig. 7). On the other hand, 4D-cell is filled with yolk granules and becomes structurally indiscernible from other yolky macromeres (Fig. 9); 4d-cell is a precursor cell of mesoblasts (mesodermal teloblasts).

Teloblastogenesis (Stages 9–12)

Mesoblasts (stage 9). About 4½ hours after the formation of 4d-cell, it is divided bilaterally into a pair of cells. These two cells are the mesoblasts, each of which subsequently produces mesodermal stem cells (msc) arranged in a row (the mesodermal germ band [MGB]) on either side of the embryo). The cleavage plane is the first landmark of the midline of the embryo; the position of these cells is nearly symmetric with respect to the midline (Fig. 10A).

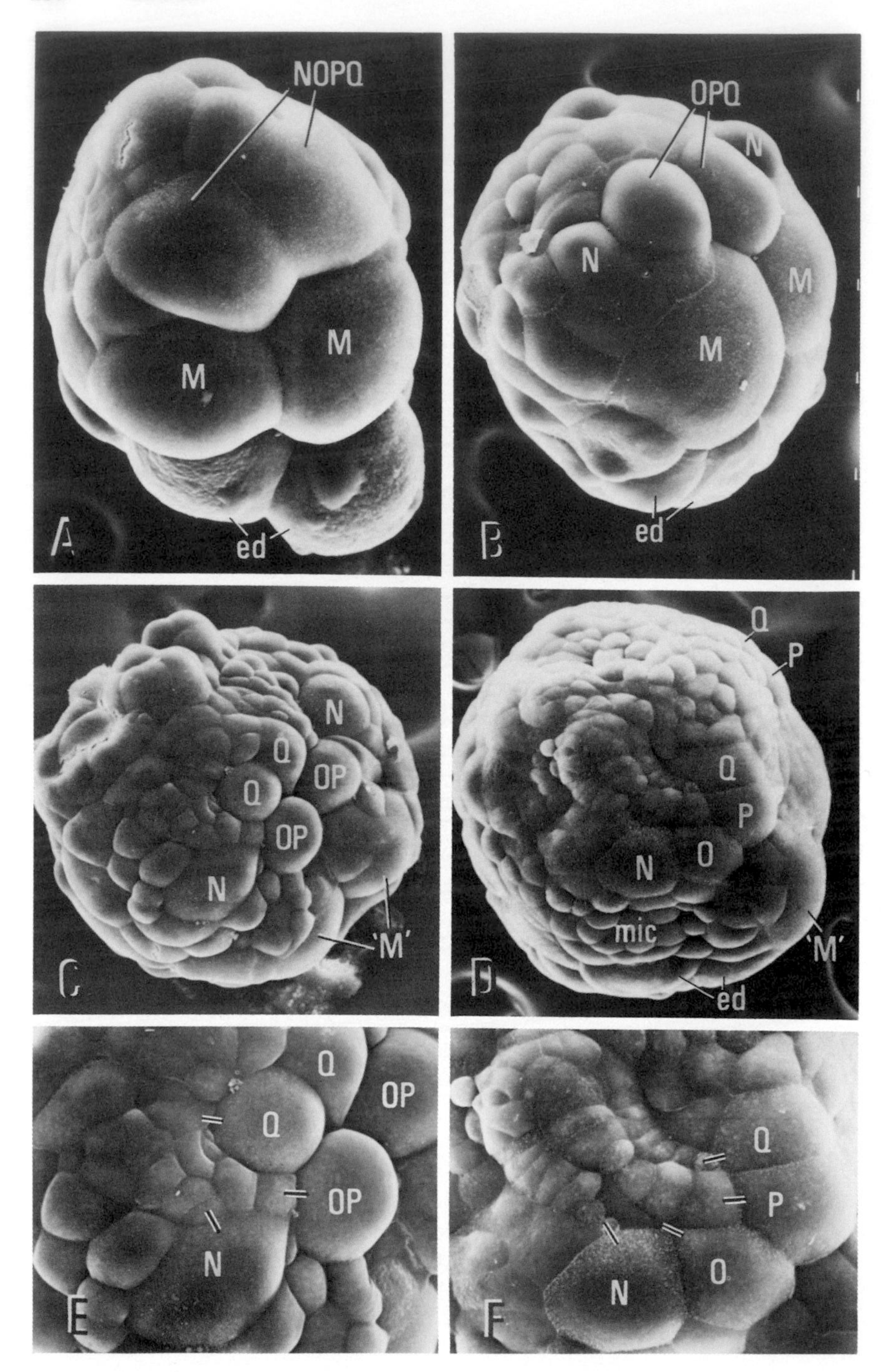
NOPQ
M
M
ed
A
OPQ
N
N
M
M
ed
B
N
Q
OP
Q
OP
N
'M'
C
Q
P
Q
P
N
O
mic
'M'
ed
D
Q
OP
Q
OP
N
E
Q
P
N
O
F

Ectoblasts (stages 10–12). During the time of division of 4d-cell, $2d^{11}$-cell divides into larger $2d^{111}$-cell and smaller $2d^{112}$-cell (Fig. 7). About 3 hours later, $2d^{111}$-cell divides bilaterally into two cells of equal size, each of which is designated NOPQ-cell in this chapter[2] (Fig. 10A).

In the next division, NOPQ-cell divides into a pair of cells, smaller *N-teloblast* and OPQ-cell (Fig. 10B). In the subsequent divisions, OPQ-cell cuts off a couple of cells behind, which straddle the mesoblasts (Penners, '22); the N-teloblast begins to bud off a row of small cells, which becomes *n-bandlet* of the ectodermal germ band (EGB). After OPQ-cell produces 4–5 small cells in a row forward, it divides into *Q-teloblast* and the OP-cell. The former is slightly smaller than and lies in front of the latter (Fig. 10C). Meanwhile, two groups of cells are separated from each other by the backward spreading of micromeres, and gradually shift from the dorsolateral to the lateral position on either side of the embryo (see Fig. 10D). Q-teloblast and OP-cell both cut off small cells in a row forward. Whereas the row from OP-cell is added to that previously formed by OPQ-cell, Q-teloblast forms a new row of small cells (*q-bandlet*). Therefore, at this developmental stage, three contiguous rows of small cells are observed on either side of the embryo (Fig. 10C,E). After OP-cell repeats the production of small cells five or six times, it divides equally into a pair of cells, *O-teloblast* and *P-teloblast*; the teloblastogenesis reaches completion. The O-teloblast lies behind the N-teloblast, and the P-teloblast behind Q-teloblast (Fig. 10D). The P-teloblast adds small cells to the row of small cells previously formed by the OPQ-cell and the OP-cell; this row is designated *p-bandlet*. On the other hand, the O-teloblast forms a new row of cells (*o-bandlet*) between the n- and p-bandlet (Fig. 10D,F). From this stage on, the EGB comprising four bandlets grows to full scale on both sides of the embryo.

Gastrulation: Germ Band Formation and Epiboly (Stages 13–15)

By the time the formation of ectoblasts reaches completion, A-, B- and C-quadrant cells have terminated the production of micromeres and already

[2]A new notational system is used to designate ectoblasts and their precursor cells, because that introduced by Penners ('22, '24a) is not only troublesome but also liable to cause misunderstanding of the fate of cells.

Fig. 10. A–D. SEM micrographs illustrating the sequence of teloblastogenesis. A, stage 11; B, stage 12a; C, stage 12b; D, stage 12c. Ectoblasts are designated N, O, P, and Q; NOPQ, and OP denote precursor cells of ectoblasts. M, mesoblast. Mesoblasts in C and D are covered with epithelium and indicated by 'M.' ed, endoderm; mic, cells derived from micromeres. × 120. E,F. Higher magnification of C and D, showing details of initiation of ectodermal germ band formation. Bars indicate relations between ectoblasts and their offsprings. × 210.

have begun to divide equally to form a endodermal germ region at the ventral side. The micromeres have divided repeatedly and spread posteriorly over the dorsal surface of the embryo to cover the mesoblasts and posteriorly located endodermal cells. Furthermore, micromere-derived cells are also found as a narrow ventrolateral band running along the EGB on either side of the embryo (Fig. 10D). Thus, shortly after the onset of formation of the EGB, three domains exposed to the exterior of the embryo are discernible: 1) short EGBs, 2) the dorsal and ventrolateral epithelial sheets derived from micromeres, and 3) the ventral yolky endoderm (presumptive midgut). On the other hand, the MGB formed by the mesoblasts lies between the EGB and the endoderm (Fig. 13A). Whereas the ectoblasts move outward from the dorsal midline, the mesoblasts remain dorsal as a contiguous pair.

Besides the elongation of the EGB and MGB on either side of the embryo, the gastrulation movement in the *Tubifex* embryo consists of two events: 1) shifting of the germ bands toward the ventral midline and final coalescence there, and 2) concurrent epibolic expansion of the epithelial sheet over large endodermal cells. The coalescence of the EGBs and MGBs is first initiated at the anterior end of the embryo at day 6 (stage 15), and progresses in an anteroposterior succession. The ventral shift of the germ bands is accompanied by changes in shape of the embryo (Fig. 11A,C; also see Fig. 3).

Concurrently with the embryo's change of shape, the dorsal epithelial sheet becomes more attenuated and increases its surface area covering the dorsal and lateral surface of the embryo. The narrow ventrolateral band of epithelial sheet on either side of the embryo (see Fig. 10D) "moves" toward the ventral midline accompanying the ventral shift of the EGBs; then they converge to cover the ventral endoderm (Fig. 11B). At the same time, the ventral epithelial sheet spreads posteriorly. As early as day 5 (stage 14), the epithelium appears to cover the whole embryo except for the EGBs and ectodermal teloblasts (Fig. 11B). Along with the coalescence of the germ bands, the epithelial cells in the ventral sheet rearrange their position and spread over the EGBs (Fig. 11D).

Later Development: Elongation and Spiralling of Embryo (Stages 16–19)

From the day 7 onward, the embryo gradually becomes elongated and curved with the ventral convexity (Fig. 12). It forms a spiral of three or four turns in the vitelline membrane. The juveniles finally hatch out passing through the cylindrical process at each end of the cocoon (see Fig. 1C). The time of hatching varies from day 15 up to day 20. Developmental stages before hatching are characterized in the live embryos as follows (at 18°C).

Eight-day. For the first time, peristaltic constrictions are observed on the anterior part of the body.

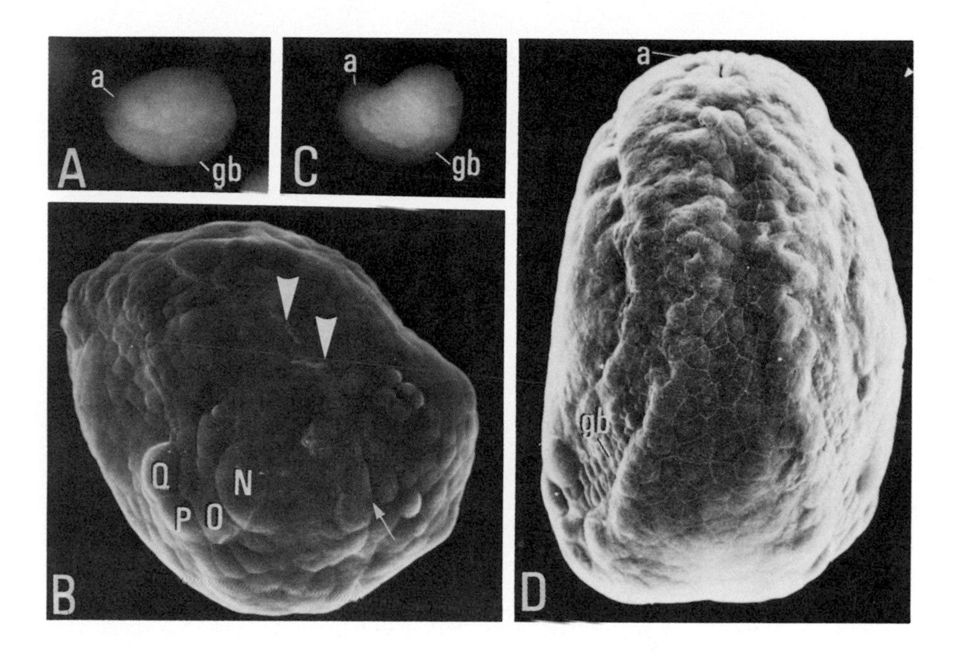

Fig. 11. Photo (A,C) and SEM (B,D) micrographs of embryos at stages 14 (A,B) and 15b (C,D). A. Side view of a live embryo. × 20. B. Posterior view. The posterior end of the embryo is indicated by arrow. Arrowheads point to the ventral midline. Note cell rows of the ectodermal germ band emanating from the ectoblasts (N, O, P, Q). × 120. C. Side view of a live embryo. × 20. D. Ventral view. The anterior half of the germ band (gb) is covered with epithelium. × 120. a, anterior end; gb, ectodermal germ band.

Nine-day. The embryo sometimes crooks at the anterior part of the body. Blood streaming is seen in the dorsal vessel running through the anterior 6–8 segments. Ciliary movement is detected in the nephridia in segments 7 and 8.

Ten-day. Although the posterior end of the body is still curved, it is much reduced in thickness. Blood is seen to be reddish under the dissecting microscope. The embryo possesses short uncinate setae, which are detected "in the body" at high magnification (× 200); however, they are not seen under the dissecting microscope.

Eleven-day. The posterior end of the body becomes slender and shows nearly the same thickness as the anterior end. However, it appears to be slightly curved and, unlike the anterior end, cannot yet move by itself. The setae expose about one fifth of the whole length to the exterior.

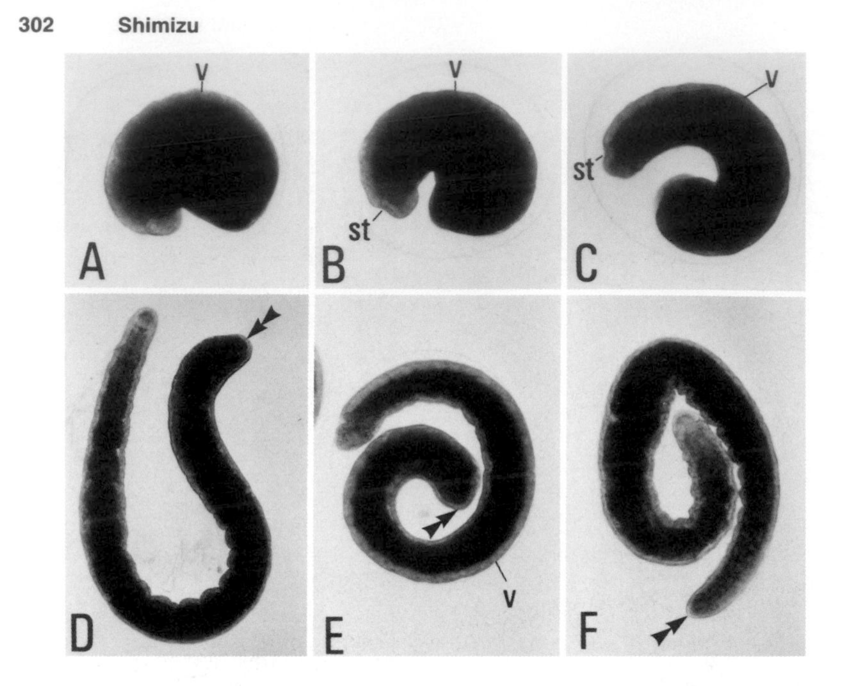

Fig. 12. Live embryos at day 7½ (A), 8 (B), 9 (C), 10 (D), 11 (E), and 12 (F). A,B,C,E. side view. D,F, dorsal view. Double arrowheads indicate the posterior end of the embryos. The light does not pass through the body of the embryo except for the epithelial cell layers. st, stomodaeum; v, ventral region. × 30.

Twelve-day. The posterior end of the embryo moves by itself and its curvature has disappeared. Rudiments of the gonad are recognizable in segments 10 and 11 when the embryo is slightly compressed.

ORGANOGENESIS

Differentiation of Mesodermal Germ Bands (MGBs)

Somite development. On both sides of the embryo, each somite is derived from one single mesodermal stem cell produced by the mesoblast (Penners, '24a; Meyer, '29). The cell undergoes a series of equal divisions; the resulting cells become aggregated into a block (somite) (Fig. 13A). Large cells in each somite produce small cells on its surface.

Along with further proliferation of the inner large cells, the coelomic cavity of the somite emerges (Fig. 13B). The first indication of the coelomic cavity is detected in the most anterior somite at day 5 (ca. 24-somite stage). Then, a pair of somites on both sides of a segment gradually extends upwards

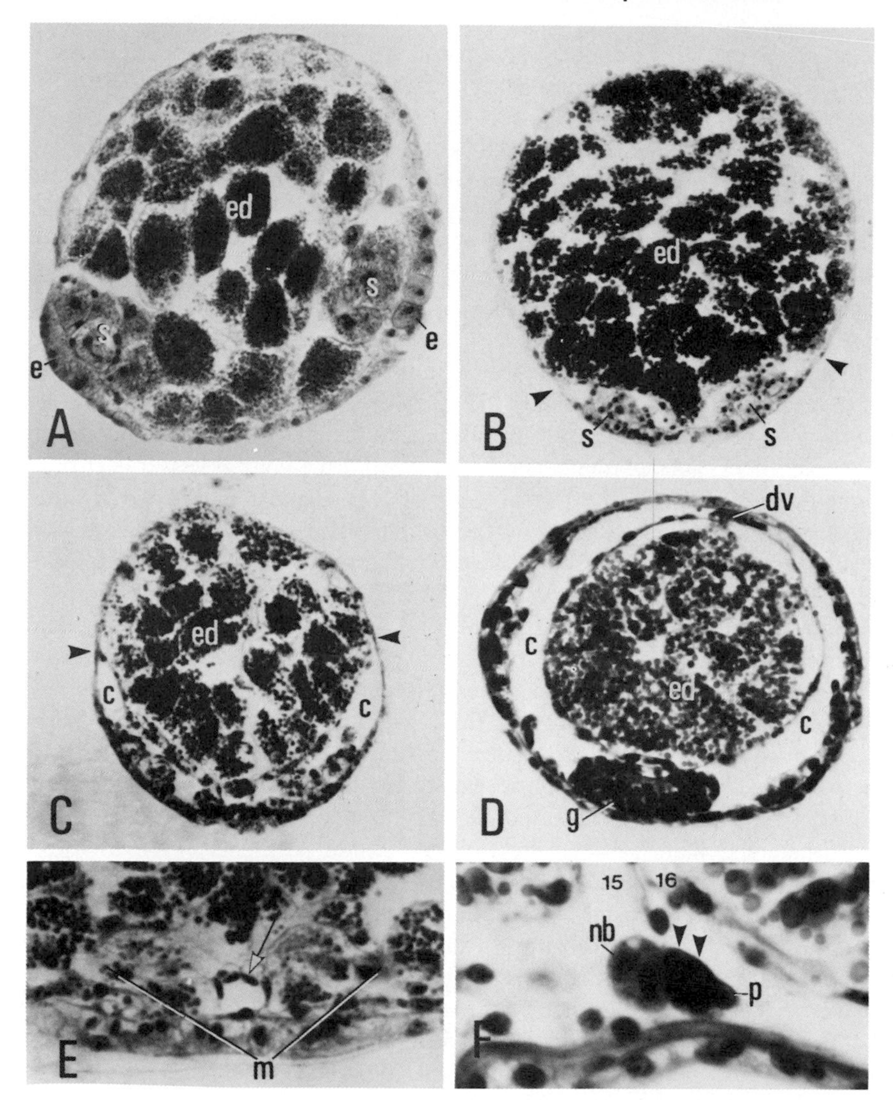

Fig. 13. Differentiation of mesoderm. A–D. Cross-sections through the mid-region of the embryos, showing the formation of the coelomic cavity (c) in the somites (s). A. stage 14. B. stage 15b. C. stage 16. D. stage 17. Arrowheads in B and C indicate the upper margin of the coelomic cavity. dv, dorsal vessel; e, ectodermal germ band; ed, endoderm; g, ganglion. × 150. E. Rudiment of ventral vessel (arrow) in an embryo at stage 16. m, mesoderm. × 400. F. Initial phase of nephridium formation in segment 16 in an embryo at stage 19. Two cells (arrowheads) have been produced by the nephridioblast (nb) located at the posterior end of segment 15. p, peritoneum. × 800.

and at day 9 encloses the endoderm forming a closed mantle of mesoderm (Fig. 13C,D). When the somites from both sides meet, their contiguous walls disappear and a continuous coelom is formed.

Coelomic walls. The coelomic walls of the somite are differentiated in the usual way into splanchnopleura, somatopleura, or septa, corresponding to their positions. The longitudinal muscle layer is differentiated from the somatic mesoderm, which lies under the ectodermal cell layer derived from the EGBs and micromeres (see below). The splanchnic mesodermal wall wraps the endoderm and is differentiated into the muscle layer and chloragogen cells.

Blood vessels. The coelomic walls definitely contribute to formation of blood vessels (Meyer, '16; Penners, '24a; Meyer, '29). The rudiments of the main vessels (dorsal and ventral) are spaces left behind between the endoderm and somite(later, splanchnopleura). The space for the ventral vessel is at first a longitudinal channel; the dorsal wall of this channel is formed by the endoderm and the ventrolateral by the somites of each side of the embryo. The channel is later closed off by the approximation of the lateral mesodermal walls (Fig. 13E). The dorsal vessel (Fig. 13D) is formed by mesodermal cells in a similar way to the ventral vessel. The perivisceral vessels by which the lumina of the dorsal and ventral vessels communicate arise by separation of the apposed walls of the intersegmental septa (Meyer, '16).

Nephridium (segmental organ). In the body of *Tubifex*, the nephridia are present in most segments, except for the first six and the genital segments (segments 9–12). Although nephridium formation in the somites of anterior segments 7 and 8 can be detected as early as day 7, it is not until shortly before hatching that segments behind the genital segments begin to form nephridia (Meyer, '29).

In the posterior segments, the nephridioblast is present on the posterior septum of each segment exposing its anterior surface to the coelomic cavity (Fig. 14A). The nephridioblast is distinguished from adjacent cells by its lenslike shape and large nucleus (Meyer, '29). This cell buds off a row of disc-shaped cells centrifugally (Figs. 13F and 14B). The row of cells is wrapped by the peritoneum and juts out into the coelomic cavity of the contiguous segment. As a result of divisions in the nephridioblast and its offspring, the row becomes longer and then the advancing margin of the row reaches the body wall (Fig. 14C). This chain of cells is a rudiment of the nephridial tube. Around this time, each cell changes its shape, becoming perforated and forming a continuous canal (Fig. 14C). Similarly, the epidermal cell that is closely associated with the forefront of the nephridial tube is also perforated and becomes a rudiment of the nephridiopore. At the same time, cilia appear in the canal. In the late stages of nephridium development, the nephridioblast divides radially into a group of four cells, which form a nephrostome.

The nephridioblast has been thought to be mesodermal in origin (Penners, '24a; Iwanoff, '28; Meyer, '29). In segment 13 and those behind it, the nephridioblast first appears in segment 13 shortly before hatching. Then, with the progress of development, similar large cells emerge gradually in more posterior segments in an anteroposterior succession. Meyer ('29) supposed that a certain cell that is located in a specific site of the septum is determined as a precursor of the nephridioblast. On the other hand, the nephridioblasts in segments 7 and 8 become recognizable in the somites only when they proliferate to some extent to form a chain of about six cells (Meyer, '29).

Differentiation of Ectodermal Germ Bands (EGBs)

Central nervous system. During gastrulation, the EGBs come to contact in the ventral midline by the n-bandlet of each side. During this process, n-bandlet cells (n-cells) undergo divisions once or twice arranging descendants into two or more longitudinal rows. These cells also divide tangentially; the superficially located cells become a part of the ordinary epidermis of the ventral body wall. The inner layer of n-cells is a rudiment of the ventral nerve cord. At day 7, those cells on each side proliferate and form a compact cell mass in each segment of the embryo (Fig. 15A). This clump of cells is the first indication of a ganglion. On the following day, the contour of the cell mass of the ganglion becomes more distinct (Fig. 15B). When the embryo develops 34–36 somites (ca. 8½ days), structures are found that longitudinally connect ganglia in the adjacent segments. Thus, the ventral nerve cord emerges as a ganglionated chain extending posteriorly from the second segment. Differentiation of ganglia advances in an anteroposterior succession.

Unlike the n-bandlets located posteriorly from segment 2, the rows of n-cells in the first segment remain apart for a short distance, leaving an interval in which a micromere-derived ectodermal cell mass is present. These bifurcated rows curve up dorsally, and meet at the most anterior portion of the

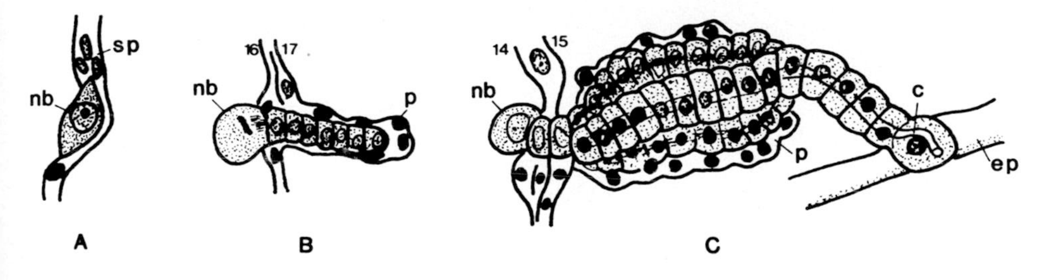

Fig. 14. Possible sequence of nephridium formation. Numerals indicate the segment number. c, canal; ep, epidermis; nb, nephridioblast; p, peritoneum; sp, septum. (Redrawn from Meyer, '29.)

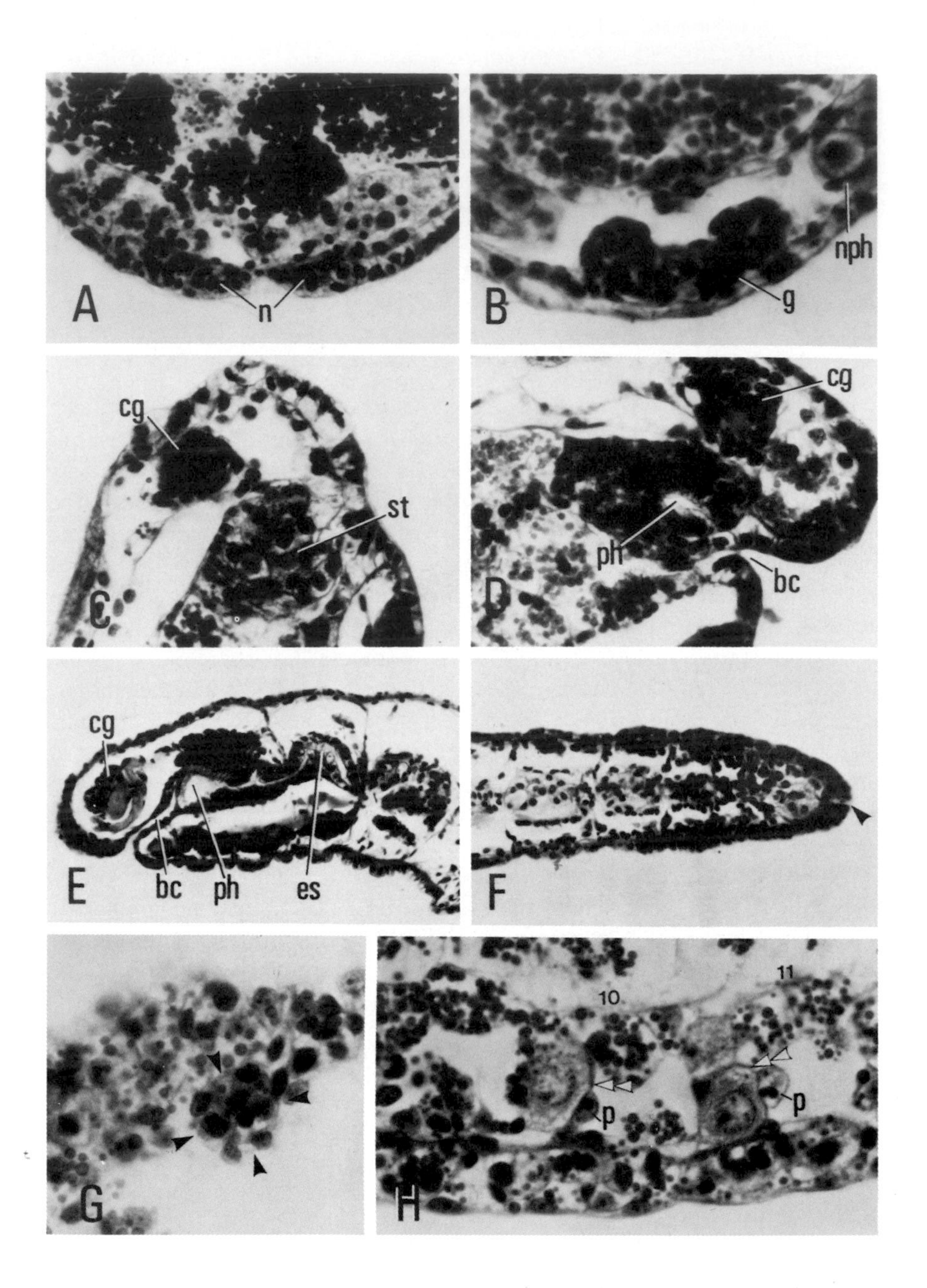
A
n
B
nph
g
cg
st
C
cg
ph
bc
D
cg
E
bc
ph
es
F
G
10
11
p
p
H

embryo. The cerebral ganglion is differentiated at this point of confluence of the neural rows (Fig. 15C,D,E).

Circular muscle layer. Cells other than n-cells in the EGBs have been thought to contribute to the formation of the circular muscle layer and a part of the epidermis. These series of cells remain each as a single row longer than the n-series. Based on the location of the circular muscle layer, Meyer ('29) suggested that o- and p-cells, but not q-cells, participate in its formation. The circular muscle layer is first observed at day 7.

Differentiation of Cells Derived From Micromeres

As described above, cells derived from the micromeres contribute to the epithelial covering of the embryo. Among organs that differentiate from the epithelial sheet, the setal sac is the only structure detectable during embryonic development. In addition to the epithelium, the stomodaeum is also derived from the micromeres (Penners, '24a).

Setal sac. At day 8, a group of 4–5 cells is found in the epidermis of both the dorsal and ventral sides of the embryo (Penners, '24a; Meyer, '29). This is the rudiment of the setal sac (Fig. 15G). Cells in each rudiment proliferate and creep into the mesoderm. It is presently unclear whether the ventral setal sac originates from micromeres or o-bandlet cells of the EGBs (Penners, '24a; Meyer, '29).

Buccal cavity and pharynx. Small cells located anteriorly in front of the EGBs divide several times to form a cell mass (stomodaeum), which later creeps inward against the endoderm (Fig. 15C). On the other hand, the buccal cavity is formed by invagination of the thickened surface layer of stomodaeal cells (Fig. 15D).

Differentiation of Endoderm

Following gastrulation, the yolky endodermal cells that fill the bulk of the interior of the embryo undergo divisions several times and gradually become smaller. At a particular point in embryonic development (ca. day 9), lumina

---

Fig. 15. A,B. Cross-section of embryos at stages 15b (A) and 17 (B). Note proliferation of neural cells (n) and formation of ganglion (g). nph, nephridium. A. × 300. B. × 600. C,D. Longitudinal sections of the anterior region of embryos at stages 17 (C) and 18 (D). Note formation of cavity in the pharynx (ph). bc, buccal cavity; cg, cerebral ganglion; st, stomodaeum. × 400. E, F. Longitudinal section of the anterior (E) and posterior (F) portion of an embryo at stage 19. Note the establishment of the anterior digestive system. Arrowhead in F points to anus. bc, buccal cavity; cg, cerebral ganglion; es, esophagus; ph, pharynx. × 200. G. Tangential section of rudiment (arrowheads) of the dorsal setal sac in an embryo at stage 16. × 600. H. Primordial germ cells (double arrowheads) in segments 10 and 11 in an embryo at stage 15b. Note large nucleus. p, peritoneum. × 600.

appear and are simultaneously surrounded by smaller cells that become arranged into a midgut epithelium (Fig. 15E). The mechanism underlying the lumen formation is not known (Penners, '24a; Iwanoff, '28).

Shortly before hatching, the esophagus first opens into the pharynx; then the continuity between the buccal cavity and pharyngeal lumen is established by a breakthrough at the base of the buccal invagination (Fig. 15E). Unlike the formation of the buccal cavity, the anus is formed not by an epidermal invagination, but by a simple perforation of the epidermis, mesoderm, and posterior end of the gut epithelium (Fig. 15F).

EXPERIMENTAL ANALYSIS OF DEVELOPMENT

Deformation Movement: A Surface Contractile Activity

Rötheli ('49, '50) found that the antimitotic reagents naphtho- and phenanthrequinone inhibit the deformation movement in the *Tubifex* egg. However, the mechanism underlying this inhibition still remains ambiguous.

Later, Shimizu ('75) reported that microfilaments are found in the cortical layer exclusively at the deforming regions, suggesting their involvement in the deformation movement as a contractile machinery. This was subsequently confirmed by experiments with cytochalasin B (50 μg/ml), which is known to prevent microfilament-related cellular phenomena. When the *Tubifex* eggs are treated with cytochalasin B, not only is the deformation movement completely inhibited, but also microfilaments disappear from the cortical layer (Shimizu, '78b). Furthermore, a groove-forming activity is elicited on the equatorial surface of the spherical eggs by Ca-ionophore A23187 (Shimizu, '78a); this activity is accompanied by the appearance of well-organized microfilaments in the cortical layer. This result indicated that the periodic appearance of surface contractile activity is correlated with changes in the intracellular Ca^{++}-level of the egg.

Although Woker ('43, '44) reported that colchicine does not affect the deformation movement, careful observations revealed that this drug inhibits the surface grooving activity on the animal hemisphere but not on the equatorial region (Shimizu, '79; see Fig. 5). As in the cleavage divisions (Lehmann, '43; Lehmann and Hadorn, '46; Brandsch and Jitariu, '71), colchicine also prevents meiotic division as a result of disruption of the meiotic apparatus. On the other hand, dinitrophenol inhibits the equatorial grooving but not the animal hemisphere grooving at the concentration (8 μg/ml) and does not disturb the meiotic events (Shimizu, '79). In normal eggs, the animal hemisphere grooving begins synchronously with the onset of meiotic telophase. These results suggest that the contractile activity of the animal hemisphere is closely related to meiotic nuclear events and that the equatorial grooving is independent of it. In this respect, Lehmann's speculation that the deformation pattern of the *Tubifex* egg is determined by the position of the meiotic apparatus (Lehmann, '46) should be reexamined, because the meiotic apparatus appears to contribute to a later portion of the deformation movement.

Pole Plasm Accumulation: An Ooplasmic Rearrangement

The pole plasm accumulation (PPA) is the most conspicuous and conceivably the important event in *Tubifex* development. As for the mechanism underlying PPA, the importance of the egg cortex has been suggested by Lehmann's centrifuge studies (for procedures, see Lehmann, '48, '56). In eggs left to themselves after centrifugation, the pole plasm material starts flowing along the surface and is redistributed in the animal and vegetal pole (Lehmann, '40, '41b). When the centrifugal force is applied to eggs that have already completed PPA, the pole plasms are hardly displaced, suggesting their firm attachment to the egg cortex (Lehmann, '38). He supposed that the PPA is accomplished through an "intracellular affinity" (or interaction) between the pole plasm material and the egg cortex at the animal and vegetal poles (Lehmann, '40, '56).

Recently, some results substantiating Lehmann's speculations on the importance of the egg cortex have been obtained by Shimizu ('82). Microfilaments are found in the cortical layer in eggs undergoing PPA and are also present in the accumulating pole plasms, linking membranous organelles. Furthermore, some filaments appear to connect membranous organelles to the filamentous cortical layer. That these microfilaments are responsible for PPA is strongly suggested by experiments with cytochalasin B: PPA is completely prevented by this drug; microfilaments are not detected in cytochalasin-treated eggs (Shimizu, '82). The actinlike nature of these filaments is also demonstrated by their decoration with heavy meromyosin from rabbit skeletal muscle (Shimizu, in preparation). Supposing the presence of bipolar myosin filaments in the *Tubifex* egg, there is a possibility that not only the egg cortex but also the pole plasm as a whole is capable of contraction. These studies show that the mechanism underlying PPA in the *Tubifex* egg can be analyzed within the context of contraction of the actomyosin system. Elucidation of the details of the mechanism awaits future studies.

Totipotency of D-cell: Significance of Pole Plasm

Penners ('24d, '25, '26, '34a, '37, '38) carried out a series of cell-ablation studies to examine the mode of development in "isolated" blastomeres. He irradiated embryos with a narrow ray of ultraviolet light and succeeded in killing given cells. The embryos were operated on in the intact cocoons and allowed to develop there.

Primordial germ cells have been thought by previous investigators to be mesodermal in origin (Iwanoff, '28; Meyer, '29, '31; Penners and Stäblein, '30). As for the site and time of their production, Meyer ('29, '31) stated that PGCs arise directly from somites in segments 10 and 11. This speculation is in part based on his impressions that cells of the embryo are arranged too compactly for PGCs to move in the embryo, and that PGCs do not seem to have the capacity for amoeboid movement. On the other hand, Penners and Stäblein ('30) reported that PGCs are precociously produced by the mesoblasts shortly before the onset of the EGB formation. According to these authors, these cells are at first located in the interstices between the endoderm and the ectoblasts at the anterior portion of the embryo; then they migrate into the somites of segments 8–12 as the embryo elongates. Finally those in genital segments 10 and 11 develop into germ cells.

Are PGCs produced directly in the somites of the genital segments, or do they migrate there from other places in the embryo? To answer this question, Penners ('34a,b) performed an interesting experiment: mesoblasts were removed by irradiation with ultraviolet light at various developmental stages on which the number of the mesodermal stem cells produced depends, so that the number of somites (and segments) could be controlled. When the embryos, which are kept alive for a long period following the operation, possess fewer than nine segments, PGC-like cells are detected in those segments located in front of the genital segments in the intact embryo. This result may be interpreted in view of Penners' speculations. However, it does not necessarily exclude Meyer's speculation, because there is a possibility that these anterior segments regulate production of PGCs in the operated-upon embryos that are defective in MGBs (cf. Penners, '37). Thus, the problem of where PGCs in the normal embryo originate seems to remain unsettled.

CONCLUDING REMARKS

In this review, I have described the developmental process of the freshwater oligochaete *Tubifex*, and pointed out some unresolved problems. As in other annelids, the most important problem in *Tubifex* development is the relationship between the localization of pole plasm and the determination of cells. At present, characterization as to the nature of certain determinants in the pole plasms is beyond our knowledge. In view of the idea that the fate of a cell depends on its cleavage history, it would be useful to determine whether the pole plasms regulate rhythms and patterns of cleavage in the *Tubifex* embryo.

LITERATURE CITED

Braidotti, P., M. Ferraguti, and T.P. Fleming (1980) Cell junctions between spermatozoa flagella within the spermatozeugmata of *Tubifex tubifex* (Annelida: Oligochaeta). J. Ultrastruct. Res. *73:*299–309.

Brandsch, R., and P. Jitariu (1971) Influence of colchicine and theophilin combined with an electromagnetic field on the first division of *Tubifex* eggs. Rev. Roum. Biol.-Zool. *16:*215–220.

Carrano, F., and F. Palazzo (1955) Localizzazione di alcuni emzimi nello sviluppo dell'uovo di Tubifex rivulorum. Riv. Biol. (Rome) *47:*193–202.

Dixon, G.C. (1915) Tubifex. In L.M.B.C. Memoirs, No. 23. London, England: Williams and Norgate, pp. 1–100.

Ferraguti, M., and G. Lanzavecchia (1971) Morphogenetic effects of microtubules. I. Spermiogenesis in annelid Tubificidae. J. Submicro. Cytol. *3:*121–137.

Fleming, T.P. (1981) The ultrastructure and histochemistry of the spermathecae of *Tubifex tubifex* (Annelida: Oligochaeta). J. Zool. (London) *193:*129–145.

Gatenby, J.B. (1916) The development of the sperm duct, oviduct and spermatheca in Tubifex rivulorum. Q. J. Microsc. Sci. *61:*317–336.

Gathy, E. (1900) Contribution à l'étude du developpement de l'oeuf et de la fecondation chez les annelides. Cellule *17:*7–62.

Henzen, M. (1966) Cytologische und mikroskopische Studien über die ooplasmatische Segregation während der Meiose de *Tubifex*-Eies. Z. Zellforsch. *71:*415–440.

Hess, O. von (1959) Phasenspezifishe Änderung im Gehalt an ungesättigten Fettsauren beim Ei von *Tubifex* während der Meiosis und der ersten Furchung. Z. Naturforsch. *14b:*342–345.

Hirao, Y. (1964) Reproductive system and oogenesis in the fresh-water oligochaete, *Tubifex hattai*. J. Fac. Sci. Hokkaido Univ. Ser. VI, Zool. *15:*439–448.

Hirao, Y. (1965a) A method for the observation of oviposition in a fresh water oligochaete, *Tubifex hattai* (in Japanese with English abstract). Zool. Mag., Tokyo *74:*283–285.

Hirao, Y. (1965b) Cocoon formation in *Tubifex* with its relation to the activity of the clitellar epithelium. J. Fac. Sci. Hokkaido Univ. Ser. VI, Zool. *15:*625–632.

Hirao, Y. (1966) The role of sand particles in the ovipository circumstance of *Tubifex*. J. Fac. Sci. Hokkaido Univ. Ser. VI, Zool. *16:*90–97.

Hirao, Y. (1968) Cytological study of fertilization in *Tubifex* egg (in Japanese with English abstract). Zool. Mag., Tokyo *77:*340–346.

Huber, W. (1946) Der normale Formwechsel des Mitoseapparates und der Zellrinde beim Ei von *Tubifex*. Rev. Suisse Zool. *53:*468–474.

Inase, M. (1960a) On the double embryo of the aquatic worm *Tubifex hattai.* Sci. Rep. Tohoku Univ. Ser. IV, Biol. *26:*59–64.

Inase, M. (1960b) The culture solution of the eggs of *Tubifex*. Sci. Rep. Tohoku Univ. Ser. IV, Biol. *26:*65–67.

Inase, M. (1967) Behavior of the pole plasm in the early development of the aquatic worm *Tubifex hattai.* Sci. Rep. Tohoku Univ. Ser. IV, Biol. *33:*223–231.

Inase, M. (1968a) The cytochemical studies on the pole plasm in the eggs of the earthworm, *Tubifex hattai.* Sci. Rep. Tohoku Univ. Ser. IV, Biol. *34:*75–80.

Inase, M. (1968b) Experimental studies on the pole plasm of *Tubifex* eggs. Sci. Rep. Tohoku Univ. Ser. IV, Biol. *34:*81–90.

Iwanoff, P. P. (1928) Die Entwicklung der Larvalsegmente bei den Anneliden. Z. Morph. Oekologie Tiere *10:*62–161.

Jaana, H. (1982a) The ultrastructure of the epithelial lining of the male genital tract and its role in spermatozeugmata formation in *Tubifex hattai* Nomura (Annelida, Oligochaeta). Zool. Anz. 209, in press.

Jaana, H. (1982b) The fine structural study of spermiogenesis in the freshwater oligochaete, *Tubifex hattai*, with a note on the histone transition in the spermatid nucleus. J. Fac. Sci. Hokkaido Univ. Ser VI, Zool. 23, in press.

Jaana, H., T. Shimizu, C. Suzutani-Shiota, and R. Yasumura (1980) Fine structure and chemical properties of the cocoon wall of the *Tubifex* (in Japanese with English abstract). Zool. Mag., Tokyo *89:*130–137.

Lehmann, F.E. (1938) Zustandsänderungen im Ei von *Tubifex* während der Reifungsteilungen. Arch. Exp. Zellforsch. *22:*271–275.

Lehmann, F.E. (1940) Polarität und Reifungsteilungen bei zentrifugierten *Tubifex*-Eiern. Rev. Suisse Zool. *47:*177–182.

Lehmann, F.E. (1941a) Die Indophenolreaktion der Polplasmen von *Tubifex*. Naturwissenschaften *29:*101.

Lehmann, F.E. (1941b) Die Lagerung der Polplasm des Tubifexeies in ihrer Abhängigkeit von der Eirinde. Naturwissenschaften *29:*101.

Lehmann, F.E. (1941c) Die Zucht von *Tubifex* fur Laboratoriumszwecke. Rev. Suisse Zool. *48:*559–561.

Lehmann, F.E. (1943) Die Zustandsänderungen der Furchungsmitosen von *Tubifex* in ihrer Abhängigkeit von chemischen Einflüssen. Rev. Suisse Zool. *50:*244–249.

Lehmann, F.E. (1946) Mitoseablauf und Bewegungsvorgänge der Zellrinde bei zentrifugierten Keimen von *Tubifex*. Rev. Suisse Zool. *53:*475–480.

Lehmann, F.E. (1948) Zur Entwicklungsphysiologie der Polplasmen des Eies von Tubifex. Rev. Suisse Zool. *55:*1–43.

Lehmann, F.E. (1956) Plasmatische Eiorganisation und Entwicklungsleistung beim Keim vom Tubifex (Spiralia). Naturwissenschaften *43:*289–296.

Lehmann, F.E., and H. Hadorn (1946) Vergleichende Wirkungsanalyse von zwei antimitotischen Stoffen, Colchicin und Benzochinon, am Tubifex-Ei. Helv. Physiol. Acta *4:*11–42.

Lehmann, F.E., and M. Henzen (1963) Zur Mikrocytologie der Meiose und Mitose von *Tubifex*. Rev. Suisse Zool. *70:*298–304.

Lehmann, F.E., and V. Mancuso (1958) Verschiedenheiten in der submikroskopischen Struktur der Somatoblasten des Embryos von *Tubifex*. Arch. Klaus-Stift. Vererb. Forsch. *32:*482–493.

Lehmann, F.E., and H.R. Wahli (1954) Histochemische und elektronenmikroskopische Unterschiede im Cytoplasma der beiden Somatoblasten des Tubifexkeimes. Z. Zellforsch. *39:*618–629.

Matsumoto, M., and M. Kusa (1966) Time-lapse cinematographic recording of *Tubifex*-eggs during maturation and early cleavage (in Japanese with English abstract). Zool. Mag., Tokyo *75:*270–275.

Matsumoto, M., and G. Yamamoto (1966) On the seasonal rhythmicity of oviposition in the aquatic oligochaete, *Tubifex hattai* Nomura (in Japanese with English abstract). Jpn. J. Ecol. *16:*134–139.

Meyer, A. (1929) Die Entwicklung der Nephridien und Gonoblasten bei *Tubifex rivulorum* Lam., nebst Bemerkungen zum natürlichen System der Oligochäten. Z. Wiss. Zool. *133:*517–562.

Meyer, A. (1931) Cytologische Studien über die Gonoblasten und andere ähnliche Zellen in der Entwicklung von *Tubifex*. Z. Morph. Oekol. Tiere *22:*269–286.

Meyer, F. (1916) Untersuchungen über den Bau und die Entwicklung des Blutgefäßsystems bei *Tubifex tubifex* (Müll.). Jena Z. Naturwiss. *54:*203–244.

Nomura, E. (1926) On the aquatic oligochaete, *Tubifex hattai*, n.sp. Sci. Rep. Tohoku Univ. Ser. IV, Biol. *1:*193–228.

Penners, A. (1922) Die Furchung von *Tubifex rivulorum* Lam. Zool. Jb. Abt. Anat. Ontog. *43:*323–367.

Penners, A. (1924a) Die Entwicklung des Keimstreifs und die Organbildung bei *Tubifex rivulorum* Lam. Zool. Jb. Abt. Anat. Ontog. *45:*251–308.

Penners, A. (1924b) Doppelbildung bei *Tubifex rivulorum* Lam. Zool. Jb. Abt. Allgem. Zool. *41:*91–120.

Penners, A. (1924c) Experimentalle Untersuchungen zum Determinationsproblem an Keim vom *Tubifex rivulorum* Lam. I. Die Duplicitas cruciata und Organbildende Keimbezirke. Arch. Mikrosk. Abt. Entwick. Mechan. *102:*51–100.

Penners, A. (1924d) Über die Entwicklung teilweise abgetöteter Eier von *Tubifex rivulorum*. Verh. Deutsch Zool. Ges. *29:*69–73.

Penners, A. (1925) Regulationserscheinungen und determinative Entwicklung nach Untersuchungen am Keim von *Tubifex*. Verh. Physik.-Med. Ges. Wurzburg N.F. *50:*198–211.

Penners, A. (1926) Experimentelle Untersuchungen zum Determinationsproblem am Keim von *Tubifex rivulorum* Lam. II. Die Entwicklung teilweise abgetöteter Keime. Z. Wiss. Zool. *127:*1–140.

Penners, A. (1933) Über Unterschiede der Kokons einiger Tubificiden. Zool. Anz. *103:*93–95.

Penners, A. (1934a) Experimentelle Untersuchungen zum Determinationsproblem am Keim von *Tubifex rivulorum* Lam. III. Abtötung der Teloblasten auf verschiedenen Entwicklungsstadien des Keimstreifs. Z. Wiss. Zool. *145:*220–260.

Penners, A. (1934b) Die Herkunft der Urkeimzellen bei *Tubifex*. Z. Wiss. Zool. *145:*389–398.

Penners, A. (1937) Regulation am Keim von *Tubifex rivulorum* Lam. nach Ausschaltung des ektodermalen Keimstreifs. Z. Wiss. Zool. *149:*86–130.

Penners, A. (1938) Abhängigkeit der Formbildung vom Mesoderm im Tubifex-Embryo. Z. Wiss. Zool. *150:*305–357.

Penners, A., and A. Stäblein (1930) Über die Urkeimzellen bei Tubificiden (*Tubifex rivulorum* Lam. und *Limnodrilus udekemianus* Claparède). Z. Wiss. Zool. *137:*606–626.

Poddubnaya, T.L. (1980) Life cycle of mass species of Tubificidae (Oligochaeta). In R.O. Brinkhurst and D.G. Cook (eds): Aquatic Oligochaete Biology. New York: Plenum Press, pp. 175–184.

Rötheli, A. (1949) Auflössung und Neubildung der Meiosespindel von *Tubifex* nach chemischer Behandlung. Rev. Suisse Zool. *56:*322–326.

Rötheli, A. (1950) Chemische Beeinflussung plasmatischer Vorgänge bei der Meiose des Tubifexeies. Z. Zellforsch. *35:*69–109.

Shimizu, T. (1975) Occurrence of microfilaments in the *Tubifex* egg undergoing the deformation movement. J. Fac. Sci. Hokkaido Univ. Ser. VI, Zool. *20:*1–8.

Shimizu, T. (1976a) The staining property of cortical cytoplasm and the appearance of pole plasm in *Tubifex* egg (in Japanese with English abstract.). Zool. Mag., Tokyo *85:*32–39.

Shimizu, T. (1976b) The fine structure of the *Tubifex* egg before and after fertilization. J. Fac. Sci. Hokkaido Univ. Ser. VI, Zool. *20:*253–262.

Shimizu, T. (1978a) Deformation movement induced by divalent ionophore A23187 in the *Tubifex* egg. Dev., Growth Diff. *20:*27–33.

Shimizu, T. (1978b) Mode of microfilament-arrangement in normal and cytochalasin-treated eggs of *Tubifex* (Annelida, Oligochaeta). Acta Embryol. Exp. *1:*59–74.

Shimizu, T. (1979) Surface contractile activity of the *Tubifex* egg: Its relationship to the meiotic apparatus functions. J. Exp. Zool. *208:*361–378.

Shimizu, T. (1981a) Cyclic changes in shape of a non-nucleate egg fragment of *Tubifex* (Annelida, Oligochaeta). Dev., Growth Differ. *23:*101–109.

Shimizu, T. (1981b) Cortical differentiation of the animal pole during maturation division in fertilized eggs of *Tubifex* (Annelida, Oligochaeta). I. Meiotic apparatus formation. Dev. Biol. *85:*65–76.

Shimizu, T. (1981c) Cortical differentiation of the animal pole during maturation division in fertilized eggs of *Tubifex* (Annelida, Oligochaeta). II. Polar body formation. Dev. Biol. *85:*77–88.

Shimizu, T. (1982) Ooplasmic segregation in the *Tubifex* egg: Mode of pole plasm accumulation and possible involvement of microfilaments. Roux Arch., in press.

Specht, W. (1961) Bildung, Bau und Funktion des sog. achromatischen Teilungsapparates der Zelle, erläutert am Beispiel der Reifungsspindel im Ei von *Tubifex*. Z. Anat. Entwick. *122:*266–288.

Stephenson, J. (1930) The Oligochaeta. Oxford: The Clarendon Press.

Suzutani, C. (1977) Light and electron microscopical observations on the clitellar epithelium of *Tubifex*. J. Fac. Sci. Hokkaido Univ. Ser. VI, Zool. *21:*1–11.

Suzutani-Shiota, C. (1980) Ultrastructural study on cocoon formation in the freshwater oligochaete, *Tubifex hattai*. J. Morphol. *164:*25–38.

Timm, T. (1980) Distribution of aquatic oligochaetes. In R.O. Brinkhurst and D.G. Cook (eds): Aquatic Oligochaete Biology. New York: Plenum Press, pp. 55–77.

Weber, R. (1956) Zur Verteilung der Mitochondrien in frühen Entwicklungsstadien von *Tubifex*. Rev. Suisse Zool. *63:*277–288.

Weber, R. (1958) Über die submikroskopische Organisation und die biochemische Kennzeichung embryonaler Entwicklungsstadien von *Tubifex*. Roux Arch. *150:*542–580.

Woker, H. (1943) Phasenspezifische Wirkung des Colchicins auf die ersten Furchungsteilungen von *Tubifex*. Rev. Suisse Zool. *50:*237–243.

Woker, H. (1944) Die Wirkung des Colchicins auf Furchungsmitosen und Entwicklungsleistungen des *Tubifex*-Eies. Rev. Suisse Zool. *51:*109–170.

Developmental Biology of Freshwater Invertebrates, pages 317–361

Embryonic Development of Glossiphoniid Leeches

Juan Fernández and Nancy Olea

ABSTRACT The embryonic development of glossiphoniid leeches, from the uncleaved egg to the completion of the gut, is described in terms of ten stages. Characterization of these stages includes analysis of the developmental mechanisms in operation and also a description of the structure of the cells and of their cell lineage relationships. The egg includes three types of cytoplasm: teloplasm, or pole plasm; perinuclear plasm; and vitelloplasm. The teloplasm, rich in cell organelles, forms throughout stage 1. During spiral cleavage, the teloplasm is partitioned in an orderly fashion as it passes into successive precursor cells that give rise to five pairs of mother cells, or teloblasts. The teloplasm of teloblasts is partly used in the construction of ectodermal and mesodermal stem cells. Different teloblasts initiate and terminate the formation of stem cells at different stages of development. Stem cells and their progeny arrange themselves into five paired longitudinal rows, or bandlets, that in turn associate with one another to form the right and left germinal bands. The germinal bands form the germinal plate, which, following segmentation, gives rise to the ectodermal and mesodermal components of the 32 body segments. Finally, segmentation of the endoderm during stage 10 leads to the formation of gut caeca. The time of development of three glossiphoniids—*Helobdella triserialis, Theromyzon rude,* and *Haementeria depressa*—is provided. Techniques for the preparation of embryos for microscopic examination as well as procedures for marking and deleting cells are also presented.

Leeches have been widely used in the past not only in medical practice but also in descriptive studies of embryonic development (see Schleip, '36) and of neural organization (see Fernández, '78). Whitman (1878, 1887), using the embryo of glossiphoniid leeches, first emphasized the importance of establishing cell pedigrees in understanding cell fates during ontogeny. The works of Retzius (1891), Cajal ('08), and Sánchez ('08), on the other hand, laid the basis for understanding the structure and relationships of nerve cells in relatively simple nervous systems.

Following the classic studies on glossiphoniid embryonic development published in the early 1930s (see Müller, '32; and Mori, '32), there were few studies of leech ontogeny for almost 50 years. Interest in leech embryology has recently been revived, partly because of the organism's value in studies of the development of the nervous system. The nervous system of the leech is an optimal combination of relative simplicity and experimental accessibility, and for this reason it has been extensively used in neurobiological research during the last 20 years.

There are several reasons why the leech embryo is an excellent organism for studies of developmental processes. First, leeches are abundant and have a wide geographical distribution. Some of them breed continously and have a life cycle of a few weeks' duration. Furthermore, many species of leeches are easy to breed in the laboratory and their embryos may be cultured outside the cocoon. In the case of leeches of the family Glossiphoniidae, embryos may reach several millimeters in diameter and their large cells can be subjected to different types of experimental manipulation. Simple procedures for the visualization of embryonic cells and techniques for marking and deleting them are now available. Leech development is simple and straightforward because most organs and tissues seem to arise from a small number of founder cells, or stem cells. Stem cells and their descendants form paired superficial structures, or germinal bands, in which the distribution of cells presents a strikingly high degree of spatial orderliness.

GENERAL CHARACTERISTICS OF LEECHES

Leeches belong to the phylum Annelida, class Hirudinea. The leech's body consists of 32 segments plus a nonsegmental prostomium. The prostomium and the first four body segments form the head region, which includes paired eyes on the dorsal surface and the anterior sucker with the mouth aperture on the ventral surface. The last seven body segments form the tail region, which includes the posterior sucker. The intervening 21 body segments, those forming the abdominal region, show a highly stereotyped metameric iteration of visceral organs such as circulatory vessels, nerve ganglia, testes, and nephridia. Body segments appear superficially subdivided into a number of annuli, some of which bear collections of sensory cells or sensilla. The annulus bearing sensilla lies at the level of the nerve ganglia and the available embryological evidence indicates that it corresponds to the first annulus of a body segment (Bathia, '70; Fernández, '80). Nephridiopores open along the ventro-lateral aspect of the abdominal region whereas the anal aperture opens on the dorsal region of the last abdominal segment. Leeches are hermaphrodites and their gonopores are found within the clitellar region, which extends along the ventral part of abdominal segments 4 to 7. The clitellar region includes numerous glands.

Glossiphoniids are freshwater leeches with a wide geographic distribution. They are found in lakes, ponds, rivers, and streams where they usually lie attached to the surface of various hard objects. Glossiphoniids have flattened bodies with a small anterior sucker and a large disc-shaped posterior sucker. Head segments carry one or more pairs of eyes and the typical midbody segment has three annuli. Glossiphoniids, like all leeches, are protandrous hermaphrodites that, at the time of mating during early spring, usually correspond to males. At this time the sperm donor ejaculates one spermatophore, which is implanted in the body wall of the recipient. Some glossiphoniids have a specialized tissue, or vector tissue, along which spermatozoa reach the ovary from the site of implantation of the spermatophore. Fertilization probably takes place in the ovary. Fertilized eggs are laid enclosed in a membranous sac, or cocoon, which contains a small amount of fluid. Cocoons usually remain attached to the ventral body wall, which forms a brooding pouch. Hatching embryos become attached to the walls of the pouch by means of a sticky secretion, whereas juveniles become attached by means of the posterior sucker. Brooding parents carry juveniles for varying lengths of time (see Sawyer, '71). Glossiphoniids feed on tissue fluid or blood withdrawn from invertebrates or from cold- or warm-blooded vertebrates. For more details on the structure of body segments and on the reproduction of leeches consult Harant and Grassé ('59), Moore ('59), and Mann ('62).

GENERAL FEATURES OF DEVELOPMENT OF LEECHES

Leeches, as well as other annelids, molluscs, and some flatworms, belong to the spiralian invertebrates. Spiralia exhibit spiral cleavage characterized by the inclination of the mitotic spindle and also by the alternation of its direction in successive cell divisions (see Costello and Henley, '76). Two other general features of leech development are the high frequency of unequal cell divisions and the mosaic character of embryogenesis. Unequal cell divisions are particularly common in early stages of development and their occurrence is not always associated with spiral division of the cells. Mosaic or determinate development refers to the property of specific blastomeres giving rise to precise structures of the embryo. Therefore, culture of isolated blastomeres usually leads to the formation of incomplete embryos (see Mori, '32).

Leeches and Oligochaetes constitute the group of Clitellate annelids whose development presents the following characteristics: 1) the D blastomere plays a crucial role in development because most tissues of the embryos and in some cases the entire embryo arise from cells descended from that blastomere; 2) eggs contain small or large yolk supplies and, accordingly, their development is either direct or passes through a larval stage; 3) ectodermal and mesodermal stem cells arise teloblastically, that is, by highly unequal divisions of large mother cells, or teloblasts, lying at the rear end of the embryo; 4) ectodermal and mesodermal stem cells, and their blast cell progeny, are assembled in an orderly manner, throughout paired structures called germinal bands; and 5) body segments are founded by a discrete number of stem cells.

Leeches of the family Glossiphoniidae differ from leeches of other families in that their eggs include large yolk reserves and consequently exhibit direct development. Furthermore, glossiphoniid eggs present prominent polar accumulations of cytoplasm rich in organelles. This cytoplasm, which constitutes the teloplasm or pole plasm, will participate in the formation of stem cells by the teloblasts. Another interesting feature of glossiphoniid embryos is that the A, B, and C macromeres do not degenerate but rather give rise to most of the endoderm of the digestive tube. For more details on leech ontogeny consult Schleip ('36), Dawydoff ('59), Mann ('62), Reverberi ('71), and Anderson ('73).

NATURAL HISTORY OF SOME GLOSSIPHONIID LEECHES

Leech development may be conveniently studied in three species of glossiphoniids: *Helobdella triserialis, Theromyzon rude,* and *Haementeria depressa. H. triserialis* is a small cosmopolitan leech, 10 to 25 mm in length, that lives under stones or hard objects immersed in shallow, slowly flowing, streams. It feeds on small snails and breeds continuously during spring and summer. A clutch may consists of as many as 50 eggs, about 400 μm in diameter, enclosed within four or five cocoons attached to the walls of the brood pouch of the parent. *T. rude* is a midsize leech that may reach 40 mm in length. Its geographic distribution corresponds to the Rocky Mountain duck migration pathways of Canada and the United States (Sawyer, '72; Davies, '73). It is found on the undersurface of hard objects immersed in lakes, ponds, and sloughs frequented by water fowl. It feeds on blood sucked from the eyes, nares, and other body parts of water fowl. Breeding occurs from May to August and a clutch consists of three to eight cocoons each containing about 60 eggs. Eggs are 700 to 800 μm in diameter. Unlike other glossiphoniids, *T. rude* lays cocoons on hard smooth surfaces and covers them until hatching. Breeding is initiated when animals weigh at least 150 mg. *T. rude* breeds once and then dies three to 19 days after release of the young (see Davies and Wilkialis, '80; Wilkialis and Davies, '80). *H. depressa* is a large leech that, when fully extended, may reach more than 200 mm in length. Its geographic distribution seems to be restricted to the Neotropical and temperate zones of South America. This glossiphoniid is usually found in the muddy bottoms of streams, ponds, and marshes visited by people and large mammals such as cattle. The leech feeds on these mammals and also on frogs. *H. depressa* breeds continously during the spring and summer and lays a variable number of cocoons that attach to the walls of the brood pouch. Each cocoon encloses one to 12 eggs, each about 2 mm in diameter. The number of eggs per laying increases with body weight. Small individuals, weighing 0.7 to 1 gm lay 20 to 30 eggs; whereas large individuals, weighing about 5 gm lay 200 to 250 eggs. For more details on the growth and reproduction of a leech of the genus *Haementeria*, consult Sawyer et al. ('81).

MAINTENANCE OF ADULTS

Animals are maintained in aquaria or glass jars dimensioned in accordance with animal number and size. Since glossiphoniids survive better in media of low oxygen content, no bubbling of the water is recommended. Good maintenance of adult leeches, including gravid ones, has been achieved with filtered spring water or with a saline solution of similar composition (see Table I). Any of these fluids must be replaced every 3 to 5 days. Immediately after feeding, water must be replaced every day for 1 to 2 weeks because the excreted ammonia may reach toxic levels. Sick animals must be quickly removed and isolated in individual small jars. They are easily detected because they usually remain motionless after agitation of the water or because they show body constrictions. Recovery of sick animals may be improved by keeping them for a while in 2% NaCl in spring water.

H. triserialis is maintained at 20° to 22°C and is fed every 2 to 3 weeks on small or midsize snails of the genus *Chilina.* Large specimens of *T. rude* collected in early spring have usually performed the last blood feeding and are thus sexually mature. These animals are maintained at 14° to 22°C, allowing control of the rate at which gonads mature and eggs are laid. Thus, at 14°C it is possible to retard egg laying for several weeks. Feeding of *T. rude* in the laboratory presents several difficulties. First, attempts to feed animals with duck's liver, blood clots, or through artificial membranes have failed (also see Davies and Wilkialis, '80). Second, feeding of juveniles on whole animals such as ducklings is usually followed by numerous losses, probably because the leeches are eaten by the birds. *H. depressa* can be maintained at 24° to 26°C and fed on unanesthesized rabbits whose ventral fur has been shaved. Other mammals such as rats, mice, and frogs may also

CHARACTERIZATION OF DEVELOPMENTAL STAGES

Development of glossiphoniid leech embryos from the uncleaved egg to the completion of the gut has been divided into ten stages (Fernández, '80). A stage is defined as that point in time at which the process diagnostic for the stage has come to completion. A diagrammatic representation of different developmental stages is given in Figure 1, and time of development for different glossiphoniids is shown in Table II. It will be noticed that the time of development is directly proportional to the size of the egg: the larger the egg, the longer it takes to complete its development. To compare the size of embryonic cells of different glossiphoniids see Table III.

TABLE II. Time of Development of Glossiphoniid Embryos at the Indicated Temperature

	H. triserialis 20°C	*T. rude*[a] 14°C	*H. depressa* 20°C
		hours	
Stage 1	0–6	0–12	0–12
Stage 2	5–6	12–15	12–15
Stage 3	6–10	15–25	15–25
Stage 4	10–15	25–40	30–50
Stage 5	15–20	35–45	40–60
Stage 6	20–55	40–90	80–120
Stage 7	55–85	90–160	160–210
Stage 8	85–140	160–230	220–280
Stage 9	140–180	200–320	320–380
Stage 10	180–300	350–850	600–800

[a]Taken from Fernández ('80).

Fig. 1. Schematic representation of the structure of *T. rude* embryos at different stages of development. a) Stage 1; b) stage 2; c) stage 3; d) early stage 4; e) late stage 4; f) late stage 5; g) early stage 6; h) late stage 6; i) early stage 7; j) late stage 7, U-shaped germinal bands; k) early stage 8; l) late stage 8; m) late stage 9; n) late stage 10. Diagrams a) to j) and m) represent dorsal views, whereas diagrams k), l), and n) represent ventral views of the embryos. Diagram m) shows the structure of the germinal plate after removal of the yolky macromeres and teloblasts. as, developing anterior sucker; gb, germinal band; gc, developing gut caeca; gp, ganglionic primordium; is, intersegmental septum; mc, micromere cap; np, nephridial primordium; nu, nuclei of macromere; pp, primordial prostomium; ps, developing posterior sucker; s, somite.

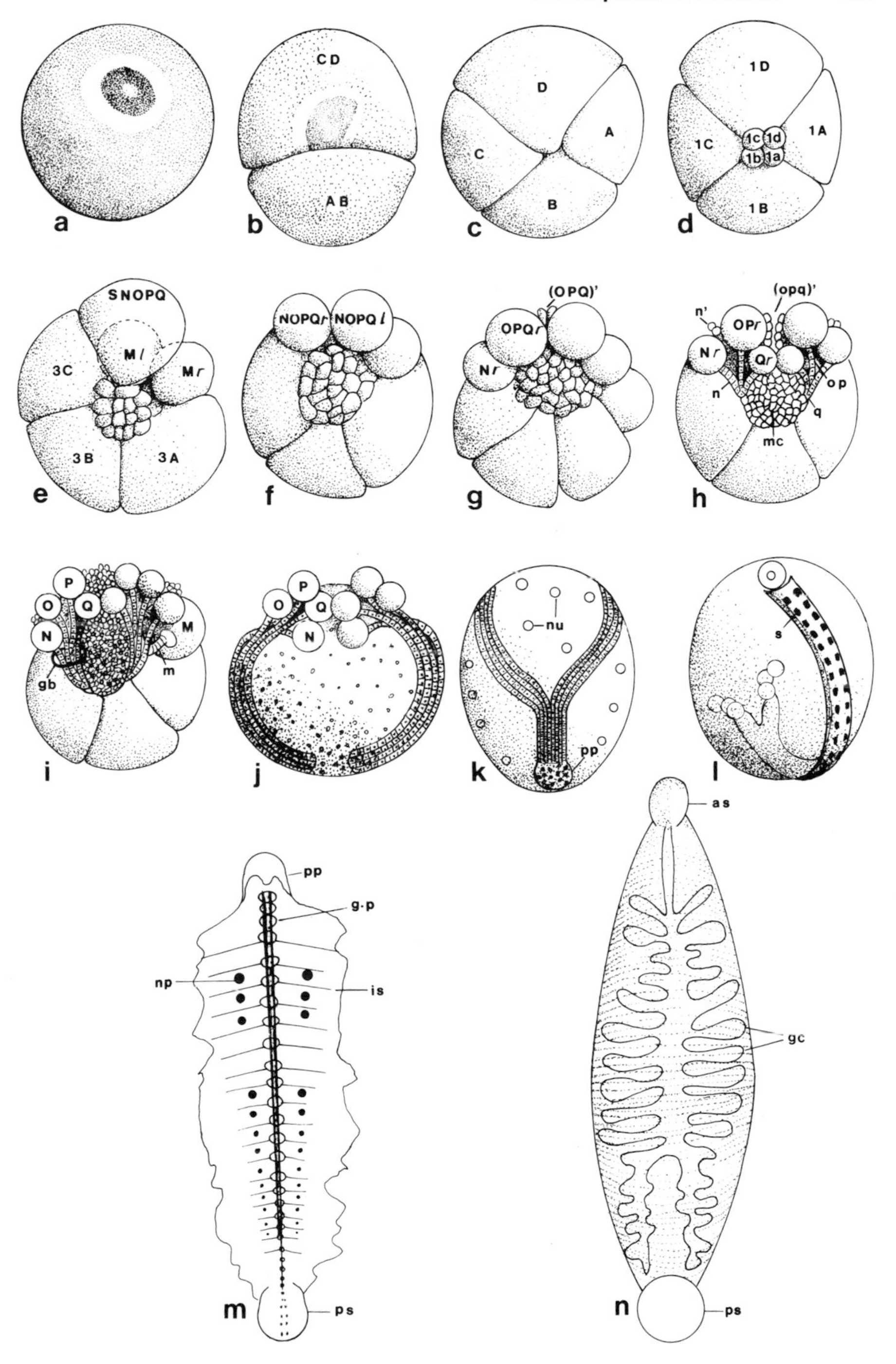
a
CD
AB
b
D
C
A
B
c
1D
1C
1c
1d
1b
1a
1A
1B
d
SNOPQ
M l
M r
3C
3B
3A
e
NOPQ r
NOPQ l
f
(OPQ)'
OPQ r
N r
g
(opq)'
n'
OP r
N r
Q r
n
op
q
mc
h
P
O
Q
N
M
m
gb
i
P
O
Q
N
j
nu
pp
k
s
l
as
pp
g.p
np
is
gc
m
ps
n
ps

1

TABLE III. Size of Embryonic Cells in Micrometers

Cell	*H. triserialis*	*T. rude*	*H. depressa*
Egg	340–400	700–800	1,800–2,200
AB	130–160 × 260–300	300–350 × 650–750	900–1,000 × 1,600–1,900
CD	200–260 × 300–400	400–500 × 650–750	1,000–1,300 × 1,600–1,900
D	180–200 × 240–300	300–400 × 500–550	800–1,000 × 1,000–1,300
First quartet micromeres	50–70	80–150	250–450
SNOPQ	160–180 × 200–240	200–300 × 350–400	500–650 × 650–800
SM	130-160 × 160-240	250-300 × 400-450	400-600 × 500-800
Mr	100–130	200–300	650–750
Ml	150–180	250–350	450–650
NOPQ	120–160	200–300	260–400
OPQ	100–130	200–300	200–400
N	80–100	150–250	180–400
(OPQ)′	20 × 50	50 × 100	80-120 × 150-180
OP	80–100	150–200	150–300
Q	70–80	100–150	180–300
O	70–80	100–120	130–230
P	70–80	100–120	160–260
m	50–70	100	100–180
n	25	60	50–75
o	20	30–50	50–60
p	20	30–40	50–60
q	20	30–60	40–50

Stage 1: The Uncleaved Egg

This stage extends from oviposition to the initiation of the first cleavage furrow and is characterized by important changes in the structure of the egg. These changes are related to the completion of meiotic maturation and to the formation of the teloplasm, or pole plasm. Eggs are laid surrounded by an acellular perivitelline membrane that forms a chamber filled with fluid in which the egg floats. Sometimes this fluid contains yolk platelets that have leaked out of the egg. The leakage of yolk platelets does not appear to harm eggs as they usually complete their development normally. The yolk is pink in *H. triserialis* and yellow in *T. rude* and *H. depressa.*

Eggs are laid during metaphase of the first meiotic division, and the first polar body is thus released during the first hours of development. Emission of the second polar body takes place by the middle of stage 1. Formation of the teloplasm takes place during the second half of stage 1, following the migration and concentration of ooplasm at the egg's poles. This phenomenon is initiated in the dorsal, or animal, hemisphere of the egg by the appearance of a superficial ring of ooplasm. A similar structure appears later in the ventral, or vegetal, hemisphere of the egg (Figs. 7, 8). Migration of the rings of ooplasm toward the animal or vegetal pole is accompanied by deformation movements that affect the structure of the cell surface. Meridional bands of contraction form grooves that give the egg a pumpkinlike appearance. By the end of stage 1, a mass of whitish ooplasm is present at each pole of the egg, forming the animal and vegetal teloplasms. A flat sheet of ooplasm is also present between the teloplasms at the center of the egg. This ooplasm surrounds the egg nucleus and will be called perinuclear plasm (Figs. 9, 10). The rest of the egg cytoplasm corresponds to the colored vitelloplasm. The teloplasms and the perinuclear plasm include large numbers of organelles, particularly mitochondria, and few yolk platelets (Figs. 10, 11). The vitelloplasm, in contrast, includes many yolk platelets and few organelles.

Stage 2: First Cleavage

The first cleavage furrow is meridional and divides the egg into two cells of unequal size: a smaller cell AB and a larger cell CD. While the perinuclear plasm subdivides into two approximately equal parts, one going into cell AB and the other going into cell CD, most of the teloplasm remains in cell CD (Fig. 12).

Stage 3: Formation of the A, B, C, and D Blastomeres

The second cleavage is also meridional and gives rise to blastomeres A, B, C, and D. The cleavage furrow divides cell CD first into a smaller blastomere C and a larger blastomere D. Division of cell AB gives rise to blastomeres A and B of about the same size (Fig. 13). Sometimes the first

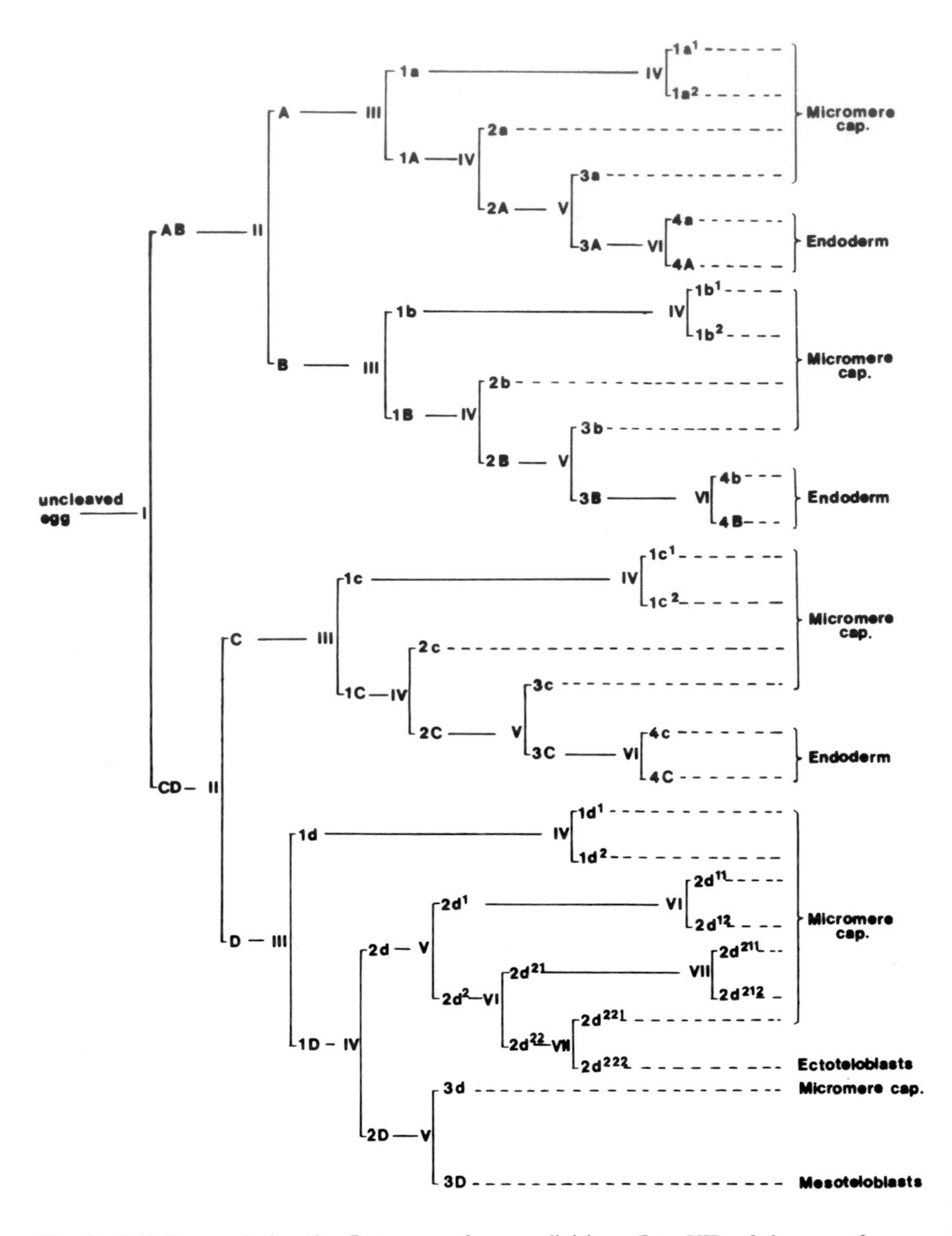

Fig. 2. Cell lineage during the first seven cleavage divisions (I to VII) of the egg of a glossiphoniid leech. Cells are designated according to the Wilsonian notational system. Modified from Müller ('32).

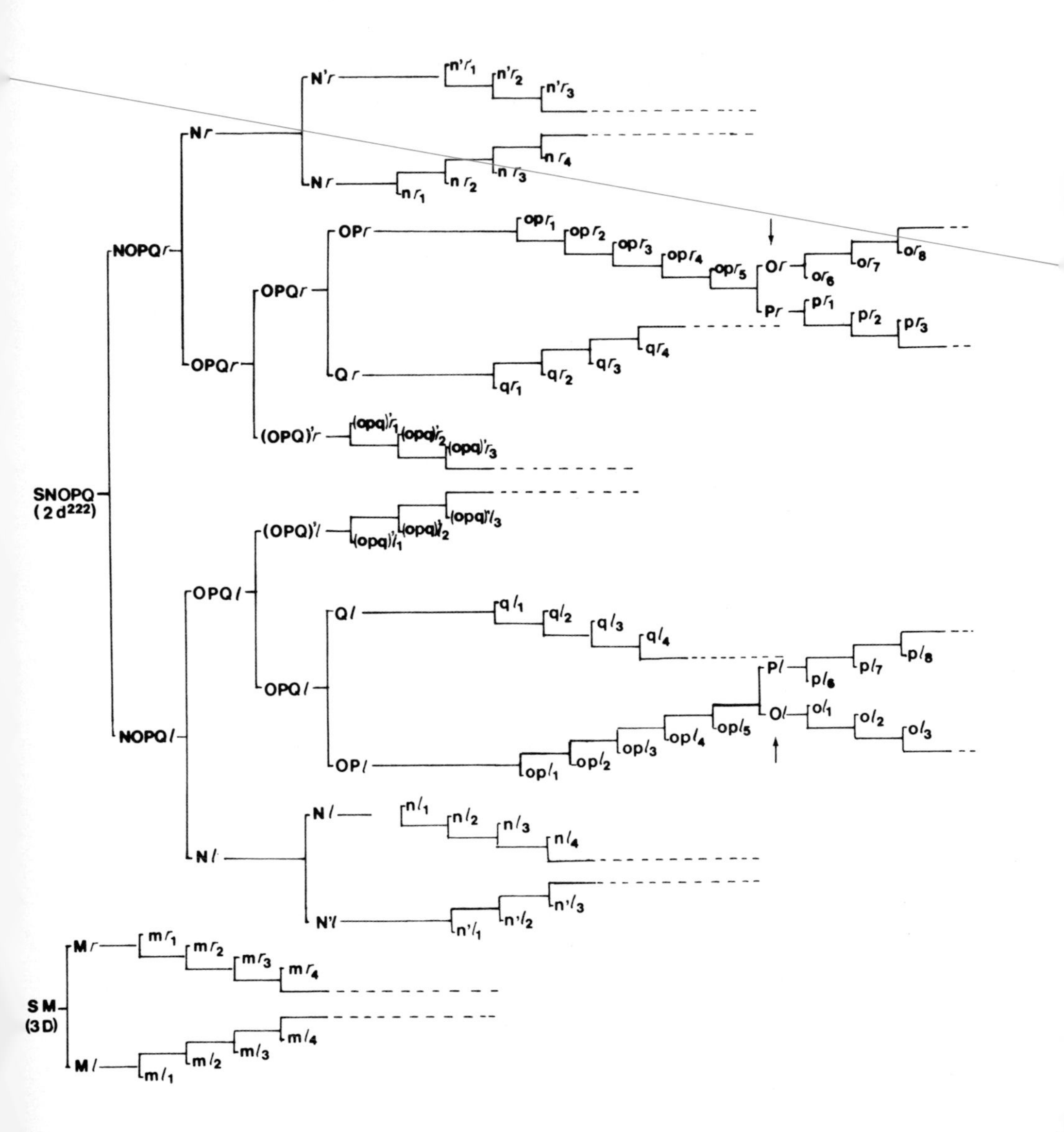

Fig. 3. Cell lineage during formation of the teloblasts and of the early germinal bandlets. Notice that upon division of the OP proteloblasts, to form the O and P teloblasts (arrows), the op bandlet becomes the o or the p bandlet. Cells are designated according to the notational system of Fernández ('80).

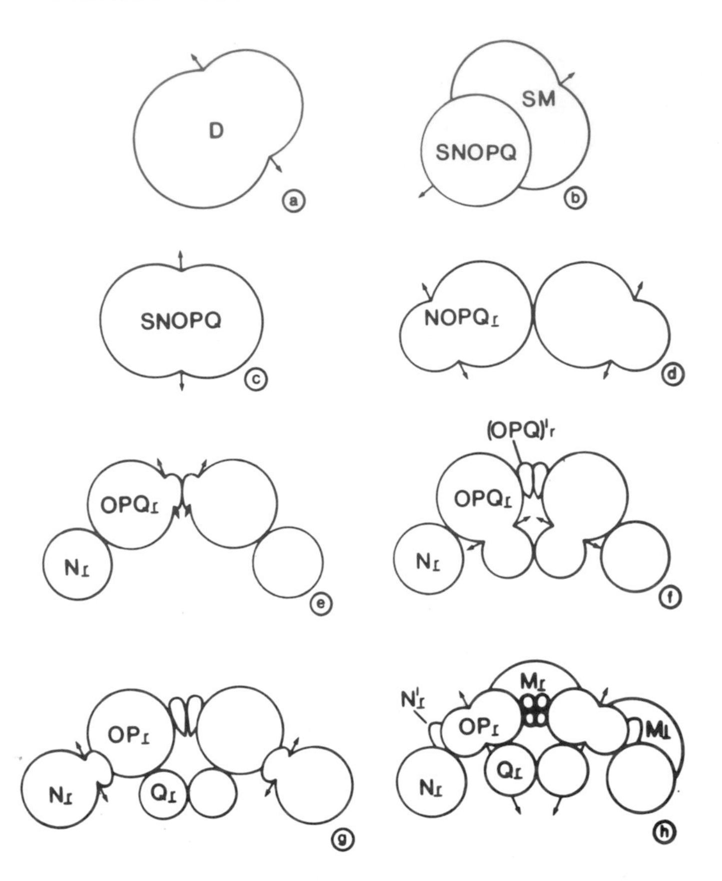

Fig. 4. Diagrammatic representation of teloblast formation from precursor cells, or proteloblasts, as seen from the animal pole of *T. rude* embryos. In all diagrams, the rostral end of the embryo is directed toward the bottom of the figure. The approximate planes of cleavage of proteloblasts, with respect to the rostrocaudal axis of the embryo, are indicated by arrows. [From Fernández ('80), reproduced by permission of Academic Press Inc.]

four blastomeres are of about the same size. Less frequently blastomeres D and C are smaller than blastomeres A and B. Unequal division of cell CD determines that most of the teloplasm remains concentrated in blastomere D. Due to migration of the animal and vegetal teloplasm to the interior of blastomere D, little of this type of cytoplasm remains at the cell surface.

Stage 4: Formation of the M Teloblast Pair

The third cleavage is equatorial and highly unequal, dividing each blastomere into a large ventral cell, or macromere, and into a smaller dorsal cell, or micromere. The cleavage furrow divides cell D first, then cell C, and finally cells B and A (Fig. 14). The first quartet of micromeres, or cells 1a, 1b, 1c, and 1d, thus form in a counterclockwise direction. Since the third cleavage furrow is oblique to the second cleavage furrow, each micromere comes to straddle two macromeres. Furthermore, the fact that the third cleavage is dexiotropic determines that cell 1d straddles macromeres 1D and 1A; cell 1a, macromeres 1A and 1B; cell 1b, macromeres 1B and 1C; and cell 1c, macromeres 1C and 1D. Hence, the third cleavage division of the egg is spiral. The macromeres undergo a number of further spiral divisions to give rise to additional micromeres that eventually form the micromere cap at the animal pole of the embryo (Fig. 2).

Division of the 1D macromere differs from the division of the other macromeres in that 2d and 2D daughter cells are of slightly different sizes. Cell 2d, the first somatoblast, lies dorsally and gives rise to the ectoteloblasts. The teloplasm of cell 2d probably comes from the animal pole of the egg. Cell 2D lies ventrally and gives rise to the mesoteloblasts. Its teloplasm probably comes from the vegetal pole of the egg. The available evidence indicates that cell 2d cuts off three micromeres to become cell $2d^{222}$ and cell 2D cuts off one micromere to become cell 3D, or second somatoblast (see Fig. 2).

Cells derived from the D blastomere will be designated by a new notational system recently proposed by Fernãndez ('80). This notational system is simpler and more straightforward than the previous system of cell designation (see Table IV). It uses the capital letters M, N, O, P, and Q and the lower case letters m, n, o, p, and q (see Fig. 3). Cell $2d^{222}$ is designated as cell SNOPQ and cell 3D as cell SM (see Fig. 15). Cell SM divides first into two daughter cells called right and left M teloblasts, or mesoteloblasts. These teloblasts, designated as cells M*r* and M*l*, are usually of different sizes and lack a bilaterally symmetrical distribution (see Fig. 17).

Stage 5: Formation of the NOPQ Cell Pair

Cell SNOPQ divides into two daughter cells of equal size designated as cells NOPQ*r* and NOPQ*l*. These cells correspond to the proteloectoblasts described by previous workers. The plane of cleavage of cell SNOPQ is meridional and defines the rostrocaudal axis and hence the midline of the embryo. The M teloblasts and the NOPQ cells lie at the caudal end of the embryo, whereas the C macromere lies at the front of the embryo. The A and B macromeres lie at the left and at the right sides of the embryo (Figs. 16, 17).

Stage 6: Formation of the N, O, P, and Q Teloblast Pairs

The NOPQ cell pair divides nearly synchronously to produce two pairs of daughter cells of different size: a larger caudo-medial pair of cells designated

TABLE IV. Designation of Cells Descending From the D Macromere[a]

Recent cell designation	Previous cell designation
SM	3D
SNOPQ	$2d^{222}$
M*r*, M*l*	Myr, Myl
NOPQ*r*, NOPQ*l*	Tr, Tl
N*r*,N*l*	Nr, Nl
N′*r*,N′*l*	nr_1, nl_1?
OPQ*r*, OPQ*l*	Mr, Ml
(OPQ)′*r*,(OPQ)′*l*	mr_1, ml_1?
Q*r*, Q*l*	Mr^3, Ml^3
OP*r*, OP*l*	Mr^{1+2}, Ml^{1+2}
O*r*,O*l*	Mr^1, Ml^1
P*r*, P*l*	Mr^2, Ml^2
m*r*, m*l*	rmK, lmK
n*r*, n*l*	nr, nl
op*r*, op*l*	mr^{1+2}, ml^{1+2}
o*r*, o*l*	mr^1, ml^1
p*r*, p*l*	mr^2, ml^2
q*r*, q*l*	mr^3, ml^3

[a]Comparison between the recent (Fernández, ’80) and the previous notational system (Müller, ’32).

as cells OPQ*r* and OPQ*l* and a smaller rostro-lateral pair of cells designated as teloblasts N*r* and N*l* (Fig. 18). Division of the nucleus and of the perinuclear plasm in the A, B, and C macromeres usually starts during or immediately after cleavage of the NOPQ cell pair. Since karyokinesis is not followed by cytokinesis, macromeres become multinucleated cells (see Figs. 36, 37). This situation persists until the late stages of development.

The OPQ cell pair divides into a smaller caudo-medial pair of cells, designated as (OPQ)′*r* and (OPQ)′*l* cells, and a larger rostro-lateral pair of cells for which the designations OPQ*r* and OPQ*l* cells are retained (Fig. 19). Shortly after formation of the (OPQ)′ cells, the OPQ cells divide again to produce a larger latero-caudal pair of cells, designated as cells OP*r* and OP*l*, and a smaller rostro-medial pair of cells, designated as teloblasts, Q*r* and Q*l* (Fig. 20). Meanwhile the N teloblasts divide to produce a pair of small caudo-medial cells, designated as cells N′*r* and N′*l* , and a pair of large latero-rostral cells for which the designations of N*r* and N*l* teloblasts are retained. Finally, toward the end of stage 6 the OP cells divide to produce two pairs of cells of similar size designated as O*r*, O*l* and P*r*, P*l* teloblasts (Figs. 21, 26).

The N′ and (OPQ)′ cells may be considered to be teloblastlike cells because they include abundant teloplasm and because their division gives rise to rows of smaller cells designated as n′*r* and n′*l* and (opq)′*r* and (opq)′*l*, respectively (Figs. 21, 28). These rows meet and blend with cells of the micromere cap. As shown in Figure 4, formation of teloblasts and of teloblastlike cells from proteloblasts occurs according to a modified spiral cleavage pattern. This determines, among other things, the stereotypic archlike arrangement of teloblasts at the embryonic surface. The orderly arrangement of teloblasts is particularly striking in embryos of *T. rude* (Fig. 28).

Teloblasts are large spherical cells that measure 70 to 750 μm in diameter. They consist of white teloplasm lying at one pole of the cell and yellow vitelloplasm that fills the remainder of the cell. The teloplasm usually forms a spherical mass that carries the multilobulated nucleus and includes large amounts of cell organelles, particularly mitochondria. Therefore, the teloplasm of teloblasts is structurally similar to the teloplasm of the egg (Figs. 29–32). Shortly after teloblasts are formed, these cells undergo a sequence of unequal divisions by which the mass of teloplasm is parceled out in the formation of disc-shaped stem cells (Figs. 30–32). Stem cells produced by each teloblast form a row called the germinal bandlet. The size of the stem cells generally depends on the size of the teloblast from which they formed. Thus, the largest stem cells come from the M teloblasts, while the smallest ones come from the O and P teloblasts. Division of stem cells gives rise to blast cells, and thus the bandlets consist of both types of cells. The five pairs of germinal bandlets are designated by lower case letters corresponding to the capital letters designating their teloblast of origin, i.e., m*r* and m*l*, n*r* and n*l*, o*r* and o*l*, p*r* and p*l*, q*r* and q*l* (Figs. 22, 23, 28, 29). As shown in Table V and in Figures 5 and 6, different teloblasts initiate the formation of stem cells at different stages of development. The arrangement of the nascent bandlets is shown in Figures 22, 24–26. Stem and blast cells of the bandlets are structurally similar. They carry a multilobulated nucleus with one or more prominent nucleoli and their cytoplasm is undistinguishable from the teloplasm of teloblasts (Fig. 32). Specialized cell junctions, such as desmosomes and gap junctions, link not only cells of a given bandlet but also cells of adjacent bandlets (Fig. 33).

By the end of stage 6, the embryo includes five pairs of teloblasts but only four pairs of germinal bandlets: the m, n, op, and q bandlets. The bandlets of either side of the embryo associate with one another to form paired structures called germinal bands (Fig. 28).

Stage 7: Formation of the U-Shaped Germinal Bands

With the formation of the last two pairs of teloblasts, the op bandlet remains attached to the O or to the P teloblast. Therefore, the op bandlet becomes the o or the p bandlet. Furthermore, with some frequency the

TABLE V. Stage at Which Teloblasts Initiate (Turn-on) and Terminate (Turn-off) the Formation of Stem Cells in *T. rude* Embryos[a]

Teloblast	Turn-on	Turn-off	Approximate time elapsed in hours at 14°C
M	late 5	mid 8	150
N	early 6	late 9	270
O	mid 6 or early 7	late 8	150
P	mid 6 or early 7	late 8	150
Q	mid 6	mid 9	200

[a]Data taken from Fernández and Stent ('80).

Fig. 5. Schematic representation of the arrangement of the M teloblast, the position of its teloplasmic pole, and the development of the m bandlets in embryos of *T. rude* at successive stages of development. a) Late stage 4. The M teloblast pair has just formed by cleavage of the SM proteloblast. Notice that prior to the formation of stem cells, the teloplasmic poles (tp) of the mesoteloblasts face each other. b) Stage 5. As a consequence of the arrangement of the mesoteloblasts, the first m stem cell pair lies in close contact with one another and with their parental cells. c) Early stage 6. The mesoteloblasts have released several stem cells and thus the m bandlets have grown in length to form a bridge that crosses the embryonic midline. Concomitantly, the teloplasmic poles of the teloblasts have moved away from each other due to lateral rotation of the cells. d) Late stage 6. The anterior region of the m bandlets is closely associated with the lower surface of the micromere cap (mc). e,f) early and late stage 7. The m bandlets have grown in length and thickness and form part of the germinal bands. The anterior regions of the m bandlets remain associated with one another and with a cluster of cells that constitutes the primordial prostomium. The available evidence indicates that the anterior regions of the m bandlets probably give rise to the head mesoderm in glossiphoniid embryos [From Fernández and Stent ('80), reproduced by permission of Academic Press Inc.]

Fig. 6. Schematic representation of the arrangement of the N, O, P, and Q teloblasts; the position of their teloplasmic pole; and the development of the n, o, p, and q bandlets in embryos of *T. rude* at successive stages of development. a,b), midstage 6. Formation of stem cells is initiated in the N teloblasts and is soon followed by the Q teloblasts and then by the OP proteloblasts. Thus, the n, q, and op bandlets are formed. Since the teloplasmic pole of the N and Q teloblasts face each other, the nascent n and q bandlets become associated with one another. c) Late stage 6. The OP proteloblasts have cleaved into the O and P teloblasts and the op bandlets of the diagram become the p bandlets. d) Early stage 7. The belatedly initiated pair of bandlets (in the diagram the o bandlets) is shorter than other bandlet pairs because it includes fewer stem cells. e) Late stage 7. The N teloblasts have moved toward the midline and lie next to the Q teloblasts. Accordingly the proximal segment of the n bandlet has crossed under the q, p, and o bandlets, forming a bandlet chiasma. f) Late stage 8. The N teloblasts have rotated by about 250°, the O and P teloblasts by about 180°, and the Q teloblasts by about 90° from the orientation that each cell had at the time it initiated stem cell production. Rotation of the O and P teloblasts is accompanied by their translation toward the midline. As a result of coalescence of the germinal bands, the n bandlet, formerly the most lateral, becomes the most medial bandlet. [From Fernández and Stent ('80), reproduced by permission of Academic Press Inc.]

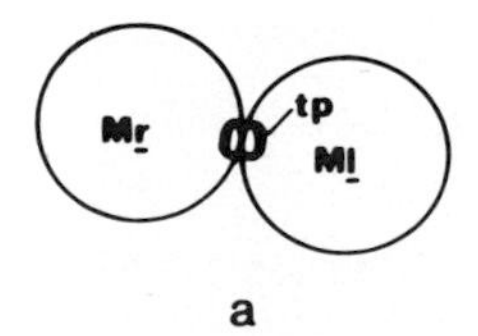

a

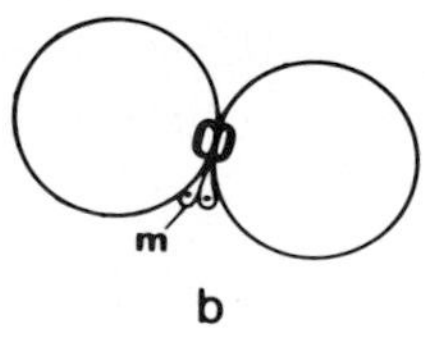

b

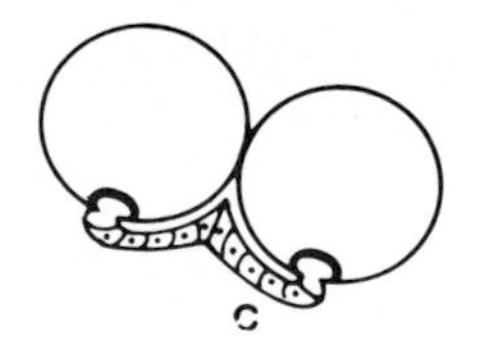
c

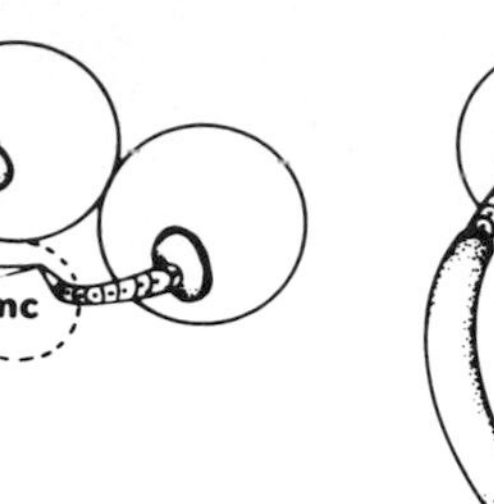

d

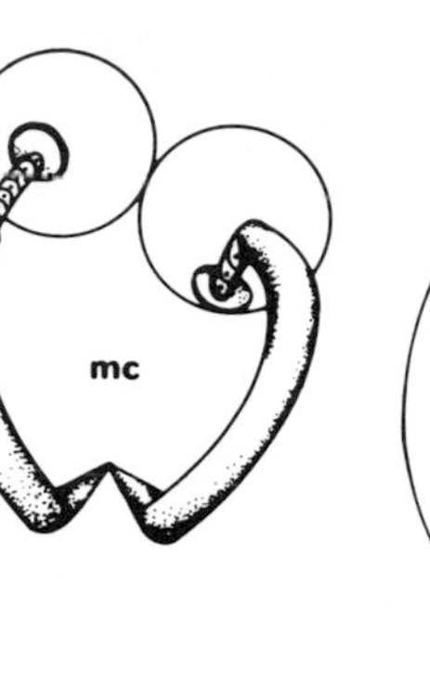

e

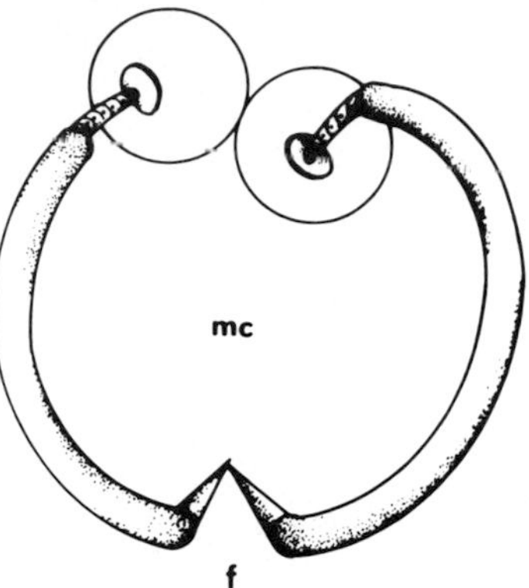

f

5

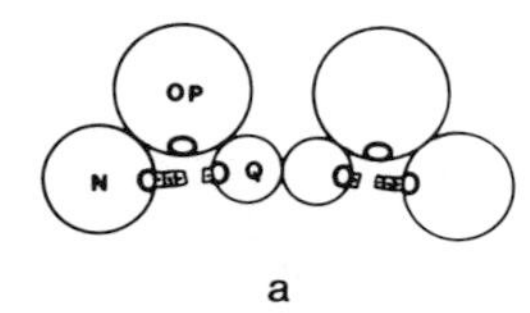

a

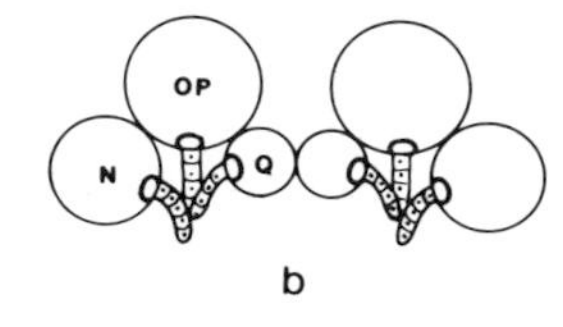

b

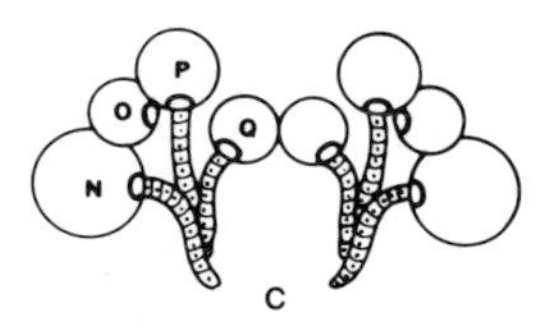

c

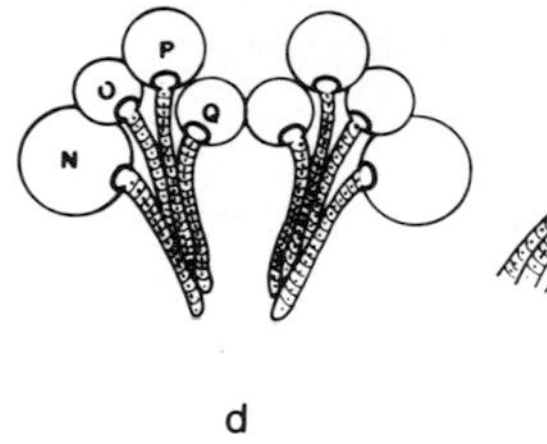

d

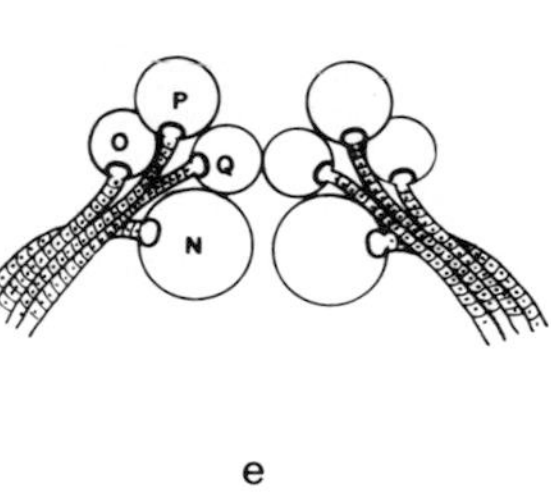

e

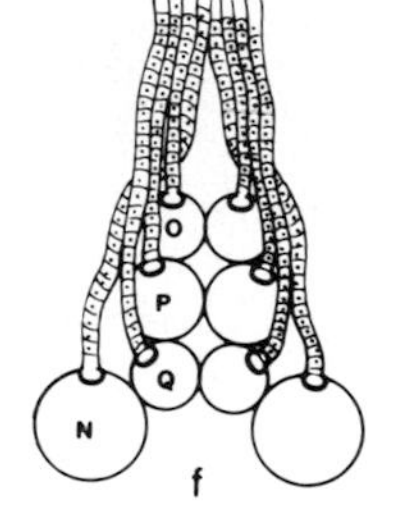

f

6

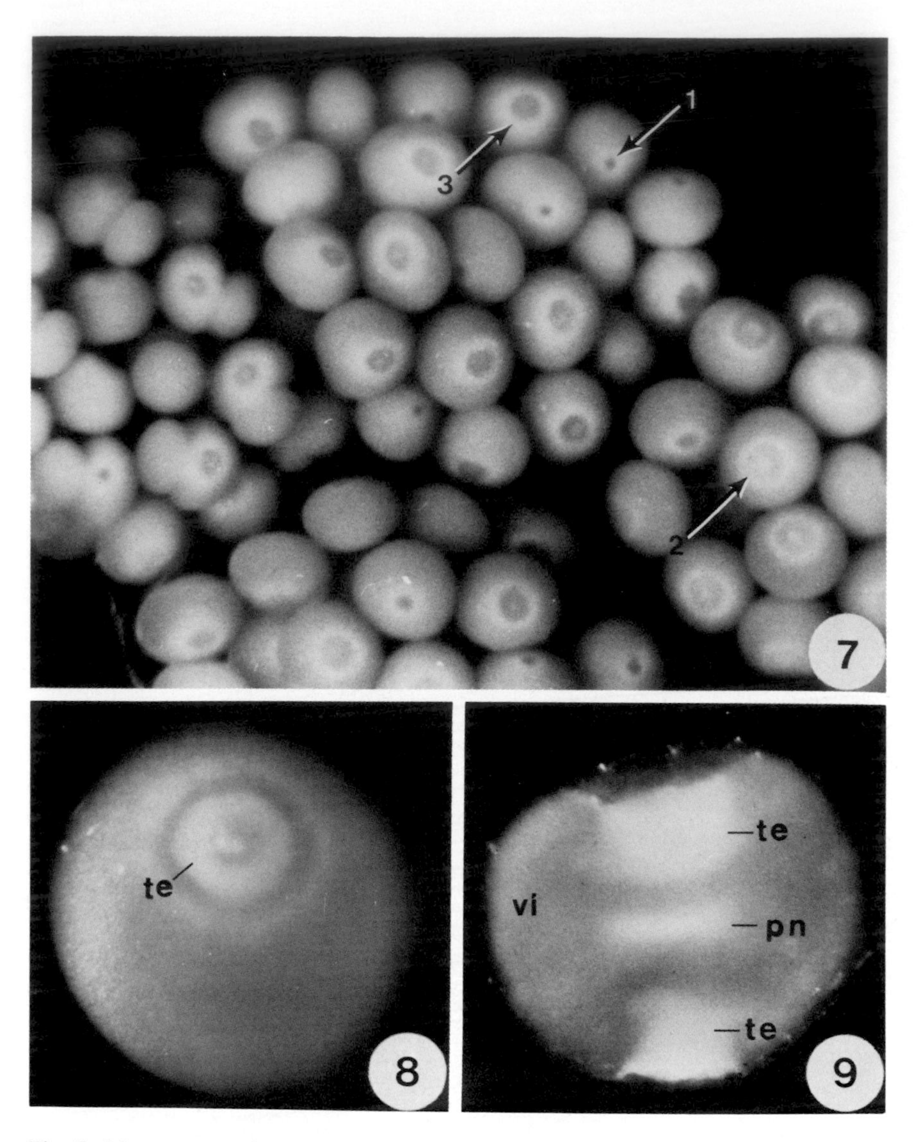

Fig. 7. Live cocoons of *T. rude* show uncleaved and cleaving eggs. The arrows 1 to 3 point to the gradual concentration of white teloplasm at the animal pole of uncleaved eggs. Cleaved eggs show that the CD blastomere is larger than the AB blastomere and includes most of the teloplasm. × 19. [Fron Fernández ('80), reproduced by permission of Academic Press Inc.]

Fig. 8. Live uncleaved egg of *T. rude* that shows the ring of teloplasm (te) of the animal pole. × 67. [From Fernández ('80), reproduced by permission of Academic Press Inc.]

Fig. 9. Transparented uncleaved egg of *T. rude* that shows the distribution of the vitelloplasm (vi), teloplasm (te), and perinuclear plasm (pn). The animal pole of the egg is at the upper part of the figure. ALFAC and methyl benzoate. × 63.

belatedly initiated bandlet crosses its sister bandlet in one or both sides forming a chiasma (Fig. 28). Each germinal band of early stage-7 embryos thus includes five bandlets. With the ongoing production of stem cells and their division into blast cells, the germinal bands advance rostro-laterally and become a pair of crescent-shaped structures lying at the dorso-lateral surface of the embryo. The rostral ends of the germinal bands are separated from each other by a cluster of cells derived from the micromere cap. This cell cluster corresponds to the primordial prostomium. The micromere cap of midstage-7 embryos consists of small rapidly proliferating cells that form a sheet bulging out the surface of the embryo. Because of the sustained proliferation of bandlet cells, the germinal bands grow throughout the embryonic surface. In small and midsize glossiphoniid embryos, such as those of *H. triserialis* and *T. rude*, expansion of the germinal bands and of the micromere cap is such that, by the end of stage 7, most of the dorsal embryonic surface is covered by these structures (Fig. 35). The germinal bands have reached the lateral edge of the dorsal embryonic surface and together constitute the U-shaped germinal bands. In large glossiphoniid embryos, such as those of *H. depressa*, the germinal bands and the micromere cap expand relatively less and, by the end of stage 7, occupy only part of the dorsal embryonic surface (Fig. 34). The germinal bands are 70–90 μm in width in *H. triserialis* and 120–150 μm in width in *T. rude* and *H. depressa*.

Stage 8: Coalescence of the Germinal Bands

As the U-shaped germinal bands of small and midsize glossiphoniid embryos continue their rostro-lateral expansion, they reach the ventral surface of the embryo and finally meet at the ventral midline. Coalescence of right and left germinal bands begins at the rostral end of the embryo just behind the primordial prostomium (Fig. 36). At this stage of development the combined germinal bands may be referred to as the V-shaped germinal bands. As the process of coalescence continues in a rostro-caudal direction, the combined germinal bands may be called the Y-shaped germinal bands (Figs. 37–39). Rostrolateral expansion of the U-shaped germinal bands of large glossiphoniid embryos, however, is not necessarily accompanied by migration of the germinal bands throughout the ventral embryonic surface. Thus, in many cases the coalesced germinal bands remain at the dorsal embryonic surface (Fig. 38). When coalescence of the germinal bands is completed, their tissues constitute the germinal plate. The first signs of peristaltic movements are detected in embryos in which more than one half of their germinal bands have already coalesced.

When coalescence of the germinal bands is completed, a thin membranous envelope is left around the yolky macromeres (Figs. 37, 46, 49). Contraction of cells present in this envelope is responsible for the early peristaltic

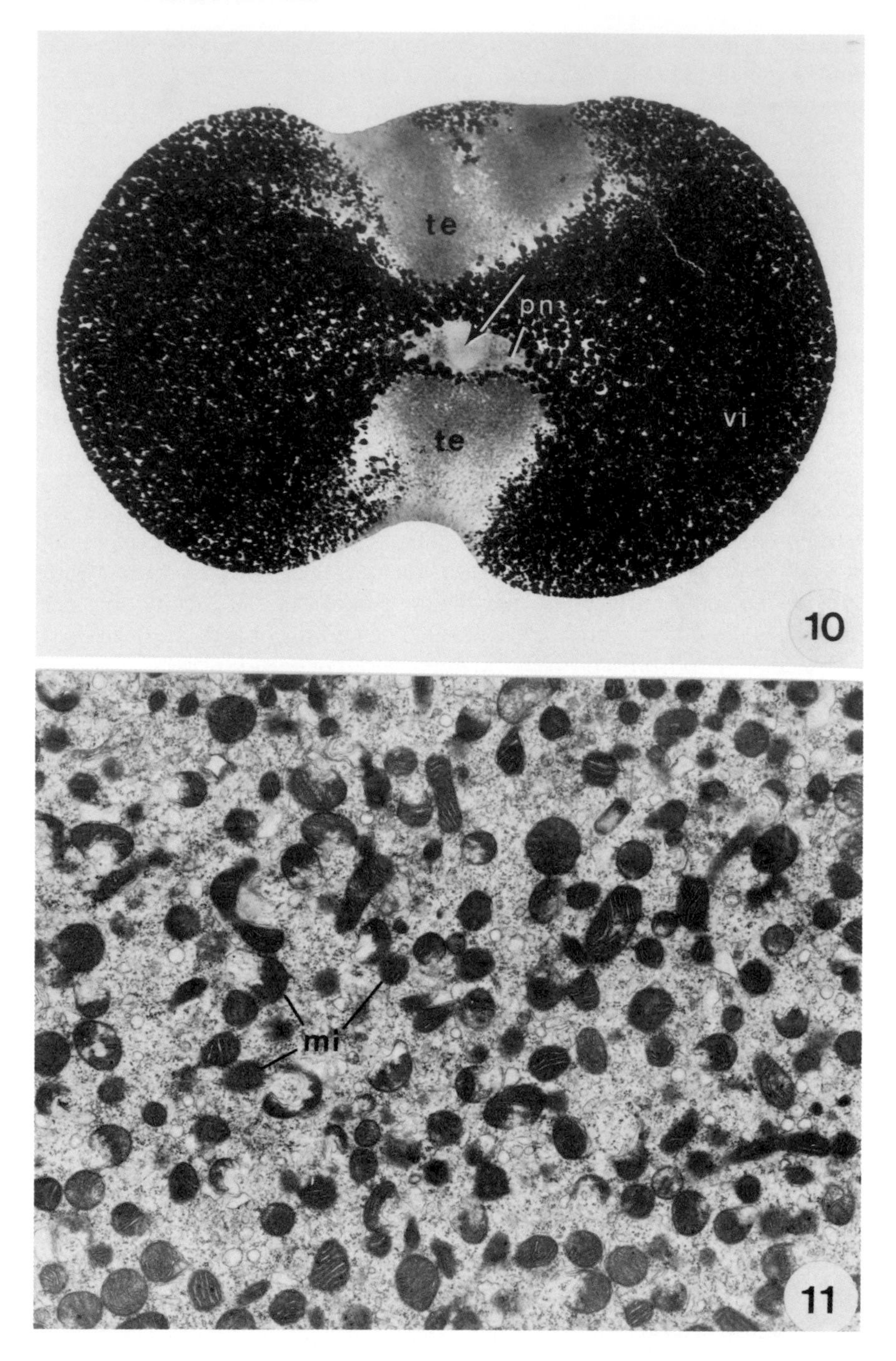
te
pn
te
vi
10
mi
11

movements of the embryo. Although the structure, origin, and fate of this membranous envelope need to be studied in detail, the available evidence indicates the following: a) it consists of more than one cell type, b) its cells probably originate from the micromere cap as well as from tissue of the germinal bands, and c) its cells probably degenerate and thus do not appear to contribute to the formation of embryonic tissue.

Formation and displacement of the germinal bands are accompanied by important changes in the position of the teloblasts. These changes have been studied in some detail in *T. rude* embryos (Fernāndez and Stent, '80). Teloblasts undergo both rotational and translational movements which change both their relative position at the embryonic surface and the position of their teloplasmic pole (Figs. 5, 6). Teloblasts rotate about axes that are approximately perpendicular to the dorsal surface of the embryo. In this manner, when coalescence of the germinal bands is completed, the teloplasmic pole of the teloblasts faces to the rear rather than to the front of the embryo, as was initially the case. Translational movements of the teloblasts lead to changes in their original distribution and also in the topography of the proximal segments of the bandlets. Thus, by the end of stage 9 the arrangement of teloblasts presents the following characteristics: a) the latero-medial cell order is N, Q, P, O, instead of N, O, P, Q; and b) the P and O teloblast pairs meet each other at the ventral midline of the embryo. Because of the relocation of the N teloblast pair, the n bandlet of either side passes below the ipsilateral o, p, and q bandlets to reach the midventral region of the germinal plate. Expansion of the micromere cap and coalescence of the germinal bands may be considered as the major gastrulation movements present in glossiphoniid embryos (see also Anderson, '73). Therefore, when coalescence of the germinal bands is completed and the yolky macromeres and teloblasts become entirely covered by the provisional membranous envelope, the gastrulation process probably reaches an end.

Fig. 10. Sectioned late uncleaved egg of *T. rude* that shows the structure of the egg cytoplasm. The vitelloplasm (vi) includes a large number of darkly stained yolk platelets, whereas the teloplasm (te) and the perinuclear plasm (pn) contain many lightly stained mitochondria. The arrow indicates the region where the egg nucleus is found. The animal pole of the egg is directed toward the top of the figure. Epon-embedded section stained with buffered toluidine blue. × 145.

Fig. 11. Low magnification electron micrograph of the animal teloplasm of a late uncleaved egg of *T. rude*. The teloplasm consists of many mitochondria (mi) that are scattered throughout a membranous-granular matrix. × 5,800.

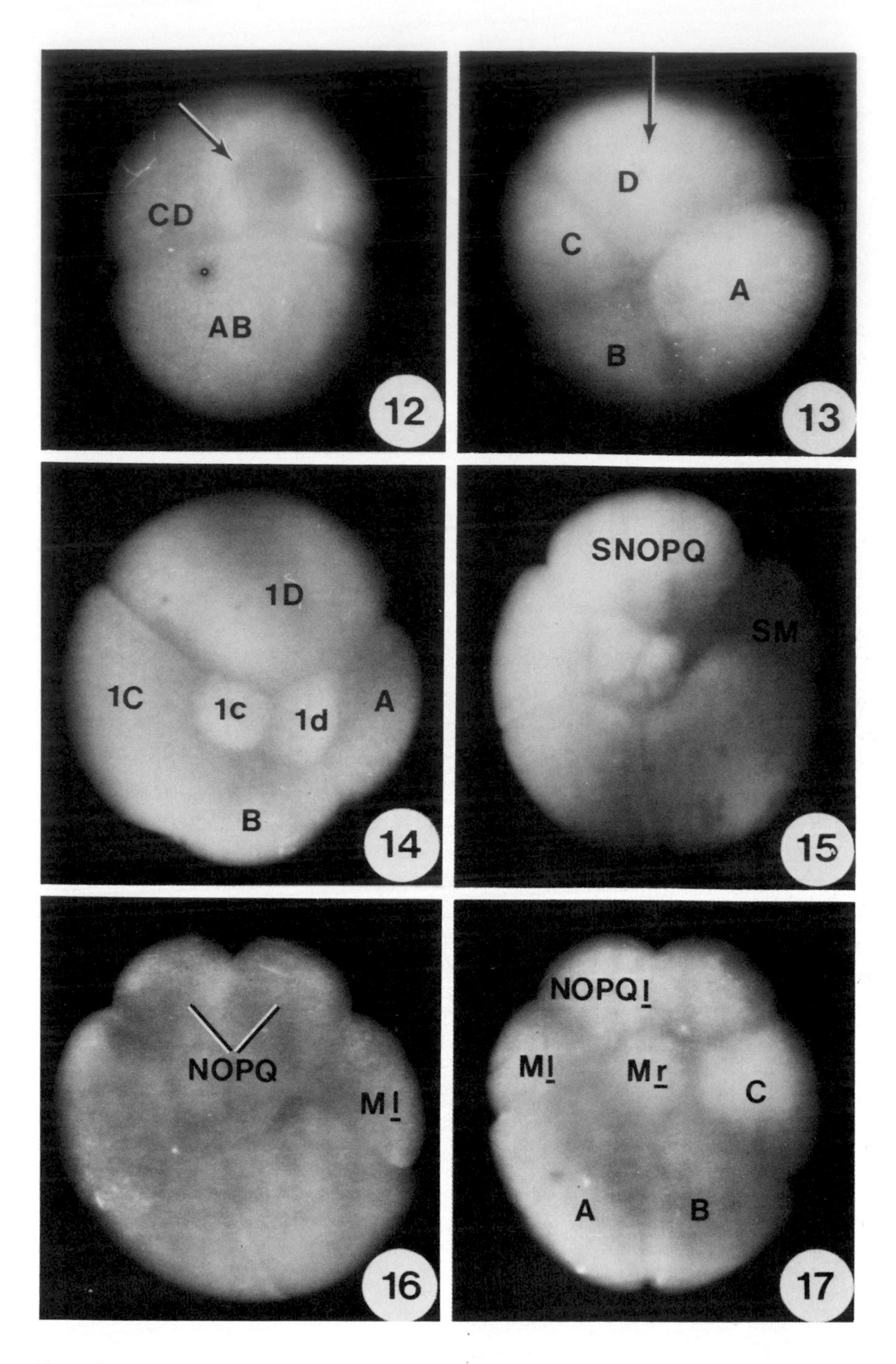

Figs. 12–17. Live embryos of *T. rude* viewed from their animal (Figs. 12–16) or vegetal (Fig. 17) pole. Arrows indicate the position of the teloplasm. Fig. 12, stage 2, × 57. Fig. 13, stage 3, × 65. Fig. 14, early stage 4, × 69. Fig. 15, mid stage 4, × 67. Fig. 16, late stage 5, × 68. Fig. 17, late stage 5, × 65. [From Fernández ('80), reproduced by permission of Academic Press Inc.]

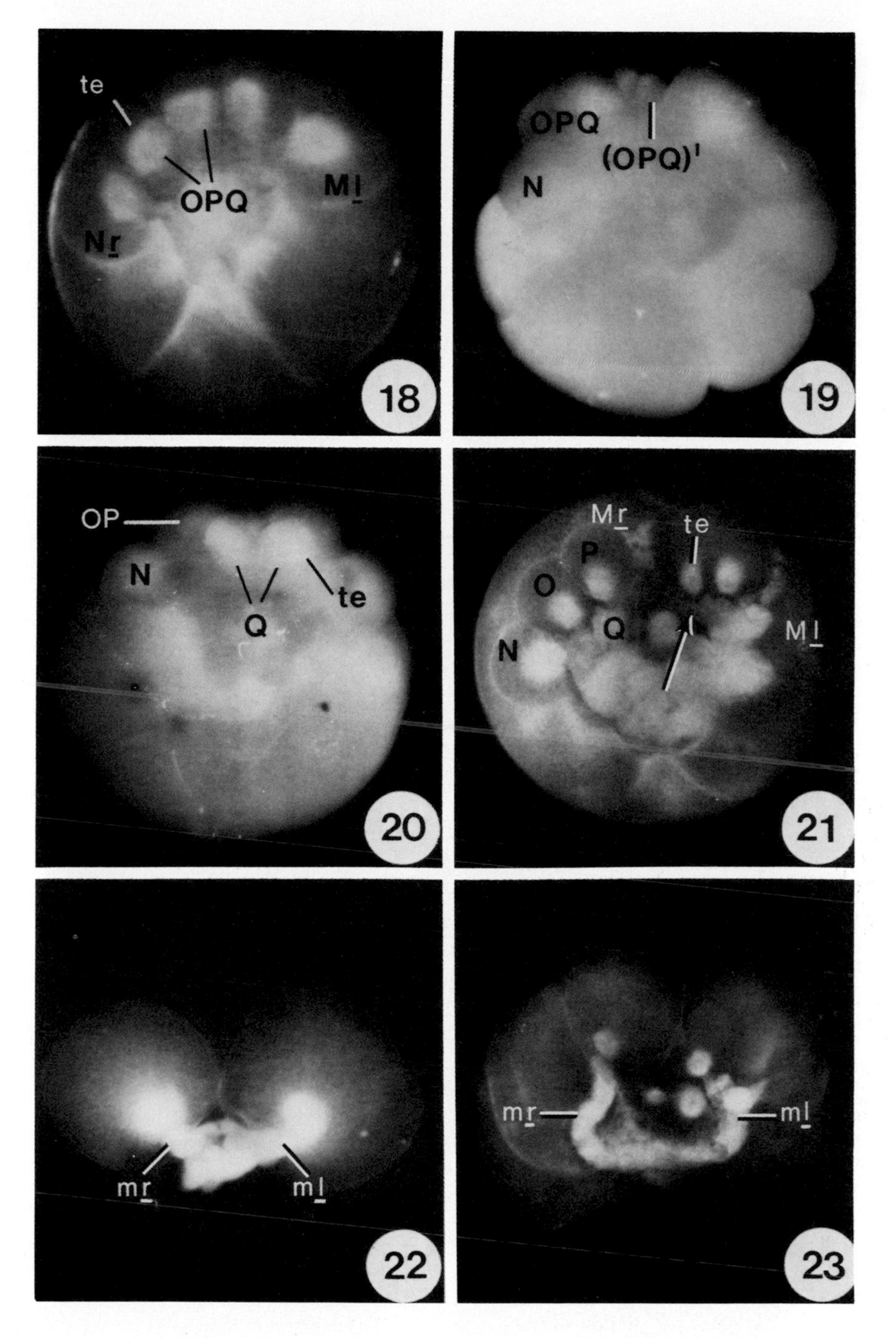

Figs. 18–21. Acid-cleared (Figs. 18, 21) and live embryos (Figs. 19, 20) of *T. rude* viewed from their animal pole. Notice the accumulation of whitish teloplasm (te) in teloblasts and proteloblasts. Fig. 18, early stage 6, × 47. Fig. 19, midstage 6, × 69. Fig. 20, late stage 6, × 64. Fig. 21, early stage 7, × 41. [Figures 19 to 21 from Fernández ('80), reproduced by permission of Academic Press Inc.]

Figs. 22, 23. These micrographs show the m teloblasts and the m bandlets dissociated from a midstage-6 (Fig. 22) and from a late stage-6 (Fig. 23) embryo of *T. rude*. Notice that the right and left m bandlets form an arch extending between the teloplasmic pole of the mesoteloblasts. Fig. 22, × 39. Fig. 23, × 28.

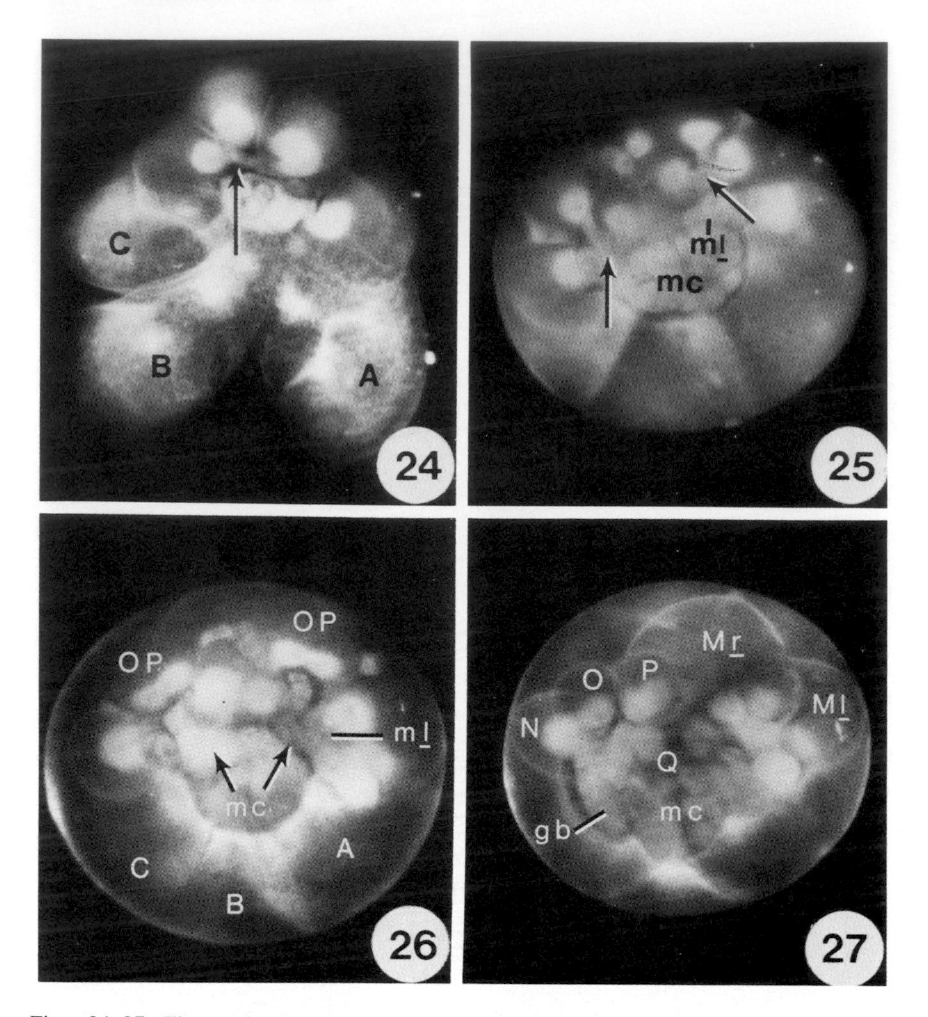

Figs. 24–27. These micrographs show the manner of growth and the arrangement of the nascent n, o, p, and q bandlets of *T. rude* embryos viewed from their dorsal aspect. [From Fernández and Stent ('80), reproduced by permission of Academic Press Inc.] Fig. 24. Partially dissected embryo of midstage 6. On the left side the nascent n and q bandlets have grown toward each other and form a bridge between the N*l* and Q*l* teloblasts (arrow). × 29. Fig. 25. Acid-cleared embryo of late stage 6 showing the association of the n, op, and q bandlets of both sides of the midline (arrows). The proximal segment of the m*l* bandlet reaches the edge of the micromere cap (mc) and joins the left ectodermal bandlets. This is the first sign of formation of the germinal bands. × 42. Fig. 26. Acid-cleared embryo of late stage 6 showing cleavage of the OP proteloblasts. Notice that the teloplasm of these cells has a figure eight shape. The n, op, and q bandlets of either side of the embryo have joined the ipsilateral m bandlet (arrows). mc, micromere cap. × 44. Fig. 27. Acid-cleared embryo or early stage 7 showing the right germinal band (gb) growing around the border of the micromere cap (mc). The M*r* and M*l* teloblasts are seen. The N, O, P, and Q teloblasts form an archlike arrangement. × 42.

Stage 9: Formation of 32 Neuromeres

The first sign of segmentation of the embryo corresponds to the agglomeration of mesodermal tissue in the form of paired rounded structures called prosomites. Prosomites are first detected in the rostral ends of the germinal bands of early stage-8 embryos (Fig. 36). Therefore, segmentation of the mesoderm is initiated sometime before the germinal bands begin to coalesce. For this reason, many somites are already present in embryos whose germinal bands have not yet completed their coalescence (Figs. 39, 42, 44). The next sign of segmentation corresponds to the agglomeration of ectodermal tissue in the form of a rostro-caudal series of paired interconnected cell masses that lie on either side of the midline of the germinal plate. These cell masses correspond to ganglionic primordia or neuromeres (Figs. 45, 46). Neuromeres are first detected in the rostral region of embryos in which approximately two thirds of the germinal bands have already coalesced. Therefore, segmentation of the neuroectoderm, and thus formation of the nerve cord, is initiated in midstage-8 embryos. The fact that segmentation of the mesoderm is initiated ahead of segmentation of the ectoderm determines that formation of 32 somites is completed earlier than formation of 32 neuromeres.

The first four ganglionic primordia develop very short connectives and thus remain close to one another to form the subesophageal ganglion. The next 21 ganglionic primordia develop long connectives and form the abdominal nerve cord. The last seven ganglionic primordia also develop short connectives and form the caudal, or tail, ganglion. The boundaries between successive embryonic body segments are formed by thin transverse sheets of tissue, or intersegmental septa. Since somites and neuromeres are not in register, septa form at the level of developing ganglia. Thus, nerve ganglia attain an intermetameric distribution (Figs. 46–49, 56). The supraesophageal ganglion arises from cells of the primordial prostomium and therefore its origin is different from that of the 32 ganglia of the ventral nerve cord (Weisblat et al., '80a).

By the end of stage 9, all teloblasts have ceased forming stem cells. As is shown in Table V, different teloblasts terminate the formation of stem cells at different stages of development. Thus, the M cells are the first teloblasts to cease forming stem cells at midstage 8, whereas the N cells are the last teloblasts to interrupt formation of stem cells at late stage 9. Since the M and N teloblasts seem to divide at about the same rate, more n than m stem cells are probably formed. Indirect estimates indicate that, in *T. rude* embryos, the M teloblasts produce 30 to 60 stem cells, whereas the N teloblasts produce 50 to 100 stem cells (Fernãndez and Stent, '80). According to these data, each body segment might be founded by one or two pairs of m stem cells. However, there is good evidence that some of the mesodermal stem cells do

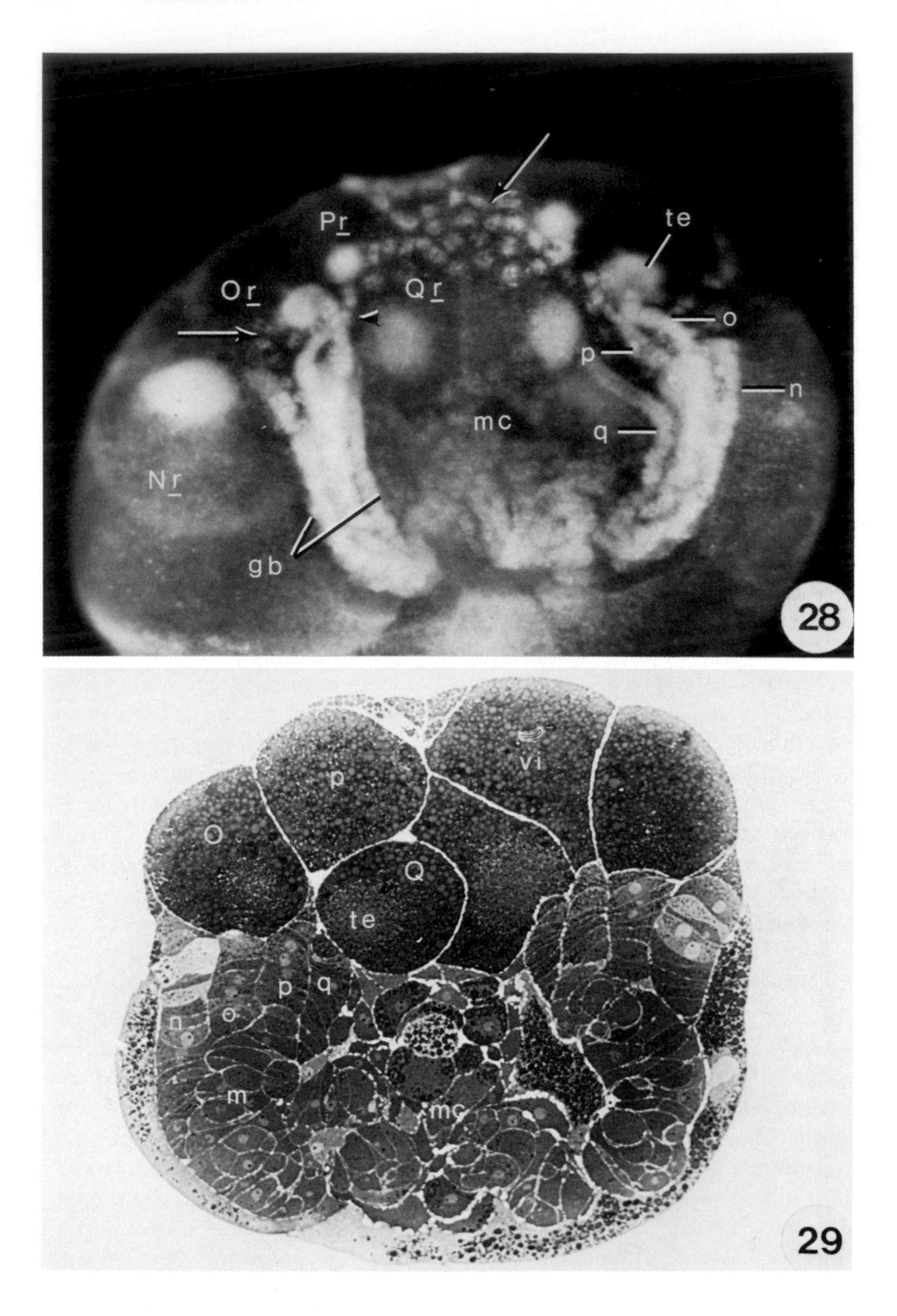
Pr
Or
Qr
te
o
p
n
mc
q
Nr
gb
28
p
vi
O
Q
te
p
q
n
o
m
mc
29

not contribute to the formation of the 32 body segments. Thus, the first m stem cells produced become associated with the primordial prostomium to form the head mesoderm (Fig. 5). The last m stem cells produced, on the other hand, fail to join the germinal bands and some of them probably degenerate. Therefore, it seems likely that each somite is founded by one mesodermal stem cell. Formation of ganglionic primordia is more complicated because several stem cells of different origin contribute to the foundation of each neuromere. The available evidence indicates that the progeny of two pairs of n stem cells and part of the progeny of a yet undetermined number of m, o, p, and q stem cells contribute to the formation of each neuromere (see Fernández and Stent, '80; Weisblat et al., '80a). The origin of the founder cells for the supraesophageal ganglion may be traced back to the A, B, and C macromeres (Weisblat et al., '80a).

Only part of the teloplasm originally present in the teloblasts is used up in the manufacture of stem cells. Thus, interruption of stem cell formation is accompanied by subdivision of the teloplasm remaining in the teloblast. This process is a consequence of division of the teloblast nucleus without subsequent cytokinesis. In this manner, teloblasts become syncytial (Figs. 36, 37).

Hatching of embryos from the perivitelline envelope occurs toward the end of stage 9 or beginning of stage 10. At that time embryos come out of the cocoon and adhere to the walls of the brooding pouch by means of a sticky secretion produced by the embryo.

Stage 10: Segmentation of the Gut

The first pair of nephridial primordia is detected in the sixth segment of midstage-9 embryos. Therefore, formation of nephridia is initiated when approximately half of the neuromeres have already been formed. Further

---

Fig. 28. Early stage 7 embryo of *T. rude*. The N, O, P, and Q teloblasts form an archlike arrangement and the n, o, p, and q bandlets converge to form the germinal bands (gb). Neither the M teloblasts nor the m bandlets are visible because they lie beneath the surface of the embryo. The N*l* teloblast is also not seen because it has already initiated its translational movement toward the midline of the embryo and partly lies below the left germinal band. The (opq)′ and n′ cells (arrows) have proliferated and joined the micromere cap (mc). The arrowhead indicates the region where the o*r* and p*r* bandlets form a chiasma. te, teloplasm. Animal pole aspect of an embryo treated with a mixture of acetic and oxalic acids and cleared with glycerol. × 75. [From Fernández ('80), reproduced by permission of Academic Press Inc.]

Fig. 29. Horizontal section just below the dorsal surface of a midstage-7 embryo of *T. rude*. The O, P, and Q teloblasts and the m, n, o, p, and q bandlets are seen on both sides of the embryo. The n, o, p, and q bandlets are contiguous and form a unicellular epithelial sheet. The m bandlets consist of several layers of cells. Teloplasm (te) is visible in the Q teloblast pair. mc, micromere cap; vi, vitelloplasm. Epon-embedded section stained with toluidine blue. × 160. [From Fernández and Stent ('80), reproduced by permission of Academic Press Inc.]

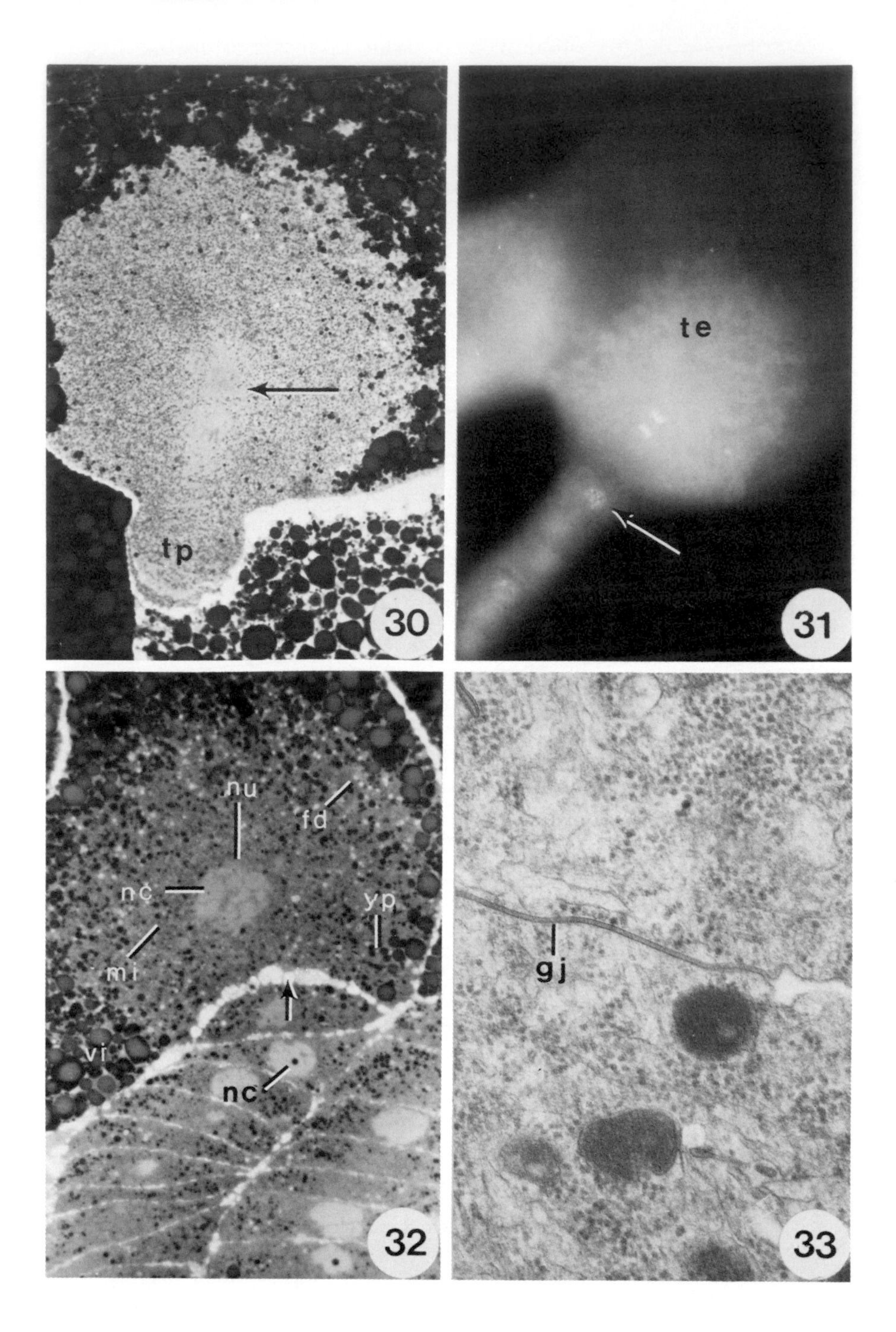
tp
30
te
31
nu
fd
nc
yp
mi
vi
nc
32
gj
33

nephridial primordia arise in a rostro-caudal direction in body segments 7 and 8 and 12 to 22 (Figs. 45, 46). Thus, glossiphoniid leeches form a total of 14 pairs of nephridia. Each nephridial primordium seems to arise from the progeny of one mesodermal precursor cell called the nephridioblast (see Bychowsky, '21). Nephridioblasts form an orderly arrangement in the anterior region of somites and thus their distribution is metameric and bilaterally symmetrical (Figs. 43, 51). Proliferation of a nephridioblast gives rise to a rounded cluster of cells that constitutes an early nephridial primordium (Fig. 47). Growth and differentiation of the tissue of the nephridial primordium occur within the substance of the intersegmental septum (Figs. 49, 51). Completion of nephridial primordia takes place in early stage-10 embryos. At this time, the primitive gut appears as a long cylinder filled with yolk platelets (Fig. 52). This cylinder arises by elongation and fusion of the A, B, and C macromeres and of the teloblasts (see Schmidt, '39; Schuster, '40). Therefore, these cells are the source of the endoderm in glossiphoniid embryos. Elongation of the macromeres is initiated in stage-8 embryos. Segmentation of the primitive gut begins by a series of constrictions of its ventral region that are in register with the intersegmental septa (Figs. 53,

---

Fig. 30. Dividing teloblast of a *T. rude* embryo. Notice that stem cells form as a result of evagination of the teloplasmic pole (tp) of the teloblast. The arrow indicates the position of the mitotic apparatus. Lightly stained granules in the teloplasm correspond to mitochondria. Epon-embedded section stained with toluidine blue. × 835.

Fig. 31. Early anaphase in a dividing teloblast of *H. triserialis* stained with Hoechst dye. Chromosomes present stronger fluorescence than the chromatin of interphase nuclei (arrow). Diffuse fluorescence of the teloplasm (te) is probably caused by the DNA present in the mitochondria. Whole-mounted embryo viewed under epi-illumination with ultraviolet light. × 430.

Fig. 32. The *Ql* teloblast and the *ql* bandlet of the embryo of Figure 29. The teloplasm includes the multiobulated nucleus (nu), which encloses tiny nucleoli (nc). Mitochondria (mi) are abundant and stain light blue, whereas yolk platelets (yp) are few in number and stain dark blue. Lipid droplets (fd) stain light green and are mostly present at the teloplasm–vitelloplasm (vi) interface. The cytoplasm of cells of the bandlets is morphologically similar to teloplasm. However, the nuclei are smaller and the nucleoli are larger in bandlet cells than in teloblasts. The arrow points to the teloplasmic pole of the teloblast, that is, the region that evaginates to form stem cells. The contact region between adjacent cells of the bandlet has a serrated aspect. × 600. [From Fernández and Stent ('80), reproduced by permission of Academic Press Inc.]

Fig. 33. Electron micrograph of a gap junction (gj) between two cells of the n bandlet. Early stage-7 embryo of *H. triserialis.* × 40,600.

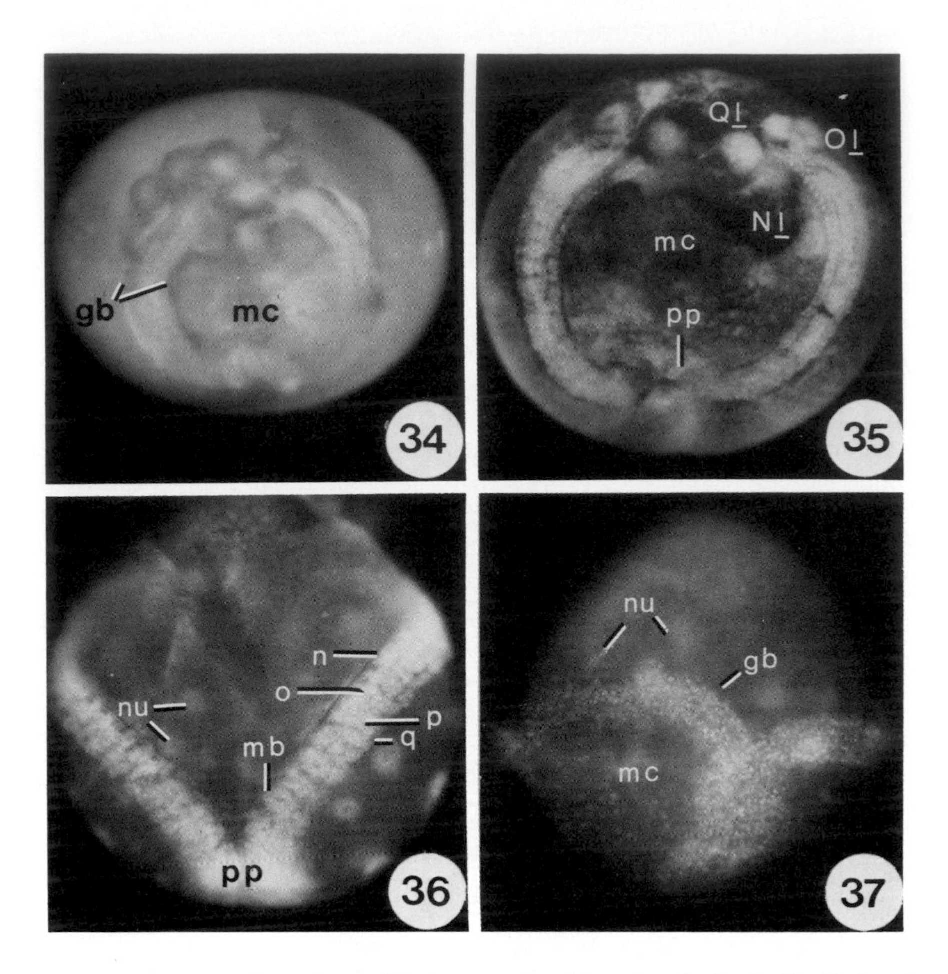

Fig. 34. Late stage-7 live embryo of *H. depressa* viewed from its animal pole. The U-shaped germinal bands (gb) and the micromere cap (mc) occupy most of the dorsal embryonic surface. Notice how clearly one can distinguish different cell types at the surface of a live embryo. × 24.

Fig. 35. Late stage-7 acid-cleared embryo of *T. rude* viewed from its animal pole. The U-shaped germinal bands and the micromere cap (mc) occupy most of the dorsal embryonic surface. The N*r* teloblast is not visible because in its medialward displacement it has been covered by the right germinal band. The N*l* teloblast, however, is visible and has changed its relative position with respect to the O*l* and Q*l* teloblasts (compare with Fig. 28). The primordial prostomium (pp) is present at the rostral pole of the embryo and separates the right and left germinal bands. Embryo treated with a mixture of acetic and oxalic acids and cleared with glycerol. × 43. [From Fernández ('80), reproduced by permission of Academic Press Inc.]

54). As these constrictions grow and reach the dorsal surface of the primitive gut, a series of annular folds of endoderm becomes established. These folds are segmental and vary both in number and morphology in different glossiphoniids. Lateral expansion of the folds is accompanied by intricate branching of the endoderm leading to the formation of the gut caeca (Fig. 55). For the reasons given above, these caeca are segmental structures (Fig. 56). Meanwhile, the lateral borders of the germinal plate have met middorsally and body closure is thus completed. Annulation of body segments and development of paired eyes is usually completed long before morphogenesis of the digestive tract has ended.

PREPARATION OF EMBRYOS FOR MICROSCOPICAL EXAMINATION

Preparation of Whole Mounts

Glossiphoniid leech embryos include a considerable amount of yolk. This material renders cells too opaque for visual examination and also causes poor preservation of the embryos when routine histological techniques are applied. To solve these problems, we have devised very simple and quick techniques that allow detailed study of glossiphoniid embryos. It has been found that some carboxylic acids, such as glacial acetic or formic acids, turn the yolk transparent while the teloplasm and perinuclear plasm remain opaque. This allows for examination of the structure of the cells in whole-mounted embryos.

Preparation of whole mounts of eggs and of early embryos (stages 1 to 3) requires fixation in ALFAC (95% ethyl alcohol, 85 ml; 40% formaldehyde, 10 ml; and glacial acetic acid, 5 ml) for 6 to 12 hours. After dehydration in

---

Fig. 36. Early stage 8 acid-cleared embryo of *T. rude* viewed from its vegetal pole. The germinal bands are now V-shaped. The primordial prostomium (pp) lies rostral to the point of convergence of the right and left germinal bands. Segmentation of the mesoderm and formation of mesodermal blocks (mb), or prosomites, has already been initiated behind the point of convergence of the germinal bands. The relative position of the n, o, p, and q bandlets has changed with respect to the embryonic midline. nu, nuclei in macromere. Embryo treated with a mixture of acetic and oxalic acids and cleared with glycerol. × 43. [From Fernández ('80), reproduced by permission of Academic Press Inc.]

Fig. 37. Midstage-8 acid-cleared embryo of *H. triserialis* stained with Hoechst dye and viewed from its vegetal pole. Notice the accumulation of fluorescent nuclei throughout the Y-shaped germinal bands (gb). Fluorescent nuclei are also present in the expanding micromere cap (mc), which will form the provisional membranous envelope of the embryo. nu, nuclei in macromere. Embryo viewed under epi-illumination with ultraviolet light. × 97.

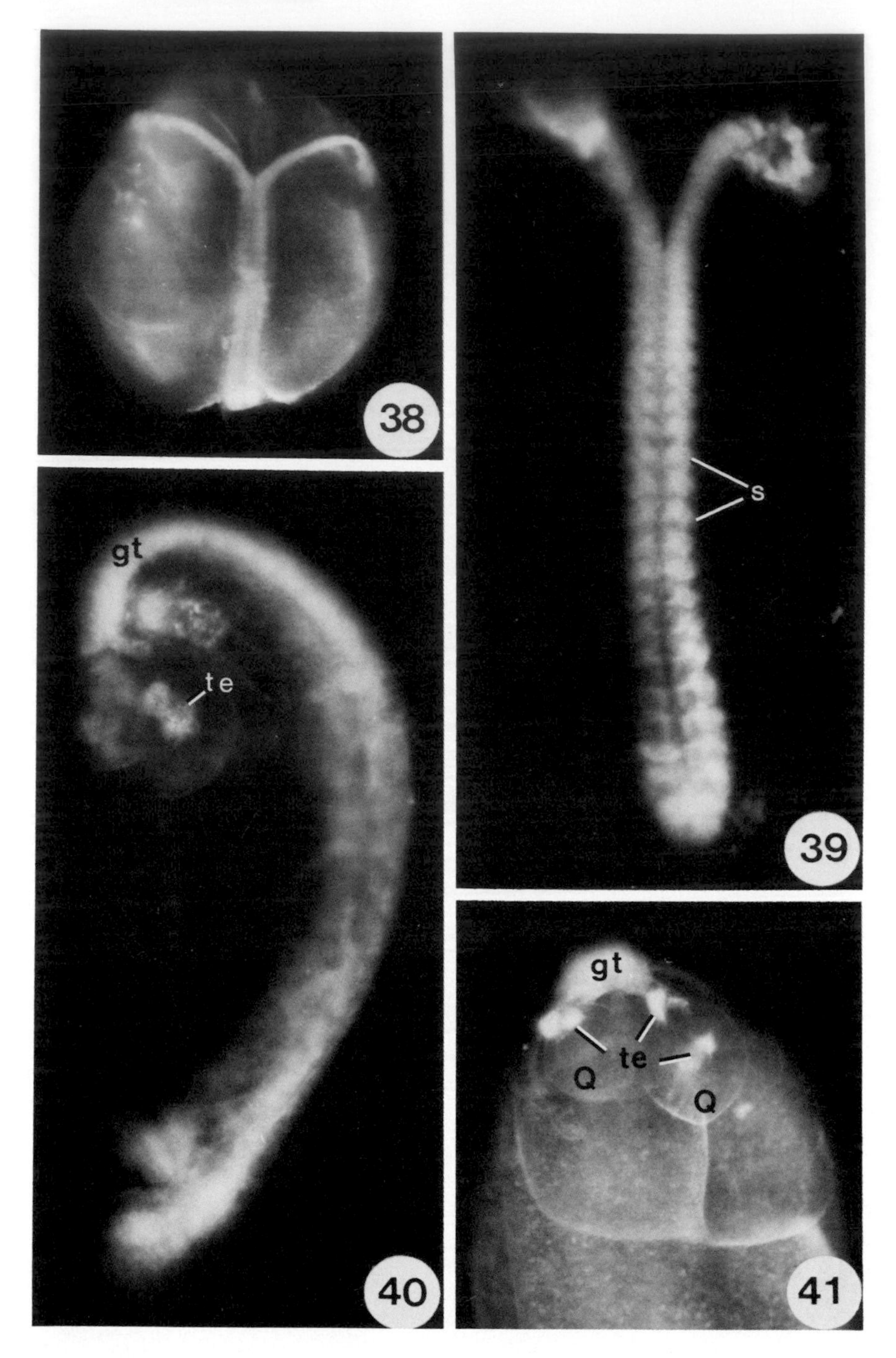
38
39
s
gt
te
40
gt
te
Q
Q
41

graded ethyl alcohol, embryos are cleared and stored in methyl benzoate (Fig. 9). Whole mounts of older embryos (stages 4 to 8) were prepared by fixation in 5% to 10% formaldehyde or Karnovsky solution to which drops of glacial acetic acid were added. Fixation in acid formaldehyde is performed as follows. Living embryos are transferred to a Petri dish containing about 15 ml of the fixative solution. After a few seconds, drops of acid are added to the fixative solution at some distance from the embryo. The transparency achieved by the embryo depends on the amount of acid added to the fixative solution and on the rate at which the dish is shaken. In general, older and larger embryos require greater amounts of glacial acetic acid. Thus, *H. triserialis* and *T. rude* require 3 to 6 drops of acid, whereas *H. depressa* requires 15 to 25 drops of acid (Figs. 18, 21, 25–27, 31, 37, 38, 41). Whole mounts of embryos at stages 4 to 8 may also be prepared by using a combination of glacial acetic and saturated oxalic acids. For this purpose, 3 to 10 drops of glacial acetic acid and then 30 drops of saturated oxalic acid are added to the fixative solution (Figs. 28, 35, 36).

Acid treatment produces swelling of the embryonic cells. The extent of swelling depends on the amount of acid used and on the amount of yolk present in the cells. Thus, cells heavily loaded with yolk, such as macromeres and teloblasts, undergo extensive swelling. Cells that include few yolk platelets, such as those of the germinal bands, may not swell at all. Acid-cleared embryos may be stored for months in the acidified fixative solution. In the case of embryos treated with oxalic acid, it is advantageous to age the

---

Fig. 38. Late stage-8 cleared embryo of *H. depressa* viewed from its animal pole. The germinal bands have not moved toward the vegetal pole and thus their coalescence is taking place in the dorsal embryonic surface. Embryo treated with glacial acetic acid. × 15.

Fig. 39. Germinal plate and germinal bands dissociated from a late stage-8 embryo of *H. depressa*. About four fifths of the germinal bands have already coalesced to form the germinal plate. Segmentation of the mesoderm tissue has so far formed about two dozen somites (s) and prosomites. × 34.

Fig. 40. Side view of the caudal end of a partially dissected *T. rude* embryo at late stage 9. The N teloblast pair has just ceased formation of stem cells and its teloplasm (te) is subdividing into two rounded masses. gt, germinal plate. × 70. [From Fernández and Stent ('80), reproduced by permission of Academic Press Inc.]

Fig. 41. Dorsal view of the caudal end of an acid-cleared *T. rude* embryo of midstage 9. The Q teloblast pair has already ceased formation of stem cells and their teloplasm (te) is subdivided into several fragments. gt, germinal plate. Embryo treated with glacial acetic acid. × 46. [From Fernández and Stent ('80), reproduced by permission of Academic Press Inc.]

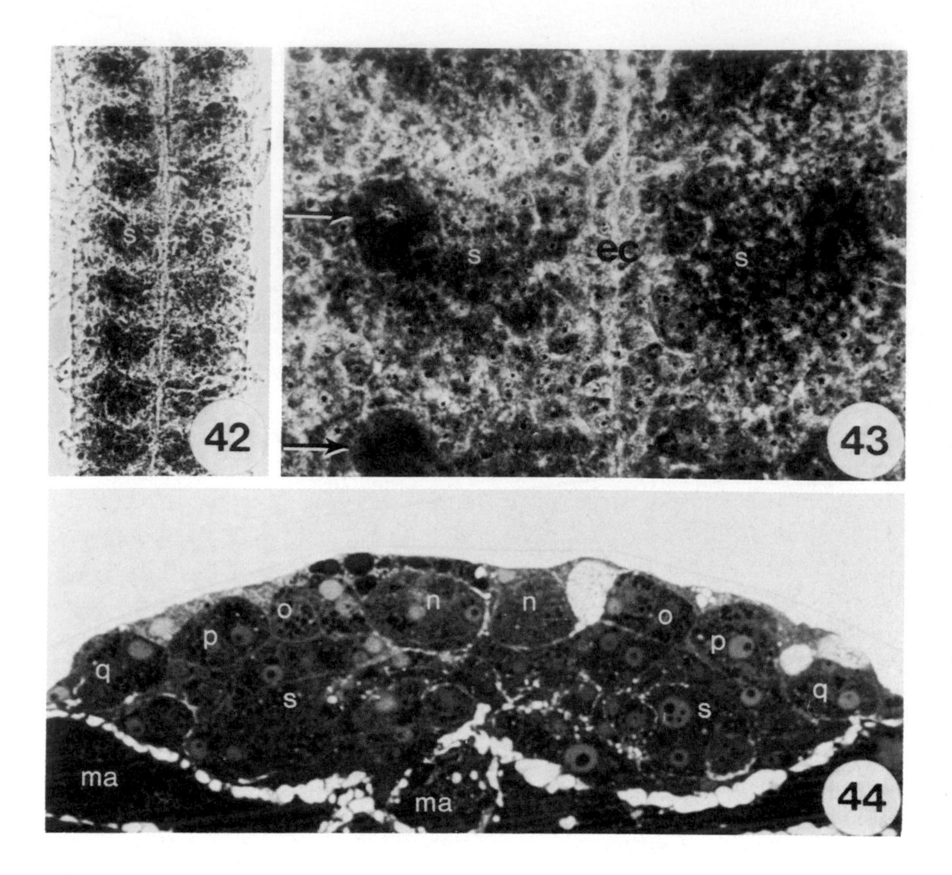

Fig. 42. Whole mount of the germinal plate of Figure 39, which was postfixed in osmium tetroxide and cleared in xylene. Somites (s) appear as rounded clusters of mesodermal cells. Phase contrast. × 74.

Fig. 43. Higher magnification of the germinal plate of Figure 42, showing the structure of the germinal plate. Somites (s) consist of cells that are larger and stain darker than those of the ectoderm. (ec). The latter has not yet segmented and forms an epithelial sheet. The large cell present in the anterior region of each somite corresponds to the nephridioblast (arrows). × 314.

Fig. 44. Transverse section of the germinal plate of a late stage-8 embryo of *T. rude*. The ectoderm is one cell thick and consists of the paired n, o, p, and q bandlets. ma, macromeres; s, somites. Epon-embedded section stained with toluidine blue. × 500.

Fig. 45. Germinal plate dissociated from a midstage-9 embryo of *T. rude*, in which about half of the ganglionic primordia (gp) have been formed. is, intersegmental septa; np, nephridial primordium; pp, primordial prostomium; s, somites. × 56.

Fig. 46. Germinal plate dissociated from an early stage-10 embryo of *H. depressa*. The germinal plate includes 32 neuromeres and 14 pairs of nephridial primordia (np). The first and the last embryonic ganglion of the abdominal nerve cord are marked with arrow-heads. The provisional membranous envelope of the embryo is marked with arrows. ps, developing posterior sucker. × 22.

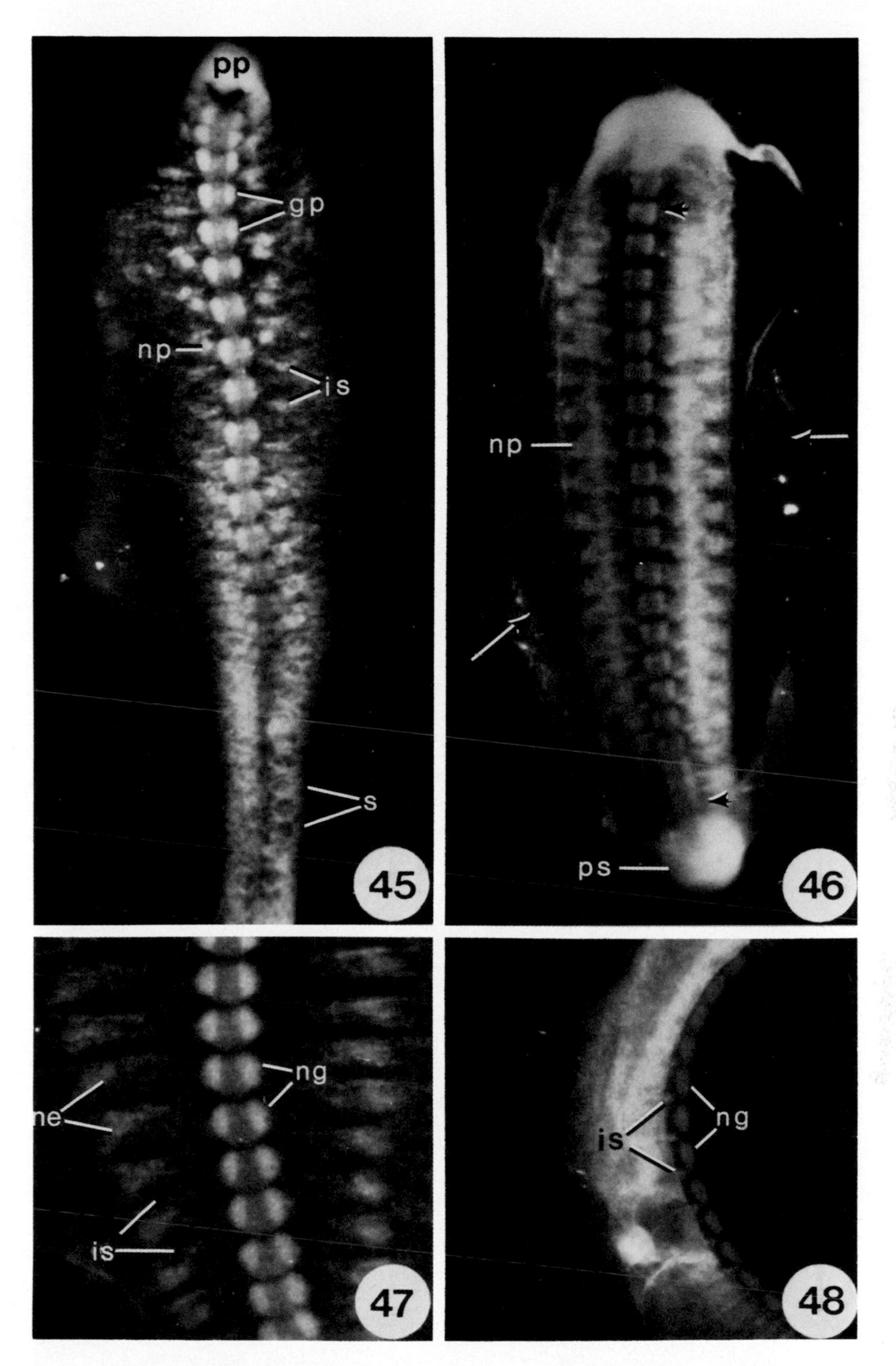

Fig. 47. High magnification of part of the germinal plate dissociated from an early stage-10 embryo of *T. rude*. It shows the distribution of embryonic nerve ganglia (ng) and developing nephridia (ne) with respect to the position of the intersegmental septa (is). × 81.

Fig. 48. Lateral view of an acid-cleared embryo, similar to that of Figure 47, which shows the relationships between embryonic nerve ganglia (ng) and intersegmental septa (is). Embryo cleared with glacial acetic acid. × 38.

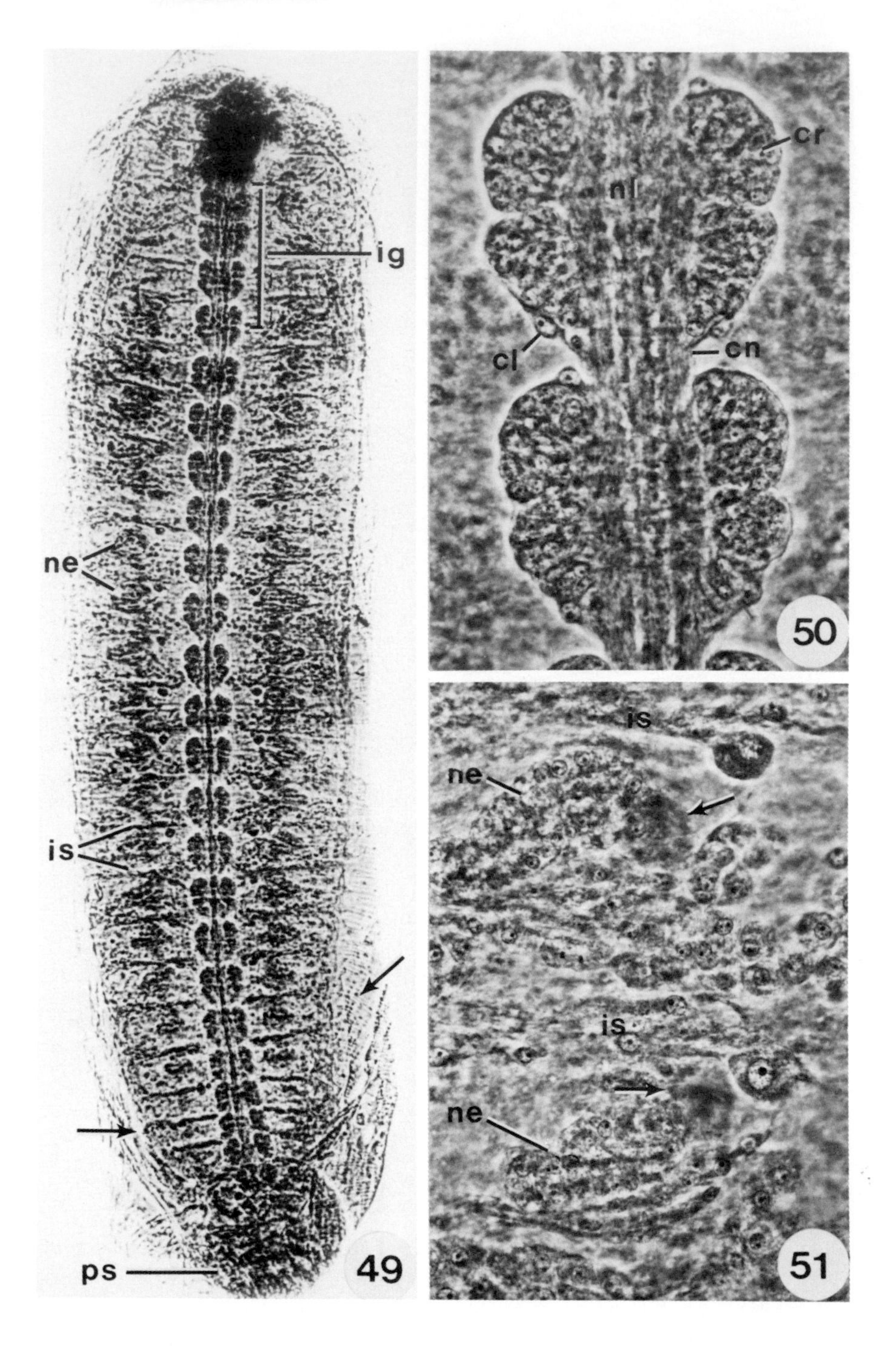
ig
ne
is
ps
49
cr
nl
cl
cn
50
is
ne
is
ne
51

preparation at room temperature or at 4°C. Aging turns the embryos less transparent but apparently increases the deposition of oxalic acid on the teloplasm. Treatment of aged embryos with 30% to 70% aqueous glycerol returns the desired degree of transparency (Figs. 28, 35, 36). Observation of transparented embryos under the dissecting microscope does not require their transfer to glass slides and is done using a dark background and reflected light.

Preparation of Dissociated Embryonic Cells

Treatment of embryos with acidified fixative solution may also be used to dissociate cells. For this purpose, larger amounts of glacial acetic acid are used. Dissociation of embryonic cells is completed with fine needles (Figs. 22–24).

Preparation of Whole-Mounted Germinal Plate

Embryos at stages 9 or 10 are anesthetized in 8% ethyl alcohol in spring water. They are then transferred to 10% formaldehyde solution to which 5 to 10 drops of glacial acetic acid are added. In this solution, embryos are opened along their dorsal region with tweezers or insect pins and their macromeres and teloblasts are dissected out. The germinal plate thus prepared is then transferred to a plastic culture dish containing nonacidified formaldehyde solution. In this solution, the germinal plate becomes sticky and can be stretched out on the bottom of the dish (Figs. 39, 45–48). The stretched germinal plate, with its dorsal region facing up, can then be examined directly under the dissecting microscope using a dark background and reflected light or may be processed for study under the light microscope. In the latter case,

---

Fig. 49. Whole mount of the germinal plate dissociated from an early stage-10 embryo of *H. depressa,* postfixed in osmium tetroxide, and cleared in xylene. The first four embryonic nerve ganglia will constitute the intraesophageal ganglion (ig), whereas the next 21 embryonic ganglia (5 to 25) will constitute the abdominal nerve cord. The intersegmental septa (is) are associated with the postero-lateral region of the embryonic nerve ganglia, which in this manner become intermetameric. Developing nephridia (ne), on the contrary, have a metameric distribution. Arrows point to the provisional membranous envelope of the embryo. ps, developing posterior sucker. Phase contrast. × 61.

Fig. 50. Higher magnification of the abdominal embryonic cord of Figure 49. The neuropile (nl), cell rind (cr), coelomic lining (cl), and connective (cn) of a developing nerve ganglion are indicated. Phase contrast. × 375.

Fig. 51. Higher magnification of two abdominal body segments of the left side of the germinal plate of Figure 49. Each developing nephridium (ne) includes a nephridioblast (arrows). is, intersegmental septa. Phase contrast. × 345.

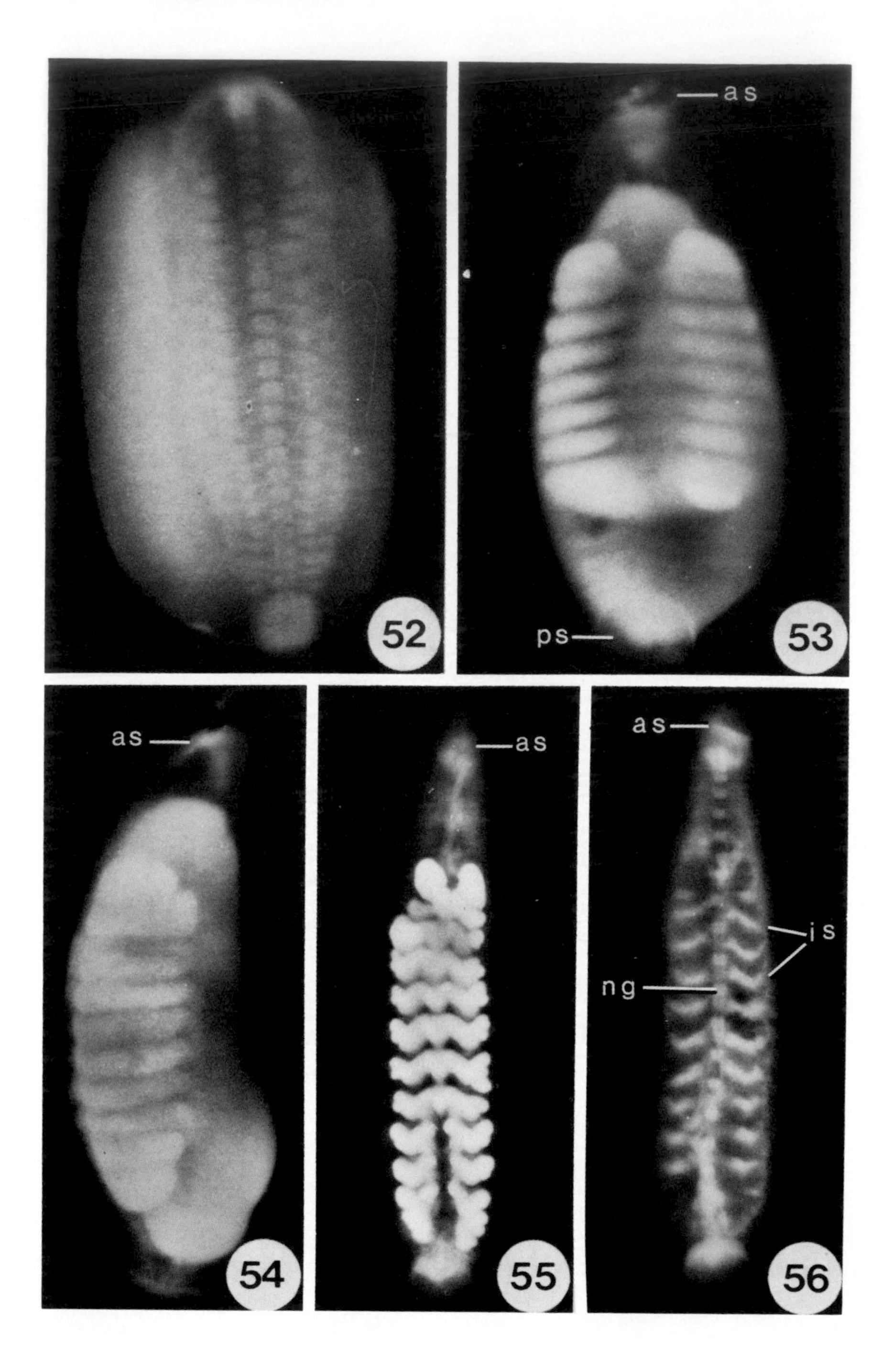
as
ps
53
52
as
54
as
55
as
is
ng
56

the germinal plate is rinsed in several changes of distilled water, postfixed in 1% OsO_4 for 1 hour in the dark, dehydrated in graded ethyl alcohol, cleared in xylene, and mounted in Permount with or without cover slips. If embryos do not remain too long in the acidified solution, very good preservation of cells is achieved. These whole mounts may be advantageously studied under the phase contrast microscope (Figs. 42, 43, 49–51).

Preparation of Embryos for Light- and Electron-Microscope Studies

Embryos are fixed at 4°C for 1 to 2 hours in Karnovsky solution (pH 7.4) diluted with an equal volume of double-distilled water. After fixation, embryos are rinsed in three changes of 0.05 M cacodylate buffer, at the same pH, and then postfixed in 1% OsO_4 for 1 to 2 hours in the dark at room temperature. The large eggs and embryos of *H. depressa* must be sliced or reduced in size during buffer rinsing. After postfixation embryos are dehydrated in graded ethyl alcohol and embedded in Epon 812. To allow reasonable penetration of the resin, the embryos are infiltrated in graded concentrations of Epon in propylene oxide in a vacuum. For light microscopy 1-μm-thick plastic sections are stained with 1% toluidine blue in borate buffer (Figs. 10, 29, 30, 32, 44). For electron micrsocopy thin sections are double stained with alcoholic uranyl acetate and lead citrate (Figs. 11, 33).

Staining of Cell Nuclei With the Fluorescent Dye Hoechst 33258

The blue fluorescing dye Hoechst 33258 (Calbiochem), a DNA-specific stain, may be used for studying the distribution of interphase and mitotic

---

Figs. 52–56. Segmentation of the gut in live anesthesized embryos of *H. depressa* at different stages of development. Fig. 52. Early stage-10 embryo viewed from the ventral aspect of the future leech. The germinal plate, whose structure is similar to that of Figure 49, rests on a huge mass of yolk contained within the primitive gut. Segmentation of the primitive gut, or endoderm, has not yet started. × 42. Fig. 53. Stage-10 embryo viewed from the ventral aspect of the future leech. The first sign of segmentation of the endoderm is the appearance of seven segmental transverse folds in the ventral surface of the primitive gut. as, developing anterior sucker; ps, developing posterior sucker. × 24. Fig. 54. Stage-10 embryo viewed from its lateral aspect. Further progress in the segmentation of the endoderm has increased the number of transverse folds to nine. Moreover, these folds have grown laterally and thus they are closer to the embryonic dorsal midline. as, developing anterior sucker. × 22. Fig. 55. Late stage-10 embryo viewed from the ventral aspect of the future leech. Each of the ten folds of the primitive gut is evolving into one bilateral pair of segmental gut caeca. The opacity of the developing caeca is due to the accumulation of yolk in their interior. as, developing anterior sucker. × 12. Fig. 56. Embryo of Figure 55 viewed from the dorsal aspect of the future leech, after removal of the developing gut caeca. This preparation demonstrates that gut caeca are segmental structures because they form from the endoderm lying between successive intersegmental septa (is). as, developing anterior sucker; ng, embryonic nerve ganglion. × 12.

cells. Provided that the ultraviolet microscope is equipped with epi-illumination, whole embryos, dissociated embryonic cells, or whole germinal plates may be used. Embryos are cleared in formaldehyde–glacial acetic acid using 2.5 μg of the dye per ml of fixative solution. When clearing of the embryos is satisfactory, they are transferred to 75% glycerol containing the same amount of dye (Weisblat et al., '80b). This solution renders embryos more transparent and also allows their storage at low temperatures and darkness. For observation, embryos are transferred to a glass slide and mounted under a cover slip using the glycerol–dye solution (Figs. 31, 37).

MARKING OF EMBRYONIC CELLS

Embryonic cells may be labeled by intracellular injection of tracer molecules that do not cross gap junctions. In this manner, the tracer molecule only passes to the progeny of the injected cell and thus cell lineage determination can be accomplished. Two such tracer molecules have recently been used: horseradish peroxidase and fluorescent dyes coupled to peptides. Tracer molecules are pressure injected within the cells by means of 1- to 3-μm-tip micropipettes. For microinjection, embryos are immobilized by suction at the tip of a hollow plastic tube connected with a syringe or by placing the embryo against a V-shaped device made of cover slips glued to the surface of a glass slide. In general the volume of tracer injected should not be higher than 5% to 10% of the volume of the injected cell. Excessive swelling of the injected cell is usually accompanied by failure to develop normally. The concentration of the tracer must be high enough so that it is not diluted excessively during succesive divisions of the cells. After injection of the tracer, the embryos are allowed to develop to a later stage.

Horseradish Peroxidase (HRP) Injection

A 2% solution of highly purified HRP (Sigma, type IX) in 0.2 M KCl is used (Weisblat et al., '78). The enzyme remains catalytically active within embryonic cells for several days after injection. Embryos are fixed in 2.5% glutaraldehyde in 0.1 M phosphate buffer pH 7.4, or in diluted Karnovsky solution, for 2 hours at 4°C. After several rinses in the buffer, the enzyme activity is demonstrated in the complete embryo by the method of Graham and Karnovsky ('66). The distribution of labeled cells may be studied in whole-mounted embryos or in thick serial sections of embryos embedded in plastic and stained with buffered toluidine blue.

Injection of Fluorescent Tracers

Fluorescent dyes that are coupled to a carrier molecule of some molecular weight do not cross gap junctions. That is the case of fluorescent dyes coupled to oligopeptides. Two such oligopeptide-fluorescent dye complexes have been successfully synthesized and tested by Weisblat et al. ('80b). These are the red fluorescing rhodamine-D-peptide (RDP) and the yellow fluorescing fluorescein-D-peptide (FDP). Use of these fluorescent tracers has the following advantages: a) it is possible to determine whether labeled cells are in interphase or in a mitotic state (labeled embryos may also be incubated in Hoechst 33258 to stain nuclear DNA), b) both fluorescent dyes can be used in combination for double label experiments, and c) presence of fluorescent dyes may be detected in living embryos. After injection of the tracer, whole-mounted embryos are prepared by the acid clearing technique already described. Embryos are mounted with glycerol and examined under the ultraviolet microscope equipped with epi-illumination and the required filters.

DELETION OF EMBRYONIC CELLS

Deletion of embryonic cells may be performed by mechanical or chemical procedures. After piercing the perivitelline membrane, cells may be pulled out of the embryo by means of a small hook prepared from an insect pin. Cells can also be destroyed by pinching their surfaces, and allowing the cytoplasm to leak out. Chemical ablation of cells is performed by microinjection of enzymes such as pronase or DNase. Pronasc (Calbiochem B grade) at a concentration of 0.5% in 50 mM KCl is coinjected with 0.2% Fast green (see Parnas and Bowling, '77; Weisblat et al., '80b). DNase I (Sigma type III) is used at a concentration of 1% in 0.15 M NaCl (S.S. Blair, personal communication). Embryos that have been subjected to this type of intervention are likely to become infected and die. For this reason, embryos that have been operated on are cultured in media containing 50 μg/ml of gentamycin sulfate.

Acknowledgments. Preparation of this chapter was supported by grants B-257-803 and B-1223-822-3 from Servicio de Desarrollo Científico, Artístico y de Cooparación Internacional de la Universidad de Chile.

The authors thank Ildina Cerda, Eugenia Díaz, Víctor Guzmán, and Lilio Yañez for technical assistance and Serena Mann for preparation of the illustrations.

LITERATURE CITED

Anderson, D.T. (1973) Embryology and Phylogeny in Annelids and Arthropods. Oxford: Pergamon Press.

Bathia, M.L. (1970) The segmentation of the Gnathobdellid leeches with special reference to the Indian leech *Hirudinaria* and medicinal leech *Hirudo*. J. Morphol. *132:*361–376.

Bychowsky, A. (1921) Ueber die Entwicklung der Nephridien von *Clepsine sexoculata* Bergmann. Rev. Suisse Zool. *29:*41–131.

Cajal, S.R. (1908) L'hypothèse de la continuité d'Apathy. Trab. Lab. Rech. Biol. Univ. Madrid *6:*21–89.

Costello, D.P., and C. Henley (1976) Spiralian development: A perspective. Am. Zool. *16:*277–291.

Davies, R.W. (1973) The geographic distribution of freshwater Hirudinoidea in Canada. Can. J. Zool. *51:*531–545.

Davies, R.W., and J. Wilkialis (1980) The population ecology of the leech (Hirudinoidea: Glossiphoniidae) *Theromyzon rude*. Can. J. Zool. *58:*913–916.

Dawydoff, C. (1959) Ontogenèse des Annélides. In P.P. Grassé (ed): Traité de Zoologie. Paris: Masson, pp. 594–686.

Fernández, J. (1978) Structure of the leech nerve cord: Distribution of neurons and organization of fiber pathways. J. Comp. Neurol. *180:*165–191.

Fernández, J. (1980) Embryonic development of the glossiphoniid leech *Theromyzon rude:* Characterization of developmental stages. Dev. Biol. *76:*245–262.

Fernández, J., and G.S. Stent (1980) Embryonic development of the gloassiphoniid leech *Theromyzon rude:* Structure and development of the germinal bands. Dev. Biol. *78:*407–434.

Graham, R.C., and M.J. Karnovsky (1966) The early stages of absorption of injected horseradish peroxidase in the proximal tubules of mouse kidney: Ultrastructural cytochemistry by a new technique. J. Histochem. Cytochem. *14:*291–302.

Harant, H., and Grassé, P.P. (1959) Classe des Annélides Achetes ou Hirudinées ou Sangsues. In P.P. Grassé (ed): Traité de Zoologie. Paris: Masson, pp. 471–593.

Inase, M. (1960) The culture solution of the eggs of *Tubifex*. Sci. Rep. Tôhoku Univ. (Biol). *26:*65–67.

Mann, H.K. (1962) Leeches (Hirudinea). Oxford: Pergamon Press

Moore, J. (1959) Hirudinea. In W.T. Edmonson (ed): Fresh Water Biology. New York: John Wiley and Sons, Inc., pp. 542–557.

Mori, Y. (1932) Entwicklung isolierter Blastomeren und teilweise abgetöteter älterer Keime von *Clepsine sexoculata*. Z. Wiss. Zool. *141:*399–431.

Müller, K.J. (1932) Uber normale Entwicklung, inverse Asymmetrie und Doppelbildungen bei *Clepsine sexoculata*. Z. Wiss. Zool. *142:*425–490.

Parnas, I., and D. Bowling (1977) Killing of single neurones by intracellular injection of proteolytic enzymes. Nature (Lond.) *270:*626–628.

Retzius, G. (1891) Biologische untersuchungen. Neue Folge II. Stockholm: Sampson and Wallin.

Reverberi, G. (1971) Annelids. In G. Reverberi (ed): Experimental Embryology of Marine and Fresh-Water Invertebrates. Amsterdam–London: North-Holland Publishing Company, pp. 126–163.

Sánchez, D. (1908) El sistema nervioso de los hirudineos. Trab. Lab. Invest. Biol. Univ. Madrid *7:*31–187.

Sawyer, R.T. (1971) The phylogenetic development of brooding behavior in the Hirudinea. Hydrobiologia *37:*197–204.

Sawyer, R.T. (1972) North American freshwater leeches, exclusive of the Piscicolidae, with a key to all species. Biol. Monogr. *46:*1–155.

Sawyer, R.T., F. Lepont, D.K. Stuart, and A.P. Kramer (1981) Growth and reproduction of the giant glossiphoniid leech *Haementeria ghilianii*. Biol. Bull. *160:*322–331.

Schleip, W. (1936) Ontogenie der Hirudineen. In H.G. Bronn (ed): Klassen und Ordnungen des Tierreichs. Leipzig. Vol. 4, Div. III, Book 4, Part 2. Frankfurt/Main: Akad. Verlagsgesellschaft, pp. 2–121.

Schmidt, G.A. (1939) Dégénérescence phylogénétique des modes de développement des organes. Arch. Zool. Exp. Gén. *81:*317–370.

Schuster, M. (1940) Die Entwicklungsgeschichtliche Bedeutung der Teloblasten bei *Glossiphonia complanata* L.Z. Wiss. Zool. Leipzig. *153:*393–461.

Weisblat, D.A., R.T. Sawyer, and G.S. Stent (1978) Cell lineage analysis by intracellular injection of a tracer enzyme. Science *202:*1295–1298.

Weisblat, D.A., G. Harper, G.S. Stent, and R.T. Sawyer (1980a) Embryonic cell lineages in the nervous system of the glossiphoniid leech *Helobdella triserialis.* Dev. Biol. *76:*58–78.

Weisblat, D.A., S.L. Zackson, S.S. Blair, and J.D. Young (1980b) Cell lineage analysis by intracellular injection of fluorescent tracers. Science *209:*1538–1541.

Whitman, C.O. (1878) The embryology of *Clepsine.* J. Microsc. Sci. *18:*215–315.

Whitman, C.O. (1887) A contribution to the history of the germ-layers in *Clepsine.* J. Morphol. *1:*105–182.

Wilkialis, J., and R.W. Davies (1980) The reproductive biology of *Theromyzon tessulatum* (Glossiphoniidae: Hirudinoidea), with comments on *Theromyzon rude.* J. Zool. Lond. *192:*421–429.

Developmental Biology of Freshwater Invertebrates, pages 363–398

Developmental Biology of the Tardigrada

Diane R. Nelson

ABSTRACT A review of the literature indicates a vast deficit in our knowledge of the biology of the Tardigrada. Discovered more than 200 years ago, these hydrophilous micrometazoans occupy a diversity of niches in freshwater, terrestrial, and marine environments throughout the world. They are metabolically active only when a film of water surrounds the body; upon desiccation, they undergo anhydrobiosis (cryptobiosis). The phylogenetic position of the group is uncertain since they possess a combination of morphological characters similar to the aschelminth complex and the arthropods. The morphology, natural history, and systematics of tardigrades have been studied the most thoroughly, but should be reevaluated with modern techniques. Collection and culturing procedures are described to enable the researcher to pursue other aspects of tardigrade biology, which will require extensive classical as well as current methods of investigation. Reproductive modes in tardigrades are reviewed, including sexual reproduction, parthenogenesis, polyploidy, and hermaphroditism. Recent studies of spermiogenesis and oogenesis and the early observations of mating and fertilization are discussed. Embryological development as described by Marcus with the light microscope is outlined; however, ultrastructural studies with electron microscopy are needed to elucidate the process. Postembryological development, including the process of periodic molting throughout the life of the tardigrade, is described based on classical life-history studies and recent electron-microscopic investigations. The nascent state of tardigrade biology provides numerous opportunities for productive biological investigation.

INTRODUCTION

General Characteristics

The phylum Tardigrada is composed of hydrophilous micrometazoans that occupy a diversity of freshwater, marine, and terrestrial habitats. Commonly called "water bears," these bilaterally symmetrical animals have a cylindrical body with four pairs of legs usually terminating in claws of various numbers and shapes. The body is generally flattened on the ventral side, convex on the dorsal side, and somewhat indistinctly divided into five body segments: a head (cephalic) segment, three trunk segments bearing the first three pairs of

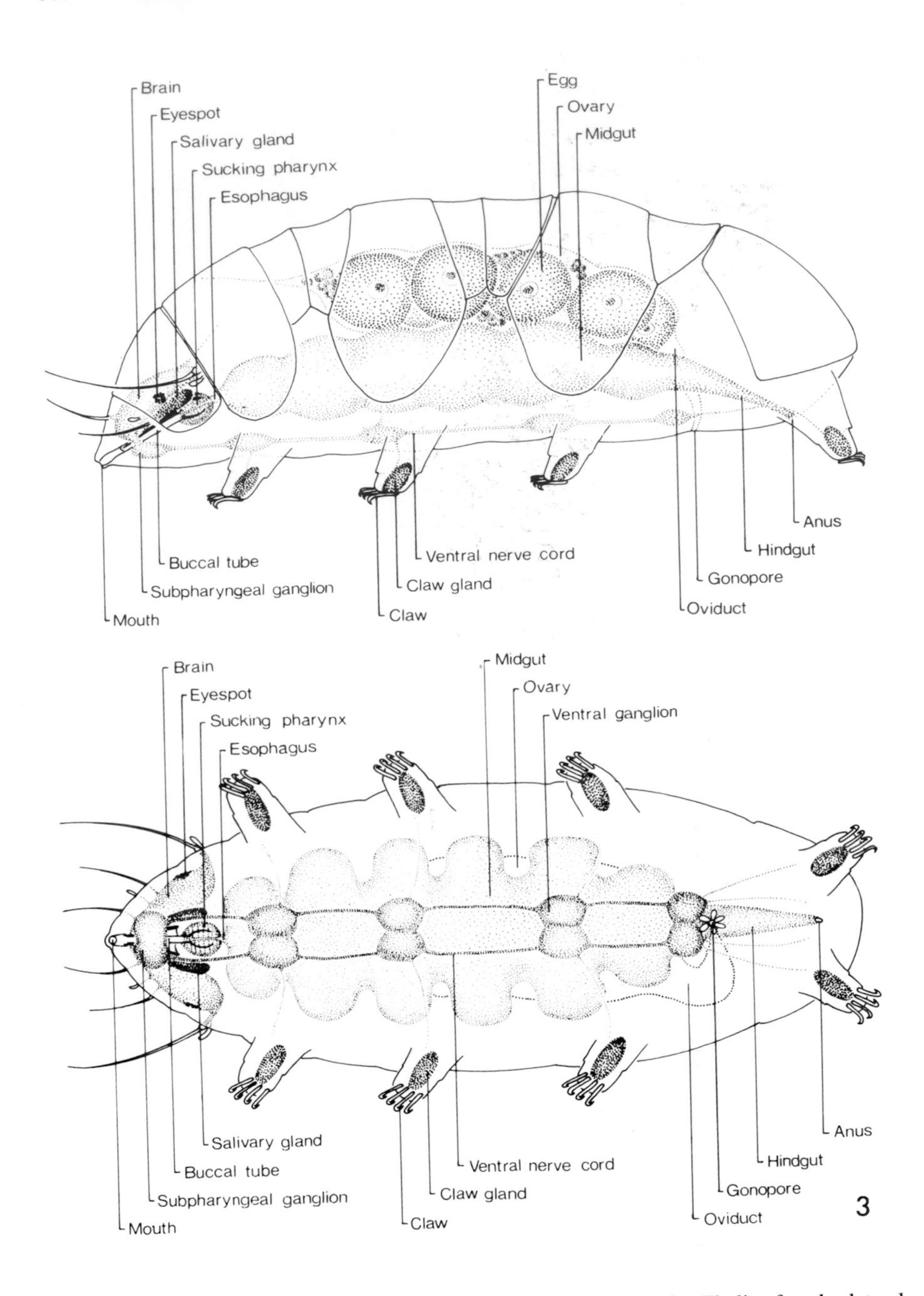

Fig. 3. Internal anatomy of a heterotardigrade, *Bryodelphax parvulus* Thulin, female; lateral view (above) and ventral view (below). (Courtesy of R.P. Higgins.)

Fig. 4. A eutardigrade, *Macrobiotus hufelandi* Schultze, from Roan Mountain, Tennessee, lateral view. × 380.

Fig. 5. *Macrobiotus hufelandi*, anterior view. × 700.

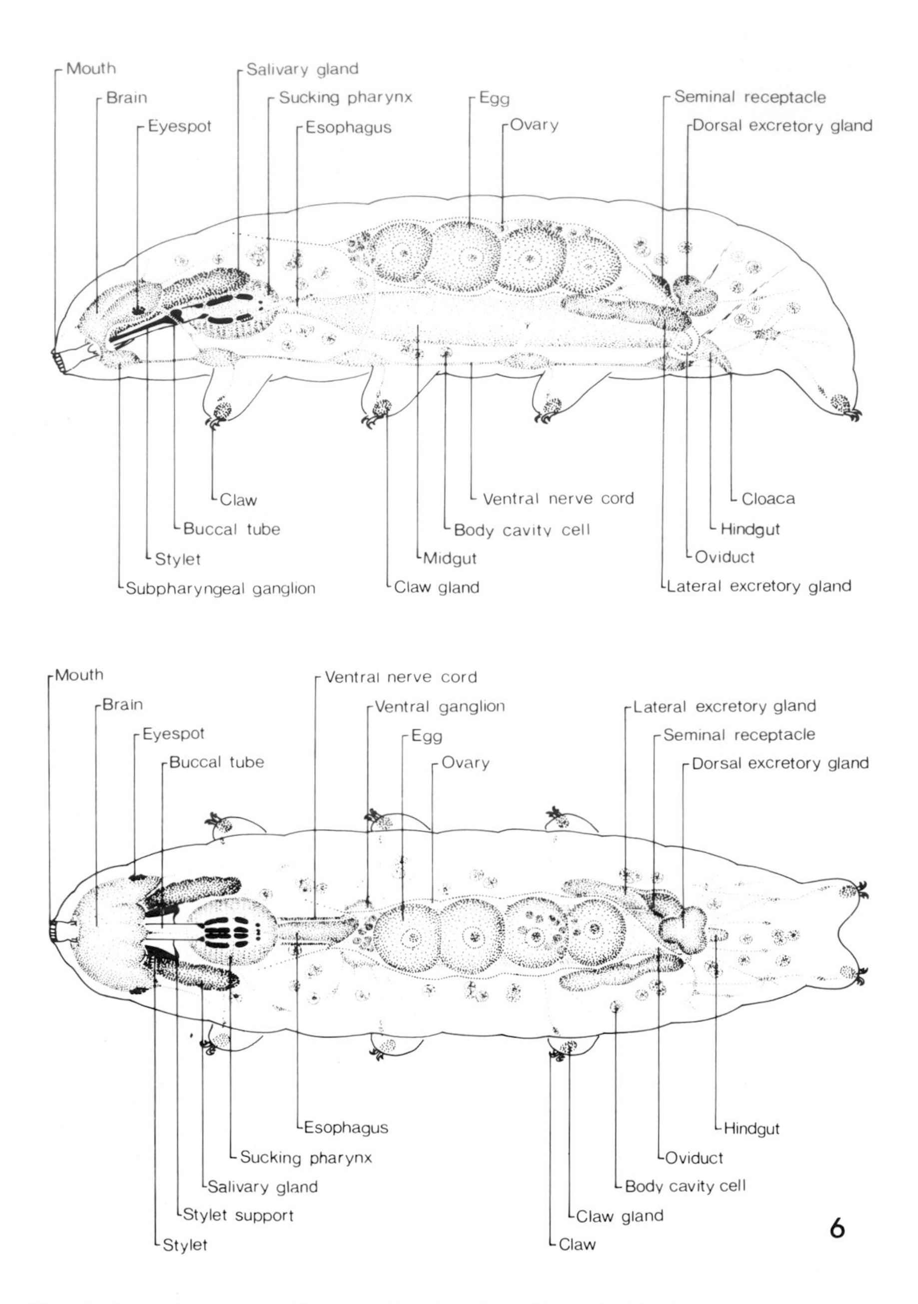

Fig. 6. Internal anatomy of a eutardigrade, *Macrobiotus hufelandi,* female; lateral view (above) and dorsal view (below). (Courtesy of R.P. Higgins.)

In addition to cuticular structures, tardigrade systematics is based on the morphology of the claws, buccal (feeding) apparatus, and eggs (Schuster et al., '80b). The size, shape, and number of claws are diagnostic of certain genera. The eutardigrades usually possess two double claws per leg, which may be similar or dissimilar in size, shape, and symmetry (Fig. 5). Members of the terrestrial heterotardigrade family Echiniscidae have four single claws per leg (Fig. 1). The complex buccal apparatus (Figs. 3, 6) consists of a buccal tube, a muscular sucking pharynx, and a pair of piercing stylets that can be extended through the mouth opening. Some variations in the feeding apparati may be correlated with food sources, which include organic detritus, bacteria, algae, other plant cells, protozoans, rotifers, nematodes, or even other tardigrades (Fig. 8). Sexual reproduction, parthenogenesis, and hermaphroditism occur in tardigrades, and the eggs, which are often essential for the determination of freshwater and terrestrial species, may be smooth or ornamented with processes. Eggs may be deposited singly outside the body of the female, in groups attached to the substrate, or inside the molted cuticle (exuvium). Development is direct, and an increase in size is the primary change that occurs during the successive molts.

The morphology of the body systems is thoroughly discussed in other sources (Ramazzotti, '72; Pennak, '78; Greven, '80; Nelson and Higgins, '83). The digestive tract is divided into a foregut, midgut, and hindgut (Figs. 3, 6). Both the foregut (buccal apparatus and esophagus) and hindgut are presumably of ectodermal origin and are lined with cuticle, which is shed prior to molting. In the Eutardigrada the hindgut is a true cloaca which receives the contents of the midgut, Malpighian tubules, and genital ducts; the cloacal opening is a transverse slit that opens ventrally just anterior to the fourth pair of legs (Figs. 9, 10). In the Heterotardigrada, the systems are separate. The oviduct leads to a rosette-shaped pre-anal gonopore, whereas the anus is a small longitudinal slit between the fourth pair of legs (Figs. 11, 12). First-instar heterotardigrades in the family Echiniscidae have a mouth but no gonopore or anus.

Four possible excretory methods were discussed by Ramazzotti ('72): 1) Through the salivary glands when the buccal apparatus is shed prior to ecdysis; 2) through ecdysis, when the old cuticle, which accumulates excretory granules, is shed; 3) through the wall of the midgut into the body fluid; and 4) through excretory glands (Malpighian tubules), which are present in eutardigrades (Fig. 6).

The nervous system (Figs. 3, 6) includes a dorsal brain (cerebral ganglion), a circumesophageal connective, a subesophageal ganglion, a circumpharyngeal connective, and four ventral trunk ganglia joined by paired nerve tracts. A cerebral eye is located in the lateral lobes of the brain in eutardigrades and some heterotardigrades. The muscular system consists of pharyn-

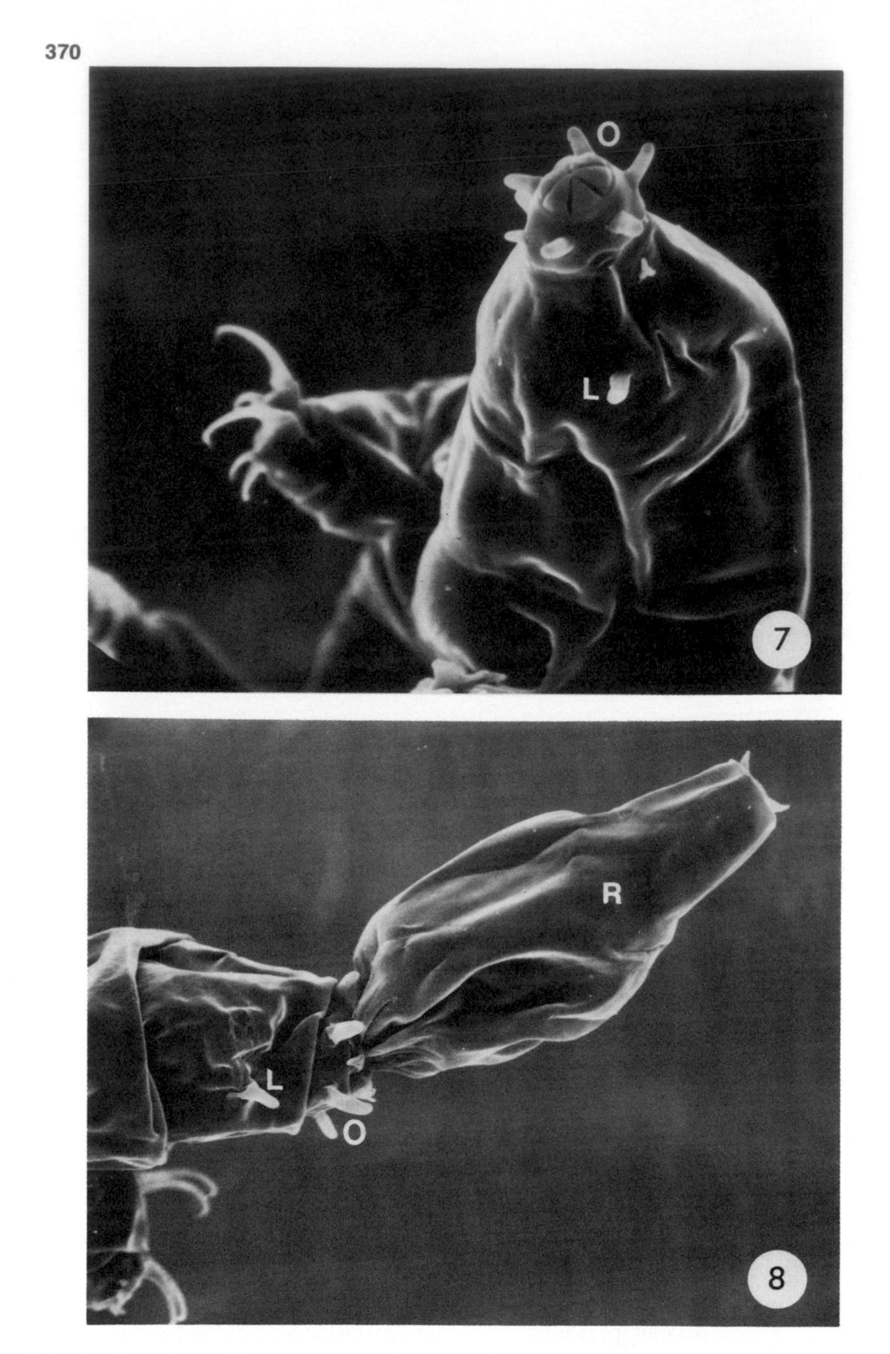

Fig. 7. Cephalic papillae of the eutardigrade *Milnesium tardigradum* Doyère, family Milnesiidai, × 1,070. Abbreviations: O, oral papilla; L, lateral papilla.

Fig. 8. *Milnesium tardigradum* feeding on a bdelloid rotifer (R), from Roan Mountain, Tennessee, × 630. Abbreviations: O, oral papilla; L, lateral papilla.

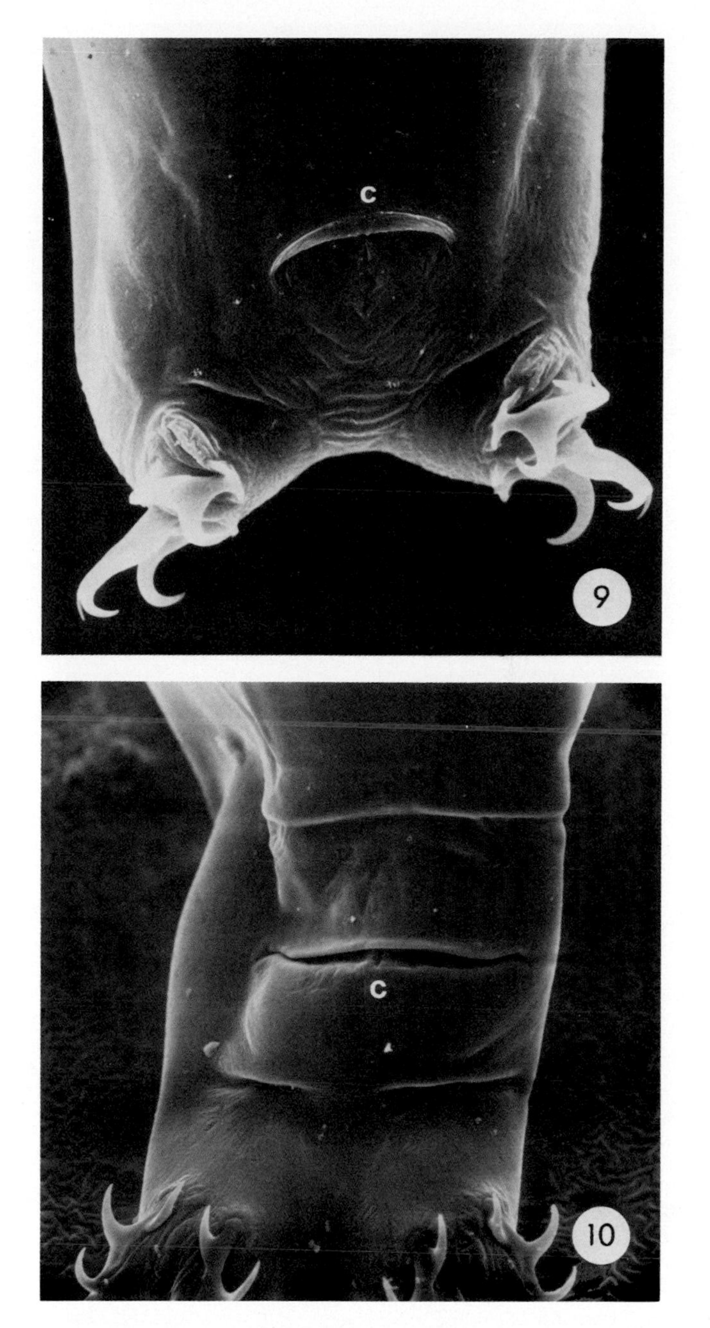

Fig. 9. Cloacal opening (c) of a eutardigrade, genus *Hypsibius,* from New Zealand. × 2,800.

Fig. 10. Cloacal opening (c) of a eutardigrade, *Milnesium tardigradum.* × 960. (Courtesy of Robert O. Schuster.)

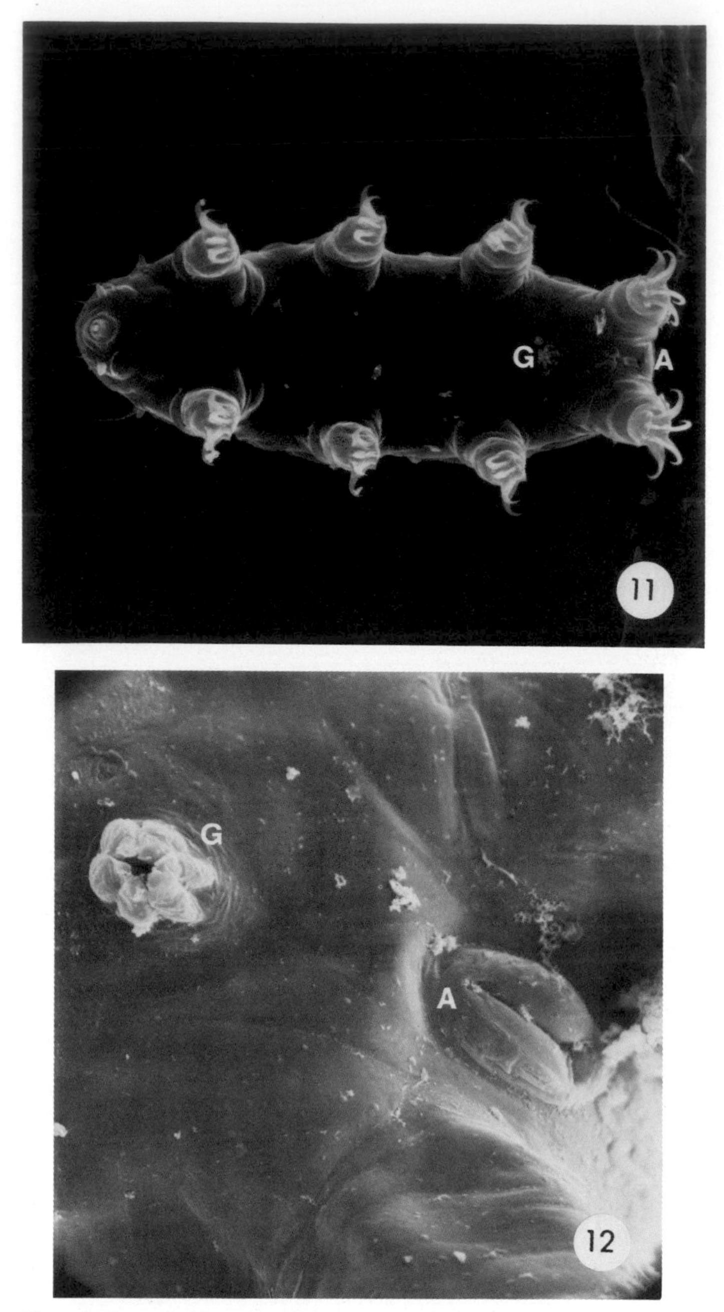

Fig. 11. Ventral view of a heterotardigrade, genus *Echiniscus*. × 300. Abbreviations: G, gonopore; A, anus.

Fig. 12. Ventral view of a heterotardigrade, genus *Echiniscoides*. × 2,220. Abbreviations: G, gonopore; A, anus. (Courtesy of Robert O. Schuster.)

geal, stylet, visceral, and somatic muscles. The somatic muscles are single, elongate, smooth muscle cells with some characteristics of obliquely striated muscle (Walz, '74, '75).

No specialized circulatory or respiratory systems are present. The body cavity is filled with a fluid and body cavity cells (coelomocytes), which circulate continually when the animal is in motion. Respiration occurs through the epidermis.

Since the discovery of tardigrades in 1773, more than 500 species have been described from a diversity of microhabitats throughout the world. Despite the cosmopolitan distribution of the phylum Tardigrada, few zoogeographic distributions have been established. The ecological requirements of some species appear to be quite broad, resulting in a worldwide distribution. Other species, with narrower tolerances, are endemic and rarely encountered. Dissemination of active tardigrades, eggs, cysts, and anhydrobiotic specimens occurs through the action of wind, water, and other animals.

All tardigrades are aquatic regardless of their specific habitat (marine, freshwater, or terrestrial) since activity (metabolism) depends on the presence of a film of water covering the body surface. As the water evaporates in terrestrial, semiterrestrial, or temporary aquatic habitats, tardigrades undergo anhydrobiosis (cryptobiosis), which enables them to survive variable periods of desiccation (Crowe, '75). Obligate freshwater and marine tardigrades may have some limited capacity for anhydrobiosis, but this has not been proved experimentally.

Marine and brackish water environments are inhabited by a single species of the class Eutardigrada, *Isohypsibius stenostomus* (Richters). The other species found in these habitats are all members of the class Heterotardigrada. Rare cases of parasitism and ectocommensalism have also been reported in heterotardigrades (Ramazzotti, '72).

Tardigrades that are distinctly aquatic (hydrophilous) live exclusively in freshwater habitats. They are not planktonic, but are occasionally found in plankton samples. Typically, these benthic invertebrates live in the interstitial space of sediments or crawl on the substrate of aquatic plants in ponds, lakes, rivers, and streams. Most species are littoral, but some specimens have been collected from lake bottoms at depths up to 150 m (Ramazzotti, '72). The aquatic habitat may also yield hygrophilous tardigrades, which typically inhabit continually moist mosses, and eurytopic species, which can tolerate a wide range of moisture conditions. Freshwater species are almost exclusively eutardigrades. Among the heterotardigrades, only *Echinursellus longiunguis* Iharos, *Carphania fluviatilis* Binda, and a few species of *Pseudechiniscus* occur in freshwater habitats.

The majority of known tardigrade species inhabit terrestrial environments: lichens, mosses, liverworts, rosette angiosperms, other plants, leaf litter, bark, and soil. Large populations of both eutardigrades and heterotardigrades are frequently encountered in the same microhabitat. Species diversity within a habitat generally ranges from two to six, although ten or more species may be found in a single sample.

The phylogenetic position of the Tardigrada is still highly debatable. Today they are regarded as a phylum with affinities to the aschelminth complex and/or the onychophoran-arthropod complex. Early researchers placed them in various systematic categories: Infusoria, Gastrotricha, Rotifera, Nematoda, Annelida, Onychophora, Linguatulida, and Arthropoda (Acarina, Crustacea, Insecta) (Riggin, '62; Ramazzotti, '72; Nelson and Higgins, '83). Additional studies of morphology and embryology with electron microscopy will undoubtedly help to elucidate their affinities.

Collection

Collection methods for aquatic tardigrades are essentially the same as those used for other benthic and interstitial microfauna (Hulings and Gray, '71). Littoral habitats are the most productive to sample. Nets with a fine mesh are dragged along the bottom; dredges or grabs may pick up sediments and vegetation; aquatic vegetation (algae, mosses, etc.) is washed and scraped. Qualitative collections of tardigrades from aquatic habitats with sandy bottoms are made by stirring the sand in a container of water and decanting immediately after the sand settles. The decanted water is passed through a sieve (No. 325; pore size 44 μm), and the specimens are rinsed from the screen to a sample jar. Sediment samples are also collected using a coring syringe; the tardigrades are separated from the sediment by elutriation. Samples with live tardigrades are examined with a dissecting microscope at a magnification of 30× or greater, and specimens are removed with an Irwin loop or a small pipette. Boiling water or boiling alcohol added to specimens in small amounts of water is adequate for tardigrade fixation. Tardigrades in large volumes of water are preserved in buffered formalin (5–10%) or glutaraldehyde (Nelson and Schuster, '81).

Collections of tardigrades from vegetation are made by removing the plant (moss, lichen, liverwort, etc.) from the substrate (tree, rock, soil) and placing the sample in a container of water for several hours to reverse anhydrobiosis. The sample is agitated and squeezed to remove specimens, eggs, and cysts, and the excess water is decanted after settling. The remaining sediment is examined for live tardigrades with a dissecting microscope. The sample can also be washed through sieves to remove particles smaller then 50 μm or larger than 1,200 μm.

Culturing

Many attempts have been made to culture tardigrades in the laboratory, but the majority have met with little success. A few species of eutardigrades have been maintained for various periods of time. Contrarily, the heterotardigrades have proved extremely difficult to culture.

Some workers have successfully reared eutardigrades through several generations on filamentous algae or on wet moss before the eventual deterioration of the culture (Oti, '56; Baumann, '61, '64, '66, '70; Dougherty et al., '61; Dougherty, '64; Ammermann, '62, '67; Crowe et al., '71b). Baumann ('61) raised *Hypsibius convergens* (Urbanowicz) in small crucibles on an algal mat of *Chlorella pyrenoidosa,* plant nutrient solution, a small piece of *Brassica rapa,* and rotifers (as supplementary food and to control bacterial growth). Crowe et al. ('71b) and Baumann ('61) reported that young eutardigrades survived on plant material alone (moss protonema, algae), but the older individuals required animal material (rotifers, nematodes) in addition as food. Ammermann ('62, '67) raised *Hypsibius dujardini* (Doyère) for more than 6 years using a substrate of sessile algae of the order Chlorococcales and a culture liquid of diluted soil decoction as prescribed by Hammerling. Dougherty ('64) cultured an Antarctic species, *Hypsibius arcticus* (Murray), on a blue-green alga in a refrigerator, but, despite initial success, the stock eventually deteriorated.

A simple method for culturing predaceous tardigrades was developed by Sayre ('69). Large numbers of a predaceous *Isohypsibius* species, originally collected from a sample of soil and alligator weed near a pond in South Carolina, were reared on the nematode *Panagrellus redivivus* (L.). The nematodes were cultured in Petri dishes on moist precooked oatmeal to which a few granules of dried yeast had been added. Large numbers of nematodes were collected and added to culture dishes containing tardigrades. To provide an inert physical support for the organisms, milled sphagnum was autoclaved, washed in demineralized water, and layered 1 cm deep in a culture dish. The moss was saturated with water after the addition of the initial population of tardigrades and nematodes. As the nematode population declined because of predation from the increasing tardigrade population, more nematodes were added. From an initial population of 50 tardigrades, as many as 5,900 were recovered 3 months later. Sayre's culture method may be adaptable to other tardigrade species as well.

Commercial Source

An excellent culture of aquatic tardigrades can be obtained from Ward's Natural Science Establishment, Inc., P.O. Box 1712, Rochester, NY 14603. The order number is 87-W-3300 (Catalog 82-83 Biology, p. 29); the cost is $4.75. The culture contains tardigrades, gastrotrichs, rotifers, and protozoans, and can be maintained for several weeks by the addition of fresh water. Tardigrades in the culture deposit eggs in the exuvium; egg development and hatching can be observed. This source is recommended for those interested in studying the developmental biology of the freshwater eutardigrades.

Three films by P. Schmidt ('71 a–c) were produced by the Institut für den Wissenschaftlichen Film, D-44 Göttingen, Nonnenstieg 72, Federal Republic of Germany, as part of "Encyclopaedia Cinematographica," a collection of documentary films on specific scientific subjects for research and teaching. These films, which may be obtained on loan from Pennsylvania State University, illustrate feeding, molting, egg laying and development, and the hatching of juveniles of *Milnesium tardigradum* Doyère and *Hypsibius dujardini.*

Literature

The definitive monograph on tardigrades, published by Ramazzotti ('72) in Italian, contains thorough sections on the systematics and biology (morphology, physiology, ecology) of the organisms, as well as keys and illustrated diagnoses of the known species, and an extensive bibliography. The supplement to this treatise (Ramazzotti, '74) introduced a modification of the systematics of the families and genera of the Eutardigrada. A revision of the monograph is currently in preparation by Ramazzotti and Maucci. Earlier monographs by Marcus ('28, '29a, b, '36) in German and by Cuénot ('32, '49) in French formed the basis of our knowledge of the embryology, physiology, morphology, taxonomy, and natural history of the group. The most complete general reference in English is Pennak's ('78) text on freshwater invertebrates. Other useful references in English include those by Riggin ('62), Nelson ('75), Nelson and Schuster ('81), and Nelson and Higgins ('83). Greven ('80) recently published a comprehensive review of tardigrade biology in German. The three international symposia on the Tardigrada included papers on various aspects of the biology and systematics of the organisms (Higgins, '75; Weglarska, '79b; Nelson, '82), and these references should be consulted for current research developments.

REPRODUCTION

Sexual Reproduction

Reproductive modes in the Tardigrada include sexual reproduction, parthenogenesis, and hermaphroditism (Ramazzotti, '72; Bertolani, '79a, b). Bisexuality is common in marine tardigrades, and sexual dimorphism is marked in the gonopore structure (Pollock, '70a, '75). In many eutardigrades, the nonmarine species, the sexes are separate but there is little external sexual dimorphism. Males of *Milnesium tardigradum* have modified claws on the first pair of legs, which may be an adaptation associated with mating. Other species may have slight variations in the size and shape of claws, especially those on the first pair of legs. In *Hypsibius oberhaeuseri* (Doyère) males, however, a flattened bump is present on the outer side of the fourth pair of legs; its function is unknown. Males are also generally smaller than females; however, since sexual maturity is reached prior to attainment of maximum length, mature males may be larger than immature females in the same population. Bisexuality in tardigrades is accompanied by meiotic development of the gametes.

In populations in which males can be identified by internal reproductive morphology, males are reported to reach peak abundance during the winter or early spring. Females, however, comprise the majority of the population throughout the year (Ramazzotti, '72). In bisexual populations of some *Macrobiotus* species, the sex ratio is approximately 1:1 (Bertolani, '75). An atypical sex ratio of 1:4 is found in some species owing to the existence of sympatric populations of bisexual (diploid) and parthenogenetic (triploid) biotypes (Bertolani, '72a, b, '75). According to Bertolani ('75, '79b, '82), the absence of males in many populations, as well as recent karyological data, indicates that polyploidy and ameiotic and meiotic parthenogenesis are common phenomena in numerous species of tardigrades.

Reproductive Apparatus

Both males and females have a single (unpaired) gonad located dorsal to the intestine (Figs. 3, 6). The sack-shaped gonad is attached to the body wall by two anterior ligaments in adult eutardigrades or by a single median ligament in heterotardigrades (Ramazzotti, '72; Pollock, '70a). Morphological variations in the gonad are associated with age, sex, species, and reproductive activity. Prior to egg deposition, the gonad of the sexually mature female is a swollen sack that covers the midgut and rectum. In males of the large eutardigrade *Milnesium tardigradum,* the testis is a small, flattened spherical structure dorsal to the posterior part of the Malpighian tubules.

In the Eutardigrada, the male has two sperm ducts which curve around each side of the intestine and open into the hindgut (cloaca). A small swollen portion of each sperm duct acts as a seminal vesicle containing mature spermatozoa. The female eutardigrade has only one oviduct on either the right or the left side of the intestine. The cloacal opening is a ventral transverse slit just anterior to the fourth pair of legs (Figs. 9, 10). In the Heterotardigrada, the genital ducts terminate in a ventral rosette-shaped gonopore anterior to the anus between the fourth pair of legs (Figs. 11, 12).

A seminal receptacle is present during the autumn and winter in some female eutardigrades in the genera *Macrobiotus* and *Hypsibius* (Marcus, '28). The seminal receptacle is a small sack, pointed at its distal end, which opens ventrally into the hindgut next to the oviduct opening. Spermatozoa, apparently agglutinated and immobile, have been observed in the receptacle, which is attached distally to the dorsal epidermis by a thin ligament (Ramazzotti, '72).

Spermatozoa

The morphology of tardigrade spermatozoa has been described for only a few species. Early workers (Henneke, '11; Baumann, '20; Marcus, '28, '29a, b) reported the length and head shape of the sperm of some eutardigrades. Spermatozoa of *Hypsibius dujardini* and *Macrobiotus hufelandi* Schultze are very thin and about 80–90 μm in length; the spiral-shaped head with 7–8 coils is about 11–12 μm long and occasionally up to 16 μm. *Isohypsibius nodusus* (Murray) also has spermatozoa of about 80 μm in length. In the marine heterotardigrade *Tetrakentron synaptae* Cuénot, one male with a body length of 105 μm was observed with spermatozoa 35 μm long in the testis (Marcus, '28). The simultaneous presence of two kinds of sperm (mature and immature?) was first observed by Henneke ('11) in the freshwater eutardigrade *Macrobiotus macronyx* Dujardin.

The first investigation of the ultrastructure of tardigrade spermatozoa was reported by Baccetti et al. ('71). The mature spermatozoan of the cosmopolitan eutardigrade *Macrobiotus hufelandi* is about 90 μm long and has three distinct regions: The cylindrical acrosome is about 3 μm long; the helical nucleus, with 14–15 coils, is about 6.5 μm long; the flagellum is about 80 μm long and tapers distally. In some specimens, however, the flagellum is 6–7 μm long and terminates in a tuft of ten fibrous units. The flagellum has the classical 9+2 arrangement of microtubular components; a very short mitochondrial sheath surrounds the anterior region of the flagellum. Specialization of spermatozoa may be correlated with a specialized mode of sperm transfer. The thread-shaped modified spermatozoan has been associated with internal fertilization (Kristensen, '79; Wolburg-Buchholz and Greven, '79). In *Macrobiotus hufelandi,* the female has a seminal receptacle and fertilization is internal. The spermatozoa are specialized with modified mitochondria (Kristensen, '79).

The sperm of the freshwater eutardigrade *Isohypsibius granulifer* Thulin is about 100 μm in length and consists of three regions: 1) A small acrosome (0.45 μm); 2) a helical nucleus (about 5 μm) with about 15 coils and a lateral "condensed body"; and 3) a flagellum about 95 μm long, which ends in a tuft and has the classical 9+2 arrangement. In contrast to *Macrobiotus hufelandi* sperm, there is no mitochondrial sheath around the flagellum, but the nucleus is partially surrounded on one side by an elongated "condensed body" of unknown origin and function. Sperm transfer occurs by copulation in *I. granulifer,* and spermatozoa are found in the ovary of the female (Wolburg-Buchholz and Greven, '79).

In the heterotardigrade *Batillipes noerrevangi* Kristensen, the sperm is only 14–17 μm in total length and is less specialized than that of eutardigrades. The sperm consists of 1) an asymmetrical rounded head (2 μm) with a "snout," a nonhelical nucleus, and an aberrant acrosome; 2) unmodified ("free") mitochondria (2.7 μm); and 3) a flagellum (12–15 μm) without a tuft. The female lacks a seminal receptacle, and fertilization is external. The male stimulates the female with his sense organs, and the female deposits her eggs on a sand grain next to the male (Kristensen, '79). Kristensen ('79) also observed other heterotardigrades in the genera *Echiniscoides* and *Pseudechiniscus* which have a "true" acrosome like the eutardigrades, but the spermatozoa are also less specialized. These "primitive" spermatozoa have a short head, unmodified mitochondria, and a primitive flagellum. The female has no seminal receptacle, and fertilization is external.

Oocytes

Little information is available on oogenesis in tardigrades. Marcus ('28, '29a, b) reported that the female gonad develops from the remnants of two posterior mesodermal coelomic pouches that fuse to form a single unpaired ovary. In immature tardigrades the ovary is filled with cells containing large nuclei and nucleoli. A single layer of epithelial cells forms the wall of the ovary. Walz ('75) proposed the presence of myofilaments in ovarian endothelial cells that may be involved in egg laying. During oogenesis the size of the ovary increases considerably. According to Weglarska ('79a), Marcus defined the type of oogenesis in tardigrades as "alimentary and nutrimentary." The ovary is "of the meriostic and polytrophic type."

Weglarska ('75) described the oocytes of the cosmopolitan eutardigrade *Macrobiotus richtersi* Murray, and later ('79a) she investigated previtellogenesis and vitellogenesis in the same species. Prior to oogenesis, the ovary averages 100 μm in length and 20 μm in width in specimens with a body length of 650 μm. The diameter of the oogonium averages 4 μm.

Cleavage with unequal cytokinesis results in several clusters of cells that initially completely fill the gonad. Each cluster consists of seven trophocytes (nurse cells) and one oocyte joined by intercellular bridges (fusomes). As the oocytes develop, the trophocytes gradually retract and disintegrate, and the oolemma is modified to form conical processes. The diameter of the mature egg averages 80 μm. The ovary containing fully developed eggs averages 300 μm in length and 110 μm in width in specimens with a body length of 640 μm.

In previtellogenesis the diameter of the oocytes ranges from 8 to 10 μm. Initially, the nuclei of the oocytes and nurse cells are similar in appearance. Yolk bodies appear in both trophocytes and oocytes.

As the oocytes enlarge to about 15 μm in diameter, vitellogenesis increases. Two types of yolk bodies are produced: One is associated with rough endoplasmic reticulum, the other with smooth. Weglarska stated that according to Nørrevang's ('68) classification, "oogenesis in *Macrobiotus richtersi* may be defined as heteronomous, because the accumulation of yolk is the result of the autosynthetic property both of the oocyte and of the nurse-cells attached to it by means of fusomes."

Weglarska ('82) also investigated the ultrastructure of the formation of the egg envelopes in *Macrobiotus richtersi.* The egg is covered with two envelopes: the vitelline membrane and the "chorion," which is not homologous to the chorion of arthropod eggs. The ovary of *M. richtersi* lacks follicle cells, which produce egg membranes in some organisms; the oocyte is active in the formation of both envelopes. During previtellogenesis and early vitellogenesis, the oocyte is surrounded by trophocytes or cells of the ovarian epithelium, and oolemma is smooth. As the trophocytes disintegrate, microvilli form on the oolemma and the surface of the trophocytes. When the oocyte diameter is about 40 μm, "chorionic bladders" resulting from exocytotic activity of the oocyte form over protrusions of the ooplasm. The isolated bladders subsequently fuse and cover the entire surface of the oocyte. The number of chorionic bladders produced by the oocyte corresponds to the number of conical processes characteristic of *M. richtersi.* The processes are flattened while the oocytes are developing within the ovary, but are extended after deposition of the egg.

The vitelline envelope is produced by the oocyte after the formation of the chorion (Weglarska, '82). The oolemma thickens and separates from the ooplasm, and a new oolemma is produced that forms numerous microvilli. A secretory product from the oocyte is released via the microvilli and forms the vitelline membrane. After oviposition, the oolemma becomes smooth and adheres closely to the vitelline membrane.

In cytological studies, Bertolani ('75) correlated the stages of meiosis in the oocytes with the molting process. The chromosomes are not visible when yolk production in the oocyte begins. During yolk production, the cuticular lining of the foregut (buccal apparatus) is shed, and the animal enters the simplex state of molting. Chromosomes become evident at the end of prophase I, which coincides approximately with the reconstruction of the foregut lining. In metaphase I, the chromosomes in the oocytes are larger than those in somatic or gonial mitoses, but the degree of difference varies with the species (Bertolani, '72a, '82). Meiosis is halted in metaphase I until oviposition, which occurs either during molting in species that deposit eggs in the exuvium or shortly after molting in those that deposit free eggs. Somatic mitoses also increase during the molt (Bertolani, '70).

Eggs

The eggs of tardigrades have high taxonomic value, especially in the eutardigrade genus *Macrobiotus* (Figs. 13–18). In some genera, smooth eggs, which are usually oval in shape and have a thin shell, are deposited in the exuvium during ecdysis (Figs. 19–21). These eggs range in number from two to 60, and are characteristic of the armored heterotardigrades *(Parechiniscus, Echiniscus, Pseudechiniscus)*, most eutardigrades in the genera *Pseudobi-*

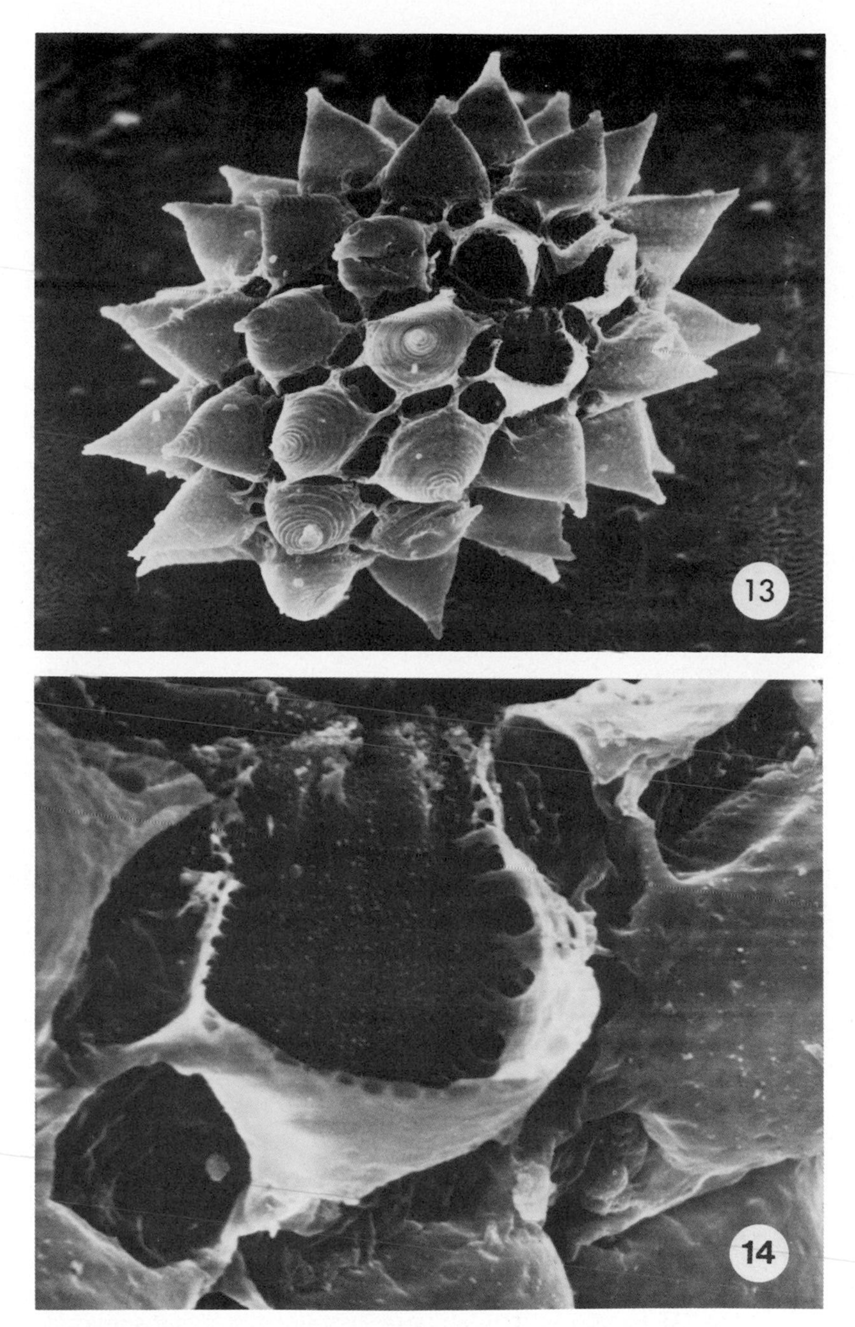

Fig. 13. Egg of *Macrobiotus areolatus* Murray, from Roan Mountain, Tennessee. Note the area where the projection had been torn away. × 1,180.

Fig. 14. Higher magnification of the egg with the torn projection. × 5,800.

Fig. 15. Egg of *Macrobiotus tonollii* Ramazzotti, from Roan Mountain, Tennessee. × 540.

Fig. 16. Egg of *Macrobiotus psephus* Bois-Reymond Marcus, from Venezuela. × 534.

Fig. 17. Egg of *Macrobiotus harmsworthi coronatus* Barros, from New Zealand. × 750.

Fig. 18. Egg of *Macrobiotus hufelandi* Schultze, from Roan Mountain, Tennessee. × 760.

otus, Hypsibius, Isohypsibius, and *Milnesium,* and only a few of the *Macrobiotus* species. The size of the oval egg varies from about 42 by 30 μm *(Hypsibius)* to 135 by 90 μm *(Milnesium)*. Smooth eggs, either solitary or in small groups, are deposited in the molted cuticle or outside the body of the female. In nearly all species of *Macrobiotus* and *Dactylobiotus,* eggs ornamented with various processes, pores, or reticulations are deposited freely

(Figs. 13–18, 22–24). As stated earlier, the projections are extended only after oviposition. The eggs are usually spherical, with a diameter up to 235 μm; the number varies from one to 15. Species identifications can be made solely on the basis of egg morphology. Toftner et al. ('75) and Grigarick et al. ('73) combined the capabilities of the scanning electron microscope (SEM) and the image-analyzing computer (IAC) to quantify characteristics of the eggs of different species in the *hufelandi* group of *Macrobiotus*. Ramazzotti ('72) and Greven ('80) have discussed and illustrated the various types of egg ornamentations and emphasized their taxonomic significance. A few species of other genera (family Hypsibiidae) also produce ornamented eggs that may be deposited either freely or in the exuvium. In some aquatic tardigrades, eggs are deposited, usually in groups, inside the hollow shells or exoskeletons of cladocerans, ostracods, or insects. One freshwater species, *Isohypsibius annulatus* (Murray), carries its exuvium containing eggs for some time; the marine species *Echiniscoides sigismundi* (Schultze) often carries one or two freely deposited eggs attached to its caudal extremity (Ramazzotti, '72). In general, the size of eggs is highly variable, even in a single population. The method of egg deposition (oviposition) is unknown for many species, especially in marine tardigrades.

Mating and Fertilization

Mating and fertilization have been observed for only a few species of tardigrades. As described by Von Erlanger (1894, 1895a, b), one or more males of some freshwater species deposit spermatozoa into the cloacal opening of the old cuticle of the female prior to or during a molt. External fertilization occurs as the female releases eggs into the exuvium (Von Wenck, '14). Henneke ('11) erroneously proposed that the males introduce sperm into the exuvium by piercing it with their stylets.

Sperm have been observed within the ovary of the female in several species. Marcus ('29a) observed mating in *Isohypsibius nodosus.* About 10 minutes after sperm were deposited, few remained in the old cuticle but many were observed in the female's ovary. The molt was completed about 24 hours later; however, no eggs were deposited in the exuvium. Internal fertilization may also occur independently of molting. Sperm deposited in the cloaca may be stored in the seminal receptacle or may fertilize the eggs in the ovary.

Hermaphroditism

Doyère (1840) proposed that tardigrades were hermaphroditic, as he interpreted incorrectly that structures near the hindgut were two testes and a seminal receptacle. Plate (1889) correctly identified the structures as the two lateral and one dorsal excretory glands (Malpighian tubules) which empty

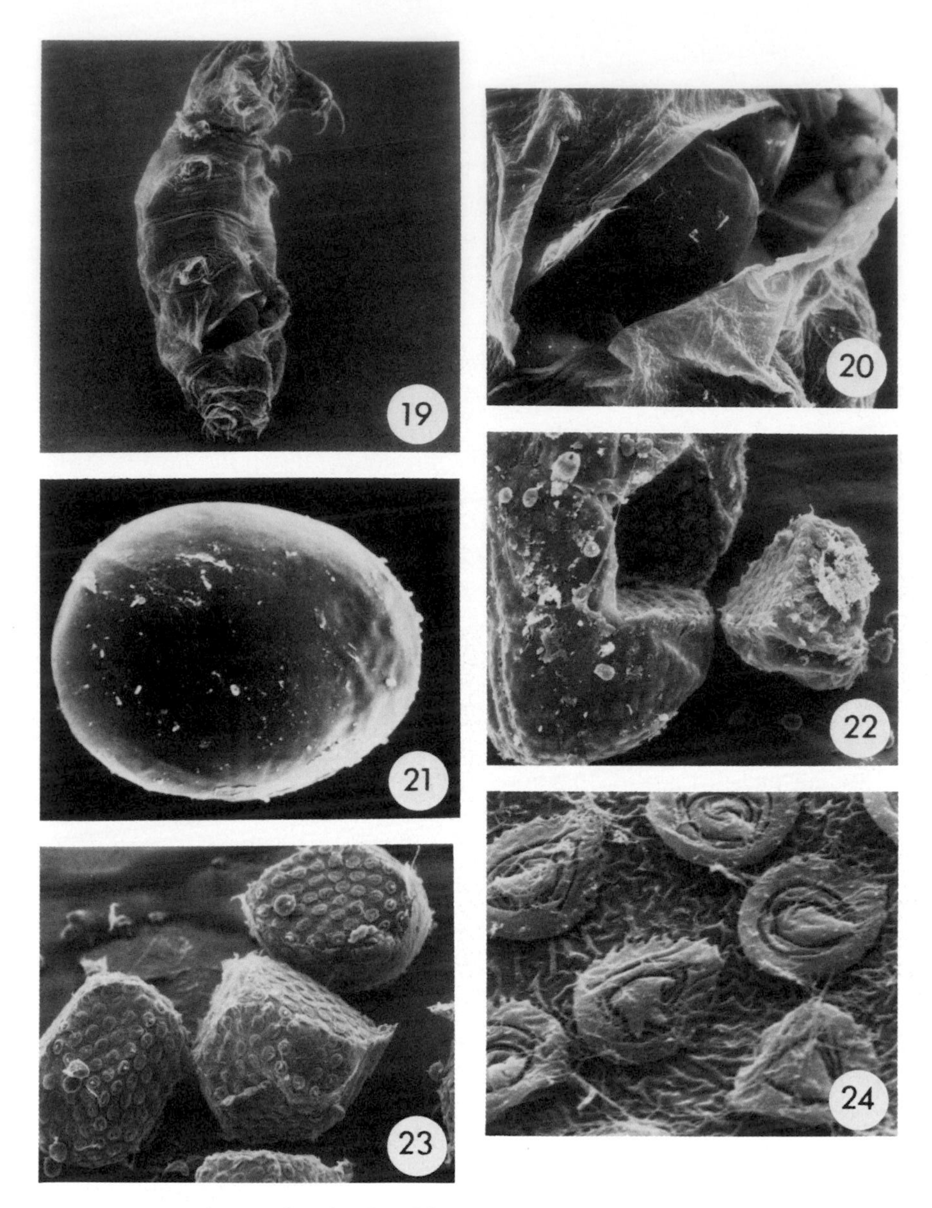

Fig. 19. An aquatic eutardigrade, *Pseudobiotus augusti,* from the Ocoee River, Tennessee, shedding the cuticle containing smooth eggs. Note the eggs through the torn cuticle. Anterior portion of the animal is abnormal owing to a fungal infection. × 116.

Fig. 20. Higher magnification of the smooth eggs of *Pseudobiotus augusti* deposited inside the exuvium. × 530.

Fig. 21. An egg of *Pseudobiotus augusti* removed from the exuvium. × 790.

into the junction of the midgut and hindgut. Since Plate's discovery, tardigrades have been considered gonochoristic (dioecious) with both bisexual and unisexual populations. Recently, however, Bertolani ('79a) reported hermaphroditic populations of aquatic eutardigrades in the genus *Isohypsibius*. In *Isohypsibius granulifer*, populations were found in which the gonad was largely female in appearance but had a small caudal portion consisting of small undifferentiated cells and male germ cells in various stages of spermiogenesis. Bisexual populations of the species were also found. In a population of *Isohypsibius baicalensis* Ramazzotti, some specimens had male germ cells interspersed with oocytes at different stages of maturation in the gonad (ovotestis). Whether self-fertilization or cross-fertilization occurs is unknown. Hermaphroditism may also exist in some *Macrobiotus* species, but has not been observed in any marine heterotardigrades. Further investigation of hermaphroditism in natural populations of tardigrades is needed.

Parthenogenesis and Polyploidy

Parthenogenesis is a common mode of reproduction in tardigrades found in leaf letter, mosses, and freshwater habitats (Ammermann, '67; Bertolani, '71a, b, '73a, b, '75, '76, '79b, '82; Bertolani and Buonagurelli, '75). In the heterotardigrade family Echiniscidae, males are absent (in the genus *Echiniscus*) or very rare (in other genera) and parthenogenesis is common, although the existence of polyploidy has not been established (Bertolani, '82). In the eutardigrades, the chromosome number (except for cases of polyploidy) is usually 5 or 6 (2n = 10 or 2n = 12). An unusual chromosome number (16) is found in the eutardigrade *Itaquascon trinacriae* Arcidiacono, which undergoes ameiotic parthenogenesis, but the degree of ploidy has not been determined (Bertolani, '79b, '82.)

Polyploidy is often associated with parthenogenesis, but neither is a requisite for the other. Some parthenogenetic populations are diploid (Ammermann, '67; Bertolani, '75, '79b; Bertolani and Buonagurelli, '75), and Bertolani ('79a) has reported a tetraploid population of a hermaphroditic

---

Fig. 22. Mass of eggs removed from the ovary of the aquatic eutardigrade *Dactylobiotus dispar* (Murray), from a temporary pond in Jefferson County, Tennessee. Note the shape of the eggs, which are tightly packed in the ovary. × 300.

Fig. 23. Eggs of *Dactylobiotus dispar* separated from the mass. Note the ornamentation on the surface of the eggs. × 310.

Fig. 24. Higher magnification of the egg of *Dactylobiotus dispar*. These eggs are laid singly outside the body of the female. Note the folded projections which become extended after oviposition and form the egg ornamentation. × 3,100.

species of *Isohypsibius*. Polyploidy is frequently found in eutardigrades in various genera and families that inhabit different microenvironments. Although polyploidy generally leads to an increase in the somatic size of an organism, a large tardigrade is not necessarily polyploid (Bertolani, '71b, '73b, '82; Bertolani and Mambrini, '77).

The freshwater eutardigrades *Marcobiotus dispar* Murray and *Hypsibius dujardini* are diploid and undergo meiotic (automictic) parthenogenesis (Ammermann, '62, '67; Bertolani, '82; Bertolani and Buonagurelli, '75). Restoration of the diploid number occurs after the first division of the oocyte; the dyads replicate and then separate. Ameiotic (apomictic) parthenogenesis occurs in triploid and tetraploid cytotypes of some freshwater and terrestrial eutardigrades *(Macrobiotus, Hypsibius, Pseudobiotus)*. According to Bertolani ('82), cytotypes are populations with the same taxonomic characteristics but with different degrees of ploidy and different modes of reproduction. Different cytotypes (diploid, triploid, and tetraploid) have been found in nearby habitats and even in the same moss. Many species have both diploid bisexual and polyploid cytotypes with ameiotic parthenogenesis. They undergo sexual reproduction when males are present and parthenogenesis when males are absent. Males are sometimes frequent in the spring but always absent in other seasons. Baumann ('64) confirmed parthenogenesis in laboratory cultures of *Milnesium tardigradum,* which also has diploid bisexual cytotypes, but excluded parthenogenesis as a mode of reproduction in *Hypsibius convergens* (Baumann, '61). The reproductive mode of the Echiniscidae remains an unsolved problem because of difficulties in rearing the organisms.

DEVELOPMENT

Embryological Development

Marcus ('28, '29a, b) described the development of the embryo of eutardigrades from observations with a light microscope. Early cleavage is holoblastic and nearly equal. The blastula is a coeloblastula with a small central cavity (blastocoel) surrounded by one layer of cells with peripheral nuclei. Gastrulation occurs by delamination. The cells of the blastula wall divide: The inner cells migrate to fill the blastocoel and form the entoderm, and the embryo becomes solid; the outer layer of cells forms the ectoderm. An internal cavity appears in the entoderm, forming the archenteron—the definitive lumen of the gut. The embryo curls up and a proctodeum forms on the ventral side and connects with the archenteron. Later a stomodeum develops and also connects with the archenteron to form a complete gut. (Note: in heterotardigrades, there is a mouth but no gonopore or anus in the first instar. This apparent contradiction has not been resolved!) Primordial egg cells develop from a particular type of entodermal cell.

Mesoderm production occurs by the formation of five pairs of archenteric pouches. The most anterior pair of pouches, located in the head, is ectomesodermal, being formed from the stomodeum; the other pairs are true enterocoelic pouches located in the body (trunk). The most posterior pair of pouches move dorsally and fuse to form the gonad. The single oviduct or two sperm ducts arise ventrally from the gonad. The walls of the remaining four pairs of pouches break down and form the muscle cells and large sperical body cavity cells floating in the coelomic fluid. The adult body cavity is technically a persistent blastocoel (Brusca, '75) or a mixocoel (Greven, '80). The debate whether the body cavity is a pseudocoelom or true coelom (hemocoel) has also not been resolved.

The length of time between oviposition and hatching of young tardigrades varies within the species, between species, and with environmental conditions, such as temperature, atmospheric humidity, and season. Greven ('80) summarized the data from Baumann ('70) for four species of eutardigrades maintained in culture at 20°C (Table II). Egg development varied from 4 to 10 days; embryological development, 5–10 to 26–31 days. Marcus ('29b) reported a minimum development time of 5 days (average time at 18°C for 100 eggs taken from ten exuvia) for *Hypsibius convergens* and a maximum of 30–40 days for eggs of *Hypsibius oberhaeuseri* and several *Macrobiotus* species. In *Milnesium tardigradum* he found that the length of time for

TABLE II. Development of Eutardigrades in Culture at 20°C (summarized by Greven, '80, from Baumann, '70)

	Macrobiotus hufelandi	*Hypsibius convergens*	*Hypsibius oberhaeuseri*	*Milnesium tardigradum*
Egg development	10 days	5–7 days	4–5 days	4 days
Embryological development	26–31 days	5–10 days	8–11 days	7 days
Body length after hatching	225–250 μm	100 μm	100–125 μm	200 μm
Duration of the simplex stage and molting	4–5 days	1–3 days	1–3 days	4 days
Sexual maturity (by second molt)	21 days	12 days	12 days	24 days
Body length at sexual maturity	350 μm	175–200 μm	200–250 μm	420–450 μm
Succession of the first four simplex stages	6, 16, 34, 52 days	4, 9, 17, 25 days	6, 9, 12, 25 days	10, 20, 34, 46 days

Echiniscus trisetosus Cuénot, the smallest size class showed more then three times as many pores per area, smaller pores, and less distance between them. Considerable variation was found in the pores of the larger size classes; there were no significant differences between the pores of these classes. In the other species, later described as a new species, *Echiniscus laterculus* (Schuster et al., '80a), there were no marked differences in the scapular plate pores in the four size classes. In both species the first instar had a mouth, but no anus or gonopore. The anus appeared in the second size class as a longitudinal slitlike opening between the fourth pair of legs. The gonopore appeared in the second size class as a simple pore anterior to the anus. In subsequent instars the gonopore formed a rosette with six oval lobes. In one species of *Echiniscus* two types of gonopores were observed. Whether the difference is due to an alteration following oviposition or to the presence of males in the population is unknown. If males were indeed present, this would be the first observation of males in any *Echiniscus* species, thus altering the view that the genus is entirely parthenogenetic. As noted for many *Echiniscus* species, only two claws (the internal claws) were present in the first size class, whereas two additional claws (the external claws) were present in all other size classes.

Other studies have described variability in taxonomic characters of adults and juveniles in species of the genus *Echiniscus.* Lattes and Tobia Gallelli ('72) reported variations in the number and size of the spines on the dentate collar of leg IV, the sculpture of the third intersegmental plate, and the cuticular appendages. Lattes ('75) compared the cuticular sculpture of adults and juveniles of *Echiniscus quadrispinosus* Richters. Maucci ('72) noted the absence of some lateral filaments on juveniles of *Echiniscus trisetosus,* although Ramazzotti ('72) reported juveniles with the same complement of appendages as adults. Contrarily, juveniles of the genus *Mopsechiniscus* have a larger number of appendages than adults; the number and length of the spines and filaments are highly variable but decrease as the animal increases in length (Ramazzotti, '72).

Molting

Periodic molting occurs throughout the life of the tardigrade. Prior to molting the cuticular lining of the foregut (buccal tube, pharynx, and esophagus) and the stylets and stylet supports are ejected (Fig. 27). The mouth opening is closed and the individual is not able to feed. This is known as the simplex stage ("Simplexstadium"—Marcus, '29a). The cuticular structures of the buccal apparatus are reformed by the salivary glands concomitantly with the synthesis of the new body cuticle from the underlying epidermis and the production of claws by the claw glands in the legs (Walz, '82). During the molt, the old body cuticle is shed, including the claws and the lining of the hindgut (Fig. 28). Usually body length increases with each molt until a maximum size is attained, although under adverse environmental conditions some individuals may be smaller after molting. Growth is more rapid during the earlier molts and, as noted previously, sexual maturity may be obtained before the tardigrade reaches its maximum size (Ramazzotti, '72). Hallas ('72) noted that some of the smaller individuals in a population may have a relatively longer buccal apparatus than the larger ones as a result of variation in the size of the eggs.

Ramazzotti ('72) and Walz ('82) reviewed studies of the life histories of certain species of tardigrades. Frequency distributions of body lengths and buccal tube lengths have indicated four to 12 molts; however, numerous problems, which have been discussed by Baumann ('61) and Ramazzotti ('72), are inherent in this method. The duration of the molting process is dependent on the species and the surrounding environmental conditions. Marcus ('29a) noted approximately 12 molts for *Macrobiotus hufelandi*, the aquatic *Macrobiotus dispar*, and *Hypsibius oberhaeuseri*. For these three species maintained in artificial cultures, he established a period of 3–12 days for the complete molting process. The length of time for each stage in the process was as follows: a) expulsion of buccal apparatus: 6 days prior to molting; b) simplex stage: 1–3 days; c) synthesis of new buccal apparatus: 2–4 days; d) period from synthesis to molting: 0–2 days; e) time needed to shed old cuticle: 0–3 days. Twelve molts were reported for *Macrobiotus areolatus* Murray (Ramazzotti, '72), *Hypsibius convergens* (Baumann, '61), and *Macrobiotus hufelandi* (Franceschi-Crippa and Lattes, '67, '68–'69).

Baumann described the life history of *Hypsibius convergens* ('61), *Hypsibius oberhaeuseri* ('66), and *Milnesium tardigradum* ('64). In all three species, females were mature after the second molt. The number of days from hatching until the first molt varied from 3 to 6 for *H. convergens,* 4 to 7 for *H. oberhaeuseri,* and 6 to 15 for *M. tardigradum.* The second molt occurred on day 8 to day 10 for the *Hypsibius* species, but on day 20 for *M. tardigradum.* The number of days between molts in older animals was 21 days in *H. convergens* but varied from 12 to 19 days in *M. tardigradum.*

Based on his observations with transmission electron microscopy, Walz ('82) recently described cuticle formation in *Macrobiotus hufelandi,* which has a cuticular structure identical to that of *M. areolatus,* studied earlier by Crowe et al. ('71a, b). Walz, however, disagreed with the model of cuticle formation proposed by Crowe et al. ('71b). According to Walz, the body wall consists of a cuticle and a single layer of epidermal cells. The cuticle is composed of three layers: 1) the epicuticle (exocuticle), subdivided into an inner and outer layer; 2) the intracuticle (mesocuticle); and 3) the procuticle (endocuticle), the innermost layer. The underlying epidermal cells secrete the thin outer epicuticle, which becomes a continuous structure at the base of

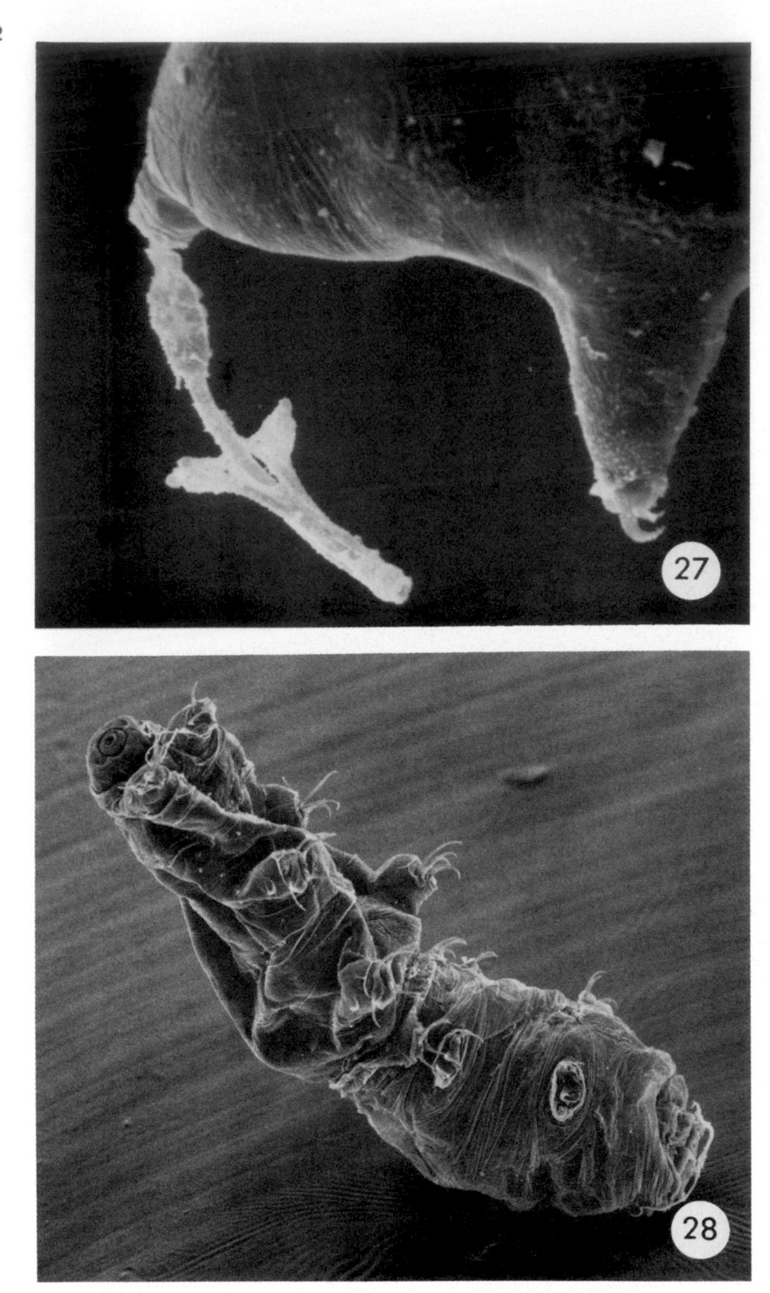

Fig. 27. A eutardigrade ejecting the buccal apparatus prior to ecdysis. × 2,040. (Courtesy of Robert O. Schuster.)

Fig. 28. An aquatic eutardigrade, *Pseudobiotus augusti* (Murray), from the Ocoee River, Tennessee, shedding the exoskeleton during molting. × 166.

the old cuticle. The inner epicuticle is then secreted, followed by the intracuticle, and finally the procuticle. During the formation of the new cuticle, the somatic muscles remain attached to the old cuticle. The muscle attachments have been described by Shaw ('74) and Crowe et al. ('71a, b).

Four cephalic sense organs of *M. hufelandi* have been described by Walz ('78, '79). These are called the a) anterolateral sensory field (ALSF), b) circum-oral sensory field (COSF), c) suboral sensory region (SOSR), and d) pharyngeal organ (PO). Changes in the ALSF, COSF, and the pharyngeal organ during molting were also analyzed by Walz ('82). The receptor lymph cavities of the COSF and the pharyngeal organ are completely absent in the molting stages. Epidermal cells that fill the cuticular pockets of the pharyngeal organ secrete the cuticular lining of the buccal tube and completely surround the outer dendritic segments of the sensory cells.

With a light microscope Marcus ('29a, b) described the formation of the solid cuticular claws by the claw glands. Observations with electron microscopy were made by Kristensen ('76), who studied claw formation in the heterotardigrade *Batillipes noerrevangi* Kristensen, and by Walz ('82), who analyzed the process in the eutardigrade *Macrobiotus hufelandi.* In both species claws are preformed by slender cell processes of the epidermal claw gland cells. These cell processes secrete the cuticular material that forms the claws.

Life-span

The life-span of tardigrades is uncertain, and its determination is complicated by the process of anhydrobiosis, which allows the organism to withstand variable periods of desiccation or other unfavorable environmental conditions in an inactive state. The active life-span is generally from a few months to 1 or 2 years, but the total life-span, the period from hatching until death, may be extended greatly by anhydrobiosis and/or encystment (Nelson and Higgins, '83; Ramazzotti, '72).

Aquatic tardigrades may possess limited abilities to undergo cryptobiosis during periods of fluctuation in water levels, but generally their active life-span is considered to be equivalent to the total life-span. Schuster et al. ('77), who studied the life history of an aquatic species, *Dactylobiotus grandipes* (Schuster et al.), in Lake Tahoe, California, considered the life cycle completed in 1 year, although the animals were essentially inactive for almost 6 months owing to cold water temperatures. Reproduction, determined by the presence of individuals containing oocytes, occurred during August, September, and October. Individuals were also kept alive in the laboratory for 6 weeks, and eggs were laid that hatched into first-instar juveniles but these died without molting. Studies by Pollock ('70b) on marine tardigrades and by Franceschi et al. ('62–'63) on moss-dwelling species suggest that life-spans may be of much shorter duration, about 3–4 months, which is more consistent with life-spans recorded for microfauna of similar size and biology (McIntyre, '69).

Marcus ('29a, '36) concluded that the active life-span of the terrestrial eutardigrades *Macrobiotus harmsworthi* Murray and *Hypsibius convergens* ranges from 18 to 30 months and that total life-span could reach 67 years. Contrariwise, Ramazzotti ('72) considered that total life-span in moss-dwelling tardigrades in "average conditions" could vary from 4 to 12 years, with exceptions in extreme conditions. The number of anhydrobiotic periods an individual can undergo in the natural habitat without damage is unknown.

CONCLUSIONS

Despite their overall abundance and cosmopolitan distribution, the Tardigrada have been relatively neglected by invertebrate zoologists. Frequently categorized as one of the "minor phyla," they are more appropriately termed one of the "lesser-known phyla." Because of difficulties in collecting and culturing the organisms and their apparent lack of economic importance to humans, our knowledge of tardigrades has remained in the nascent state since their discovery over 200 years ago. The lack of basic biological information has hindered the growth of interdisciplinary studies, which are just beginning to develop.

Early European workers described the majority of tardigrade species from temporary wet mounts of live specimens observed with the light microscope, often resulting in inadequate, incorrect, or incomplete descriptions and illustrations. Morphology, systematics, and natural history were focal points of their studies. Today, new insights into the biology of the Tardigrada are evolving from more comprehensive, contemporary investigations utilizing transmission and scanning electron microscopy as well as histochemistry, cytological techniques, and phase- and interference-contrast microscopy. The "state of the art," though still primitive, provides a vast array of opportunities for classical and contemporary biological investigations. Within the last decade, an upsurge in interest and activity has led to a boom in tardigrade research.

LITERATURE CITED

Ammermann, D. (1962) Parthenogenese bei dem Tardigraden *Hypsibius dujardini* (Doy.). Naturwiss. *49:*115.

Ammermann, D. (1967) Die Cytologie der Parthenogenese bei dem Tardigraden *Hypsibius dujardini*. Chromosoma *23:*203–216.

Baccetti, B., F. Rosati, and G. Selmi (1971) Electron microscopy of tardigrades. IV. The spermatozoan. Monitore Zool. Ital. *5:*231–240.

Baumann, H. (1920) Mitteilungen zum feineren Bau der Tardigraden. Zool. Anz. *52:*56–67.

Baumann, H. (1961) Der Lebensablauf von *Hypsibius (H.) convergens* Urbanowicz (Tardigrada). Zool. Anz. *167:*362–381.

Baumann, H. (1964) Über den Lebenslauf und die Lebensweise von *Milnesium tardigradum* Doyère (Tardigrada). Veröff. Überseemus. Bremen *3:*161–171.

Baumann, H. (1966) Lebenslauf und Lebensweise von *Hypsibius (H.) oberhaeuseri* Doyère (Tardigrada). Veröff. Überseemus. Bremen. *3:*245–258.

Baumann, H. (1970) Lebenslauf und Lebensweise von *Macrobiotus hufelandii* Schultze (Tardigrada). Veröff. Überseemus. Bremen. *4:*29–43.

Bertolani, R. (1970) Mitosi somatiche e costanza cellulare numerica nei Tardigradi. Rend. Acc. Naz. Lincei, Ser. VIII, *48:*739–742.

Bertolani, R. (1971a) Rapporto-sessi e dimorfismo sessuale in *Macrobiotus* (Tardigrada). Rend. Acc. Naz. Lincei, Ser. VIII, *50:*377–382.

Bertolani, R. (1971b) Partenogenesi geografica triploide in un Tardigrado *(Macrobiotus richtersi).* Rend. Acc. Naz. Lincei, Ser. VIII, *50:*487–489.

Bertolani, R. (1972a) Osservazioni cariologiche su alcuni *Macrobiotus* (Tardigrada). Rend. Acc. Naz. Lincei, Ser. VIII, *52:*220–224.

Bertolani, R. (1972b) La partenogenesi nei Tardigradi. Boll. Zool. *39:*577–581.

Bertolani, R. (1973a) Presenza di un biotipo partenogenetico e suo effetto sul rapporto-sessi in *Macrobiotus hufelandi* (Tardigrada). Rend. Acc. Naz. Lincei, Ser. VIII, *54:*469–473.

Bertolani, R. (1973b) Primo caso di una popolazione tetraploide nei Tardigrada. Rend. Acc. Naz. Lincei, Ser. VIII, *55:*571–574.

Bertolani, R. (1975) Cytology and systematics in Tardigrada. In R.P. Higgins (ed): International Symposium on Tardigrades, Pallanza, Italy, June 17-19, 1974. Mem. Ist. Ital. Idrobiol. *32*(Suppl.):17–35.

Bertolani, R. (1976) Osservazioni cariologiche su *Isohypsibius augusti* (Murray, 1907) e *I. megalonyx* Thulin, 1928 (Tardigrada) e ridescrizione delle due specie. Boll. Zool. *43:*221–234.

Bertolani, R. (1979a) Hermaphroditism in tardigrades. Int. J. Invert. Reprod. *1:*67–71.

Bertolani, R. (1979b) Parthenogenesis and cytotaxonomy in Itaquasconinae (Tardigrada). In B. Weglarska (ed): Second International Symposium on Tardigrades, Crakow, Poland, July 28–30, 1977. Zeszyty Naukowe Uniw. Jagiellońskiego, Prace Zoologiczne *25:*9–18.

Bertolani, R. (1982) Cytology and reproductive mechanisms in tardigrades. In D. Nelson (ed): Proceedings of the Third International Symposium on the Tardigrada, Johnson City, Tennessee, August 3–6, 1980. Johnson City: East Tennessee State University Press, pp. 93–114.

Bertolani, R., and G.P. Buonagurelli (1975) Osservazioni cariologiche sulla partenogenesi meiotica di *Macrobiotus dispar* (Tardigrada). Rend. Acc. Naz. Lincei, Ser. VIII, *58:*782–786.

Bertolani, R., and V. Mambrini (1977) Analisi cariologica e morfologica di alcune popolazioni di *Macrobiotus hufelandi* (Tardigrada). Rend. Acc. Naz. Lincei, Ser. VIII, *62:*239–245.

Brusca, G.J. (1975) General Patterns of Invertebrate Development. Eureka, California: Mad River Press, pp. 72, 74.

Crowe, J.H. (1975) The physiology of cryptobiosis in tardigrades. In R.P. Higgins (ed): International Symposium on Tardigrades, Pallanza, Italy, June 17–19, 1974. Mem. Ist. Ital. Idrobiol. *32*(Suppl.):37–59.

Crowe, J.H., I.M. Newell, and W.W. Thomson (1971a) Fine structure and chemical composition of the cuticle of the tardigrade, *Macrobiotus areolatus* Murray. J. Microsc. *11:*107–120.

Crowe, J.H., I.M. Newell, and W.W. Thomson (1971b) Cuticle formation in the tardigrade, *Macrobiotus areolatus* Murray. J. Microsc. *11*:121–132.
Cuénot, L. (1932) Tardigrades. Faune de France *24*:1–96. Paris: Paul Lechevalier.
Cuénot, L. (1949) Les Tardigrades. Traité de Zoologie *6*:39–59.
Dougherty, E.C. (1964) Cultivation and nutrition of micrometazoa. II. An Antarctic strain of the tardigrade *Hypsibius arcticus* (Murray, 1907) Marcus, 1928. Trans. Am. Microsc. Soc. *83*:7–11.
Dougherty, E.C., D.J. Ferral, and B. Solberg (1961) Xenic cultivation of Antarctic micrometazoa. I. The tardigrade *Hypsibius arcticus* (Murray, 1907) Marcus, 1928. Am. Zool. *1*:350–351.
Doyère, L. (1840) Mémoire sur les Tardigrades. Ann. Sci. Nat., Zool., Ser. 2, *14*:269–362.
Franceschi, T., M.L. Loi, and R. Pierantoni (1962–63) Risultati di una prima indagine ecologica condotta su popolazioni di Tardigrada. Boll. Mus. Ist. Biol. Univ. Genova *32*:69–93.
Franceschi-Crippa, T., and A. Lattes (1967) Analisi della variazione della lunghezza degli esemplari di una popolazione di *Macrobiotus hufelandii* Schultze in rapporto con l'esistenza di mute. Boll. Mus. Ist. Biol. Univ. Genova *35*:45–54.
Franceschi-Crippa, T., and A. Lattes (1968–69) Ulteriore contributo allo studio della variazione della lunghezza individuale di *Macrobiotus hufelandii* Schultze in rapporto alle mute. Boll. Mus. Ist. Biol. Univ. Genova *36*:41–45.
Greven, H. (1980) Die Bärtierchen. Die Neue Brehm-Bücherei. Wittenberg Lutherstadt: A. Ziemsen Verlagg, 101 pp.
Grigarick, A.A., R.O. Schuster, and E.C. Toftner (1973) Descriptive morphology of eggs of some species in the *Macrobiotus hufelandii* group (Tardigrada: Macrobiotidae). Pan-Pacific Entomol. *49(3)*:258–263.
Grigarick, A.A., R.O. Schuster, and E.C. Toftner (1975) Morphogenesis of two species of *Echiniscus.* In R.P. Higgins (ed): International Symposium on Tardigrades, Pallanza, Italy, June 17–19, 1974. Mem. Ist. Ital. Idrobiol. *32*(Suppl.):133–151.
Hallas, T.E. (1972) Some consequences of varying egg-size in Eutardigrada. Vidensk. Meddr. Dansk Naturh. Foren. *135*:21–31.
Henneke, J. (1911) Beitrage zur Kenntnis der Biologie und Anatomie der Tardigraden (*Macrobiotus macronyx* Duj.) Zeitschr. Wiss. Zool. *97*:721–752.
Higgins, R.P. (ed) (1975) International Symposium on Tardigrades, Pallanza, Italy, June 17–19, 1974. Mem. Ist. Ital. Idrobiol. *32*(Suppl.):469 pp.
Hulings, N.C., and J.S. Gray (eds) (1971) A manual for the study of meiofauna. Smithsonian Contr. Zool. *78*:1–83.
Kristensen, R.M. (1976) On the fine structure of *Batillipes noerrevangi* Kristensen 1976. l. Tegument and moulting cycle. Zool. Anz. *197*:129–150.
Kristensen, R.M. (1979) On the fine structure of *Batillipes noerrevangi* Kristensen, 1976 (Heterotardigrada). 3. Spermiogenesis. In B. Weglarska (ed): Second International Symposium on Tardigrades, Crakow, Poland, July 28–30, 1977. Zeszyty Naukowe Uniw. Jagiellońskiego, Prace Zoologiczne *25*:97–105.
Lattes, A. (1975) Differences in the sculpture between adults and juveniles of *Echiniscus quadrispinosus.* In R.P. Higgins (ed): International Symposium on Tardigrades, Pallanza, Italy, June 17–19, 1974. Mem. Ist. Ital. Idrobiol. *32*(Suppl.):171–176.
Lattes, A., and F. Tobia Gallelli (1972) Variabilità intraspecifica di *Echiniscus (E.) quadrispinosus* Richters e differenziazione di questa specie da *Echiniscus (E.) merokensis* Richters. Boll. Musei Ist. Biol. Univ. Genova *40*:137–152.
Marcus, E. (1928) Zur Embryologie der Tardigraden. Verhandl. Deutsch. Zool. Ges. *32*:132–146.

Marcus, E. (1929a) Tardigrada. In H.G. Bronn (ed): Klassen und Ordnungen des Tierreichs, Bd. 5, Abtlg. IV, Buch *3:*1–603. Leipzig: Akademische Verlagsgesellschaft.

Marcus, E. (1929b) Zur Embryologie der Tardigraden. Zool. Jahrb. Anat. *50:*333–384.

Marcus, E. (1936) Tardigrada. In F. Schultze (ed): Das Tierreich *66:*1–340. Berlin and Leipzig: Walter de Gruyter.

Maucci, W. (1972) Tardigradi muscicoli della Turchia. Mem. Mus. Civ. St. Nat. Verona *20:*169–221.

McIntyre, A.D. (1969) Ecology of marine meiobenthos. Biol. Rev. *44:*245–290.

Nelson, D.R. (1975) The hundred-year hibernation of the water bear. Natural History *84(7):*62–65.

Nelson, D.R. (ed) (1982) Proceedings of the Third International Symposium on the Tardigrada, Johnson City, Tennessee, Aug. 3–6, 1980. Johnson City: East Tennessee State University Press, 236 pp.

Nelson, D.R., and R.P. Higgins (1983) Tardigrada. In D. Dindal (ed): Soil Biology Guide. New York: John Wiley (in press).

Nelson, D.R., and R.O. Schuster (1981) Tardigrada. In S.H. Hurlbert, G. Rodriguez, and N.D. Santos (eds): Aquatic Biota of Tropical South America, Part 2: Anarthropoda. San Diego, California: San Diego State University Press, pp. 161–166.

Nørrevang, A. (1968) Electron microscopic morphology of oogenesis. Int. Rev. Cytol. *23:*113–186.

Oti, K. (1956) Cultural experiments with terrestrial mosses submerged in water. I. Modifications of leaves. Jpn. J. Ecol. *5:*103–111.

Pennak, R.W. (1978) Tardigrada. In: Fresh-water Invertebrates of the United States. New York: John Wiley, pp. 239–253.

Plate, L. (1889) Beiträge zur Naturgeschichte der Tardigraden. Zool. Jahrb. Anat. *3:*487–550.

Pollock, L.W. (1970a) Reproductive anatomy of some marine Heterotardigrada. Trans. Am. Microsc. Soc. *89(2):*308–316.

Pollock, L.W. (1970b) Distribution and dynamics of interstitial Tardigrada at Woods Hole, Massachusetts, USA. Ophelia *7:*145–165.

Pollock, L.W. (1975) Tardigrada. In A.C. Giese and J.S. Pearse (eds): Reproduction of Marine Invertebrates, Vol. II. New York: Academic Press, pp. 43–54.

Ramazzotti, G. (1972) Il Phylum Tardigrada, 2nd ed. Mem. Ist. Ital. Idrobiol. *28:*1–732.

Ramazzotti, G. (1974) Supplemento A, Il Phylum Tardigrada, Seconda Edizione, 1972. Mem. Ist. Ital. Idrobiol. *31:*69–179.

Riggin, G.T. (1962) Tardigrada of southwest Virginia with the addition of a description of a new marine species from Florida. Va. Agr. Exp. Sta. Tech. Bull. *152:*1–145.

Sayre, R.M. (1969) A method for culturing a predaceous tardigrade on the nematode *Panagrellus redivivus.* Trans. Am. Microsc. Soc. *88:*266–274.

Schmidt, P. (1971a) Organisation und Fortpflanzung von Tardigraden. Film C 1062/1971, Encyclopaedia Cinematographica. Göttingen: Institut für den Wissenschaftlichen Film, pp. 1–20.

Schmidt, P. (1971b) *Hypsibius dujardini* (Tardigrada), Organisation und Fortpflanzung. Film E 1658/1971, Encyclopaedia Cinematographica. Göttingen: Institut für den Wissenschaftlichen Film, pp. 1–16.

Schmidt, P. (1971c) *Milnesium tardigradum* (Tardigrada), Organisation und Fortpflanzung. Film E 1659/1971, Encyclopaedia Cinematographica. Göttingen: Institut für den Wissenschaftlichen Film, pp. 1–13.

Schuster, R.O., A.A. Grigarick, and E.C. Toftner (1980a) A new species of *Echiniscus* from California (Tardigrada: Echiniscidae). Pan-Pacific Entomol. *56(4):*265–267.

Schuster, R.O., D.R. Nelson, A.A. Grigarick, and D. Christenberry (1980b) Systematic criteria of the Eutardigrada. Trans. Am. Microsc. Soc. *99(3)*:284–303.

Schuster, R.O., E.C. Toftner, and A.A. Grigarick (1977) Tardigrada of Pope Beach, Lake Tahoe, California. Wasmann J. Biol. *35(1)*:115–136.

Shaw, K. (1974) The fine structure of muscle cells and their attachments in the tardigrade *Macrobiotus hufelandi*. Tissue Cell *6(3)*:431–445.

Von Erlanger, R. (1894) Zur Morphologie und Embryologie eines Tardigraden (*Macrobiotus macronyx* Duj.) Vorl. Mitt. I. Biol. Ctrbl. *14(16)*:582–585.

Von Erlanger, R. (1895a) Zur Morphologie und Embryologie eines Tardigraden (*Macrobiotus macronyx* Duj.) Vorl. Mitt. II. Biol. Ctrbl. *15(21)*:722–777.

Von Erlanger, R. (1895b) Beitrage zur Morphologie der Tardigraden. I. Zur Embryologie eines Tardigraden: *Macrobiotus macronyx* Dujardin. Morphol. Jahrb. *22(4)*:491–513.

Toftner, E.C., A.A. Grigarick, and R.O. Schuster (1975) Analysis of scanning electron microscope images of *Macrobiotus* eggs. In R.P. Higgins (ed): International Symposium on the Tardigrades, Pallanza, Italy, June 17–19, 1974. Mem. Ist. Ital. Idrobiol. *32*(Suppl.):393–411.

Von Wenck, W. (1914) Entwicklungsgeschichtliche Untersuchungen an Tardigraden (*Macrobiotus lacustris* Duj.) Zool. Jahrb. Abt. Anat. *37*:465–514.

Walz, B. (1974) The fine structure of somatic muscles of Tardigrada. Cell Tissue Res. *149*:81–89.

Walz, B. (1975) Ultrastructure of muscle cells in *Macrobiotus hufelandi*. In R.P. Higgins (ed): International Symposium on the Tardigrades, Pallanza, Italy, June 17–19, 1974. Mem. Ist. Ital. Idrobiol. *32*(Suppl.):425–443.

Walz, B. (1978) Electron microscopic investigation of cephalic sense organs of the tardigrade *Macrobiotus hufelandi* C.A.S. Schultze. Zoomorphologie *89*:1–19.

Walz, B. (1979) Cephalic sense organs of Tardigrada. Current results and problems. In B. Weglarska (ed): Second International Symposium on Tardigrades, Crakow, Poland, July 28–30, 1977. Zeszyty Naukowe Uniw. Jagiellońskiego, Prace Zoologiczne *25*:161–168.

Walz, B. (1982) Molting in Tardigrada. A review including new results on cuticle formation in *Macrobiotus hufelandi*. In D.R. Nelson (ed): Proceedings of the Third International Symposium on the Tardigrada, Johnson City, Tennessee, Aug. 3–6, 1980. Johnson City: East Tennessee State University Press, pp. 129–148.

Weglarska, B. (1975) Studies on the morphology of *Macrobiotus richtersi* Murray, 1911. In R.P. Higgins (ed): International Symposium on the Tardigrades, Pallanza, Italy, June 17–19, 1974. Mem. Ist. Ital. Idrobiol. *32*(Suppl.):445–464.

Weglarska, B. (1979a) Electron microscope study on previtellogenesis and vitellogenesis in *Macrobiotus richtersi* J. Murr. (Eutardigrada). In B. Weglarska (ed): Second International Symposium on Tardigrades, Crakow, Poland, July 28–30, 1977. Zeszyty Naukowe Uniw. Jagiellońskiego, Prace Zoologiczne *25*:169–189.

Weglarska, B. (ed) (1979b) Second International Symposium on Tardigrades, Crakow, Poland, July 28–30, 1977. Zeszyty Naukowe Uniw. Jagiellońskiego, Prace Zoologiczne *25*:1–197.

Weglarska, B. (1982) Ultrastructural study of the formation of egg envelopes in *Macrobiotus richtersi* (Eutardigrada). In D.R. Nelson (ed): Proceedings of the Third International Symposium on the Tardigrada, Johnson City, Tennessee, Aug. 3–6, 1980. Johnson City: East Tennessee State University Press, pp. 115–128.

Wolburg-Buchholz, K., and H. Greven (1979) On the fine structure of the spermatozoan of *Isohypsibius granulifer* Thulin 1928 (Eutardigrada) with reference to its differentiation. In B. Weglarska (ed): Second International Symposium on Tardigrades, Crakow, Poland, July 28–30, 1977. Zeszyty Naukowe Uniw. Jagiellońskiego, Prace Zoologiczne *25*:191–197.

Developmental Biology of Freshwater Invertebrates, pages 399–483

Development of the Pulmonate Gastropod, *Lymnaea*

John B. Morrill

ABSTRACT This paper reviews the salient features of the natural history of freshwater pond snails, *Lymnaea,* and other freshwater pulmonates, with respect to their utility for experimental analyses of embryonic development. Methods for maintaining breeding colonies of adults and for obtaining and culturing embryos are outlined. Both general and specific features of development from gametogenesis to the juvenile snail stage are described. Standard timetables of the developmental morphology are provided for gametogenesis, oviposition to first cleavage, the cleavage period, gastrulation, and the postgastrula to hatching juvenile snail stage. The physical and chemical aspects of the egg capsule, embryonic nutrition, and developmental chemistry are summarized. Developmental abnormalities are reviewed. Because of the mosaic nature of their developmental patterns, freshwater pulmonate embryos are exceptional models for the analysis of positional information, pattern formation, and morphogenesis.

INTRODUCTION

Freshwater pond snails are gastropod molluscs belonging to the subclass Pulmonata, order Basommatophora. Within this order are 13 families of which eight occur in freshwater habitats (Hubendick, '78). The embryology of only a relatively few species has been studied to any extent in three families—Lymnaeidae (34 genera), Planorbidae (84 genera), and Physidae (7 genera)—that are commonly distinguished from one another by the general shape of the adult shell (Fig. 1). Beneath the shell is a highly differentiated animal (Fig. 2) that develops from an egg 80 to 150 μm in diameter that is laid or oviposited in a capsule filled with albuminous fluid. The fertilized egg passes its entire development from oviposition until the juvenile snail stage in the capsule.

The extensive literature on the natural history, reproductive biology, and embryology of pond snails has been reviewed in major treatises and mono-

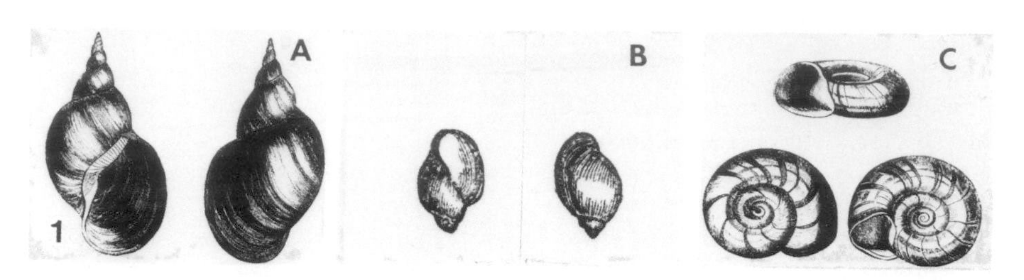

Fig. 1. Shapes of shells of adult freshwater pulmonates commonly used in embryological studies. A. *Lymnaea* sp. (dextral). B. *Physa* sp. (sinistral). C. *Planorbis* sp. (sinistral).

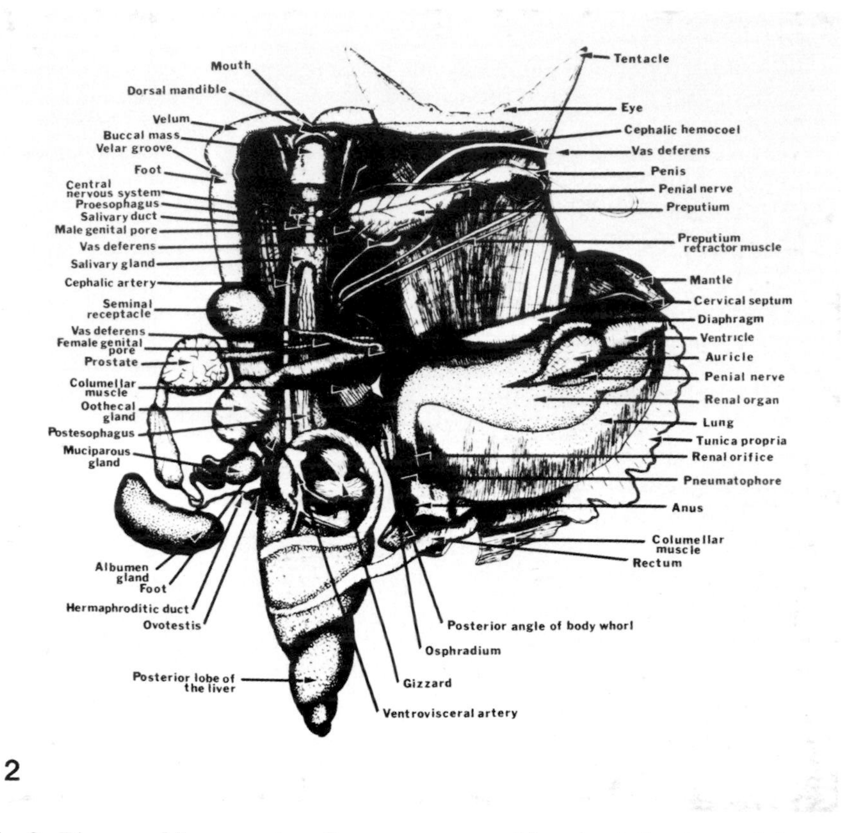

Fig. 2. Diagram of *Lymnaea stagnalis appressa* removed from its shell and opened along the left side. The dorsum is flattened to the right and the organs are spread for clarity. In their natural position the reproductive organs overlie the postesophagus; the lung covers the reproductive organs. Courtesy of Dr. Carriker (Carriker, '46).

graphs. Yet there is no guide or manual to introduce and orient the professional biologist or the teacher and student of comparative embryology to various apsects of development in this group of invertebrates. Accordingly, this chapter includes guidelines and brief summaries of information on the natural history and life cycles of adult snails and laboratory breeding colonies, methods for rearing embryos, and overviews of normal and abnormal development. Because this chapter complements and supplements what has been published previously, I have cited those references that are most pertinent with respect to their contents and completeness of their bibliographies.

The general developmental patterns of basommatophoran pulmonates are similar. Because of the many descriptive and experimental studies on the development of *Lymnaea stagnalis* by Professor C.P. Raven and his students and co-workers (Raven, '66), the development of *Lymnaea* will be emphasized. My own investigations have been on the North American *Lymnaea palustris,* so most of the illustrations are of this species and include previously unpublished data.

GENERAL FEATURES OF DEVELOPMENT

Pulmonates are hermaphrodites, thus both cross- and self-fertilization may occur; the latter thereby allows for the establishment of genetic strains. Fertilization is internal in a region where the common hermaphroditic duct bifurcates into the male and female reproductive tracts. Following fertilization the egg is surrounded by nutritive secretions (albumen) from the albumen gland. An individual egg and its surrounding albumen is then encapsuled by a set of membranes as it passes down the oviduct. Depending upon the species and the breeding conditions, a few to many egg capsules are further surrounded by additional secretions and laid in a common jelly mass in lymnaids and physids or in a capsule in planorbids (Bondesen, '50).

The general features of the development of the egg from oviposition until the juvenile snail hatches from its egg capsule 7 to 20 days later can be easily observed at low magnifications with either the stereo or compound microscope. Among the earlier accounts, that of Lankester (1874) is one of the most readable. Raven ('48, '75) gives more formal and yet concise reviews of pulmonate development as well as a detailed description (Raven, '66). Hence only the major morphological features and temporal aspects of the development (Fig. 3) of the *Lymnaea* egg from oviposition until hatching from the capsule are described in this section.

During the first 2 to 3 hours following oviposition the fertilized egg undergoes two maturation divisions to form the first and second polar bodies at the animal pole (Fig. 4A). At the end of the second maturation division the chromosomes at the animal pole of the egg swell into small vesicular

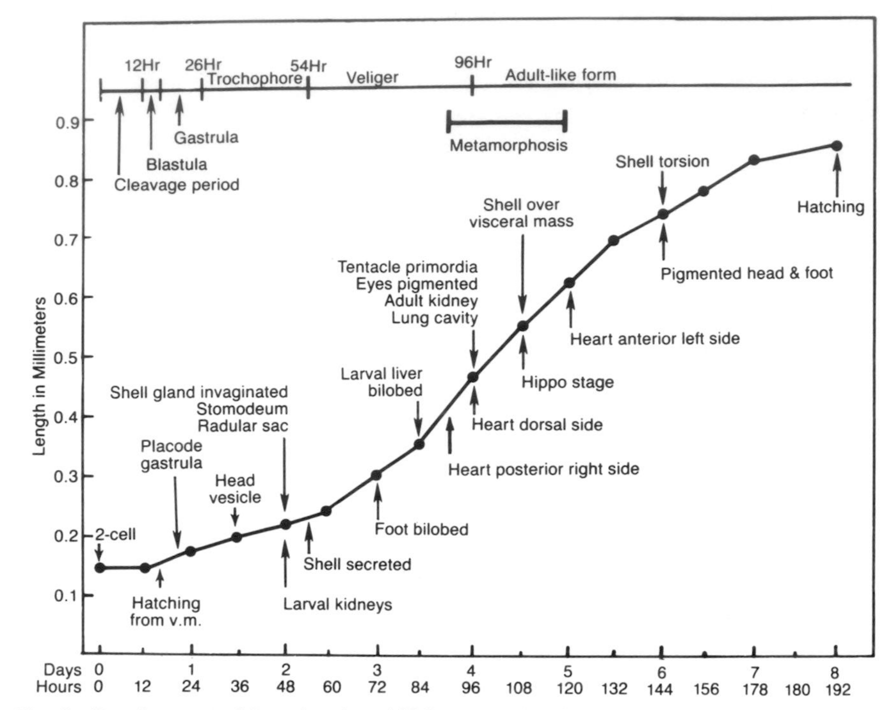

Fig. 3. Development of *L. palustris* at 25°C. versus time in days and hours beginning at the 2-cell stage including the length of the embryo, the six major developmental stages, and major morphologic criteria used to stage embryos.

karyomeres, which coalesce to form the female pronucleus that remains at the animal pole after the second maturation. The male pronucleus migrates to the animal pole. At the onset of first cleavage a central spindle forms between the two pronuclei and the egg divides into two cells of equal size by unipolar cytokinesis. Because of their attachment to the vitelline membrane and the attachment of the vitelline membrane to the tips of microvilli on the surface of the egg, the polar bodies mark the animal pole of the egg from the

---

Fig. 4. Scanning electron micrographs of the egg of *L. palustris* during the early cleavage period. A. 1-cell. B. 2-cell, stage 8. C. 4-cell, stage 13. D. 8-cell, stage 17. E. 12-cell, stage 20a. F. early 24-cell, stage 23. Note polar bodies lost from eggs in (D) and (F). Bar (A–F), 40 μm. b, membrane blisters, a fixation artifact. cf, capsule fluid. d, dimples in surface of cells in vegetal hemisphere of egg and blastomeres resulting from shrinkage of vacuolated gamma granules beneath the oolemma. m, micromere. M, macromere. p1, first polar body. p2, second polar body. vm, vitelline membrane.

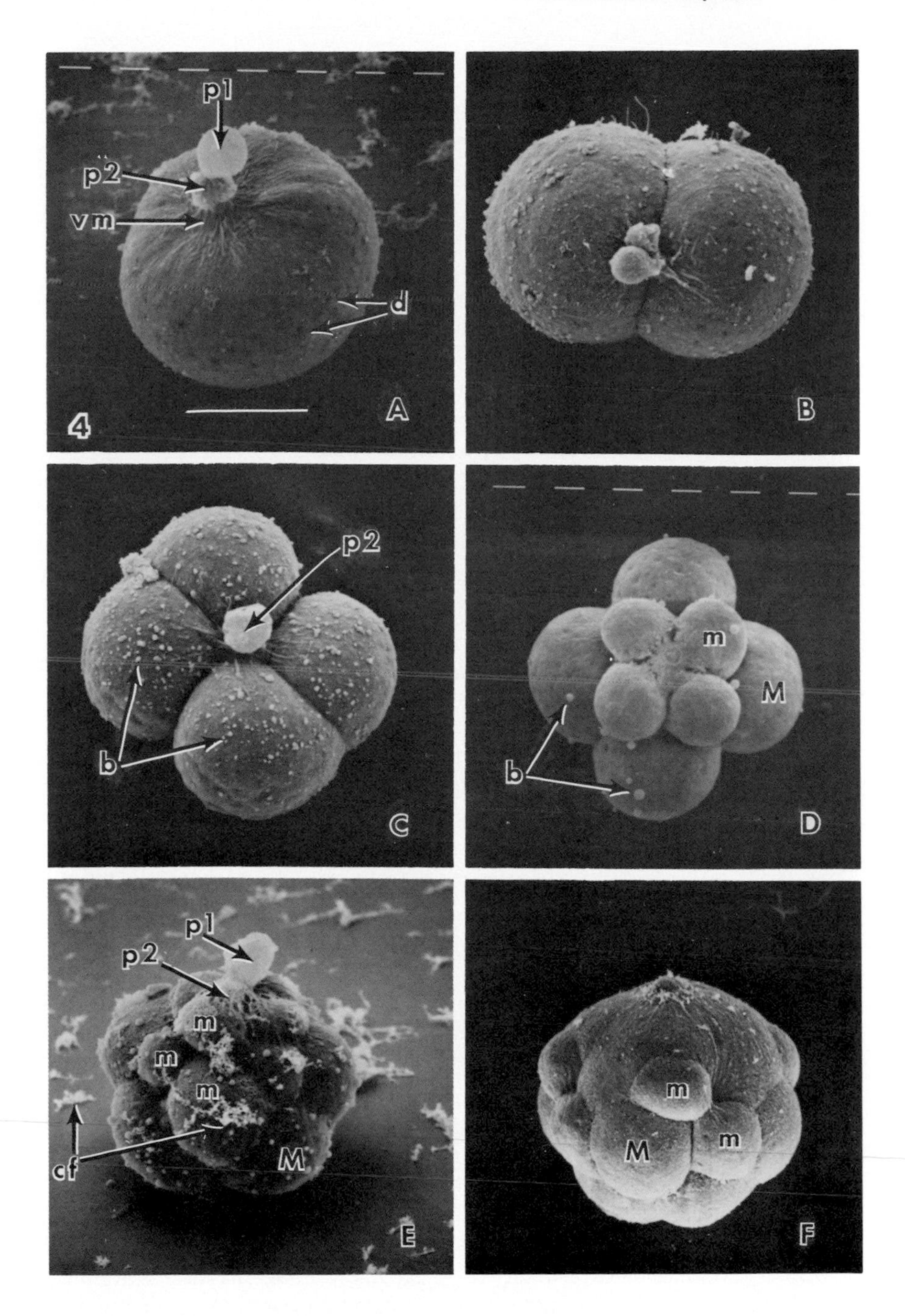
p1
p2
vm
d
4
A
B
p2
b
C
m
M
b
D
p1
p2
m
m
m
cf
M
E
m
M
m
F

time of their formation until the early gastrula stage 16 hours after first cleavage when the vitelline membrane and polar bodies are cast off together.

In addition to the maturation divisions between oviposition and first cleavage, there is an ooplasmic segregation of cytoplasmic organelles that results in the formation of a mitochondria-rich animal pole plasm at the animal pole, a continuous layer of subcortical plasm beneath the oolemma, and a more central endoplasmic region with numerous vacuolated gamma yolk granules.

With first cleavage the egg enters the cleavage period, which I have divided into the early cleavage period (2- to 24-cell stage), the 24-cell stage resting period and the late cleavage period (25- to 49-cell stage). In the early period cleavage follows the spiral molluscan pattern (Fig. 4B–F) in which three quartets of micromeres are formed by alternating dexiotropic and laeotropic supraequatorial divisions of the four macromeres. The egg exhibits a radial symmetry as a result of synchronous cleavage. The size of the cells is similar in the four quadrants of the macromeres and the tiers of micromeres. Thus there is no morphologically distinct D quadrant. Following each cleavage the rounded cells flatten, the embryo undergoes compaction, and a fluid-filled cleavage cavity forms and disappears shortly before the next cleavage. The formation and disappearance of the cleavage cavity recur throughout the early cleavage period.

Up to the 24-cell stage the egg divides at the rate of one cleavage every 70 to 85 minutes. At the 24-cell stage the egg stops dividing for 2.5 to 3 hours. During this "resting period" the inward cytoplasmic extensions of the micromeres and macromeres occlude the cleavage cavity. The tips of 14 micromeres converge and make contact with one vegetal cross-furrow macromere, the presumptive 3D macromere. From this stage on the recurrent cleavage cavity is reduced to a number of lenticular, intercellular spaces. Thus there is no blastocoelic cavity in *Lymnaea* and perhaps no blastula stage.

The 24-cell stage is also characterized by the migration of subcortical accumulations (SCA) of RNA-rich cytoplasm from the vegetal cross-furrow region of the macromeres to their inner tips and by the presence of gap junctions between adjoining micromeres and between micromeres and macromeres. Also during this stage functional nucleoli and RNA synthesis are first detected (Biggelaar, '71a). In *L. palustris* individual cells begin to imbibe the capsule fluid by endocytosis. Morphologically the 24-cell stage marks the beginning of the bilateral asynchronous late cleavage period and the appearance of the future dorso-ventrality of the embryo.

The 24-cell stage ends and the late cleavage period begins when the outwardly bulging 3D macromere divides to form the 4d micromere (M cell or primary mesentoblast) and the 4D macromere. This division is followed by a series of asynchronous and bilaterally symmetrical divisions in the other quartets of cells during the next 3 hours. The late cleavage period ends approximately 12 hours after first cleavage with the bilateral division of the (M) micromere into the right and left M^1 and M^2 cells and the division of the 3A-3C macromeres to form the other micromeres (4a-4c) of the fourth quartet.

Over the next 4 hours the descendants of the first and second quartet of micromeres divide to form a crosslike arrangement of cells—the molluscan cross—in the animal hemisphere of the egg. The development of the molluscan cross continues into the gastrula period, which begins approximately 16 hours after first cleavage when the embryo prepares to "hatch" from or cast off the vitelline membrane and attached polar bodies. In the process of hatching, the vegetal end of the embryo exhibits a series of amoeboidlike bulges, the vitelline membrane ruptures at the vegetal end of the egg, and the embryo appears to slide out of the ruptured vitelline membrane. A few minutes later the vegetal end of the embryo flattens as the macromeres (5A-5D) and micromeres (4a-4c, 5a-5d) begin to invaginate, forming a shallow depression. During the gastrula period, which may be divided into six phases (G1–G6), the initial invagination at the vegetal pole deepens, the animal pole region flattens and then is depressed, and the future posterior end of the embryo enlarges relative to the anterior end of the embryo. This results in a placode-shaped gastrula (Fig. 5A) when viewed from the side. The flattened placode gastrula occurs approximately 24 hours after first cleavage; gastrulation ends 2 hours later when the stomodeum forms by the convergence and union of the lateral lips of the gastropore. Differentiation of larval organs begins during the gastrula period with the formation of cilia on the surfaces of the primary velar cells.

Following gastrulation the embryo differentiates first into a trochophore-like larva with conspicuous larval liver cells forming a cuplike layer around the endodermal lumen (Fig. 5B); other larval organs and adult organ primordia appear. Thus the 2-day (48-hour) trochophore larva has a pair of larval kidneys (protonephridia), an invaginating larval shell gland, ciliated apical plate cells, a full complement of ciliated velar cells, a radular pouch in the ventral wall of the pharynx, and a bulging foot primordium ventral to the stomodeum.

By the third day of development the embryo has become a veliger larva with a recognizable bilobed liver (Fig. 5C). The foot is bilobed, the larval shell (protoconch) has begun to be secreted, and the embryo has begun to elongate as well as to exhibit the first signs of asymmetry. Twenty-four hours later (4-day stage) the veliger begins to metamorphose into a juvenile adult (Fig. 5D). The adult tentacle and eye primordia appear; the shell covers the posterior half of the body or visceral mass; the adult mantle fold and lung primordia begin to differentiate. The heart, which first appears on the right

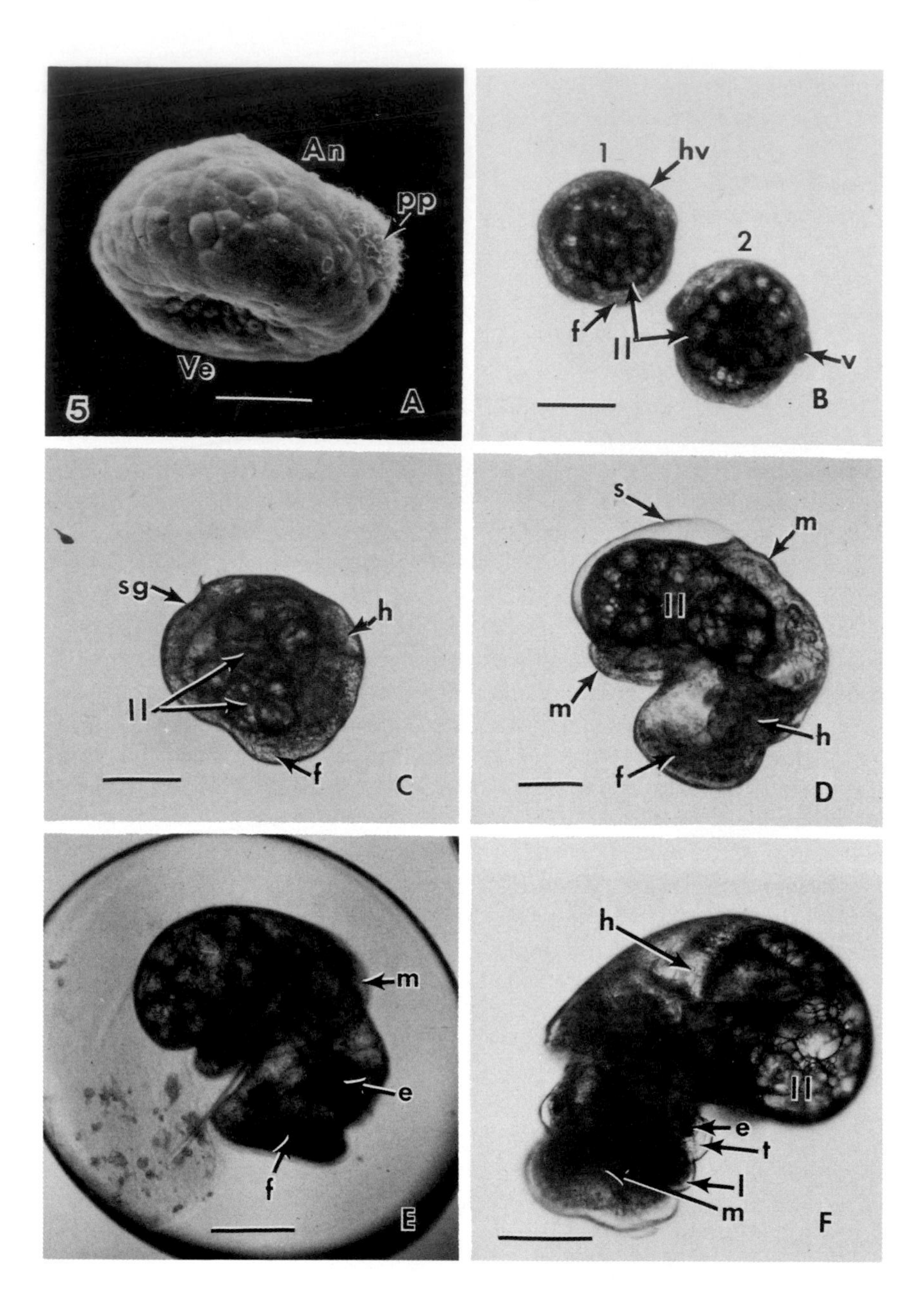
5
A
An
pp
Ve
B
1
2
hv
f
II
v
C
sg
h
II
f
D
s
m
II
m
h
f
E
m
e
f
F
h
II
e
t
l
m

side of the 3.5-day embryo, is now located on the dorsal side of the visceral mass and the bilobed foot is distinctly separate from the head region.

In the middle of the metamorphic period the 4.5-day embryo resembles a miniature hippopotamus when viewed anteriorly and is aptly referred to as the "hippo" stage (Raven, '49). By this stage the shell completely covers the visceral mass. From this stage until hatching the various adult organ primordia continue to differentiate and grow and the whole embryo grows (Fig. 5E,F) until it is nearly as large as the capsule and has exhausted the nutrients of the capsule fluid.

The temporal appearances of the larval and adult organs in *L. palustris* are summarized in Table I. This table is adapted from Cumin's ('72) study on organogenesis in *L. stagnalis*. Included in this table are Cumin's morphological stages E2–E11 and developmental stages of *L. palustris* at 25°C. Throughout this chapter postgastrula stages of *L. palustris* will be referenced both by age in days or hours at 25°C and by Cumin's stages in parentheses, i.e., 2-day trochophore (E2).

Because more than 90% of the nutrient reserves are in the capsule fluid, growth accompanies morphogenesis and differentiation of the pulmonate embryo (Fig. 3). However, the processes associated with overall growth of the embryo may be uncoupled from those associated with morphogenesis and differentiation after the trochophore stage by culturing decapsulated embryos in pond water.

In concluding this section I refer the reader to the following references for detailed descriptions and illustrations of the development of *Lymnaea* in particular and freshwater pulmonates in general: oviposition to first cleavage (Raven, '45), first cleavage to the trochophore stage (Raven, '46), trochophore stage to hatching (Cumin, '72), hatching stage to juvenile snail (Arni, '73). Several other studies provide excellent illustrations and morphological descriptions of whole embryos and histological sections of embryos sectioned in different planes (Fol, 1879; Rabl, 1879; Wierzejski, '05; Kubota, '54; Régondaud, '64; Arni, '74). Additional details of the morphogenesis and histogenesis of specific larval and adult organs may be found in the following:

Fig. 5. Normal stages of development of *L. palustris* at 25° C. A. SEM of 24-hour placode gastrula viewed from the right side showing the indented animal and vegetal poles. B. LM of 2-day (E2) trochophores (1. side view, 2. posterior view). C. LM of 3-day (E3) veliger viewed from the right side. D. LM of 4-day (E5) hippo stage viewed from right side. E. LM of 5-day (E7) juvenile snail. F. LM of 6-day (E8) juvenile snail. Bars (A) 40 μm, (B-D) 100 μm, (E,F) 200 μm. An, animal pole. e, eye. f, foot. h, head. hv, head visicle. l, lappet. ll, larval liver cells. m, mantle margin. pp, ciliated primary prototrochal cells. s, shell. sg, shell gland. t, tentacle. v, velum. Ve, vegetal pole.

TABLE I. Temporal Appearances of Larval and Adult Organs in *Lymnaea palustris* Cultured at 25°C[a]

	Stage (days of development)										
	E1 (1.5)	E2 (2.0)	E3 (2.5)	E4 (3.5)	E5 (4.0)	E6 (4.5)	E7 (5.0)	E8 (6.0)	E9 (7.0)	E10 (8.0)	E11 (9.0)
Velum	+[b]	+	+	+	+	+	+	+	+	+	+
Larval liver cells	+	+	+	+	+	+	+	+	+	+	+
Stomodeum	+	+	+	+S	+S	+	+	+	+	+	+
Protonephridia	(+)	+	+	+	+	+	D	D	−	−	−
Esophagus	P	+	+	+	+	+	+	+	+	+	+
Shell	−	(+)	+	+	+	+	+	+	+	+	+
Stomach	−	(+)	(+)	+	+	+	+	+	+	+	+
Intestine	−	P	(+)	+	+	+	+	+	+	+	+
Radular apparatus	−	U	(+)	(+)	(+)	(+)	+	+	+	+	+
Adult kidney	−	U	U	U	U	+	+	+	+	+	+
Nuchal cells	−	−	+	+	+	+	+	+	+	+	+
Cerebral ganglia	−	−P	U	U	U	+	+	+	+	+	+
Eyes	−	−	I	(+)	(+)	+	+	+	+	+	+
Tentacles	−	−	I	(+)	(+)	+	+	+	+	+	+
Heart	−	−	−	+	+	+	+	+	+	+	+
Pedal ganglia	−	−	−	P	U	U	+	+	+	+	+
Pleural ganglia	−	−	−	P	P	U	+	+	+	+	+
Parietal ganglion	−	−	−	P	P	U	+	+	+	+	+
Visceral ganglion	−	−	−	P	U	U	+	+	+	+	+
Oral lappets	−	−	−	I	(+)	+	+	+	+	+	+
Lung cavity	−	−	−	I	(+)	+	+	+	+	+	+
Buccal ganglia	−	−	−	−	(+)	+	+	+	+	+	+
Osphradial ganglion	−	−	−	−	P	U	U	+	+	+	+
Osphradium	−	−	−	−	P	I	I	+	+	+	+
Mantle cavity	−	−	−	−	I	+	+	+	+	+	+
Jaws	−	−	−	−	(+)	(+)	+	+	+	+	+
Pericardium	−	−	−	−	(+)	(+)	+	+	+	+	+
Adult digestive gland	−	−	−	−	−	(+)	(+)	(+)	(+)	(+)	(+)
Salivary glands	−	−	−	−	−	−	(+)	+	+	+	+
Lung spiracle	−	−	−	−	−	−	−	+	+	+	+
Gonad	−	−	(+)	(+)	(+)	(+)	(+)	(+)	(+)	(+)	(+)
Reproductive tract	−	−	−	−	(+)	(+)	(+)	(+)	(+)	(+)	(+)

[a]Adapted from Cumin, '72.

[b]Key: −, absent; (+), process of developing; +, fully developed; D, degenerating; P, proliferation of cells; U, undifferentiated cell complex; I, invagination; S, sinking inward.

larval protonephridia (Meisenheimer, 1899); the foregut (Raven, '58); the midgut and larval liver cells (Arni, '75); the shell gland and mantle (Raven, '52; Timmermans, '69; Kniprath, '77,'79,'81); the lung (Régondaud, '64); sensory organs, the head, and ganglia of the central nervous system (Raven, '49; Kruglyanskaya and Sakharov, '75); the nuchal cells (Régondaud, '72; Régondaud and Brisson, '76); the radula (Kerth, '79); the statocysts (Geuze, '68); the gonads and reproductive tract (Fraser, '46; Luchtel, '72; Brisson and Besse, '75; Brisson and Régondaud, '77). Finally, Fioroni ('66) and Fioroni and Schmekel ('76) summarize the early literature in a comparative manner for prosobranch and pulmonate gastropods.

NATURAL HISTORY OF ADULTS

The Lymnaeidae, Planorbidae, and Physidae inhabit all types of freshwater habitats worldwide. Therefore, throughout their geographic range taxonomic species vary morphologically and physiologically. This led to the splintering of *L. stagnalis* into at least eight subspecies and *L. palustris* into three subspecies (Baker, '11). Recent studies by Hubendick ('51) on the Lymnaeidae and Te ('78) on the Physidae have eliminated the taxonomic confusion in these families. In spite of the phenotypic plasticity of adult snails, the haploid number of chromosomes is 18 in subspecies and races of *L. stagnalis* and *L. palustris* (Inaba, '69; Patterson and Burch, '78).

Two practical aspects of taxonomy need emphasis. The first is that within the geographic range of any species, even among local populations, there can be considerable phenotypic variation, not only in morphology but also in the physiology and fecundity of the adults (Forbes and Crampton, '42). The second is the scientific name of the species. The European *L. stagnalis* (L.) differs from the North American *L. stagnalis* physiologically. Currently the North American species is most often named *L. stagnalis appressa.* The taxonomic naming of *L. palustris* is more complicated. The true *L. palustris* (Muller) is an European species. While a closely related North American species is referred to as *L. palustris* in the literature, it has also been named *L. elodes, Stagnicola palustris, S. palustris elodes,* and *S. elodes.*

Hubendick ('78) has summarized the various morphological features, habitats, and geographic distribution for the families and genera in Basommatophora. His review and that of Russell-Hunter ('78) on their ecology cover nearly every facet of the natural history of freshwater pulmonates, while Baker ('11) provides the embryologist with a readable introduction to the morphology, ecology, and behavior of the lymnaeids.

Species of lymnaeids, physids, and planorbids occur in the euphotic zones of neutral to hard, calcium-rich, eutrophic bodies of fresh water. They feed on the *Aufuchs* of aquatic macrophytes, benthic algae, and detritus. Typically they occur in quiet backwater areas of vegetated shores of lakes, ponds, marshes, and drainage ditches, and between and beneath rocks of nonvegetated shorelines and slow-moving streams. In general, the larger the body of water and the more productive and diverse the types of aquatic vegetation, the larger the population of any one species and the greater the diversity of species that may be present. When the aquatic vegetation dies, the snails usually disappear from an area.

The general features of the reproductive life cycles are reviewed by Bondesen ('50), Duncan ('59, '75), and Russell-Hunter ('78). In the temperate zone most species exhibit an annual life cycle in which juveniles or adults overwinter in the benthos. With the rise in water temperature in early spring, these individuals resume their growth; egg laying commences following the development of the female reproductive system and continues into midsummer and early fall depending on the water temperature. Thus the best time to collect reproductively active adults in the temperate zone is between March and September. However, individuals collected from wild populations at any time of the year can be conditioned to breed in a few days to weeks in the laboratory.

Collected specimens can be identified to the level of family and genus on the basis of shell characters—size, shape of shell, and shape of shell aperture—and whether the shell is right handed or left handed. A high-spired, dextral shell occurs in most Lymnaeidae (Fig. 1A) although sinistral-shelled individuals and races do occur. The inheritance of sinistrality appears to have been studied only in *Lymnaea peregra* (Boycott et al., '31; Ubbels et al., '69; Raven, '72a; Freeman, '77). Physidae (Fig. 1B) and some Planorbidae have sinistral shells. In most planorbids the whorls of the spire are in a single plane producing a sinistral, discoidal type of shell (Fig. 1C).

For the embryologist wishing to study the reproductive system and unravel it from the other adult organs of a snail removed from its shell, the procedural and anatomical descriptions of Baker ('11), Carriker ('46), and McCraw ('57) are most helpful. Snails should be anesthetized before removing the shell and before subsequent dissections. Of the various agents used to relax freshwater snails the more successful include nembutal and chloretone (Carriker, '46); nembutal and methanol (McCraw, '58); and nitrogen gas, nembutal, Sandoz M.S. 222, and CO_2 (Lever et al., '64). The last method is recommended particularly for larger species, e.g., *L. stagnalis* and *L. palustris,* because it is effective in minutes instead of hours and anesthetized snails recover rapidly. Once relaxed the snail may be dissected with fine needles, forceps, and iridectomy scissors.

OBTAINING AND MAINTAINING ADULTS

Obtaining Adults

One or more species of pond snails can be found in almost any local, eutrophic body of hard water that has emergent or floating vegetation. Cattail marshes are excellent sites especially for *L. stagnalis* and *L. palustris*. "Pond" snails are available from biological supply companies, wholesale aquarium supply companies, and local pet shops. Usually such snails are species of Physidae and Planorbidae. If a particular species is not available locally or commercially, enough snails or egg masses to start a breeding colony may be requested from individual researchers. Juvenile or adult snails can be shipped like other aquatic invertebrates via air express; individual egg masses may be shipped in vials of pond water. Juveniles of young adults of *L. stagnalis* are less likely to die in transit than mature adults.

Maintaining Adults

General conditions for rearing freshwater pulmonates in the laboratory are outlined in Galtsoff et al. ('37). More specific conditions for maintaining breeding colonies of *L. stagnalis* are described by Noland and Carriker ('46), Steen ('67, '68), Steen et al. ('69), and Mooij-Vogelaar et al. ('70). These papers also detail and review the literature on the life history in the laboratory, including growth, life span, sexual maturation, rhythms and stimulation of oviposition activity, frequency and duration of egg laying, total egg production per snail, and rate of embryological development, as well as the effects of water, temperature, light, atmospheric pressure, food, overcrowding, changing of water, copulation, and isolation of adults.

Following the methods of Noland and Carriker, I have maintained for 15 years a laboratory breeding colony of *L. palustris* that began with snails collected from wild populations in Connecticut, Iowa, and Virginia. Breeding adults reside in 5-liter polypropylene circular tanks. Tanks are continuously aerated by an air pump or by compressed air from a laboratory air line with a charcoal and glass wool recirculating air filter hung on the outside of the tank. Each cylindrical tank rests on a small ball-bearing turntable to allow the tank to be rotated when oviposited egg masses are removed. The tanks are arranged in two rows on a laboratory bench (Fig. 6). The front row of tanks contains breeding adults with 20 to 25 snails per tank; the elevated posterior row of tanks serves as a nursery for restocking the breeding tanks. Since the tanks have neither vegetation nor gravel, the snails are forced to lay their eggs on the walls and bottoms of the tanks. The snails are fed *ad libitum* with the greenest outside leaves trimmed from iceberg head lettuce or the greenest leaves from leafy varieties of lettuce. Fecal matter is aspirated

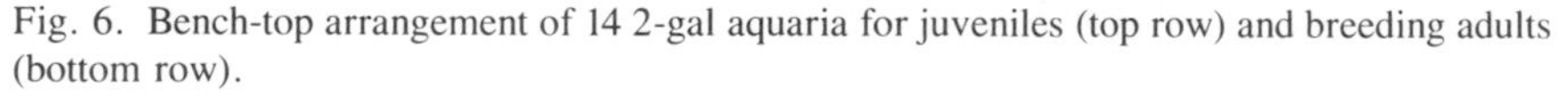
Fig. 6. Bench-top arrangement of 14 2-gal aquaria for juveniles (top row) and breeding adults (bottom row).

from the bottom of the tanks daily. A small quantity of fine sand is placed in each tank for the snails' gizzards. The tanks are illuminated with fluorescent ceiling lights or a set of lights on a stand over the tanks. Under a 16-hour light/8-hour dark cycle the snails tend to lay eggs between 0600 and 0800 hours and again between 1800 and 2000 hours. Under more or less continuous illumination, oviposition is more irregular. Even after many generations, egg laying in the laboratory decreases in the fall and then increases in the spring.

Whereas *L. stagnalis* first begin to lay eggs approximately 3 months after oviposition, when the length of the shell is 40–50 mm (Noland and Carriker, '46), *L. palustris* begin to lay eggs when the length of the shell is 15 to 16 mm, 6 to 8 weeks after oviposition. McCraw ('70) has described the changes in size and morphology of the reproductive tract of *L. palustris* relative to the increase in length of shell. Even after several generations in the laboratory many individual snails of the same generation show noticeable and measurable variations in fertility, rate of growth, and longevity. For example, in six tanks of breeding adults oviposition may occur regularly in only one or two tanks because of genetic differences (Forbes and Crampton, '42) or variations in the water quality of individual tanks.

I have used filtered local pond water, spring water, and deep well water with equal success; tap water nearly always kills the snails and the developing embryos. Since water that will support both adults and embryos is not always available, I have listed in Table II formulae for six artificial pond waters. The critical elements are pH (7.0 to 8.4); sodium, calcium and magnesium

TABLE II. Formulae for Six Artifical Pond Waters[a]

	Concentration in milligrams per liter of distilled or deionized water					
Substance	1	2[b]	3	4	5	6
NaCl	2,000	1,000	2,000	1,300	20	—
KCl	—	50	100	97	3.3	—
Na_2HPO_4	—	25	50	17	—	—
KH_2PO_4	100	—	—	—	—	—
$MgCl_2$	300	—	—	325	—	33
$MgSO_4 \cdot 7H_2O$		100	200	—	—	30
$Mg(HCO_3)_2$	—	—	—	—	59	—
$CaCl_2$	300	87	100	558	—	—
$KHCO_3$	—	—	—	—	—	10
$NaHCO_3$	2,000	120	—	1,769	—	58
$Ca(HCO_3)_2$	—	—	—	—	162	—
pH	—	8.4	7.3	7.3	—	7.3

[a]References for the formulae: 1) Carriker, M.R. ('43) Nautilus *57:*52–59; 2) Friedl, F.E. ('61) Exp. Parasitol. *15:*7–13; 3) Chernin, E. ('57) Proc. Soc. Exp. Biol. Med. *96:*204–210; 4) Basch, P.F. and J.J. Di Conza ('73) J. Trop. Med. and Hygiene *22:*805–813; 5) Greenway, P. ('70) J. Exp. Biol. *53:*147–163; 6) Taylor, H.S. ('69) J. Exp Biol. *59:*453–564.
[b]Formula 2 has in addition 25 mg disodium EDTA, 2.5 mg phenol red, 10 μg $CuSO_4 \cdot 5H_2O$, 100 μg $FeCl_3$, 100 μg H_3BO_3, 100 μg $MnSO_4 \cdot 4H_2O$, 10 μg $ZnCl_2$.

salts; temperature 15° to 25°C; and fouling of the water. Even when snails are fed regularly, growth and egg laying will decline unless the water is changed and the walls and filter system cleaned biweekly.

OBTAINING AND REARING EMBRYOS

In my laboratory four to six breeding tanks of *L. palustris* yield adequate numbers of egg masses daily for classroom demonstrations and descriptive embryological studies. When greater numbers of egg masses are needed, oviposition is stimulated by changing the water in the tanks. This induces laying of egg masses in the next 3 hours by individuals with a high oviposition activity. A sudden rise in temperature also stimulates oviposition in freshwater pulmonates (Duncan, '75). Oviposition may also be stimulated in *L. palustris* by transferring snails from a breeding tank at 20°C to illuminated tanks of well-aerated water with or without submerged aquatic plants, e.g., *Elodea sp.*, *Vallisenaria sp.*, at 25°C. Frequently the snails must be entrained to this regime of temperature shock over a period of days before the stimulated snails will lay eggs in the first 3 to 4 hours.

When a *Lymnaea* egg mass is first laid, it appears opaque (Fig. 8A). Within a few hours the egg mass becomes transparent and jellylike (Fig. 8B) as the granules in the tunica interna surrounding the egg capsules dissolve.

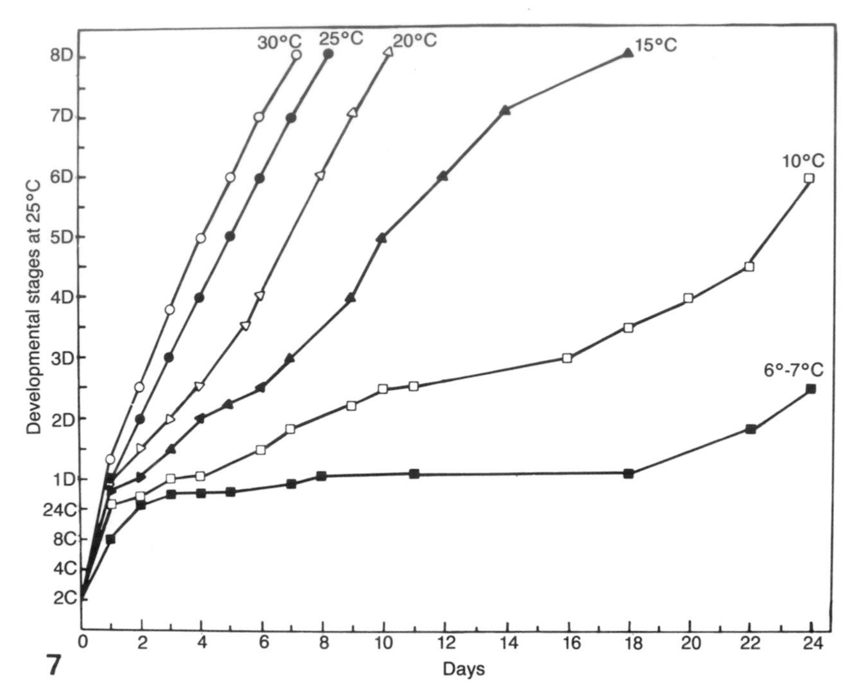

Fig. 7. Development of *L. palustris* at six temperatures with respect to developmental stages at 25°C. Ordinate scale: 2C-24C, 2-cell through 24-cell stage; 1D-8D, 1-day gastrula through hatching from egg capsule on 8th day.

Fig. 8. Egg masses and egg capsules of *L. palustris*. A. Opaque egg mass immediately after oviposition. B. Same egg mass 2 hours later after dissolution of granules in tunica interna. C. Distal end of egg mass. D–H. Egg mass and egg capsules treated with 0.1% sodium oxalate to localize calcium binding sites. D. Margin of egg capsule. E. Teased egg mass with calcium oxalate crystals in tunica interna. F. Egg capsule with calcium oxalate crystals throughout the capsule fluid. G. Calcium oxalate crystals in lighter layer of partially stratified capsule fluid. H. Egg capsule centrifuged at 2,500 g to stratify the capsule fluid into three layers and to fragment the 1-cell egg. The centripetal egg fragment is in the centripetal layer of fluid. Calcium oxalate crystals are in both the centripetal and centrifugal layers of fluid. I. Egg capsule fixed in Bouin's fixative to coagulate the capsule fluid immediately surrounding the embryo. Bars (A,B) 4mm, (C–I) 250 μm. c, viscous component of capsule fluid coagulated with Bouin's fixative. cf, capsule fluid. co, calcium oxalate crystals. e, embryo. lm, limiting membrane. me, membrana externa. mi, membrana interna. o, ovum. tc, tunica capsularis. ti, tunica interna. y, yolk fragment.

Fig. 9. Egg mass of *Planorbis* sp. Bar 5 mm.

Fig. 10. SDS-PAGE patterns of *L. palustris* capsule fluid. A. Bands stained with Coommassie blue. B. Bands stained with PAS. A and B are from two separate gels and two egg masses run under similar conditions.

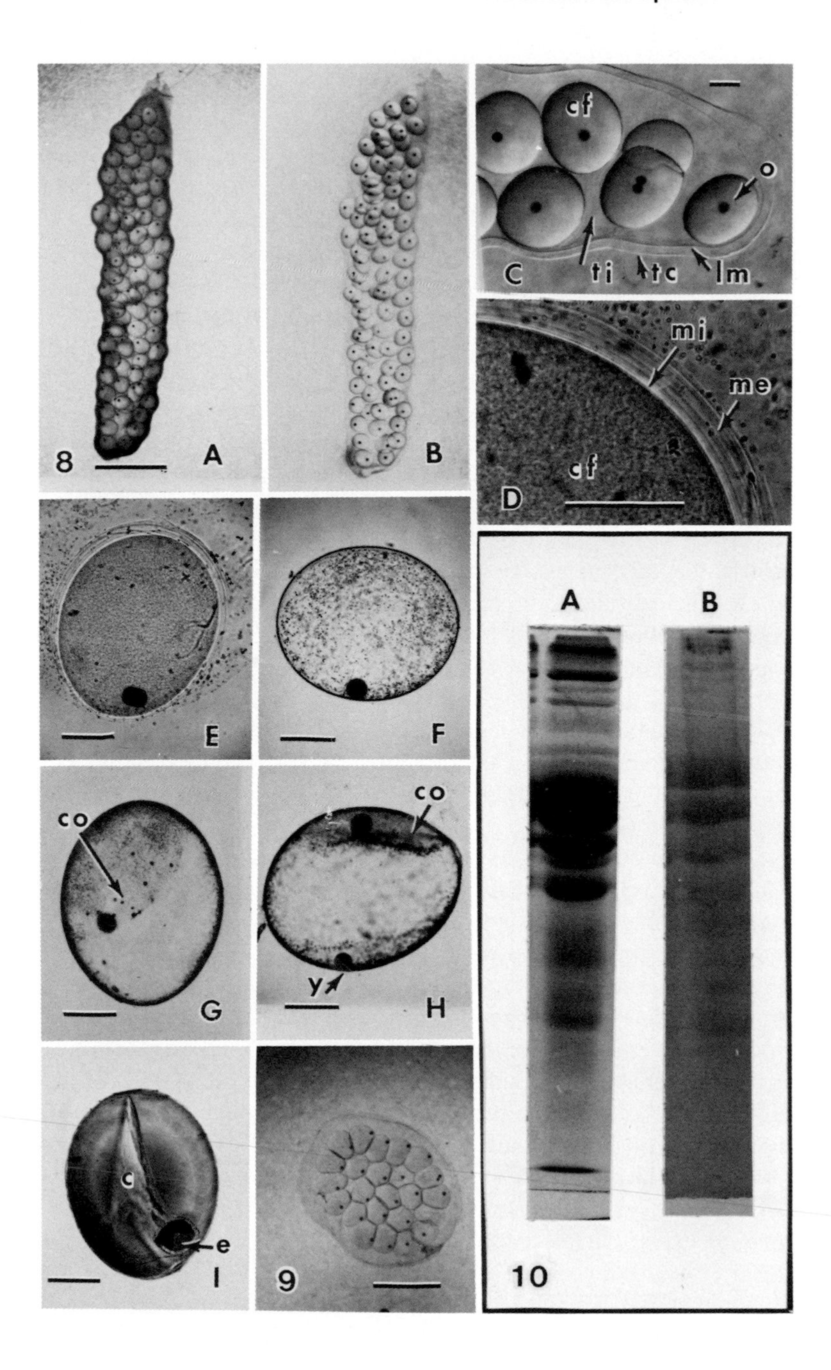
8
A
B
cf
o
C
ti
tc
lm
mi
me
cf
D
E
F
co
G
co
y
H
c
e
I
9
A
B
10

Oviposited egg masses can be dislodged from the surfaces of the tank and from shells of snails by pressing one's finger against the egg mass. The freed egg mass is then drawn into a large bore pipette and transferred to a shallow culture dish (60–90 mm in diameter). The egg masses in the collecting dish are then sorted into a series of smaller dishes on the basis of their stages of development with the aid of a low power stereomicroscope.

The egg capsules are isolated from the jelly mass and cultured in millipore-filtered pond water (MFPW) for both descriptive and experimental studies. While there are several methods for sterilizing egg masses (Beadle, '69) and egg capsules (Friedl, '64), MFPW is satisfactory except for those experiments where bacteria may contaminate the experimental culture medium. To isolate egg capsules, an egg mass is transferred with forceps to a piece of filter paper on a glass plate where the egg mass is slit lengthwise with a curved scalpel blade. The eggs are gently teased and rolled free from the egg jelly composed of the tunica capsularis and tunica interna (Fig. 8C). The freed capsules are gently rolled across the moistened filter paper. In *L. palustris* the inner capsule membrane (membrana interna) is sufficiently tough for the capsules to be rolled across dry filter paper, which removes the outer capsule membrane (membrana externa, Fig. 8D). The isolated capsules are then scraped up on a clean scalpel blade and transferred to 30-mm covered plastic culture dishes for further observation. This isolation procedure can be used with other species of lymnaeids and physids but not with planorbids. Planorbid egg capsules must be cut or teased individually from their surrounding investments. Thus planorbids are usually cultured as an entire egg mass (Fig. 9).

Because the egg passes its entire development in a relatively large ovoid capsule, the development and response to experimental treatments of individual eggs can be followed by culturing single egg capsules in depression slides, depression plates, or microtiter test plates. Egg capsules may also be pipetted onto filter paper and then transferred with a scalpel to 2% agar plates where they are arranged in rows with the aid of a small hair loop. One or more grooves are cut in the agar plate and filled with water to maintain sufficient moisture. Individual egg capsules may be transferred from the plate to slides for detailed observations and photographing and then returned to the plate via pipetting the capsule onto filter paper and from there transferring the capsule with a scalpel to its original position on the plate. This filter paper transfer procedure is also used when it is necessary to transfer eggs from one experimental solution to another in order to minimize potential dilution effects and enhance the effectiveness of rinsing procedures, particularly when eggs are treated with small volumes of expensive solutes.

For both the European *L. stagnalis* and *L. palustris*, 25°C has become the standard temperature for culturing embryos and for establishing timetables of development. When the room temperature is below 25°C a small bench-top water bath makes an excellent incubator. However, 25°C may not be an optimal temperature for normal development of other species. For each species it is important to establish a timetable of development at a standard temperature. It is also important to determine the rate of normal development at different temperatures (Fig. 7), particularly when it is desirable to slow down the several events and extend the length of a developmental stage for descriptive and experimental studies. For example, *L. palustris* eggs cultured at 6°C develop from the 2-cell to 24-cell stage in 48 hours, whereas it takes 6 hours at 25°C.

To prevent abnormalities caused by overcrowding, one should limit the number of egg capsules per unit volume of culture medium. With *L. palustris* four egg capsules per 1 ml is the standard for rearing eggs from the early cleavage stages to hatching. One egg capsule per 0.1 ml is standard for short-term exposures to experimental solutions. Because each egg is, in effect, surrounded by an unstirred layer of capsule fluid, encapsuled eggs should be cultured on a shaking table to optimize uniform response to experimental solutions.

Eggs in an egg mass do not develop synchronously; the onset of first cleavage may differ by 40 to 60 minutes. To synchronize small numbers of eggs, eggs are sorted into groups at 1- to 5-minute intervals as they begin first or second cleavage. Large numbers of *L. palustris* eggs from several egg masses may be synchronized by transferring groups of synchronized eggs to a 6°C water bath and then returning the pooled eggs to 25°C.

Eggs and embryos may be isolated from their capsules by puncturing and tearing the capsule membrane with two pairs of well-sharpened jeweler's forceps. Isolated embryos may be cultured in hanging drops of capsule fluid or pond water or rinsed free of capsule fluid and capsule membranes by gently swirling the shallow plastic culture dish and aspirating off the capsule fluid and debris. One can isolate and rinse 50 to 100 embryos in 15 minutes. Larger numbers of embryos for cell cultures or biochemical analyses may be isolated by the method of Basch and DiConza ('74a).

Eggs or embryos isolated from their capsules before the trochophore stage usually do not develop normally regardless of whether they are cultured in pond water or in capsule fluid. Post-trochophore embryos, however, will develop normally even when cultured in pond water. Several methods for adult organ and embryonic cell culture have been developed (Flandre, '72; Basch and DiConza, '73, '74b; Bayne, '76; Gomot, '77) but have rarely been used by embryologists.

THE NATURAL HISTORY OF THE EGG CAPSULE

The egg capsule membrane and capsule fluid have both protective and nutritive functions, which are quite complex. The size of the capsule and the capsule fluid composition vary from species to species, egg mass to egg mass, and even between egg capsules in an egg mass. Thus variations in rates of normal development, time of hatching from the capsule, and responses of individual eggs to experimental treatments are partly due to variations in the capsule fluid. While encapsuled eggs develop in freshwater and even in distilled water for 24 hours, the immediate chemical and physical environment of the egg is the relatively large unstirred layer of capsule fluid.

The literature on the composition and osmotic properties of the capsule fluid of *Lymnaea* has been reviewed by Morrill et al. ('64, '76), Bayne ('68), Beadle ('69), and Taylor ('73). The *L. stagnalis* egg capsule has a volume of 0.64–0.85 μl, an average wet weight of 638 μg, and an average dry weight of 107 μg. Of the dry weight, 35% is galactogen (2.2×10^6 daltons, Fleitz and Horstmann, '67); the remainder is proteinaceous. The galactogen content may range from 18 to 60 μg per capsule (Horstmann, '56, '58). In *L. palustris* the average volume of the egg capsule is 0.16 μl, the galactogen content is 24 μg, and the protein content is 16 μg. The capsule fluid of *L. palustris* has been separated into 18 electrophoretically distinct protein bands that also develop with the periodic acid-Schiff (PAS) stain for carbohydrates (Fig. 10). This indicates that at least some of the carbohydrate moiety of the capsule fluid may be associated with one or more glycoproteins.

In *L. stagnalis,* Taylor ('73) measured the ionic composition of the capsule fluid and found the following milliequivalent values: calcium (12–15), sodium (1.2–2.7), magnesium (7), and potassium (0.5). Because of the semipermeable nature of the capsule membrane, a Donnan equilibrium exists across the capsule membrane with a potential of 23 mV, inside negative. Yet there is no evidence that any of the sodium or calcium in the capsule fluid is nonexchangeable or "bound" (Taylor, '73). However, the membranes and fluid of the capsule do function as cation buffers. Calcium leaves the capsule fluid slowly in distilled water. With the histochemical sodium oxalate test for calcium, calcium oxalate crystals are localized in the tunica externa, tunica interna, membrana externa, membrana interna, and the capsule fluid (Fig. 8D–F). The number and size of the calcium oxalate crystals in the capsule fluid may vary between capsules from different egg masses. Furthermore, in capsules where the capsule fluid partially stratifies under normal gravity or is stratified into three layers by centrifuging (2500 g), the calcium oxalate crystals are localized in the centripetal layer (Fig. 8G,H). While the several species of macromolecules that function as calcium buffers are yet to be identified, the simple oxalate test is a useful qualitative method to detect the competitive interaction between calcium and other cations for calcium-binding sites in the capsule fluid.

An additional aspect of the heterogeneous nature of the capsule fluid is the presence of a viscous component that immediately surrounds the *L. palustris* egg (Fig. 8I). Whether this component is secreted by the albumen gland and whether it plays a causal role in the early stages of normal development are not known.

The inner proteinaceous membrane (membrana interna) of the egg capsule is approximately 0.1 μm thick. Its porosity varies permitting entry of molecular weights up to 342 in *Biomphalaria sudanica,* 504 in *L. stagnalis* (Beadle, '69), and 1,500 to 2,000 in *L. palustris.* The capsule fluid has an internal colloidal, osmotic pressure of 1.5 mM/1 (25 mm Hg) in *Biomphalaria,* 4–5 mM/1 (75 mm Hg) in *L. stagnalis* , and 0.25 mM in *L. palustris.* The maintenance of a constant volume of the egg capsule is regulated by hydrostatic back pressure because of the inelastic membrana interna. Therefore, particularly during the first 2 days of development, the encapsuled egg develops under a positive hydrostatic pressure. The newly laid *Lymnaea* egg, when isolated from its capsule, is in osmotic equilibrium with 93 mM (2.1 atmospheres) nonelectrolyte solutions (Raven '66). Thus the *Lymnaea* egg may require the positive hydrostatic pressure of the capsule for normal development.

Newly laid and early cleavage stage *L. palustris* eggs do not develop beyond the gastrula stage when: 1) they are isolated from their capsules and cultured in hanging drops of capsule fluid; 2) the capsules are punctured at 2–4-hour intervals to relieve the hydrostatic pressure; or 3) the capsules are cultured in nonelectrolyte solutions, e.g., carbowax or Ficoll, 0.25 mM, in which the capsule loses it turgor and becomes flaccid.

In conclusion, the capsular fluid is not only the major complex nutrient for the developing embryo, it is a both chemically and physically complex colloidal environment whose ion buffering and physical properties merit further study.

MICROTECHNIQUES FOR THE STUDY OF DEVELOPMENT

Because of the opacity of the freshly laid egg (120–150 μm, *L. stagnalis* ; 108–126 μm, *L. palustris;* 80–150 μm in other species) and the developing embryo, only the grossest morphological features are discernible in the living embryo. Therefore, for detailed cell lineage analyses and histological, cytological, cytochemical, and ultrastructural studies, eggs and embryos must be fixed and prepared as whole mounts or embedded and sectioned.

Historically, in nearly all the light microscopical (LM) studies, eggs and embryos have been decapsulated and fixed in Bouin's fixative or other "classical" LM fixatives. While they preserve the general shapes of cells and tissues and are suitable for differential histological staining reactions, these fixatives may alter the normal shapes of cells as well as destroy the integrity of the intracellular structure and cellular organelles. The morphological features of the egg and embryo are preserved in a more "natural" state with fixatives used for ultrastructural analyses. The marked contrast between the degree of cytoplasmic integrity in *Lymnaea* eggs fixed in Bouin's fixative versus eggs fixed in glutaraldehyde and osmium is shown in Figure 11A–D.

Therefore, for routine LM histological and cytological studies as well as transmission electron-microscopical (TEM) and scanning electron-microscopical (SEM) analyses we find the following protocol quite satisfactory: fix in 2% glutaraldehyde in 0.05 M phosphate or 0.05 M cacodylate buffer (pH 7.2) for 30 to 90 minutes at 22–25°C, rinse in 0.05 M buffer, postfix in 0.5% osmium tetroxide in 0.05 buffer for 60 minutes at 4°C, rinse in distilled water, dehydrate in an ethanol or acetone series, and embed in Lockwood's Dow 732/332 epoxy resin for LM and TEM sections or prepare the material for SEM analysis by the critical point drying method. Both eggs and embryos should be gently rinsed free of residual capsule fluid, particularly for SEM analyses. Small polyethylene rather than glass containers should be used since the cleavage stage embryos tend to stick to glass surfaces. Continuous gradient exchange dehydration is superior to the traditional step gradient dehydration and will minimize both overall and differential shrinkage of embryos and organelles. To minimize damage to cells and surfaces of the embryos, embryos are pipetted into polyethylene tubes (Beem capsules with the tapered end cut off) that are covered at both ends with 40-mesh Nitex held in place by the rims of Beem capsule caps. Initially the tube with its lower end capped is stood upright in a small fixing dehydration container half filled with pond water; washed embryos are pipetted into the tube; the top of the tube is capped; and the series of fixatives, rinses, and dehydrating agents is added and withdrawn from the container. This way the embryos are neither lost nor injured. Once the embryos are in 100% ethanol or acetone the capped tube can be transferred to a critical point drier for SEM analyses or the embryos can be carefully pipetted into a second container and processed for embedding in resin. The above procedure is especially useful for

---

Fig. 11. LM and TEM micrographs illustrating the effects of Bouin's and glutaraldehyde/osmium fixation on the structure of the cytoplasm. A. LM of 1-cell egg fixed in Bouin's at first maturation division. B. LM of 2-cell egg fixed in glutaraldehyde/osmium. C. TEM of the peripheral equatorial region of a 1-cell stage egg fixed in Bouin's at the first maturation division (from Luchtel, '76). D. TEM of the cortical region in the animal pole hemisphere of one blastomere of a 2-cell stage egg fixed in glutaraldehyde/osmium. Bars (A,B) 50 μm, (C,D) 1 μm. ap, animal pole plasm. b, beta yolk granules. bt, beta yolk granules transforming into vacuolated gamma yolk granules. ev, endocytotic vesicles. g, Golgi complex. l, lipid droplets. m, mitochondria. ma, meiotic aster. n, nucleus with karyomeres and nucleolarlike bodies. o, oolemma. pb, polar body. vg, vacuolated gamma yolk granule.

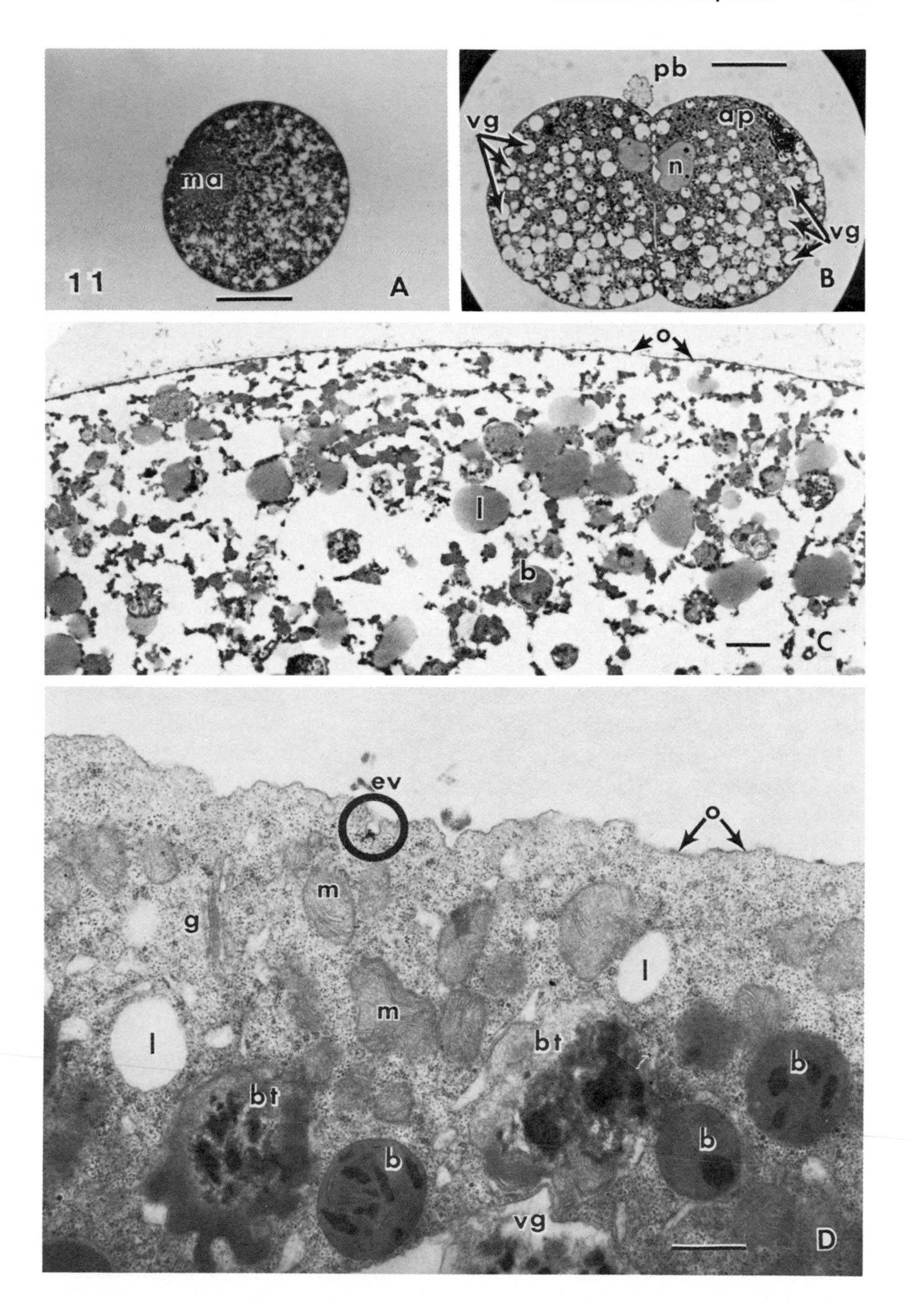
ma
11
A
pb
vg
ap
n
vg
B
o
l
b
C
ev
o
m
g
l
m
l
bt
bt
b
b
b
vg
D

preparing individual or small numbers (10–20) of eggs for correlative LM, TEM, and SEM morphological analyses.

NORMAL DEVELOPMENT

Origins of Gametes to Oviposition

Origins. Embryonic development begins in the hermaphroditic gonad in pulmonates where the male and female cells differentiate from the germinal epithelial cells. In *Lymnaea* the gonad originates at the late trochophore veliger stage (E3) in the ventral mesoderm where two large, central cells (the germ line) are surrounded by a group of smaller cells (the nutritive cell line). The presumptive female tract and most of the male tract originate from the proliferation of ectodermal cells in the ventral body wall anterior to the gonad (Brisson and Régondaud, '77). The further development of the reproductive system has been described by Archie ('41), Fraser ('46), Cumin ('72), and Arni ('73) and reviewed by Martoja ('64). Following hatching the growth and maturation of the reproductive system parallel the general growth of the snail (McCraw, '70). In *Lymnaea* the gonad or ovotestis becomes embedded in the adult digestive gland. As in other eukaryotes, the maturation and release of the gametes as well as the activities of the accessory organs associated with reproduction are under neuroendocrine control (Geraerts and Bohlken, '78; Jong-Brink et al., '79; Geraerts and Mohamed, '81). Since the experimental approaches to examination of neuroendocrine control are similar to those used with echinoderms and amphibians, the pulmonates offer another potentially interesting system for the embryologist interested in the causal events associated with gametogenesis and gamete release (Geraerts and Joosse, '75; Geraerts and Bohlken, 76; Geraerts and Algera, '76).

Ovotestis. The general morphological and histological aspects of the ovotestes have been described by Archie ('41), Joosse and Reitz ('69), and Jong-Brink et al. ('76) and reviewed by Duncan ('75). The ovotestis is located near the tip of the shell along the columellar surface of the digestive gland; branches of the ovotestis ramify between the lobes of the digestive gland and end in blind sacs called acini. Figure 12 shows a general scheme of an acinus. An outer layer of connective tissue and a basal lamina surround the germinal epithelium, which is continuous with the epithelium of the vas efferens. Both sex and nurse cells arise from the germinal epithelium, which is composed of ciliated cells and cells with microvilli. A narrow germinal band or ring of epithelial cells separates the proximal region of the acinus, where male and female sex cells develop, from the distal region of the acinus. Both oogenesis and spermiogenesis appear concurrently in an acinus in two concentric areas; the central vitellogenic area, where the acinus wall is next to the digestive gland, is surrounded by a spermatogenic area (Joosse and Reitz, '69).

Spermatogenesis. Jong-Brink et al., ('77) provide an excellent modern review and ultrastructural analysis of spermatogenesis, while Anderson and Personne ('76) review the morphology and physiology of the mature sperm. Additional recent ultrastructural studies include SEM and freeze-fracture analyses (Maxwell, '77), acrosome formation (Takaichi and Dan, '77), flagellum formation (Dan and Takaichi, '79), and nuclear maturation (Terakado, '81).

Oogenesis. In additional to the detailed reviews of oogenesis by Raven ('61, '66, '75), there are several original descriptive studies (Hartung, '47; Bretschneider and Raven, '51; Recourt, '61; Ubbels, '68; Jong-Brink et al., '76; Rigby, '79) that detail the morphological and cytoplasmic differentiation of the oocyte. Additional information on the cytochemistry of the nucleoli and the time course of RNA synthesis during oogenesis is given by Kielbówna and Koscielski ('74), Tapaswi ('74), and Bolognari and Carmignani ('76).

During oogenesis the oogonium originates in the germinal epithelial ring and migrates with a presumptive follicle cell along the basal lamina to the distal part of the acinus. During early development each oocyte becomes surrounded by two layers of follicle cells.

The growth and maturation of the oocyte have been divided into five successive stages (Fig. 13, stages D–H). In stage 1 (D) there are no follicle cells and yolk platelets are absent. In stage 2 (E) follicle cells attach to the oocyte and yolk platelets first appear. In stage 3 (F) the follicle cells completely surround the oocyte. In stage 4 (G) an apical cleft begins to form between the follicle cells at the luminal side of the oocyte. In stage 5 (H) the apical cleft enlarges and the apical area of the oocyte is populated with short (0.3–0.5-μm) microvilli embedded in the vitelline membrane. In pulmonates vitellogenesis occurs throughout oogenesis and is controlled by the dorsal body hormone of the dorsal body cells of the cerebral ganglia (Geraerts and Joosse, '75; Geraerts and Mohamed, '81). The origins and development of the yolk platelets described by Recourt ('61), Terakado ('74), and Jong-Brink et al. ('76) complement the LM studies of Bretschneider and Raven ('51) and Ubbels ('68).

According to Rigby ('79), the fully developed oocyte (90–105 μm in diameter) has three broad zones: the apical area, the basal area, and the intermediate zone bordered by the follicle cells. The apical area (the future animal pole) is exposed to the lumen of the acinus. The basal area (the future vegetal pole) is next to the basal lamina adjoining the intercellular matrix between the digestive gland and the gonad. Root-like "rhizoids" project from the base of the oocyte and form tight junctions, gap junctions, and desmosomes with branching processes from the bases of the follicle cells. Lipid droplets first accumulate in this region of the oocyte. The intermediate zone of the oocyte is apposed by the follicle cells and characterized by tight

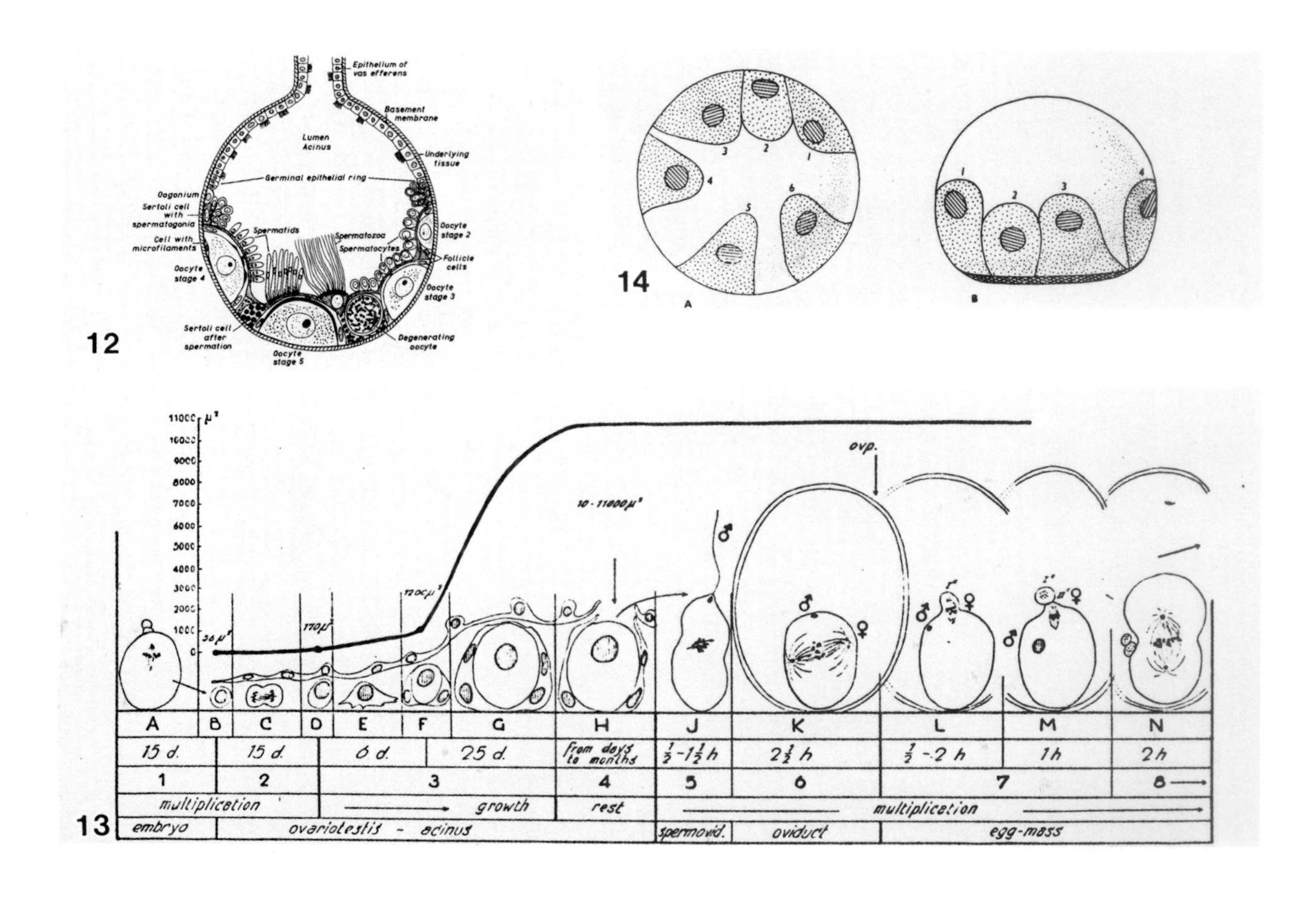
12
Epithelium of vas efferens
Basement membrane
Lumen Acinus
Underlying tissue
Germinal epithelial ring
Oogonium
Sertoli cell with spermatogonia
Cell with microfilaments
Oocyte stage 4
Spermatids
Spermatozoa
Spermatocytes
Oocyte stage 2
Follicle cells
Oocyte stage 3
Sertoli cell after spermation
Oocyte stage 5
Degenerating oocyte
14
A
B
13
ovp.
10 - 11000 μ²
A B C D E F G H J K L M N
15 d. 15 d. 6 d. 25 d. from days to months 1/2-1 1/2 h 2 1/2 h 1/2-2 h 1h 2h
1 2 3 4 5 6 7 8
multiplication growth rest multiplication
embryo ovariotestis - acinus spermovid. oviduct egg-mass

junctions, gap junctions, and desmosomes that alternate with intercellular clefts between the inner layer of follicle cells and the oocyte. Cytoplasmic projections from the follicle cells may penetrate the oocyte and be "ripped off." Raven ('63, '67) has shown that the inner follicle cells are asymmetrically arranged around the oocyte (Fig. 14). Thus the future axes of symmetry of the fertilized egg and embryo appear to originate while the oocyte is still in the follicle.

During oogenesis the volume of the oocyte increases 223-fold and nuclear volume increases 162-fold over a period of at least 30 days. During this period there is considerable activity of the nucleoli and synthesis of two kinds of yolk platelets—the beta and gamma yolk granules. The mature oocyte may not be released for days or months. In starved snails the oocytes may be resorbed by the follicle cells (Joosse et al., '68).

Ovulation. Ovulation is regulated by a hormone produced by the caudodorsal, neurosecretory cells of the cerebral ganglia in *Lymnaea* (Geraerts and Bohlken, '76). Within minutes after the injection of extracts of these cells into the hemolymph, oocytes appear in the spermoviduct. The release of the oocyte from its follicle may involve both local intercellular muscular contractions and enlargement of the clefts between the oocyte and surrounding follicle cells. Once released from their follicles, mature oocytes pass from the lumina of the acini to the spermoviduct (hermaphroditic duct) by ciliary activity of the endothelial cells lining the acinus, the vas efferentia, and the spermoviduct. Usually upon release from the follicle the germinal vesicle breaks down and the first maturation division begins and then is arrested at metaphase.

Fertilization. The egg is fertilized while arrested at metaphase of the first maturation division. In the course of fertilization one or more sperm may

---

Fig. 12. Schematic, longitudinal section through an acinus of the ovotestis, of *Biomphalaria glabrata*. Courtesy of Dr. Jong-Brink (Jong-Brink et al., '76).

Fig. 13. Phases of the life cycle of the oocyte of *L. stagnalis* from formation of primordial germ cells (B). C. Multiplication of primary sex cells. D. Oogonium. E. Amoeboid phase of oogonium. F. Formation of follicle. G. Growth phase. H. Resting phase before release from follicle. J. Insemination or fertilization phase. K. Prematurity phase. L. First maturation division after oviposition. M. Second maturation division. N. First cleavage. Ordinate: cross-sectional area of oocyte in microns. Courtesy of Dr. Raven (Bretschneider and Raven, '51).

Fig. 14. Diagrams of the asymmetric arrangement of inner follicle cells around oocyte of *L. stagnalis*. A. Viewed from luminal, presumptive animal pole end of oocyte. B. Lateral view showing cross-hatched, presumptive vegetal pole end area that is in contact with the basement lamina of ovotestis' wall. Courtesy of Dr. Raven (Raven, '63).

enter any region of the egg's surface. Whole sperm enter the egg but only one sperm forms a pronucleus. Thus the eggs are the physiologically polyspermic type and no cortical reaction occurs.

While self-fertilization occurs in isolated individuals of *Lymnaea,* ordinarily cross-fertilization follows copulation. During copulation the ejaculation of sperm and seminal plasm into the lower vaginal region of the female tract also affects the activity of the female tract as a whole (Rudolph, '79). Sperm from the copulating partner are passed along a ciliated furrow of the female tract (Fig. 15B) (Bretschneider, '48a) as far as the junction of the male and female tracts where eggs leaving the spermoviduct are fertilized in the fertilization pocket or carrefour (Crabb, '27; Bretschneider, '48a; Duncan, '75). No one knows whether the fertilization pocket provides a special environment for fertilization.

Passage of egg in female tract. Figure 15A shows the general features of the reproductive tracts. Detailed descriptions of the genital system and the functional anatomy of the reproductive tracts are given by Bretschneider ('48b), Plesch et al. ('71), Duncan ('75), and Jong-Brink et al. ('79).

Following fertilization the egg passes down the glandular female tract (Fig. 15C) becoming surrounded first with secretions from the albumen gland. Usually a single egg and some albumen are encapsuled by the proteinaceous membrana interna secreted by the convoluted posterior pars contorta of the oviduct. Here the specific size and shape of the egg capsule are determined. The outer, polysaccharide-rich membrane (membrana externa) of the capsule is secreted by cells in the anterior pars contorta region of the oviduct. Egg capsules accumulate in the pars recta where they are collectively "cemented" or enveloped by secretions from the muciparous gland that forms the tunica interna. The tunica interna is covered next with a thin layer of material secreted by the pars recta region. This layer forms the limiting membrane between the tunica interna and the tunica capsulis secreted by the oothecal gland region of the tract. The outermost layer of material of the egg

---

Fig. 15. Schematic diagrams of the reproductive tract and accessory glands of *Lymnaea.* A. External anatomical relationships. (A) Courtesy of Dr. Plesch et al. (Plesch et al., '71). B. Path of sperm in cross-fertilization. C. Path of ovum. (B, C) Courtesy of Dr. Bretschneider (Bretschneider, '48b). ag, albumen gland. apc, anterior pars contorta. bc, bursa copulatrix. c, carrefour. cil. furr., ciliated furrow. ec, egg capsule. fp, fertilized pocket. gp, gynopore. me, membrana externa. mg, muciparous gland. mi, membrana interna. og, oothecal gland with three zones of glandular epithelium. pc, pars contorta. pf, perivitelline fluid. pg, prostate gland. ppc, posterior pars contorta. pr, pars recta oviductus and praeputium of penial complex. sp, sperm balls and sperm duct. sp. gen., spermatogenesis in ovotestis. spo, spermoviduct. tc, tunica capsularis. ti, tunica interna. u, uterus. ud, ureter or vas deferens. v, vagina. vd, vas deferens. ve, verge. vs, vesicular seminales of spermoviduct.

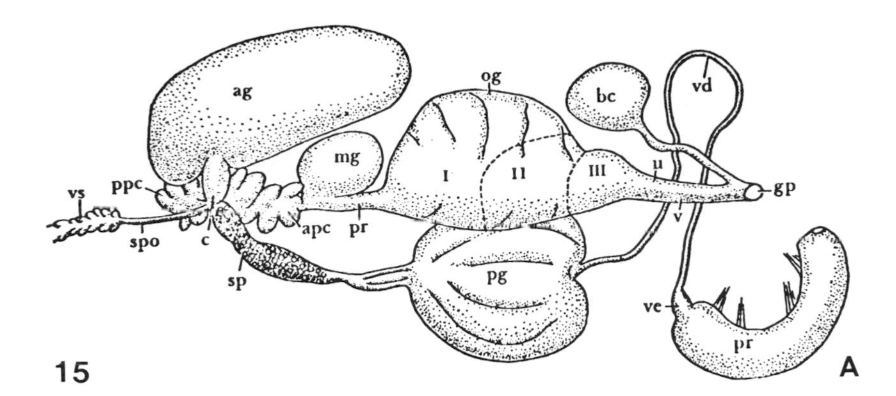
ag
og
bc
vd
mg
I
II
III
u
vs
ppc
gp
spo
c
apc
pr
v
sp
pg
ve
pr

15 A

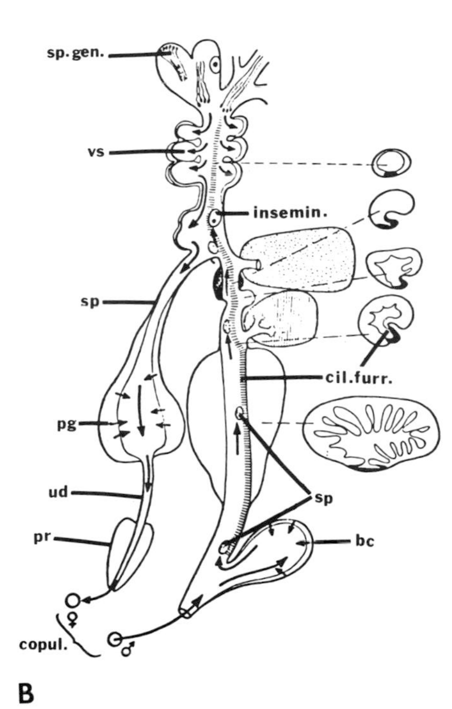
sp. gen.
vs
insemin.
sp
cil.furr.
pg
sp
ud
pr
bc
copul.

B

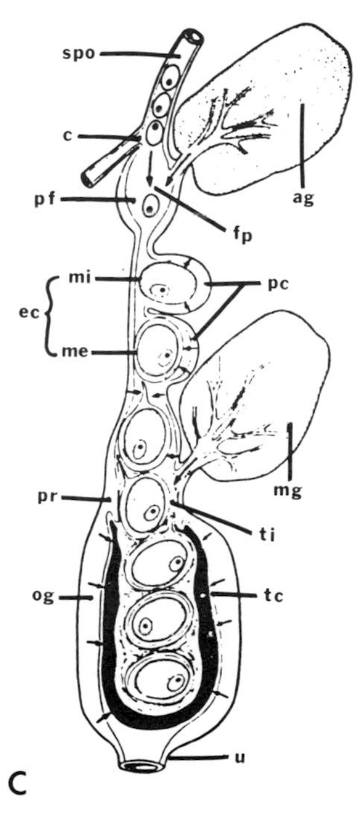
spo
c
ag
pf
fp
mi
pc
ec
me
mg
pr
ti
og
tc
u

C

mass—the pallium gelatinosum—may be secreted either by zone III of the oothecal gland or by mucous cells around the gynopore.

The specific and comparative aspects of the actual oviposition of strings of eggs in egg masses and the arrangement of eggs in a common capsule have been described in detail by Bondesen ('50). Whereas the numbers of eggs oviposited may vary with individuals of a particular species, the shape of the egg mass and the arrangement of the egg capsules within the egg mass are species specific. In *Lymnaea* the distal end of the egg mass is blunt and the proximal end tapered (Fig. 8A).

Oviposition to First Cleavage

Between oviposition and first cleavage the maturation divisions are completed; ooplasmic segregation of cytoplasmic organelles occurs; the viscosity and surface tension increase; the egg swells slowly (40%–50% increase in volume); and more and more gamma yolk platelets become vacuolated. The phenomena occurring during this period have been studied and reviewed by Raven ('45, '48, '66). Given the recent interest in the regulation of meiosis and cytoplasmic differentiation during the maturation period in other eggs, the earlier studies of Raven and his colleagues provide an excellent departure point for future experimental analyses of the *Lymnaea* egg.

Maturation divisions. At 25°C the first polar body in *Lymnaea* forms 30–70 minutes after oviposition; the second polar body forms 40–50 minutes after the first polar body; and first cleavage occurs 70–80 minutes after the second polar body (2.5 to 3 hours after oviposition). The temporal variations between eggs from the same egg mass are most probably due to the eggs having already begun their maturation process as they pass down the female tract prior to oviposition. The clear polar bodies formed at the animal pole average 10 μm in diameter and remain attached to the egg's surface via the vitelline membrane and possibly a cytoplasmic connection (midbody) in the case of the second polar body. Typically, the first polar body lies on the outer surface of the vitelline membrane and the second polar body is beneath the vitelline membrane. Raven's ('45) timetable (Table III) summarizes the nuclear events during the maturation period.

In *L. stagnalis* and *L. palustris* maturation begins when the egg is in the spermoviduct. The oocyte's centrioles move to the opposite ends of the germinal vesicle, small asters appear around them, and the germinal vesicle regresses. Then the maturation spindle forms and migrates with the enlarging asters to the animal pole where the peripheral aster "connects" with the oolemma at the animal pole. Following the first meiotic division and the formation of the first polar body, the deep centrosphere surrounding the undivided centriole migrates as a whole toward the animal pole where it becomes the peripheral centrosphere. There the centrosphere is transformed into an asymmetric, egg-shaped spindle and an aster forms around the

TABLE III. Timetable of Maturation Events Between Oviposition and First Cleavage[a]

	Egg nucleus	Sperm nucleus	Sperm aster
60 minutes before 1st p.b.	1st maturation spindle, metaphase	Subcortical sperm-head	
30 minutes before 1st p.b.	End of mat. sp. reaches egg cortex	Subcortical sperm-head	
20 minutes before 1st p.b.	Anaphase of 1st mat. division	Subcortical sperm-head	
Extrusion of 1st p.b.	Telophase of 1st mat. division; central aster large	Subcortical sperm-head	Small
40 minutes before 2d p.b.	2d mat. spindle, metaphase	Subcortical sperm-head	Growing
20 minutes before 2d p.b.	Anaphase of 2d mat. division	Subcortical sperm-head	Growing
Extrusion of 2d p.b.	Telophase of 2d mat. division; central aster large	Subcortical sperm-head	Growing
20 minutes after 2d p.b.	Karyomeres in large aster	♂ pronucleus moves to sperm aster	Sperm aster with clear central area
60 minutes after 2d p.b.	Karyomeres at animal pole under cortex	♂ pronucleus in sperm aster	Sperm aster with clear central area
40 minutes before 1st cleavage	Pronuclei united		Sperm aster disappears
20 minutes before 1st cleavage	Cleavage spindle, metaphase		

[a]After Raven, '45.

peripheral centriole. The sperm aster and nucleus migrate toward the spindle where the sperm aster fuses with the blunt, deep end of the spindle. Thus the sperm aster becomes the inner aster of the second meiotic spindle. Raven has described the cytological details and variations of this remarkable process in *Lymnaea* and other pulmonates (Raven et al., '58; Raven, '64a). Immediately after the formation of the second polar body the chromosomes remaining in the aster swell to form karyomeres that coalesce to form the female pronucleus. The female pronucleus remains at the animal pole; the sperm aster moves towards the animal pole; and the sperm nucleus enlarges to become the male pronucleus that comes to rest beside the female pronucleus about 40 minutes before first cleavage. About 20 minutes later the nuclear membranes disappear, the cleavage spindle forms, and the zygote begins to divide into two cells.

Ooplasmic segregation. During meiosis, mitochondria first accumulate around the maturation spindles and the asters and then accumulate beneath the animal pole cortex in the region of the animal pole plasm. In *L. palustris* and *L. stagnalis* the animal pole plasm appears shortly after the formation of the second polar body. However, the cytoplasmic redistribution and localization of other organelles around the animal vegetal axis occurs earlier. Between ovulation and oviposition a differentially stained plasm rich in beta yolk platelets forms in a conelike region at the vegetal pole. During the maturation period this plasm spreads toward the animal pole beneath the oolemma to form the subcortical plasm (Raven, '45). By the 4-cell stage the subcortical plasm is differentially concentrated in the cells of the animal hemisphere (Raven, '46). At the LM level of resolution the central endoplasmic region is characterized by the presence of many mitochondria (alpha granules) and vacuolated gamma granules. In the equatorial region beneath the oolemma in *Lymnaea* are six asymmetrically arranged patches of RNA-rich plasm called the subcortical accumulations (SCA).

The normal distributional patterns of the organelles and stainable plasms may be altered by centrifugal force. The degree of stratification and the redistribution of substances following centrifugation have been studied by Raven and Brunnekreeft ('51) and Raven and Wal ('64). The redistribution of substances involves first the destratification and homogenous distribution followed by resegregation of the organelles and plasms similar to that in the normal egg.

Cortical contractions may be associated with ooplasmic segregation. Such contractions are normally not visible except at the time of the formation of the second polar body, when at early anaphase the surface of the egg at the animal pole exhibits a wavelike series of blebs or undulations. At the same time, an asymmetrical bulge occurs at the vegetal pole. This suggests that ooplasmic segregation and the cortical contractions during the maturation period are associated with the organization and activity of both cytoskeletal and cytocontractile elements as well as polarized ion fluxes. TEM studies (Morrill and Perkins, '73; Luchtel, '76) show that cytoplasmic microtubules are present in both the cortical and endoplasmic regions of the egg. However, the organization of the cytoskeleton remains to be determined. Meshcheryakov and Filatova ('81) have prepared sealed myosin-extracted plasma membrane ghosts of *L. stagnalis* eggs. After being irrigated with heavy meromyosin these ghosts undergo contractions on the addition of ATP and have fibrils of varying diameters (Meshcheryakov, '81). Interestingly, isolated cytoplasm treated with heavy meromyosin does not contract; furthermore, cytochalasin B prevents neither the binding of heavy meromyosin nor the contractile reaction in the cortical membrane ghosts. Finally, preliminary measurements of the intracellular electro-potential across the oolemma in different regions of *Physa acuta* after the formation of the second polar body show that the membrane potential is 67.4 mV inside negative, is the same for different regions of the egg, and is potassium dependent (Meshcheryakov and Vorobeva, '73).

Vitelline membrane. Normally, transitory changes in the shape of the egg are restricted by the surrounding vitelline membrane that is attached to the tips of microvilli on the egg's surface (Morrill and Perkins, '73). Bluemink ('67) and Morrill and Perkins ('73) have described the morphology of the vitelline membrane. At the ultrastructural level the vitelline membrane consists of two layers of flaky material 0.1–0.8 μm thick. However, it is poorly characterized both morphologically and biochemically. Its several properties appear to alter between oviposition and first cleavage (Hudig, '46). In eggs isolated from their capsules the vitelline membrane can be detached in sucrose solutions and in calcium-free media or digested by treatment with trypsin (10 mg per ml for 10–35 minutes) (Mescheryakov and Beloussov, '75). Preliminary experiments with *L. palustris* indicate that the vitelline membrane may be removed from encapsuled eggs with a solution of dithiothreitol (DTT); encapsuled eggs without a vitelline membrane can develop into normal embryos. The vitelline membrane is removed more rapidly in eggs treated with DTT before the formation of the first polar body than in eggs treated at the 2-cell stage. This again suggests that the vitelline membrane becomes "toughened" after oviposition. Once procedures are perfected for removing the vitelline membrane from the egg while it is still in the capsule, further analysis will be possible of ooplasmic segregation and other apsects of early embryonic development that heretofore have been restricted to decapsuled eggs that may divide normally but fail to develop beyond the gastrula stage.

Early Cleavage Period

The morphology, nomenclature, and cell lineages during the early and late cleavage periods in pulmonates have been described in detail for *Physa* (Wierzejski, '05), *Planobis* (Holmes, 1900), *Lymnaea* (Verdonk, '65), and *Biomphalaria* (Camey and Verdonk, '70). Although the study of Camey and Verdonk is on a sinistral species, it is an excellent introduction to the morphology of the cleavage patterns, giving the normal sequences of cleavages and the cell lineages of cells up to the 130-cell stage. These studies are complemented by Raven's ('46) definitive histochemical and cytochemical study on the development of the egg from first cleavage to the trochophore stage. Several ultrastructural studies (Bluemink, '67; Elbers, '69; Morrill and Perkins, '73; Luchtel, '76; Wal, '76a,b; Morrill and Macey, '79; Dorresteijn et al., '81) provide additional information on the nature of the cell surface, cytoplasmic matrix and organelles, cellular junctions, and mobilization of the nutrient reserves in the yolk platelets.

During the early cleavage period pulmonates exhibit spiral cleavage that is both radially symmetrical and nearly synchronous in the four quadrants along the animal–vegetal axis. Thus, not until the 24-cell stage is there an overt morphological distinction between the classic A, B, C, and D quadrants. During the early cleavage period, metabolic processes associated with the cell cycle, cell division, the mobilization of nutrient reserves, and osmoregulation parallel the specific processes associated with ooplasmic segregation of morphogens and cytoplasmic organelles, spatial and temporal patterns of the precocious determination of embryonic axes of symmetry, and the developmental fates of individual cells and cell lines. The mosaic nature of the cleavage pattern and the cellular determination of the future morphological patterns of larval and adult organs result from an interaction between the cortical and subcortical cytoplasm (Raven, '70).

To interpret experimentally induced alterations in cleavage patterns and normal development, it is important to know not only the cleavage stage but the phase when a particular experimental treatment is begun and terminated relative to recurrent cellular activities within a cleavage stage. Accordingly, the external morphology of the *Lymnaea* egg during the early cleavage period is shown in Figure 16. Stages 1 through 19 have been described by Raven ('46); I have extended Raven's landmarks and numerical stages to the 24-cell stage. While the temporal length of a particular Raven stage may vary according to temperature and other conditions, the external morphology of the stage is a useful standard reference. Raven's stages are complemented by Biggelaar's ('71b,c) analyses of DNA synthesis and the duration of cell cycles from the 2-cell to the 49-cell stage (Table IV). During the cleavage period the lengths of the cell cycles are due to variations in length of the S and G_2 phases. The M phase is constant, and there is no G_1 phase.

Historically, few experimental studies have used either closely synchronized eggs or standard temperatuures and stages of development. All too frequently the response of the eggs varies from normal development to death following a particular treatment regime. With respect to eggs treated with substances that have "specific" molecular targets, Geilenkirchen's ('67) treatment procedure has increased the uniformity of alterations in developmental patterns. Briefly, synchronized control and experimental eggs of *L. stagnalis* or *L. palustris* are transferred from 25°C to their respective solutions at 6°C, treated at 6°C for 2 hours, washed in pond water for 2 hours at 6°C, and then reared in pond water at 25°C. So treated, the control eggs

---

Fig. 16. Normal stages according to Raven of the external appearance of the *Lymnaea* egg during the early cleavage period from 2nd polar body formation to the 24-cell stage. Stage O, 2nd polar body. Stages 1–9, 2-cells; Stages 10–15, 4-cells. Stages 16–19, 8-cells. Stages 20a and b, 12-cells. Stages 21a and b, 16-cells. Stage 22, 20-cells (formation of 3rd quartet of micromeres). Stage 23, division of 2nd quartet of micromeres, 24-cells. Stage 24a, compaction, 24-cell stage. Stage 24b, 24-cell stage plus 120 minutes. Numbers in lower left corner of each photo are stages; numbers in upper right corner are minutes before or after stage 1 of the first cleavage for *L. palustris* at 25°C. Photos are of *L. palustris*. Stages 1–19 are from Raven ('46).

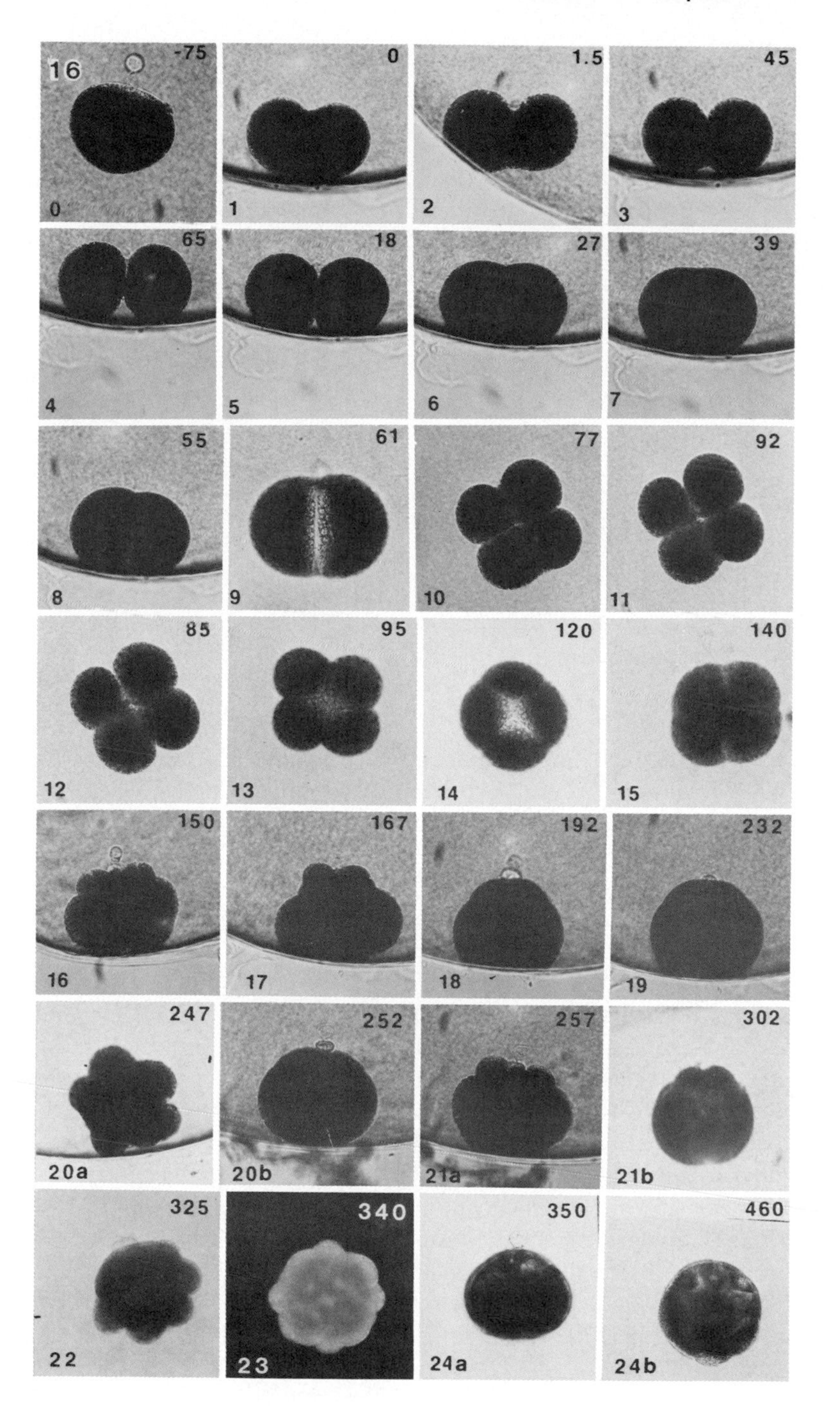
16
-75
0
0
1
1.5
2
45
3
65
4
18
5
27
6
39
7
55
8
61
9
77
10
92
11
85
12
95
13
120
14
140
15
150
16
167
17
192
18
232
19
247
20a
252
20b
257
21a
302
21b
325
22
340
23
350
24a
460
24b

TABLE IV. Duration of Cell Cycles in Early Cleavages of *Lymnaea stagnalis* in Minutes[a]

Cells	G_1	S	G_2	M	Total time (Minutes)
Macromeres					
AB-CD	—	30	20	34	84
1A-1D	—	24	20	34	78
2A-2D	—	22	19	34	75
3D	—	32	134	34	200
3A-3C	—	32	224	34	290
1st quartet					
1a-1d	—	24	37	35	96
$1a^1$-$1c^1$	—	50	216	34	300
$1d^1$	—	50	246	34	330
$1a^2$-$1b^2$	—	100	206	34	340
$1c^2$-$1d^2$	—	110	256	34	400
2nd quartet					
2a-2d	—	23	28	34	85
$2a^1$-$2d^1$	—	50	106	34	190
$2a^2$-$2d^2$	—	50	146	34	230
3rd quartet					
3a-3d	—	29	207	34	270

[a]Adapted from Biggelaar ('71b, c).

develop normally and the experimentally treated eggs respond uniformly (70%–90%). This procedure has proven particularly effective when eggs are treated with ions such as lithium or antimicrotubule and antimicrofilament agents.

Because the first cleavage spindle originates near the animal pole, the first cleavage begins as an indentation at the animal pole and continues via initial unipolar ingression. An extension of the groove around the egg results in two nearly equal, spherical cells connected by a residual midbody. This is followed first by a flattening of the cells against each other. Then they are separated by the formation of fluid-filled intercellular spaces that fuse to form a lenticular cleavage cavity. When eggs are treated with Cytochalasin or compressed with a glass rod during the ingression phase, the furrow regresses resulting in a binucleate egg (Mulherkar et al., '77; Meshcheryakov and Veryasova, '79b). Regression of the cleavage furrow also occurs when encapsulated eggs are mechanically agitated in a test tube vibrated with a vortex mixer. Cytochalasin causes an elongation of the blastomeres along the polar axis in eggs treated at metaphase of the second to fourth cleavage and lobe-like protrusions may occur at the vegetal ends of the blastomeres and macromeres. Further, the spiral bending of the contact zones of the cells is eliminated.

Descriptive and experimental studies have been directed to a number of other aspects of cellular division including the dynamics of the cell surface during early cleavage (Mescheryakov, '78b), the orientation of cleavage spindles and rotation of blastomeres relative to the spiral cleavage pattern and cortical organization (Guerrier, '71; Meshcherayakov and Beloussov, '73, '75; Meshcheryakov, '78a,b,'79; Meshcheryakov and Veryasova, '79a,b), the determinative significance of cellular arrangements (Biggelaar, '78; Bezem et al., '81), DNA synthesis (Biggelaar, '71b,c; Bajaj et al., '73), RNA synthesis (Biggelaar, '71a; Brahamachary et al., '71a; Boon-Niermeijer, '74, '75), and protein synthesis (Jockusch, '68; Wal, '76b; Morrill et al., '76).

The cleavage cavity. During each cleavage cycle up to the 24-cell stage, the intercellular space enlarges to form a fluid-filled cleavage cavity (Fig. 17A-C). This recurrent cleavage cavity (Raven, '66) is not only involved in the regulation of water (Beadle, '69; Taylor, '77) but may also influence the shape of the mitotic apparatus (Meshcheryakov and Veryasova, '79b). Because the cleavage cavity may occur in eggs in which cytokinesis is inhibited, the cleavage cavity may be considered a "water pumping" organelle. That a cleavage cavity forms at all is due to the precocious formation of septate desmosomes (Fig. 18) between the cells beginning at the 2-cell stage. Thus Taylor ('77) has shown that with each cleavage cycle the egg functions as a several-compartment, leaky epithelium with respect to the movement of water and ions and that water movement is coupled to sodium fluxes.

During a given cleavage cycle tight junctions and septate junctions form between the cells in their peripheral regions of contact. The cleavage cavity fills slowly with fluid via the coalescence of intercellular clefts. Raven ('46) originally described the presence of secretion cones along the luminal margins of the cells. These cones may be vacuolated gamma granules (Fig. 17H) that dishcarge not only water and ions but also complex macromolecules such as mucopolysaccharides and glycosaminoglycans. Figure 17C,G shows a network of fibrous material in the cleavage cavity, while Figure 17E,F shows the as yet unidentified polygranules that resemble hyaluronic acid morphologically.

The vacuolated gamma granules may be involved in the metabolically dependent osmoregulation of water in the cleaving egg as well as in the 1-cell stage egg (Raven, '66; Taylor, '77). LM sections of cleaving eggs show that vacuolated gamma granules are not only localized along the cleavage cavity margins of the cells but also along the intercellular margins and in the cortical regions of cells (Fig. 11B; Fig. 17A,B; Fig. 23C,D). Brahmachary et al. ('71b) and Brahmachary and Ghosal ('76) observed a rhythmic incorporation of ^{35}S during the cleavage period into a fraction precipitable with cetylpyridium chloride. Future studies on water regulation and its association

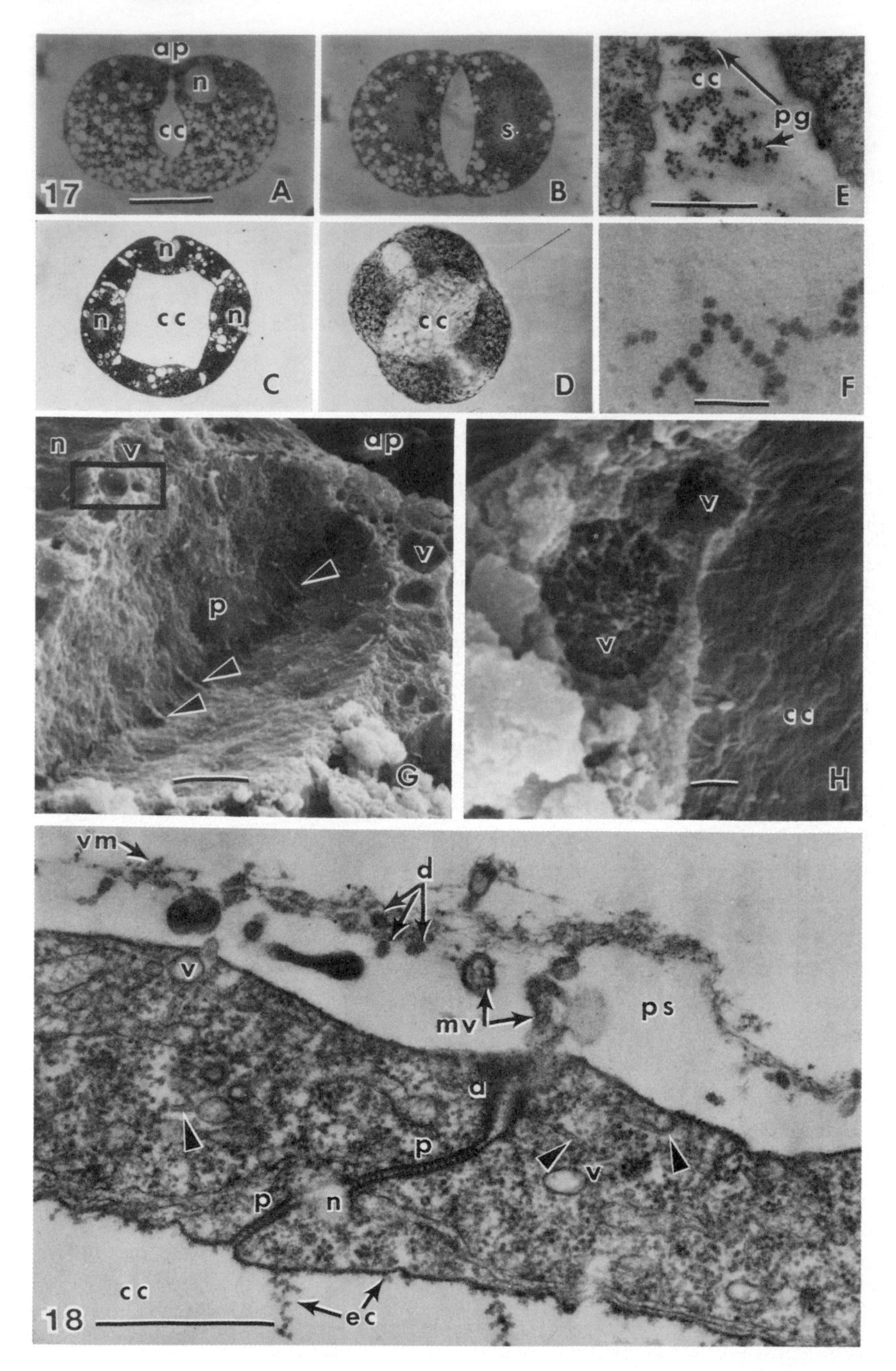
ap
n
cc
17
A
s
B
cc
pg
E
n
n
cc
n
C
cc
D
F
n
v
ap
v
p
G
v
v
cc
H
vm
d
v
mv
ps
a
p
p
n
v
cc
ec
18

with formation of the cleavage cavity formation should clarify the various mechanisms involved and the roles played by various components of the intravacuolar and extracellular matrices in water regulation.

In addition to acting as a permeability barrier in the formation of the cleavage cavity, the septate desmosomes may be involved in the rapid contraction of the cleavage cavity and the release of fluid. In thin sections of 2-cell stage eggs and older embryos, the profiles of the septate desmosomes are frequently interrupted by an intercellular space or "nonfunctional" region (Fig. 18) (Luchtel, '76) of varying diameter that actually may be involved in the pulsatory expulsion of cleavage cavity fluid described by Comadon and deFonbrune ('35). Since the optical properties of the expelled intercellular fluid of embryos differ from the surrounding capsule fluid, one can follow the process microscopically. The appearance of septate desmosomes may also assist the cleavage furrow microvilli (5 to 8 μm long) and filopodic processes that extend from one cell to another (Fig. 17G) to maintain the positional relationships of the cells between sucessive cleavages.

Centrifuged eggs. The numerous descriptive and experimental analyses of the *Lymnaea* egg from oviposition to the 24-cell stage have led to several theories regarding the organization of the egg and the mechanisms of cellular patterns, determination, and differentiation. However, analyses of what is essential for normal development have been confounded by the presence of the nonessential. Accordingly, in this section I will outline the procedures for using the centrifuge as an aid in quantitative microscopy and as an instrument to remove the nonessential cytoplasmic organelles from 1-cell and early cleavage stage eggs, as well as to localize what may be essential for normal development in pulmonate eggs.

Fig. 17. Cleavage cavity of *L. palustris* during the early cleavage period. A. LM sagittal section of 2-cell egg. B. LM sagittal section of 2-cell egg (stage 9). C. LM equatorial section of 16-cell egg. D. LM of whole mount of 4-cell egg (stage 14) fixed and stained with silver nitrate. E. TEM of 2-cell egg with polygranules (arrows) in cleavage cavity. F. TEM of branching chains of polygranules. G. SEM of cleavage cavity of fractured 2-cell egg (stage 8) where cells are joined. H. Enlargement of boxed area in (G) showing fibrous material in one vacuole. Bars (A-D) 50 μm, (E) 0.5 μm, (F) 0.1 μm, (G) 5 μm, (H) 1μm. ap, animal pole. cc, cleavage cavity. n, nucleus. s, anaphase mitotic spindle. p, pit-like area. pg, polygranules. v, intracellular vacuole.

Fig. 18. TEM of a septate junctional complex between two micromeres of a 24-cell stage embryo of *L. palustris*. Bar 0.5 μm. a, zonula adherens. cc, cleavage cavity. d, dense bodies at tips of microvilli. ec, extracellular coat on cleavage cavity side of cells. mv, microvilli. n, nonfunctional region. p, pleated septate junction. ps, perivitelline space. v, vacuole. vm, vitelline membrane. arrowheads, microtubules.

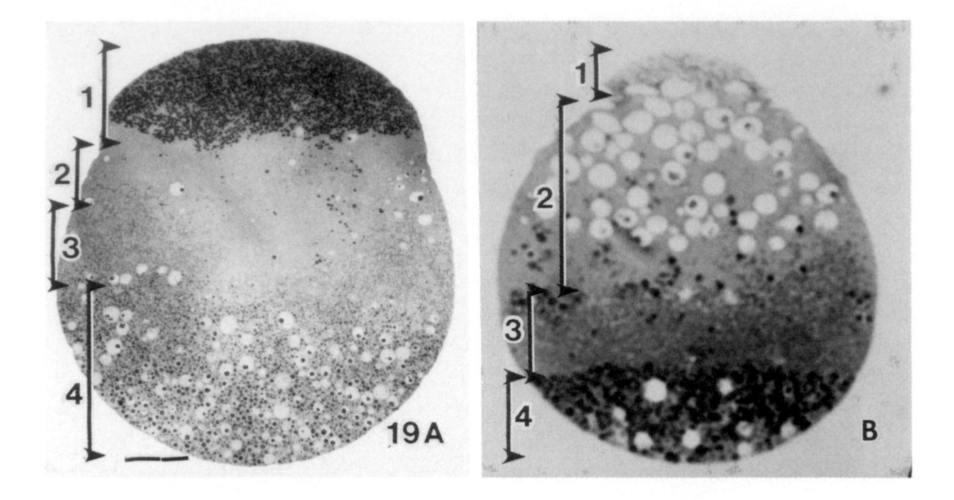

Fig. 19. Longitudinal sections through centrifuged 1-cell stage eggs to show variations in degrees of stratification of cytoplasmic organelles relative to species and stage of development. A. Low-power TEM micrograph of an egg of *L. stagnalis* centrifuged at 2,000 g for 5 minutes within 20 minutes after oviposition and before formation of the first polar body. Lipid droplets (zone 1) are at the centripetal end of the egg, followed by a layer relatively free of large organelles (zone 2), then a layer of mitochondria (zone 3), and a centrifugal layer (zone 4) of beta and vacuolated gamma yolk granules and mitochondria. Original photograph courtesy Dr. Daniel L. Luchtel. B. LM micrograph of an egg of *L. palustris* centrifuged at 2,500 g for 10 minutes, 20 minutes after the formation of the first polar body. At this stage the majority of the vacuolated gamma granules are in zone 2. Bar 10 μm.

To estimate the numbers of and characterize the morphology of different types of organelles, particularly relatively rare organelles such as multivesicular bodies, encapsuled eggs are placed in small centrifuge tubes and centrifuged at various gravitational forces for different lengths of time. When 1-cell stage pulmonate eggs are centrifuged for 10 minutes at 1,250 to 6,300 g at 20° to 25°C, the cytoplasmic organelles stratify into four to six zones or layers. Clement ('38), Raven and Bretschneider ('42), and Raven ('45) found that between oviposition and first cleavage the composition of the zones changed particularly with respect to the distribution and number of vacuolated gamma granules. Figure 19A,B illustrates this phenomenon as well as illustrating the resolution of a low-magnification TEM micrograph of a thin section compared to a high magnification LM micrograph of a section 0.5 μm thick. Even though both eggs were decapsuled and fixed within 20 minutes after centrifugation, destratification of the zones is evident. Therefore, to optimize the degree of stratification and segregation of organelles, encapsuled eggs should be centrifuged at 4°–10°C at a centrifugal force that will stratify but not fragment the egg. Lower centrifugal forces for long periods are more effective than high forces for short periods. Encapsuled eggs are then quick-fixed while still on the centrifuge, using either a plankton centrifuge or a centrifuge with a continuous flow head. After fixation the eggs are dissected from their capsules, embedded, and sectioned perpendicular to the plane of stratification.

Whereas most studies have been directed toward the possible functions of different organelles and their redistribution in uncleaved and early cleavage stage pulmonate eggs, Hartung ('47) and Bretschneider and Raven ('51) centrifuged whole snails to stratify organelles in developing oocytes. Thus centrifugation provides the embryologist with an alternate approach to quantitative ultrastructural, cytochemical, and autoradiographic analyses.

For the embryologist interested in the organization and behavior of the cytoskeletal and cytoplasmic contractile systems, the yolk, lipid droplets, and vacuolated gamma granules may be removed from the encapsuled egg via centrifugal forces that cause the 1-cell egg to be pulled into two to three fragments (Morrill, '64; Clement, '38). Whether an encapsuled egg is fragmented by centrifuging depends partly on where it comes to rest relative to the interfaces between the zones of stratified capsule fluid and partly on the relative size of the egg compared to diameter of the capsule. Clement ('38) fragmented the 1-cell egg of *Physa heterostropha* into three fragments (Fig. 21K) of which the nucleated hyaline fragment developed into a normal snail. In *L. stagnalis* and *L. palustris* with larger capsules the egg usually fragments into a centrifugal yolk fragment and a larger centripetal, nucleated fragment, which contains the lipid droplets and vacuolated granules. However, if the *Lymnaea* egg is centrifuged shortly before the formation of the first or second polar body, the lipid droplets and varying amounts of vacuoles may be removed via the formation of giant polar bodies (Fig. 21A-D; Morrill, '63b). The actual fragmentation of a centrifuged egg and the rounding up of the egg fragments is a remarkable process visible only with a centrifuge microscope (Fig. 20).

When one is interested in the developmental potential of nucleated fragments of centrifuged eggs, the orientiation of the egg's animal–vegetal (A–V) axis relative to the centrifugational axis at the time of centrifuging must be determined. Again the encapsulated pulmonate egg offers several fortuituous opportunities. When first laid, the egg is heavier than the capsule fluid. With the uptake of water and increasing numbers of vacuolated yolk granules, the specific gravity of the egg decreases and the egg floats to the top of the capsule. At the same time, the egg's center of gravity shifts toward the animal end of the egg during ooplasmic segregation and the egg rotates with its animal pole lowermost by the time of first cleavage. In *L. palustris* and *P.*

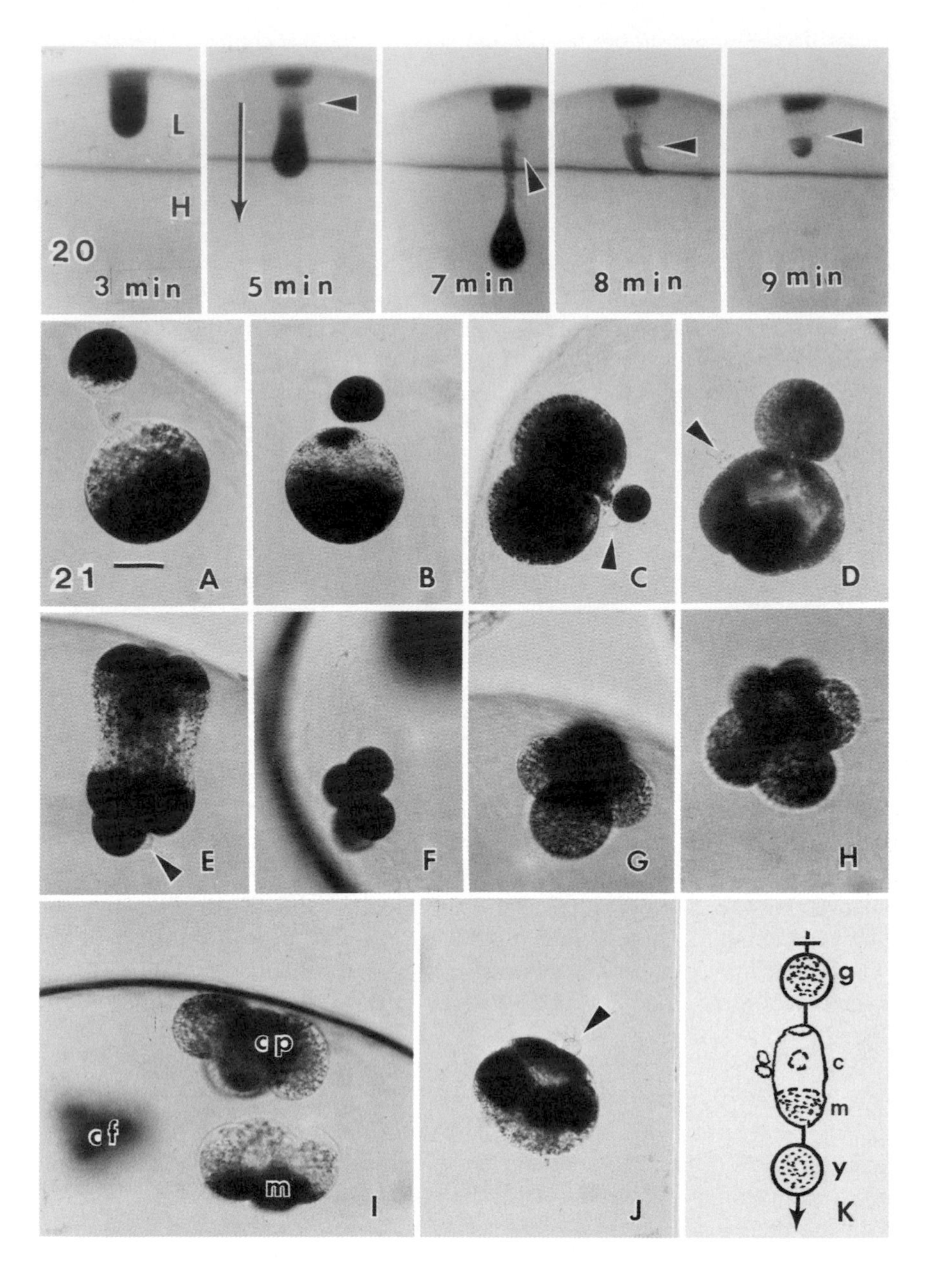
L
H
20
3 min
5 min
7 min
8 min
9 min
21
A
B
C
D
E
F
G
H
cp
cf
m
I
J
g
c
m
y
K

heterostropha the egg remains inverted near the top of the capsule through the 4- to 8-cell stage, after which its orientation and position in the capsule may undergo several changes during the subsequent cleavages. Figure 22A–L illustrates these phenomena originally described by Comadon and deFonbrune ('35) and Raven ('45).

In both *L. palustris* and *P. heterostropha,* fragmenting the 1-cell egg shortly after the formation of the second polar body is optimal for subsequent normal cleavage and development of the egg. At this time the eggs tend to orient with their A–V axis perpendicular to the centrifugal axis. In general, eggs centrifuged at the time of cell division and during prophase develop abnormally (Geilenkirchen, '64b). In *L. palustris* 2-cell stage eggs rarely recover from centrifugation because they are compressed against the top of the capsule. However, early 4-cell stage eggs withstand centrifugation (Fig. 21E). At this stage the egg may be fragmented with the centrifugal, yolk fragment containing the animal pole region of each of the four cells. The

---

Fig. 20. Stratification, elongation, and fragmentation of a 1-cell stage egg of *L. palustris* centrifuged 10 minutes after the formation of the 2nd polar body (p.b.) viewed through a centrifuge microscope. The light micrographs were taken at 3, 5, 7, 8, and 9 minutes (min) after the centrifuge had attained a centrifugal force of 2,500 g. Arrow indicates direction of centrifugal force; arrowheads mark position of polar bodies and animal pole of the egg. L, light, and H, heavy, layers of capsule fluid.

Fig. 21. Light micrographs of eggs of *L. palustris* centrifuged at the 1-, 2-, 4-, or 8-cell stage at 2,500 g for 10 minutes. A. Initial formation of giant 2nd p.b. of an egg centrifuged 5 minutes before formation of the 2nd p.b. and photographed 2 minutes after removal from centrifuge. B. Same egg as in (A) 10 minutes later. C. 2-cell stage egg initially centrifuged at time of formation of 1st p.b. showing a large, dark 1st p.b. and a normal, clear 2nd p.b. (arrow). D. 8-cell stage egg centrifuged at time of formation of 2nd p.b. with swollen, giant 2nd p.b. and normal 1st p.b. (arrow). E. 4-cell stage egg 2 minutes after removal from centrifuge, p.b. (arrow). F–H. 4-cell stage that oriented on the centrifuge as in (E) and then fragmented into a yellow, centrifugal fragment (F) in which the animal pole region of each cell contained beta yolk granules and a nucleated, centripetal fragment (G) that cleaved into 8 cells (H) consisting of 4 macromeres filled with vacuolated gamma granules and 4 micromeres filled with mitochondria. I. 4-cell stage egg broken into three fragments of which the centripetal (cp) and middle (m) fragments are in focus and the smaller yolk-filled centrifugal (cf) fragment is out of focus. J. Centrifugal fragment of an egg centrifuged at the 8-cell stage. In this egg the nonnucleated centripetal fragment contained the vegetal pole region of each of the 4 macromeres. Arrow points to polar body. K. Schematic diagram of 1-cell stage egg of *Physa heterostropha* pulled into three fragments by a centrifugal force of 6,300 g 10 minutes after the formation of the 2nd p.b. The nucleated clear (c) fragment contains the polar bodies and mitochondria (m); the centripetal grey (g) fragment contains the lipid droplets and vacuolated gamma yolk granules; and the centrifugal, yellow (y) fragment contains the beta yolk granules. Bar (A–J) 50 μm. Courtesy of Dr. Clement (Clement, '38).

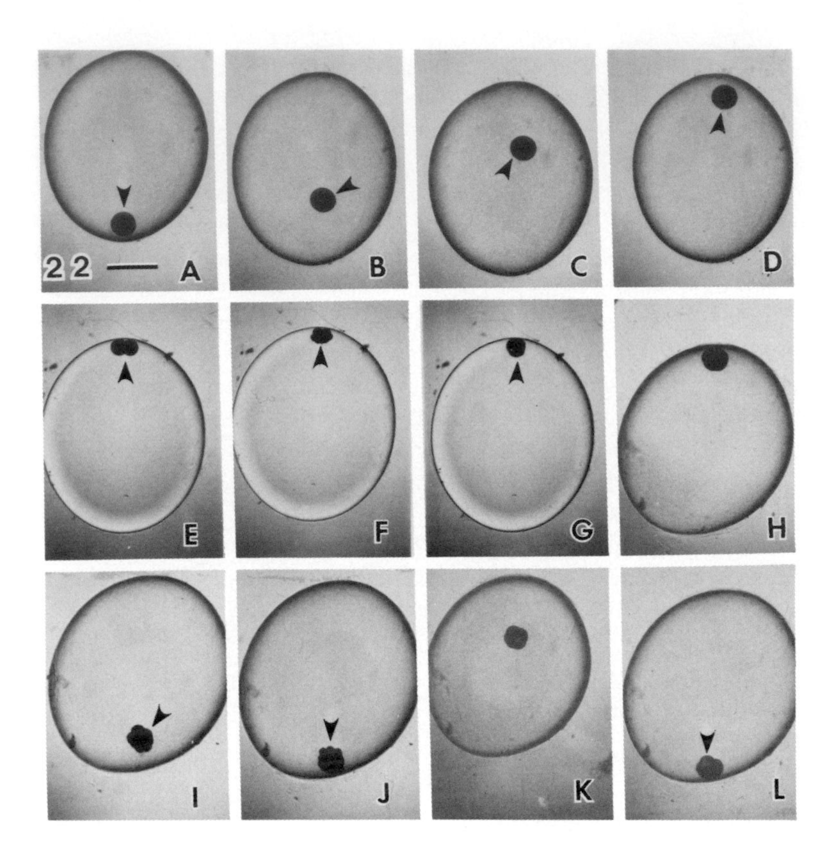

Fig. 22. Rise and fall of egg of *L. palustris* in the egg capsule and orientation of the animal vegetal (A–V) axis caused by changes in specific gravity of the whole egg and the center of gravity within the egg from time of 1st p.b. formation until the 24-cell stage; viewed with a horizontal microscope. Arrows point to the location of the animal pole. A. 1st p.b., egg at bottom of capsule and animal pole up. B. 2nd p.b., egg rising and beginning to rotate on its A–V axis. C. 10 minutes after 2nd p.b., egg inverted near top of capsule. E. 2-cell stage, egg inverted. F. 8-cell stage. G. 12-cell stage. H. 16-cell stage. I. Egg descending and rotating with animal pole up. J. 12-cell. K. 20-cell, egg descending. L. Early 24-cell stage, egg animal pole up. Bar (A–L) 200 μm.

nucleated, centripetal fragment (Fig. 21G,H) may cleave normally and develop into a normal snail. Four-cell stage eggs may also fragment into three fragments (Fig. 21I). The middle nucleated fragment may cleave normally until the 24-cell stage but is arrested at the gastrula stage. In eggs centrifuged at the 8-cell stage, fragmentation results in the removal of the vegetal ends of the four macromeres; the nucleated fragment (Fig. 21J) may cleave normally until the 24-cell stage but is arrested at the gastrula stage.

In conclusion, fragmenting the encapsuled egg by centrifugal force not only "clears" the cytoplasm of nonessential cytoplasmic constituents but it also removes various regions of the egg and undetermined amounts of the original egg cortex. Thus the normal development of a nucleated fragment will depend in part on the orientation of the egg on the centrifuge; nucleated fragments lacking the original vegetal pole region may divide normally during the early cleavage period but stop at or before the gastrula stage. Centrifuged eggs and nucleated fragments of centrifuged eggs that develop beyond the gastrula stage may exhibit a variety of morphological abnormalities. The most intriguing of these abnormalities are the dislocation or multiplication of organ primordia, which may result from the initial displacement of morphogens in the subcortical cytoplasm and subsequent relocation in cells other than those originally fated to give rise to the normal patterns.

Twenty-four Cell Stage

Until the 24-cell stage the overt mosaic nature of the spiral cleavages exhibits a radial symmetry acccompanied by the recurrent formation and disappearance of the cleavage cavity. There is an orderly embryonic determination with respect to the tempo of division and presumptive fates of individual cells and postcleavage period morphogenesis beginning at the 4-cell stage. This is shown by blastomere deletion experiments (Morrill et al., '73; Cather et al., '76) and lithium pulse experiments (Verdonk, '65, '68; Biggelaar and Boon-Niermeijer, '73; Biggelaar, '77). This determination is first coupled with a specific positioning of cells relative to one another and then, during the 24-cell stage, with the positioning of the micromeres relative to the presumptive 3D macromere (Guerrier and Biggelaar, '79; Biggelaar et al., '81).

In *L. palustris,* shortly after the completion of the sixth cleavage and the beginning of the 24-cell stage, the initial cleavage cavity (Fig. 23A) becomes occluded by the cells elongating centrally. In particular the elongated extensions of the derivatives ($1q^1$ and $1q^2$) of the first quartet of micromeres extend along one side of the cleavage cavity (Fig. 23B) and contact the inner tip (Fig. 23D) of the ovoid presumptive 3D macromere, which ultimately makes contact with 20 of the other 23 cells. As early as 20 minutes after initial compaction of the 24 cells, one of the cross-furrow macromeres, the 3D cell, can be distinguished from the 3A, 3B, and 3C cells in serial cross-sections by its oval shape (Fig. 23C) and by its having extended further into the cleavage cavity prior to contacting the $1q^1$ and $1q^2$ micromeres. In *L. stagnalis,* on the other hand, Biggelaar ('76) observed that both cross-furrow macromeres initially make contact with the projections of the animal pole

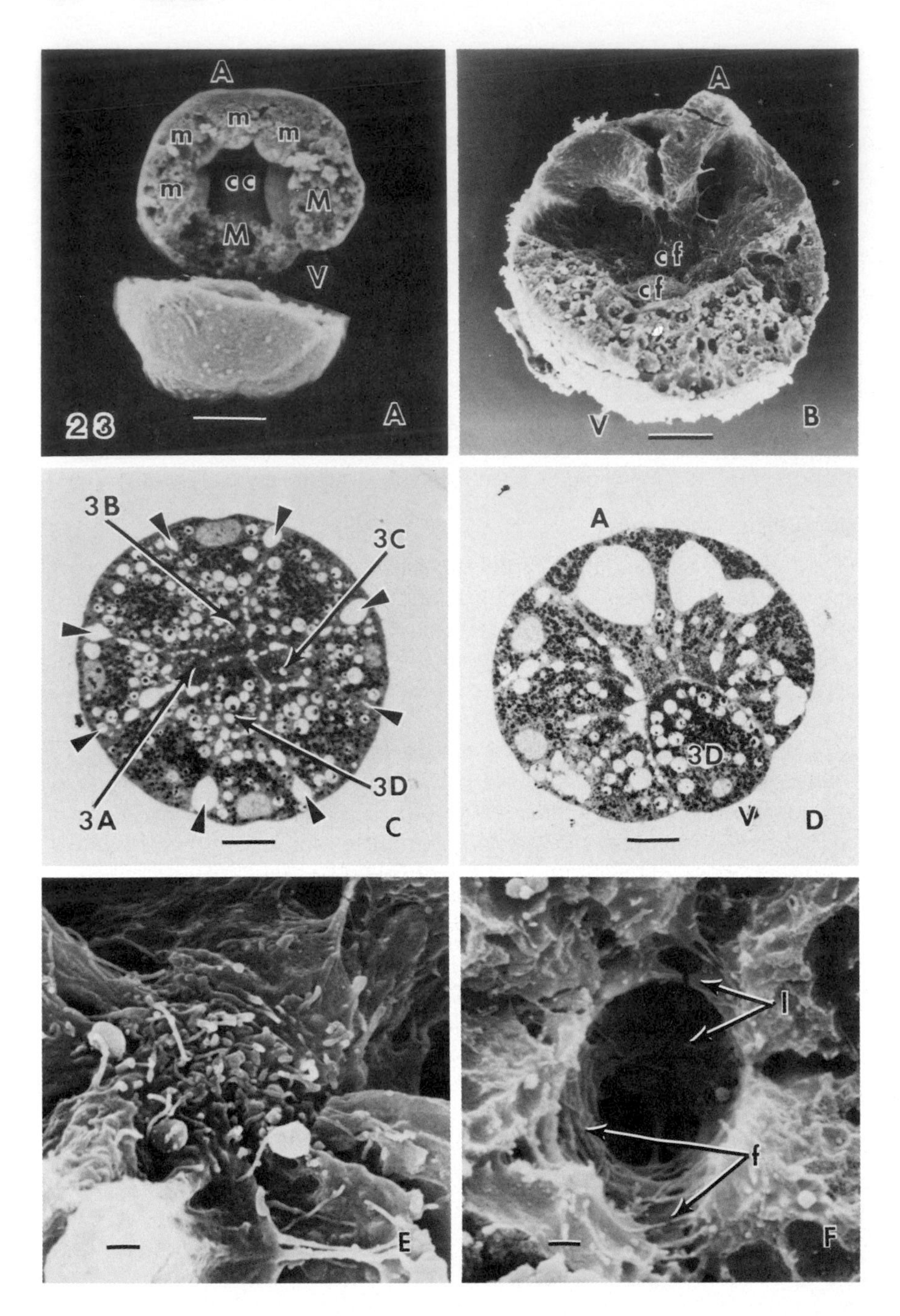
A
m
m
m
m
cc
M
M
V
23
A
A
cf
cf
V
B
3B
3C
3A
3D
C
A
3D
V
D
E
l
f
F

micromeres and that subsequently the cellular contacts are restricted to one or the other of the apparently equivalent macromeres.

In *L. palustris* numerous filopodia are produced by the elongating micromeres at their tips (Fig. 23E) and along their lateral margins,, where broad sheetlike lamellipodia of adjacent cells are in contact (Fig. 23F). The filopodia and lamellipodia form interlacing networks between adjacent cells. Well-developed septate desmosomes join adjacent cells at their outer periphery. While microtubule profiles are common in all regions of the cells, they are especially abundant in the peripheral cortical regions (Fig. 24A), in the elongated $1q^1$ and $1q^2$ micromeres, and at the inner terminal tips of these cells (Fig. 24B). Just proximal to the inner tips is a region densely populated with mitochondria (Fig. 24C). During the 24-cell stage the differential segregation of visible plasms and organelles continues and is most pronounced in the polarized movement of the SCAs or ectosomes to the inner tips of the macromeres (Raven, '74) and the accumulation of mitochondria at the inner tips of all the cells.

In addition to the peripheral intercellular septate and intermediate junctions and adhesive cellular contacts of filopodia and lamellipodia of neighboring cells, morphologically distinct gap junctions occur between most of the cells. While putative gap junctions first appear at the 4-cell stage and well-developed gap junctions occur at the 8- and 16-cell stages, it is not until the 24-cell stage that the gap junctions (Fig. 24C,D,E) assume a degree of permanency (Dorresteijn et al., '81). Thus during the 24-cell stage the previously

---

Fig. 23. Micrographs of the 24-cell stage egg of *L. palustris*. A. SEM of an egg fixed at the moment of compaction (24-cell plus 5 minutes) and fractured along the animal–vegetal axis showing the hemispherical shapes of the cleavage cavity ends of the macromeres and micromeres. B. SEM of an egg fixed at the 24-cell plus 10 minutes stage fractured along the animal–vegetal axis showing the occlusion of the cleavage cavity by the elongating first quartet of micromeres that extend preferentially along one side of the cleavage cavity and contact one of the crossfurrow macromeres. C. LM of a cross-section through the equatorial region of an egg fixed at the 24-cell plus 30 minutes stage. Note that the 3-D macromere is distinctly oval as compared to the 3A, 3B, and 3C macromeres. The original cleavage cavity is reduced to lenticular intercellular spaces between the blastomeres. D. LM of a longitudinal section through the 3D macromere of an egg fixed at 24-cells plus 60 minutes. E. SEM of the tips of the first quartet of micromeres ($1q^1$ and $1q^2$) that contact the inner end of the 3D macromere. This egg was fixed at the 24-cell plus 20 minutes stage and then fractured in a plane perpendicular to the animal-vegetal axis. F. SEM of a 24-cell plus 60 minute stage egg fractured at the level of the second quartet of micromeres ($2q^2$) to show the complexity of the filopodia and lamellopodia. Bars (A) 20 μm, (B–D) 10 μm, (E,F) 1 μm. A, animal pole. cc, cleavage cavity. cf, cross furrow macromere. f, filopodia. l, lamellopodia. m, micromere. M, macromere. V, vegetal pole. arrowheads, lenticular intercellular spaces.

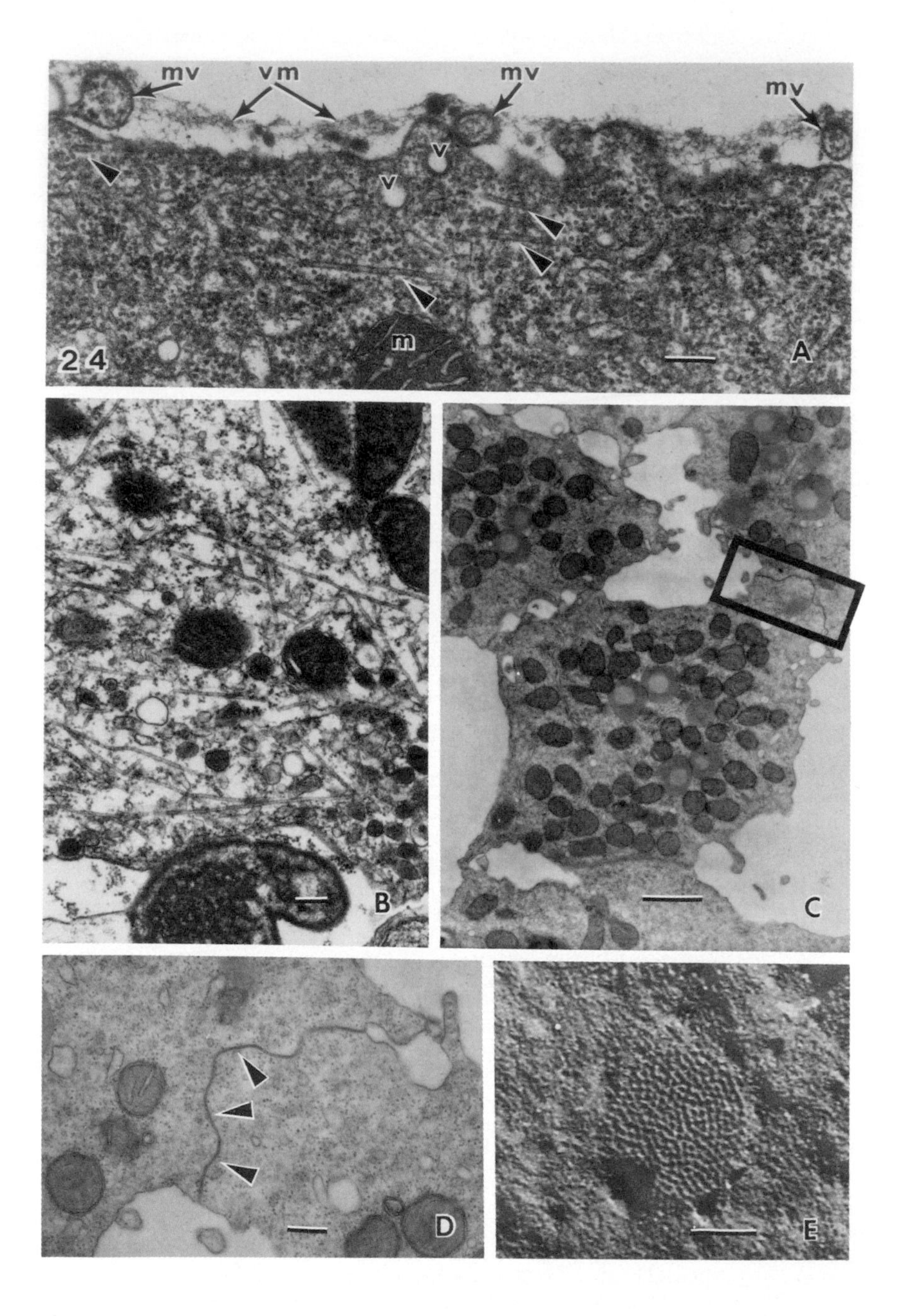
mv
vm
mv
mv
v
v
m
24
A
B
C
D
E

recurrent cleavage cavity is more or less permanently occupied by cellular elongations and the cells are morphologically "coupled" by gap junctions. In addition the cells in each tier of micromeres assume characteristic shapes and are positioned relative to the 3D macromere that marks the morphological appearance of dorso-ventrality.

It is intuitively tempting to suggest that the cells at the 24-cell stage are physiologically coupled and that morphogenetic information in the 3D macromere is transmitted to the cells with which it establishes contact. If this is true, then the positional pattern of the tips of the micromeres around the tip of the 3D macromere may be critical with respect to the subsequent determination and differentiation of individual lines of cells. Normally the positional pattern of elongation of the cells into the cleavage cavity and the establishment of normal cellular contacts during the 24-cell stage is temporally coupled to other cellular activities. However, Biggelaar ('77) uncoupled several cellular activities by treating eggs with lithium chloride (Li) during the 4-cell stage. Li-embryos cleaved in a normal manner until the 16-cell stage. Then the division of the 2A-2D macromeres and the onset of the 24-cell stage was delayed by 120 minutes relative to the controls. At the end of the 24-cell stage the 3D macromere did not divide precociously. Instead the 3D macromere divided at the same time as the 3A-3C macromeres (190 minutes after the 24-cell stage). While in Li-embryos the normal, common cell borders between the presumptive 3D macromere and the micromeres did not occur, the ectosomes migrated on schedule. Morphogenetically, Li-embryos exhibited a radial cleavage pattern after the 24-cell stage and developed into radialized exogastrulae in which all the primary prototrochal cells were ciliated instead of only the anterior prototrochal cells.

The analyses of the causal relationships between morphogenetic gradients in the *Lymnaea* egg and the cellular pattern during the cleavage period come to focus more and more on the 24-cell stage as the critical period in early development, a point when the cellular basis of either normal or abnormal

---

Fig. 24. TEMs of selected regions of cells of the 24-cell stage egg. A. Cortical region of a micromere showing microtubules (arrowheads), coated vesicles, a mitochondrion, cross-sections of three microvilli, and the vitelline membrane. B. Longitudinal section through the distal tip of an elongated $1q^1$ micromere in contact with the 3D macromere showing the numerous microtubules and relatively few mitochondria typical of the distal-most tips of the micromeres. C. Cross-section through the distal end of two $1q^1$ micromeres proximal to the distal tip region. Note the relative density of mitochondria and lipid droplets. D. Cellular junction between a $1q^1$ and $1q^2$ micromere in the box in (C) is a putative gap junction (arrows). E. Freeze-fracture replica of a macular gap junction. Bars (A,B) 200 nm, (C,D) 0.1 μm, (E) 100 nm. m, mitochondria. mv, microvilli. v, coated vesicles. vm, vitelline membrane.

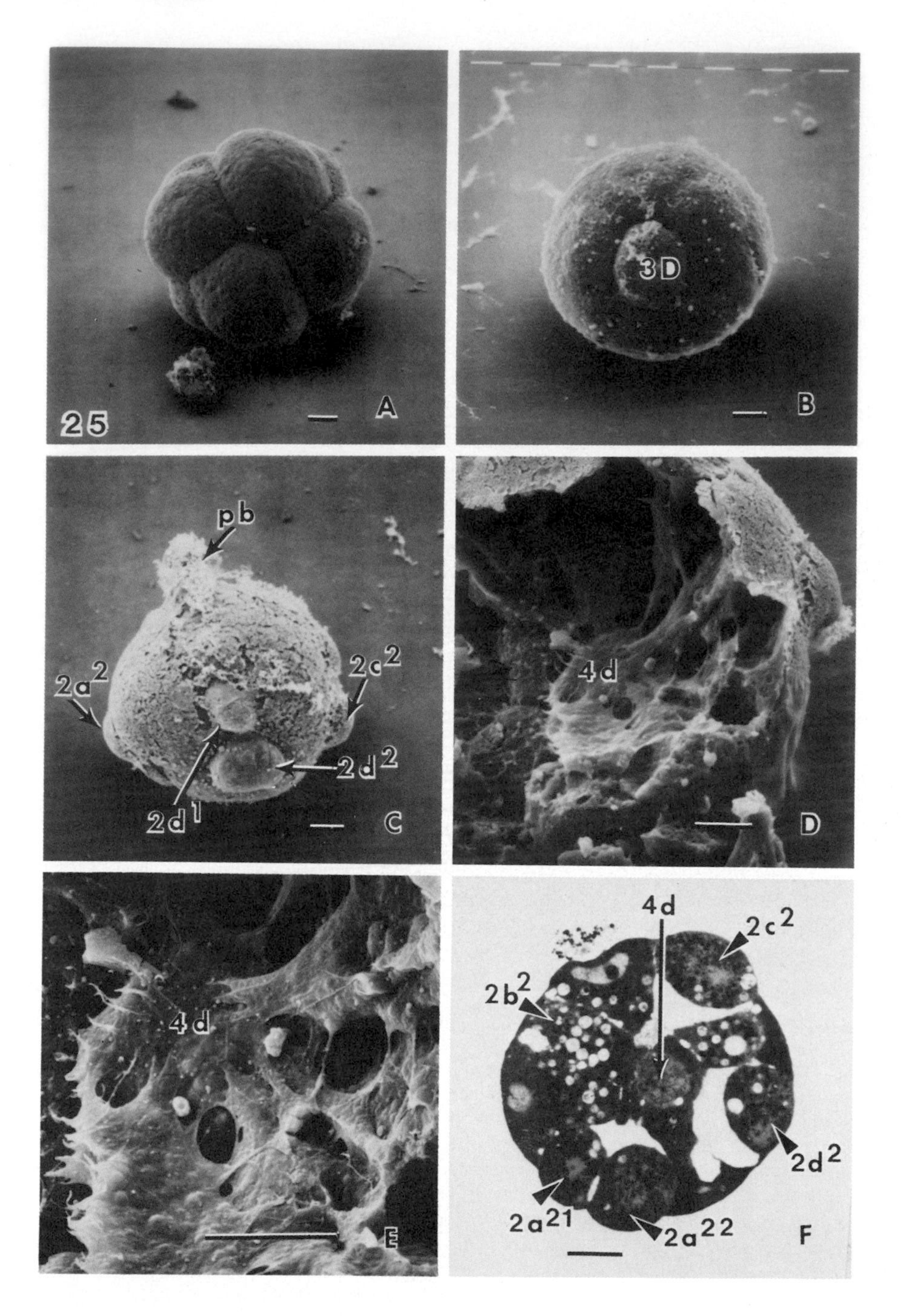
25
A
3D
B
pb
2a^2
2c^2
2d^2
2d^1
C
4d
D
4d
E
4d
2c^2
2b^2
2d^2
2a^{21}
2a^{22}
F

organ patterns is all but finalized. According to Raven ('76) the bilateral patterns of larval head organs and dorso-ventrality in particular originate by the interaction of the original axial gradient and the dorso-ventral concentration gradient when the 3D macromere contacts the first quartet of micromeres during the 24-cell stage. Utilizing the ordered patterns of determinate cleavage in normal eggs, Bezem and Raven ('76) have developed a computer program to simulate the normal development of *Lymnaea*. This program has been successfully used to model the cellular patterns leading to the normal larval head and to Li-induced radialized larvae (Raven and Bezem, '76).

Late Cleavage Period

The precocious division of the 3D macromere (Fig. 25B) into the 4D macromere and the primary mesentoblast (M cell or 4d micromere) marks the end of the 24-cell stage and the beginning of the period of bilateral, asynchronous cleavages (Fig. 25C) where the dorso-ventral polarity is morphologically manifested (Biggelaar, '71c, '76; Camey and Verdonk, '70). Figure 26 summarizes the timetable of cell divisions during this period, which arbitrarily ends at the 49-cell stage with the bilateral division of the M cell into the M_1 and M_2 cells and the formation of the remaining fourth quartet of micromeres (4a-4c). The exact timing of cellular divisions during this period may differ between species.

Cells that are not dividing retain their cleavage cavity extensions. In particular the derivatives of the first quartet of micromeres remain in contact with the inner tip of the 4d micromere (Fig. 25D,E). The lateral margins of the 4d micromere remain in contact with other micromeres except when the

---

Fig. 25. Photomicrographs of the egg of *L. palustris* during the early phases of the late cleavage period to illustrate the morphological appearance of bilateral symmetry and cellular contacts with the 4d micromere (primary mesentoblast cell). A. SEM vegetal pole view of a 24-cell plus 10 minute egg showing the arrangement of the cross-furrow macromeres and non-crossfurrow macromeres. B. SEM vegetal pole view of a 24-cell plus 150 minute egg when the bulging 3D macromere is preparing to divide. C. SEM of egg fixed 1.5 hours after the division of the 3D macromere viewed from the future posterior end of the embryo. The micromeres $2d^1$, $2d^2$, $2c^2$, and $2a^2$ have rounded up and are dividing; the other cells and polar bodies are hidden beneath the enveloping vitelline membrane. D. SEM of the egg in (C) fractured to show the convergence of the elongated micromeres on the inner end of the 4d micromere. E. A higher magnification view of the arrangement of protoplasmic extensions of the micromeres where they contact the 4d micromere. F. LM of a cross-section at the level of the $2q^2$ micromeres of an egg fixed 1.5 hours after the division of the 3D macromere. The micromeres $2c^2$ and $2d^2$ are in anaphase of mitosis; $2a^2$ has divided into $2a^{21}$ and $2a^{22}$; the nucleus of $2b^2$ is in interphase. The cytoplasm of the 4d micromere has many mitochondria as well as darkly staining granules. Bars (A–F) 10 μm.

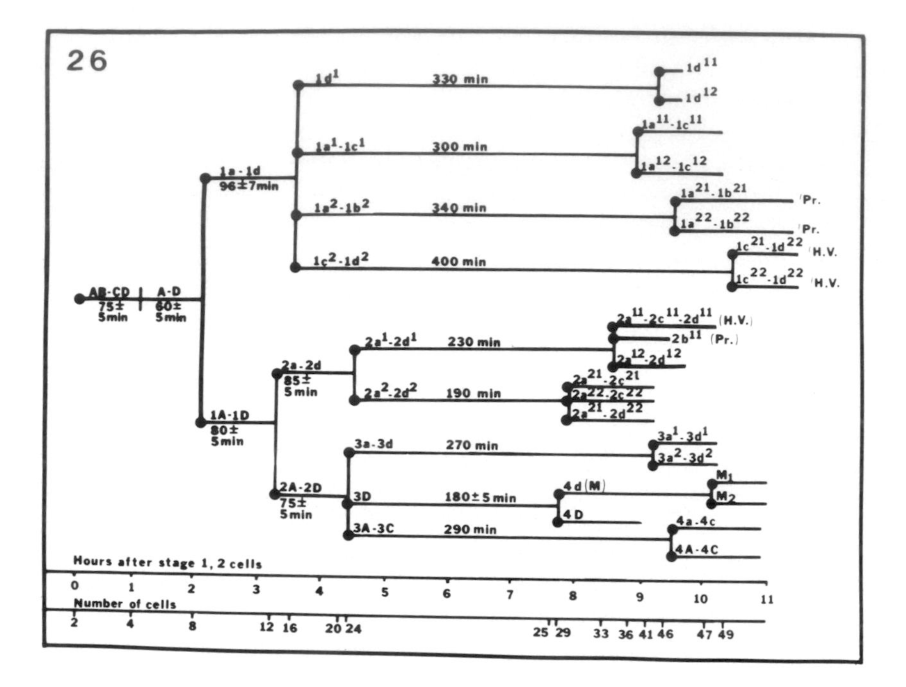

Fig. 26. Cell lineage and timetable of cell divisions of the *Lymnaea* egg from the 2-cell to 49-cell stage including the approximate lengths of cell cycles in minutes where they are known. Note the asynchronous divisions in the descendants of the first and second quartets of micromeres and the precocious division of 3D into 4d and 4D at the end of the 24-cell "resting" stage. Cells of the head vesicle (H.V.) and the prototroch (Pr.) stop dividing after their formation (modified from Biggelaar, '71c).

latter round up to divide (Fig. 25F). The various cellular events and cellular interactions during this late cleavage period have not received critical attention beyond the cell lineage analysis of Camey and Verdonk and the cytochemical study of Raven ('46).

Cell lineages and cell fates. During the late cleavage period certain cells destined to form the primary ciliated prototrochal (velar) cells and the head vesicle cells (Fig. 26) cease to divide. The divisions of other cell lines destined to form larval and adult organs continue, but the number of divisions, their tempo, and cleavage planes differ. Cell lines destined to form larval organs stop dividing before those destined to form adult organ primordia.

Historically, cell lineage analyses or genealogies and cellular configurations involved the differential staining of whole embryos with silver nitrate (Holmes, 1900; Verdonk, '65; Camey and Verdonk, '70; Arnolds, '79).

TABLE V. Larval and Adult Organs Derived From the First Five Quartets of Micromeres and the Macromeres

	Larval organs	Adult organ Primordia
Micromeres		
1a-1d	Apical plate	Paired cephalic plates
1a, 1b, 2a, 2b	Ciliated prototroch/ velum	
1a, 1c, 1d, 2a, 2c, 2d	Head vesicle	
2c, 2d	Shell gland	Mantle
2a, 2c		Stomodeum
3c, 3d		Foot
3a, 3b		Secondary mesenchyme
		Ectoderm
4d	Primary Mesenchyme	Pericardium
	Paired protonephridia	Primordial germ cells
4a-4c, 5a-5c	Large celled larval liver	
Macromeres		
4D, 5A-5C		Esophagus, stomach, intestine

Now, with procedures for removing the vitelline membrane of pregastrula embryos, coupled with SEM analyses of individual whole embryos before and after dry fracturing, cell lineage analyses of normal and experimentally treated embryos should yield additional information and insights into the mosaic cellular patterns in pulmonate embryos.

Table V outlines those larval organs and adult organ primordia known to originate from descendants of the first five quartets of micromeres and the macromeres. The histories of the micromeres and macromeres have been detailed by Holmes (1900) Wierzejski ('05), Verdonk ('65, '68), and Camey and Verdonk ('70). By the gastrula stage the descendants of the first two quartets of micromeres in the animal hemisphere form the molluscan cross (Fig. 27A,B). Figures 28A–F and 29A–C show the development of the molluscan cross and the emergence of the cellular arrangements leading to the larval and adult organ primordia of the head region of the trochophore and veliger larvae. Table VI summarizes the cell lineages of the three major ectodermal organs of the head region—the apical plate between the paired cephalic plates, the velum, and the head vesicle, which lies posterior to the apical plate and cephalic plates. Figure 29D shows one of several types of head malformations produced by lithium treatment during the early cleavage (Verdonk, '68) to illustrate the marked morphogenetic effects that result from

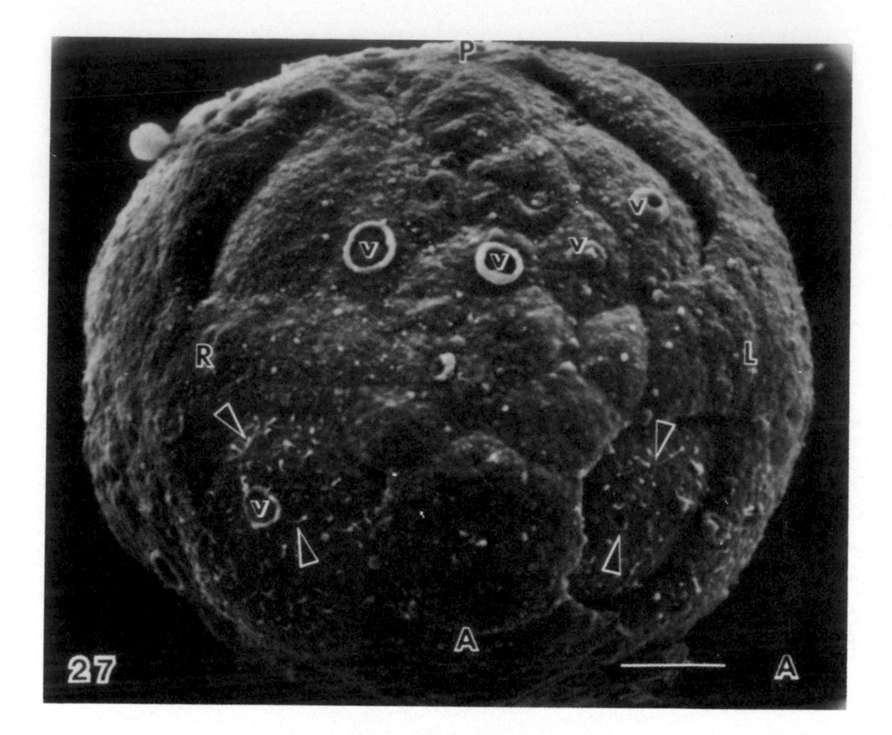
P
V
V
V
V
R
L
V
A
27
A

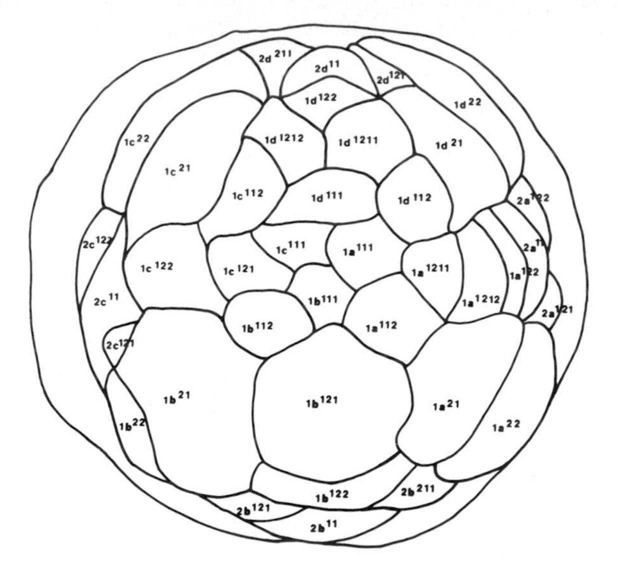
2d211
2d11
2d121
1d122
1d22
1c22
1d1212
1d1211
1d21
1c21
1c112
1d111
1d112
2a122
2c122
1c111
1a111
2a11
1c122
1c121
1a1211
1a122
2c11
1b111
1a1212
2a121
1b112
1a112
2c121
1b21
1b121
1a21
1a22
1b22
1b122
2b211
2b121
2b11

B

the alteration of cleavage patterns of even a few cells, in this case the apical plate cells.

Gastrula Period

Gastrulation has been described by Rabl (1879), Holmes (1900), Wierzejski ('05), and Camey and Verdonk ('70).

In *Lymnaea* the onset of gastrulation is signaled by the hatching of the embryo from the vitelline membrane. Little is known about the hatching process other than that the vitelline membrane appears to rupture at the vegetal end of the embryo which bulges outward at approximately the 64-cell stage (15–16 hours post 2-cell). Frequently the remnant of the vitelline membrane with the attached polar bodies is cast off as the embryo first flattens and then invaginates at the vegetal pole. How the original attachments of the tips of microvilli to the vitelline membrane are severed is not known.

In *L. palustris* gastrulation occurs over a period of 8 hours at 25°C. I have divided this period into six morphological phases (G-1 to G-6), shown in Figure 30A–F. Morphogenetically the initial invagination at the vegetal pole involves an inward sinking of the macromeres (4A-4D) and the fourth quartet of micromeres. Subsequent to the initial invagination both the macromeres and the fourth quartet of micromeres divide. The initial horseshoe-shaped, bilaterally symmetrical, invaginating vegetal pole region deepens as the endomeres become flask-shaped and contact the inner ends of the animal pole micromeres. Presumably contraction of the inner extensions of the endomeres results in a flattening of the animal pole followed by an inward depression of the animal pole to form the placode-shaped gastrula (Fig. 5A). The invagination of the endomeres is accompanied by cellular proliferation of ectomeres and the convergence of the right and left foot primordia toward the midline of the vegetal pole (Fig. 30D). The fusion of the foot primordia along the midline begins at the posterior end of the embryo and proceeds anteriorly (Fig. 30E) until the original gastropore is transformed into the stomodeal primoridum located at the anterior end of the embryo ventral to

---

Fig. 27. Animal pole view of the arrangement of cells at the molluscan cross stage. A. SEM of *L. palustris* embryo 19 hours after first cleavage. B. Map of identifiable cells in (A). In the process of preparing the embryo, the interradial cells (primary trochoblasts) between the arms of the cross sank inward thereby defining the right (R) and left (L) lateral arms, the anterior ventral (A) arm and posterior dorsal (D) arm of the cross. The anterior, right and left prototrochal cells ($1a^{22}$, $1a^{21}$, $1b^{21}$, $1b^{22}$) have begun to form cilia (arrowheads). The ringlets and conical protrusions on individual cells are macropinocytic albumen vacuoles. Bar (A) 10 μm.

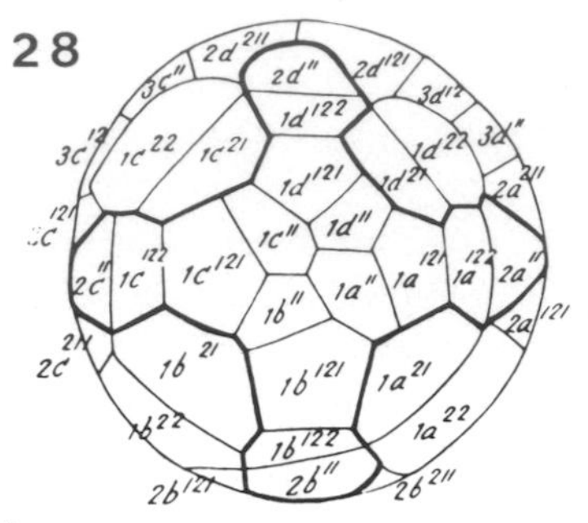

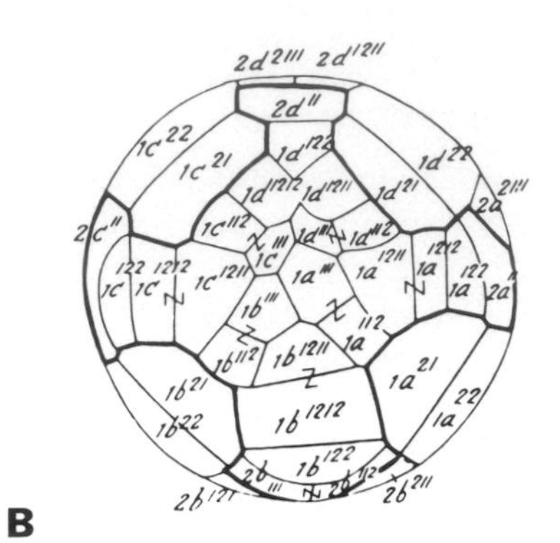

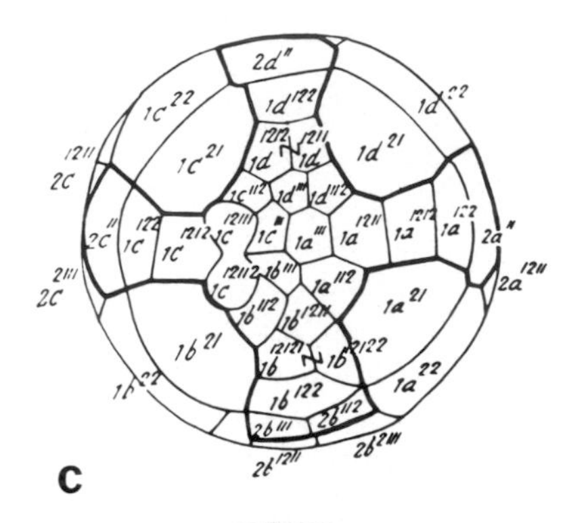

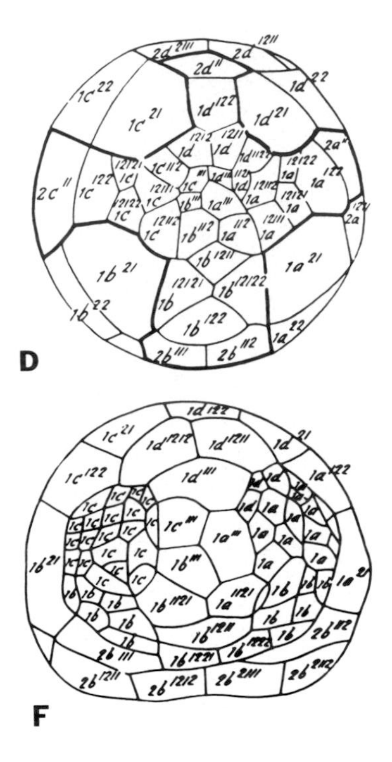

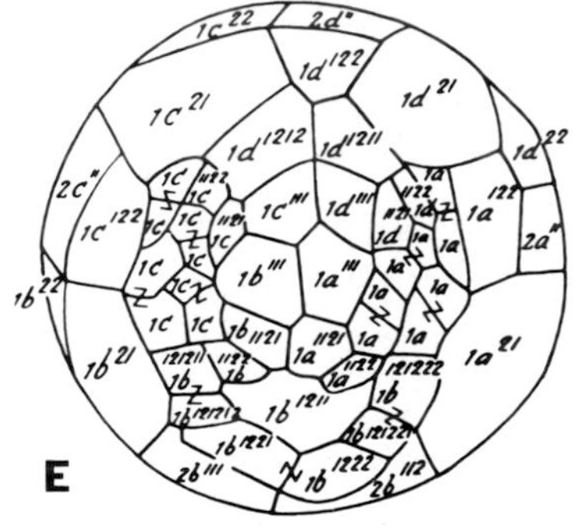

Fig. 28. Diagrams of cell lineages of the animal hemisphere and development of the "molluscan cross" from the early gastrula to trochophore stage of *L. stagnalis* cultured at 25°C. Hours after first cleavage: A: 15; B: 19; C: 21; D: 24; E: 28; F: 40. Surface views of embryos stained with silver nitrate. Courtesy of Dr. Verdonk (Verdonk, '65).

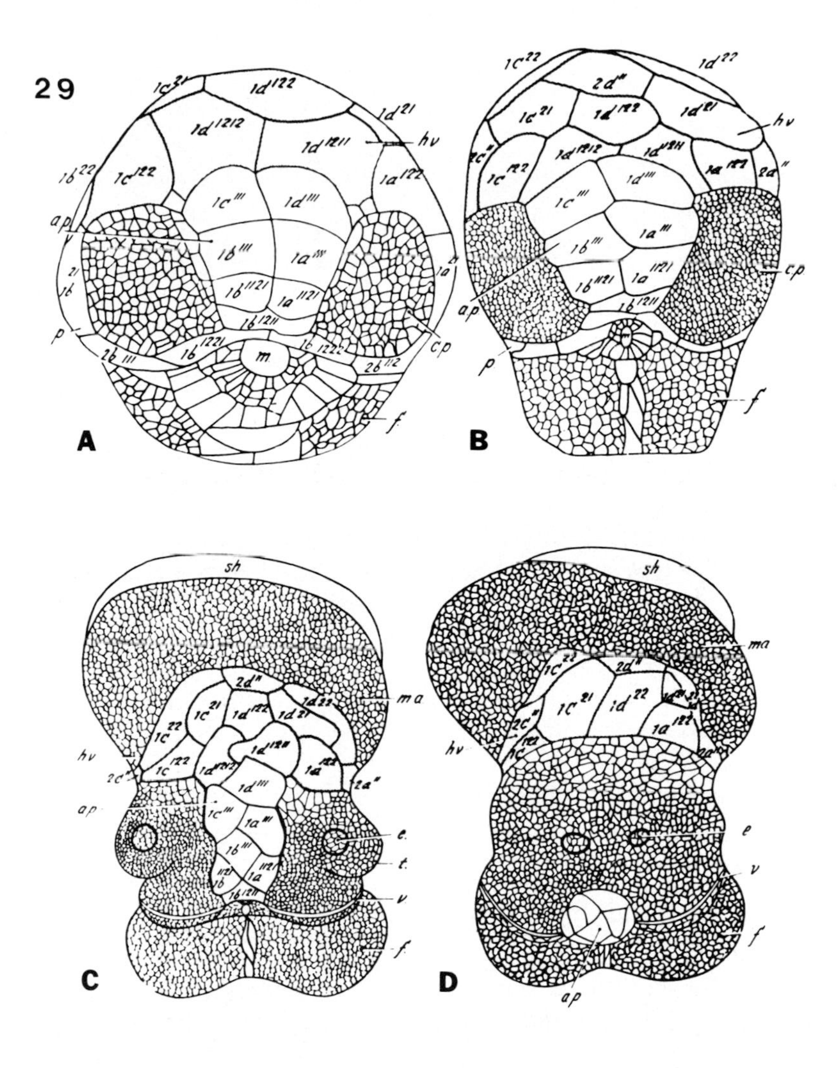

Fig. 29. Diagrams of the developing head region of: A) the normal trochophore; B) Early veliger; C) Normal "hippo" stage veliger larva; D) A synophthalmic embryo with two eyes in the bridge of cells connecting the tentacle fields in *L. stagnalis.* The embryo in (D) was produced by treatment of a 2-cell stage egg with 2.5×10^{-5} M lithium chloride for 24 hours. Courtesy of Dr. Verdonk (Verdonk, '65). ap, apical plate. cp, cephalic plate. e, eye. f, foot. hv, head vesicle. m, mouth. ma, mantle. p, prototroch. sh, shell. t, tentacle. v, velum.

TABLE VI. Cell Lineage of Larval Ectodermal Organs[a]

Larval Organ	Cells
Apical plate	$1a^{111}$, $1b^{111}$, $1c^{111}$, $1d^{111}$ $1a^{1121}$, $1b^{1121}$, $1b^{1211}$
Velum	$1a^{21}$, $1a^{22}$, $1b^{21}$, $1b^{22}$ $2b^{111}$, $2b^{112}$ $2b^{121}$, $1b^{122}$, $2b^{211}$, $2b^{212}$
Head vesicle	$1a^{122}$, $1c^{122}$, $1c^{21}$, $1c^{22}$ $1d^{1211}$, $1d^{1212}$, $1d^{122}$ $1d^{21}$, $1d^{22}$, $2a^{11}$, $2c^{11}$ $2d^{11}$

[a]Adapted from Verdonk, '65; Camey and Verdonk, '70.

the velum (Fig. 30F) and dorsal to the crescentlike arrangement of the stomoblasts.

The morphogenetic movements during the gastrula period are accompanied by an increase in the number of cells from 64 at the beginning to 130 at the end, and by the appearance of cilia (Fig. 27A) on the two anterior pairs of primary velar cells ($1a^{21}$, $1a^{22}$, $1b^{21}$, $1b^{22}$). Raven ('46) has described the histological and cytochemical changes at this stage. The presence of many macroendocytotic vesicles on the surfaces of the cells reflects an exceptionally active uptake of capsule fluid during this period. Whereas gastrulation is under genomic control (Verdonk, '73; Morrill et al., '76), formation of albumen vesicles and the appearance of cilia on the primary velar cells may occur in actinomycin-D treated embryos that are arrested at the gastrula stage.

Embryogenesis and Organogenesis

Morphogenesis of the postgastrula embryo of *Lymnaea* has been described in detail by Raven ('46), Régondaud ('64), and Cumin ('72) as well as by other workers cited above under "General Features of Development." There-

---

Fig. 30. SEMs of the vegetal and anterior ends of the embryo of *L. palustris* during the period of gastrulation at 25°C. A. G-1 phase, 16 hours. B. G-2 phase, 20 hours. C. G-3 phase, 22 hours. D. G-4 phase, 23 hours. E. G-5 phase (placode gastrula), 24 hours. F. G-6 (head vesicle), 26 hours after the 2-cell stage. Bar (A–F) 10 μm. a, anterior end. d, dorsal side. f, foot primordium. ma, macromeres. gp, gastropore. p, posterior end. st, stomodeum, v, ventral side.

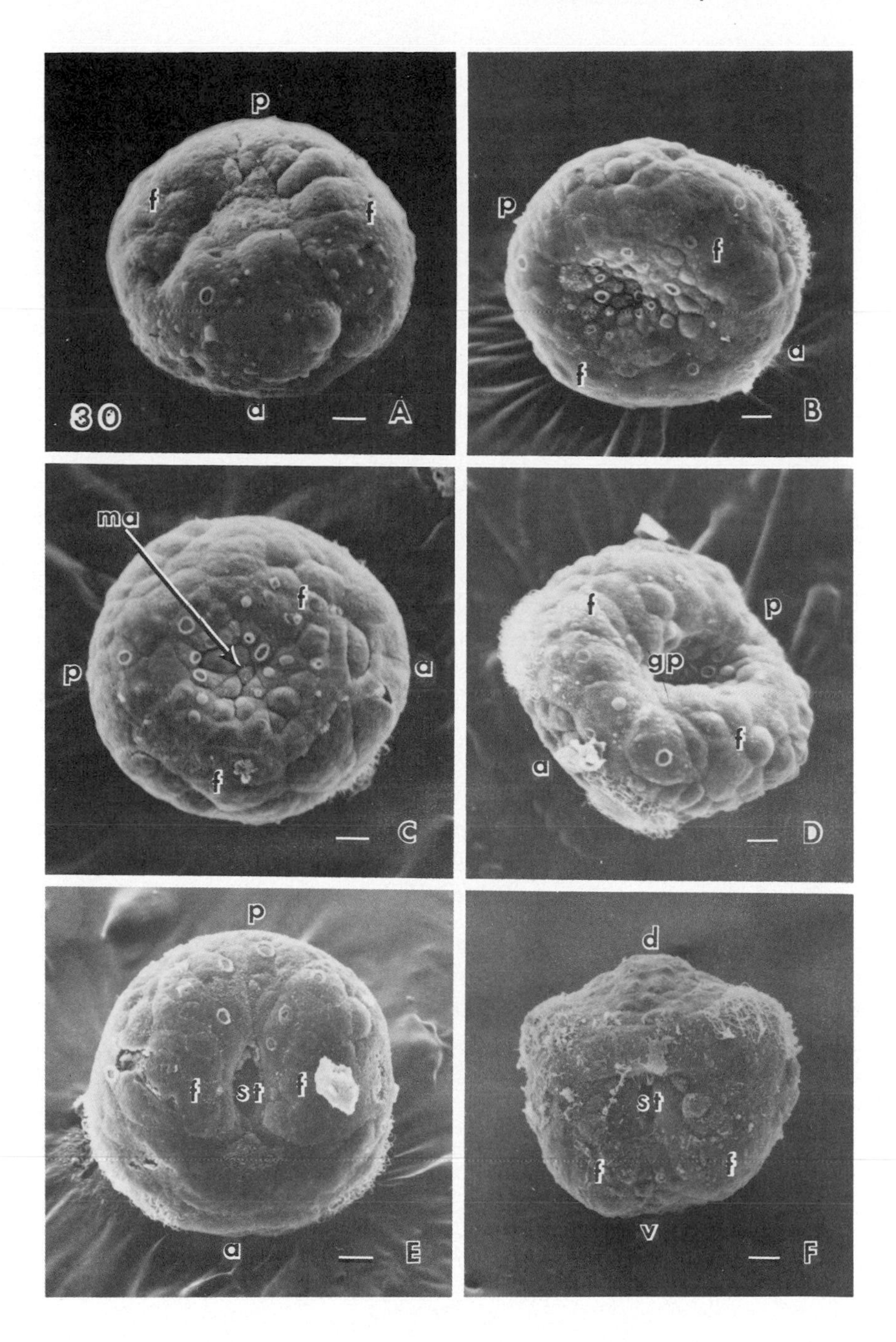
30
p
f
f
a
A
p
f
f
a
B
ma
f
p
a
f
C
f
p
gp
a
f
D
p
f
st
f
a
E
d
st
f
f
v
F

fore, I will restrict myself to certain aspects of embryogenesis and organogenesis that are of potential interest to the experimental embryologist.

By the end of the gastrula period (G-6) the embryo begins to assume the form of the trochophore larva (Fig. 30F). The dorsal posterior region of the head may become swollen as the large, flattened, larval head vesicle cells are distended by the accumulation of fluid in the body cavity in this region. In some pulmonates, e.g. *Physa,* there is a morphologically distinct head vesicle stage. At this stage only the primary trochoblasts or velar cells are ciliated and the embryo swims slowly.

Ciliogenesis. In *L. palustris* we have observed a consistent pattern of swimming behavior in which the embryo rotates backwards, turning upside down and then right side up. At the same time the embryo rotates slowly in a clockwise direction. In other words, rotational movement of the embryo is analogous to a diver's back flip with a slow side twist. This swimming behavior continues until the 4-day (E5) veliger stage when the rotary movement is supplemented by a forward swimming movement. By the E7 stage, when the foot is functionally developed, swimming movements cease and the juvenile snail creeps along the inner surface of the capsule. It is interesting that the inherent asymmetrical organization of the egg is manifested not only in the morphogenesis of the various organs but also in the swimming behavior of the embryo.

The developmental patterns of the ciliated ectodermal cells are best studied with SEM. The times of appearance of cilia on individual ectodermal cells of *L. palustris* are summarized in Table VII. Figure 31A–C shows the appearance of the cilia on the cells of the velum, apical plate, and midline (pedal plate) of the foot. By the 2-day (E2) definitive trochophore (Fig. 31C) the ciliation of the larval ciliated cells is completed. During the veliger phase (E3-E5) additional ectodermal cells become ciliated along the future ventral, lateral margins of the mantle (Fig. 31E), in the area of the future mantle cavity (Fig. 34A) and at the bases of the tentacle primordia (Fig. 34B). The ciliated cells of the apical plate and velum persist through metamorphosis and are present in the juvenile snail (Fig. 31F).

Preliminary ultrastructural analyses of the ectodermal cilia show that electron-dense material is attached to the surface of the ciliary membrane at periodic intervals (Fig. 32A,B). In addition, the shafts of the velar cilia appear connected to one another by a sheetlike layer of extracellular material near the bases of the cilia. Whether this material coordinates the ciliary movements is not known.

In the course of our SEM analysis of ciliogenesis in *L. palustris,* we discovered ventral to the ciliated velar cells a row of rounded cells (Fig. 31B) that appear shortly after gastrulation and persist at least through the 4-day

TABLE VII. Developmental Patterns of Ciliated Ectodermal Cells in *L. palustris* During the First Four Days of Development

Region of ectoderm	Time of appearance in hours after first cleavage											
	18	24	30	36	42	48	54	60	72	84	90	96
Velum cells												
$1a^{21}$	X											
$1a^{22}$	X											
$1b^{21}$	X											
$1b^{22}$	X											
$2b^{111}$				X								
$2b^{112}$				X								
Apical plate cells												
$1a^{111}$				X								
$1b^{111}$				X								
$1c^{111}$				X								
$1d^{111}$				X								
$1b^{1211}$				X								
$1b^{1121}$					X							
$1a^{1121}$					X							
Foot												
Midventral line					X							
Anterior transverse line						X						
Mantle margin									X			
Presumptive mantle cavity									X			

(E5) stage. These cells are bulbous and densely covered with conspicuous microvilli. Neither their origins nor their functions are known.

Shell gland and mantle formation. The shell gland is derived from the descendants of the $2d^{12}$ and $2d^{21}$ cells (Holmes, 1900). SEM of *L. palustris* embryos shows an initial, rosettelike arrangement of cells at the posterior end of the embryo at the gastruula stage (Fig. 33A) that becomes a discrete cluster of cells at the midtrochophore (E1) stage (Fig. 33B). By the definitive trochophore (E2) stage the shell gland is characterized by a central, shallow, invaginated region encircled by a plate of presumptive secretory cells just ventral to the posterior margin of the head vesicle (Fig. 33C). By the E3 stage the invagination has deepened and the encircling plate of cells has begun to secrete the initial pellicle (Fig. 33D). By the veliger stage the shell gland has enlarged (Fig. 33E) but still maintains its medial, posterior position relative to the head vesicle. The margin of the shell gland begins to be depressed relative to an encircling marginal area of ectoderm, thereby creating the periostracal groove (Fig. 33E,F). The subsequent development of the

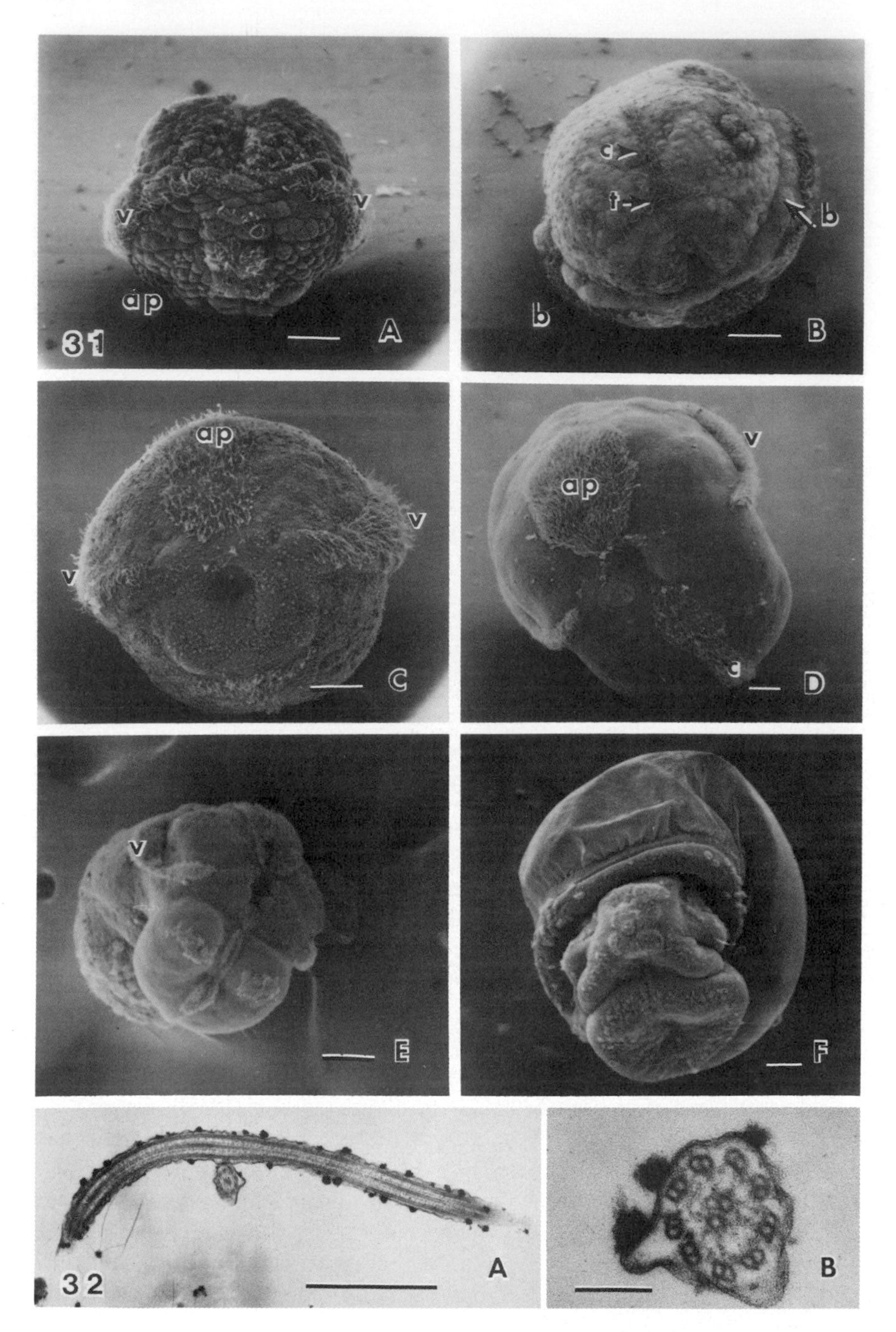
31
A
v
v
ap
B
c
t
b
b
C
ap
v
v
D
v
ap
c
E
v
F
32
A
B

mantle fold, the invagination of the lung cavity, and the formation of the mantle cavity are shown in Figure 34A–D. By the E7 stage the juvenile snail has a fully differentiated mantle margin (Fig. 35).

Other organs. A full description of the histogenesis and organogenesis of the other organs is beyond the scope of this chapter. However, Figure 36A–D illustrates the complex nature of the embryo from the E2 trochophore stage to the E5 metamorphosing veliger stage. The studies of Régondaud ('64) and Cumin ('72) provide excellent stage-by-stage descriptions of the morphogenesis of the embryo, and are complemented by Raven's ('66) descriptions of the development of individual organs.

BIOCHEMISTRY OF DEVELOPMENT

Compared to the detailed analyses of morphogenesis, our knowledge of the molecular aspects of *Lymnaea* development is rudimentary (Raven, '72b) and limited mainly to analyses of carbohydrate metabolism (Horstmann, '64; Goudsmit, '64, '72, '76), uric acid synthesis (Conway et al., '69), RNA synthesis (Brahmachary, '73), and protein synthesis and differentiation (Morrill et al., '76; Moon and Morrill, '79) of whole embryos.

Both the biochemical and experimental (Verdonk, '73) evidence to date indicate that the development of *Lymnaea* is not under genomic control until gastrulation and that most genes directly involved with differentiation, morphogenesis, and growth are activated at the trochophore (E2) stage when there is an abrupt increase in total RNA and protein synthesis relative to DNA synthesis (Fig. 37A–C).

Fig. 31. SEM micrographs of the anterior end of postgastrula embryos of *L. palustris* showing the changes in general morphology and the patterns of the ciliated ectodermal cells of larval and adult organs. A. 36-hour trochophore oriented with its ventral, anterior side up, showing the ciliated cells of the velum and the apical plate. B. 42-hour trochophore oriented upside down to show the double row of ciliated cells between the right and left halves of the foot primordium and the short transverse row of ciliated cells at the anterior end of the foot. Note a row of bulbous cells covered with microvilli located immediately ventral to the ciliated cells of the velum. C. Frontal view of a 48-hour trochophore. D. Frontal view of a 72-hour (3-day) veliger. E. 4-day "hippo" stage. F. 5-day juvenile snail. Bars (A–D) 20 μm; (E,F) 40 μm. ap, apical plate. b, bulbous cells. c, ciliated cells (pedal plate) of foot. t, transverse row of ciliated cells. v, velum.

Fig. 32. TEM micrographs of cilia of the primary velar cells of a gastrula stage embryo. A. Longitudinal section of a cilium. B. Cross-section of a cilium. Glutaraldehyde–tannic acid, post osmication (GTO) fixation. Bars (A) 1 μm; (B) 0.1 μm.

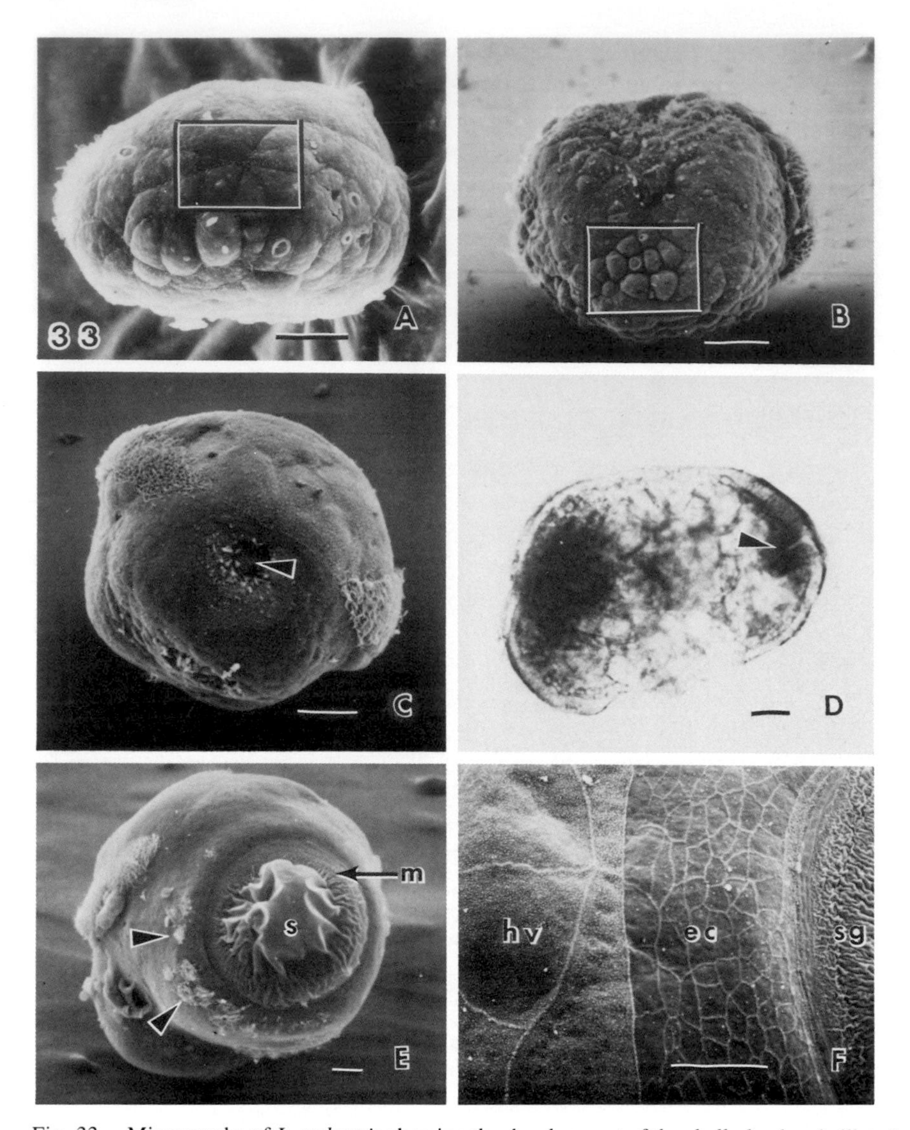

Fig. 33. Micrographs of *L. palustris* showing the development of the shell gland and ciliated ectodermal cells from the gastrula to the veliger stage. A. SEM of 20-hour (G-3 stage) gastrula; the presumptive shell gland area is boxed. B. SEM of 36-hour trochophore with presumptive shell gland area boxed. C. SEM of 42-hour trochophore with an invaginating shell gland (arrow). D. LM of a whole mount of a 60-hour veliger showing the invaginated shell gland (arrow). E. SEM of a 72-hour veliger showing the circular margin of the larval shell gland, the larval shell wrinkled in preparing the specimen, and ciliated ectodermal cells (arrows) along the future ventro-lateral margin of the mantle fold. F. SEM of the dorsal ectoderm of the body or visceral mass of a 72-hour veliger. Bars (A–F) 20 μm. ec, dorsal ectoderm. hv, head vesicle cells with microvilli. m, margin of shell gland. s, shell, sg, shell gland.

Because the pulmonate embryo grows rapidly after gastrulation, it is important to convert biochemical data on whole embryos to a per unit embryo protein base. Furthermore, since a certain percentage of the total measurable embryo protein following gastrulation is ingested capsule fluid (Morrill et al., '76), this component should be subtracted from the total embryo protein to obtain adjusted values for embryonic protein at different stages (Fig. 37D).

The amount of ingested capsule fluid protein may be estimated by measuring the amount of galactogen in the embryos at each stage of development and determining the ratio of the initial concentration of capsule fluid protein to capsule fluid galactogen in newly laid egg capsules. Figure 38 shows the average concentrations of total protein and galactogen in the capsule fluid and embryo and hatched juvenile snail of *L. palustris.* The time of hatching of the juvenile snail appears to be correlated with the embryo's having consumed the nutrients in the capsule fluid. Preliminary experiments with *Physa fontinalis* indicate that certain terminal stages of morphogenesis may be delayed in a prehatching, juvenile snail embryo that is cultured in a hanging drop culture of capsule fluid. Thus exhaustion of the capsule fluid may play a causal role in morphogenesis and differentation as well as stimulating hatching activity.

Starved embryos. When trochophore stage (E2) and older embryos are decapsulated and isolated from the capsule fluid and cultured in pond water, they continue to differentiate. The degree of differentiation and length of survival is greater in postmetamorphic (E5–E11) than in premetamorphic (E2–E4) embryos. In isolated embryos the large larval liver cells regress and the fluid-filled body cavity swells. In addition, whitish granules appear in the cells of the paired protonephridia, indicating a precocious development of nitrogen catabolism. Similar alterations in morphogeneis may result from treating embryos with actinomycin D (Morrill et al., '76).

The degree to which normal development depends on the utilization of galactogen and proteins of the capsule fluid is not known; however, circumstantial evidence indicates that the uptake and catabolism of the capsule fluid are essential for normal development. Goudsmit ('76) measured the metabolism of ^{14}C-galactogen in whole embryos and embryo homogenates of *Bulimnaea megasoma* embryos and found that a measurable amount of the ^{14}C was incorporated into lipid, proteins and nucleic acids. Interestingly, neither of the two β-galactosidases in the *L. stagnalis* embryo cleaved galactogen. The isotopic labeling of the constituents of the capsule fluid (Goudsmit, '76) combined with *in vitro* culturing of embryos in different components of the capsule fluid could further our understanding of the role embryonic nutrition may play in morphogenesis, differentiation, and growth of the pulmonate embryo. The development of the pulmonate embryo may depend on the normal metabolism of galactogen.

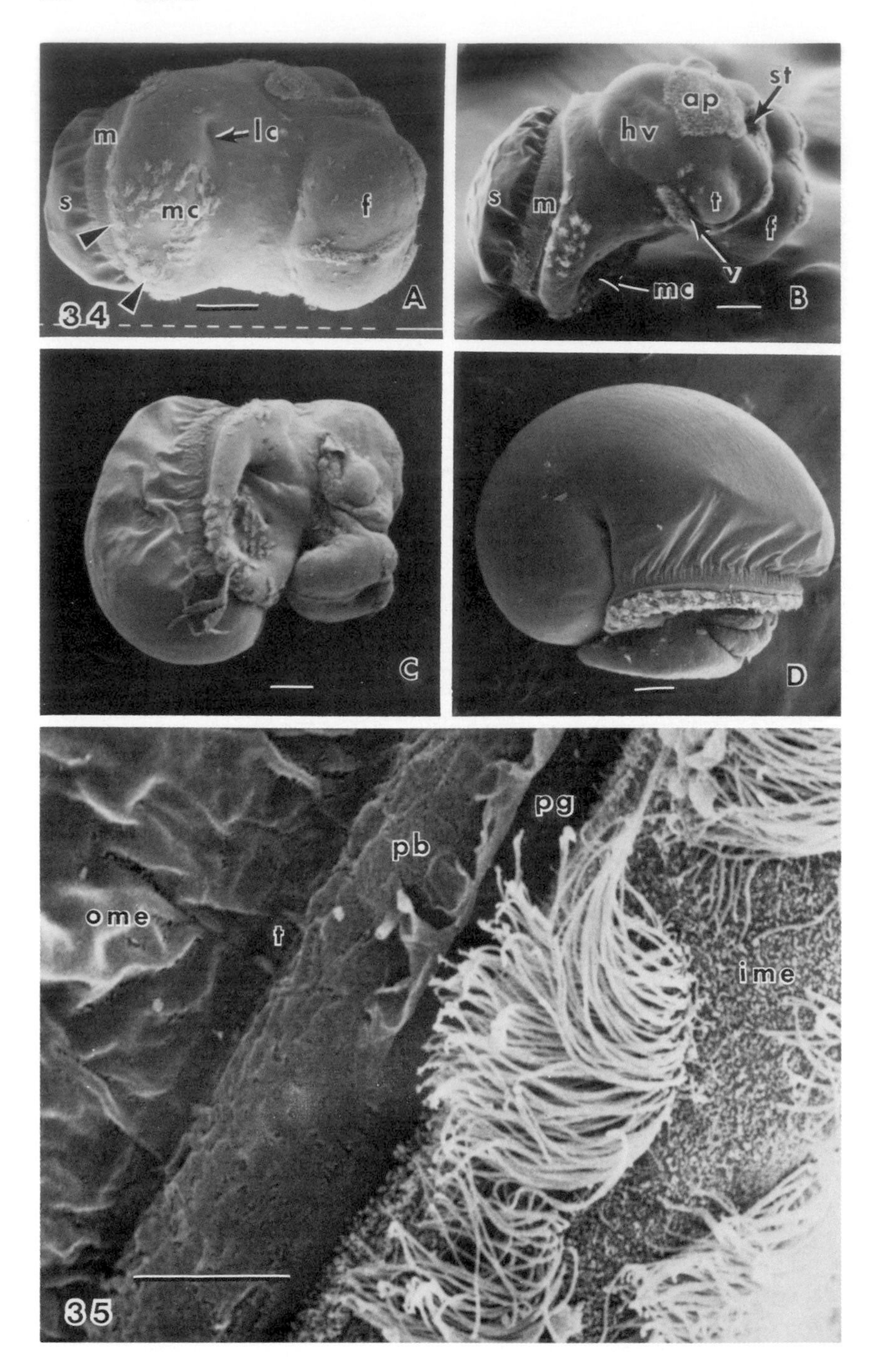
m
lc
s
mc
f
34
A
ap
st
hv
s
m
t
f
v
mc
B
C
D
pg
pb
ome
t
ime
35

MOBILIZATION OF NUTRIENT RESERVES

The two major sources of nutrient reserves are the intraovum yolk platelets and glycogen granules and the extraovum albuminous capsule fluid (Fioroni and Schmekel, '75). The mobilization of the yolk has been studied in detail (Bluemink, '67, '69; Arni, '74; Wal, '76a,b), but it probably is not essential for normal development since the hyaline fragment of the 1-cell stage egg may develop into a normal juvenile snail (Clement, '38). In contrast, the capsule fluid is essential for normal development.

Beginning during the early cleavage stages, capsule fluid is ingested by both micropinocytosis and macropinocytosis at the surfaces of cells exposed to the capsule fluid (Raven, '75). The onset of macropinocytosis of capsule fluid may vary; in *L. palustris* it begins at the 24-cell stage when an occasional conical protrusion 0.4 to 0.5 μm in diameter arises on the outer surfaces of one or more micromeres (Fig. 39A). The albumen vacuole first appears as an elevated ringlet of the cell membrane on the surface of a cell (Fig. 39E,F). In section the ringlet appears as a pair of pseudopodia (Fig. 39B). A network of microfilaments occurs at the base of the ringlet (Fig. 39C). As the ringlet extends outward from the cell, the peripheral edges converge centrally forming a conical protuberance (Fig. 39E). When the margins of the ringlet fuse, the membrane on the luminal side of the protuberance surrounds a droplet of capsule fluid to form the albumen vacuole (Fig. 39D), which migrates from the surface of the cell into the ectoplasmic region (Fig. 39B). Bluemink ('67) and Arni ('74, '75) have described the subsequent fates of the albumen vacuole and its contents.

The macropinocytosis of capsule fluid and the formation of albumen vacuoles occur at the peripheral surfaces of the outer, ectodermal cells of the

---

Fig. 34. SEMs of *L. palustris* during the period of metamorphosis (E4–E7). A. Ventro-lateral view of the right side of a 3.5-day veliger. B. Dorso-lateral view of the right side of a 4-day "hippo" stage. C. 4.5-day embryo viewed from the right side showing the general morphology of the mantle fold, the mantle cavity, head region and foot. D. 5-day juvenile snail viewed from the right side showing initial torsion of the shell. Bars (A–D) 50 μm. ap, apical plate. f, foot. hv, head vesicle. lc, lung cavity primordium. m, mantle. mc, mantle cavity. s, shell. st, stomodeum. t, tentacle. v, velum. arrowheads, row of ciliated cells along presumptive mantle margin.

Fig. 35. SEM micrograph of the margin of the mantle of a 5-day embryo. Bar 10 μm. ime, inner mantle epithelium. ome, outer mantle epithelial region overlain by non-calcified periostracum. pb, periostracal belt covered with newly secreted periostracum. pg, periostracal groove. t, transitional region.

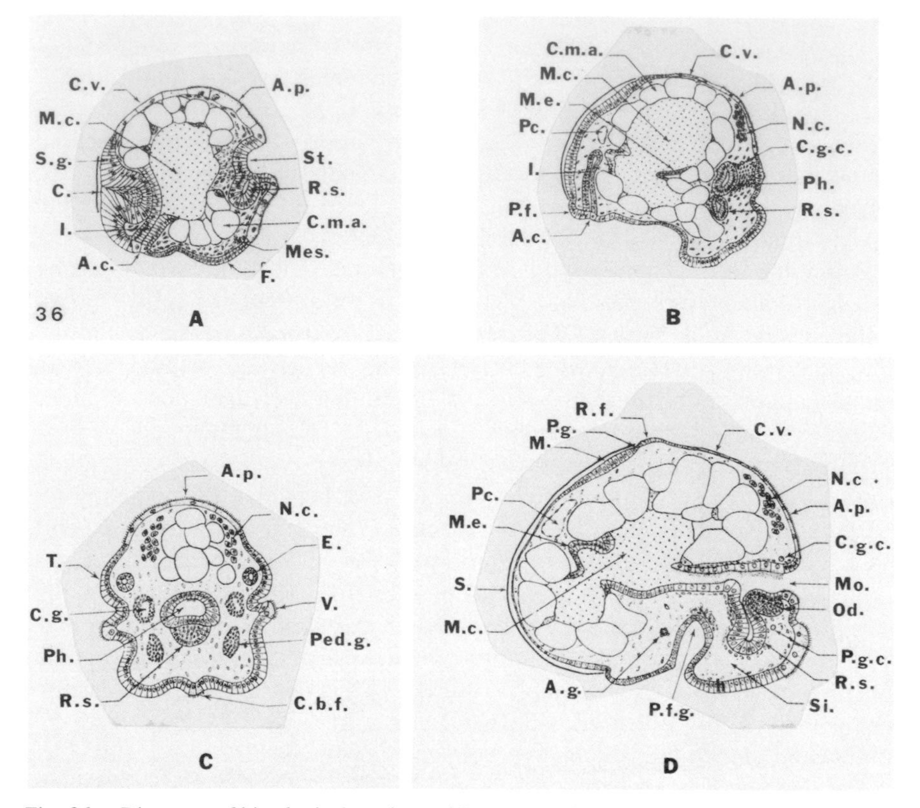

Fig. 36. Diagrams of histological sections of *Lymnaea* embryos. A. 2.5-day embryo, sagittal section. B. 3–3.5-day embryo, sagittal section. C. 3.5–4.0-day embryo, cross-section through head-foot region. D. 4-day embryo, medial sagittal section. Courtesy of Dr. Régondaud ('64). Ac, anal cells. Ag, abdominal ganglion. Ap, apical plate. C, protoconch shell. Cbf, ciliated band of foot. Cg, cerebral ganglion. Cgc, cerebral ganglion commissure. Cma, albumenous larval liver cells. Cv, cephalic vesicle. E, eye. F, foot. I, intestine. M, mantle. Mc, mesenteron cavity. Me, mesenteron endothelium. Mes, mesenchyme. mf, mantle fold. Mo, mouth. Nc, nucal cells. Od, odontophore. Pc, pericardium. Pf, pallial fold. Pfg, posterior foot groove. Pg, pallial groove. Pgc, pedal ganglion commissure. Ped.g, pedal ganglion. Ph, pharynx. Rs, radular sac. Sg, shell gland. Si, hemocoel sinus. T, tentacle. V, velum.

embryo through the early trochophore stage. Macropinocytic activity is particularly pronounced during the gastrula period (Figs. 27A, 30, 39E). With the development of the digestive tract during the trochophore period, the capsule fluid is ingested orally and incorporated by macropinocytosis into the larval liver cells of the midgut (Arni, '75). Other endodermal cells and the ectodermal cells may continue to exhibit micropinocytotic activity (Arni,

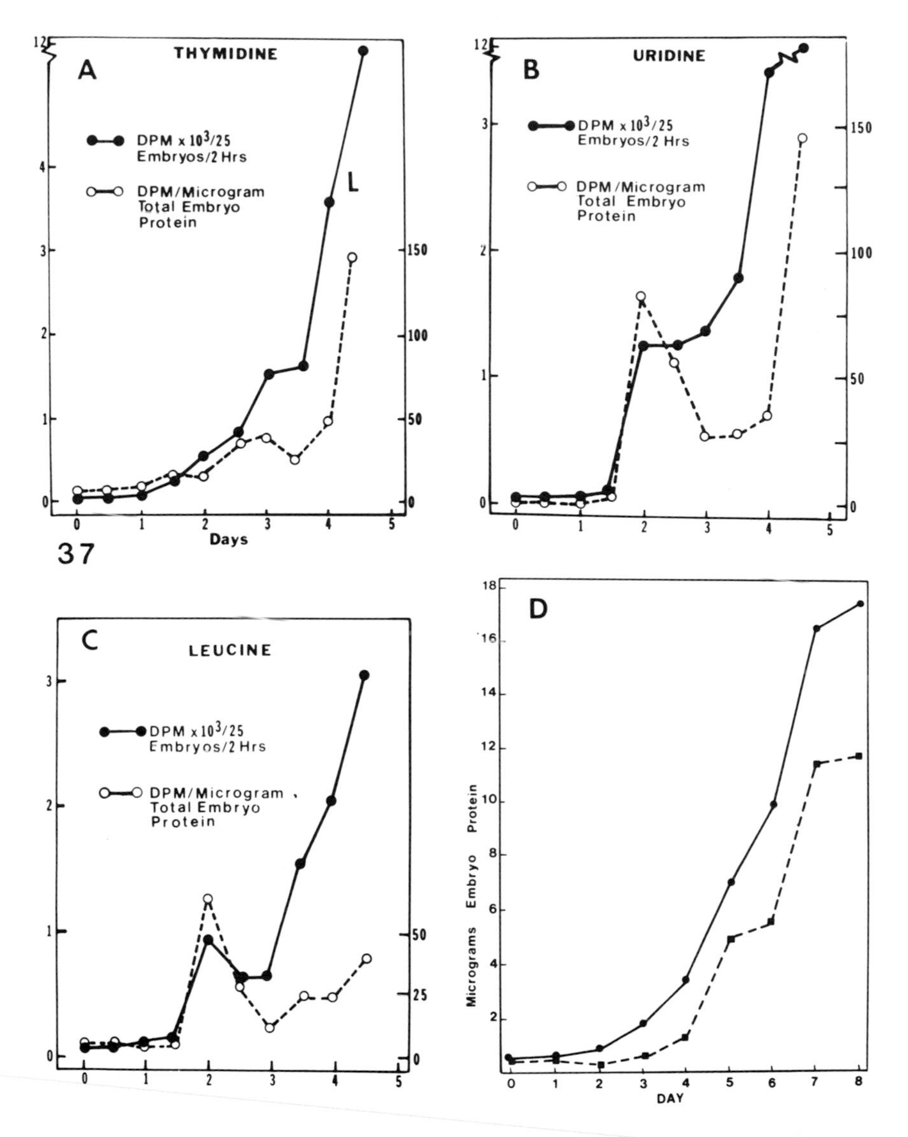

Fig. 37. Nucleic acid and protein synthesis of the *L. palustris* embryo during the first 4 days of development at 24°C. A. incorporation of thymidine H^3 into total DNA. B. Incorporation of uridine H^3 into total RNA. C. Incorporation of leucine H^3 into total protein. Units: x axis, day of development; left y axis, DPM $\times$ 10^3 per 25 embryos per 2 hours; right y axis, DPM per microgram of total embryo protein. D. Comparison between the total embryo protein (μg protein per embryo) (solid line) and total embryo protein less the amount of ingested capsule fluid protein (dashed line) during *L. palustris* development at 25°C.

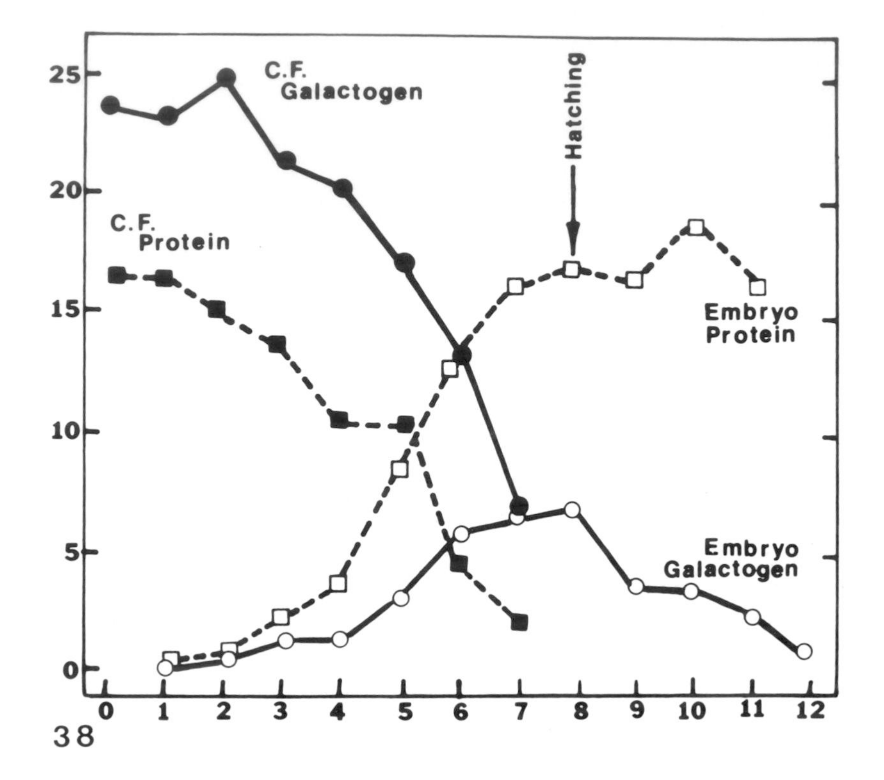

Fig. 38. Total protein and galactogen of embryo and capsule fluid (C.F.) of *L. palustris*. Units: x axis, days of development at 25°C; y axis, micrograms per embryo or capsule fluid per capsule (Morrill et al., '76).

Fig. 39. Micrographs of endocytotic albumen vesicles in *L. palustris*. A. LM of an equatorial section of a 49-cell stage egg showing two endocytotic vesicles (arrowheads) in the process of formation. B. TEM of the cortical region of a micromere of a 24-cell stage egg. C. TEM of a developing endocytotic vesicle of a 24-cell stage egg, GTO fixation. D. TEM of 24-cell stage egg. E. SEM of an animal pole view of a G1 stage gastrula. F. SEM of the central region of the molluscan cross of a G1 stage gastrula showing four phases in the formation of the endocytotic vesicle. Bars (A) 10 μm; (B–D) 1 μm;(E–F) 10 μm. av, albumen vacuole. c, cone phase. e, post enclosion phase. ev, endocytotic vesicle. l, lipid droplets. m, mitochondria. mf, microfilament network. pg, polygranules. r, ringlet phase. rm, ruffled membrane phase. sc, surface coat. vm, vitelline membrane. y, yolk platelet.

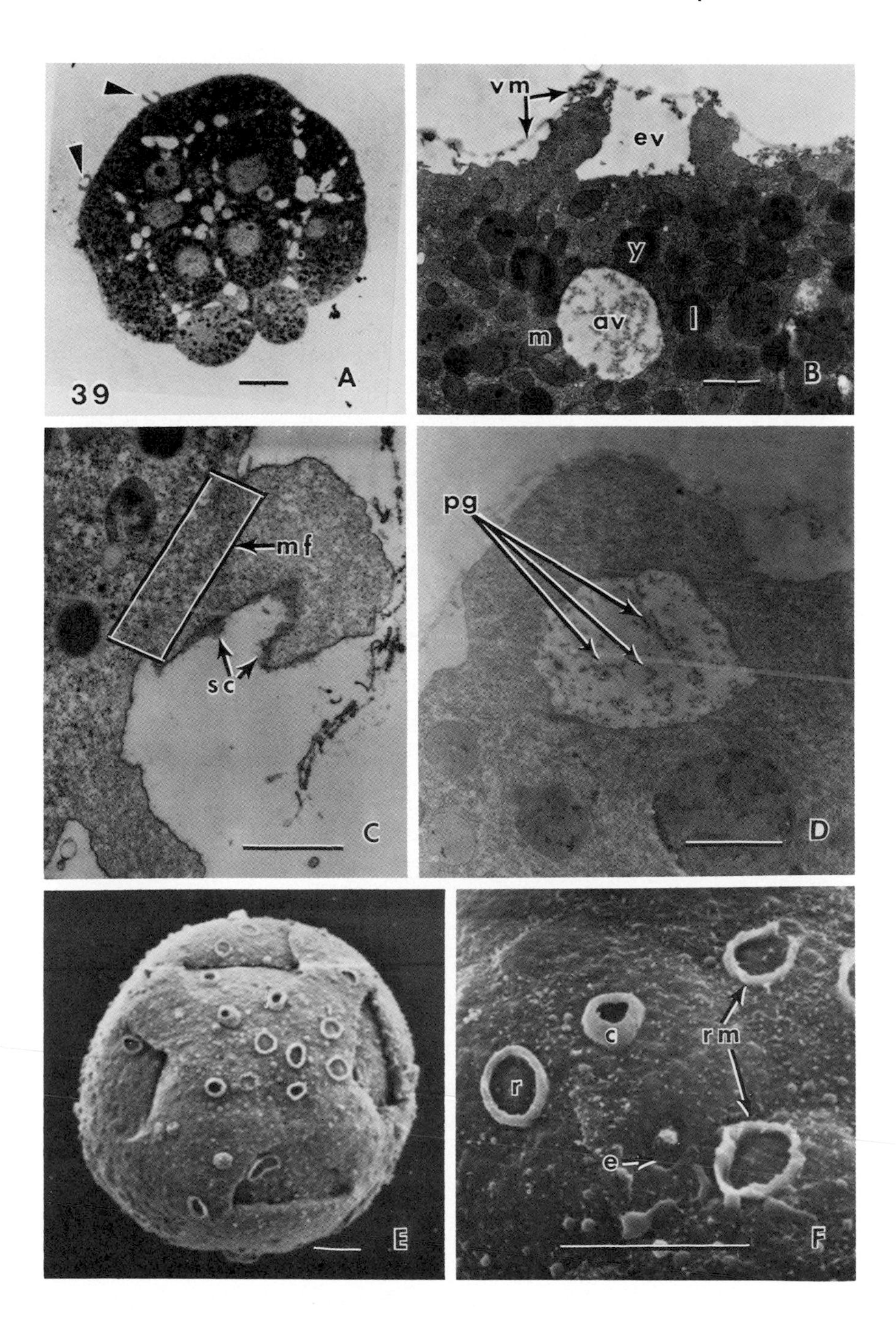
39
A
vm
ev
y
av
l
m
B
mf
sc
C
pg
D
E
c
rm
r
e
F

'74). Thus the frequency of appearance and the distribution of the macropinosomes parallel other morphogenetic events.

There are several aspects of macropinocytosis that deserve further comment. During the formation of the macropinosome the total surface area of the cell increases. Whether this is restricted to the elongating ringlet is not known. Whether the site and initiation of the formation of a ringlet is mediated by the distribution of cell surface "albumen" receptors is not known. Whether macropinocytosis is stimulated extracellularly or intracellulary is not known. Bluemink's ('67) experiments indicate that an albumen vacuole may form in less than 2 minutes. Both the rate and frequency of pinocytotic activity may vary from one cell to the next. Figure 39B shows one cell at the 24-cell stage with an albumen vacuole still in the peripheral ectoplasmic region while a second macropinosome is beginning to form. SEM photographs of older embryos show that two or three ringlets of pinosomes may occur simultaneously on the outer surface of a cell.

In addition to the macropinosomes in pulmonates being peculiar cellular organelles, their formation results in the internalization of the original oolemma during the cleavage period. Thus, morphogenetic determinates that might be associated with the original oolemma may no longer be necessary in particular cells once they begin to form albumen vacuoles.

ABNORMAL DEVELOPMENT

Teratological Experiments

The eggs and embryos of *Lymnaea* and other freshwater pulmonates have been subjected to numerous kinds of physical and experimental treatments to alter the normal course of development. The various treatments may be classified as to their effects on the egg from oviposition through the early cleavage period or their morphogenetic effects on postcleavage period development. The earlier experimental analyses have been reviewed by Raven ('48, '64b, '66, '70) and related to his general cortical field or cortical gradient theory of development. Nearly all the experimentally induced abnormalities are also observed to occur spontaneously in response to unknown environmental conditions. Such "spontaneously" occurring abnormalities have been recorded by Pelseneer ('20) in his treatise on the variations in molluscs.

Prior to 1964 many of the experimental studies on *Lymnaea* failed to pay strict attention to the degree of synchrony of the treated eggs; the specific moment at which the treatment began and ended; or the temperature, tempo, and pattern of cleavages of the treated eggs. Subsequent to Geilenkirchen's ('64b) analysis of the periodic sensitivity of eggs centrifuged at different moments during the cell cycle and Verdonk's ('65) cell lineage analysis of lithium-induced head malformations, the importance of taking into account the temporal and spatial mosaicism of the early embryo with respect to any particular treatment regime has been evident. Thus the morphogenetic effects of centrifugation (Geilenkirchen, '64b), lithium (Geilenkirchen, '67; Biggelaar, '71d), ultraviolet light (Labordus, '70) puromycin (Boon-Niermeijer, '76a), and heat shock (Boon-Niermeijer, '76b) have in each instance been related in part to the lengthening of one or more cleavage cycles relative to the normal events associated with cellular determination during the early cleavage period and possibly the normal positioning of cells prior to and during the 24-cell stage. The third cleavage and the formation of the first quartet of micromeres is a particularly sensitive period with respect to cellular patterning and normal development.

Ordinarily the development of each egg subjected to experimental treatments is recorded at daily intervals until it dies or reaches the early juvenile stage (E6–E7). The various developmental disturbances may be classified as follows according to the convention of Geilenkirchen ('64a).

First period death. Embryos may cleave normally or abnormally and die prior to the ciliated gastrula stage without exhibiting any visible morphological abnormality.

Second period death. Embryos develop beyond the gastrula stage and die but without showing specific abnormalities.

Arrested gastrulae. Embryos are arrested at the gastrula stage and may develop into dumbbell-shaped exogastrulae (Fig. 40A,B), irregularly-shaped gastrulae (Fig. 40C), or vesicular, hydropic gastrulae (Fig. 40D). The development of exogastrulae has been analyzed by Raven ('52, '66).

Unspecific malformations. This group includes those embryos whose development is arrested at the trochophore or veliger stage (E2–E4) but do not exhibit visible head, shell, or foot malformations. Frequently in these arrested embryos the body becomes hydropic; the larval liver cells regress; and the protonephridial cells contain white deposits (Fig. 40E). The foot or head may also become hydropic; in some cases two or all three regions of the embryo may become hydropic (Fig. 40F). Hydropia most probably results from the impairment of the metabolically dependent mechanisms for controlling the fluid in the extracellular spaces (Beadle, '69). However, hydropia may also result from the abnormal development and function of the primary mesoderm cells derived from the M_1 and M_2 cells and the secondary mesoderm cells derived from the $3a^{21}$, $3a^{22}$, $3b^{21}$, and $3b^{22}$ cells (Morrill, '63a).

Head malformations. This group includes embryos in which the eyes and tentacles are displaced, reduplicated (Fig. 40G), fused, or reduced (Raven, '49, '66).

Shell malformations. This group includes embryos in which the shell rudiment may be lacking or limited to the area of the original shell gland; the

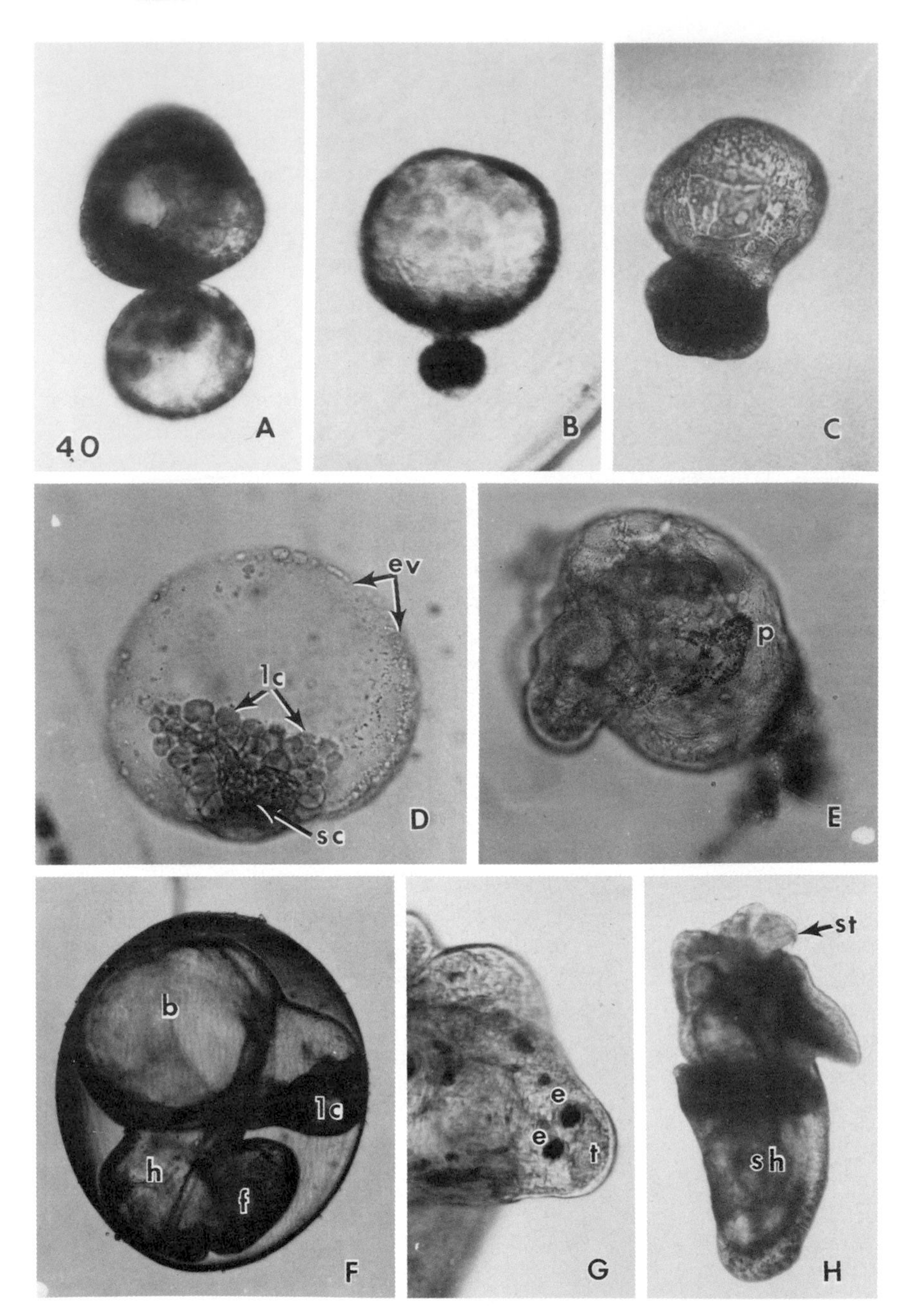
40
A
B
C
ev
lc
sc
D
p
E
b
lc
h
f
F
e
e
t
G
st
sh
H

shell is everted, flattened, helmet-shaped, or cone-shaped (Fig. 40H). Raven ('52) has analyzed the abnormal morphogenesis.

Foot malformations. This group includes embryos with an asymmetric foot of which only one side develops normally, a supernumerary foot, a split foot in which each lateral foot primordium develops into a foot, or a foot that is oriented at an angle relative to the midline of the embryo.

Stomodeum malformations. The foregut may evert either asymmetrically or symmetrically (Fig. 40H) to form a conical outgrowth between the head and foot. Raven ('58) has analyzed the development of this malformation.

Normal. Embryos that develop normally to the juvenile snail stage (E7—E8) are classified as normal even though their development may be retarded and their size reduced.

In most of the teratological studies on *Lymnaea,* the percentage of embryos exhibiting a specific morphological abnormality is too low to allow one to sample a population of treated eggs at intervals to determine the causal mechanisms and cellular lineages leading to the abnormality. Only in the case of lithium-treated embryos has a detailed cell lineage analysis been undertaken to explain the origins of the several types of head malformations (Verdonk, '65, '68). The analysis of the cellular patterns of head malformations caused by cell deletions at the 8-cell stage (Cather et al., '76) and by puromycin (Boon-Neirmeijer, '76a) represent a compromise approach to the probable orgins of a particular morphological abnormality once the cellular genealogy is known. Since lithium treatment of early cleavage stage eggs can result in 80% to 100% of the eggs developing into exogastrulae, the analysis of the developmental origins of Li-exogastrulae is a promising subject for future study.

Polyvitelliny

Occasionally two or more eggs may occur in single egg capsules, which are generally at the posterior end of the egg mass in *Lymnaea.* Such eggs may develop independently of one another or else fuse to form conjoined

---

Fig. 40. LM micrographs of common types of developmental abnormalities of *Lymnaea* embryos. A. Dumbbell exogastrula. B. Dumbbell exogastrula with one half swollen or hydropic. C. Arrested irregularly shaped gastrula. D. Hydropic gastrula. E. Arrested, early veliger with hydropic body. F. Swollen hydropic veliger filling the egg capsule. G. An asymmetric triophthalamic embryo with two eyes at the base of the right tentacle. H. 5-day juvenile stage embryo with cone-shaped shell and everted stomoderum. b, hydropic body. e, eye. ev, ectodermal vesicle. f, foot. h, hydropic head. lc, large-celled endoderm. p, protonephridium with intracellular deposits. sc, small-celled endoderm. sh, shell. st, everted stomodeum. t, tentacle.

monsters (Pelseneer, '20; Crabb, '31; Winsor and Winsor, '32; George, '58). The most developmentally interesting monsters are those produced by the fusion of only two embryos to form conjoined twins. Fusion occurs only at the gastrula stage between the time of hatching from the vitelline membrane and the active movement of the ciliated, placode gastrula. Although the exact mechanism of fusion is not known, all surfaces of a gastrula except the animal pole region appear capable of fusing with the surfaces of a second gastrula. If the anterior–posterior axes of the fused embryos are reversed the two embryos tend to develop independently of one another. However, if the two embryos fuse with their anterior–posterior axes in register, then one of several types of monsters may result. The conjoined snail may have two feet and two heads but only one body covered by a single mantle and shell; or it may have two bodies, two shells, and two heads but only one foot. Apparently neither the presumptive foot nor the mantle are irreversibly determined at the gastrula stage. Given the possibility of removing the vitelline membrane at earlier stages of development, the experimental fusion of embryos could be a novel way of analyzing cellular determination in the pulmonate egg.

CONCLUDING REMARKS

The numerous morphological analyses of nearly every facet of the development of freshwater pulmonate molluscs, especially species of *Lymnaea,* provide an excellent basis for future correlative LM, TEM, and SEM analyses of the cellular basis of morphogenesis. Because of the development of both intracellular and extracellular patterns, as well as the determination of cell lines during the cleavage period, both normal and abnormal cleavage patterns offer unusual opportunities for analyzing positional information associated with morphogenesis. Future experimental analyses of the development of pulmonates will require renewed attention not only to the correlative temporal and spatial aspects of cell lineages but also the potential morphogenetic roles of intercellular communication, cellular shapes, specication of cellular adhesion, and extracellular matrices.

LITERATURE CITED

Anderson, W.A., and P. Personne (1976) The molluscan spermatzoon: Dynamic aspects of its structure and function. Am. Zool. *16:*293–313.

Archie, V.E. (1941) Histology and developmental histology of the ovotestis of *Lymnaea stagnalis lillianae.* Thesis, University of Wisconsin.

Arni, P. (1973) Vergleichende Untersuchungen an Schlupfstadien von neun Pulmonaten-Arten (Mollusca, Gastopoda). Rev. Suisse Zool. *80:*323–402.

Arni, P. (1974) Licht- und Elektronenmikroskopische Untersuchungen an Embryonen von *Lymnaea stagnalis* L. (Gastropoda, Pulmonta) mit besonderer berucksichtigung der fruhembryonalen ernahrung. Z. Morphol. Tiere. *78:*299–323.

Arni, P. (1975) Licht- und Elektronenmikroskopische Untersuchungen zur Entwicklung und Degeneration transitorischer Speicherzellen der Mitteldarmregion von *Lymnae stagnalis* L. (Gastropoda, Pulmonata). Z. Morphol. Tiere. *81:*221–240.

Arnolds, W.J.A. (1979) Silver staining methods for the demarcation of superficial cell boundaries in whole mounts of embryos. Mikroskopie *35:*202–206.

Bajaj, S.K., K. Rao, and C.M.S. Dass (1973) Synthesis of DNA in snail embryos during the early cleavage. Indian J. Exp. Biol. *11:*277–280.

Baker, F.C. (1911) The Lymnaeidae of North and Middle America, recent and fossil. Special Publ. 3, Chicago Acad. Sci., p. 539.

Basch, P.F., and J.J. DiConza (1973) Primary cultures of embryonic cells from the snail *Biomphalaria glabrata.* Am. J. Trop. Med. Hyg. *22:*805–813.

Basch, P.F., and J.J. DiConza (1974a) *Biomphalaria glabrata:* Procedures for obtaining large numbers of embryos for cell cultures. J. Invertebr. Pathol. *24:*122–123.

Basch, P.F., and J.J DiConza (1974b) *Biomphalaria glabrata:* Growth and differentiation of embryo fragments in extended culture. J. Invertebr. Pathol. *24:*125–126.

Bayne, C.J. (1968) Histochemical studies on the egg capsules of eight gastropod molluscs. Proc. Malac. Soc. Lond. *38:*199–212.

Bayne, C.J. (1976) Culture of molluscan organs: A review. In K. Maramorasch (ed): Invertebrate Tissue Culture, Research Applications. New York: Academic Press, pp. 57–59.

Beadle, L.C. (1969) Salt and water regulation in the embryos of freshwater pulmonate molluscs. J. Exp. Biol. *50:*475–479.

Bezem, J.J., H.A. Wagemaker, and J.A.M. Biggelaar (1981) Relative cell volumes of the blastomeres in embryos of *Lymnaea stagnalis* in relation to bilateral symmetry and dorsoventral polarity. Proc. K. Ned. Akad. Wet. *79:*9–20.

Bezem, J.J., and C.P. Raven (1976) Some algorithms used in the simulation of embryonic development. In A. Lindenmayer and G. Rozenberg (eds): Automata, Languages and Development. Amsterdam: North-Holland Publishing Co., pp. 147–158.

Biggelaar, J.A.M. van den (1971a) RNA synthesis during cleavage of the *Lymnaea* egg. Exp. Cell Res. *67:*207–210.

Biggelaar, J.A.M. van den (1971b) Timing of the phases of the cell cycle with tritiated thymidine and Fuelgen cytophotometry during the period of synchronous division in *Lymnaea.* J. Embryol. Exp. Morphol. *26:*351–366.

Biggelaar, J.A.M. van den (1971c) Timing of the phases of cell cycle during the period of asynchronous division up to the 49-cell stage in *Lymnaea.* J. Embryol. Exp. Morphol. *26:*367–391.

Biggelaar, J.A.M. van den (1971d) Development of division asynchrony and bilateral symmetry in the first quartet of micromeres in eggs of *Lymnaea.* J. Embryol. Exp. Morphol. *26:*393–399.

Biggelaar, J.A.M. van den (1976) Development of dorsoventral polarity preceding the formation of the mesentoblast in *Lymnaea stagnalis.* Proc. K. Ned. Akad. Wet. *79:*111–126.

Biggelaar, J.A.M. van den (1977) Significance of cellular interaction for the differentation of the macromeres prior to the formation of the mesentoblast in *Lymnaea stagnalis.* Proc. K. Ned. Akad. Wet. *80:*1–12.

Biggelaar, J.A.M. van den (1978) The determinative significance of the geometry of the cell contacts in early molluscan development. Biol. Cellularie *32:*151–161.

Biggelaar, J.A.M. van den, and E.K. Boon-Niermeijer (1973) Origin and prospective significance of division asynchrony during early molluscan development. In M. Balls and F.S. Billett (eds): The Cell Cycle in Development and Differentiation. New York: Cambridge University Press, pp. 215–228.

Biggelaar, J.A.M. van den, A.W.C. Dorresteijn, S.W. deLaat, and J.G. Bluemink (1981) The role of topographical factors in cell interaction and determination of cell lines in molluscan development. In H.G. Schweiger (ed): International Cell Biology. Berlin: Springer-Verlag, pp. 526–538.

Bluemink, J.G. (1967) The subcellular structure of the blastula of *Lymnaea stagnalis* L. (Mollusca) and the mobilization of the nutrient reserve. Thesis, University of Utrecht.

Bluemink J.G. (1969) Are yolk granules related to lysosomes. Zeiss Inf. *73:*95–99.

Bolognari, A., and M.P.A. Carmignani (1976) Aspetti del nucleolo scondo considerazion: Antiche e recenti. Riv. Biol. Norm. Patol. *2:*27–68.

Bondesen, P. (1950) A comparative morphological-biological analysis of the egg capsules of freshwater pulmonate gastropods. Hydrophila, Basommatophora, Pulmonata, Natura Jutl. *3:*1–201.

Boon-Niermeijer, E.K. (1974) The effect of puromycin on the progression of the cleavage cycle of *Lymnaea stagnalis.* Proc. K. Ned. Akad. Wet. *77:*16–28.

Boon-Niermeijer, E.K. (1975) The effect of puromycin on the early cleavage cycles and morphogenesis of the pond snail *Lymnaea stagnalis.* Wilhelm Roux Arch. Dev. Biol. *177:*29–40.

Boon-Niermeijer, E.K. (1976a) Head malformations arising after puromycin treatment in *Lymnaea stagnalis.* Proc. K. Ned. Akad. Wet. *79:*34–41.

Boon-Niermeijer, E.K. (1976b) Morphogenesis after heat shock during the cell cycle of *Lymnaea.* A new interpretation. Wilhelm Roux Arch. Dev. Biol. *180:*241–252.

Boycott, A.E., D. Diver, S.L. Garstang, and F.M. Turner (1931) The inheritance of sinistrality in *Limnaea peregra* (Mollusca, Pulmonata). Phil. Trans. R. Soc. B *219:*51–131.

Brahmachary, R.L. (1973) Molecular embryology of invertebrates. In M. Abercrombie, J. Brachet, and T.J. Kind (eds): Advances in Morphogenesis. New York: Academic Press, pp. 115–173.

Brahmachary, R.L., and D. Ghosal (1976) Sulfated macromolecules in early embryos of *Limnaea.* Z. Naturfursch. *31:*488.

Brahmachary, R.L., D. Ghosal, and P.K. Tapaswi (1971a) Rhythmic incorporation of P^{32} and C^{14}-uracil in early mitotic cycles in *Limnaea* (Mollusc) eggs. Z. Naturfursch. *26:*822–824.

Brahmachary, R.L., D. Ghosal, P.K. Tapaswi, and T.K. Basu (1971b) Rhythmic incorporation of ^{35}S during early embryonic mitosis in *Limnaea* (Mollusc). Exp. Cell Res. *65:*325–328.

Bretschneider, L.H. (1948a) Insemination in *Limnaea stagnalis* L. Proc. K. Ned. Akad. Wet. *51:*357–362.

Bretschneider, L.H. (1948b) The mechanism of oviposition in *Limnaea stagnalis.* Proc. K. Ned. Akad. Wet. *51:*616–626.

Bretschneider, L.H., and C.P. Raven (1951) Structural and topochemical changes in the egg cells of *Limnaea stagnalis* L. during oogenesis. Arch. Neerl. Zool. *10:*1–31.

Brisson, P., and C. Besse (1975) Etude ultrastructurale de l'ébauche gonadique chez l'embryon de *Lymnaea stagnalis* L. (Gasteropode Pulmone Basommatophore). Bull. Soc. Zool. France *100:*345–349.

Brisson, P., and J. Régondaud (1977) Origine et structure de l'ébauche de la gonade chez les gasteropodes pulmones basommatophores. Malacologia *16:*457–466.

Camey, T., and N.H. Verdonk (1970) The early development of the snail *Biomphalaria glabrata* (Say) and the origin of the head organs. Neth. J. Zool. *20:*93–121.

Carriker, M.R. (1946) Morphology of the alimentary system of the snail (*Lymnaea stagnalis appressa* say). Trans. Wisconsin Acad. Sci. *29:*209–236.

Cather, J.N., N.H. Verdonk, and G. Zwaan (1976) Cellular interactions in the early development of the gastropod eye, as determined by deletion experiments. Malacol. Rev. *9:*77–84.

Clement, A.C. (1938) The structure and development of centrifuged eggs and egg fragments of *Physa heterostropha.* J. Exp. Zool. *79:*435–460.

Comadon, J., and P. deFonbrune (1935) Recherches effectuées aux premiers stades du développement d'oeufs de Gasteropodes et d'un ver à l'aide de la Cinématographie. Arch. Anat. Microsc. *31:*79–100.

Conway, A.F., R.E. Black, and J.B. Morrill (1969) Uric acid synthesis in embryos of the pulmonate pond snail, *Limnaea palustris:* Evidence for a unique pathway. Comp. Biochem. Physiol. *30:*793–802.

Crabb, E.D. (1927) The fertilization process in the snail *Lymnaea stagnalis appressa* Say. Biol. Bull. *53:*67–109.

Crabb, E.D. (1931) The origin of independent and of conjoined twins in fresh water snails. Wilhelm Roux' Arch. Entwicklungsmech. Org. *124:*332–356.

Cumin, R. (1972) Normentafel zur Organogenese von *Limnaea stagnalis* (Gastropoda, Pulmonata) mit besonderer berucksichtigung der mitteldarmdruse. Rev. Suisse Zool. *79:*709–774.

Dan, J.C., and S. Takaichi (1979) Spermiogenesis in the pulmonate snail, *Euhadra hickonis.* III. Flagellum formation. Dev. Growth Differ. *21:*71–86.

Dorresteijn, A.W.C., J.A.M. Biggelaar, J.C. Bluemink, and W.J. Hage (1981) Electron microscopical investigations of the intercellular contacts during the early cleavage stages of *Lymnaea stagnalis* (Mollusca, Gastropoda). Wilhelm Roux' Arch. Dev. Biol. *190:*215–220.

Duncan, C.J. (1959) The life cycle and ecology of the freshwater snail *Physa fontinalis* (L.). J. Anim. Ecol. *28:*97–117.

Duncan, C.J. (1975) Reproduction. In V. Fretter and J. Peake (eds): Pulmonates, Vol. 1. Functional Anatomy and Physiology. New York: Academic Press, pp. 309–365.

Elbers, P.F. (1969) The primary action of lithium chloride on morphogenesis in *Lymnaea stagnalis.* J. Embryol. Exp. Morphol. *22:*449–463.

Fioroni, P. (1966) Zur morphologie und embryogenese des darmtraktes und der transitorischen organe bei Prosobranchiern (Mollusca, Gastropoda). Rev. Suisse Zool. *73:*621–876.

Fioroni, P., and L. Schmekel (1975) Development and habitat dependence in gastropods—An ontogenetic comparison. Forma Functio *8:*209–252.

Fioroni, P., and L. Schmekel (1976) Ontogeneses of gastropods with rich nutritive resources. Zool. Jb. Anat. Bd. *96:*74–171.

Flandre, O. (1972) Cell culture of molluscks. In E.C. Vago (ed): Invertebrate Tissue Culture. New York: Academic Press, pp. 361–383.

Fleitz, V.H., and H.J. Horstmann (1967) Über das native Galaktogen aus den Eirn der Schlammschnecke *Lymnaea stagnalis.* Hopper-Seyler's Z. Physiol. Chem. *348:*1301–1306.

Fol, H. (1879) Etudes sur le développement des Mollusques. Troisième mémoire: Sur le développement des Gasteropodes pulmones. Arch. Zool. *8:*103–222.

Forbes, G.S., and H.E. Crampton (1942) The differentiation of geographical groups in *Lymnaea palustris.* Biol. Bull. *82:*26–46.

Fraser, L.A. (1946) The embryology of the reproductive tract of *Lymnaea stagnalis appressa* Say. Trans. Am. Microsc. Soc. *65:*279–298.

Freeman, G. (1977) The transformation of the sinistral form of the snail *Lymnaea peregra* into dextral form. Am. Zool. *17:*946 (abstract).

Friedl, F.E. (1964) A method for securing the snail, *Lymnaea stagnalis jugularis* (Say), free from bacteria. Exp. Parasitol. *15:*7–13.

Galtsoff, P.S., F.E. Lutz, P.S. Welch, and J.G. Needham (1937) Culture Methods for Invertebrate Animals. New York: Comstock Publishing Co.

Geilenkirchen, W.L.M. (1964a) The action and interaction of calcium and alkali chlorides on eggs of *Limnaea stagnalis* and their chemical interpretation. Exp. Cell Res. *34:*463–487.

Geilenkirchen, W.L.M. (1964b) Periodic sensitivity of mechanisms of cytodifferentiation in cleaving eggs of *Limnaea stagnalis.* J. Embryol. Exp. Morphol. *12:*183–195.

Geilenkirchen, W.L.M. (1967) Programming of gastrulation during the second cleavage cycle in *Limnaea stagnalis:* A study with lithium chloride and actinomycin D. J.Embryol. Exp. Morphol. *17:*367–374.

George, J.C. (1958) Experimental fusion of embryos of *Limnaea stagnalis* L. Proc. K. Ned. Acad. Wet. *61:*595–597.

Geraerts, W.P.M., and L.H. Algera (1976) The stimulating effect of the dorsal-body hormone on cell differentiation in the female accessory sex organs of the hermaphrodite freshwater snail, *Lymnaea stagnalis.* Gen. Comp. Endocrinol. *29:*109–118.

Geraerts, W.P.M., and S. Bohlken (1976) The control of ovulation in the hermaphroditic freshwater snail *Lymnaea stagnalis* by the neurohormone of the caudodorsal cells. Gen. Comp. Endocrinol. *28:*350–357.

Geraerts, W.P.M., and S. Bohlken (1978) The endocrine control of reproduction in the hermaphrodite pulmonate snail *Lymnaea stagnalis.* In P.J. Gaillard and H.H. Boer (eds): Comparative Endocrinology. Amsterdam: Elsevier/North-Holland, pp. 21–24.

Geraerts, W.P.M., and J. Joosse (1975) Control of vitellogenesis and of growth of female accessory sex organs by the dorsal body hormone (DBH) in the hermaphroditic freshwater snail *Lymnaea stagnalis.*Gen. Comp. Endocrinol. *27:*450–464.

Geraerts, W.P.M., and A.M. Mohamed (1981) Studies on the role of the lateral lobes and the ovotestis of the pulmonate snail *Bulinus truncatus* in the control of body growth and reproduction. Int. J. Invert. Reprod. *3:*297–308.

Geuze, J.J. (1968) Observations on the function and the structure of the statocysts of *Lymnaea stagnalis* (L.). Neth. J. Zool. *18:*155–204.

Gomot, L. (1977) Invertebrate organ culture media (other than insects). In M. Recheigl, Jr. (ed): CRC Handbook Series in Nutrition and Food, Section G: Vol. IV. Culture media for cells, organs and embryos. Cleveland: CRC Press, pp. 121–170.

Goudsmit, E.M. (1964) The metabolism of galactogen and glycogen by the pulmonate snails, *Bulimnaea megasoma* and *Helix pomatia.* Thesis, University of Michigan.

Goudsmit, E.M. (1972) Carbohydrates and carbohydrate metabolism in Mollusca. In M. Florkin and B.T. Scheer (eds): Chemical Zoology. New York: Academic Press, pp. 219–243.

Goudsmit, E.M. (1976) Galactogen catabolism by embryos of the freshwater snails, *Bulimnaea megasoma* and *Lymnaea stagnalis* . Comp. Biochem. Physiol. *53:*439–442.

Guerrier, P., (1971) La polarisation cellulaire et les caractères de la segmentation au cours de la morphogénèse spirale (Gasteropodes pulmones, Lamellibranches, Annélides polychètes). Ann. Biol. *10:*152–192.

Guerrier, P., and J.A.M. van den Biggelaar (1979) Intracellular activation and cell interactions in so-called mosaic embryos. In N. Le Dourain (ed): Cell Lineage, Stem Cells and Cell Determination. New York: Elsevier/North-Holland, pp. 29–36.

Hartung, E.W. (1947) Cytological and experimental studies on the oocytes of freshwater pulmonates. Biol. Bull. *92:*10–22.

Holmes, S.J. (1900) The early development of *Planorbis*. J. Morphol. *16:*369–458.

Horstmann, H.J. (1956) Der Galaktogengehalt der Eier von *Lymnaea stagnalis* L. während der Embryonalentwicklung. Biochem. Z. *328:*342–347.

Horstmann, H.J. (1958) Sauererstoffuerbranch und Trockengewicht der Embryonen von *Lymnaea stagnalis* L. Z. Vgl. Physiol. *41:*390–404.

Horstmann, H.J. (1964) Stoffwechsel während der Embryonal-und Jugendentwicklung der Lungenschnecken. Helgol. Wiss. Meeresunters. *9:*336–343.

Hubendick, B. (1951) Recent Lymnaeidae, their variation, morphology, taxonomy, nomenclature and distribution. K. Sven. Vetenskapsakad. Handl. *3:*1–223.

Hubendick, B. (1978) Systematics and comparative morphology of the basommatophora. In V. Fretter and J. Peake (eds): Pulmonates: Systematics, Evolution and Ecology. New York: Academic Press, pp. 1–49.

Hudig, O. (1946) The vitelline membrane of *Limnaea stagnalis*. Proc. K. Ned. Akad. Wet. *49:*554–564.

Inaba, A. (1969) Cytotaxonomic studies of lymnaeid snails. Malacologia *7:*143–168.

Jockusch, B. (1968) Protein synthesis during the first three cleavages in pond snail eggs (*Lymnaea stagnalis*). Z. Naturforschg. *23:*1512–1516.

Jong-Brink, M. de, H.H. Boer, J.G. Hommes, and A. Kodde (1977) Spermatogenesis and the role of Sertoli cells in the freshwater snail *Biomphalaria glabrata*. Cell Tiss. Res. *181:*37–58.

Jong-Brink, M. de J.P. ter Borg, M.J.M. Bergamin-Sassen, and H.H. Boer (1979) Histology and histochemistry of the reproductive tract of the pulmonate snail, *Bulinus truncatus*, with observations on the effects of castration on its growth and histology. Int. J. Invert. Reprod. *1:*41–56.

Jong-Brink, M. de, A. de Wit, G. Kraal, and H.H. Boer (1976) A light and electron microscope study on oogenesis in the freshwater pulmonate snail *Biomphalaria glabrata*. Cell Tiss. Res. *171:*195–219.

Joosse, J., M.H. Boer, and C.J. Cornelisse (1968) Gametogenesis and oviposition in *Lymnaea stagnalis* as influenced by uv-irradiation and hunger. Symp. Zool. Soc. Lond. *22:*213–235.

Joosse, J., and D. Reitz (1969) Functional anatomical aspects of the ovotestis of *Lymnaea stagnalis*. Malacologia *9:*101–109.

Kerth, K. (1979) Phylogenetische Aspekte der Radulamorphogenese von Gastropoden. Malacologia *19:*103–108.

Kielbówna, L., and B. Koscielski (1974) A cytochemical and autoradiographic study of oocyte nucleoli in *Limnaea stagnalis* L. Cell Tiss. Res. *152:*103–111.

Kniprath, E. (1977) Zur ontogenese des schalenfeldes von *Lymnaea stagnalis*. Wilhelm Roux' Arch. Dev. Biol. *181:*11–30.

Kniprath, E. (1979) The functional morphology of the embryonic shell-gland in the conchiferous molluscs. Malacologia *18:*549–552.

Kniprath, E. (1981) Ontogeny of the molluscan shell field: A review. Zool. Scr. *10:*61–79.

Kruglyanskaya, Z.Y., and D.A. Sakharov (1975) Appearance of biogenic amines in the developing nervous system of embryos of the mollusk *Lymnaea stagnalis*. Sov. J. Dev. Biol. *6:*160–162.

Kubota, T. (1954) The development of the egg of *Lymnaea pervia* v. Martens. Kagoshima Daigaku Bunrigakubu Reka Hoboku *3:*61–73.

Labordus, V. (1970) The effect of ultraviolet light on developing eggs of *Lymnaea stagnalis* (Mollusca, Pulmonata). IV. The interference of irradiations with morphogenesis. Proc. K. Ned. Akad. Wet. *73:*477–493.

Lankester, E.R. (1874) Observations on the development of the pond snail (*Lymnaea stagnalis*), and on the early stages of other Mollusca. Q.J. Microsc. Sci. *14:*365–391.

Lever, J., J.C. Jager, and A. Westerveld (1964) A new anaesthetization technique for freshwater snails, tested on *Lymnaea stagnalis* . Malacologia *1:*331–337.

Luchtel, D.L. (1972) Gonadal development and sex determination in pulmonate molluscs. I. *Arion circumscriptus.* Z. Zellforsch. *130:*279–301.

Luchtel, D.L. (1976) An ultrastructural study of the egg and early cleavage stages of *Lymnaea stagnalis,* a pulmonate mollusc. Am. Zool. *16:*406–419.

Martoja, M. (1964) Développement de l'appareil reproducteur chez les gasteropodes pulmones. Ann. Biol. *3:*14–232.

Maxwell, W.L. (1977) Freeze-etching studies of pulmonate spermatozoa. Veliger *20:*71–72.

McCraw, B.M. (1957) Studies on the anatomy of *Lymnaea humilis* Say. Can. J. Zool. *35:*751–768.

McCraw, B.M. (1958) Relaxation of snails before fixation. Nature *181:*575.

McCraw, B.M. (1970) Aspects of the growth of the snail *Lymnaea palustris* (Muller). Malacologia *10:*399–413.

Meisenheimer, J. (1899) Zur Morphologie der Urniere der Pulmonaten. Z. Wiss. Zool. *65:*709–724.

Meshcheryakov, V.N. (1978a) Dynamics of the cell surface during early cleavage and its relation to the polarity of the zygote in gastropoda. Z. Obshch. Biol. *39:*916–926.

Meshcheryakov, V.N. (1978b) Orientation of cleavage spindles in pulmonate mollusks. Part I. Role of blastomere shape in orientation of second division spindles. Sov. J. Dev. Biol. *9:*487–493.

Meshcheryakov, V.N. (1979) Orientation of cleavage spindles in pulmonate mollusks. II. Role of structure of intercellular contacts in orientation of the third and fourth cleavage spindles. Sov. J. Dev. Biol. *9:*495–501.

Meshcheryakov, V.N. (1981) Isolation of the egg cortical layer in pulmonate molluscs. Ontogenez *12:*177–186.

Meshcheryakov, V.N., and L.V. Beloussov (1973) Effect of trypsin on spatial organization of early cleavage of the mollusks *Limnaea stagnalis* L. and *Physa fontinalis* L. Ontogenez *4:*359–372.

Meshcheryakov, V.N., and L.V. Beloussov (1975) Asymmetrical rotations of blastomeres in early cleavage of gastropoda. Wilhelm Roux's Arch. Dev. Biol. *177:*193–203.

Meshcheryakov, V.N., and L.G. Filatova (1981) The molluscan egg ghosts as a mechanochemical model of cytokinesis. I. ATP-dependent contractile processes in the ghosts treated with fragments of myosin from rabbit skeletal muscles. Tsitol. Akad. Nauk SSSR *23:*305–311.

Meshcheryakov, V.N., and G.V. Veryasova (1979a) Orientation of cleavage spindles in pulmonate mollusks. Sov. J. Dev. Biol. *10:*22–32.

Meshcheryakov, V.N., and G.V. Veryasova (1979b) Orientation of cleavage spindles in pulmonate molluscs. III. Form and localization of mitotic apparatus in binuclear zygotes and blastomeres. Ontogenez *10:*24–35.

Meshcheryakov, V.N., and I.A. Vorobeva (1973) Preliminary results on the membrane potential of the zygote of the mollusc *Physa actua* in normal eggs and under the action of urea. Biological Sciences (Russian) *6:*47–50.

Mooij-Vogelaar, J.W., J.C. Jager, and W.J. van der Steen (1970) The effect of density changes on the reproduction of the pond snail *Lymnaea stagnalis* L. Neth. J. Zool. *20:*279–288.

Moon, R.T., and J.B. Morrill (1979) Further studies on the electophoretically mobile acid phosphatases of the developing embryo of *Lymnaea palustris* (Gastropoda, Pulmonata). Acta Embryol. Exp. *1:*3–15.

Morrill, J.B. (1963a) Morphological effects of cobaltous chloride in the development of *Limnaea stagnalis* and *Limnaea palustris*. Biol. Bull. *125:*508–522.

Morrill, J.B. (1963b) Development of centrifuged *Limnaea stagnalis* eggs with giant polar bodies. Exp. Cell Res. *31:*490–498.

Morrill, J.B. (1964) The development of fragments of *Limnaea stagnalis* eggs centrifuged before second cleavage. Acta Embryol. Morphol. Exp. *7:*5–20.

Morrill, J.B., C.A. Blair, and W.J. Larsen (1973) Regulative development in the pulmonate gastropod, *Lymnaea palustris,* as determined by blastomere deletion experiments. J. Exp. Zool. *183:*47–55.

Morrill, J.B., and L.E. Macey (1979) Cell surface changes during the early cleavages of the pulmonate snail, *Lymnaea palustris*. Amer. Zool. *19:*902 (abstract).

Morrill, J.B., E. Norris, and S.D. Smith (1964) Electro- and immunoelectrophoretic patterns of egg albumen of the pond snail *Limnaea palustris*. Acta Embryol. Morphol. Exp. *7:*155–166.

Morrill, J.B., and F.O. Perkins (1973) Microtubules in the cortical region of the egg of *Lymnaea* during cortical segregation. Dev. Biol. *33:*206–212.

Morrill, J.B., R.W. Rubin, and M. Grandi (1976) Protein synthesis and differentiation during pulmonate development. Am. Zool. *16:*547–561.

Mulherkar, L., S.C. Goel, and M.V. Joshi (1977) Effects of cytochalasin H on the cleaving eggs of *Limnaea acuminata*. Indian J. Exp. Biol. *15:*1089–1093.

Noland, L.E., and M.R. Carriker (1946) Observations on the biology of the snail *Lymnaea stagnalis apressa* during twenty generations in laboratory culture. Am. Midl. Nat. *36:*467–493.

Patterson, C.M., and J.B. Burch (1978) Chromosomes of pulmonate molluscs. In V. Fretter and J. Peake (eds): Pulmonates: Systematics, Evolution and Ecology. New York: Academic Press, pp. 1–49.

Pelseneer, P. (1920) Les variations et leur herdite chez mollusques. Mem. Ac. Roy. Belg., Clin. Sci. (in-80, 2 ser.) *5:*1–314.

Plesh, B., M. de Jong-Brink, and H.H. Boer (1971) Histological and histochemical observations on the reproductive tract of the hermaphrodite pond snail *Lymnaea stagnalis*. Neth. J. Zool. *21:*180–201.

Rabl, C. (1879) Ueber die Entwicklung der Tellerschnecke. Gegenbaurs Morphol. Jahrb. *5:*562–660.

Raven, C.P. (1945) The development of the egg of *Limnaea stagnalis* L. from oviposition till first cleavage. Arch. Neerl. Zool. *7:*91–121.

Raven, C.P. (1946) The development of the egg of *Limnaea stagnalis* L. from the first cleavage till the trochophore stage, with special reference to its "chemical embryology." Arch Neerl. Zool. *8:*353–434.

Raven, C.P. (1948) The chemical and experimental embryology of *Limnaea*. Biol. Reviews *23:*333–369.

Raven, C.P. (1949) On the structure of cyclopic, synophthalmic and anophthalmic embryos obtained by the action of lithium in *Limnaea stagnalis*. Arch. Neerl. Zool. *8:*1–32.

Raven, C.P. (1952) Morphogenesis in *Limnaea stagnalis* and its disturbance by lithium. J. Exp. Zool. *121:*1–78.

Raven, C.P. (1958) Abnormal development of the foregut in *Limnaea stagnalis*. J. Exp. Zool. *139:*189–245.

Raven, C.P. (1961) Oogenesis: The Storage of Developmental Information. New York: Pergamon Press.

Raven, C.P. (1963) The nature and origin of the cortical morphogenetic field in *Limnaea*. Dev. Biol. *7:*130–143.

Raven, C.P. (1964a) The formation of the second maturation spindle in the eggs of various Limnaeidae. J. Embryol. Exp. Morphol. *12:*805–823.

Raven, C.P. (1964b) Development. In K.M. Wilbur and C.M. Yonge (eds): Physiology of Mollusca. Vol. I. New York: Academic Press, pp. 165–195.

Raven, C.P. (1966) Morphogenesis: The Analysis of Molluscan Development. Oxford: Pergamon Press.

Raven, C.P. (1967) The distribution of special cyytoplasmic differentiations of the egg during early cleavage in *Limnaea stagnalis.* Dev. Biol. *16:*407–437.

Raven, C.P. (1970) The cortical and subcortical cytoplasm of the *Lymnaea* egg. Int. Rev. Cytol. *22:*1–44.

Raven, C.P. (1972a) Determination of the direction of spiral coiling in *Lymnaea peregra.* Acta Morphol. Neerl. Scand. *10:*165–178.

Raven, C.P. (1972b) Chemical embryology of mollusca. In M. Florkin and B.T. Scheer (eds): Chemical Zoology, Vol. VII: Mollusca. New York: Academic Press, pp. 155–185.

Raven, C.P. (1974) Further observations on the distribution of cytoplasmic substances among the cleavage cells of *Lymnaea stagnalis.* J. Embryol. Exp. Morphol. *31:*37–59.

Raven, C.P. (1975) Development. In V. Fretter and J. Peake (eds): Pulmonates: Functional Anatomy and Physiology. New York: Academic Press, pp. 367–400.

Raven, C.P. (1976) Morphogenetic analysis of spiralian development. Am. Zool. *16:*397–403.

Raven, C.P., and J.J. Bezem (1976) Analysis of pattern formation in gastropods by means of computer simulation. In A. Lindenmayer and G. Rozenberg (eds): Automata, Languages and Development. New York: North-Holland Publishing Co., pp. 139–158.

Raven, C.P., and L.H. Bretschneider (1942) The effect of centrifugal force upon the eggs of *Limnaea stagnalis* L. Arch. Neerl. Zool. *6:*255–278.

Raven, C.P., and F. Brunnekreeft (1951) The formation of the animal pole plasm in centrifuged eggs of *Limnaea stagnalis* L. Proc. K. Ned. Akad. Wet. *54:*440–450.

Raven, C.P., F.C.M. Escher, W.M. Herrebrut, and J.A. Leussink (1958) The formation of the second maturation spindle in the eggs of *Limnaea, Limax,* and *Agriolimax.* J. Embryol. Exp. Morphol. *6:*28–51.

Raven, C.P., and U.P. van der Wal (1964) Analysis of the formation of the animal pole plasm in the eggs of *Limnaea stagnalis.* J. Embryol. Exp. Morphol. *12:*123–139.

Recourt, A. (1961) Elektronenmicroscopish onderzoek naar de oogenese bij *Limnaea stagnalis* L. Thesis, Utrecht.

Régondaud, J. (1964) Origine embryonnaire de la cavité pulmonaire de *Limnaea stagnalis* L. Considérations particulières sur la morphogénèse de la commissure viscérale. Bull. Biol. Fr. Belg. *98:*433–471.

Régondaud, J.M. (1972) Observation ultrastructurale des cellules nucales de l'embryon de *Lymnaea stagnalis* L. (Gasteropode pulmone basommatophore). C.R. Acad. Sci. Paris *275:*679–682.

Régondaud, J., and P. Brisson (1976) Données radioautographiques après incorporation de leucine tritiée dans les cellules nucales de l'embryon de *Lymnaea stagnalis* L. (Gasteropode pulmone basommatophore). Bull. Soc. Zool. Fr. *101:*477–480.

Rigby, J.E. (1979) The fine structure of the oocyte and follicle cells of *Lymnaea stagnalis,* with special reference to the nutrition of the oocyte. Malacologia *18:*377–380.

Rudolph, P.H. (1979) The strategy of copulation of *Stagnicola elodes* (Say) (Basommatophore: *Lymnaeidae*). Malacologia *18:*381–389.

Russell-Hunter, W.D. (1978) Ecology of freshwater pulmonates. In V. Fretter and J. Peake (eds): Pulmonates: Systematics, Evolution and Ecology. New York: Academic Press, pp. 385–429.

Steen, W.J. van der (1967) The influence of environmental factors on the oviposition of *Lymnaea stagnalis* (L). under laboratory conditions. Arch. Neerl. Zool. *17:*403–468.

Steen, W.J. van der (1968) Note on the effect of excessive amounts of food on the reproduction of *Lymnaea stagnalis* (L.) (Gastropoda, Pulmonata). Neth. J. Zool. *18:*411–414.

Steen, W.J. van der, N.P. van den Haven, and J.C. Jager (1969) A method for breeding and studying freshwater snails under continuous water change, with some remarks on growth and reproduction in *Lymnaea stagnalis* (L.). Neth. J. Zool. *19:*131–139.

Takaichi, S., and J.C. Dan (1977) Spermiogenesis in the pulmonate snail, *Euhadra hickonis.* I. Acrosome formation. Dev. Growth Differ. *19:*1–14.

Tapaswi, P.K. (1974) Further investigation on transcription during oogenesis and immediately after activation by sperm in *Limnaea* (Mollusca) eggs. Acta Biol. Exp. *2:*191–195.

Taylor, H.H. (1973) The ionic properties of the capsular fluid bathing embryos of *Lymnaea stagnalis* and *Biomphalaria sudanica* (Mollusca: Pulmonata). J. Exp. Biol. *59:*543–564.

Taylor, H.H. (1977) The ionic and water relations of embryos of *Lymnaea stagnalis,* a freshwater pulmonate mollusc. J. Exp. Biol. *69:*143–172.

Te, G. (1978) A systematic study of the family Physidae (Basommatophora: Pulmonata). Thesis, University of Michigan.

Terakado, K. (1974) Origin of yolk granules and their development in the snail, *Physa acuta.* J. Electron Micros. *23:*99–106.

Terakado, K. (1981) Chromatin arrangement and axis formation in the spermiogenesis of a pulmonate snail. Dev. Growth Differ. *23:*381–399.

Timmermans, L.P.M. (1969) Studies on shell formation in molluscs. Neth. J. Zool. *19:*417–532.

Ubbels, G.A. (1968) A Cytochemical Study of Oogenesis in the Pond Snail *Limnaea stagnalis.* Rotterdam: Bronder Ubbels.

Ubbels, G.A., J.J. Bezem, and C.P. Raven (1969) Analysis of follicle cell patterns in dextral and sinistral *Limnaea peregra.* J. Embryol. Exp. Morphol *21:*445–466.

Verdonk, N.H. (1965) Morphogenesis of the head region in *Limnaea stagnalis* L. Thesis, Utrecht.

Verdonk, N.H. (1968) The determination of bilateral symmetry in the head region of *Limnaea stagnalis.* Acta Embryol. Morphol. Exp. *10:*211–227.

Verdonk, N.H. (1973) Gene expression in early development of *Limnaea stagnalis.* Dev. Biol. *35:*29–35.

Wal, U.P. van der (1976a) The mobilization of the yolk of *Lymnaea stagnalis* (Mollusca). I. A structural analysis of the differentiation of the yolk granules. Proc. K. Ned. Akad. Wet. *79:*393–404.

Wal, U.P. van der (1976b) The mobilization of the yolk of *Lymnaea stagnalis.* II. The localization and function of the newly synthesized proteins in the yolk granules during early embryogenesis. Proc. K. Ned. Akad. Wet. *79:*405–420.

Wierzejski, A. (1905) Embryologie von *Physa fontainalis* L. Z. Wiss. Zool. *83:*503–706.

Winsor, C.P., and A.A. Winsor (1932) Polyvitelline eggs and double monsters in the pond snail *Lymnaea columella* Say. Biol. Bull. *63:*400–404.

Developmental Biology of Freshwater Invertebrates, pages 485–533

General Biology and Cytology of Cyclopoids

Grace A. Wyngaard and C. C. Chinnappa

ABSTRACT The cyclopoid copepods are an especially interesting group from a cytological standpoint. They exhibit considerable genetic variability, and their patterns suggest some very interesting problems concerning chromosomal evolution in the group. The chapter describes the life cycle, general biology, and distribution of free-living freshwater cyclopoids. Procedures for collection and laboratory rearing methods are provided. Emphasis is given to species *Mesocyclops edax* and *Cyclops strenuus*, on which most of the existing cytological work has been done. The developmental biology has not been thoroughly examined for any one species; thus a selective overview of the existing studies of oogenesis, spermatogenesis, and embryology is reviewed.

INTRODUCTION

Cyclopoid copepods are extremely important freshwater organisms found from the Arctic to the equator. The numerous species that have been described are both herbivores and carnivores. Although the taxonomy of freshwater copepods appears to be on a solid footing considering the numerous species list and keys (e.g., Gurney, '33; Rylov, '48; Yeatman, '59; Dussart, '69), especially for representatives from temperate zones, there are many examples of variant populations. Interest in this group has recently been renewed because of their potential role in the dynamics of secondary production (Monakov, '76). In this context, researchers have long been hampered by their inability to accurately identify individual copepod species from diverse localities since their morphological features can be highly polymorphic. Nilssen ('79), in his recent paper on problems of subspecies recognition in freshwater cyclopoids, summarizes the systematics in this group in an historical viewpoint: "These arthropods are characterized by considerable phenotypic plasticity, many sibling-species, polymorphism and polyphenism. The taxonomy within freshwater copepods is mostly based upon the morphology of the animals and splitting of number of species into subspecies is often based on very scarce material. Seldom have taxonomists working on freshwater copepods used adequate taxonomic theory or taken advantage of tools of modern taxonomy."

In the past these diverse cyclopoids have conveniently been assigned *sensu lato* to a well-documented species. In the absence of other criteria the procedure is satisfatory and may be the only reasonable course of action available. As a consequence of this approach, however, many cyclopoid species are now considered cosmopolitan or, at least very widely distributed (e.g., *Mesocyclops leuckarti* (Claus), *Ectocyclops phaleratus* (Koch), *Eucyclops serrulatus* (Fischer), and *Thermocyclops crassus* (Fisher)). The question arises whether the same broadly defined morphospecies, when it is apparently found in several geographically isolated sites, is indeed a single species or is possibly a complex of sibling species. It is of further interest to ascertain the degree of genetic isolation if it exists among various populations of cyclopoids that have been defined as the same species. Cytogenetic investigations with other organisms have offered a direct method for resolving various taxonomic conundrums (e.g., Sonneborn, '57; Rothfels, '56; White, '71; Cleland, '62). In recent years Harding ('50), S. Beermann ('59, '66), Einsle ('75), and Chinnappa and Victor ('79b) have stressed cytogenetics as part of systematics to delimit species of Copepods. Banarescu ('74) claimed that research works without good knowledge of evolution, and taxonomy can drive this part of systematics into pure typology. Most research workers include some ecological data in their systematic work, and the studies on the ecology of zooplankton are being conducted extensively. Arriving at correct identification of taxa depends on the understanding of their biology. Then, perhaps, the ecological investigations become more meaningful.

Cyclopoid copepods have received little study to date from developmental biologists. This lack of attention largely reflects a sparse knowledge of how to rear these animals in the laboratory. Much progress has recently been made in culture techniques, and the first portion of this chapter emphasizes the general biology and laboratory culture of freshwater cyclopoids. Particular emphasis is given to *Mesocyclops edax* Forbes and *Cyclops strenuus* Fish., on which some cytological work has been done. The developmental biology has not been thoroughly examined for any one species. Thus, a selective overview of developmental studies is presented. The latter portion of this chapter focuses on cytological studies of cyclopoids and points out some problems associated with chromosomal evolution.

Life Cycle

Freshwater cyclopoids inhabit a variety of habitats including pitcher plants, bromeliads, springs, rivers, temporary and permanent ponds, and lakes. A typical cyclopoid life cycle consists of an embryonic ("egg") stage, five or six larval (naupliar: NI–NVI) stages (Fig. 1), five copepodid (juvenile: CI–CV) stages, and an adult stage (Figs. 2–4) which does not molt. There is considerable sexual dimorphism, and reproduction is always sexual. Development from egg to adult may proceed uninterrupted or include a diapause phase. The total life-span may vary from several weeks to over a year, according to species, temperature, and food.

General Features of Postembryonic Development

The greater portion of cyclopoid development is postembryonic. The life cycle consists of an embryonic ("egg") stage, five or six larval (naupliar) stages, five copepodid stages, and an adult stage. Ewers ('30) depicts the naupliar stages of several cyclopoid species. There are conflicting reports in the literature regarding the number of naupliar instars. Elgmork and Langeland ('70) review these studies and propose that free-living cyclopoids normally have six naupliar instars. Eggs are carried in thin-walled sacs by the female until they hatch into free-swimming nauplii. The NI nauplius is unsegmented, and has a median eye and three pairs of appendages: antennules, antennae, and mandibles (Fig. 1).

Copepodid stages are segmented and resemble the adult form. The first copepodid stage (CI) has two pairs of swimming legs and one urosome segment. The remaining legs and urosome segments are added during succeeding molts. Recognition of copepodid stages is based upon segmentation

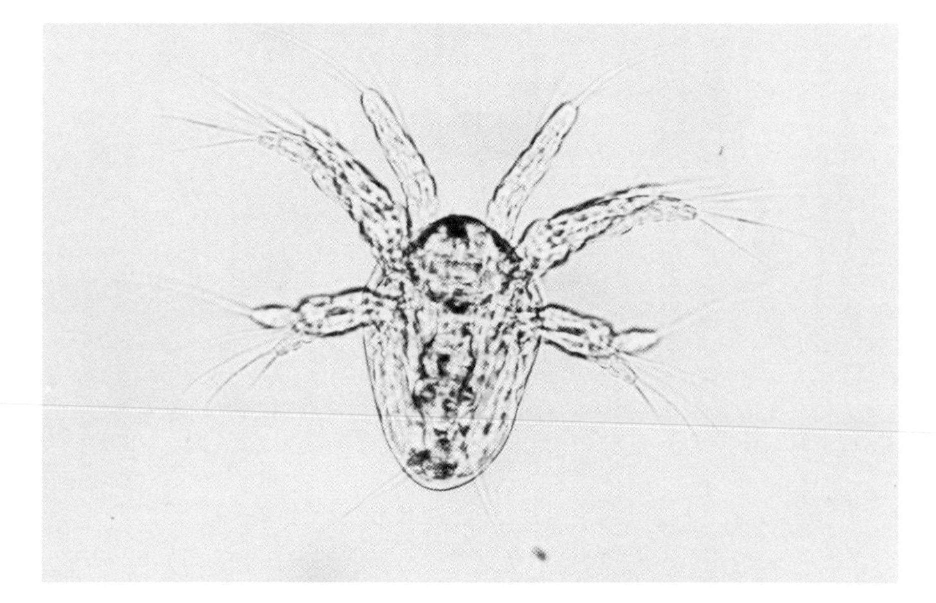

Fig. 1. Nauplius larva of *Mesocyclops edax*.

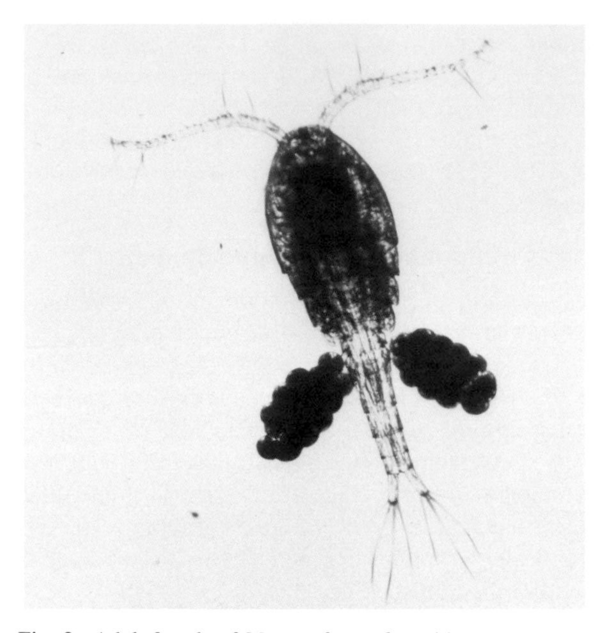

Fig. 2. Adult female of *Mesocyclops edax* with two egg sacs.

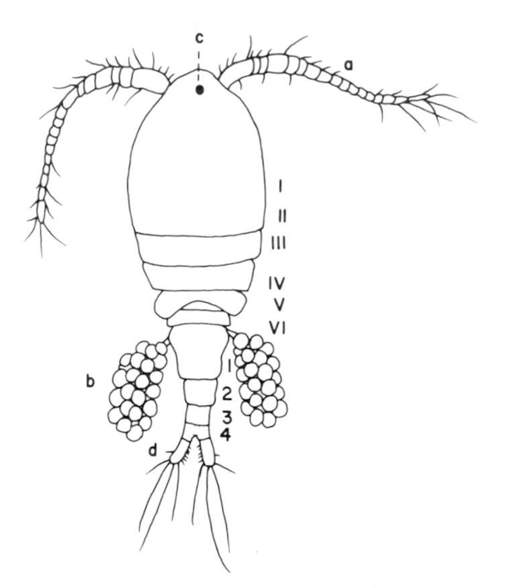

Fig. 3. Generalized adult female cyclopoid. a—First antennae; b—egg; c—eye; d—caudal rami; I—cephalothorax (head fused with first thoracic segments); II to V — second, third, fourth, and fifth thoracic segments; 1 to 4 — abdominal (urosome) segments; 1—genital segment.

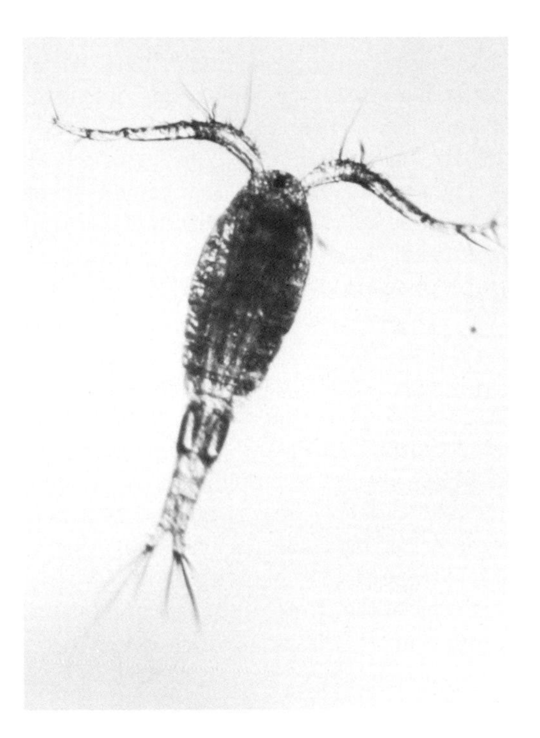

Fig. 4. Adult male of *Mesocyclops edax*.

of body and details of appendages, which is exemplified by Ravera's ('53) description of the morphological stages of *C. strenuus*. The molting process occurs rapidly. The copepodids and adults slip out of the anterior end of the carapace, which has split along the dorsal surface. Growth is determinant, and the adult does not molt.

The characteristic features of adult external morphology are shown in Figures 2 and 4 using *M. edax*. The adult body is divided into 1) the cephalothorax (head plus thoracic segments), 2) the urosome (abdomen), and 3) the caudal rami (two setose projections attached to the last urosome segment) (Fig. 3). The head contains the single median eye (hence the name *Cyclops*) and five appendages: first antennae, second antennae, mandibles, first maxillae, and second maxillae. The first two thoracic segments are fused to the head, the first bearing maxillipeds and the second bearing the first pair of legs. The adult has five pairs of swimming legs. The fifth pair is reduced and is important in taxonomic identification. Thoracic segments 3–6 are distinct, and each bears a pair of legs. The genital pore lies in the first urosome segment. In females this genital segment is formed by fusion of the seventh thoracic segment yielding only four distinct urosome segments in the adult. Males have five distinct urosome segments (Fig. 4), as these segments are not fused. The seventh thoracic segment of the male usually bears a sixth pair of very rudimentary legs. Thus the female adult body consists of nine distinct segments and the male adult body consists of ten distinct segments. Females carry two egg sacs laterally on the genital segment. Adults are sexually dimorphic, and the sexes may be distinguished in some species as early as CIV or CV stages. Adult female size varies according to species, locale, food, and temperature. Body lengths of female freshwater cyclopoids range from 0.50 mm (*Tropocyclops prasinus*) to 4.0 mm (*Macrocyclops fuscus*).

The adult male (Fig. 4) is characteristically smaller than the adult female, and its body form is more slender. The pair of geniculate first antennae aid in grasping the females during mating and in transferring spermatophores to the genital segment. Males are easily recognized by the geniculate antennae and swollen first-urosome segment, which contains two spermatophore sacs.

Adult females and males of most species are readily distinguished with the aid of a dissecting microscope at 25 × magnification. Maturity can be quickly ascertained by examination of the shape of the last urosome segment to which the caudal rami are attached. If it is longer than it is wide, the individual is immature; otherwise, it is an adult.

Internal Anatomy

Rylov ('48) presents a detailed review of the internal anatomy of cyclopoids. The male genital organis of *C. strenuus* (Heberer, '26) and *A. viridis* (Heberer, '26; Rousset et al., '81) have been studied in detail. The unpaired testis is situated in the first thoracic segment. Bilobed and unpaired testes are found in some parasitic cyclopoids (Rousset et al., '81). From the anterior portion of the testis in *A. viridis* arise a pair of symmetrically situated genital ducts (vasa deferentia) that are directed anteriorally and then posteriorally. They widen to form the seminal vesicles that contain the spermatophores (first-urosome segment).

The ovary may be either paired (*Macrocyclops, Ectocyclops, Eucyclops serrulatus, Tropocyclops prasinus,* and *Paracyclops affinis*) or unlobed (*C. strenuus, C. insiginis, Acanthocyclops bicuspidatus, A. vernalis, Microcyclops gracilis, M. leuckarti,* and *M. dybowskii*) (Matschek, '10). It is positioned between the intestine and dorsal body integument, and may fill a large space when it becomes distended with eggs. Eggs are often visible through the translucent exoskeleton of adult females.

Near the terminal portion of the oviducts are glandular cells that produce a secretion for the egg shell. Circular muscles around the opening of the oviduct eject the mature eggs into the receptaculum seminis, where the spermatophores are stored and fertilization occurs. The spermatozoa from one spermatophore may inseminate multiple batches of eggs. Secretions for the egg-sac membrane are produced by skin glands near the genital opening (Walter, '22). Their adherence to one another forms the egg sacs. At this stage the "eggs" can be teased apart and will remain viable.

Other studies have examined the digestive system of *A. viridis* (Farkas, '23); ganglionic mass or "brain" of *A. viridis, Macrocyclops fuscus,* and *C. strenuus* (Richard, 1891) and *Mesocyclops oithonoides* (Hanström, '24); and eye of *C. strenuus* and *M. fuscus* (Gicklhorn, '30). The marine calanoid *Calanus finmarchicus* Gunnerus (Marshall and Orr, '72) is the only copepod for which a great deal is known of the internal anatomy. Comparable intensive study of a freshwater cyclopoid remains to be done.

Reproduction

Cyclopoids reproduce sexually. No instance of parthenogenesis has ever been reported. Males and females may mate more than once. The mating process may vary in duration from minutes to hours (Hill and Coker, '30). Cannibalism of males by females after mating occurs in some species, and may vary in extent among populations of the same species (Wyngaard and Allan, in preparation). Females of *M. edax* store sperm and produce multiple clutches from one insemination. This is probably common among cyclopoids. It is not known if adult cyclopoids can store sperm during diapause.

Distribution

Freshwater cyclopoids are distributed throughout much of the world. Hutchinson ('67) reviews their distribution, which is incompletely known owing to sparse sampling in areas. Distributional information is available for North America (Wilson and Yeatman, '59), Mexico (Comita, '51), southeastern Canada (Carter et al., '80), southern and western Canada (Anderson, '74), Arctic and sub-Arctic Canada (Reed, '63), Old World (Rylov, '48), Europe (Kiefer and Fryer, '78), Great Britain (Gurney, '33), France (Dussart, '69), Morocco (Dumont and Decraemer, '77), Southeast Asia (Fernando, '80a), Ceylon (Fernando, '80b), Mali (Dumont et al., '81), Egypt (Kiefer and Fryer, '78), Africa (Lindberg, '58), Cuba (Smith and Fernando, '78a), Java and Bali (Kiefer, '33), and Guam (Watkins and Belk, '75). The distribution of the genus *Cyclops* is Palaearctic, and this genus has a high percentage of cold-water forms (Rylov, '48). Warm-water stenotherms are absent. *Cyclops strenuus* is widespread in the Palaearctic, though there are many varieties of this species. It is often confused with *Cyclops abyssorum.* Nilssen ('79) compares the ecology and morphology of the two species, and Einsle ('64) uses cytological characters to distinguish them. Wilson and Yeatman ('59) report *C. strenuus* to be rare in North America, but it has been reported in Alaska (Rylov, '48), northern Canada (Patalas, '75), and Ontario, Canada (Smith and Fernando, '78b). Most records of this species in North America are probably *C. scutifer* (Yeatman, '44; Wilson and Yeatman, '59). *Cyclops* is not prevalent in southern latitudes (Rylov, '48).

In contrast to *Cyclops,* the genus *Mesocyclops* (including *Thermocyclops*) is exclusively warm-water stenothermal and is found primarily in southern latitudes. It is believed to have a subtropical origin and radiated northward. It is absent in the high Arctic regions. *Mesocyclops leuckarti* Claus is the only true cosmopolitan freshwater cyclopoid, as it is found in all zoogeographical areas. It is not as common in North America as in Europe, and most reports of *M. leuckarti* in North America before the 1940s are probably *M. edax.* Coker ('43) definitively separated these two species. *M. edax* is one of the commonest cyclopoids in eastern North America, ranging from Central America (Deevey et al., '80) to southern Canada (Carter et al., '80).

The taxonomy of cyclopoids is in need of much revision. Einsle ('75) has revised the European *Cyclops,* and Kiefer and Fryer ('78) review the Copepoda as a whole. Dumont et al. ('81) point out that *M. leuckarti,* among other species, is not cosmopolitan, but instead consists of numerous morphologically stable taxa. Subspecies recognition of these variable cyclopoids is one current problem (Nilssen, '79) that emphasizes the need for more breeding (e.g., Price, '58; Allan, in preparation) and cytological (Einsle, '64, '80; Chinnappa and Victor, '79a) studies.

Seasonal Cycles and Habitat Type

Many north-temperate cyclopoid populations pass through only one, two, or a few generations per year with individuals in each generation developing more or less synchronously, as a cohort. Depending upon the species and locale, one of the generations may undergo a summer or winter diapause, in the mud, or overwinter in the plankton. Diapause is characterized by arrested development, a metabolic slowdown, empty gut, and large oil globules that are visible through the exoskeleton on the cephalothorax.

Most studies that have examined the causal determinants of diapause have focused on proximal factors such as photoperiod and temperature. Physiological studies such as that examining oxidative metabolism of diapausing *Diacyclops navus* (Watson and Smallman, '71) are few. Epp and Lewis ('80) have compared oxygen consumption rates of nondiapausing nauplii and copepodids, and report an abrupt change in the logarithmic metabolism-weight relationship between the two life stages.

Cyclopoids usually overwinter or diapause as CIVs, CVs, or adult females. It is questionable whether cyclopoids have diapausing eggs, as is common in diaptomid copepods (Hutchinson, '67); however, Roy ('32) reports resistance to desiccation of eggs, CV, and adults of *C. furcifer* in the mud. Because of the relative synchrony of maturation of each generation, reproducing females may be present for only a few weeks to a month, two or three times a year. Laboratory propagation can circumvent this problem.

The variable diapause behavior in cyclopoids is reviewed by Sarvala ('79). *C. strenuus* undergoes a summer dormancy in the CIV stage in the mud (Elgmork, '55). These CVs reappear in the plankton during autumn or winter (Elgmork, '55, '59; Einsle, '64, '67). In some instances a small fraction of the populations breeds in the summer (Einsle, '64, '69; Elbourn, '66).

In Minnesota, *M. edax* has spring and summer generations that complete maturation, and a third generation that diapauses predominantly as CIVs through the winter and reproduces the following spring (Comita, '72). Some Connecticut populations diapause as CIV and adults (Elgmork, '64).

In contrast, subtropical populations of *M. edax* (Wyngaard et al., '82) have reproducing individuals and all developmental stages present throughout the year. The few existing studies of other subtropical (Gophen, '78) and tropical (Burgis, '71; Lewis, '79) cyclopoids suggest that this pattern may be general.

Most cyclopoids live in the bottom or shoreward portion of lakes or ponds, although *M. edax* inhabits the open-water zone. *Mesocyclops edax* predominates in productive lakes, but inhabits a wide variety of water bodies and is tolerant of a wide range of pH. *Cyclops strenuus* inhabits chiefly ponds and littoral portions of lakes (Elgmork, '59; Einsle, '75, '80).

Collection of Field Animals

Collection of planktonic cyclopoids is most simply carried out by towing a planton net vertically or obliquely through the water. A mesh size of 158 μm will suffice to capture adults of most species, but a 76-μm mesh is needed if the naupliar stages are to be captured. The cone-shaped net may be attached to either a detachable bucket with a mesh window (e.g., Wisconsin net by Wildco) or a glass vial. If diapausing plankton are to be captured from the mud, a benthic sampler such as the Ekman grab should be used. The net, benthic grab, and similar sampling devices are discussed in detail in Edmondson and Winberg ('71) and Lind ('74). Sampling littoral areas may be accomplished by towing a small plankton net along the bottom or in the vegetation. Small mesh dip nets may also be used.

The extent of vertical migration varies among species and locale. The late juvenile and adult stages of many cyclopoids remain in the bottom mud or deeper waters during the day and migrate vertically to the surface waters at night. Collection of migrating populations necessitates night plankton collections or benthic samples during the day if reproductive adults are desired (Woodmansee and Grantham, '61). Where species of *Mesocyclops* and *Cyclops* coexist in stratified lakes, one might find *Mesocyclops* inhabiting the warm top layer and *Cyclops* inhabiting the cooler, deeper waters (Smyly, '78; Stücke, '81).

Laboratory Culture

Successful laboratory culture of freshwater cyclopoids depends on having the appropriate diet, simulated freshwater medium, and appropriate temperature and light intensity. Culture techniques presented in this chapter have yielded high survivorship and fecundity for *Mesocyclops edax* and *Cyclops bicuspidatus thomasi*, as well as certain diaptomid copepodids, cladocerans, and freshwater rotifers. The success obtained thus far with a variety of species suggests that these methods may be useful for many other cyclopoid species as well.

This section outlines procedures for preparing animal and algal media and culturing animal and algae prey for laboratory maintenance of breeding populations of *M. edax*. *Mesocyclops edax* is omnivorous, as are many cyclopoids. Algae are consumed throughout the life-span, though they play a declining role in the diet of late copepodid and adult stages which consume animal prey.

Space limitations do not allow mention of all aspects of general laboratory culture methods. Many essential details are provided by Guilliard's ('75) excellent discourse on preparation of media and algae culture, comments on lights, glassware, etc.

Preparation of Animal Media

A simulated freshwater medium is recommended to insure constant, reproducible quality. The quality of spring water from biological suppply companies can vary considerably, and one inferior batch could result in loss of laboratory cultures. Pond or lake water is variable in quality and may not be convenient.

The simulated freshwater medium given here is a modification of D'Agostino and Provasoli ('70) (see Table I for formula). To prepare the medium:

1. Add to double-distilled water:
   a. 7 × Salt solutions (omit $NaSiO_3 \cdot H_2O$ unless diatoms are being cultured)
   b. Hutner's trace elements (See Table II for formula)
   c. Vitamins
2. Fill Pyrex Erlenmeyer flasks three-fourths full with the above mixture. Plug flasks with cotton wrapped in cheesecloth and cap with foil. Autoclave at 20 psi, slow exhaust, for only 8–10 min so as not to denature vitamins.
3. When medium has completely cooled, adjust pH to 7.0 with NaH_2CO_3.

TABLE I. Animal Medium Modified After D'Agostino and Provosoli ('70)

Salt	Final stock (gm/l)	Add to 1 liter
KCL	6.25 gm/250 ml	2 ml
$MgSo_4 \cdot 7H_2O$	5.00 gm/250 ml	2 ml
$CaCl_2 \cdot 2H_2O$	3.31 gm/250 ml	2 ml
K_2HPO_4	0.75 gm/250 ml	2 ml
KH_2PO_4	0.75 gm/250 ml	2 ml
$NaNo_3$	6.25 gm/150 ml	2 ml
$NaSiO_3 \cdot 9H_2O$	2.50 gm/250 ml	2 ml
Vitamin B_{12}	0.05 gm/l	0.1 ml
Thiamine HCl	0.02 gm/l	0.1 ml
Ca-pantothenate	0.05 gm/l	0.1 ml
Hutner's trace elements		0.3 ml

Preparation of Algal Medium

The preparation of algal medium and subsequent culture of algae are the most time-consuming aspects of laboratory culture of freshwater cyclopoids. The formula for the algal growth medium is that used by Stemberger ('81), which is a modification of Woods Hole, MBL medium (Nichols, '73) and Guillard ('75) WC medium (see Table III for formula). To prepare medium:

1. Add to distilled water:
   a. 13 × Salt solutions (remember $FeCl_3$ and EDTA are combined, so add twice the amount of this solution as of the other 12 solutions). Omit $Na_2SiO_3 \cdot 9H_2O$ unless diatoms are being cultured.
   b. Vitamins
2. Filter medium through 5-μm filter, using a filtering apparatus attached to a vacuum pump.
3. Filter medium through a .22-μm filter. This removes most bacteria. This slow process is made efficient if the 5 μm filtered medium is placed in a carboy above the .22μm filter and allowed to trickle through the filter. The filter paper may have to be changed during the process, depending on the level of water contamination.
4. Pour medium into storage flasks, taking care to leave space at top of flask to prevent expulsion of the plug. Medium may also be autoclaved in the containers in which the algae are to be cultured. For example, fill 2-liter Erlenmeyer flasks with 1.4 liter of medium, 125 ml flasks with 75 ml of medium, and test tubes halfway to the top.
5. Cap flasks with cheesecloth-covered cotton plug and foil. Screw caps on test tubes should be slightly loosened.

TABLE II. Algal Medium Modified After Nichols ('73)

Salt	Final stock (gm/l)	Add to 1 liter
a. Macronutrients—use 1 ml per liter		
$CaCl_2 \cdot 2H_2O$	36.76	1 ml
$MgSO_4 \cdot 7H_2O$	36.97	1 ml
$NaHCO_3$	12.60	1 ml
K_2HPO_4	4.355	1 ml
$NaNO_3$	42.505	1 ml
$Na_2SiO_3 \cdot 9H_2O$[a]	28.42	1 ml
b. Micronutrients—use 1 ml per liter		
Na_2EDTA[b]	4.36	1 ml
$FeCl_3 \cdot 6H_2O$[b]	3.15	1 ml
$CuSO_4 \cdot 5H_2O$	0.01	1 ml
$ZnSO_4 \cdot 7H_2O$	0.022	1 ml
$CoCl_2 \cdot 6H_2O$	0.01	1 ml
$MnCl_2 \cdot 4H_2O$	0.18	1 ml
$Na_2MoO_4 \cdot 2H_2O$	0.006	1 ml
H_3Bo_3	130 mg/l	1 ml
Thiamine HCl (B1)	0.1 mg/l	1 ml
Biotin	0.5 μg/l	1 ml
Cyanocobalamin (B12)	0.5 μg/1	1 ml

[a]Omit $Na_2SiO_3 \cdot 9H_2O$ unless diatoms are being cultured.
[b]Mix together Na_2EDTA and $FeCl_3 \cdot 6H_2O$ and store and dispense as one solution.
Reproduced with permission, from the Annual Review of Microbiology, Volume 26, © 1972 by Annual Reviews Inc.

TABLE III. Hutner's Trace Elements (after Hutner, '72)

Element	Final medium (Mg/liter)	Salt	Gravimetric weight	
			Factor	Salt (gm)
Fe	6.0	$Fe(NH_4)_2(SO_4)_2 \cdot 6H_2O$	7.0	42.0
Mn	5.0	$MnSO_4 \cdot H_2O$	3.0	15.5
Zn	5.0	$ZnSO_4 \cdot 7H_2O$	4.4	22.0
Mo	2.0	$(NH_4)_6MoO_7O_{24} \cdot 4H_2O$	1.8	3.6
Cu	0.4	$CuSO_4$(anhydrous)	2.5	1.0
V	0.2	NH_4VO_3	2.3	0.46
Co	0.1	$CoSO_4 \cdot 7H_2O$	4.8	0.48
B	0.1	H_3BO_3	5.7	0.57
Ni	0.1	$NiSO_4 \cdot 6H_2O$	4.5	0.45
Cr	0.1	$CrK(SO_4)_2 \cdot 12H_2O$	9.6	0.96
			Total:	87.02

6. Autoclave at 20 psi, slow exhaust for 8–10 min.
7. Allow medium to cool at least 24 hours, or longer, if cloudiness in medium has not disappeared. The pH should be 7.0–7.2 and may not need adjustment. If necessary, adjust pH by adding 1 N HCl drop by drop. The buffer in the medium is a weak one. The Tris buffer is omitted as it is toxic to zooplankton.

General Comments on Preparation of Media

Considerable efficiency and convenience can be attained if several suggestions are adopted. Vitamins can be premeasured and mixed together in advance, then stored frozen in small plastic autoclavable screw cap vials. This helps maintain sterility and does not require measuring of vitamins each time the algal medium is prepared. An automatic pipette (e.g., Gilson Pipetteman by Ranin Instrument Co.) with disposable pipette tips ensures accuracy in dispensing solutions and saves time in preparation and dishwashing. Media can be made in large batches and will remain usable for several months if stored in the dark at cold temperatures, preferably 4 °C.

Salts for stock solutions should be weighed out on an electrobalance and mixed with distilled water in plastic bottles. Glass vessels may take up silicate. Stocks of salt solutions and Hunter's trace elements should be stored in the dark at low temperatures and replenished every 6 months.

Culture of Algae for Feeding to Cyclopoids

Probably the most important factor affecting culture success of freshwater cyclopoids is the food on which they are reared. A variety of factors affect adequacy of a particular food, among which capturablity, size, and nutritional content are important. All cyclopoids are raptorial, in contrast to filter-feeding calanoid copepods, and the relative size of cyclopoid to prey item may affect feeding efficiency considerably. The alga used is especially important. The flagellated yellow-brown alga *Cryptomonas ozolini* Sküja (class Cryptophyceae) has proved to be an excellent food source for *M. edax*. It was isolated from a Michigan lake in which *M. edax* occurs, and is obtainable from the Starr Collection at the University of Texas at Austin (culture number UTEX-LB2194). The algal culture methods presented here will probably suffice for many other algae.

Algae may be raised in large batches, but pure stock cultures should be maintained separately in screw-cap test tubes (ca. 18 × 150 mm), lipless tubes plugged with cotton, or 125-ml Erlenmeyer flasks. Screw caps on test tubes should be tightened and then loosened one-quarter turn to facilitate gas exchange. Algal stocks should be transferred to fresh medium weekly using a sterile Pasteur pipette or by flaming the open ends of both tubes and pouring a portion of the algal stock into the tube containing fresh medium. Numerous tubes of stocks should be maintained if many large amounts of algae are needed. If stock cultures become contaminated, isolate single cells and place individually into sterile algal medium repeatedly, using sterile pipette technique (Guillard, '73).

Large quantities of algae can be grown in 2-liter Erlenmeyer flasks filled with 1.4 liters of algal medium and innoculated with week-old stock cultures. The risk of contamination increases with increasing size of the culture vessel. These culture methods will not provide axenic culture conditions. Sterile technique should be used whenever possible to avoid high levels of bacterial contamination, as bacteria will compete with algae for nutrients. In nature, cyclopoids most likely consume bacteria, but little is known of the bacterial species preferred and the extent to which this occurs. All algal cultures should be shaken by hand daily to maintain cells in suspension, if a shaker table is unavailable.

Algae for feeding should be harvested in early log-phase growth to ensure nutritional constancy. *Cryptomonas* cultures generally reach this stage in 8–10 days. The algal culture should not be fed directly to zooplankton, because it contains algal wastes and metabolites. To concentrate algae and remove the medium, centrifuge the culture in 15-ml centrifuge tubes for 15 min at 1,000 rpm, using a desk-top centrifuge. Pour off the supernatant and then resuspend the pellet in several milliliters of fresh medium. The concentrated algae from numerous centrifuge tubes can be pooled and set aside for feeding. The suspension should be dark brown. A reddish tint indicates lysed cells that result from centrifugation at speeds too high or for too long a duration. A quick examination under a compound microscope will indicate whether cells are alive and motile.

Temperature and Light

Cryptomonas ozolini grows very well at 10–15°C and a light intensity of about 2,000–4,300 lux. A light:dark cycle of 14:10 is preferable to continuous illumination, which may disrupt the natural algal physiology. Stock cultures should be maintained under these conditions. Temperature constancy may be difficult to maintain in small incubators. In such cases, a walk-in chamber is preferable. Algae cultured in large batches may be grown at room temperature using 40-watt "cool-white" fluorescent bulbs.

Dishwashing

Clean glassware that is free of contaminants is essential to successful laboratory culture. Dirty glassware should be soaked in hot soapy water (Luiqui-Nox detergent is preferable), rinsed in tap water, soaked for 5 minutes in a 10–30% dilution of HNO_3, and rinsed several times first in tap water and then in deionized water. The acid rids the glass of soap residue, which is toxic to zooplankton. Plastic Petri dishes may receive this treatment also, though certain polycarbonates used to make carboys and centrifuge bottles can not withstand frequent acid washing. Glassware to be used for algal culture should be autoclaved as an additional precaution.

Culture of Animal Prey

Animal prey is very important in the diet of *M. edax*. Reproduction will be low or nonexistent if animal food is absent from the diet. In nature cyclopoids are omnivorous and consume other zooplankton, such as cladocerans, rotifers, diaptomid copepodids, ciliates, and other cyclopoids. Monakov ('72) and Brandl and Fernando ('78) review cyclopoid feeding studies. These natural prey are time-consuming to culture and compete for algae in the copepod culture, making laborious the long-term maintenance of *M. edax* cultures.

An excellent alternative to zooplankton prey is newly hatched *Artemia salina* nauplii. *Mesocyclops leuckarti* (Gophen, '76), and *M. edax* (Allan, in preparation), *C. abyssorum* (Whitehouse and Lewis, '73), and *A. viridis* (Smyly, '70) have been reared successfully with brine shrimp nauplii. Dried eggs of *Artemia* are inexpensive, may be purchased from most aquarium stores or biological supply companies, and remain viable for several years. To hatch eggs, place them in seawater (or Instant Ocean mixed according to directions) in a small Erlenmeyer flask with bubbled air. Eggs hatch in 24–30 hours at room temperature. Egg viability is variable. Do not wait until all eggs have hatched before feeding nauplii to copepods. It is extremely important that only newly hatched (less than 1 day old) nauplii are used as food. These young are still small enough to be ingested, and possess high nutritional content. Older brine shrimp will have used much of the lipid reserves in growth and thus will provide less nutrition. Rinse *Artemia* nauplii with simulated lake-water medium or distilled water to avoid contamination of cultures with seawater. *M. edax* usually ingests *Artemia* nauplii whole, though very small cyclopoid species may not be able to consume even the smallest *Artemia* nauplii. In addition, the smaller size of males relative to females may keep males of some species from efficiently ingesting prey that are suitable for females.

Alternatives

Freshwater cyclopoids have been cultured on *Scenedesmus* (Smyly, '70), *Chlorella* (Gophen, '76; Epp and Lewis, '80), *Chlamydomonas* (Wyngaard, unpublished), *Haematococcus pluvialis* (Epp and Lewis, '80), and *Euglena* (Whitehouse and Lewis, '73; Jamison, '80a) or a mixture of several of these algae (Munro, '74), but it is not known how the survivorship and fecundity compares with those reared on cryptomonads. Other successful diets include trout chow and dried alfalfa (Hunt and Robertson, '77), and a rice-bacteria mixture (Gophen, '76), both of which promote growth of microorganisms on which the cyclopoids feed. Most of these studies utilized mixed algae and protozoan species and filtered lake or pond water, which are variable in quality and unrepeatable. Brandl ('73) provides an excellent description of mass rearing techniques for *A. vernalis, A. americanus* Marsh, *A. robustus* Sars, *A. viridis, C. vicinus* Uljanin, *C. strenuus, C. tatricus* Kozminski, *Diacyclops bicuspidatus* Claus, *M. leuckarti, Eucyclops serralatus* Fisch., and *Macrocyclops albidus* Jurine. These species were maintained in the laboratory for at least ten generations and bred well when fed a green alga (e.g., *Chlorella* sp., *Scenedesmus* sp., *Chlamydomonas reinhardtii,* or *C. moewussi*) and protozoans (e.g., *Paramecium caudatum* Ehrenberg, *P. aurelia* Ehrenberg, or *P. bursaria* Ehrenbert et Focke). These *Paramecium* are easily cultured. *Ankistrodesmus falcatus* is a green alga often used for growing crustaceans, especially cladocerans. Nauplii of *M. edax* die when reared on it, which may be due either to starvation or a toxic effect. Stein ('73) lists biological collections from which many alga strains may be obtained. Strains vary among collections, and caution should be exercised in choosing the appropriate one.

Guillard ('75) describes a continuous culture apparatus for algae that considerably minimizes daily and weekly maintenance. This method is more likely to be successful with rapidly growing algae such as *Chlamydomonas* and *Scenedesmus* than is the cryptomonad, which is highly susceptible to contamination from airborne green algae such as *Chorella* or *Ankistrodesmus*. Commercial supply companies such as Carolina Biological Supply Co. distribute soil extract media, which may be used instead of the algal medium presented here (Table II) for some algae.

Calanoid nauplii (Jamison, '80b) and diaptomid copepodids (Wyngaard, unpublished) have been used as food for late-stage copepodids and adults. The recent interest in the role of protozoans (Porter et al., '79) and bacteria

(Gophen et al., '74) in zooplankton diets may provide additional food sources for freshwater laboratory cultures.

Mass Culture of *M. edax*

M. edax can be mass-cultured at room temperature in 250-ml beakers containing animal medium and algae. Many other containers are appropriate. A Petri dish can serve as a lid to minimize contamination. The *Cryptomonas* density should be approximately 10^4 cells/ml, and the culture should have a light yellow-brown tint. If smaller cells (e.g., 5 μm diameter) such as *Chlamydomonas* and *Scenedesmus* are used rather than *C. ozolini* (8–12 μm diameter), an alga concentration of 10^5 cells/ml will provide good growth. It is important not to overfeed. Estimates of algal density can be made using a Sedgewick Rafter cell (see Lind, '74, for details). With experience, a quick examination of a 10-ml subsample under a dissecting microscope will suffice. Cultures need only minimum light (one fluorescent 40-watt tube on a 14:10 L:D cycle will suffice). Low illumination helps prevent overgrowth of algae. Every 2 or 3 days, half of the culture medium should be replaced with fresh medium. A strainer of plankton netting prevents loss of animals. At low temperatures (ca. 15°C), this need be done only weekly. Certain combinations of photoperiod and temperature induce diapause, so the appropriate photoperiod and temperature will vary according to species. *Mesocyclops edax* and *M. leuckarti* (Jamison, '80b) commence ingestion of large animal prey such as *Artemia* or diaptomid nauplii and copepodids during the CIII or CIV stage, and become increasingly carnivorous as they mature. An animal food ratio of 4–5 *Artemia* nauplii per CIV or adult per day should be added daily. Uneaten nauplii should be removed the following day.

M. edax (Wyngaard, personal observations) and probably most other omnivorous cyclopoids are cannibalistic. The extent to which large individuals prey upon smaller individuals can be minimized by separately rearing cohorts of different ages. This can be accomplished by isolating ovigorous females and, shortly after hatching, separating the newly hatched nauplii from the mother (e.g., using 20 × 60 mm Petri dishes). The sibs are reared together in small (100 ml) or large (250 ml) beakers (depending on clutch size) until the late copepodid stages. At this time they are separated and placed in large beakers with unrelated individuals. The nature of the study will determine the permissible degree of inbreeding.

Individuals may be handled with disposable Pasteur pipettes with the opening enlarged to a diameter of about 5 mm. No mortality should occur from such handling. Eggs can be removed from the sacs of live females using a glass rod probe to which a tungsten filament is attached. The tungsten may be drawn out to a very fine, sturdy point by dipping it into molten sodium nitrite.

The culture techniques described in this chapter yield approximately 95% survivorship and high fecundity for *M. edax*. Development rate and body size are largely influenced by food and temperature. Maturation times from NI to adult female for a *M. edax* population from Florida are a little over 1 week at 30°C, about 2 weeks at 25°C, and 1.5–2 months at 15°C. Other populations of *M. edax* exhibit slightly slower or faster rates at the same temperature (Wyngaard, unpublished). Males reach maturity several days sooner than females and also have a shorter physiological life-span. Maturation rates of *M. leuckarti* (Gophen, '76), *C. vicinus* (Munro, '74), and *C. vernalis* (Hunt and Robertson, '77) have been determined.

Clutch size of *M. edax* varies considerably. A Florida population averaged 40 eggs/clutch, whereas a Michigan population averaged 20 eggs/clutch in the lab (Wyngaard, in preparation). Up to 15 clutches may be produced by one female from a single insemination. The interclutch duration varies from about a day at 30°C to a week or so at 15°C, depending largely on temperature and food.

More is known about embryonic development time than postembryonic development time for most cyclopoids. Embryonic duration is almost entirely temperature-dependent. Elapsed time from "egg" appearance in the sac to hatching for *M. edax* is 28.0, 48.9, and 98 hours at 30, 25, and 15°C, respectively (Wyngaard et al., '82; Wyngaard, unpublished). Embryonic duration times have been determined for *Thermocyclops neglectus* (Burgis, '70); *C. strenuus, M. leuckarti,* and *A. viridis* (Smyly, '74); *T. oithoinoides* (Frenzel, '77); *C. scutifer* (Taube and Nauwerck, '67), and *C. vernalis* (Hunt and Robertson, '77), to name a few. Cooley and Minns ('78) review estimates of these duration times for a number of cyclopoids and conclude that different populations of the same species may exhibit different temperature responses.

Considerable differences in developmental rates and hatching success at various temperatures may exist among species. Eggs of *C. b. thomasi* in Lake Ontario exhibit a high hatching success at 4°C whereas *M. edax* incurs excessive egg mortality. Carter ('74) reported no hatching of *M. edax* at 5°C. Eggs of *M. leuckarti* require considerably longer to develop at 4°C than do those of *C. strenuus* and *A. viridis*. Taube ('66) reports *M. leuckarti* cannot develop at 2°C but that *C. scutifer* eggs have a high hatching success at this temperature. Gophen reports egg mortality for *M. leuckarti* above 27°C, but Wyngaard (in preparation) found 100% hatching success of *M. edax* at 30°C. Eggs of *Thermocyclops neglectus* will hatch at 35°C, but developmental rate decreases and egg mortality increases at 32.5°C (Burgis, '70).

Spermatogenesis and Oogenesis

Spermatogenesis of cyclopoids has been studied by Lérat ('05), Chambers ('12), and especially Heberer ('24, '32). More recently, the genital tract and spermatogenesis of *A. viridis* have been studied in detail (Rousset et al., '81). While in the testis, the young spermatocyte contains a large, subspherical nucleus with clusters of strongly osmophilic, condensed chromatin along the periphery of the nucleoplasm. The nucleus releases mitochondrial RNA into the cytoplasm when it positions itself against the nuclear membrane. The globular spermatids then elongate, flatten, and become more deformed in shape when they are cramped together and pass through the opening into the vas deferens. The spermatid transforms into a spermatozoan in the vas deferens. The proximal portion of this duct secretes a substance containing osmophilic, ovoid granules, which later become the core of the spermatophore. The midsection secretes the peripheral substance and coat of the spermatophore. At this point the spermatozoan is immobile and disc-shaped. Secretions of the seminal vesicle facilitate swelling of the bean-shaped spermatophore, and also permit adhesion onto the female genital segment. Rousset et al. ('81) compare the morphology of the sperm and male genital tract of *A. viridis* with that of *C. strenuus* and several parasitic cyclopoids in the context of phylogenetic relationships.

The comments below describing oogenesis were obtained from Raven ('61). The main emphasis of investigations of oogenesis has been placed on the changes in nuclear structure and the role of the nucleolus during oocyte growth. Most of these references predate a more sophisticated understanding of the role of nucleoli in synthesis and assembly of ribosomal RNA. During the major phase of vitellogenesis, the nucleolus of *Cyclops* is connected with the nuclear membrane by a kind of funnel (Fautrez, '58, 59). The nucleolar products are extruded into the cytoplasm of oocytes when the fluid or semifluid nucleolus connects with the nuclear membrane and flows out into the cytoplasm (Fautrez, '58). Intranucleolar vacuoles break through the cortex and pour out their contents into the nucleoplasm (Frautrez and Fautrez-Firlefijn, '51). This conception of nucleolar function is not entirely credible in the light of current understanding of cell biology, and the problem could stand reinvestigation by electron microscopy and more advanced methods of histochemistry. Terpilowska ('71) described the polylecithal egg of *A. viridis* and its pronucleii just prior to, during, and following the copulation phase.

Early Development

Few embryological studies of cyclopoids exist. Green ('71), Rylov ('48), Terpilowska ('77), and Beerman ('77) review much of the existing work. Fuchs ('14) described the segregation of entodermal, mesodermal, and ectodermal layers of *C. viridis* and compared it to other entomostracans. Cleavage is determinate in all copepods and does not begin until the egg leaves the oviduct. Stich ('50a) compared the arrangement of plasma and basophilic granula of normal and centrifuged oocytes of two- and four-cell embryos of *C. viridis.* He concluded from the centrifugation studies that cleavage was determinant in the very early cleavage stages. Oocytes, however, could develop normally following mild centrifugation. Additional centrifugation studies demonstrated that the nucleus of the eggs could be split into karyomeres that reunited following centrifugation (Stich, '50b). Terpilowska ('77) provides an excellent, detailed account of cleavage and gastrulation in the egg of *A. viridis (C. viridis).* Cleavage is complete, equal, and initially synchronous. The germ line cells are differentiated at the two-cell stage, and are characterized by 1) ectosomes (agglomerations of RNA) and 2) commencement of delayed mitotic divisions of granular cells compared to the remaining blastomeres. She points out that a study of the ultrastructure of these ectosomes during embyrogenesis is needed to understand the structural and molecular basis of determination of germ line cells in the Copepoda. The successive cleavage divisions leading to the beginning of formation of the blastocoel at the 16-cell stage, two primordial germ cells, and gastrulation at the ninth cleavage division are depicted. Measurement of nucleic acid content throughout embryogenesis indicated that RNA appears in the nucleoli as late as the blastula stage, indicating the ability to synthesize RNA at this stage. The cyclic distribution of glycogen (an important source of energy) within the nucleus during various cleavage stages is also documented.

A timetable of cleavage divisions of *Cyclops divulsus, C. furcifer,* and *C. strenuus* is given by Beerman ('77). She reports that the first three cleavage divisions in all three species parallel that described by Rückert (1895) and Amma ('11) for "*C. strenuus.*" The blastomeres of *C. divulsus* divide synchronously until the fourth cleavage division (16-cell stage) when one of the 16 cells divides to form the progenitors of the two definitive primordial germ cells and what Amma ('11) believes to be the stem cell of the entoderm. The remaining 15 cells undergo fifth-cleavage division during which diminution mitosis takes place (Beerman, '59) (Table IV). When the 30 cells formed from the fifth cleavage undergo a sixth cleavage, the stem cell of the entoderm undergoes diminution. The two definitive primordial germ cells

TABLE IV. Timetable of Cleavage Division in *C. divuslus* at 18–20°C (from S. Beermann, '77, Chromosoma by permission)

Cleavage interval	Duration of interval in min	Number of nuclei at beginning of division	Number of nuclei at end of division	Special features
0–1	45–60	1	2	No differentiation between eu- and heterochromatin
1–2	45–60	2	4	No differentiation between eu- and heterochromatin
2–3	45–60	4	8	No differentiation between eu- and heterochromatin
3–4	45–60	8	16	No differentiation between eu- and heterochromatin
4–5	210–240	15 + 1 PGC + 1 PEC	30 + 1 PGC + 1 PEC	Differentiation into eu- and heterochromatin. After 60 min division of 1 cell, formation of nucleoli. Diminution at 5th cleavage in 15 cells.
5–6	45–60	30 + 2 PGC	62 + 2 PGC	Division of the PGC between 5th and 6th cleavage without diminution. Diminution division of the PEC. Heterochromatin only in 2 PGC's

PGC = primordial germ cell.
PEC = primordial entodermal cell.

are formed from the stem cell of the germ line that divides without diminution (Fig 5).

Cleavage in *C. furcifer* resembles that of *C. divulsus* up to the 16-cell stage. Just prior to the fifth cleavage, one cell divides and forms the primordial germ cell and stem cell of the entoderm. The remaining 15 cells divide to form 30 cells, which undergo chromatin diminution until just prior to the sixth cleavage. During the seventh cleavage, the cells that complete the first diminution undergo a second diminution of chromatin. Two divisions, later the primordial entodern and its two daughter cells undergo chromosomal diminution.

Cyclops strenuus differs from the above two species in that the division yielding the stem cell of the germ line has occurred by the third cleavage division (Amma, '11; Häcker, '03). The eighth blastomere divides, yielding the precursor of the primordial germ cell and a stem cell of the entoderm or its precursor. Chromatin diminution occurs in the remaining seven cells in the fourth cleavage division. The primordial germ cell possibly undergoes diminution during the fifth cleavage. Of special interest are differences in rates of development among sister embryos that are homozygous or heterozygous for heterochromatin-rich chromosomes, which Beermann noted at the third cleavage. These differences become more pronounced in succeeding cleavages.

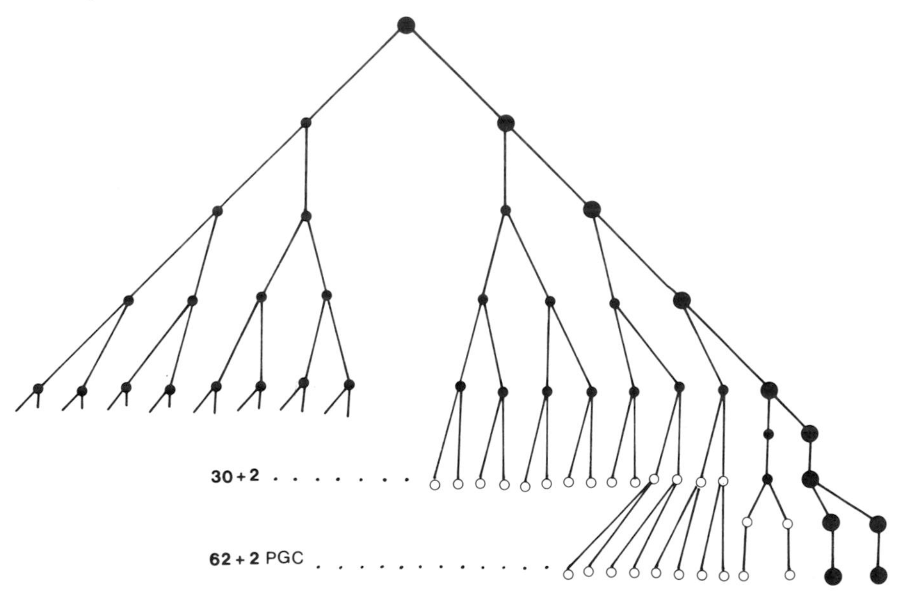

Fig. 5. Cleavage and diminution program in *Cyclops divulsus*. Open circles represent diminuted, small closed circles represent not diminuted (germ line large circles). PGC — primordial germ cell. (Redrawn from S. Beermann, '77, Chromosoma, with permission.)

The embryos are contained in an egg membrane and held within a sac until hatching. These eggs are spheroid in shape and vary considerably in size and color. The color may vary within a single species and appears to be due to variation in the carotenoid-protein links within plant food (Dupraw, '58). The size varies around 100 μm in diameter, according to species and locale. Larger clutches tend to have smaller eggs. Smaller eggs hatch into smaller nauplii (Wyngaard, unpublished). Hatching occurs when the permeability of the inner membrane changes, and the osmotic entry of water causes the inner membrane to rupture (Davis, '59).

CYTOGENETIC STUDIES ON CYCLOPOIDS: BACKGROUND AND CURRENT STATUS

In 1912, Chambers reported the chromosome numbers of eight North American cyclopoids: *Macrocyclops albidus* (Jurine), *Diacyclops bicuspidatus thomasi* (S. A. Forbes), *Cyclops fuscus* (Jurine), *Orthocyclops modestus* (Herrick), *Acanthocyclops viridis* (Jurine), *A. americanus* (Marsh, 1893), *A. viridis robustus* (Sars), and *A. vernalis* (Fischer). By 1951 Makino was able to list the chromosome numbers of 22 freshwater cyclopoid copepods reported from Europe and North America. Since then the bulk of the cytogenetic information of copepods has been based on studies of European species of *Cyclops* (O. F. Müller), *Ectocyclops* (Brady), *Acanthocyclops* (Kiefer), *Eucyclops* (Claus), and *Macrocyclops* (Claus). The cytology of copepods is remarkable in several aspects. This chromosomal diversity may provide useful taxonomic indications at the species level. In particular, the unusual chromosomal features that have been found among copepods include 1) achiasmate meiosis in females (Matschek, '10; W. Beermann, '54; S. Beermann, '59; Rüsch, '60; Chinnappa, '80); 2) chiasmate meiosis in males (S. Beermann, '77); 3) diverse sex chromosome mechanisms such as OO, XO, XY types (W. Beermann, '54; Rüsch, '60; Ar-Rushdi, '63; S. Beermann, '77; Chinnappa and Victor, '79b); 4) complex heterozygosity (Chinnappa and Victor, '79a); and 5) variations in both heterochromatin content and size of chromosome sets at the diploid level, as well as size differences in haploid sets (S. Beermann, '77). An interesting drawback to a complete cytogenetic analysis of copepods is that it is often difficult to obtain usable chromosomal preparations from males (S. Beermann, '77; Chinnappa and Victor, '79b). It is also common that cyclopoids are often heavily infested with protozoan parasites to the extent that chromosomal preparation from gonads is not possible. Intensive parasitization of this type has also been recorded by Michajlow ('72). In addition to cytogenetic studies, genetic studies in the future may also aid in delimiting "morphospecies." The chromosome numbers so far recorded for cyclopoids by previous authors are presented in

TABLE V. Known Chromosome Numbers of Cyclopoid Copepods

Species	2n	n	Authority
Macrocyclops albidus (Jurine, 1820)	14	7X	Matscheck, 1909
			Braun, 1909
			Chambers, 1912
			Chinnappa and Victor, 1979b
Macrocyclops ater (Herrick, 1882)	22	7X	Chinnappa and Victor, 1979b
Macrocyclops fuscus (Jurine, 1820)	14	7X	Braun, 1909
			Matscheck, 1910
			Chambers, 1912
			Stella, 1931[a]
			Rüsch, 1960
Macrocyclops fuscus distinctus	11		Braun, 1909
			Amma, 1911[a]
	14	7X	Stella, 1931[a]
Eucyclops serrulatus (Fischer, 1851)	14	6+2XX	Matscheck, 1910
	13	6+XX	Rüsch, 1960
			Chinnappa and Victor, 1979b
Encyclops speratus (Lilljoborg, 1901)	13	6+XX	Chinnappa and Victor, 1979b
Tropocyclops prasinus (Fischer, 1860)	11	5+XX	Braun, 1909
			Matscheck, 1910
Tropocyclops prasinus mexicanus (Kiefer, 1938)	11	5+XX	Chinnappa and Victor, 1979b
Paracyclops affinis (Sars, 1863)	14	7X	Matscheck, 1910
Ectocyclops strenzkei Herbst	11	5+XX	W. Beermann, 1954
	12	5+XΞ	W. Beermann, 1954
Ectocyclops phaleratus (Koch, 1838)	13	6+X	Braun, 1909
			Matscheck, 1910
Mesocyclops edax (S.A. Forbes, 1891)	14	7X	Chinnappa and Victor, 1979a, b
Mesocyclops leuckarti (Claus, 1857)	14	7X	Braun, 1909
			Matscheck, 1910
	12	6X	Chinnappa and Victor, 1979b
Cyclops insignis Claus, (1857)	22	11X	Amma, 1911
			Braun, 1909
			Matscheck, 1910
Cyclops fuscus	8	4	Häcker, 1872[a]
Cyclops strenuus Fischer, (1851)	22	11X	Braun, 1909
			Häcker, 1892[a], 1896[a]
			Lérat, 1905
			Matscheck, 1910
			Amma, 1911
Cyclops abyssorum Sars, (1863)	22	11X	S. Beermann, 1977
Cyclops furcifer	22	11X	S. Beermann, 1977

Continued on next page.

TABLE V. Known Chromosome Numbers of Cyclopoid Copepods, continued

Species	2n	n	Authority
Claus, (1857)			
Acanthocyclops vernalis (Fischer, 1853)	10	5X	Braun, 1909
	11	5+X	Matscheck, 1910
	10	4+XYX	Rüsch, 1960
	10	4+XXX	Rüsch, 1960
	8	4X	Chinnappa and Victor, 1979b
Acanthocyclops viridis (Jurine, 1820)	12	6X	Amma, 1911 Chambers, 1912 Braun, 1909 Matscheck, 1910 Stella, 1931
	12	5+X+X	Rüsch, 1960
	12	5+XXΞ	Rüsch, 1960
Acanthocyclops viridis	24	12X	Häcker, 1895, 1897, 1903, 1904[a]
Acanthocyclops americanus (Marsh, 1893)	10	5XΞ	Chambers, 1912
A. robustus (Sars, 1863)	4	2X	Chambers, 1912
A. vernalis	6	3X	Chambers, 1912
A. robustus	6	3XΞ	Rüsch, 1960
Diacyclops bicuspidatus (Claus, 1857)	18	9X	Braun, 1909 Matscheck, 1909
D. b. thomasi (S.A. Forbes, 1882)		9X	Chambers, 1912 Chinnappa and Victor, 1979b
D. b. odessanus Schmankevitch	18	9X	Braun, 1909
Metacyclops gracilis (Lilljoborg, 1853)	6	3X	Braun, 1909 Matscheck, 1910
Cryptocyclops bicolor (Sars, 1863)	12	6X	Braun, 1909
Thermocyclops dybowskii (Landi, 1890)	18	9X	Braun, 1909 Matscheck, 1910
Orthocyclops modestus (Herrick, 1853)	8	—	Chambers, 1912

Published nomenclature has been updated according to Dussart ('69) and Yeatman ('59).
[a]See Makino ('51).

Table V. Some additions of chromosome numbers for material studied from North America, Asia, Europe, and the Caribbean are given in Table VI.

From recent cytogenetic studies (Chinnappa and Victor, '79a, b) three categories of cyclopoids can be separated.

TABLE VI. Chromosome Numbers of Cylopoid Copepoda Examined in the Present Study[a]

Species	nX	Location
Thermocyclops crassus (Fischer)	7	Sri Lanka
Mesocyclops leuckarti (Claus)	7	Sri Lanka
Acanthocyclops latipes (Lowndes)	3	Austria
Macrocyclops sp.	7	Austria
Cyclops strenuus (Fischer)	11	Austria
Eucyclops serrulatus (Fischer)	6 + x	Austria
Macrocyclops cf. albidus (Jurine)	7	Haiti
Mesocyclops leuckarti (Claus)	7	Lake District, Britain
Acanthyocyclops viridis (Jurine)	6	Czechoslovakia
Cyclops vicinus Uljanin	11	Czechoslovakia
Acanthocyclops americanus (Marsh)	3	Czechoslovakia
Macrocyclops fuscus (Jurine)	7	United States
M. albidus (Jurine)	7	United States
Eucyclops agilis (Koch)	6 + x	United States
Tropocyclops prasinus mexicanus Kiefer	5 + x	United States
Microcyclops viricans rubellus (Lillj.)	3	United States
Mesocyclops edax (Forbes)	7	United States
Acanthocyclops vernalis (Fischer)	4	United States

[a]Some species have only been tentatively identified.

Category 1: Holartic species

Taxonomically defined North American copepods whose chromosome number (2n) differs from the European forms of the same "species." For example, the North American forms of *Acanthocyclops vernalis* (Fischer) and *Mesocyclops leuckarti* (Claus) have diploid chromosome constitutions of 8 and 12 respectively. By contrast, the European forms of *M. leuckarti* have 2n=14 (Braun, '09; Matschek, '10) and the 2n for the European *A. vernalis* is 10 (Rüsch, '60). Another pattern occurs with *A. americanus* (Marsh). In North America the 2n value for this species is 10 whereas in Europe it is 6.

Generally this "vernalis" group is considered as a single species, *Acanthocyclops vernalis* (Price, '58; Yeatman, '59; Smith, '77) with *Cyclops parcus* Herrick, 1882, *C. robustus* Sars, 1863, *C. brevispinosus* Herrick, 1884, and *C. americanus* Marsh, 1893, as synonyms. However, breeding experiments suggested that there are, at least, seven reproductive isolates of *A. vernalis* (Price, '58). Moreover, Price ('58) concluded that isolate C was actually *C. brevispinosus*. Currently, *C. viridis* is considered equivalent to *A. vernalis* (Gurney, '33; Price, '58; Yeatman, '59). Thus under the heading "*A. vernalis*" one finds *C. brevispinosus, C. parcus,* and *C. americanus,* which have 2n values of 4, 6, and 10, respectively, whereas *A. vernalis* as defined by conventional taxonomic criteria has a 2n value of 8. Consequently, in this vernalis group alone one finds that the various forms have two, three, four, five, and six chromosome pairs. Indeed, the vernalis group is a species complex! Sex chromosome differentiation is a feature of certain European copepods (W. Beermann, '54; Rüsch, '60). *A. vernalis*, for example, shows female structural digamety of the XY type (Rüsch, '60). By contrast, not all of the members of the North American vernalis complex have sex chromosomes in the female. In fact, of ten species of North American copepods examined to date only two have sex chromosome differentiation (Chinnappa and Victor, '79b).

Category 2: Holarctic Species With the Same Diploid Value as Other Closely Related Species Reported Previously

For example, the North American form *Eucyclops speratus* Lilljoborg (2n=12+X) has the same chromosome constitution as *C. serrulatus* Herrick (Chinnappa and Victor, '79b). Taxonomically, Yeatman ('59) placed Herrick's subspecies *elegans* as a synonym of the European *Eucyclops speratus*. The Ontario specimens of *E. speratus* differ from the European forms with respect to spinules that occur along the entire length of the ramus (Smith, '77). *E. serrulatus* is a smaller organism with caudal rami of variable length (4–5 times the breadth), and differs from *E. speratus* in that it has longer rami (6–8 times the breadth). Moreover, these "species" often are found together (Smith, '77). Since the chromosomal complement between these two forms does not differ, the morphological variations may represent a continuum and the designation *E. speratus* may be questionable. Whether these two forms are reproductively isolated remains to be tested.

Category 3: Species of Copepods That May Be Uniquely North American

An example of such an endemic species is *Mesocyclops edax* S. A. Forbes. In this species the ring association of chromosomes during meiosis that includes all 14 chromosomes and other variations (Chinnappa and Victor, '79a; present report) is a chromosomal configuration that, to date, is unique among animals.

Certainly, the application of cytogenetic techniques to help elucidate taxonomic problems is not novel; but such an approach to the study of copepods could be of some help in classifying whether current species designations are legitimate. As well, it is anticipated that such investigations will uncover systems that would, in the short term, require that operational definitions of biological species be used. In the latter instances, the cytogenetic evidence alone may be either insufficient or inadequate to be able to delimit precisely a particular species. Studies so far have, however, been instructive. Strains of copepods with "identical" morphology have been found to have divergent chromosomal complements. By contrast, strains with differing morphologies, as judged by taxonomic criteria, have similar chromosome complements. Finally, copepods have been discovered with unique chromosome complements. That some widely separated copepod strains have retained morphological constancy while revealing considerable chromosomal plasticity presents an intriguing problem from an evolutionary viewpoint.

Future studies in cytotaxonomy can be used to answer specific questions about the specific status of the following copepods: 1) Is the *strenuus* group—i.e., *Cyclops strenuus, C. strenuus abyssorum, C. prealpinus, C. vicinus,* and *C. furcifer*—a species complex? 2) Does the North American *Diacyclops bicuspidatus thomasi* and its European relative *D. bicuspidatus* share the same chromosome complement? 3) Is the *Acanthocyclops vernalis* complex an artificial lumping together of diverse species? 4) Does the homogeneity among widely distributed species—e.g., *Mesocyclops leuckarti* and *Thermocyclops crassus*—signify chromosomal identity?

CYTOLOGY OF *MESOCYCLOPS EDAX* AND *CYCLOPS S. STR.*

Mesocyclops edax, 2n=14 and *Cyclops s. str. (C. strenuus, C. furcifer, C. divulsus)* 2n=22, show similarities in several cytological features. These are with respect to their chromosome morphology, heterozygosity, and chromatin diminution of heterochromatin early in embryogenesis. Studies on the cytology of the genus *Cyclops* are extensive (Häcker, 1895, Rückert, 1895; Braun, '09; Matschek, '10; Lérat, '05; Amma, '11; S. Beermann, '59, '66, '77; Einsle, '64, '68), whereas studies on *M. edax* have been restricted (Chinnappa and Victor, '79; Chinnappa, '80). Oogenesis and spermatogenesis essentially follow the same pattern in the two taxa. Meiosis in females is achiasmate as in all cyclopoids, whereas male meiosis is probably chiasmate. Extreme stickiness and clumping of male meiotic chromosomes, however, present problems in obtaining good squash preparations (S. Beermann, '77; Chinnappa, '80).

Oogenesis in Bivalent Forming Races of *Mesocyclops edax*

Early prophase nuclei characteristically show darkly stained spheres, formed by the heterochromatic regions of the chromosomes, whereas the euchromatic regions are threadlike (Fig. 6). During the growth phase of the oocyte, homologous chromosomes become distinctly separated and soon the distance-parallel pairing is established (Fig. 7). Parallel pairing of the homologous chromosomes is restricted to the median euchromatic regions whereas the heterochromatic ends do not show pairing affinities (Figs. 8, 9). Though the distance-parallel pairing is clear, it appears as though the homologous euchromatic regions are often interconnected by certain fine fibrillar structures (Fig. 8). Whether these are synaptonemal complexes, or merely the chromatin fibrils radiating from the chromosomes, is difficult to ascertain at this stage. The presence of synaptonemal complexes has, however, been suggested to occur, and serves as a mechanism for proper disjunction of chromosomes at anaphase I in achiasmate organisms (Gassner, '69; Rasmussen, '77). Synaptonemal complexes have also been observed in the chiasmate organisms, and their function has been interpreted in maintaining and regulating bivalent structure, chiasma maintenance and meiotic chromosome disjunction (Maguire, '78, '79; Moses, '68; Moens and Church, '79). In the ring-forming races of *M. edax* the euchromatic regions of adjacent chromosomes are connected by interchromosomal fibrils (Fig. 11). By diakinesis the chromosomes become shorter (Fig. 9), and during metaphase I the bivalents orient themselves in the equatorial plane and the unusual bending of the H segments of the chromosomes becomes established. By early anaphase I the H segments of seven bivalent chromosomes are bent and are directed away from the metaphase plate (Fig. 10). The underlying mechanism of the unusual orientation of metaphase I chromosomes in *M. edax* oocytes cannot be explained at this time. Anaphase I disjunction of 7/7 chromosomes is regularly observed. The second meiotic division could not be observed due to technical difficulties.

Oogenesis in Ring-Forming Races (Female Meiosis)

Initial stages of oocyte development resemble that of bivalent forming races. The chromosome configurations, however, cannot be clearly observed until diplotene. At this stage the chromosomes arrange themselves in a closed chain configuration, and all fourteen chromosomes can be identified (Fig. 11). This ringlike configuration is attained by the participation of adjacent chromosomes. The euchromatic regions of individual chromosomes tend to pair with those of two different adjacent chromosomes. There is no indication of chiasma formation at the region of pairing. The chromosomes are held together by fibrillar connections of unknown derivation.

In the ring the chromsomes are designated as chromosomes 1 to 14. The sequence of the chromosomes is represented in Figure 15. The pattern of chromosome association and the position of each chromosome in the ring is constant. The chromosomes in the figure are labeled in accordance with Belling's segmental interchange theory, under which a normal chromosome would be designated Aa, whereas an interchange chromosome resulting from reciprocal translocation between chromosomes Aa and Bb would be desig-

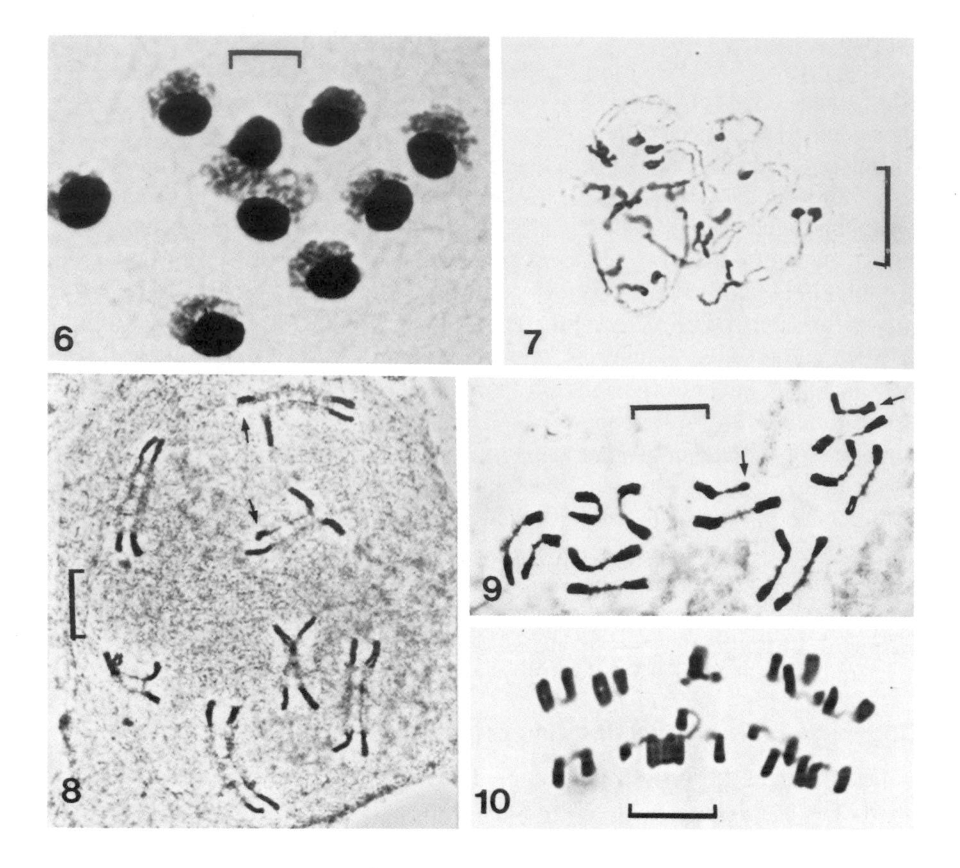

Figs. 6–10. *Mesocyclops edax*. Oogenesis. In all figures the bar represents 10 μm. (Figures 7, 8, and 10 are taken from Chinnappa, '80, with permission.)

Fig. 6. Early pachytene stage. Terminal H segments of chromosomes condensed into darkly stained blocks.

Fig. 7. Late pachytene showing seven bivalents.

Fig. 8. Diplotene bivalents showing distance-parallel pairing. Euchromatic regions of the homologues show interchromosomal connections. Two chromosomes are with shorter H segments (arrows).

Fig. 9. Early metaphase I. Seven bivalents. Arrows indicate chromosomes with short H segments.

Fig. 10. Early anaphase I chromosomes showing characteristic bending of H segments.

nated as aB. The mode of distribution of the chromosomes is that adjacent chromosomes pass to opposite poles. This feature is demonstrated in Figure 12. Thus two groups of chromosomes (maternal and paternal) are considered as two complexes. One complex is composed of Aa, Bb, Cc, Dd, Ee, Ff, and Gg (Fig. 12, inner circle). The other complex is composed of aB, bC, cD, dE, eF, fG, and gA (Fig. 12, outer circle). Within any individual female the pattern of association of chromosomes in the ring appears constant. The ring association of chromosomes continues to persist through diakinesis and metaphase 1. At anaphase 1 alternate disjunction of the chromosomes results in a 7:7 distribution (Figs. 13, 14).

The ring association of chromosomes in *Mesocyclops edax* was interpreted as a result of structural heterozygosity resulting from multiple sequential interchanges (Chinnappa and Victor, '79b). This type of "complex heterozygosity" can be compared to that known in the plant genus *Oenothera* 2n=14. The main contributions on the cytogenetics of *Oenothera* by Renner and his school ('17, '19, '25, '43) and Cleland and his co-workers ('22, '23, '57, '62) in understanding the evolution of the system are classics today.

The principal features of Oenothera genetics can be summarized as follows: Most Oenotheras are "complex heterozygotes," as Renner called them. Each race has two different complexes or permanent genomes. This results from 1) the inclusion of all genes within a single linkage group for which the plant is heterozygous, 2) the presence of one or more lethals in each genome, and 3) self-pollination. Hybrids derived from crossing of true-breeding complex heterozygotes, however, may have two or more linkage groups. Genes linked in the parents are not necessarily linked in the hybrids (Cleland, '62). The presence of extensive chromosome linkage in *Oenothera* was related to the cytological basis for the extensive genetic linkage that was found to exist in various races. The plants that form only bivalents lack lethals, and have large, open-pollinated flowers. The large ring-forming race possess balanced lethals and small self-pollinating flowers. The other major group of races have large, open-pollinated flowers and an absence of lethals. These have developed greater heterogeneity with regard to chromosome configurations. Several cytotypes can be encountered in a local population, mostly intermediate between large rings of chromosomes and the all bivalent condition. Plants that possess one or more rings of chromosomes do not ordinarily breed true for these ring types, since they lack lethal factors. Since outcrossing is the rule, however, crosses between plants whose gametes have diverse segmental arrangements are constantly producing plants with rings, so that a balance tends to be maintained between plants with and without ring configurations (Cleland, '62).

Studies on the North American Oenotheras have shown that practically all races found from the Rocky Mountains eastward are complex heterozygotes,

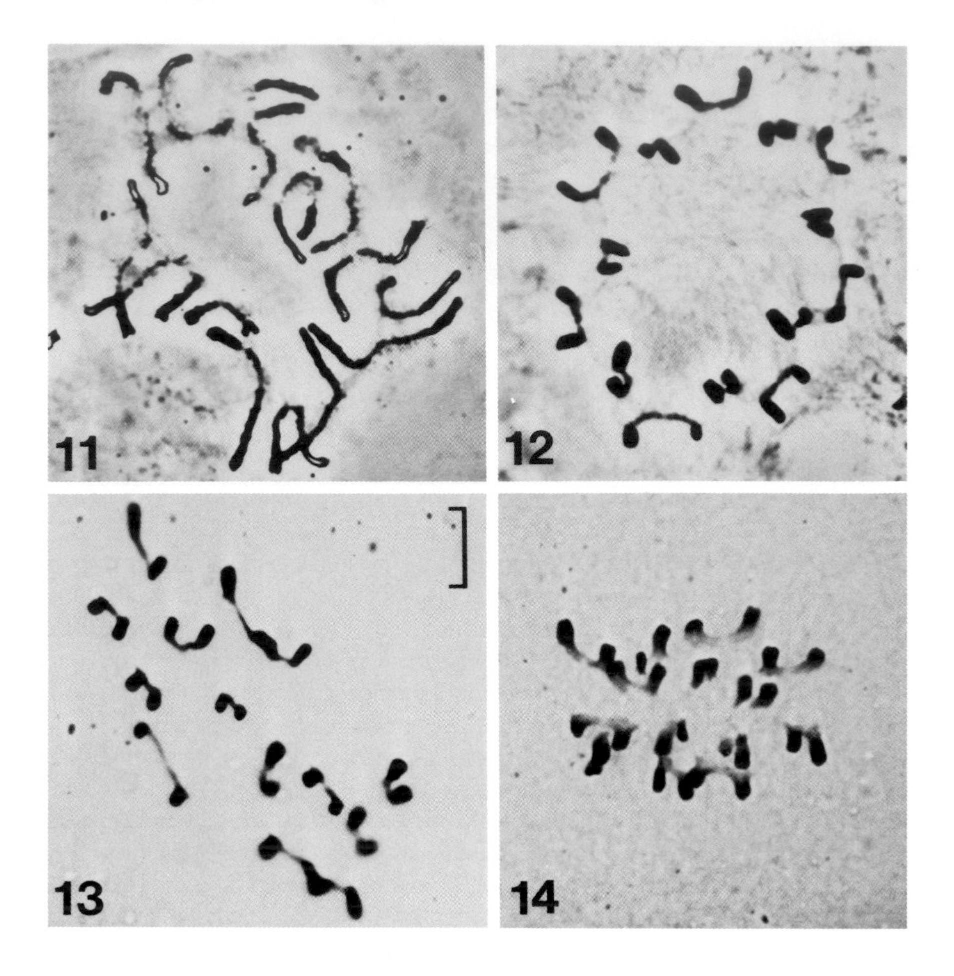

Figs. 11–14. *Mesocyclops edax.* Oogonial meiosis in ring-forming races.

Figs. 11, 12. Diplotene and late diakinesis stages with rings of 14 chromosomes in each. Note the interchromosomal connections between the euchromatic regions of adjacent chromosomes.

Figs. 13, 14. Metaphase I and early anaphase I showing alternate disjunction of chromosomes. (From Chinnappa and Victor, '79a, with permission.)

with balanced lethals and self-pollination. By contrast, the Oenotheras found in the southwestern and western parts of the United States and northern Mexico are largely complex homozygotes, open-pollinated and free of lethals. At present, the ring-formers can be divided into five groups that are sufficiently distinct in their cytological, morphological, and ecological char-

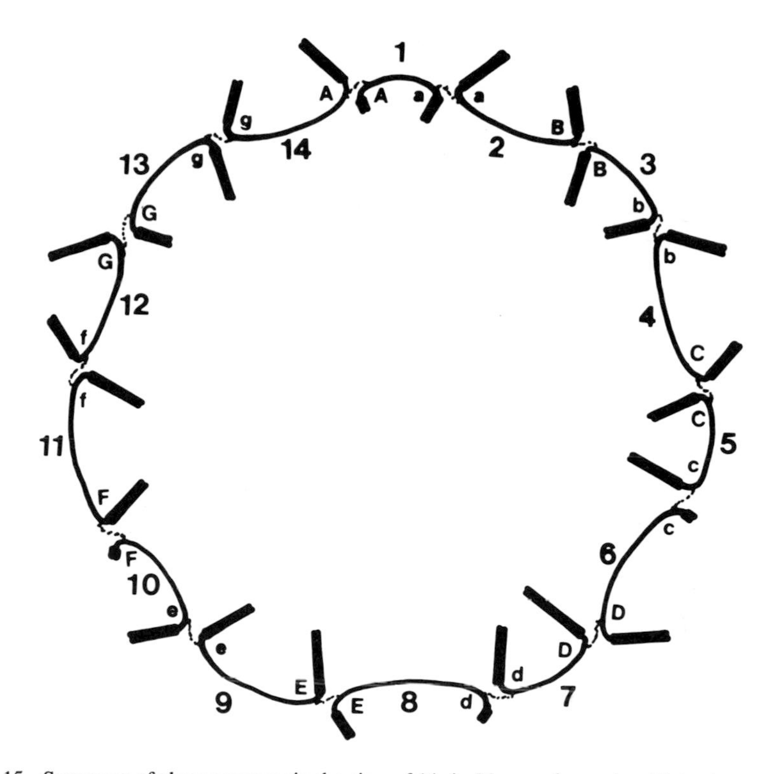

Fig. 15. Sequence of chromosomes in the ring of 14, in *Mesocyclops edax*. Note chromosomes 1, 6, 8, and 10 carry smaller H segments in one of the arms. The chromosomes are labeled in accordance with the segmental interchange theory. (Redrawn from Chinnappa and Victor, '79a, with permission.)

acteristics. North American Oenotheras, on the one hand, show a great variety of phenotypic expression. On the other hand, these variations often grade into one another so intricately that the systematists find it very difficult to discover clear-cut boundary lines separating one taxon from another. The unusual genetic system evolved in *Oenothera* has produced through much of its range a unique type of population structure.

Whereas bivalent forming races of 2n = 14 are homozygous for the seven-chromosome genome, the races with a ring of 14 chromosomes are complex heterozygotes. Theoretically, assuming that a ring always has an even number of chromosomes in it, 15 arrangements of 14 chromosomes into rings and pairs are possible. All these configurations have been encountered in *Mesocyclops edax* in individual females analyzed from limited numbers of populations. The possible configurations and their distribution are presented in

TABLE VII. Possible Configurations of 14 Chromosomes Into Rings and Pairs, and Their Frequency Occurrences in Various Populations of *Mesocyclops edax*

	Banister Lake, Onatario	Guelph Lake, Ontario	Long Lake, Ontario	Heart Lake, Ontario	Valens Lake, Ontario	Maplehurst Lake, Ontario	Nipissing Lake, Ontario	Douglas Lake, Michigan	Lake Thonotosassa, Florida	Fairy Lake, Florida
1. Ring of 14	16	10	7	—	-	2	3	10	6	—
2. Ring of 10 + ring of 4	—	1	—	—	1	—	—	—	—	—
3. Ring of 8 + ring of 6	2	7	—	—	—	—	—	—	—	—
4. Ring of 6 + two rings of 4	—	1	—	—	—	—	—	—	—	—
5. Ring of 12 + 1 pair	21	6	4	11	2	3	—	9	—	—
6. Ring of 8 + ring of 4 + 1 pair	—	1	—	4	—	—	—	—	—	—
7. Two rings of 6 + 1 pair	1	—	—	—	—	—	—	—	—	—
8. Three rings of 4 + 1 pair	—	—	—	5	1	—	—	—	—	—
9. Ring of 10 + 2 pairs	13	1	—	4	1	—	—	5	—	—
10. Ring of 6 + ring of 4 + 2 pairs		1	—	8	—	—	—	—	—	—
11. Ring of 8 + 3 pairs	—	1	1	1	—	—	—	5	—	—
12. Two rings of 4 + 3 pairs	1	—	—	—	—	—	—	—	—	—
13. Ring of 6 + 4 pairs	—	—	—	3	—	1	—	1	—	—
14. Ring of 4 + 5 pairs	—	—	—	—	—	—	2	—	1	26
15. Seven pairs	—	—	27	—	—	3	2	3	35	10
Total number of animals analyzed	54	29	39	36	5	9	7	33	42	36

Table VII. Figures 16–23 are examples of some of the configurations from various heterozygote females. Heterozygosity of a different nature has been reported in *Cyclops* (S. Beermann, '77). Females of *Cyclops strenuus* (2n=22), from one population, are heterozygous for an entire set of large and a second set of small chromosomes in their germ lines. In metaphase I, the 11 bivalents of heterozygous oocytes regularly arrange themselves in such a manner that all of the large dyads move to one pole and the small ones to the other. This extreme genome dimorphism in the females is suggested to persist, on the basis of female heterogameticity and coordinated orientation and segregation of all the two types of chromosomes during meiosis I. Males are homozygous for the large set.

The evolution of such "similar" cytogenetic systems in plant and animal species is of interest. Taking into consideration the biological differences between these two unrelated organisms, it is important to discover the features and strategies of population structures in *Mesocyclops edax* and their mode of evolution and what the possibilities may be for future evolutionary development in the cylopoid species.

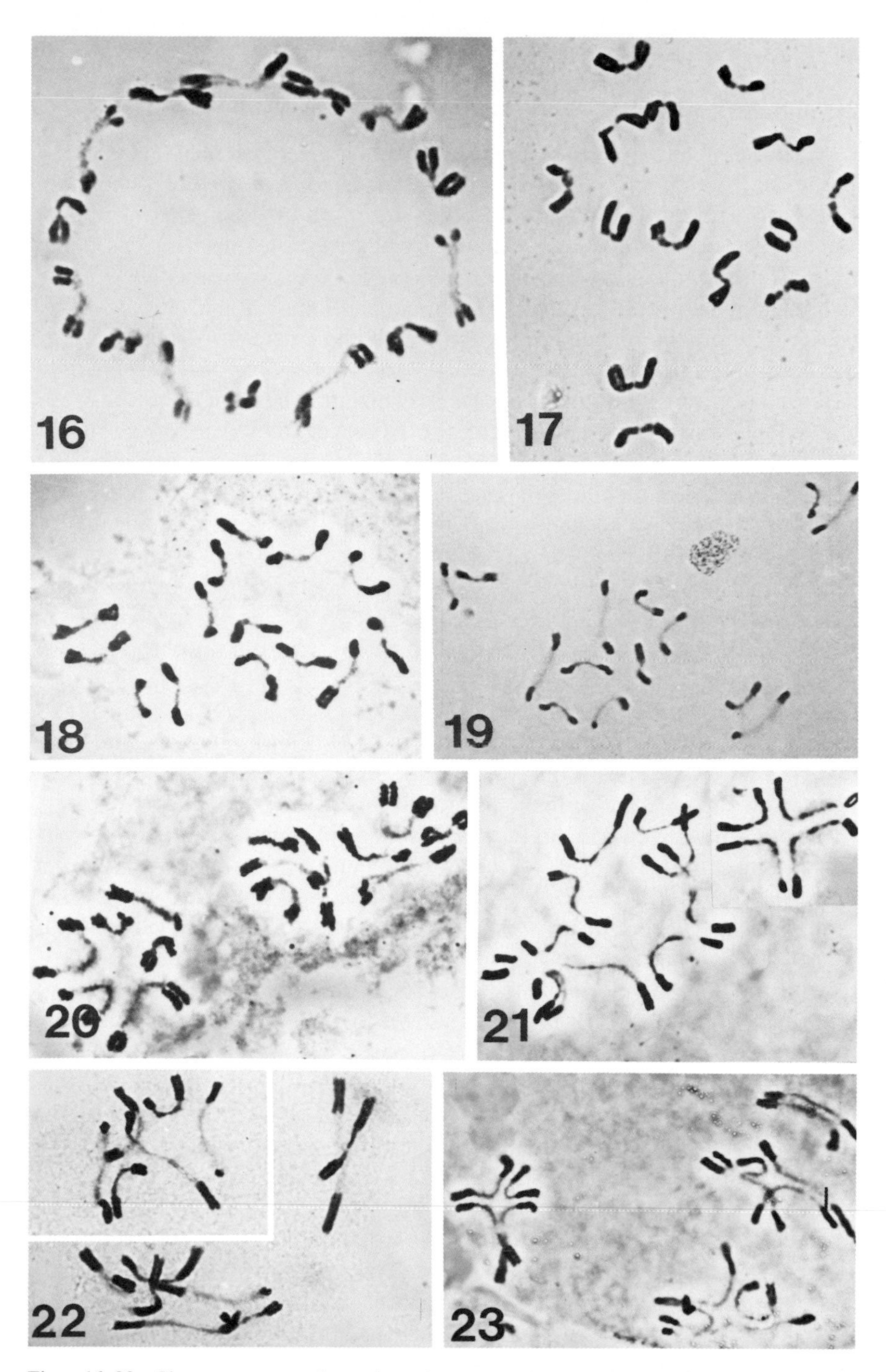

Figs. 16–23. Chromosome configurations in *Mesocyclops edax* from various populations. Fig. 16—ring of 14; Fig. 17—ring of 12 + 1 pair; Fig. 18—ring of 10 + 2 pairs; Fig. 19—ring of 8 + 3 pairs; Fig. 20—rings of 6, 4, 4; Fig. 21—rings of 10, 4; Fig. 22—rings of 6,6 + 1 pair; Fig. 23—rings of 4, 4, 4 + 1 pair.

Chromosome Morphology and Chromatin Diminution

Structure of chromosomes can be studied during oogonial mitosis and oogenesis. The chromosome complement for *Mesocyclops edax* is shown in Figures 24 and 25. The 14 chromosomes are more or less metacentric, and the monocentric nature is demonstrated by anaphase flexture (Fig. 24). The chromosomes in the germ line are characteristic in having a median euchromatic region and large terminal H segments. Some of the chromosomes clearly show smaller intercalary H segments at the kinetochoric regions (Fig. 25). Two chromosomes show secondary constrictions on the H segments, and they are presumably the nucleolar organizers. These NOR regions lose their identity during diplotene-diakinesis. However, they can be recognized in some cells where they bear satellite-like structures. The measurements of the chromosomes at diplotene-diakinesis stages from a bivalent forming race suggest that there are no major size differences in the chromosomes; they could however, be arranged in decreasing order of size. There are two chromosomes that consistently show short H segments (Fig. 9, arrows). The differences in the size of H segments reflect in the occurrence of hetermorphic bivalents. If, however, the euchromatic regions of the chromosomes are taken into consideration, they appear to be constant for a given homologous pair. The diploid complement can then be arranged in accordance with the size and be designated as chromosomes 1–14. On the basis of measurements the karyotype arrangement of seven homologous pairs from one animal is shown in Figure 26. The two chromosomes that have smaller H segments are chromosomes 2 and 14. In the case of ring-forming races there are four chromosomes with shorter H segments. The constancy in the lengths of the euchromatic regions and variation in the lengths of H segments suggest that the morphology of the chromosomes appears to be affected by the change in the amounts of heterochromatin on the distal ends of the chromosomes. Heterochromatic polymorphism has also been reported to occur in *Cyclops* species (Beermann, '77). It has been shown that some species of the genus *Cyclops* (2n=22) have several homologous chromosomes with different amounts of germ-line-limited heterochromatin. In *Cyclops strenuus* this heteromorphism is generally expressed only in sizes of the chromosomes. In *C. furcifer* and *C. divulus*, which have chromosome morphology similar to that of *M. edax*, homologous H segments show length differences. This kind of polymorphism, which was found in the females from widely separated populations, showed striking similarities in the type of heterochromatic diplotene bivalents. It has been suggested (S. Beermann, '77) that such observed variability of homologous H segments is maintained as a component of a balanced polymorphism in the populations.

Whether the changes in chromosome morphology and complex heterozygosity seen in *Mesocyclops edax* has any effect on the physiology and

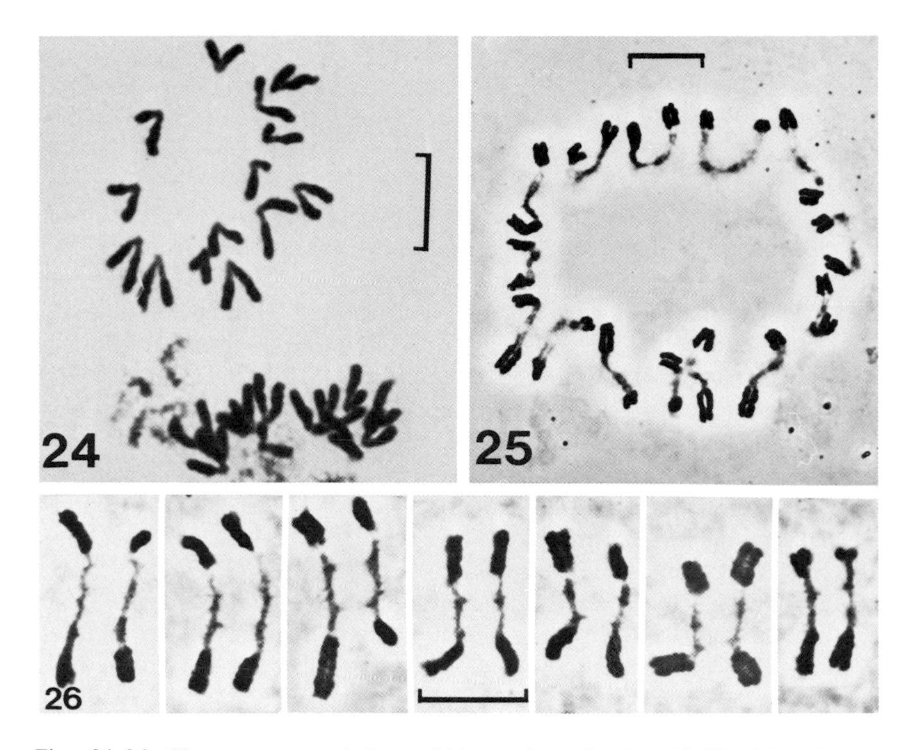

Figs. 24–26. Chromosome morphology of *Mesocyclops edax,* 2n=14. Fig. 24. Miotic anaphase chromosomes in oocyte. Fig. 25. Chromosomes at diplotene in oocyte. Fig. 26. Diplotene bivalents of homozygous female. Most bivalents show interchromosomal connections. Bivalents 1 and 7 are heteromorphic for H-segments. (Figures 24 and 25 are from Chinnappa and Victor, '79a, with permission.)

morphology of the animal is not known. Such investigations are worth pursuing to understand the cytogenetic effects on various populations of *M. edax.* The Long Lake population with seven bivalents is presumed to be homozygous for a seven-chromosome genome. The existence of such races is of considerable importance particularly to establish the basic chromosome morphology and to study the consequences of interchanges observed in various populations. The origin and evolution of complex heterozygosity coupled with achiasmate meiosis in the cyclopoids and their possible cytogenetic implications are important aspects of future studies.

Some cyclopoids are known to have an unusual feature closely related to that described in some Ascarids—that is, chromatin diminution taking place during cleavage divisions of the embryonic egg. The early development of several copepods is characterized by an elimination of up to 50% or more of chromosomal DNA from the nuclei of future somatic cells (S. Beermann, '77). Details of diminution division in three species of *Cyclops* have been well documented by S. Beermann ('59, '77). The diminution in the three species basically follows the same pattern, but they differ as to the time and duration of the diminution process. In *C. divulsus,* elimination is completed in one step and is confined to the fifth cleavage division (Fig. 5; Table IV). In *C. furcifer,* diminution begins in the sixth cleavage division and is not completed until after the seventh cleavage. In the interphase before the sixth cleavage, the eliminated chromatin first becomes visible in the form of numerous very small particles dispersed throughout the nucleus in the early prophase of elimination division. Only as the division proceeds do the particles merge into larger more conspicuous ones that fill the spindle during meta- and anaphase. It was suggested that the elimination of chromatin was in the form of smaller fragments and that the entire H segments were being totally eliminated. In the case of *Mesocyclops edax*, however, the chromatin diminution during early cleavage stages suggests that the elimination of H segments is completed in one step (Chinnappa, '80). Early stages until third cleavage show chromosome morphology unchanged (Fig. 27). At fourth cleavage division, chromatin elimination is observed. During anaphase, up to 26 darkly stained large chromatin bodies could be counted at the equatorial plate (Fig. 28). There are also smaller chromatin particles. At the poles the chromosomes are euchromatic since having lost the heterochromatin.

S. Beermann's studies ('59, '66, '77) indicate that the eliminated chromatin derives from the disintegration of distinct heterochromatic segments. It has been suggested further that during diminution the interstitial as well as terminal DNA segments are removed without disrupting the chromosome morphology in *Cyclops furcifer*. A model was presented (S. Beermann, '66) to demonstrate that the DNA to be eliminated will first arrange as a series of loops which will then be released from the chromosome in the form of rings. This feature of chromatin ring formation as a product of chromatin diminution has been further demonstrated by means of electron microscopy using a microspreading procedure (S. Beermann and Meyer, '80). The differentiation into soma and germ line of Copepods and Ascarids is an example of cell-differentiation phenomena on the basis of chromatin elimination. In both species the heterochromatin of the chromosomes of somatic cells is eliminated during early cleavage divisions, whereas it is preserved in the germ cells. In the three species of *Cyclops* studied by S. Beermann, there is twice as much DNA in the germ line genome as in the somatic genome, and heterochromatin-rich H segments of the germ line chromosomes mainly contain satellite DNA with highly repetitive, simple sequences. Moritz and Roth ('76) also noted that the germ-line-limited DNA in *Ascaris* consists entirely of highly repeated DNA sequences. Boveri ('10) and Mortiz ('67)

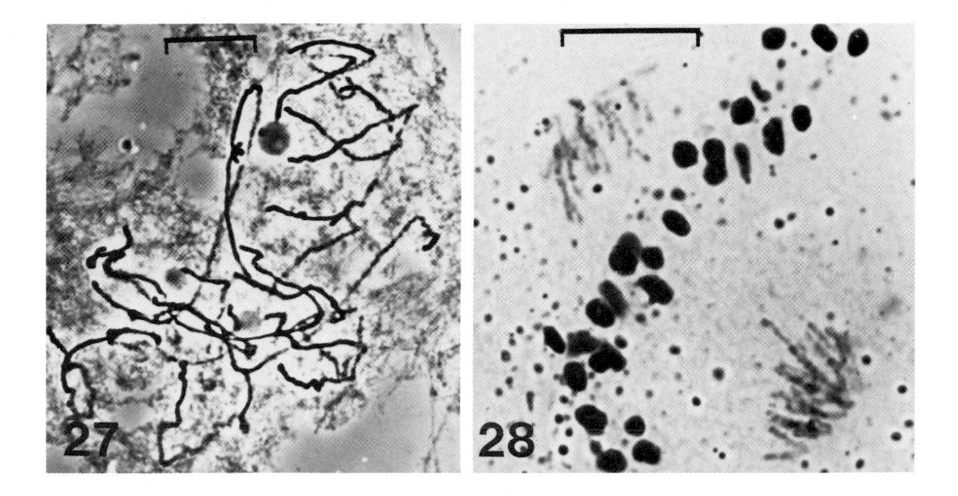

Figs. 27, 28. *Mesocyclops edax.* Cleavage mitosis. Fig. 27. Phrophase of second cleavage showing heteropychosis of terminal H segments. Fig. 28. Late anaphase at fourth cleavage division. Note that the eliminated chromatin at the equatorial plane and the chromosomes at the poles are euchromatic.

demonstrated in *Ascaris* that the persistence of germ-line-limited chromatin is a necessary criterion for germ-line quality of blastomeres. However, S. Beermann ('77) concluded that the elimination of extra chromatin does not serve as a basic mechanism controlling germ-line differentiation. An additional study of chromatin diminution in *C. strenuus* (Kiknadze, '72, cited in Terpilowska, '71) suggests that eliminated chromatin contains genes for a "nondifferentiated state" of blastomeres.

Variation in DNA content during embryogenesis has also been studied in the marine calanoid *Pseudocalanus* (Robins and McLaren, '82), *C. strenuus* (Beermann, '77), and *Acanthocyclops gigas* (Terpilowska, '71). The loss of chromatin in *Pseudocalanus* and *A. gigas* does not appear to involve chromatin diminution. Chromatin is lost during the first cleavage division in *Pseudocalanus*. Terpilowska ('71) found increased DNA content in the 16-cell blastula relative to the 8-cell and gastrula stages, as well as a gradual heterochromatization of chromosomes during embryogenesis (Terpilowska, '74). She hypothesizes that a mechanism other than those proposed by Stich ('62) and S. Beermann ('59, '66) must be involved in elimination of DNA, and that it is related to the process of differentiation.

It would be of considerable interest to know whether the germ-line genome in cyclopoids contains sequences not found in the somatic cells, and whether the highly repetitive satellite DNAs are responsible for homologous chromosome recognition in the case of organisms with achiasmate meiosis. As Roth ('79) states, dealing with the DNA properties of germ-line-limited DNA and organization of the somatic genomes in *Ascaris suum* and *Parascaris equorum*: "In relation to the DNA composition in germ line and soma of *Ascaris* there are two questions which remain to be answered. First, are there some 'originals' of the germ line satellite DNA sequences in the somatic genome, possibly at the telomeres of the somatic chromosomes and secondly soma? On the basis of the technology of DNA restrictions and gene cloning it is now possible to perform experiments to resolve both problems."

METHODS FOR CYTOLOGICAL STUDIES

Light Microscopy (Feulgen and Acetic-orcein Staining for Chromosomes)

Whole animals are fixed in a mixture of ethyl alcohol and glacial acetic acid (3:1) for 6 to 24 hours. They may be stored in 70% ethyl alcohol for about 15 days, if necessary. Storage for longer periods may not yield good results. To observe mitotic and meiotic divisions, the slides are prepared according to one of the following procedures: 1) The fixed animals are subjected to hydrolysis in 1N HCl at 55°C for 5–7 min. The animals are transferred to Schiff's reagent for 60 min, washed in SO_2 water, and transferred to distilled water. The animals are placed on a slide and the gonads are dissected out in 45% acetic acid. The tissue is gently teased apart, covered with a coverslip, and squashed firmly. 2) The gonads are dissected in 45% acetic acid and squash preparations are made with 2% acetic-orcein stain.

To analyze the cleavage divisions at appropriate stages, a few eggs are separated from freshly deposited egg batches after various intervals, and then fixed and stained either by the Feulgen method or with 2% acetic-orcein. To make permanent slides, place the freshly prepared slides on dry ice for at least 1 hour, pry off the coverslip, quickly dip the slide and coverslip in 9% ethanol.

C-Banding for Heterochromatic Segments of Chromosomes

Fix the material (gonads, testis, eggs) in 3:1 ethanol and acetic acid for at least 6 hours. To soften the tissue place the tissue in 45% acetic acid for 10 minutes. Then squash the tissue on a slide under a coverslip. The slides are frozen on dry ice, the coverslips are removed, and the slides are transferred to 95% ethanol for 5–10 minutes and then air-dried for 1 to several days. The air-dried slides are incubated for 10 minutes in a saturated solution of barium hydroxide (8 gm/100 ml water at 40°C), rinsed quickly in 0.1 N HCl and then in distilled water, dehydrated quickly through 50%, 70%, and 95%

and absolute ethanol, and air-dried. The slides are again incubated in 2 × SSC (SSC = 0.15 M NaCl, 0.015 M Na citrate, pH 7) at 60°C for 1 hour, rinsed in distilled water, and stained with 5% Geimsa (Gurr's improved R66, buffered to pH 6.8 with M/15 sodium phosphate buffer). Staining time varies up to 1 hour, depending upon the batch of slides. The slides are then rinsed in the same buffer, air-dried, and mounted using either Gurr's DEPEX or PERMOUNT.

N-Banding for Location of Nucleolar Organizers

The material is squashed according to the procedure mentioned for C-banding. The air-dried slides are incubated at 95°C for 5–10 minutes in 1 M NaH_2PO_4 solution (pH 4.2, adjusted with 1 N NaOH). After thorough rinsing in distilled water, they are stained for 20 minutes with Geimsa (diluted 1:2 with 1/15 phosphate buffer, pH 7), and rinsed in tap water, air-dried, and mounted using either DEPEX or PERMOUNT. Details are provided by Matsui and Sasaki ('73), Lin and Uchida ('73), and Webb ('76).

Feulgen Cytophotometry to Determine Nuclear DNA Content

Fix the tissue in 3:1 ethanol and acetic acid for 6 hours and squash on a thin slide with 50% acetic acid. The coverslips are pried off after the slides are frozen on dry ice. They are placed in isopropanol for 8–10 days at 4°C. Hen blood was diluted with 4% sodium citrate and smeared next to the cyclopoid tissue (these nuclei act as standards). The slides are subjected to a second fixation treatment in 4:1 ethanol and 40% formaldehyde for 30 minutes. They are then exposed to 5 N HCl at 25°C for 60 minutes. The slides are Feulgen-stained by immersing in Schiff's reagent for 60 minutes. They are washed in distilled water and SO_2 water, passed through the alcohol series to xylol, and mounted in DEPEX. The slides are stored in the dark to prevent fading. The stained preparations are scanned either with a microspectrophotometer (UMSP Carl Zeiss; S. Beermann, '77) or with a microdensetometer (John and Hewitt, '66). The amount of stain per nucleus is measured at a wavelength of 560 nm. In any given experiment, the fixation, hydrolysis, and staining should be carried out at the same time and with the same batch of reagents and the same equipment.

Electron Microscopy

Material is fixed in 4% cold phosphate-buffered (pH 7) glutaraldehyde. If eggs are to be processed they are punctured with a very fine needle and fixed at 4°C for several days. They are washed in phosphate buffer for several hours and are osmicated in a 2% phosphate-buffered OsO_4 solution for at least 24 hours at room temperature. They are then washed in buffer followed by distilled water. At this stage they may be subjected to a prestaining fixation in 2% aqueous uranyl acetate solution at 60°C for 1–3 hours. Finally, they are dehydrated through an alcohol series followed by propylene oxide and embedded in Epon. Polymerization in gelatin capsules takes place at 60°C for 24 hours. Ribbons of sections are collected with a loop and deposited on Formvar-covered, single-hole grids. Sections are stained with 2% uranyl acetate at 45°C for 30 minutes, washed, and then stained briefly with lead citrate (Reynolds, '63).

Whole-mount Electron Microscopy-Mirospreading Technique for Isolation of Nuclear Components Under Optical Control

The developmental stages in the eggs are checked on control eggs by fixing and staining at regular intervals. Thus one can select the eggs for observation at certain stages of cleavage division. Place double-distilled water in a black embryonic watch glass. While the egg is held with fine forceps close to the surface of water, it is quickly made to burst open to release the contents on the water surface. The spread of the material can be followed under the dissecting microscope with proper illumination. Spread nuclear material is picked up by gently touching a carbon coated 200-mesh copper grid to the surface. The material is fixed by floating grids face down on either a 10% formalin solution in 0.1% sucrose (pH 8.5) or glutaraldehyde (6.5%, pH 7.4) for 2–10 mins. Grids are then drained and washed gently for 30 seconds in 0.4% "Fotoflo" (Kodak) adjusted to pH 8 with borate buffer, drained, and air-dried on filter paper in a closed petri dish. Dry grids are immersed surface upward in alcoholic phospotungstic acid (PTA) made up just before use by adding 6 drops of 95% ethanol to 2 drops of 4% aqueous PTA in a deep depression slide. The staining is done for 30 seconds to one minute and the grids are washed for 30 sec in 95% ethanol and air-dried. The grids are then ready for observation in the EM. Details are provided by Counce and Meyer ('73), Meyer and Hennig ('74), and Moses ('77).

Acknowledgments. We thank Dr. C. H. Fernando, University of Waterloo, Canada, and Dr. Z. Brandl, Hydrobiological Laboratory of the Botanical Institute, Prague, for collecting some of the species of cyclopoids for cytological studies. Chromosome numbers of the species from the United States were determined by Mr. Robert Kraft, Duke University, Durham, North Carolina. Comments by J. D. Allan, D. Bonar, H. J. Dumont, J. L. Elmore, J. Green, R. Kraft, S. Twombly, and H. C. Yeatman improved the manuscript.

LITERATURE CITED

Amma, K. (1911) Über die Differenzierung der Keimbahnzellen bei den Copepoden. Arch. Zellforsch. *6*:497–576.

Anderson, R.S. (1974) Crustacean plankton communities of 340 lakes and ponds in and near the national parks of the Canadian Rocky Mountains. J. Fish. Res. Board Can. *31:*855–869.

Ar-Rushdi, A.H. (1963) The cytology of achiasmatic meiosis in the female *Tigriopus* (Copepoda). Chromosoma *13:*526–539.

Banarescu, P. (1974) The typological species concept and modern methods of taxonomy. Z. Zool. Syst. Evol. Forsch. *12:*295–299.

Beermann, S. (1959) Chromatin-Diminution bei Copepoden. Chromosoma (Berl.) *10:*504–514.

Beermann, S. (1966) A quantitative study of chromatin diminution in embryonic mitosis of *Cyclops furcifer*. Genetics *54:*567–576.

Beermann, S. (1977) The diminution of heterochromatic chromosomal segments in *Cyclops* (Crustacea, Copepoda). Chromosoma (Berl.) *60:*297–344.

Beermann, S., and G.F. Meyer (1980) Chromatin rings as products of chromatin diminution in *Cyclops.* Chromosoma *77:*277–283.

Beermann, W. (1954) Weibliche Heterogametie bei Copepoden. Chromosoma *6:*381–390.

Boveri, T. (1910) Die Potenzen der Ascaris-Blastomeren be abgeänderter Furchung. Zugleich ein Beitrag zur Frag qualitativ ungleicher Chromosomenteilung. In: Festscher. f.R. Hertwig (Jena) *3:*131–214.

Brandl, Z. (1973) Laboratory culture of cyclopoid copepods on a definite food. Vestnik Cs. Spol. Zool. *37:*81–88.

Brandl, Z., and C.M. Fernando (1978) Prey selection by the cyclopoid copepods *Mesocyclops edax* and *Cyclops vicinus.* Verh. Int. Verein. Limnol. *20:*2505–2510.

Braun, H. (1909) Die spezifischen Chromosomenzahlen der einheimischen Arten der Gattung *Cyclops.* Arch. Zellforsch. *3:*449–482.

Burgis, M.J. (1970) The effect of temperature on the development time of eggs of *Thermocyclops* sp., a tropical cyclopoid copepod from Lake George, Uganda. Limnol. Oceanogr. *15:*742–747.

Burgis, M.J. (1971) The ecology and production of copepods, particularly *Thermocyclops hyalinus*, in the tropical Lake George, Uganda. Freshwater Biol. *1:*169–192.

Carter, J.C.H. (1974) Life cycles of three limnetic copepods in a beaver pond. J. Fish. Res. Board Can. *31:*421–434.

Carter, J.C.H., M.J. Dadswell, J.C. Roff, and W.G. Sprules (1980) Distribution and zoogeography of planktonic crustaceans and dipterans in glaciated eastern North American. Can. J. Zool. *58:*1355–1357.

Chambers, R. (1912) Egg maturation, chromosomes, and spermatogenesis in *Cyclops.* Univ. Toronto Studies, Biol. Ser. No. *14:*5–37.

Chinnappa, C.C. (1980) Bivalent forming race of *Mesocyclops edax* (Copepoda, Crustacea). Can. J. Genet. Cytol. *22:*427–431.

Chinnappa, C.C., and R. Victor (1979a) Cytotaxonomic studies on some cyclopoid copepods (Copepoda, Crustacea) from Ontario, Canada. Can. J. Zool. *57:*1597–1604.

Chinnappa, C.C., and R. Victor (1979b) Achiasmatic meiosis and complex heterozygosity in female cyclopoid copepods (Copepoda, Crustacea). Chromosoma *71:*227–236.

Cleland, R.E. (1922) The reduction divisions in the pollen mother cells of *Oenothera franciscana.* Am. J. Bot. *9:*391–413.

Cleland, R.E. (1923) Chromosomal arrangements during meiosis in certain Oenotheras. Am. Nat. *57:*562–566.

Cleland, R.E. (1957) Chromosome structure in *Oenothera* and its effect on the evolution of the genus. Cytologia (Proc. Int. Genet. Symposia) *1956:*5–19.

Cleland, R.E. (1962) The cytogenetics of *Oenthera*. Adv. Genet. *11:*147–237.

Coker, R.E. (1943) *Mesocyclops edax* (S.A. Forbes), *M. leuckarti* (Claus) and related species in America. J. Elisha Mitchell Sci. Soc. *59:*181–200.

Comita, G.W. (1951) Studies on Mexican copepods. Trans. Am. Microsc. Soc. *70:*367–379.
Comita, G.W. (1972) The seasonal zooplankton cycles, production and transformation of energy in Severson Lake, Minnesota. Arch. Hydrobiol. *70:*14–66.
Cooley, J.M., and C.K. Minns (1978) Prediction of egg development times of freshwater copepods. J. Fish. Res. Can. *35:*1322–1329.
Counce, S.J., and G.F. Meyer (1973) Differentiation of the synaptonemal complex and the kinetophore in *Locusta* spermatocytes studied by whole mount electron microscopy. Chromosoma *44:*231–253.
D'Agostino, A.S., and L. Provasoli (1970) Dixenic culture of *Daphnia magna* Straus. Biol. Bull. *139:*485–494.
Davis, C.C. (1959) Osmotic hatching in the eggs of some freshwater copepods. Biol. Bull. *116:*15–29.
Deevey, E.S. Jr., G.B. Deevey, and M. Brenner (1980) Structure of zooplankton communities in the Peten Lake District, Guatemala. In W.C. Kerfoot (ed): Evolution and Ecology of Zooplankton Communities. Spec. Symp. III. Am. Soc. Limnol. Oceanogr., New England Press, pp. 669–678.
Dumont, H.J., J. Pensaert, and I. Van de Velde (1981) The crustacean zooplankton of Mali (West Africa). Hydrobiologia *80:*161–187.
Dumont, H.J., and W. Decraemer (1977) On the continental copepod fauna of Morocco. Hydrobiologia *52:*257–278.
Dupraw, E.J. (1958) Analysis of egg color variations in *Cyclops vernalis.* J. Morphol. *103:*31–63.
Dussart, B. (1969) Les Copépodes des Eaux Continentales. Cyclopoidis et Biologie. Paris: N. Boubée, 282 pp.
Edmondson, W.T., and G.G. Winberg (1971) A Manual on Methods for the Assessment of Secondary Productivity in Fresh Waters. IBP Handbook No. 17, Oxford: Blackwell Scientific Publications.
Einsle, U. (1964) Die Gattung *Cyclops s. str.*, im Bodensee. Arch. Hydrobiol. *60:*133–199.
Einsle, U. (1967) Die äusseren Bedingungen der Diapause planktisch lebender *Cyclops*—Arten. Arch. Hydrobiol. *63:*387–403.
Einsle, U. (1968) Cytologisch-taxonomische studien an *Cyclops*—Population Schleswig-Holsteins. Gewasser Abwasser *47:*31–40.
Einsle, U. (1969) Populations dynamische and synökologische Studien am Crustaceen-Plankton zweier Kleimseen. Beitr. Naturk. Forsch. SW Deutschl. *28:*53–73.
Einsle, U. (1975) Revision der Gattung *Cyclops s. str.* speziell der *abyssorum*. Gruppe. Mem. Ist. Ital. Idrobiol. *32:*57–200.
Einsle, U. (1980) Systematic problems and zoogeography in cyclopoids. In W.C. Kerfoot (ed): Evolution and Ecology of Zooplankton Communities. Spec. Symp. III. Am. Soc. Limnol. Oceanogr., New England Press, pp. 669–678.
Elbourn, C.A. (1966) The life cycle of *Cyclops strenuus strenuus* Fischer in a small pond. J. Anim. Ecol. *355:*333–347.
Elgmork, K. (1955) A resting stage without encycstment in the annual cycle of the freshwater copepod *Cyclops strenuus strenuus.* Ecology *36:*739–743.
Elgmork, K. (1959) Seasonal occurrence of *Cyclops strenuus* in relation to environment in small water bodies in Southern Norway. Folia Limnol. Scand. *11:*1–196.
Elgmork, K. (1964) Dynamics of zooplankton communities in some small inundated ponds. Folia Limnol. Scand. *12:*1–83.
Elgmork, K., and A.L. Langeland (1970) The number of naupliar instars in Cyclopoida (Copepoda). Crustaceana *18:*275–282.
Epp, R.W., and W.M. Lewis, Jr. (1980) The nature and ecological significance of metabolic changes during the life history of copepods. Ecology *61:*259–264.

Ewers, L. (1930) The larval development of freshwater copepods. Contr. No. 3, The Franz Theodore Stone Laboratory of the State University. Ohio State University, pp. 1–43.

Farkas, B. (1923) Beiträge zur Kenntnis der Anatomie und Histologie des Darmakanals der Copepoden. Acta Lit. Acad. Sci. Reg. Univ. Francisco-Josephine, Sect. Sc. Nat. *1:*47–76.

Fautrez, J. (1958) Etude histochimique de l'oocyte de premier order. C.R. Assoc. Anat. *45:*1.

Fautrez, J. (1959) Le nucléole et l'anabolisme protéique. Biol. Jaarb *27:*17.

Fautrez, J., and N. Fautrez-Firlefign (1951) A propos de la chromatine et des nucléoles dans la vesicule germinative de l'oocyte de quelques crustaces. Biol. Jaarb. *18:*27.

Fernando, C.H. (1980a) The species and size composition of tropical freshwater zooplankton with special reference to the Oriental Region (South East Asia). Int. Revue Ges. Hydrobiol. *65:*411–426.

Fernando, C.H. (1980b) The freshwater zooplankton of Sri Lanka, with a discussion of tropical freshwater zooplankton composition. Int. Rev. Ges. Hydrobiol. *65:*85–125.

Frenzel, P. (1977) Zur Bionomie von *Thermocyclops oithonoides.* Arch. Hydrobiol. *80:*108–130.

Fuchs, K. (1914) Die Kiemblätterontwicklung von *Cyclops viridis* Jurine. Zool. Jahrb. *38:*103–156.

Gassner, G. (1969) Synoptonemal complexes in the achiasmatic spermatogenesis of *Bolbe nigra* Giglio-Tos (Mantoidea). Chromosoma *26:*22–34.

Gicklhorn, J. (1930) Notiz über die sagen. Cornealinsen von *Cyclops strenuus* Zool. Anz. *90:*250–258.

Gophen, M. (1976) Temperature effect on lifespan, metabolism and development time of *Mesocyclops leuckarti* (Claus). Oecologia (Berl.) *25:*271–277.

Gophen, M. (1978) Errors in the estimation of recruitment of early stages of *Mesocyclops leuckarti* (Claus) caused by the diurnal periodicity of egg-production. Hydrobiologia *57:*59–64.

Gophen, M., B. Cavari, and T. Berman (1974) Zooplankton feeding on differentially labelled algae and bacteria. Nature *247:*393–394.

Green, J. (1971) Crustaceans. In G. Reverberi (ed): Experimental Embryology of Marine and Fresh-water Invertebrates. Amsterdam: North-Holland, pp. 313–362.

Guillard, R.R.L. (1973) Methods for microflagellates and nannoplankton. In J.R. Stein (ed): Handbook of Phycological Methods. New York: Cambridge University Press.

Guillard, R.R.L., (1975) Culture of phytoplankton for feeding marine invertebrates. In W.L. Smith and M.H. Chanley (ed): Culture of Marine Invertebrate Animals. New York: Plenum Press, pp. 29–60.

Gurney, R. (1933) British Fresh-water Copopoda III. Cyclopoida. London: Ray Society.

Häcker, V. (1895) Über die selbständigkeit der väterlichen und mütterlichen kerbestandteile während der Embryonalentwicklung von *Cyclops.* Aech. Mikro. Anat. *46:*579–618.

Häcker, V. (1903) Über das Schicksal der elterlichen und grosselterlichen Kernunteile. Jena A. Naturw. *37:*297–400.

Hanström, B. (1924) Beitrag zur Kenntnis des zentralen Nervensystems der Ostracoden und Copepoden. Zool. Anz. *61:*31–38.

Harding, J.P. (1950) Cytology, genetics and classification. Nature *166:*769–771.

Heberer, G. (1924) Die Spermatogenese der Copepoden. I.Z. Zool. *123:*555–646.

Heberer, G. (1926) Zur Kenntnis der männlichen Generationsorgane der Cyclopiden. Verh. Deutsche Zool. Ges. *31:*114–148.

Heberer, G. (1932) Die Spermatogenese der Copepoden II. Z. Zool. *142:*141–253.

Hill, L.L., and R.E. Coker (1930) Observations on mating habits of *Cylops.* J. Elisha Mitchell Soc. *45:*206–220.

Hunt, G.W., and A. Robertson (1977) The effect of temperature on reproduction of *Cyclops vernalis* Fischer (Copepoda, Cyclopoida). Crustaceana *32:*169–177.

Hutchinson, G.E. (1967) A Treatise on Limnology II. Introduction to Lake Biology and the Limnoplankton. New York: John Wiley.

Hutner, S.H. (1972) Inorganic nutrition. Ann. Rev. Microbiol. *26:*313–346.

Jamieson, C.D. (1980a) Observations on the effect of diet and temperature on rate of development of *Mesocyclops leuckarti* (Claus) (Copepoda, Cyclopoida). Crustaceana *38:*145–154.

Jamieson, C.D. (1980b) The predatory feeding of copepodid stages III to adult *Mesocyclops leuckarti* (Claus). In W.C. Kerfoot (ed): Evolution and Ecology of Zooplankton Communities. Spec. Sym. III Am. Soc. Limnol. Oceanogr. University Press of New England, pp. 518–537.

John, B., and G.M. Hewitt (1966) Karyotype stability and DNA variability in the Acrididae. Chromosoma *20:*155–172.

Kiefer, F. (1933) Die freilebenden Copepoden der Binnengwässer von Insulinde. Arch. Hydrobiol. (Suppl.) *12:*519–621.

Kiefer, F. (1978) Zur Kenntnis der Copepoden faun ägyptischer Binnengewässer. Arch. Hydrobiol. *84:*480–499.

Kiefer, F., and G. Fryer (1978) Das Zooplankton der Binnegewässer. Teil 2. Stuttgart: E. Schweizerbart'sche Verlag.

Kiknadze, I.I. (1972) Funkcional'naja Organizacija Chromosom. Leningrad: Nauka, 221 pp.

Lérat, P. (1905) Les phénomènes de maturation dans l'ovogénèse et la spermatogénèse du *Cyclops strenuus.* Cellule *22:*163–202.

Lewis, W.M. Jr. (1979) Zooplankton Community Analysis. Studies on a Tropical System. New York: Springer-Verlag.

Linn, C.C., and I.A. Uchida (1973) Fluorescent banding of chromosomes (Q-bands). In P.F. Kruse and M.K. Patterson (eds): Methods and Applications of Tissue Culture. New York: Academic Press, pp. 778–781.

Lind, O.T. (1974) Handbook of Common Methods in Limnology. St. Louis: C.V. Mosby.

Lindberg, K. (1958) Cyclopoides du Soudan (A.O.F.). Bull. IFAN. A *20:*115–116.

Maguire, M.P. (1978) A possible role of synaptonemal complex in chiasma maintenance. Exp. Cell Res. *112:*297–308.

Maguire, M.P. (1979) An indirect test for a role of the synaptonemal complex in the establishment of sister chromatid cohesiveness. Chromosoma *70:*313–321.

Makino, S. (1951) Chromosome numbers in animals. Ames: Iowa State College Press, 290 pp.

Marshall, S.M., and A.P. Orr (1972) The Biology of a Marine Copepod. New York: Oliver and Boyd.

Matschek, H. (1909) Zur Kenntnis der Eireifung und Eiablage bei Copepoden. Zool Anz. *34*:42–54.

Matschek, H. (1910) Ueber Eireitung und Eioblage bei Copepoden. Arch. Zellforsch. *5:*36–119.

Matsui, S., and M. Sasaki (1973) Differential staining of nucleolus organizers in mammalian chromosomes. Nature *246:*148–150.

Meyer, G.F., and W. Hennig (1974) The nucleolus in primary spermatocytes of *Drosophila heydei.* Chromosoma *46:*121–144.

Michajlow, W. (1972) Euglenoidina parasitic in Copepoda. Warsaw: PWN—Polish Scientific Publishers. 224 pp.

Moens, P.B., and K. Church (1979) The distribution of synaptonemal complex material in metaphase I bivalents of *Locusta* and *Chloealtis* (Orthoptera: Arcrididae). Chromosoma *73:*247–254.

Monakov, A.V. (1972) Review of studies of feeding of aquatic invertebrates conducted at the Institute of Biology of Inland Waters, Academy of Science, USSR. Jr. Fish. Res. Board Can. *24:*363–384.

Monakov, A.V. (1976) Nutrition and trophic relations of fresh water Copepoda (in Russian). Leningrad: Nauka, 170 pp.

Moritz, K.B. (1967) die Blastomerenddifferenzierung für Soma and Keimbahn bei *Parascaris equorum*. Unterschungen mittel UV-Bestrahlung und Zentrifugierung. Wilhelm Roux's Arch. Entwickl. Mech. Org. *159:*203–266.

Moritz, K.B., and G.E. Roth (1976) Complexity of germ line and somatic DNA in *Ascaris.* Nature *259:*55–57.

Moses, M.J. (1968) The synaptonemal complex and meiosis. In K.S. Sparks et al. (eds): Molecular Human Cytogenetics. New York: Academic Press.

Munro, I.G. (1974) The effect of temperature on the development of eggs, naupliar, and copepodite stages of two species of copepods, *Cyclops vicinus* Uljanin and *Eudiaptomus gracilis* Sars. Oecologia *16:*355–367.

Nichols, H.W. (1973) Growth media—Freshwater. In J.R. Stein (ed): Handbook of Phycological Methods. New York: Cambridge University Press.

Nilssen, J.P. (1979) Problems of subspecies recognition in freshwater cyclopoid copepods. Z. Zool. Syst. Evolut. Forsch. *17:*285–295.

Patalas, K. (1975) The crustacean plankton communities of fourteen North American great lakes. Verh. Int. Verein. Limnol. *19:*504–511.

Porter, J.G., M.L. Pace, and J.F. Battey (1979) Ciliate protozoans as links in freshwater planktonic food chains. Nature *277:*563–565.

Price, J.L. (1958) Cryptic speciation in the vernalis group of Cyclopidae. Can. J. Zool. *36:*285–303.

Rasmussen, S.W. (1977) The transformation of the synaptonemal complex into the "elimination chromatin" of *Bombyx mori* oocyte. Chromosoma *60:*205–221.

Raven, C.P. (1961) Oogenesis: The Storage of Developmental Information. New York: Pergamon.

Ravera, O. (1953) Gli stadi di sviluppo dei copepodi pelagici del Lago Maggiore Mem. Ist. Ital. Idrobiol. *7:*129–159.

Reed, E.B. (1963) Records of freshwater Crustacea from arctic and subarctic Canada. Can. Nat. Mus. Bull. *199:*29–59.

Renner, O. (1917) Versuche über die genetische Konstitution der Oenotheren. Z. Vererbungslehre *18:*121–294.

Renner, O. (1919) Zur Biologie und Morphologie der männlichen der Haplonten einiger Oenotheren. Z. Botan. *11:*305–380.

Renner, O. (1925) Unterschungen über die faktorielle Konstitution einiger Komplex-heterozytischer Oenotheren. Bibliotheca Genet. *9:*1–168.

Renner, O. (1943) Ueber die Entstehung homozygotischer Formen aus heterozygotischen Oenotheren. II. Die Translocationshomozygoten. Z. Botan. *39:*49–105.

Reynolds, E.S. (1963) The use of lead citrate at high pH as an electron-opaque stain in electron microscopy. J. Cell Biol. *17:*208–213.

Richard, J. (1891) Recherches sur le système glandulaire et sur le système nerveux des Copepodes libres d'eau douce. Ann. Sc. Nat. Zool. *12:*113–270.

Robins, J.H., and I.A. McLaren (1982) Unusual variations in nuclear DNA contents in the marine copepod *Pseudocalanus*. Can. J. Genet. Cytol. *24* (in press).

Roth, G.E. (1979) Satellite DNA properties of germ line limited DNA and the organization of the somatic genomes in the Nematodes *Ascaris suum* and *Parascaris equorum*. Chromosoma *74:*335–371.

Rothfels, K.H. (1956) Blackflies: Siblings, sex and species grouping. J. Hered. *36:*113–122.

Rousset, V., F. Coste, J.F. Manier, and A. Raibaut (1981) L'Appareil génital male et la spermatogénèse d'un copépode cyclopoïde libre: *Acanthocyclops (Megacyclops) viridis* (Jurine, 1820) (Copepoda, Cyclopoida). Crustaceana *40:*65–78.

Roy, J. (1932) Copépodes et Cladocères de l'ouest de la France. Recherches biologiques et faunistiques sur le plancton d'eau douce des vallées du Loir et de la Sarthe. Thèse. Faculte des Sciences, Univ. Paris, No. 2207, Sér. A. No. *1338:*1–226.

Rückert, J. (1895) Zur Befruchtung von *Cyclops strenuus*. Anat. Anz. *10:*708–725.

Rüsch, M.E. (1960) Untersuchungen über Geschlechtsbestimmungsmechanismen bie Copepoden. Chromosoma *11:*619–632.

Rylov, V.M. (1948) Freshwater Cyclopoida—Fauna of U.S.S.R. Crustacea Vol. 3 No. 3, Israel Program for Sci. Transl., Jerusalem.

Sarvala, J. (1979) Benthic resting periods of pelagic cyclopoids in an oligotrophic lake. Holartic Ecol. *2:*88–100.

Smith, K. (1977) The taxonomy and the distribution of Ontario copepod Crustacea (Calanoida and Cyclopoida) with keys to the species. M.Sc. thesis, University of Waterloo.

Smith, K., and C.H. Fernando (1978a) The freshwater calanoid and cyclopoid Crustacea of Cuba. Can. J. Zool. *56:*2015–2023.

Smith, K., and C.H. Fernando (1978b) A guide to the freshwater calanoid and cyclopoid copepod crustacea of Ontario. University of Waterloo Biology Series No. 18.

Smyly, W.J.P. (1970) Observations on rate of development, longevity and fecundity of *Acanthocyclops viridis* (Jurine) (Copepoda, Cyclopoida) in relation to type of prey. Crustaceana *18:*21–36.

Smyly, W.J.P. (1974) The effect of temperature on the development time of the eggs of three freshwater cyclopoid copepods from the English Lake District. Crustaceana *27:*278–284.

Smyly, W.J.P. (1978) Strategies for co-existence in two limnetic cyclopoid copepods. Verh. Int. Verein. Limnol. *20:*2501–2504.

Sonneborn, T.M. (1957) Breeding systems, reproductive methods, and species problems in Protozoa. In: The Species Problem. Washington: American Association for the Advancement of Science, pp. 155–324.

Stein, J.R. (1973) Handbook of Phycological Methods. Culture Methods and Growth Measurements. New York: Cambridge University Press.

Stemberger, R.S. (1981) A general approach to the culture of planktonic rotifers. J. Can. Fish Aquat. Sci. *38:*721–724.

Stich, H. (1950a) Histochemische Untersuchungen frühembryonaler Sonderungsprozesse normaler und zentrifugierter Eier von *Cyclops viridis*. J. Arch. Entwicklungsmech. Organ. *144:*364–380.

Stich, H. (1950b) Karyologische Untersuchungen an zentrifugierten Keimen von *Cyclops viridis* J. Biol. Zentralbl. *69:*197–209.

Stich, H. (1962) Variations of the deoxyribonucleic acid (DNA) content in embryonal cells of *Cyclops strenuus*. Exp. Cell Res. *26:*136–143.

Stücke, D.E. (1981) Seasonality and distribution of two limnetic cyclopoid copepods, *Diacyclops bicuspidatus thomasi* (S.A. Forbes) 1881 and *Mesocyclops edax* (S.A. Forbes) 1881 in relation to lake thermal and oxygen structure. M.S. thesis, Case Western Reserve University, Cleveland, Ohio.

Taube, I. (1966) The temperature dependence of the development of the embryo in *Mesocyclops leuckarti* (Claus) and *Cyclops scutifer* Sars. Thesis, Limnol Institute, Uppsala, Sweden.

Taube, I., and A. Nauwerck (1967) Zur populationsdynamik von *Cyclops scutifer* Sar. I. Die Temperatureabhängigkeit der Embryonalentwicklung von *Cyclops scutifer* Sars im Vergleich zu *Mesocyclops leukcarti* (Claus) Rep. Inst. Freshwater Res. Drottiningholm *47:*76–86.

Terpilowska, B. (1971) Quantitative investigations of DNA in early embryogenesis of *Acanthocyclops gigas* (Claus). Zool. Pol. *21:*163–177.

Terpilowska, B. (1974) The structure and behavior of chromosomes in embryonic development of *Acanthocyclops gigas* (Claus). Zool. Pol. *24:*11–28.

Terpilowska, B. (1977) Cytological and cytochemical investigations of the early stages of *Acanthocyclops viridis* (Jurine) embryogenesis. Zool. Pol. *21:*483–516.

Walter, E. (1922) Ueber die Lebensdauer dev freilebenden Süsswawser-Cyclopiden und andere Fragen ihrer Biologie. Zool. Zahrb. Syst. *44:*375–420.

Watkins, A.L., and D. Belk (1975) The Copepoda of Guam. Crustaceana *28:*302–304.

Watson, N.H.F., and B.N. Smallman (1971) The role of photoperiod and temperature in the induction and termination of an arrested development in two species of freshwater cyclopoid species. Can. J. Zool. *9:*855–862.

Webb, G.C. (1976) Chromosome organization in the Australian plague locust, *Chortoicetes terminifera.* I. Banding relationships of the normal and supernumerary chromosomes. Chromosoma *55:*229–249.

White, M.J.D. (1971) The value of cytology in taxonomic research in Orthoptera. Proc. Int. Study Conf. on Current and Future Problems of Acridology, 1970. London.

Whitehouse, J.W., and B.G. Lewis (1973) The effect of diet and density on development, size and egg production in *Cyclops abyssorium* Sars, 1863 (Copepoda, Cyclopoida). Crustaceana *25:*15–236.

Wilson, M.S., and H.C. Yeatman (1959) Free living copepods. In W.T. Edmondson (ed): Freshwater Biology. New York: John Wiley, pp. 735–861.

Woodmansee, R.A., and B.J. Grantham (1961) Diel vertical migrations of two zooplankters *(Mesocyclops* and *Chaoborus)* in a Mississippi Lake. Ecology *42:*619–628.

Wyngaard, G.A., J.L. Elmore, and B.C. Cowell (1982) Dynamics of a subtropical plankton community, with emphasis on the copepod *Mesocyclops edax.* Hydrobiologia *89:*39–48.

Yeatman, H.C. (1944) American cyclopoid copepods of the *viridis-vernalis* group (including a description of *Cyclops carolinianus* n. sp.) Am. Midl. Nat. *32:*1–90.

Developmental Biology of Freshwater Invertebrates, pages 535–576

Development of Freshwater Bryozoans (Phylactolaemata)

Hideo Mukai

ABSTRACT The Phylactolaemata are a small group, occurring exclusively in fresh water. They reproduce by means of external buds, statoblasts, and the production of larvae. The phylactolaemate colonies may be divided into roughly two types—branching and massive. The budding pattern is essentially similar in both types: each zooid produces less than four buds successively on the ventral cystid wall. Statoblasts can be divided morphologically into three types: floatoblasts, sessoblasts, and piptoblasts. The processes involved in sessoblast formation are described here for the first time. Sessoblasts, like other types of statoblasts, grow initially in the funiculus; then, they are cemented to the ectocyst, and thus to the substratum, by their dorsal side. This cementing is accomplished through the participation of both the statoblast and the cystid. The first sign of germination of the statoblasts is the thickening of the epidermal layer on the ventral side, where an invagination appears and grows into the primary polypide. In the leptoblast, a kind of floatoblast peculiar to *Plumatella casmiana*, the formation of the primary polypide has already begun by the later phases of statoblast formation. Much of our knowledge on the germination capacities of statoblasts has been obtained with the so-called spinoblasts. Biochemical aspects or endogenous controlling mechanisms of dormancy, quiescence, and germination of statoblasts have scarcely been elucidated.

Phylactolaemates are hermaphroditic and viviparous; a fertilized egg is taken into the embryo sac, where it grows into a ciliated larva that is nearly equivalent to a juvenile colony. Entoderm formation is suppressed. Much more study is needed concerning various aspects of sexual reproduction and embryogenesis. Other developmental aspects, such as the growth of the colony and the degeneration, regeneration, and malformation of the zooid, have been reviewed. Regulation among the different modes of reproduction in the colony presents many fascinating problems which await further study.

Dedicated to the memory of the late Professor Emeritus Hidemiti Oka with cordial thanks for introducing me to this field of work.

INTRODUCTION

The phylum Bryozoa or Ectoprocta is usually divided into three distinct classes: Phylactolaemata, Stenolaemata, and Gymnolaemata (Ryland, '70). The Phylactolaemata are restricted to fresh water, the Stenolaemata are exclusively marine, and the Gymnolaemata are chiefly marine. The present article deals only with the Phylactolaemata. They constitute a small group, comprising about 29 (Lacourt, '68) to 39 species (Bushnell, '73). The general biology of the Phylactolaemata has been reviewed by Hyman ('59) and Brien ('60).

The Phylactolaemata are sessile colonial organisms. They are widely distributed—both ecologically and geographically—in almost all lakes, ponds, reservoirs, streams, rivers, and similar freshwater bodies. Colonies may be found attached to submerged logs, twigs, water plants, the underside of rocks, and similar substrates. *Fredericella* is generally found in lakes at considerable depths; the other genera can be found in shallow water. Phylactolaemate colonies show two general types of construction: the plumatellid and the lophopodid (Hyman, '59). In the plumatellid type, exemplified by *Fredericella*, *Plumatella*, *Hyalinella*, and others, the colony is highly branched (Fig. 1A); in the lophopodid type, exemplified by *Lophopus*, *Lophopodella*, *Pectinatella*, and *Cristatella*, the colony is a compact gelatinous mass (Fig. 1B). In *Pectinatella*, many colonies are attached together to form a huge gelatinous mass. *Cristatella mucedo*, which is considered the most advanced species, forms bilateral vermiform colonies. Slow locomotion is a characteristic of young colonies of lophopodids, but *Cristatella* colonies move about throughout life.

Phylactolaemates reproduce both sexually and asexually. Asexual reproduction is accomplished in two ways, that is, by formation of ordinary external buds and by formation of an internal bud known as the statoblast.

ZOOID STRUCTURE, BUDDING PATTERN, AND BLASTOGENESIS

Structure of the Zooid

The colony of bryozoans is often called a zoarium. Each complete unit of the colony is a zooid, which consists of a polypide and a cystid. The polypide is an evaginable entity composed of the tentaculated lophophore and viscera, and the cystid represents the body wall that accomodates the polypide. Phylactolaemates lack polymorphism, and the structure of the zooid is essentially similar in all species (Fig. 2).

The anterior end of the polypide is formed of the lophophore, which is horseshoe-shaped except in *Fredericella* where it is almost circular. The lophophore is fringed with a single row of ciliated tentacles, which are

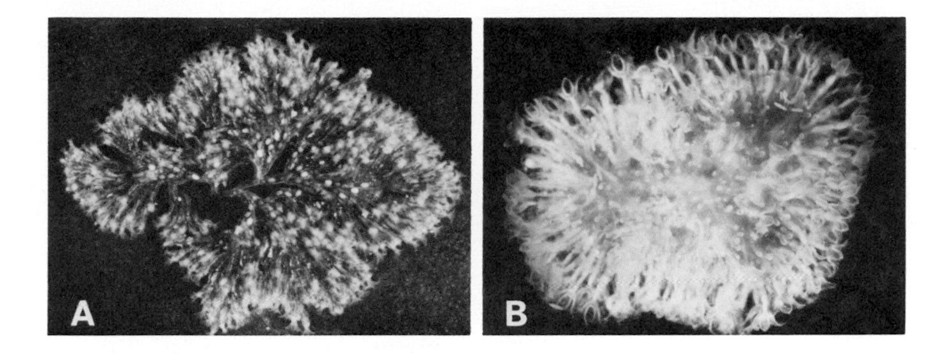

Fig. 1. Phylactolaemate colonies. A. Branching colony of *Plumatella casmiana*. × 2.7. B. Massive colony of *Pectinatella gelatinosa*. × 2.5. (Part B is from Mukai et al., '79, with permission of the Faculty of Education, Gunma University.)

connected basally by a thin intertentacular membrane. The mouth is situated in the bend of, and is embraced by, the lophophore. The mouth leads to the V-shaped digestive tract, which is differentiated into a pharynx, esophagus, stomach, and intestine, while the anus opens near the mouth outside the lophophore. By convention the oral side of the zooid is considered ventral and the anal side dorsal.

The mouth is overhung dorsally by a projecting flap, the epistome. This organ is peculiar to the phylactolaemates, and is believed to represent the protosome, its lumen therefore being the protocoel. The space within the lophophore constitutes the mesocoel, which also extends into individual tentacles. The mesocoel is separated from the metacoel, or the main coelom between the digestive tract and the cystid, by an incomplete diaphragm. The brain ganglion is located on the diaphragm between the mouth and the anus.

The cystid is composed of two parts, the inner living and the outer nonliving. In the terminology of bryozoans, the former is often referred to as the endocyst and the latter as the ectocyst. The term "zooecium" is usually applied to the ectocyst (Hyman, '59; Ryland, '70), but sometimes it indicates the whole cystid. The ectocyst varies according to species from a thin, hard cuticle to a thick, soft gelatinous deposit and usually covers the whole endocyst, but in *Pectinatella magnifica* and *Cristatella mucedo* it is confined to the basal wall of the colony. The endocyst is composed of several distinct layers. The outermost is the epidermis, which secretes the ectocyst. The epidermal cells are largely cylindrical or cuboidal; some of them have basophilic cytoplasm and some others are vacuolated or contain various types of granules. Typically the epidermis is underlain by a circular muscle layer,

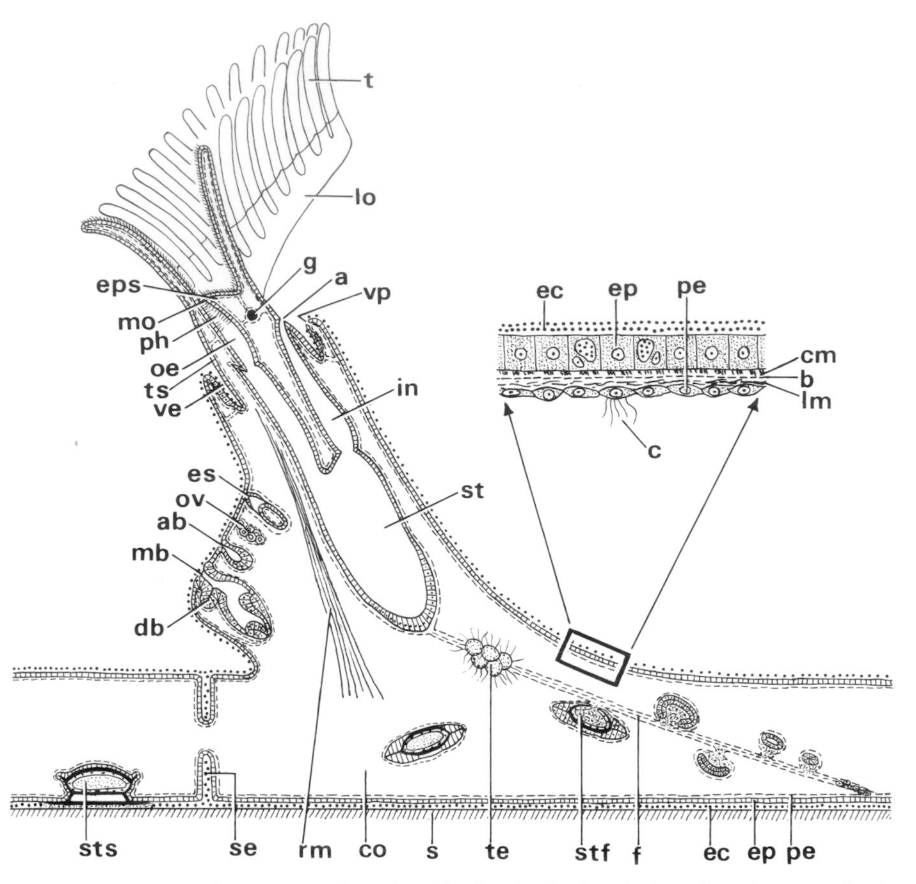

Fig. 2. Diagrammatic representation of median longitudinal optical section of a generalized phylactolaemate zooid. ab, adventitious bud; a, anus; b, basement membrane; c, cilia; cm, circular muscle layer; co, coelom (metacoel); db, duplicate bud; ec, ectocyst; ep, epidermis; eps, epistome; es, embryo sac; f, funiculus; g, ganglion; in, intestine; lm, longitudinal muscle layer; lo, lophophore; mb, main bud; mo, mouth; oe, oesophagus; ov, ovary; pe, peritoneum; ph, pharynx; rm, retractor muscle; s, substratum; se, septum; st, stomach; stf, statoblast (floatoblast); sts, statoblast (sessoblast); t, tentacle; te, testis; ts, tentacle sheath; ve, vestibule; vp, vestibular pore.

a basement membrane and a longitudinal muscle layer. The innermost layer of the endocyst is a peritoneum bearing bunches of long cilia at intervals, which serve to circulate the body fluid containing several kinds of coelomocytes.

Apically or anteriorly the endocyst is permanently invaginated to form a vestibule. At the basal end of the vestibule, the invaginated endocyst or the vestibular wall is continuous with the tentacle sheath. This structure constitutes a cylindrical eversible thin-walled organ and envelops the tentacles when the polypide is retracted. The polyide is connected ventrolaterally to the cystid by powerful retractor muscles. The posterior end of the stomach is linked to the ventral wall of the cystid by the funiculus, a hollow cord lined by a basement membrane accompanied by longitudinal muscle fibers and covered externally by the peritoneum.

At least in some species, and presumably in all phylactolaemates, a pore is present near or at the apical end of the dorsal vestibular wall. Through this vestibular pore the metacoel communicates with the exterior. In *Fredericella* and in some *Plumatella* species, metacoels of a colony are separated by imperfect septa, but in other species such septa are vestigial or completely lacking and the digestive tracts of many zooids are suspended in a common coelom.

Budding Pattern

The colony expands by producing more zooids through budding. In all phylactolaemates except *Stephanella hina* (see below), the budding zone is confined to the ventral cystid wall of the zooid. The budding pattern of phylactolaemates was studied by many earlier workers (e.g., Braem, 1890), and a simple pattern of budding has been established by Brien ('36, '53).

In the plumatellid type, each grown zooid or polypide (represented by the letter *A*) is accompanied by three buds arranged linearly on the ventral side (Fig. 2). They are a main bud *B*, a duplicate bud *C* developed on the ventral side of *B*, and an adventitious bud *B'* on the dorsal side of *B*. In other words, each polypide has simultaneously a combination of buds *B–C* and an adventitious bud *B'*. When the main bud *B* grows out into a fully formed polypide, the duplicate bud *C* separates away and becomes the main bud of *B*, producing ventrally its own duplicate bud, while a new adventitious bud arises between *B* and *C*. Meanwhile the adventitious bud *B'* becomes in turn the main bud of the maternal polypide *A*, developing ventrally a duplicate bud, while the maternal polypide *A* may produce an adventitious bud of the next order between *A* and *B'*. These processes continue to form a branching colony. However, it is important to note that not all the buds produced grow into functional zooids. A zooid produces up to four, but more frequently one or two, daughters (Wayss, '68; Wood, '73). (For rather exceptional cases of *Fredericella sultana* and *Plumatella fruticosa*, see "Growth of the Colony" below.) As the zooids regress and die after giving off their buds, active zooids are found only in the distal parts of the colony.

In the lophopodid type, budding proceeds in basically the same way, but with some variations. In this type of colony, buds appear in a bilateral arrangement with the *B–C* combination to one side of the median sagittal plane of the maternal polypide *A* and the adventitious bud *B'* to the other. A zooid born on the right side of the maternal zooid carries the *B–C* combination on the left side, and *vice versa* (Oka and Oda, '48; Oda, '60). Thus each zooid generally produces two daughters, not simultaneously but successively;

sometimes, however, a third bud may also develop exceptionally late. These daughter zooids repeat the process of budding, thereby producing a compact colony. As the older, more central zooids undergo a continual process of disintegration, actively budding zooids are confined to the peripheral areas of the colony.

Stephanella hina, in which zooids protrude from branched stolon-like creeping tubes, displays a special mode of budding. In this species, the terminal part of a creeping tube forms buds one by one through which the tube grows and extends. Moreover, a pair of buds can arise, at least in young colonies, at either the front or the rear part of a polypide (Oka, '08; Toriumi, '55b). Further details of the budding pattern of this species await future study.

Blastogenesis

The blastogenesis of phylactolaemates was studied by Davenport (1890), Braem (1890), Oka (1891), Kraepelin (1892), and Brien ('36, '53), among others. Round basophilic cells at the base of the epidermis constitute blastogenic cells. They proliferate and push their way inward, following the disappearance of the underlying muscular layers and basement membrane, in the form of a solid knob covered by the peritoneum (Fig. 3A). Braem (1890) believed that these basal epidermal cells constitute a reserve of embryonic cells that are transmitted from generation to generation. This view seems to have been supported by Marcus ('34), but Brien ('36, '53) denied all possibility of the existence of a germinal line, not only for asexual, but also for sexual reproduction. According to this author, the basal epidermal cells are derived simply by the dedifferentiation of columnar epidermal cells.

The primordium elongates and hollows to a double-layered closed vesicle (Fig. 3B), which soon extends to communicate with the exterior (Fig. 3C,D). The lumen of the vesicle puts out two tubular outgrowths into the inner layer on the median plane (Fig. 3E). These two tubes ultimately unite to form a continuous digestive tract. The tube on the future dorsal side is essential and becomes the stomach and intestine, while the other on the ventral side gives rise to the pharynx and esophagus. This account is accepted by most authors (Davenport, 1890; Braem, 1890; Kraepelin, 1892). Oka (1891), who worked with *Pectinatella gelatinosa*, presented a somewhat different story of the formation of the digestive tract. A constriction is formed at about the middle of the vesicle (Fig. 3F). The constricted opening is the mouth of the future polypide, and the lower chamber differentiates into the digestive tract, producing the intestine as an outgrowth (Fig. 3G). In either case, the brain ganglion arises as an evagination of the dorsal wall of the future pharynx. The lophophore, epistome, and tentacle sheath develop from the upper part of the bud (Fig. 3G,H).

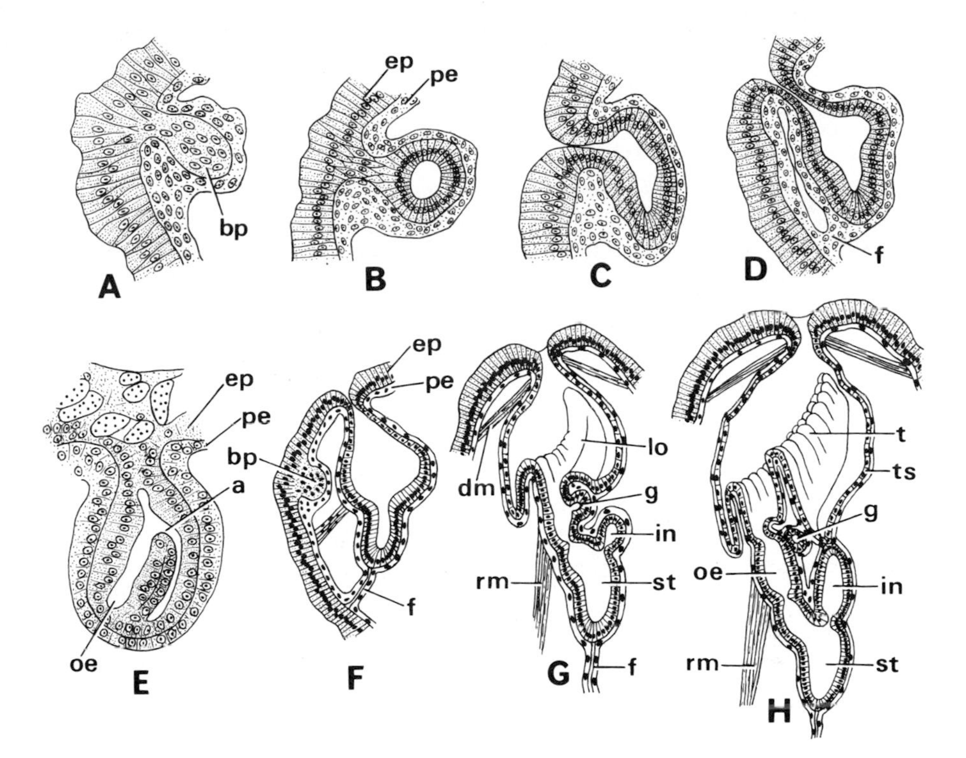

Fig. 3. Blastogenesis. A–D, F–H. Successive stages of blastogenesis in *Pectinatella gelatinosa*. E. Bud of *Cristatella mucedo*, with two outgrowths of lumen to form digestive tract. (A–D, F–H, after Oka, 1891, with permission of the Faculty of Science, University of Tokyo; E, after Davenport, 1890, with permission of Museum of Comparative Zoology, Harvard University.) a, anus of future polypide; bp, bud primordium; dm, dilator muscle of vestibule; ep, epidermis; f, funiculus; g, ganglion; in, intestine; lo, lophophore; oe, oesophagus; pe, peritoneum; rm, retractor muscle of lophophore; st, stomach; t, tentacle; ts, tentacle sheath.

The funiculus is formed precociously at the closed vesicle stage. According to Braem (1890), Oka (1891), and Brien ('53), the funiculus arises as a longitudinal ridge of the peritoneal layer on the ventral side of the vesicle, then the ridge is separated off as the funiculus. The view, advanced by Davenport (1890), Kraepelin (1892), and von Buddenbrock ('10), that the funiculus is formed by a proliferation of peritoneal cells from the free end of the bud, making a secondary connection with the ventral cystid wall, is probably incorrect. The funiculus is at first solid, but later becomes hollow. As the bud grows, the point of attachment of the funiculus to the cystid increasingly separates from the polypide.

TYPES OF STATOBLASTS

Statoblasts are asexually produced internal buds that are unique to the phylactolaemates; they develop in the funiculus and detach from it when completed. A mature statoblast is either oval or almost circular and consists of a yolky germinal mass enclosed in a protective sclerotized shell. The side on which the statoblast is attached to the funiculus is the ventral or deutoplasmic side, and the opposite is the dorsal or cystigenic side. The ventral side is easily distinguished from the dorsal side by the presence of a faint concentric annular pattern at the center. Upon germination, the shell separates along the equatorial suture into two valves, a dorsal and a ventral.

Various morphological features of the statoblast differ with different genera or species and have been considered reliable diagnostic characters (Toriumi, '56; Lacourt, '68; Mukai and Oda, '80). The possible phylogenetic relations within the phylactolaemates are presented in Figure 4, which may facilitate an understanding of the following description. Several classification systems of statoblasts have been proposed (Brown, '33; Rogick, '43b; du Bois-Reymond Marcus, '53; Wiebach, '63). In this article, statoblasts are grouped morphologically into three major types: floatoblasts, sessoblasts, and piptoblasts. Photographs of some statoblasts are shown in Figure 5; and scanning electron micrographs of various statoblasts are presented in Bushnell and Rao ('74, '79), Wiebach ('75), and Mundy ('80).

Floatoblasts

Statoblasts equipped with a float or a well-developed equatorial annulus of empty cells are called floatoblasts. They are produced in all genera except *Fredericella*. When released from the colony, floatoblasts of most species rise to the water surface, but those of the genera *Hyalinella*, *Lophopus*, *Lophopodella*, and *Pectinatella gelatinosa* sink to the bottom. Once dried, however, floatoblasts of the latter species also float. Each valve of a floatoblast includes a central capsular portion and a peripheral annular portion. Generally the annulus covers the capsule more extensively on the dorsal side. In *Cristatella mucedo*, the float is entirely attached to the dorsal valve. The floatoblasts usually float with the dorsal side up.

Floatoblasts of *Lophopus* are pointed at both ends, and those of *Lophopodella*, *Pectinatella*, and *Cristatella* are ornamented with hooked spines. Such statoblasts (including or excluding floatoblasts of *Lophopus*) are often named spinoblasts, which are, however, not a natural group. There is considerable intraspecific variation in the number of spines of spinoblasts. This has been studied by a number of workers chiefly from the taxonomical point of view (see Mukai and Oda, '80, for a review of the literature).

Most species produce only a single kind of floatoblast; however, *Plumatella casmiana* produces three morphologically distinct kinds of floatoblasts (Wiebach, '63; Wood, '73). One is an extremely fragile, minimally sclerotized form, termed a leptoblast; another is an ordinary, heavily sclerotized form, called a pycnoblast; and the third form is an intermediate floatoblast.[1]

Sessoblasts

Statoblasts cemented through the ectocyst to the substratum are here called sessoblasts. The side facing the substratum is considered dorsal. The sessoblast has a thin peripheral annulus or lamella, which is homologous to the float of the floatoblast but is never pneumatic. The dorsal valve is equipped with a peculiar attaching apparatus (see "Formation of Sessoblasts," below). Sessoblasts are produced by the genera *Fredericella* (see below), *Stephanella*, *Plumatella*, *Gelatinella*, and *Stolella*, but are completely lacking in the genera *Lophopus*, *Lophopodella*, *Pectinatella*, and *Cristatella*. In *Hyalinella*, sessoblasts may be absent, or very limited in number, if present (Toriumi, '56; Lacourt, '68).

Piptoblasts

The statoblasts of *Fredericella* have no float, so that they are often classified in the literature as sessoblasts. Actually, however, only some of

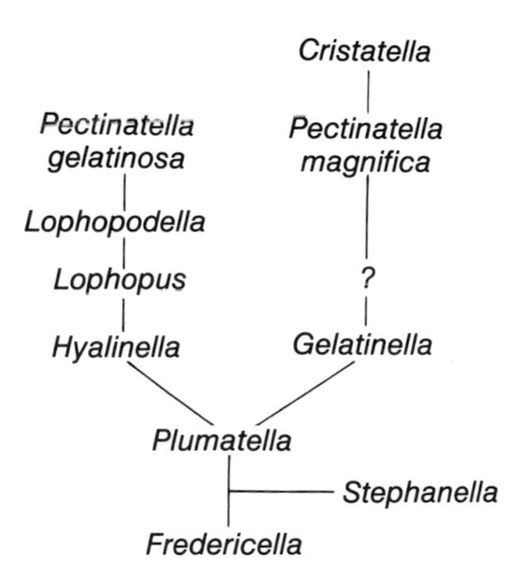

Fig. 4. Tentative representation of the phylogenetic relationships within the Phylactolaemata. A few genera are not shown. (Based chiefly on Mukai and Oda, '80, and unpublished observations.)

[1]Both Rogick ('43a) and Toriumi ('55a) distinguished only two types of floatoblasts in *P. casmiana*: a "thin-walled" type and a "capsuled" or "thick-walled" type. The thick-walled type is known to be highly variable in shape (Viganò, '68). I have been studying *P casmiana* in Central Japan, and have obtained only the above two types of floatoblasts. The thin-walled type is clearly the leptoblast. The thick-walled type seems to be identical with the intermediate floatoblast of Wiebach ('63) and Wood ('73).

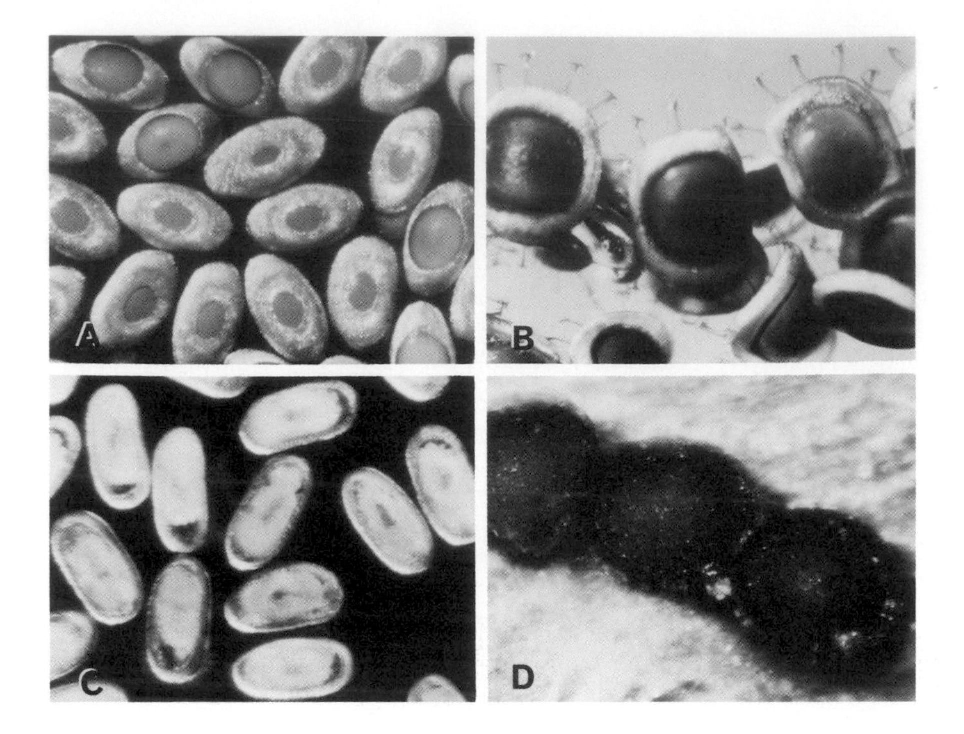

Fig. 5. Statoblasts. A. Floatoblasts of *Plumatella emarginata*. × 40. B. Floatoblasts (spinoblasts) of *Pectinatella magnifica*. × 20. C. Floatoblasts (leptoblasts) of *Plumatella casmiana*. × 40. D. Sessoblasts of *Plumatella repens*. × 40.

them are fixed to the substratum (hence true sessoblasts defined above), but the majority of them are free in the coelom. Du Bois-Reymond Marcus ('53) called the latter type of statoblasts piptoblasts. According to this author, in *F. australiensis* the sessoblasts and the piptoblasts are nearly equal in shape, the former being only a little smaller. In *F. sultana*, according to Toriumi ('51), out of 33 statoblasts examined, the attaching apparatus was distinct in 9, recognizable in 6, absent in 3, and indistinct in the rest, suggesting the presence of some intermediate forms between the sessoblast and the piptoblast. On the other hand, Wood ('73) asserts that the piptoblast resembles the floatoblast more closely than the sessoblast in certain features. In the piptoblast, the annulus is absent or, if present at all, is very incomplete (Lacourt, '68; Bushnell and Rao, '79; Wood, '79; Mundy, '80). The piptoblast may represent the most primitive type of statoblast, from which both the sessoblast and the floatoblast have evolved.

MORPHOLOGY OF THE FORMATION AND GERMINATION OF STATOBLASTS

Formation of Floatoblasts and Piptoblasts

Early stages of the formation of floatoblasts and piptoblasts have been studied by a number of workers (Braem, 1890; Oka, 1891; Kraepelin, 1892; von Buddenbrock, '10; Brien, '53). The primordium of a statoblast is composed of two sorts of cells, peritoneal and epidermal. The peritoneal cells are those of the funiculus; the epidermal cells are derived by migration from the cystid during a short period following the appearance of the funiculus in a young bud. These migrating cells have derivations similar to the blastogenic cells that give rise to the inner layer of a bud (Brien, '53). Some of the ingressed epidermal cells clump together as a ball, which soon becomes hollow, turning into a vesicle. The epidermal vesicle creates a bulge in the peritoneal covering of the funiculus as it grows (Fig. 6A). Meanwhile, peritoneal cells accumulate beneath the vesicle and later begin to store reserve food in the form of yolk granules; therefore, we may speak of these peritoneal cells as yolk cells. The epidermal vesicle, now occupying the dorsal side of the statoblast, flattens into a two-layered disc, then grows around the mass of yolk cells turning into a two-layered cup (Fig. 6B), the rim of which is to be closed later at the center of the ventral side. The two epidermal layers are of the same thickness at first, with columnar basophilic cells. Subsequently, the outer layer grows thicker as the inner layer becomes thinner. Between the two layers, a thin sclerotized sheet, which is known to be chitinous, is secreted by the outer epidermal layer (Tajima, '80). It subsequently attains considerable thickness and becomes a capsule enclosing the germinal mass, which is composed of the inner epidermal layer and the yolk cells.

The capsule soon develops an equatorial suture that delimits two valves. Then, the outer epidermal cells bordering the suture begin to project as a crest. This process is lacking or very limited in piptoblasts. In floatoblasts, the epidermal crest becomes more and more extensive and forms an annular float (Fig. 6C). Each cell of the float secretes chitinous walls, first basally and then laterally; the cytoplasm becomes less basophilic and eventually the cell contents disappear, leaving a hollow interior. The lateral walls thus formed are provided with one or more small pores (Braem, 1890; Lacourt, '68; Mukai and Oda, '80; Mundy, '80). Meanwhile, the outer epidermal cells resting on both sides of the capsule remain basophilic and migrate peripherally in the form of thin layers that cover the growing annular float (Fig. 6D). The hooked spines of *Pectinatella magnifica* and *Cristatella mucedo* are elaborated by these epidermal layers, but those of *Lophopodella carteri* and *Pectinatella gelatinosa* are formed by marginal cells from the float (Mukai and Oda, '80). The expanding epidermal layers continuously

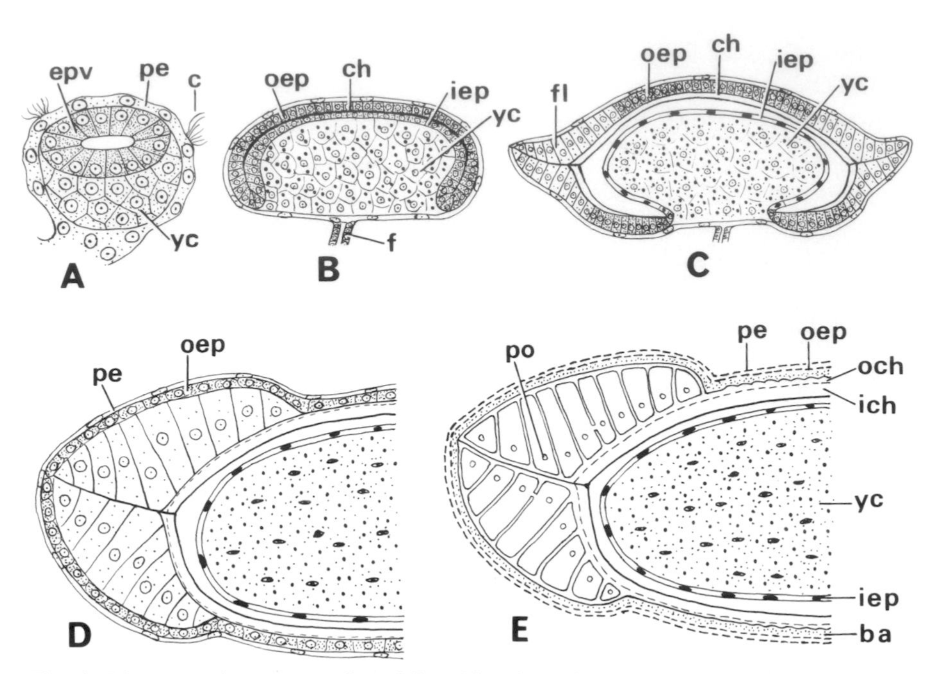

Fig. 6. Diagrammatic representation of floatoblast formation. The dorsal side is upward. A. Epidermal-vesicle stage. B. Two-layered epidermal cup with a chitinous layer between the two layers is enclosing yolk cells, which are accumulating yolk granules. C. Float is protruding. D. Float is covered with outer epidermal cells migrated from the capsular region. E. Mature stage. ba, basophilic layer; c, cilia; ch, chitinous layer; epv, epidermal vesicle; f, funiculus; fl, float; ich, inner chitinous layer (capsule); iep, inner epidermal layer; och, outer chitinous layer (periblast); oep, outer epidermal layer; po, pore; pe, peritoneal layer; yc, yolk cells.

secrete the chitinous substance, with which the opening ends of the float cells are closed. This is followed in some species by the secretion of a soft or hard gelatinous basophilic layer (Toriumi, '56; Mukai and Oda, '80). In *P. magnifica* and *C. mucedo*, the basophilic layer is soft and gelatinous and decays after the statoblast is released into the water.

In histological sections, the shell of a mature statoblast is demarcated not only into a dorsal and a ventral valve by the equatorial suture line, but also, in general, into an inner and an outer chitinous layer by a clear line (Fig. 6E). The inner layer constitutes the capsule proper; the outer layer is known as the periblast (Wood, '79), includes the float and spines when these are present, and in some species is overlain by the basophilic layer. Brief exposure to hot 10% potassium hydroxide permits such statoblasts to be separated into four basic parts: two capsule valves and two periblast valves

(Wood, '79; Mundy, '80). In the statoblasts of *P. magnifica* and *C. mucedo*, no demarcating line between the capsule and the periblast is visible (Mukai and Oda, '80); when treated with a hot potassium hydroxide solution, their shell only separates into two valves, dorsal and ventral (Mundy, '80). Here, it may be added that when the statoblasts of *L. carteri* and *P. gelatinosa* are treated with 10 M KOH at room temperature for about 3 hours, only the periblast separates into two valves while the capsule remains intact and viable with the germinal mass (Mukai, '77; Mukai and Oda, '80).

Returning to an early stage, when the capsule closes at the center of the ventral side, the inner epidermal layer with pale cytoplasm and condensed nuclei becomes a thin epithelial covering of the yolk cells. (For the only exception, see "Formation and Germination of Leptoblasts," below.) This epidermal layer contains a substantial amount of glycogen (Tajima, '80). In most species, the yolk cells initially produce large spherical yolk granules (ca. 4.5 μm in diameter); then, toward the terminal phase of vitellogenesis, much smaller yolk granules appear. However, in *Lophopodella carteri* and *Pectinatella gelatinosa*, yolk granules are uniformly small, fusiform, and about 2 μm long (Mukai and Oda, '80). The formation of yolk granules of *P. gelatinosa* was studied through the electron microscope by Tajima and Mukai ('75) and Terakado and Mukai ('78). The yolk granules are elaborated within Golgi vesicles. Two types of yolk cells have been described by the latter authors; one is an ordinary mononucleate cell with less dense cytoplasm and the other is a more rarely encountered irregularly-shaped multinucleate cell with dense cytoplasm. The significance of these two different types of yolk cells is not clear. Yolk cells contain lipid droplets and glycogen in addition to numerous glycoprotein yolk granules. As yolk cells accumulate these nutrients, their cytoplasm is gradually confined to the cell periphery and the central perinuclear region. Subsequently, the cell boundaries eventually disappear and, in the mature statoblast, small condensed nuclei are scattered within the yolk mass. Histochemical studies of the germinal mass have been performed by Lacourt and Willighagen, ('66) and Mukai ('73).

In *Fredericella sultana*, *Lophopus crystallinus*, *Pectinatella magnifica*, and *Cristatella mucedo*, each zooid usually produces only one or a few statoblasts; in many other species, however, several or more statoblasts per zooid are usual. When more than two statoblasts are formed on a funiculus, they are arranged in orderly succession from the oldest, which is next to the stomach end, to the youngest, which is nearest the cystid. The shell is initially light yellow, then darkens to the characteristic hue of the mature statoblast. A rudiment may develop into a mature statoblast within several days. Mature floatoblasts and piptoblasts accumulate in the coelom where they continue to be enclosed within a couple of degenerated cellular envelopes, the outer one originated from the funicular peritoneum and the inner one from the outer epidermal layer of the statoblast. Some of these floatoblasts may be released when the colony divides or may be released from a disintegrated portion of the colony. However, small spineless floatoblasts are often expelled through the vestibular pore (Marcus, '41; '42; Wiebach, '52, '53; Wayss, '68).

Formation of Sessoblasts

So far as I know, only Kraepelin (1892) gives a brief account of the formation of sessoblasts. Much of this section is based on my own study with *Plumatella casmiana*.

Rudiments of sessoblasts develop, as do those of floatoblasts and piptoblasts, within the funiculus. This was determined using both living colonies and histological sections. When colonies that are producing only leptoblasts and sessoblasts are examined histologically, these two types of statoblasts can easily be distinguished at the initial stage of capsule formation, for the outer epidermal cells are far taller in sessoblasts than in leptoblasts. At this stage, a growing sessoblast lies on the basal cystid wall by its dorsal side (Fig. 7A,B).

The ectocyst of this species is originally hard, gelatinous, and translucent. When overlain by the sessoblast, the ectocyst becomes dark brown and opaque. This is due to the secretion of pigmented material from the epidermis of the endocyst. In fact, the epidermal cells under and around the sessoblast contain fine pigmented granules in their cytoplasm, and the inner portion of the ectocyst, which has been newly secreted by these epidermal cells, is heavily colored. In addition, the secreted pigmented material invades the outer, colorless, older portion of the ectocyst, turning it dark brown. This pigmented material may act as a cementing substance, giving the ectocyst a leatherlike appearance and stiffness.

Meanwhile the outer epidermal layer of the sessoblast produces two chitinous layers projecting from the capsular portion (Fig. 7C,D). One is the equatorial annulus; the other is borne on the dorsal valve along its subequatorial zone and gives rise to the so-called attaching apparatus. Concurrently, within the endocyst a pigmented layer is secreted as a protrusion of the ectocyst; this layer is oval in outline, corresponding to that of the attaching apparatus, and may be called the "cementing layer." Then, the free end of the attaching apparatus and that of the cementing layer are united to form a continuous wall holding the sessoblast. Within the chamber delimited by this wall, the enclosed endocyst and the cellular envelope of the statoblast are united. Outside the wall also, a similar union is attained between the endocyst and the cellular envelope of the statoblast. Thus, the sessoblast now lies within the body wall, between the ectocyst and the "endocyst" modified from the statoblast covering. Kraepelin (1892) has illustrated a sessoblast at about this stage.

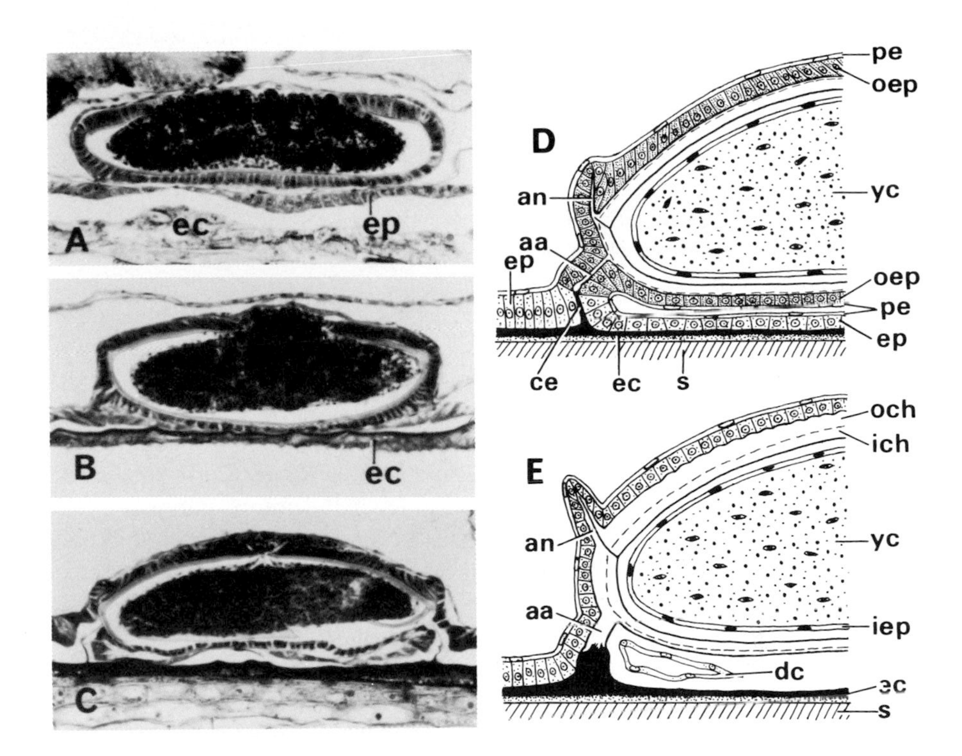

Fig. 7. Sessoblast formation in *Plumatella casmiana*. The ventral side is upward. A. Young sessoblast lying on the basal cystid wall. Separation between endocyst and ectocyst is due to fixation. × 160. B. The dorsal side of the sessoblast is more closely attached to the cystid. Substratum is removed. × 140. C. Beginning of annulus, attaching apparatus and cementing layer. Dorsal chitinous layer was broken during sectioning. × 140. D. Part of section C, diagrammatic drawing. E. Part of a mature sessoblast. aa, attaching apparatus; an, annulus; ce, cementing layer; dc, degenerating cell layers; ec, ectocyst; ep, epidermis of cystid; ich, inner chitinous layer (capsule); iep, inner epidermal layer; och, outer chitinous layer (periblast); oep, outer epidermal layer; pe, peritoneal layer; s, substratum; yc, yolk cells.

The cell layers lining the substatoblast chamber soon degenerate. In the mature sessoblast, the ventral valve is much thicker than the dorsal valve (Fig. 7E). In both valves, especially in the former, the dermarcating line between the capsule and the periblast is clear. The germinal mass within the capsule is formed through essentially the same processes as those in the floatoblasts and piptoblasts described above. The fine structure of the attaching apparatus of several sessoblasts has been studied by Wiebach ('75). The sessoblasts are exposed after the disintegration of the colony but remain attached to the substratum until germination.

Germination of Statoblasts

The main articles on the germination of statoblasts are those of Braem (1890, '13), Oka (1891), and Kraepelin (1892). Although these authors worked with floatoblasts, the morphogenetic process of germination may be essentially the same in all types of statoblasts. As already indicated, the germinal mass of a mature statoblast consists of a flattened epithelium of seemingly inert epidermal cells and an enclosed, nucleated yolk mass of peritoneal origin. According to Braem (1890, '13), the first sign of germination is a circular disklike thickening of the epidermal layer on the ventral side (Fig. 8A). The thickened epidermal layer is now composed of columnar cells that contain basophilic cytoplasm and a large nucleus, and it is lined by a peritoneal layer developing from the yolk mass. Along the submarginal zone of the blastogenic disk thus formed, a circular fold or groove develops within the epidermal layer (Fig. 8B). Soon the inner portion becomes encircled by the groove, which becomes oval, displays bilateral symmetry, and penetrates into the yolk more deeply on the ventral side of the future polypide. The outer margin of the groove gradually closes to form a double-layered closed vesicle (Fig. 8C), which gives rise to a polypide through essentially the same processes as in budding from the zooid as has already been described. During this period, the epidermal cells not engaged in producing the bud also reassume their basophilic appearance and the yolk cells arrange themselves as a peritoneum beneath the epidermis. In this way a cystid arises. As development proceeds, the attachment site of the bud to the cystid is gradually shifted to the equatorial region of the statoblast. Oka (1891) claims that in *Pectinatella gelatinosa* the rudiment of a polypide is formed from the begin-

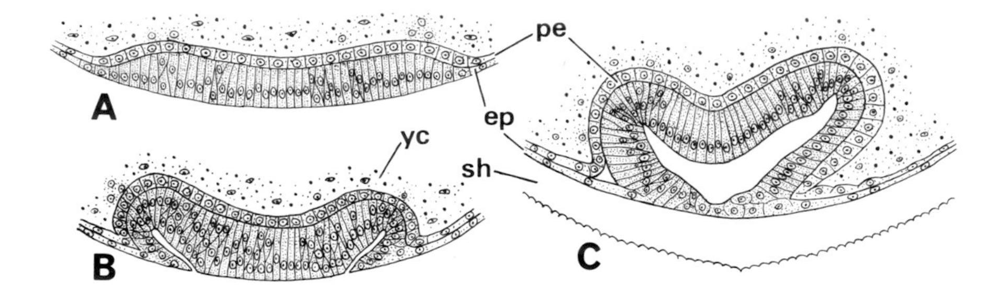

Fig. 8. Diagrammatic representation of early stages of germination in the statoblast of *Cristatella mucedo*. A. Blastogenic disk on the ventral side. The shell is not shown. B. Deep fold develops in the epidermis of the blastogenic disc. The shell is not shown. C. Closed-vesicle stage. The oral or ventral side of the future polypide is to the right. (A-C, redrawn from Braem, '13, with permission of E. Schweizerbart'sche Verlagsbuchhandlung.) ep, epidermis; pe, peritoneum; sh, shell; yc, yolk cells.

ning at a site on the equatorial plane. He probably missed the initial stage of germination. In fact, I have observed the formation of the blastogenic disk in this species in the central area of the ventral side.

At an early stage of germination, prior to the formation of the closed vesicle, the shell separates along the suture into two valves, and the expanded germinal mass becomes visible (Fig. 9A). The separation is probably attained by the secretion of a lytic enzyme, but no definite evidence to support this assumption is available. Meanwhile, a club-shaped protrusion from the cystid, which is known as the mucous pad (Brooks, '29), appears from between the valves (Fig. 9B). Soon after this event, a polypide evaginates from the opposite side (Fig. 9C). There seems to be no definite relation between the axis of the polypide and that of the statoblast (Oka, 1891; Braem, '13). Note that in the terminology employed in this article the primary polypide carries the dorsal valve on its ventral side and the ventral valve on its dorsal side.

Each statoblast gives rise to a single polypide. When evaginated, this primary polypide is already accompanied by its buds located on the ventral cystid wall. Thus, a colony is usually founded by the primary polypide and its daughters, though in *Fredericella sultana* an adventitious bud may also arise, soon after the evagination of the primary polypide, at the dorsal posterior end of the cystid, independently from the primary polypide (Braem, '08; Brien, '36).

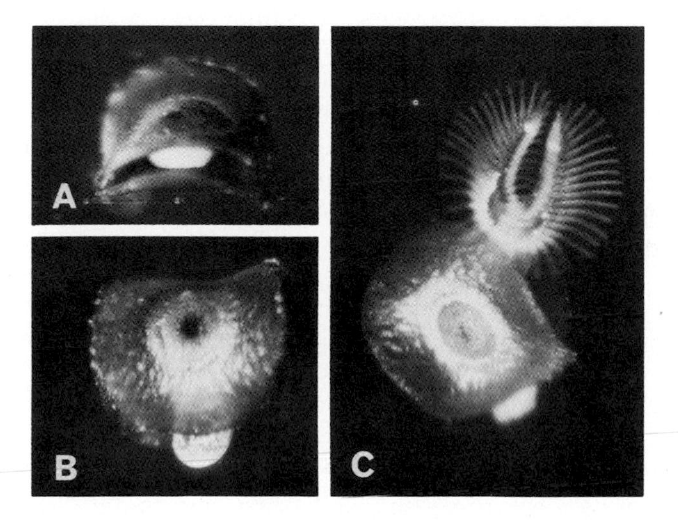

Fig. 9. Germination of statoblast in *Pectinatella gelatinosa*. A. Germinal mass is seen between two valves. × 20. B. Mucous pad is protruded. × 20. C. Ventral view of evaginated ancestrula. × 20.

Formation and Germination of Leptoblasts

The leptoblast is a floatoblast that only occurs in *Plumatella casmiana*. It has a thin translucent shell with a narrow float. All investigators (Rogick, '41a; Wiebach, '63; Viganò, '68) have recognized that the leptoblast develops a young polypide within the capsule while it is still retained by the colony, and that the separation of the valves, followed by the evagination of the polypide, takes place as soon as it is released from the colony. However, no detailed histological study of the leptoblast has yet been published. The following account is based on my own observations.

The early phases of the formation of leptoblasts are quite similar to those of ordinary floatoblasts. However, the later part of leptoblast formation is accompanied by the formation of a polypide. At a stage prior to the closure of the capsule, the inner epidermal layer may flatten as usual on the dorsal side; however, on the ventral side its cells continue to be columnar and basophilic and are accompanied by peritoneal cells that are not engaged in the formation of yolk granules. This two-layered ventral plate constitutes the so-called blastogenic disk, but it is still pierced by peritoneal cells at the center (Fig. 10A,B). Soon, a deep fold develops along the submarginal zone of the blastogenic disk, causing the invagination of a bud primordium into the yolk cell mass (Fig. 10C,D). Then, as the opening of the capsule closes, the connection of the primordium with the outer epidermal layer is cut off. Thus, a double-layered closed vesicle is formed within the capsule on the ventral side. Subsequently, the vesicle gives rise to a polypide, which is almost complete when the statoblast is freed from the funiculus (Fig. 10E).

Almost all leptoblasts are expelled through the vestibular pore. Prior to the expulsion, a free leptoblast is brought to the apical end of the metacoel on the dorsal side of the polypide, the polypide retracts completely, and slow contractions may be observed in the cystid wall covering the leptoblast. Expulsion itself is accomplished suddenly, by a single rapid contraction of the cystid wall. Sometimes, two leptoblasts are released simultaneously through a vestibular pore.

Wiebach ('63) divided the statoblasts into two groups according to their capacity for germination: opsioblasts, in which an obligate dormant period is necessary before germination, and tachoblasts, which can germinate without a dormant period. He regarded the leptoblast as an extreme case of tachoblast. This grouping seems rather arbitrary. From the developmental point of view, the leptoblast should be distinguished from all other kinds of statoblasts. Leptoblast formation overlaps with the formation of the polypide. Therefore, the leptoblast develops continuously, and for this asexual body even the term "stato"-blast seems to be unsuitable in the strict sense of the word. In all other kinds of statoblasts, their formation and germination are

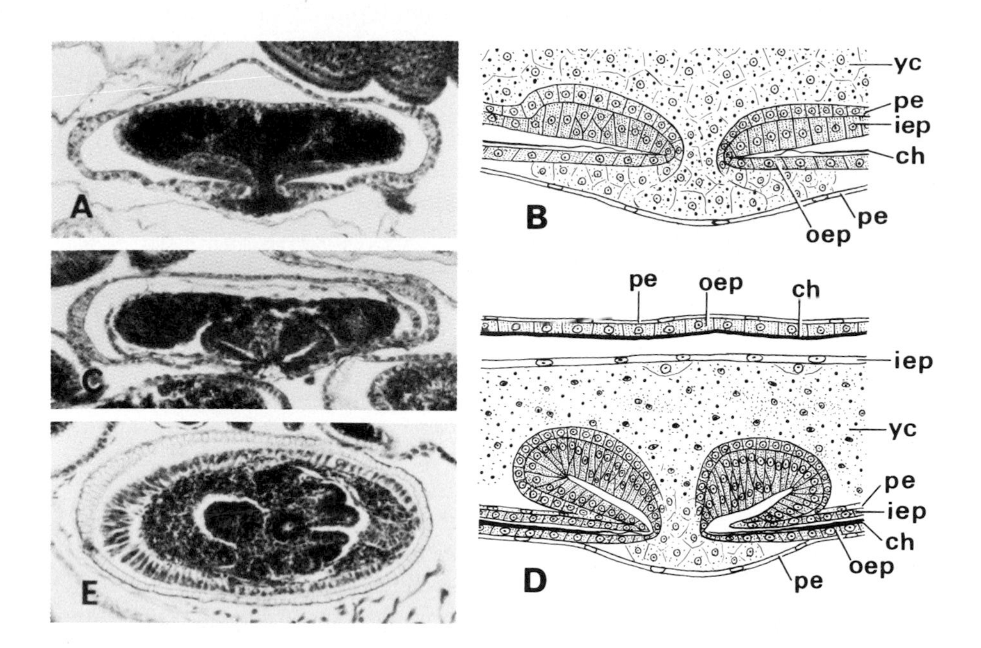

Fig. 10. Leptoblast formation. In all sections except E, the ventral side is downward. A. Blastogenic disk on the ventral side is pierced by yolk cells. × 140. B. Part of section A. C. Just before the closure of the capsule, the blastogenic disk is folded. × 120. D. Part of section C. E. Frontal section of a mature leptoblast with a polypide. The orifice of the polypide is to the right. × 120. ch, chitinous layer; iep, inner epidermal layer; oep, outer epidermal layer; pe, peritoneal layer; yc, yolk cells.

clearly separated, and they can survive adverse conditions without germination. Is the leptoblast a primitive statoblast or a specialized one derived from an ordinary floatoblast? The latter alternative seems more probable.

The shell of the leptoblast is translucent, so that the contour of the polypide developing within the statoblast, which is still attached to the funiculus, is discernible. Therefore, precise information about the relation between the body axis of the maternal polypide and that of the polypide within the statoblast could be obtained by close examination.

GERMINATION CAPACITIES OF STATOBLASTS

A comprehensive review of this topic has been presented by Bushnell and Rao ('74), and readers are referred to it for the information then available.

Before going into details, a few technical and terminological remarks may be necessary. So far, most laboratory studies have been done with floatoblasts, especially spinoblasts. Mature statoblasts can be easily obtained by natural release or artificial dissection from living colonies, or after disintegration of colonies in glass vessels. The statoblasts collected in this way are usually stored in water at low temperature (about 5°C), but for statoblasts of certain species preservation under dry conditions (Oda, '59) or in water at room temperature in the dark (Mukai, '74) is also possible. To test germination capacity, the preserved statoblasts should be placed in well-oxygenated water under appropriate environmental conditions. Oda ('59) employed zinc dental cement to attach floating statoblasts to a glass plate.

Generally, the term "germination" should refer to the whole period from the start of the formation of the primary polypide to its evagination. However, in this section the separation of the valves is referred to as germination, which is usually followed by the evagination of the primary polypide. Oda ('59) distinguished two states in resting statoblasts: quiescence and dormancy. Quiescence is the state in which viable statoblasts are inhibited from germinating because of the absence of one or more external factors necessary for germination. Dormancy is the state in which viable statoblasts do not germinate under the environmental conditions that are ordinarily favorable for rapid germination.

External Germination Factors and Quiescence

Temperature is a major factor controlling germination; it affects both the percentage of and the time until germination. According to Oda ('59), *Lophopodella carteri* statoblasts can germinate in a temperature range from 15° to 35°C, the optimum being 25°C. Information concerning the effect of temperature on germination may be found in Brown ('33) and Oda ('79) for *Pectinatella magnifica*, and in Mukai ('74) for *Pectinatella gelatinosa*.

Light is also an essential factor for germination; statoblasts usually do not germinate under conditions of continuous darkness. This has been established for *L. carteri* (Oda, '59, '80; Bergin et al., '64), *P. gelatinosa* (Mukai, '74), and *P. magnifica* (Oda, '79). However, in the last-mentioned species statoblasts are freed from light dependency by exposure to low temperatures (Oda, '79). The early observation of Brown ('33) that light has little effect on the germination of *P. magnifica* statoblasts might be understood in this context. Both in *P. gelatinosa* (Mukai, '74) and in *L. carteri* (Oda, '80), the light sensitivity of statoblasts is related to the intensity or the duration of illumination necessary for germination, and these values vary with the light and temperature conditions to which the statoblasts have been previously exposed. Oda ('80) could germinate *L. carteri* statoblasts released from colonies reared under dark conditions (at 20°C) by exposing them to 500-lux light for only 1 minute. Oda ('79, '80) demonstrated with statoblasts of *P. magnifica* and *L. carteri* that not only the alternation from dark to light but also that from light to dark is effective in inducing germination. Furthermore, it has been shown for *L. carteri* statoblasts that alternating light as opposed to continuous light tends to promote their germination (Oda, '59, '80; Bergin et al., '64). As considered by these authors, this phenomenon may indicate a simple type of photoperiodism. However, it seems more probable that alternating light may simply act as a trigger for germination, and the length of the photoperiod may not be an important aspect. With regard to wave length, visible light except for the violet and blue portion of the spectrum is effective in the induction of germination in all of the species mentioned (Mukai, '74; Oda, '79, '80).

Various external factors can inhibit germination, causing statoblasts to remain quiescent. Oda ('59) distinguished three kinds of quiescence, which depend on the absence of specific factors: thermo-, photo-, and hydro-quiescence. Mukai ('74) added the category chemo-quiescence, in which statoblasts are inhibited from germinating by the presence of certain chemicals. He reported that the water in which a number of *P. gelatinosa* statoblasts had been stored for a long time inhibits their germination. The assumed inhibitor contained in such "conditioned water" is dialyzable and heat-labile. Black and Proud ('76) failed to find such an inhibitor for *P. magnifica*, but later Oda ('79) demonstrated that the statoblasts of this species also produce a similar inhibitor. He noted that the conditioned water has the specific odor of hydrogen sulfide. As indicated before, the basophilic layer of the statoblasts of this species decays after they are released from the colony. The germination inhibitory action of the so-called conditioned water might be due to some degradation product(s) originating from the statoblasts or some other sources. Much more study is needed to determine the origin and chemical nature of the inhibitor(s). Statoblasts usually do not germinate as long as they are retained in living colonies, though rare examples of germinating floatoblasts within the coelom have been reported (Oda, '59; Bushnell, '66). Whether this also represents an example of chemo-quiescence remains an open question. Statoblasts of *P. magnifica* do not germinate in water low in or devoid of oxygen, although they do retain their viability (Brown, '33).

Biochemistry of Germination

The only available information on the biochemistry of germination is provided by Black and Proud ('76). They used thermo-quiescent statoblasts (stored at 4°C) of *Pectinatella magnifica*. When incubated at 25°C, the statoblasts germinate after a pregermination period of about 3 days. The oxygen consumption of the statoblasts increases about fourfold (from 0.33 to 1.4 μl/hr/100) during the pregermination period; then, a dramatic fourfold increase in respiration occurs between 2 and 6 hours after germination, which is followed by only a small further increment during subsequent development. During the pregermination period, there are no significant changes in DNA, RNA, or protein, though polysomes and ribosomes in the 8,000-g supernatant increase and the polysome/monosome ratio changes from 0.1 to 0.2 during this period. After germination, DNA and RNA increases steadily, but protein remains constant. Actinomycin D does not prevent germination, but blocks further development.

Oda ('59) studied the effect of X-rays on germination with dried statoblasts of *Lophopodella carteri*. The most sensitive stage occurred 12 hours after immersing the statoblasts in water at 25°C. At this time, according to Oda, the statoblasts have absorbed enough water and it is just before the epidermal layer begins to thicken, even though its cells may already have been activated. This stage precedes the maximal proliferation of cells.

Dormancy

Both *Pectinatella magnifica* (Brown, '33; Oda, '79) and *Pectinatella gelatinosa* (Mukai, '74) produce dormant statoblasts. In *Lophopodella carteri*, dormant statoblasts are only produced by midsummer colonies (Oda, '59). In nature, the dormant period of statoblasts of these species lasts for more than several weeks; the statoblasts gradually awaken from dormancy but remain in quiescence with the decrease in water temperature in late autumn. Other scattered information suggests the occurrence of dormancy in statoblasts of some other species as well; Bushnell and Rao ('74) made preliminary comments that both floatoblasts and sessoblasts of *Plumatella fruticosa* appear to have a dormant period.

Oda ('59) reported that in statoblasts of *L. carteri* dormancy could easily be broken by drying (for about 2 days) or by exposure to low temperature (for more than 3 weeks at 10°C). For the statoblasts of *P. magnifica*, exposure to air for a short period (6 hours) is somewhat more effective than low temperature in breaking dormancy (Oda, '79). But in *P. gelatinosa*, neither drying nor low temperature is effective in breaking dormancy (Mukai, '74).

Black and Proud ('76) measured the respiratory rate of *P. magnifica* dormant statoblasts. In the first 2 days after release from colonies, statoblasts consumed O_2 at 0.4 μl/hr/100 (at 25°C). The respiratory rate of statoblasts held at 25°C was 0.02 μl/hr/100 at 10–14 days after release. This decrease in the respiratory rate after release coincided with the loss of permeability to ^{14}C-uridine and phenylalanine. After 8 months at 4°C, when the statoblasts were quiescent, respiration of statoblasts returned to 25°C averaged 0.33 μl/hr/100.

Viability

Recently, I examined the germination capacities of *Pectinatella gelatinosa* statoblasts stored in water for long periods (Table I). From the results obtained, it might be safely concluded that the statoblasts of this species can be stored in water for several years without conspicuous loss of germination capacity and that approximately half of them can survive for almost 10 years. At the time of incubation, in almost all the statoblasts collected in 1970, the germinal mass had shrunk, leaving a narrow space between it and the capsule. Furthermore, in a considerable number of statoblasts collected in 1969, the germinal mass had collapsed within the capsule. Such statoblasts were not used in experiments. Most of the 1971 statoblasts appeared normal, though some of them had a shrunken germinal mass.

The viability of dried statoblasts was studied by Brown ('33), Rogick ('38, '40, '41b, '45), Oda ('59, '79), Mukai ('74), and Bushnell and Rao ('74). The evidence indicates that the statoblasts of *Pectinatella magnifica* are relatively susceptible to drying, and those of *Lophopodella carteri* have the highest resistance to desiccation. Oda ('59) used *L. carteri* statoblasts kept dry in a room for various periods. Drying for less than a year had little effect on germination capacity. Subsequently, a gradual decline in germination capacity was noted; after drying for more than 3 years only a few statoblasts

TABLE I. Viability of *Pectinatella gelatinosa* Statoblasts Stored in Water at Room Temperature in the Dark[a]

Date of collection	Percent germination (evagination)	Pregermination period in days
Sep. 30,'69	2(1)	7
Oct. 18, '69	0(0)	—
Nov. 2, '70	10(1)	6–9
Sep. 10, '71	49(48)	5–8
Oct. 29, '71	59(11)	2–11
Sep. 4, '74	44(43)	3–6
Sep. 6, '75	98(98)	3–6
Oct. 8, '76	50(49)	2–5
Sep. 30, '78	100(100)	2–3
Sep. 10, '79	100(100)	2–5

[a]Statoblasts were incubated on May 23, 1981, at 25°C under 150-lux light. For each lot, 100 statoblasts were used as a set. Through daily observations continued until June 6, germination or the splitting of the valves and the evagination of the primary polypide were recorded separately.

germinated. He could germinate two statoblasts after drying for 2,296 days. Statoblasts of the same species, after severe desiccation with $CaCl_2$ for several days, did not germinate (Oda, '59).

Generally, statoblasts tolerate low temperatures very well; however, they are comparatively intolerant of high temperatures (Brown, '33; Oda, '59, '79; Mukai, '74). Brown ('33) found with the statoblasts of *P. magnifica* and other species that a few of them were still viable after passing through the digestive tracts of animals (amphibians, turtles, and ducks). Oda ('59) showed that dried statoblasts of *L. carteri*, as opposed to undried ones, have relatively greater resistance to x-radiation. Mukai ('77), using undried statoblasts of *P. gelatinosa*, found a rather unusual viability pattern following exposure to various acids and alkalis. For example, 1 M HCl completely inhibits germination within 30 minutes at room temperature. As the concentration of HCl decreases from this level, the inhibitory effect also diminishes. On the other hand, statoblasts treated with 3 M or 5 M HCl for 4 hours are perfectly viable. Additional data on the resistance of statoblasts to various organic and inorganic chemicals can be found in Bushnell ('74), Bushnell and Rao ('74), and Mukai ('77).

SEXUAL REPRODUCTION

All phylactolaemates are hermaphroditic; gonads of both sexes develop in each zooid (Fig. 2). The gonads, together with the sex cells, are of peritoneal origin. The testis occurs on the funiculus. In *Cristatella mucedo*, testes may also develop on the imperfect partitions that connect the upper and basal body walls in the margin of the colony. The partitions are composed of the peritoneum and muscle cells. In all species, the testis and statoblasts often develop on the same funiculus. In such cases, the former is located nearer to the stomach. In the absence of statoblasts, the funiculus may become almost completely covered with the testis. The ovary is initiated on the ventral cystid wall and, as it grows, extends into the coelom. All phylactolaemates are viviparous, brooding the embryo in an embryo sac (often called ooecium) attached to the cystid wall between the maternal polypide and the ovary. Various aspects of the sexual reproduction of phylactolaemates remain obscure; conflicting descriptions are often encountered. No ultrastructural information is currently available.

Sperm, Eggs, and Fertilization

Spermatogenesis in phylactolaemates has been studied by Kraepelin (1892), Braem (1897), Retzius ('06), Marcus ('34, '41), Brien ('60), and others. Spermatogonia derived from peritoneal cells proliferate to produce clusters of spermatocytes, which give rise to spermatids through meiotic division. In each cluster of spermatids, their nuclei are positioned peripherally around a shared cytoplasmic mass or cytophore (Fig. 11A), which leads to synchronized spermiogenesis. Characteristically, the distal end of the flagellum receives a cover of cytoplasm at an early stage prior to the formation of the midpiece (Marcus, '34; Franzén, '70). As tails grow, they cover the surface of the cytophore, moving gently with the stream of coelomic fluid. Mature spermatozoa leave the cytophore and are circulated with coelomic fluid. The mature spermatozoon is rather large and is composed of three distinct parts: a short conical head, a long midpiece with a helical structure probably arising from mitochondrial material, and a long tail relatively rich in cytoplasm. No acrosomal structure has been detected in the head region. Recently, the spermatogenesis of bryozoans including the Phylactolaemata has been reviewed by Franzén ('77).

The ovary contains a substantial number of oocytes with the most advanced at the free end (Fig. 11B). Each oocyte is surrounded by a thin folliclelike layer of ovarian cells. An early oocyte contains homogeneous cytoplasm. In a full-grown oocyte, the cytoplasm is often stratified so that an outer and an inner layer surrounding the large germinal vesicle can be distinguished. The oocyte has little yolk, though some spherical, more or less irregularly shaped granules may be present in the outer cytoplasmic layer. The significance of the cytoplasmic stratification and the origin, chemical nature, function, and fate of the granules are variously described in the literature (Braem, 1897; Marcus, '34; Brien and Mordant, '55; Brien, '60).

The embryo sac is formed as an invagination of the endocyst following the disappearance of muscle layers and consists of elongated epidermal cells with slightly basophilic cytoplasm and basal nuclei covered by the peritoneum (Fig. 11C). It arises only in the presence of an adjacent ovary (Marcus, '34; Brien, '36). The opinion of some early workers, especially Braem (1897), that the embryo sac is an altered polypide bud has been rejected by Brien ('36, '53). The embryo sac receives only a single egg; the manner in which the egg is taken into the embryo sac is not known. The embryo sac is generally believed to receive an egg from the directly adjacent ovary (Braem, 1897; Brien, '60), though Marcus ('34) has claimed that in *Lophopus crystallinus* a mature egg is first discharged into the coelom and circulated about, makes contact with an embryo sac, and then enters it.

When, where, and how fertilization takes place are not yet clear. Some workers (e.g., Kraepelin, 1892; Marcus, '34) maintain that fertilization takes place in the ovary, but Braem (1897) believes that it occurs in the embryo sac. Fertilization has been assumed to be accomplished by the sperm of the same zooid, or at least of the same colony. Using a culture medium without sperm of foreign origin, Oka and Oda ('48) reared *Lophopodella carteri* colonies derived from a single statoblast and obtained swimming larvae. In a

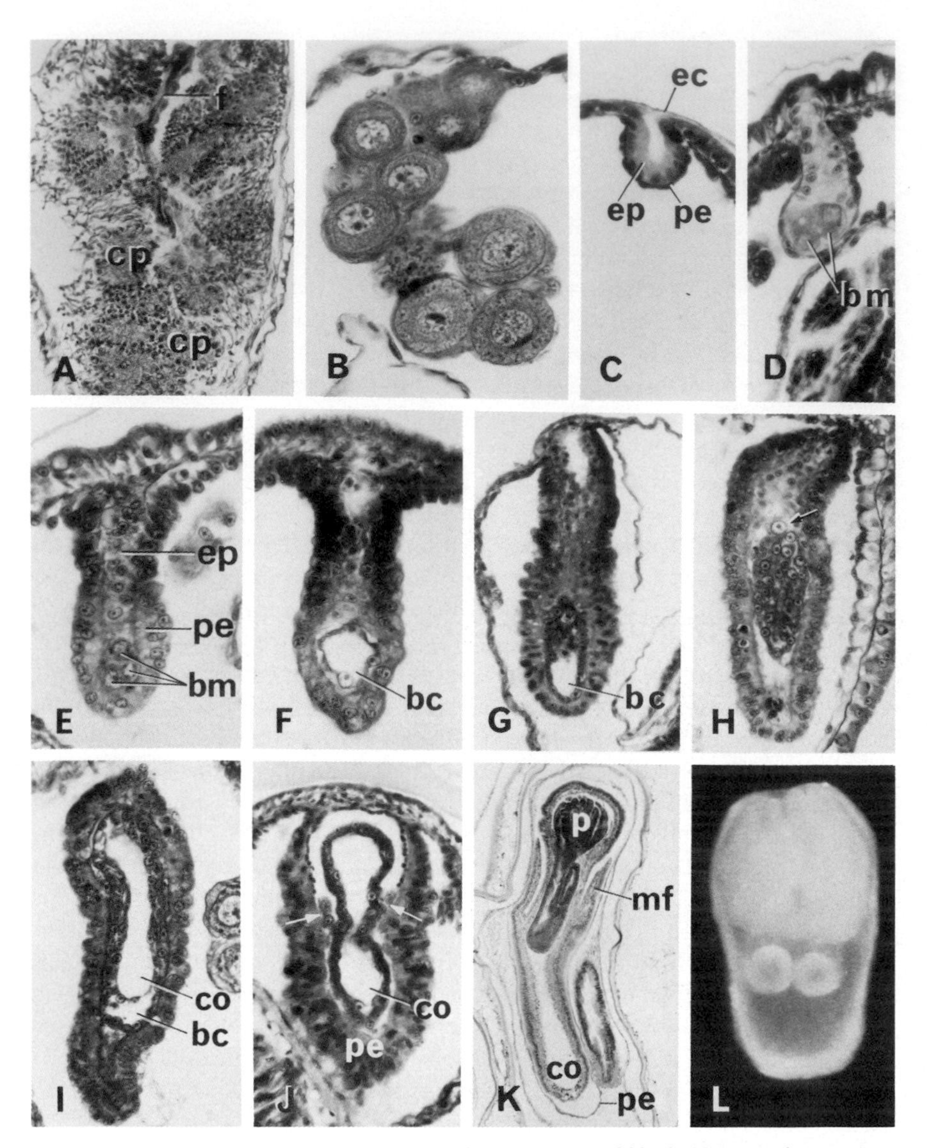

Fig. 11. Sexual reproduction. A. Testis. × 180. B. Ovary. × 300. C. Young embryo sac. × 300. D. Early cleavage stage. × 300. E. Cleavage stage. × 300. F. Blastula stage. × 300. G. Embryo that is proliferating internal mesodermal cells. × 200. H. Embryo whose blastocoel is filled with mesodermal cells. The arrow indicates a cap cell. × 200. I. Embryo with a diminishing blastocoel at the lower end. × 200. J. Two-layered embryo. Arrows indicate the placental ring. × 200. K. Nearly mature larva within the embryo sac. × 45. L. Free larva with two polypides. The leading pole is downward. × 60. A–C, E, G–I, K, L: *Plumatella casmiana*; D, F: *Plumatella emarginata*; J: *Hyalinella punctata*. bc, blastocoel; bm, blastomeres; co, coelom; cp, cytophore; ec, ectocyst; ep, epidermal layer of embryo sac; f, funiculus; mf, mantle fold; p, polypide; pe, peritoneal layer of embryo sac.

number of gymnolaemate and stenolaemate bryozoans, sperm have been observed to be released through the tips of tentacles (Silén, '66, '72; Bullivant, '67). A pore at the tip of each tentacle has been observed in some phylactolaemates (Braem, 1890; Marcus, '34), but its function is not known. The ejection of coelomic fluid containing mature sperm through the vestibular pore has been observed in *Plumatella* (Wiebach, '53), *Lophopodella*, and *Pectinatella* (Oda, '58), but what happens to the sperm after their release has not been ascertained. Thus, the possibility of cross-fertilization in phylactolaemates is still an open question.

Embryogenesis

Main accounts of the embryogenesis of phylactolaemates are those of Kraepelin (1892) for *Plumatella*, Braem (1897) for *Plumatella fungosa*, Braem ('08) for *Fredericella sultana*, Marcus ('34) for *Lophopus crystallinus*, and Brien ('53) for *P. fungosa*.

A fertilized egg is located at the bottom or lower end of the embryo sac. No reliable observations on the formation of polar bodies have been reported (but see Marcus, '34). Early cleavage produces an irregular mass of blastomeres (Fig. 11D,E), which give rise to a one-layered sac presumably equivalent to a blastula (Fig. 11F). At the upper end of the blastula, cells proliferate internally (Kraepelin, 1892; Braem, 1897) and they fill the blastocoel secondarily (Kraepelin, 1892). Figure 11G, H shows this process. Subsequently, the cells within the blastocoel give rise to a mesodermal peritoneal layer probably by rearrangement (Fig. 11I). During this process, both the embryo and the embryo sac continuously broaden and elongate. It is often found that a growing embryo has a large blastocoel in its lower part between the ectoderm and mesoderm, suggesting different growth rates for these two blastoderms. Braem (1897) believed that in *Plumatella* the lower portion of the ectoderm not lined by the mesoderm disintegrates. This seems doubtful. The same author (Braem, '08) reported that in *Fredericella* the blastocoel located in the upper part of the embryo is the last to disappear. Eventually, the embryo becomes, in essence, a cystid (Fig. 11J). There is no entoderm.[2]

In the course of embryogenesis, the epidermis of the embryo sac is separated from that of the cystid by the development of a muscle layer (Brien, '53). The epidermal lining of the embryo sac completely disappears from the region surrounding the embryo. As the embryo grows, the degradation of

[2]Braem (1897) believed that in *Plumatella* a few cells migrate inward from the upper end of the embryo during later cleavage, and that these cells represent the entoderm but quickly degenerate. He regarded the upper end as the vegetal pole, and the one-layered vesicle stage as the gastrula or pseudoblastula. Further, Braem ('08) reported that in *Fredericella* a few transitory entodermal cells are segregated outside the embryo at the vegetal pole. These findings by Braem have not been verified by later investigators.

the epidermal layer advances upward. In the uppermost part, epidermal cells with atrophied nuclei remain (Fig. 11J). In contrast to the epidermis, the outer or peritoneal layer of the embryo sac proliferates, and its cells become cylindrical and basophilic with large nuclei located near the outer ends.

While the embryo is proliferating the internal mesodermal cells, its upper surface is capped by a group of enlarged round cells (cap cells) derived from the ectoderm of the embryo (Fig. 11H). The cap cells are of extra-embryonic nature. As the embryo grows, in *Plumatella* the cap is removed and its cells seem to participate in a placental ring that is in contact with both the embryo and the embryo sac at about the middle of the former (Braem, 1897; Brien, '53). This is the case in *Hyalinella* also (Fig. 11J). The so-called placental ring may have a nutritive function, or it may act simply to anchor the embryo in place. In *Lophopus*, the cap cells seem to migrate loosely in the space between the embryo and embryo sac without forming a definite placental ring (Marcus, '34). In *Fredericella*, the cap cells remain covering the upper end of the embryo in the form of a placental disk (Braem, '08).

Meanwhile, the embryo develops one to four buds, depending on the species, on its upper end; these buds proceed to differentiate into polypides exactly as in asexual budding. While the polypides are developing, an annular fold of the cystid appears below the polypidial region or below the placental ring if present. This mantle fold (Fig. 11K) advances upward, rupturing the placental ring and enveloping the polypidial region completely except for a small terminal opening. The exposed cystid wall of the embryo develops cilia; ciliation does not occur on the inner wall of the mantle fold and on the polypidial region. The ciliated part is a temporary larval covering or mother cystid, and the nonciliated part is a permanent daughter cystid. The full-grown embryo or larva lies free in the embryo sac, whose wall is now reduced to a thin, stretched peritoneal layer. Generally, the maternal polypide degenerates before the resulting ciliated larva leaves the embryo sac.

Larva and Metamorphosis

Occasional rotation of the mature larva within the embryo sac has been observed with *Plumatella fungosa* (Marcus, '26a) and *Lophopodella carteri* (Oda, '60). The larva escapes through an opening formed at the site of attachment of the embryo sac to the maternal cystid. The cast-off embryo sac disintegrates within the coelom in *P. fungosa* (Marcus, '26a) and *Cristatella mucedo* (Mukai, unpublished observations); in *L. carteri* the embryo sac is everted and comes out at the release of the larva, then disintegrates there (Oka and Oda, '48). As to the external factors controlling the release of larvae, nothing is known. In *L. carteri*, the majority of larvae appear in the morning but some of them come out in the afternoon (Oda, '60). Larvae of

P. fungosa were reported to appear at night (Braem, 1897) or in the morning (Marcus, '26a). *Pectinatella magnifica* larvae are usually released at night (Hubschman, '70).

Released larvae swim about, usually rotating with the pole opposite the polypides forward (Fig. 11L). A nervous network with associated sensory cells is concentrated at this pole (Braem, '08; Marcus, '26a,b). Larvae of *P. magnifica* (Williams, '21) and *P. fungosa* (Marcus, '26a) respond negatively to light. This reaction seems to be photokinetic rather than phototactic (cf. Ryland, '77). Negative geotaxis and positive aerotaxis have been noted for the larvae of *P. fungosa* (Marcus, '26a). Larvae begin to metamorphose after a free-swimming period lasting for only a few hours. Little is known about the factors responsible for the induction of metamorphosis. Hubschman ('70) studied the substrate discrimination of *P. magnifica* larvae. They favor stone over glass surfaces, which are in turn preferred to sand-size particles. At metamorphosis, the larva attaches to some suitable substratum by its leading pole. After, or occasionally before, attachment the mantle fold rolls back, exposing the polypide-bearing region. Thus as soon as settlement is accomplished, the polypides evaginate. In time the degenerated tissue of the larval covering at the base of the new colony is sucked up into the coelom and the daughter cystid closes under this region completely.

The larva is in effect a juvenile colony. The number of primary polypides in the larva is somewhat variable even within the same species. Generally, there is one in *Fredericella* and there are two in *Plumatella* and *Hyalinella* and four in *Pectinatella magnifica* and *Cristatella* (cf. Rogick, '39). *Pectinatella gelatinosa* larvae have two primary polypides (Oda and Nakamura, '80). These facts may suggest that in the evolution of phylactolaemates asexual reproduction has been pushed backward to the embryonic stage, resulting in a drastic suppression of larval structures. Further detailed information of living larvae and their metamorphosis is given by Rogick ('39) for *Hyalinella punctata*, Brien ('53) for *Fredericella sultana* and *Plumatella fungosa*, and Oda ('60) for *Lophopodella carteri*. For comparative aspects of phylactolaemate larvae with those of other groups of bryozoans, see Zimmer and Woollacott ('77).

DEVELOPMENTAL BIOLOGY OF THE COLONY AND THE ZOOID

Laboratory Culture of the Colony

To study various developmental aspects of the colony or the zooid, continuous observation of living animals is required. The laboratory culture of freshwater bryozoans has been attempted by a number of workers with varying degrees of success (Otto, '21; Marcus, '26a; Brandwein, '38; Rogick, '35, '38, '45; Oka and Oda, '48; Oda, '60, '80; Wayss, '68; Wood,

'71, '73: Mukai et al., '79). Recently, Jebram ('77) reviewed culture methods for both freshwater and marine bryozoans.

It is rather difficult to culture freshwater bryozoans under laboratory conditions, particularly *Pectinatella* and *Cristatella*. Conditions indispensable for successful rearing are maintaining a sufficient supply of food and keeping the colonies clean. Since the phylactolaemates are voracious and display a low digestive efficiency (Rüsche, '38), the amount of nourishment available in the culture medium is vitally important. Their diet consists of minute organisms, such as diatoms, desmids, other unicellular algae, protozoans, and bacteria. As the culture medium most investigators simply employ fresh pond-water (preferably habitat water) or old aquarium water rich in food organisms. Aquarium scrapings, fresh yeast cake, bacterial scum from water containing cornmeal (Rogick, '45), or commercially prepared fish food (Marcus, '26a) may be added to the culture medium. According to Wood ('71), the least troublesome and most reliable food source appears to be the suspended organic matter that occurs naturally in a large thriving aquarium. By circulating the water of such an aquarium between it and a culture vessel, he was able to maintain several consecutive generations of *Fredericella* and *Plumatella*. Brandwein ('38) prepared cultures of *Chilomonas* and *Colpidium* in finger bowls containing an agar layer with embedded wheat grains overlaid with a very dilute balanced salt solution. *Pectinatella magnifica* colonies attached to the inner surfaces of small Petri dishes were reared within these finger bowls. Development was rapid, but frequent subculturing was necessary. Wayss ('68) maintained *Plumatella repens* colonies in continuous laboratory culture for 3 years; the colonies were fed on small organisms (*Chlorogonium*, *Haematococcus*, *Paramecium*, etc.) cultured in a medium mixed 1:1 from a Knop solution and a soil solution. Oda ('80) has succeeded in long-term rearing of *Lophopodella carteri* colonies by using pure-cultured *Chlamydomonas reinhardtii* as food organisms.

Cultures may be started from a statoblast, a larva, or a piece of an old colony. Usually colonies are grown on glass slides, watch-glasses, the inside bottoms of small Petri dishes, or similar vials, which are then set in a larger vessel containing a culture medium. It is important that the colonies be kept "upside down" so that all fecal pellets and debris fall away from the animals. Moreover, colonies and their environs should be kept clean by carefully removing accumulations of organic debris, daily when possible, with a pipette, fine needle, or forceps. To avoid the accumulation of debris, Mukai et al. ('79) raised juvenile colonies of *Pectinatella gelatinosa* emerged from statoblasts on a string suspended in the culture medium. Frequent changes of the culture medium or transfers of colonies to fresh cultures are generally desirable. Wood ('71) described a simple and convenient aquarium method for rearing colonies of *Fredericella* and *Plumatella*. The same author (Wood, '73) also developed a device for field culture of colonies of these genera.

Growth of the Colony

As a rule, a colony is founded either by the germination of a statoblast (statoblast colony) or by the attachment of a larva (embryo colony). In the former case, only a single polypide evaginates, and in the latter case one to four polypides appear almost simultaneously. Starting from these primary zooids or ancestrulae, colony formation (or astogeny) proceeds through asexual budding, the general pattern of which has been described previously. Oda ('60) has presented graphically the growth of both statoblast and embryo colonies of *Lophopodella carteri*. Bushnell ('66) studied the growth of *Plumatella repens* colonies in two lakes in Michigan. The growth rate measured by the increase in the number of polypides tended to be geometric, with a doubling time of 3.4 to 4.9 days in summer colonies or 4.0 to 7.4 days in spring colonies. According to the laboratory study of Wayss ('68), the optimum temperature for the growth of *P. repens* colonies was found to be between 28° and 32°C. An extremely low (12°C) or high (36°C) temperature, especially the latter, affected the order of budding within the whole colony and hence the colony shape by decreasing the branching of the colony. Wood ('73) performed intensive analytical studies both in the field (Sawhill Pond near Boulder, Colorado) and in the laboratory on the colony growth of *P. repens*, *Plumatella casmiana*, and *Fredericella sultana*. In all three species, an increase in colony age was generally accompanied by a reduction in zooid longevity, by a decrease in the percentage of zooids involved in statoblast and bud production, and by the production of statoblasts and buds at an increasingly earlier zooid age. Again, in all species the polypide mortality and budding rate tended to change together. Growth data of laboratory-reared colonies are presented by Rogick ('35, '38) for *Lophopodella carteri*, by the same author (Rogick, '45) for *Hyalinella punctata*, and by Mukai et al. ('79) for *Pectinatella gelatinosa*.

Compact gelatinous colonies of *Lophopus*, *Lophopodella*, *Pectinatella*, and *Cristatella* increase in number by vigorous divisions. Mukai et al. ('79) found in field studies that a statoblast colony of *P. gelatinosa* produces up to 398 colonies in 3 months. Svetlov ('35) studied the regulation of colony shape following artificial cutting of *Cristatella mucedo* colonies. Frequent fragmentation of colonies takes place also in branching forms such as *Fredericella sultana* (Braem, '08; Otto, '21; Wood, '73), *Stolella evelinae* (Marcus, '41), and *Hyalinella punctata* (Rogick, '45). In *F. sultana* at least, living branches are often carried away to establish new colonies elsewhere (Wood, '73).

In *Plumatella fruticosa* and *F. sultana*, especially in the former, zooids often successively bud several or more daughter zooids, most of which detach from the maternal cystid. Stumps of cast-off daughter zooids are found as a linear series of serrations on the ventral cystid wall. Such detached zooids are believed to be reproductive bodies capable of founding new colonies

(Toriumi, '51, '54; Wiebach, '54). However, Wiebach ('54) failed to see any colony formation from detached zooids of *P. fruticosa* after a month of observation in nature, and only some indications of budding were noted.

Degeneration, Regeneration, and Malformation of the Zooid

Every polypide of phylactolaemate colonies is a transitory organism and degenerates after existing for a short period. There are considerable but scattered data on the longevity of polypides (see the literature cited under "Laboratory Culture of the Colony," above). However, the conditions under which the data were collected are so varied that meaningful comparisons are necessarily limited. In natural populations of *Plumatella repens* Bushnell ('66) found the range of polypide longevity to be 4–53 days; few polypides lived for longer than 6 weeks and 80% died within 28 days (at 14°–31°C). According to the field observations of Wood ('73), the polypide longevity was 1–44 days (average 7.4 days) in *P. repens*, 2–31 days (average 6.8 days) in *Plumatella casmiana*, and 2–39 days (average 7.8 days) in *Fredericella sultana*. As was already noted, the mean polypide longevity varied considerably with colony age; the general trend was toward decreasing longevity as the colony grew older. The fact that the mortality of polypides and budding rate in a colony tend to change together (Wood, '73) strongly suggests the existence of some competition for nourishment between grown zooids and early buds. According to the laboratory study of Mukai et al. ('79), ancestrulae evaginated from statoblasts of *Pectinatella gelatinosa* survived for up to 29 days when bud growth was suppressed, but for less than 12 days when their main buds grew to polypides (at 28°C). Brooks ('29) reported that ancestrulae of *Pectinatella magnifica* lived for 6 weeks (2 weeks on yolk stored in their body cavity and a further 4 weeks on ingested food) without forming colonies in the laboratory. Temperature is an important external factor affecting the life span of polypides. Wayss ('68) found during laboratory study of *P. repens* that polypide longevity decreases as the water temperature rises; the mean polypide longevity was 74 days at 12°C but it was 7.5 days at 36°C. Oka and Oda ('48) reported that polypides of *Lophopodella carteri* when cultured in the laboratory at 25°C lived for about 15 days, but at lower temperatures they lived for up to 52 days.

Processes of polypide degeneration were described in detail by Rogick ('38) for *Lophopodella carteri*. A polypide retracts permanently, followed by the disintegration of the tentacles, lophophore, digestive tract, and other viscera to form a spherical mass, known as the "brown body." The brown body usually breaks away from the cystid and circulates in the coelom of the colony, but sometimes it is constricted off together with an apical part of the cystid embracing it (Marcus, '26a, '34; Mukai et al., '79). Brown bodies within the coelom are extruded through some part of the cystid wall, most frequently the basal portion (Rogick, '38; Oka and Oda, '48). In *Stolella evelinae* (Marcus, '41) and *Plumatella repens* (Wayss, '68), they may also be expelled through the vestibular pore. It is often said that definite brown bodies are formed only in marine bryozoans, where degeneration of a polypide is usually followed by the regeneration of a new polypide occupying the same cystid, but degenerative processes seem to be similar in all groups of bryozoans (cf. Gordon, '77). In the phylactolaemates, degeneration of a polypide is generally followed by regression of its cystid. However, Wood ('73) observed the occurrence of replacement budding in both laboratory and field colonies of *Plumatella casmiana*. In *Fredericella sultana*, similar replacement budding can be induced at least experimentally (see below).

In *Lophopus crystallinus* (Otto, '21) and *Lophopodella carteri* (Oka, '44) tentacles or distal parts of the lophophore arms will regenerate when cut off, but no regeneration follows when both lophophore arms are removed near their bases. In *Plumatella repens*, according to Wayss ('68), the polypide survives after partial amputation of the stomach but does not regenerate the lost part. When the stomach is wholly removed, the wound does not close and the polypide dies. In this species, budding activity is confined to the normal budding zone, and a healthy maternal polypide is necessary for budding. However, once a bud is formed it can develop into a polypide in the absence of the maternal polypide or even when transplanted elsewhere in the cystid wall. Such a polypide produces its buds normally.

Fredericella sultana has the highest regenerative ability. Brien ('36) used ancestrulae derived from statoblasts of this species and cut the zooid obliquely through the vestibular region into two parts. When the budding zone was included in the posterior part, the isolated anterior part died. The remaining posterior part regenerated a new anterior part from the ventral wall of the cystid so far as the digestive tract remained intact. When it degenerated, the main bud developed into a polypide but the adventitious bud, located on the dorsal side of this polypide, degenerated. When the budding zone was included in the anterior part, the isolated anterior part turned into a closed vesicle or a small cystid by the disintegration of the lophophore. The main bud developed into a complete but small polypide while the adventitious bud degenerated. The posterior part of this cut regenerated the removed anterior part from the ventral wall of the cystid as long as the digestive tract survived intact. The polypide, in turn, developed an adventitious bud on the ventral side. However, when the digestive tract degenerated a new polypide developed to replace the lost polypide from the cystid wall at the site of attachment of the old esophagus and this polypide produced an adventitious bud on its ventral side. From these results, Brien ('36) concluded that the cystid wall on the ventral side of a polypide is in a physiological state that is capable of supporting budding, whereas such a physiological condition is lacking on the dorsal side.

When cystid tubes of *F. sultana* are cut across, the endocyst closes the wounds, withdrawing from the cut ends. The newly formed cystid ends may produce adventitious buds (Braem, '08; Otto, '21). According to the latter author, the formation of such buds is more likely to occur in cuts located distally or in young portions of the colony, with less frequent occurrence toward the proximal direction. A polypide may be formed either at the distal or at the proximal end of a cystid tube, but more frequently it is formed at the former.

Otto ('21) further declared that in *Fredericella sultana*, *Plumatella fungosa*, and *Plumatella fruticosa* cutting or wounding of the cystid induces the formation of a funiculus and eggs from the cystid wall at or near the wound. The funiculus thus formed is connected to the cystid at both ends and may develop statoblasts and a testis. This finding of Otto must be regarded with reserve, since Wayss ('68) failed to verify it with *Plumatella repens*. However, Wayss ('68) revealed another aspect of *P. repens*' regenerative capacity. When a funiculus was cut at either or both ends, neither regeneration of a new funiculus nor reattachment of the cut end(s) of the old funiculus took place. Statoblasts growing in a funiculus isolated in the coelomic cavity continued their development, but somewhat more slowly than those in an intact funiculus. If a funiculus is cut near the stomach, new formation of statoblast rudiments is possible; however, if a funiculus is cut near the cystid it does not occur.

Rare examples of malformed zooids have been reported in *Hyalinella punctata*, *Lophopodella carteri, Pectinatella gelatinosa*, and *Cristatella mucedo* by Rogick ('45), Oda ('54), and especially by Oda and Nakamura ('73). According to the last-named authors, the abnormal polypides thus far reported are classified into the following four groups: polypides with an X-shaped lophophore, polypides with an E-shaped lophophore, polypides with separated lophophores, and polypides with a half lophophore. All these polypides except the last group are broadly divided into two types, oral and anal, according to the plane of duplication of body parts. In both types all intermediates toward complete twin polypides have been observed. Almost all malformations were found in ancestrulae emerged from statoblasts, and daughter zooids budded from such abnormal zooids were always normal morphologically. All abnormal zooids were found by chance, and nothing is known of the mechanism of their abnormal morphogenesis. Recently, Jebram ('78) presented a comprehensive review of abnormalities in bryozoans including the Phylactolaemata.

Life-Cycle Events

The annual cycle of phylactolaemates in temperate zones appears to be largely under the control of temperature cycles. In most species, colonies die out at the onset of winter; statoblasts survive and germinate in spring. The ability to withstand extremes of temperature varies with species (cf. Hyman, '59); for example, colonies of *Fredericella sultana* may survive during mild winters (Brown, '33; Bushnell, '66; Wood, '73). An obvious exception is found in *Stephanella hina*, whose colonies flourish only during winter and early spring (Oka, '08; Toriumi, '55b). At least in some species, colony longevity in nature is much shorter than the growing period of the species. At Sawhill Pond, colonies of *Plumatella repens* lived for as long as 59 days and those of *Plumatella casmiana* over 64 days; longevity of *F. sultana* colonies was significantly greater (Wood, '73).

Generally, colonies hatched from overwintered statoblasts rapidly proceed to gametogenesis and statoblast formation. Sexual reproduction occurs but once annually; it usually lasts 3 to 4 weeks in nature. In *Cristatella mucedo* (Braem, 1890), *Plumatella fungosa* (Brien and Mordant, '55), and possibly *Hyalinella punctata* (Rogick, '45), gamete production and statoblast production are more or less separated, and the former precedes the latter; however, in *Lophopodella carteri* (Oda, '59, '60) and many other species they occur concurrently. Complete omission or extremely limited occurrence of sexual activity has been reported for various species, both in nature and in the laboratory (Rogick, '45; Brien and Mordant, '55; Wayss, '68; Wood, '73; Mukai, '74; Mukai et al., '79; see Hyman, '59, for earlier papers). In laboratory cultures of *Plumatella repens* Wayss ('68) observed that testes mature at 28°, 32°, and 35°C, but at 18° and 24°C testes do not appear and at 12°C they appear but do not mature. Ovaries were not formed at any temperature. As a rule, statoblast production continues until colonies disintegrate. There are few analytical studies on conditions favorable for statoblast production. Oda ('80) briefly noted that in *Lophopodella carteri* more statoblasts were formed in the dark than under conditions of light.

The number of generations arising from statoblasts each year varies from one to three with different species, or even in the same species, depending upon temperature. In many species, statoblasts released in summer can germinate within that season, but those liberated during the autumn usually overwinter before germinating. Hence, most species in temperate zones are believed to have two regular statoblast generations per year (cf. Brown, '33). *Fredericella sultana* (Braem, '08), *Pectinatella magnifica* in Michigan (Brown, '33), and *Pectinatella gelatinosa* (Mukai, '74) have only one statoblast generation each year. Their statoblasts do not germinate the same year in which they are produced and remain dormant or quiescent until the following spring. Oda ('79) presented some evidence of a secondary statoblast generation in *P. magnifica* in Japan. Brown ('33) had evidence of two statoblast generations of *Plumatella repens*, but Bushnell ('66) observed three for this species. According to Oda ('59, '60), in Central Japan *L. carteri* produces statoblast colonies three times a year; dormancy occurs only in statoblasts formed in midsummer (Fig. 12).

Both Rogick ('43a) and Toriumi ('55a) have reported that in *Plumatella casmiana* thin-walled floatoblasts (leptoblasts) are more abundant during the earlier part of the growing season and thick-walled floatoblasts (see footnote 1) are more abundant later in the season. Most colonies contain only a single kind of floatoblast, but some colonies contain both kinds simultaneously. Chiefly from field observations at Sawhill Pond, Wood ('73) concluded that colonies from intermediate floatoblasts produce both leptoblasts and intermediate floatoblasts (but at different times, never simultaneously); leptoblast colonies form both more leptoblasts and pycnoblasts (these two types having never been found together in the same colony); and pycnoblast colonies produce only intermediate floatoblasts. As was previously noted, leptoblasts germinate soon after release from the colony. Pycnoblasts are fairly common, according to Wood ('73), during June and July, and they germinate within 2 to 3 weeks. Only intermediate floatoblasts overwinter to germinate the following spring.

In field colonies of *Plumatella repens* and *P. casmiana*, sessoblasts appear sporadically, but most frequently during the spring and fall months. Their appearance is usually followed shortly by the death of the zooids—a fact that was also confirmed by laboratory studies (Wood, '73). From these results, Wood ('73) considered that sessoblasts form in response to suboptimal environmental conditions, or to general physiological deterioration of individual zooids. Some results from my own preliminary observations may be worth adding here. In the early summer of 1981, I collected from a pond several large colonies of *P. casmiana* that were grown on a plastic plate and laden with numerous leptoblasts, thick-walled floatoblasts, or both. A small number of sessoblasts were found in some colonies. These colonies were brought into the laboratory, where they were reared with aquarium water containing green algae and other food organisms. In a few days, numerous sessoblasts appeared in the central area of each colony; most polypides gradually died and within a week or so active polypides were confined to the marginal zone of the colony. Under the same culture conditions, young colonies raised from leptoblasts vigorously produced leptoblasts; only some of them produced a few sessoblasts as well. Again, under the same culture conditions, *Plumatella emarginata* colonies raised from floatoblasts constantly formed sessoblasts,

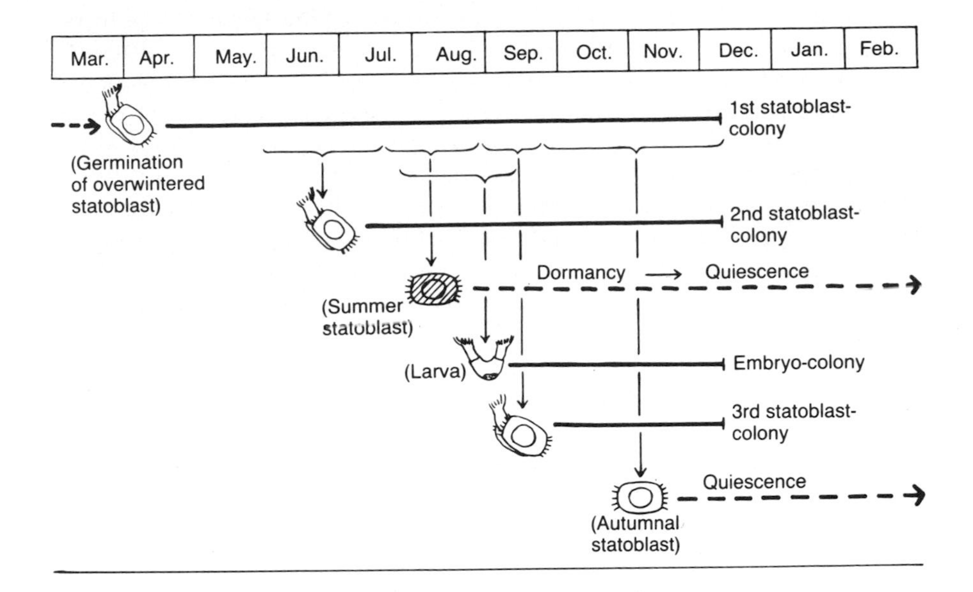

Fig. 12. Annual reproductive cycle of *Lophopodella carteri* in Central Japan. (Redrawn from Oda, '59, with permission of the author.)

with or without forming floatoblasts. It seems probable, though verification is lacking, that a statoblast rudiment can develop into either a sessoblast or a floatoblast depending on some external conditions and, by extension, on the physiological condition of the colony or of the zooid.

CONCLUDING REMARKS

It will be obvious from the preceding pages that the study of the developmental biology of freshwater bryozoans is still in its infancy. Although phylactolaemates are by no means uncommon, they cannot easily be found without a careful search because they are sessile and the growing season of colonies is rather short. Even when colonies or statoblasts are found, identification of species is often very difficult, especially so in plumatellids. Furthermore, laboratory culture of colonies is not easy. These points may explain to a large measure the neglect of this interesting group of animals in modern biology.

Phylactolaemates reproduce by means of external budding, internal budding, and embryogenesis. Our knowledge of even the basic morphological events involved in these processes is often incomplete. Sessoblast formation, for example, was first described in this article. Information available on various aspects of sexual reproduction, e.g., gametogenesis, fertilization, transfer of an egg into the embryo sac, cleavage, and embryogenesis, is limited and is often contradictory. Thus a wide variety of reproductive events in phylactolaemates require much more morphological study, not only at the light microscopic but also at the ultrastructural level.

Regulation among different modes of reproduction is a fascinating problem that awaits future study; clearly such study will depend heavily upon the development of culture methods for colonies. Many species produce both floatoblasts and sessoblasts. Moreover, *Plumatella casmiana* characteristically produces two or three kinds of floatoblasts. Factors or mechanisms that control the formation of different kinds of statoblasts have been only partially elucidated. The fact that statoblast rudiments can continue to grow even after the isolation of the funiculus in the coelom strongly suggests the possibility of the application of organ-culture methods for the study of this and related problems.

So far, most of the information on the germination capacities of statoblasts was obtained with the so-called spinoblasts; our knowledge on the germination capacities of other types of statoblasts is still fragmentary. We know very little of the biochemistry or endogenous control mechanisms of dormancy, quiescence, and germination of statoblasts. These phenomena in statoblasts are shared with various germinative systems of other organisms and present many problems that deserve modern treatment.

The phylactolaemates, like other members of the Bryozoa, are all colonial organisms. There is no doubt that various developmental events shown by each zooid are under the control of both the zooid itself and the colony as a whole. Analytical studies along this line are scarce. Experimental fusion of colonies or the exchange of the body fluid between colonies may be a promising approach for the analysis of regulatory mechanisms involved in the colony level. These organisms could also present some interesting models for research on aging.

Acknowledgments. I wish to express my thanks to Professor Shuzitu Oda of Rikkyo University for his generous help in surveying the literature and for his encouragement during the course of this work. I am also grateful to Professors Ronald R. Cowden and Frederick W. Harrison for their kind invitation to write this chapter and for their editorial efforts.

LITERATURE CITED

Bergin, E.A., W.R. Tenney, and W.S. Woolcott (1964) Photoperiodism as a factor influencing the germination of statoblasts of the bryozoan *Lophopodella carteri* (Hyatt). Va. J. Sci. *15:*287 (abstract).

Black, R.E., and V.K. Proud (1976) Biochemical changes in temperature-activated statoblasts of an ectoproct bryozoan. J. Exp. Zool. *197:*141–147.

Braem, F. (1890) Untersuchungen über die Bryozoen des süssen Wassers. Zoologica (Stuttgart) *2:*1–134.

Braem, F. (1897) Die geschlechtliche Entwicklung von *Plumatella fungosa*. Zoologica (Stuttgart) *10:*1–96.

Braem, F. (1908) Die geschlechtliche Entwicklung von *Fredericella sultana* nebst Beobachtungen über die weitere Lebensgeschichte der Kolonien. Zoologica (Stuttgart) *20:*1–38.

Braem F, (1913) Die Keimung der Statoblasten von *Pectinatella* und *Cristatella*. Zoologica (Stuttgart) *26:*35–64.

Brandwein, P.F. (1938) The culture of *Pectinatella magnifica* Leidy. Am. Nat. *72:*94–96.

Brien, P. (1936) Contribution à l'étude de la réproduction asexuée des Phylactolémates. Mém, Mus. R. Hist. Nat. Belg. (ser. 2) *3:*569–625.

Brien, P. (1953) Étude sur les Phylactolémates. Ann. Soc. R. Zool. Belg. *84:*301–440.

Brien, P. (1960) Classes de Bryozaires. In P.-P. Grassé (ed): Traité de Zoologie. Vol. 5, Part 2. Paris: Masson, pp. 1053–1335.

Brien, P., and C. Mordant (1955) Relations entre les réproductions sexuée et asexuée à propos des Phylactolémates. Ann. Soc. R. Zool. Belg. *86:*169–189.

Brooks, C.M. (1929) Notes on the statoblasts and polypids of *Pectinatella magnifica*. Proc. Acad. Nat. Sci. Philadelphia *81:*427–441.

Brown, C.J.D. (1933) A limnological study of certain fresh-water Polyzoa with special reference to their statoblasts. Trans. Am. Microsc. Soc. *52:*271–316.

Bullivant, J.S. (1967) Release of sperm by Bryozoa. Ophelia *4:*139–142.

Bushnell, J.H. (1966) Environmental relations of Michigan Ectoprocta, and dynamics of natural populations of *Plumatella repens*. Ecol. Monogr. *36:*95–123.

Bushnell, J.H. (1973) The freshwater Ectoprocta: A zoogeographical discussion. In G.P. Larwood (ed): Living and Fossil Bryozoa. London, England: Academic Press, pp. 503–521.

Bushnell, J.H. (1974) Bryozoans (Ectoprocta). In C.W. Hart, Jr., and S.L.H. Fuller (eds): Pollution Ecology of Freshwater Invertebrates. New York: Academic Press, pp. 157–194.

Bushnell, J.H., and K.S. Rao (1974) Dormant or quiescent stages and structures among the Ectoprocta: Physical and chemical factors affecting viability and germination of statoblasts. Trans. Am. Microsc. Soc. *93:*524–543.

Bushnell, J.H., and K.S. Rao (1979) Freshwater Bryozoa: Micro-architecture of statoblasts and some Aufwuchs animal associations. In G.P. Larwood and M.B. Abbott (eds): Advances in Bryozoology. London, England: Academic Press, pp. 75–92.

Davenport, C.B. (1890) *Cristatella*: The origin and development of the individual in the colony. Bull. Mus. Comp. Zool. Harvard Univ. *20:*101–151.

du Bois-Reymond Marcus, E. (1953) Bryozoa from Lake Titicaca. Bol. Fac. Fil., Ciênc. Letr. Univ. São Paulo, Zool. *18:*149–163.

Franzén, Å. (1970) Phylogenetic aspects of morphology of spermatozoa and spermiogenesis. In B. Baccetti (ed): Comparative Spermatology. New York: Academic Press, pp. 29–45.

Franzén, Å. (1977) Gametogenesis of bryozoans. In R.M. Woollacott and R.L. Zimmer (eds): Biology of Bryozoans. New York: Academic Press, pp. 1–22.

Gordon, D.P. (1977) The aging process in bryozoans. In R.M. Woollacott and R.L. Zimmer (eds): Biology of Bryozoans. New York: Academic Press, pp. 335–376.

Hubschman, J.H. (1970) Substrate discrimination in *Pectinatella magnifica* Leidy (Bryozoa). J. Exp. Biol. *52:*603–608.

Hyman, L.H. (1959) The lophophorate coelomates: Phylum Ectoprocta. The Invertebrates. New York: McGraw-Hill, *5:*275–515.

Jebram, D. (1977) Experimental techniques and culture methods. In R.M. Woollacott and R.L. Zimmer (eds): Biology of Bryozoans. New York: Academic Press, pp. 273–306.

Jebram, D. (1978) Preliminary studies on "abnormalities" in bryozoans from the point of view of experimental morphology. Zool. Jahrb. Anat. Ont. Tiere *100:*245–275.

Kraepelin, K. (1892) Die deutschen Süsswasser-Bryozoen. II. Entwicklungsgeschichtlicher Teil. Abhandl. Naturwiss. Ver. Hamburg *12:*1–67.

Lacourt, A.W. (1968) A monograph of the freshwater Bryozoa: Phylactolaemata. Zool. Verhandl. *93:*1–159.

Lacourt, A.W., and R.G.J. Willighagen (1966) Histochemical investigation of *Pectinatella magnifica* (Leidy, 1851). Koninkl. Nederl. Akad. Wetenschappen (ser. C) *69:*22–23.

Marcus, E. (1926a) Beobachtungen und Versuche an lebenden Süsswasserbryozoen. Zool. Jahrb. Syst. Oekol. Geogr. Tiere *52:*279–350.

Marcus, E. (1926b) Sinnesphysiologie und Nervensystem der Larve von *Plumatella fungosa* (Pall.). Verhandl. Dtsch. Zool. Ges. *31:*86–90.

Marcus, E. (1934) Ueber *Lophopus crystallinus* (Pall.). Zool. Jahrb. Anat. Ont. Tiere *58:*501–606.

Marcus, E. (1941) Sôbre Bryozoa do Brasil. I. Bol. Fac. Fil., Ciênc. Letr. Univ. São Paulo, Zool. *5:*3–208.

Marcus, E. (1942) Sôbre Bryozoa do Brasil. II. Bol. Fac. Fil., Ciênc. Letr. Univ. São Paulo, Zool. *6:*57–105.

Mukai, H. (1973) Histological and histochemical studies on the formation of statoblasts of a fresh-water bryozoan, *Pectinatella gelatinosa*. J. Morphol. *141:*411–426.

Mukai, H. (1974) Germination of the statoblasts of a freshwater bryozoan, *Pectinatella gelatinosa*. J. Exp. Zool. *187:*27–40.

Mukai, H. (1977) Effects of chemical pretreatment on the germination of statoblasts of the freshwater bryozoan, *Pectinatella gelatinosa*. Biol. Zentralbl. *96:*19–31.

Mukai, H., T. Karasawa, and Y. Matsumoto (1979) Field and laboratory studies on the growth of *Pectinatella gelatinosa* Oka, a freshwater bryzoan. Sci. Rep. Fac. Educ. Gunma Univ. *28:*27–57.

Mukai, H., and S. Oda (1980) Comparative studies on the statoblasts of higher phylactolaemate bryzoans. J. Morphol. *165:*131–155.

Mundy, S.P. (1980) Stereoscan studies of phylactolaemate bryozoan statoblasts including a key to the statoblasts of the British and European Phylactolaemata. J. Zool. *192:*511–530.

Oda, S. (1954) On the double monsters of polypides in freshwater Bryozoa (in Japanese with English abstract). Coll. Breed. (Tokyo) *16:*15–18.

Oda, S. (1958) On the outflow of the blood in colonies of freshwater Bryzoa (in Japanese). Kagaku (Tokyo) *28:*37.

Oda, S. (1959) Germination of the statoblasts in freshwater Bryozoa. Sci. Rep. Tokyo Kyoiku Dai. (sec. B) *9:*90–132.

Oda, S. (1960) Relation between asexual and sexual reproduction in freshwater Bryozoa. Bull. Mar. Biol. Stat. Asamushi *10:*111–116.

Oda, S. (1979) Germination of the statoblasts of *Pectinatella magnifica*, a freshwater bryozoan. In G.P. Larwood and M.B. Abbott (eds): Advances in Bryozoology. London, England: Academic Press, pp. 93–112.

Oda, S. (1980) Effects of light on the germination of statoblasts in freshwater Bryozoa. Annot. Zool. Japon. *53:*238–253.

Oda, S., and R.M. Nakamura (1973) The occurrence of double polypides in freshwater Bryozoa. In G.P. Larwood (ed): Living and Fossil Bryozoa. London, England: Academic Press, pp. 523–528.

Oda, S., and R.M. Nakamura (1980) Sexual reproduction in *Pectinatella gelatinosa*, a freshwater bryozoan. Proc. Jpn. Soc. Syst. Zool. *19:*38–44.

Oka, A. (1891) Observations on fresh-water Polyzoa. J. Coll. Sci. Imp. Univ. Tokyo *4:*89–150.

Oka, A. (1908) Ueber eine neue Gattung von Süsswasserbryozoen. Annot. Zool. Japon. *6:*277–285.

Oka, H. (1944) An experiment on the regeneration of freshwater Bryozoa (in Japanese). Zool. Mag. (Tokyo) *56:*9–11.

Oka, H., and S. Oda (1948) Observations on freshwater Bryozoa, with special reference to their reproduction (in Japanese). Coll. Breed. (Tokyo) *10:*39–48.

Otto, F. (1921) Studien über das Regulationsvermögen einiger Süsswasserbryozoen. Arch. Entw.-Mech. Org. *47:*399–442.

Retzius, G. (1906) Die Spermien der Bryozoen. Biol. Untersuch. (N.F.) *13:*45–48.

Rogick, M.D. (1935) Studies on fresh-water Bryozoa. III. The development of *Lophopodella carteri* var. *typica*. Ohio J. Sci. *35:*457–467.

Rogick, M.D. (1938) Studies on fresh-water Bryozoa. VII. On the viability of dried statoblasts of *Lophopodella carteri* var. *typica*. Trans. Am. Microsc. Soc. *57:*178–199.

Rogick, M.D. (1939) Studies on fresh-water Bryozoa. VIII. Larvae of *Hyalinella punctata* (Hancock), 1850. Trans. Am. Microsc. Soc. *58:*199–209.

Rogick, M.D. (1940) Studies on fresh-water Bryozoa. XI. The viability of dried statoblasts of several species. Growth *4:*315–322.

Rogick, M.D. (1941a) Studies on fresh-water Bryozoa. X. The occurrence of *Plumatella casmiana* in North America. Trans. Am. Microsc. Soc. *60:*211–220.

Rogick, M.D. (1941b) The resistance of fresh-water Bryozoa to desiccation. Biodynamica *3:*369–378.

Rogick, M.D. (1943a) Studies on fresh-water Bryozoa. XIII. Additional *Plumatella casmiana* data. Trans. Am. Microsc. Soc. *62:*265–270.

Rogick, M.D. (1943b) Studies on fresh-water Bryozoa. XIV. The occurrence of *Stolella indica* in North America. Ann. N.Y. Acad. Sci. *45:*163–178.

Rogick, M.D. (1945) Studies on fresh-water Bryozoa. XV. *Hyalinella punctata* growth data. Ohio J. Sci. *45:*55–79.

Rüsche, E. (1938) Hydrobiologische Untersuchungen an niederrheinischen Gewässern. X. Nahrungsaufnahme und Nahrungsauswertung bei *Plumatella fungosa* (Pallas). Arch. Hydrobiol. *33:*271–293.

Ryland, J.S. (1970) Bryozoans. London, England: Hutchinson University Press, 175pp.

Ryland, J.S. (1977) Taxes and tropisms of bryozoans. In R.M. Woollacott and R.L. Zimmer (eds): Biology of Bryozoans. New York: Academic Press, pp. 411–436.

Silén, L. (1966) On the fertilization problem in the gymnolaematous Bryozoa. Ophelia *3:*113–140.

Silén, L. (1972) Fertilization in the Bryozoa. Ophelia *10:*27–34.

Svetlov, P. (1935) Regulationserscheinungen an *Cristatella*-Kolonien. Z. Wiss. Zool. *147:*263–274.

Tajima, I. (1980) Electron microscope studies on the statoblast of a fresh-water bryozoan, *Pectinatella gelatinosa*. II. Changes in the fine structure of cystigenous cells during statoblast formation (in Japanese with English abstract). Zool. Mag. (Tokyo) *89:*26–40.

Tajima, I., and H. Mukai (1975) Electron microscope studies on the statoblast of a fresh-water bryozoan, *Pectinatella gelatinosa*. I. Vitellogenesis in the "yolk cell" during statoblast formation (in Japanese with English abstract). Zool. Mag. (Tokyo) *84:*205–216.

Terakado, K., and H. Mukai (1978) Ultrastructural studies on the formation of yolk granules in the statoblast of a fresh-water bryozoan, *Pectinatella gelatinosa*. J. Morphol. *156:*317–338.

Toriumi, M. (1951) Taxonomical study on fresh-water Bryozoa. I. *Fredericella sultana* (Blumenbach). Sci. Rep. Tohoku Imp. Univ. (ser. 4) *19:*167–177.

Toriumi, M. (1954) Taxonomical study on fresh-water Bryozoa. VIII. *Plumatella fruticosa* Allman. Sci. Rep. Tohoku Imp. Univ. (ser. 4) *20:*293–302.

Toriumi, M. (1955a) Taxonomical study on fresh-water Bryozoa. X. *Plumatella casmiana* Oka. Sci. Rep. Tohoku Imp. Univ. (ser. 4) *21:*67–77.

Toriumi, M. (1955b) Taxonomical study on fresh-water Bryozoa. XI. *Stephanella hina* Oka. Sci. Rep. Tohoku Imp. Univ. (ser. 4) *21:*131–136.

Toriumi, M. (1956) Taxonomical study on fresh-water Bryozoa. XVII. General consideration: Interspecific relation of described species and phylogenic consideration. Sci. Rep. Tohoku Imp. Univ. (ser. 4) *22:*57–88.

Viganò, A. (1968) Note su *Plumatella casmiana* Oka (Bryozoa). Riv. Idrobiol. *7:*421–468.

von Buddenbrock, W. (1910) Beiträge zur Entwicklung der Statoblasten der Bryozoen. Z. Wiss. Zool. *96:*477–524.

Wayss, K. (1968) Quantitative Untersuchungen über Wachstum und Regeneration bei *Plumatella repens* (L.). Zool. Jahrb. Anat. Ont. Tiere *85:*1–50.

Wiebach, F. (1952) Ueber den Ausstoss von Flottoblasten bei *Plumatella fruticosa* (Allman). Zool. Anz. *149:*181–185.

Wiebach, F. (1953) Ueber den Ausstoss von Flottoblasten bei Plumatellen. Zool. Anz. *151:*266–272.

Wiebach, F. (1954) Ueber *Plumatella fruticosa* (Allman.) Arch. Hydrobiol. *49:*258–272.

Wiebach, F. (1963) Studien über *Plumatella casmiana* Oka (Bryozoa). Vie Milieu *14:*579–596.

Wiebach, F. (1975) Specific structures of sessoblasts (Bryozoa, Phylactolaemata). Docum. Lab. Géol. Fac. Sci. Lyon (H.S.) *3:*149–154.

Williams, S.R. (1921) Concerning "larval" colonies of *Pectinatella*. Ohio J. Sci. *21:*123–127.

Wood, T.S. (1971) Laboratory culture of fresh-water Ectoprocta. Trans. Am. Micrsc. Soc. *90:*92–94.

Wood, T.S. (1973) Colony development in species of *Plumatella* and *Fredericella* (Ectoprocta: Phylactolaemata). In R.S. Boardman, A.H. Cheetham, and W.A. Oliver, Jr. (eds): Animal Colonies. Stroudsburg: Dowden, Hutchinson & Ross, pp. 395–432.

Wood, T.S. (1979) Significance of morphological features in bryozoan statoblasts. In G.P. Larwood and M.B. Abbott (eds): Advances in Bryozoology. London, England: Academic Press, pp. 59–74.

Zimmer, R.L., and R.M. Woollacott (1977) Metamorphosis, ancestrulae, and coloniality in bryozoan life cycle. In R.M. Woollacott and R.L. Zimmer (eds): Biology of Bryozoans. New York: Academic Press, pp. 91–142.

Index